Fundamentals of Ship Hydrodynamics

Fundamentals of Ship Hydrodynamics

Fluid Mechanics, Ship Resistance and Propulsion

Lothar Birk
School of Naval Architecture and Marine Engineering
The University of New Orleans
New Orleans, LA
United States

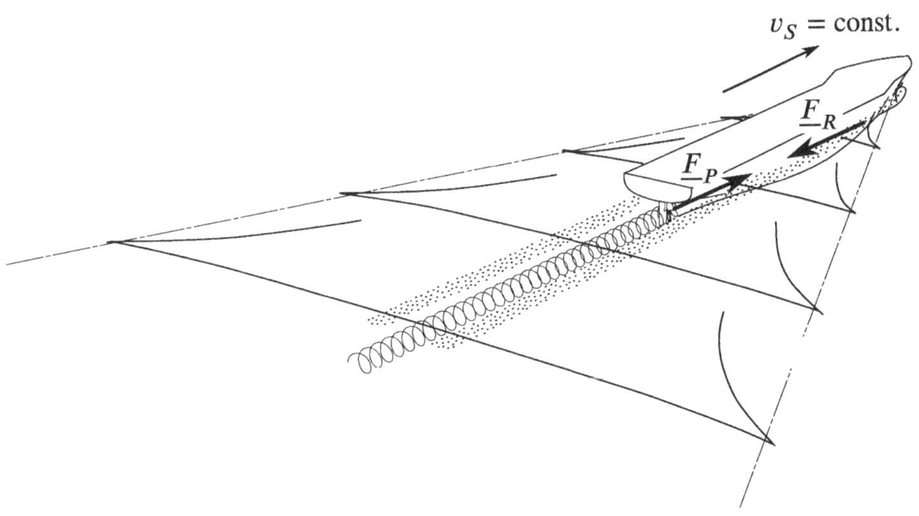

This edition first published 2019

Registered Offices
John Wiley & Sons, Inc., 111 River Street, Hoboken, NJ 07030, USA
John Wiley & Sons Ltd, The Atrium, Southern Gate, Chichester, West Sussex, PO19 8SQ, UK

Editorial Office
The Atrium, Southern Gate, Chichester, West Sussex, PO19 8SQ, UK

For details of our global editorial offices, customer services, and more information about Wiley products visit us at www.wiley.com.

Wiley also publishes its books in a variety of electronic formats and by print-on-demand. Some content that appears in standard print versions of this book may not be available in other formats.

Library of Congress Cataloging-in-Publication Data

Names: Birk, Lothar, 1963- author.
Title: Fundamentals of ship hydrodynamics : fluid mechanics, ship resistance and propulsion / Lothar Birk, University of New Orleans.
Description: Hoboken, NJ : John Wiley & Sons, Ltd, [2019] | Includes bibliographical references and index.
Identifiers: LCCN 2018060347| ISBN 9781118855485 (hardcover) | ISBN 9781118855515 (epub)
Subjects: LCSH: Ships–Hydrodynamics.
Classification: LCC VM156 .B5335 2019 | DDC 623.8/12–dc23
LC record available at https://lccn.loc.gov/2018060347

Cover Design: Wiley
Cover Image: © zennie / Getty Images

Set in 10pt Warnock Pro Regular by Lothar Birk

Printed in Great Britain by TJ International Ltd, Padstow, Cornwall

10 9 8 7 6 5 4 3 2 1

To My Family

They make everything worthwhile!

Contents

List of Figures

List of Tables

Preface

Target readership

This book has been designed as a textbook to support ship resistance and propulsion related courses at the undergraduate level. As such, its main audience is naval architecture and marine engineering students and students in related fields. However, since the book covers topics in fairly great detail, it is suited for self study for everybody with a working knowledge of calculus, statics, and dynamics. Graduate students and practicing engineers, who venture from other engineering disciplines into maritime fields of study or work, might use this book as a preparation for new tasks.

Identified needs

Over the past 20 years, I have taught ship resistance and propulsion at three different universities and consistently made the following observations:

- The foundation laid by basic fluid mechanics courses in modern engineering curricula is incomplete and does not cover everything essential to courses focused on ship resistance and propulsion.
- A wealth of excellent reference books exists covering all aspects of ship hydrodynamics. However, no matter how strongly I recommend one of them, most students find them too expensive or too intimidating and do not use them as study aids.
- Arguably, most reference books are not organized in a way which lends itself to support class work. The chapters are designed so broadly that it becomes difficult to assign specific parts to individual class periods.

Interior vs. exterior flow

In many engineering curricula in the United States, basic fluid mechanics is covered in a single course. This is just enough to cover hydrostatics and the basic equations of fluid dynamics but leaves hardly any room for boundary layer theory, potential flow, wave theory, and foil and wing theory. In addition, teaching fluid mechanics courses is often the responsibility of mechanical engineering departments, which naturally concentrate on pipe flows and turbo machinery rather than exterior flows.

Details vs. coverage

Authors of reference books assume more prerequisite knowledge than a typical undergraduate student of today actually has. After all, their target audience are practicing engineers. In addition, reference books attempt to be comprehensive and cover a broad range of topics and tend to omit a lot of detail. The gaps may be easily filled by an expert but often pose a seemingly insurmountable obstacle for students trying to understand the origins of a theory or fathom exactly how a certain method works. As a consequence, I find myself compelled to explain to students what reference books cover with statements like 'as one can easily see'. Unfortunately, covering extensive details in class distracts students from important assumptions and conclusions.

Objectives

Based on these observations, I set out to write a textbook which meets the following objectives:

- Whereas most reference books on ship hydrodynamics spend just a couple of pages to summarize results of fluid mechanics, this textbook dedicates considerable space to it. However, it is not meant as a substitution, but rather a logical continuation of a basic fluids mechanics course. The material is presented with its application to ship resistance and propulsion in mind.
- Instead of covering all possible aspects of ship hydrodynamics, a selection of topics is covered in greater detail. Sentences like 'as one can easily see' or 'after some manipulations' are kept to a minimum. The detailed coverage allows the teacher to concentrate on important assumptions and conclusions. Students will find and study the details in the associated sections.
- Each chapter covers material for one, or sometimes two, class periods, which should simplify reading assignments. As a consequence, the book is organized into an unusually large number of chapters. Margin notes are used as an additional organizational aid. There is a continuous thread throughout the book, but the chapters are relatively independent from each other. This should make it easier to skip some of them, assign them as extra reading, or rearrange their order according to the needs of a specific course.

Content overview

The junior level ship resistance and propulsion course serves a dual purpose in our naval architecture and marine engineering curriculum at the University of New Orleans. On one hand, it identifies and explains basic flow patterns around a ship sailing at constant speed. On the other hand, it prepares students to conduct basic ship design tasks like resistance and powering estimates. Starting with basic fluid mechanics and ending with powering estimates spans a wide arc. The only way to keep the page count in check was to concentrate on the immediate topics at hand rather than venturing into all variations and alternatives. The reader will notice that the book focuses on displacement type monohulls driven by marine propellers. As a consequence, multihulls, planing boats, and other propulsion systems are not covered. Fundamental analytical and experimental methods are discussed but not computational fluid dynamics.

Organization

The book is subdivided into 51 chapters organized into three parts: basic fluid mechanics, ship resistance, and propulsion. However, the boundaries are blurred as I attempt to connect basic theory with its application in ship hydrodynamics wherever possible. The first chapter specifies the calm water resistance and propulsion problem. The second chapter defines ship resistance and its major components. In Chapters 3 through 13 we develop important equations describing viscous flow around submerged bodies and use them to assess the frictional resistance of a ship. Chapters 14 through 18 analyze inviscid flow and combine it with viscous flow theory to explain viscous pressure resistance. Chapters 19 through 25 tackle wave theory and wave resistance.

Chapters 26 through 30 explain the concepts and theories which govern ship model testing and the prediction of full scale resistance. Chapter 31 provides a first look at resistance estimates for ship design purposes.

Chapter 32 marks the beginning of the ship propulsion part. Basic terminology, propulsor action, hull–propeller interaction, and propeller geometry are illustrated in Chapters 33 through 35. Chapters 36 through 40 cover the basic flow theory for lifting foils and wings. Chapters 41 through 44 deal again with model testing and discuss experiments with model propellers and self-propelled ship models.

Chapters 45 through 49 address the problem of cavitation, cavitation avoidance, and how to select a propeller for a specific ship. Finally, Chapters 50 and 51 describe in detail two methods to estimate resistance and powering requirements in early design stages.

Nomenclature

Symbols are typically explained when they are introduced. A conscious effort has been made to use the terminology and symbols according to the *Dictionary of Hydrodynamics* and the *ITTC Symbol and Terminology List* maintained and published by the International Towing Tank Conference (ITTC). Both documents are part of the quality systems manual and can be found on the ITTC's website at www.ittc.info (ITTC, 2017a,b).

Cartesian coordinate system

In most cases a Cartesian coordinate system $< x, y, z >$ is employed with its positive x-axis pointing forward (in the direction of motion), its y-axis pointing to port, and its z-axis pointing upwards.

Summary

A textbook is always a conglomerate of the combined knowledge and wisdom of all who have worked in the specific field. All the presented work has originally been developed by others and I have made every effort to point the reader to the correct sources. My job has been to illustrate and explain everything, and as such the errors are all mine. If you find any errors, please feel free to point them out to me via e-mail at lothar.birk@marine-hydrodynamics.com.

Slidell, December 2018
Lothar Birk

References

ITTC (2017a). *Dictionary of hydrodynamics – Alphabetic.* International Towing Tank Conference, Quality Systems Group.

ITTC (2017b). *ITTC symbols and terminology list – Alphabetic.* International Towing Tank Conference, Quality Systems Group.

Acknowledgments

With the encouragement of my students, who told me on several occasions I should convert my class notes into a book, I started writing this textbook roughly five years ago. When Paul Petralia from Wiley & Sons, Ltd contacted me, I had the book proposal ready to go and luckily the publishing house accepted my concept. I am grateful for the opportunity and their support.

I would like to thank my colleagues Dr. Janou Hennig, Dr. Alfred Kracht, and Dr. Walter Kühnlein for providing photos for the book. Many thanks also to Dr. Settles for his photo of the laminar–turbulent transition in the airflow above a candle.

Writing a book is a milestone in a professional career. It provides time to pause and reflect on how one got to this point. It is obvious to me that I had great teachers in high school (which is called Gymnasium in Germany) and throughout my studies at Technische Universität Berlin. Thank you, Mr. E. Jäckle, Mr. H. Riekert, Professor G.F. Clauss, Professor H. Nowacki, and Professor E. Wolf, who all sparked my curiosity and inspired my desire to learn more. I hope all readers find great teachers and mentors like I did.

Last but not least, I want to thank my children Benjamin and Kathleen, who both helped to correct their dad's basic English, and of course my love and wife Carola, who encouraged me all the way and patiently endured my prolonged occupation with this book project.

About the Companion Website

The companion website for this book is at

www.wiley.com/go/birk/hydrodynamics

The website includes:

- Python scripts
- Figures

Scan this QR code to visit the companion website.

1

Ship Hydrodynamics

The field of ship hydrodynamics considers the interaction of vessels with surrounding fluids. As the prefix 'hydro' suggests, we are most concerned with water; however, the air flow around the super structure has to be dealt with as well. In this chapter, we narrow down this broad field and define calm water hydrodynamics as the context of this book. We will also discuss the role and responsibilities of the naval architect in the analysis of ship hydrodynamics and – in broad terms – what tools we have at our disposal to solve hydrodynamic tasks in ship design.

Learning Objectives

At the end of this chapter students will be able to

- review the complexity of the ship propulsion problem
- explain the concepts of calm water, trial, service, and open water condition
- understand the role of ship resistance and propulsion in ship design
- distinguish basic tools to predict and investigate hydrodynamic performance

1.1 Calm Water Hydrodynamics

The real world

Boats and ships crossing any body of water have to negotiate the environment presented by wind, waves, currents, and the boundaries of their domain. Figure 1.1 shows the major interactions between ship and environment.

- Wind blowing over water creates a seemingly chaotic pattern of waves through which the ship sets its course. Wind and waves may come from different directions relative to the ship's path. A ship will change the shape of waves in its vicinity. This is called wave diffraction. Wind and waves exert forces on the vessel which vary with time.
- Wind and wave forces cause the ship to move. This movement creates additional waves, similar to the waves created by a stone dropped into a pond. This is known as wave radiation.

Fundamentals of Ship Hydrodynamics: Fluid Mechanics, Ship Resistance and Propulsion, First Edition.
Lothar Birk.

Companion website: www.wiley.com/go/birk/hydrodynamics

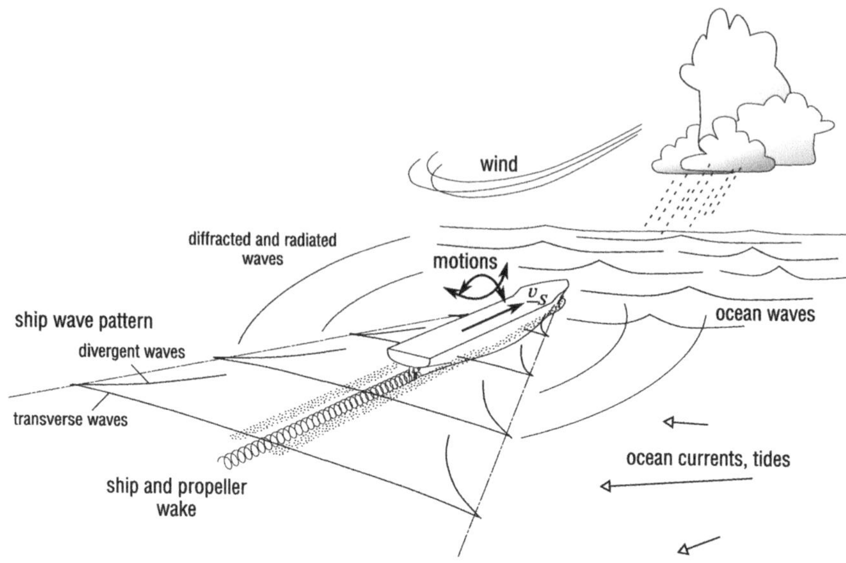

Figure 1.1 Ship sailing in its natural habitat

- Currents may cause the general flow direction to be oblique to the ship's path.
- A ship, even when moving in undisturbed water, creates a well organized wave pattern. It appears in a triangular region behind the ship and consists of divergent and transverse waves.
- Some of the waves created by the vessel will break. A mix of water and air creates a band of froth along the ship's path.
- Small and large eddies appear next to and behind the ship. They are the result of friction between hull and water. A boundary layer forms over the submerged hull surface and merges into a disturbed flow region behind the ship generally known as wake.
- A rotating propeller will generate its own twisted wake, further complicating the flow patterns.

Unsteady flow A moving vessel constantly displaces water and air molecules from their original position. As stated in Newton's laws of motion,[1] forces (Latin: actio) must act on the molecules to change magnitude or direction of their velocities. In turn, a reaction force (Latin: reactio) is exerted by water and air on the ship. Vessels are usually self-propelled, i.e. they have some means of propulsion. The most common propulsor today is the marine propeller. Sails, oars, paddle wheels, and water jets may be applied depending on purpose, size, and speed of the vessel. Propulsors create the force necessary to overcome the reaction force by water and air. Although power settings of the engine turning the propeller are kept constant, the speed of a vessel will still vary because

[1] Sir Isaac Newton (*1642 – †1726), famous English mathematician, astronomer, and physicist. The * marks the year of birth and the † marks the year of death

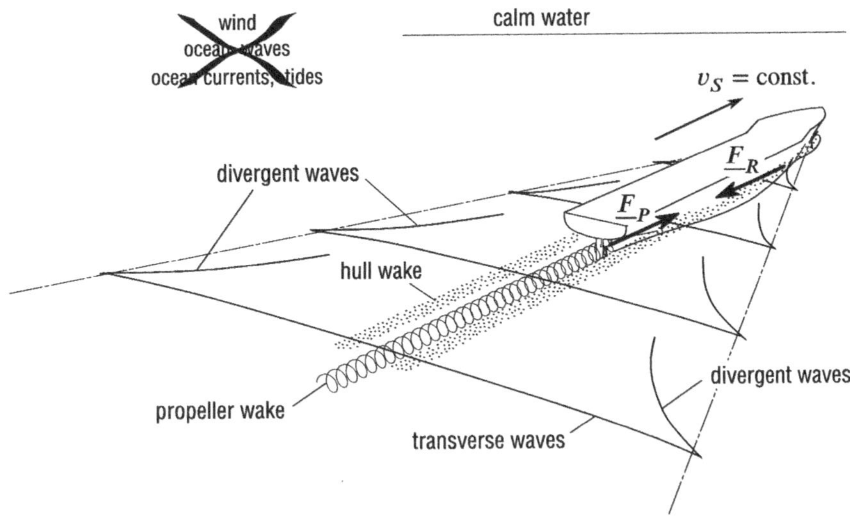

Figure 1.2 Self-propelled ship sailing in calm water with constant speed

waves and wind will alter the reaction force. In summary, the flow around a ship hull is time dependent and, as a consequence, the flow will be unsteady.

Calm water condition

In order to reduce the number of variables which influence the flow, it is often worthwhile to study phenomena separately. In this book, we will eliminate the time varying components of the flow and ignore for the most part effects of wind, waves, and currents.

- We consider only the force necessary to move the ship through air and water which are initially at rest.
- Only the waves generated by the ship itself are considered.

This is known as calm water condition (Figure 1.2).

Trial condition

A ship trial is conducted before final delivery of a new vessel. Performance is measured and compared to the contracted requirements. Trials are conducted in deep, open waters since the vicinity of the sea bottom or shore lines has a negative impact on ship performance. At the time of a trial, the hull is freshly painted and free of any marine growth. Together with the calm water condition this is referred to as trial condition.

Service condition

The contract between owner and shipbuilder typically specifies a combination of engine power and ship speed which has to be attained on the ship's trial. However, the naval architect must optimize the vessel for its intended service. A ship will encounter currents, wind, and waves during its voyages. Marine fouling over time will increase the roughness of any hull surface. These are the actual service conditions. As a result, resistance will be higher in service than at the ship's trial. Additional resistance has to be considered during selection of engine and propeller, otherwise the ship will not be able to perform as intended.

Application of calm water or trial conditions leaves us with a greatly simplified scenario depicted in Figure 1.2. The ship's velocity is assumed to be constant in direction and

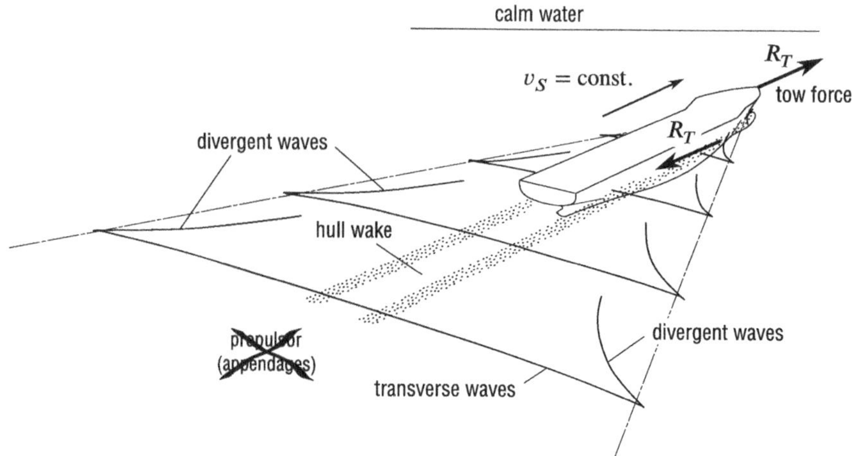

Figure 1.3 Towed bare hull (no propeller or appendages) moving in calm water

magnitude $\underline{v}_s$ = const.[2] Then all forces acting on the ship must be in equilibrium according to Newton's first law. The propulsor provides exactly the force $\underline{F}_P$ necessary to compensate the force $\underline{F}_R$ exerted by water and air on the ship.

$$\underline{F}_P + \underline{F}_R = \underline{0} \tag{1.1}$$

In ship resistance and propulsion, we are concerned with the steady forward motion. For that reason the discussion may be restricted in many cases to just the force components pointing in longitudinal direction.

Separation of hull and propeller

Simulations of the flow around a ship–propulsor system are a challenge even for today's multiprocessor computers. It is also quite difficult to make measurements in this closed system which, according to Equation (1.1), has no resultant external force. Similar problems exist in structural analysis. In order to reveal shear forces and bending moments in a beam, one side of the beam is 'removed' to reveal the internal forces. To that effect, the ship–propulsor system is split into two parts which are treated separately:

Total resistance

- Bare hull: hull without propulsor and usually without any appendages like rudder, struts, and bilge keels (Figure 1.3). We remove the propulsor, which means the bare hull has to be towed to achieve the desired speed. The required tow force is equal to the ship hull's total resistance R_T.

Open water condition

- Propulsor: the propulsor is removed from the ship and its properties are investigated in undisturbed parallel flow instead of the disturbed flow field generated by the hull (Figure 1.4). This is known as open water condition.

Advantages

Separation of the hull–propulsor system into its subsystems has advantages which are exploited in experimental studies of ship hydrodynamics:

- forces of water and air on the hull and the force generated by the propulsor are revealed,

[2] Underlined quantities, like velocity $\underline{v}$, represent vectors. See the beginning of Chapter 3 for details.

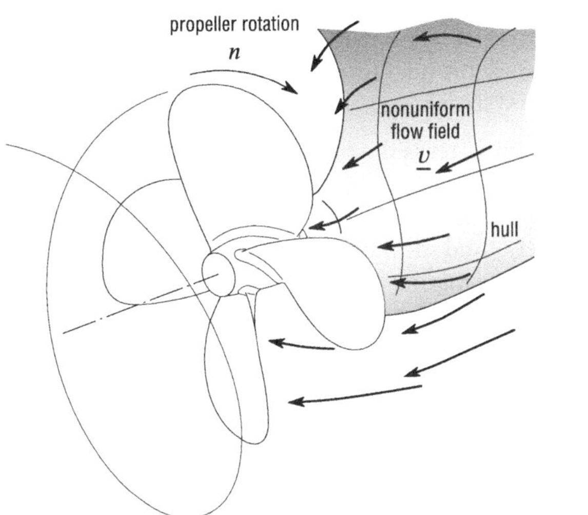

(a) Propeller working in a nonuniform flow field behind the hull (behind condition)

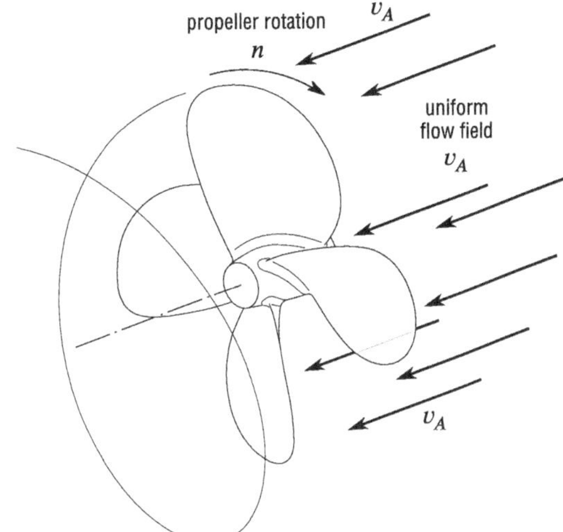

(b) Propeller working in a uniform flow (open water condition)

Figure 1.4 Comparison of inflow conditions for a propeller operating in behind and in open water condition

- hulls may be investigated without a specific propulsor, and
- propulsor performance may be determined without the influence of a ship's wake.

Disadvantages

However, separating hull and propulsor considerably changes the hydrodynamic system. As a consequence, we need to apply corrections to model test and simulation results performed for hull or propulsor alone.

- The flow around the ship hull will be different for hulls with a propulsor attached and for hulls without a propulsor. This is especially true for marine propellers. Rotating propellers accelerate fluid already upstream of the propeller. Hence, they have a direct impact on the flow around the stern of a vessel.
- Propellers, water jets, or paddle wheels do not operate without a ship attached to them. The hull will create a nonuniform flow field called wake in which the propulsor is working as indicated in Figure 1.4(a). Therefore, the open water condition with uniform inflow into the propulsor shown in Figure 1.4(b) is an unrealistic, hypothetical case.
- When performance is separately determined for hull and propulsor, the question arises as to how the results are reconciled to make a prediction for the complete hull–propeller system.

Hull-propulsor interaction

Although one may argue that it would be better to only investigate the complete hull–propulsor system, it is today's practice to perform model tests and even calculations for the separated components. Effects of the omitted part on the performance of the other part are quantified by additional hydrodynamic characteristics. They will be discussed in Chapter 34 about hull–propeller interaction. This established practice goes back many years, and naval architects have to remind themselves that careful interpretation of these characteristics is required because the quantities they describe do not exist in the original, combined ship–propulsor system.

Book topic The flow of water around the ship in calm water is the topic of this book. First, we will discuss the resistance problem, i.e. the flow around a ship hull without its propulsor and the forces associated with this flow. The resistance problem also provides the context for review of the underlying concepts of fluid mechanics. In the second half, we will study the propulsion problem. We examine marine propellers as the most widely used propulsor and finally learn how to combine the results of hull and propeller analysis to solve ship design related propulsion problems.

1.2 Ship Hydrodynamics and Ship Design

Your future job The definition of an effective hull shape is a naval architect's responsibility. Hull geometry significantly impacts the performance of a vessel and must be assessed from earliest design stages until completion of the design.

In a nutshell, the responsibilities of a naval architect charged with the design and hydrodynamic assessment of a hull include:

- developing a form with minimum powering requirements
- selection and design of an appropriate propulsor of high efficiency
- contributing information about necessary engine power and propulsor specifications to the overall ship design
- ensuring that the vessel behaves well in waves
- ensuring that the vessel is sufficiently course stable and/or maneuverable depending on its mission.

In this book, we consider the first three items related to resistance and propulsion in calm water.

Objective The main objective in this element of ship design is the development of a ship–propulsor combination which provides the most economic and ecological system to fulfill the vessel's mission. Overall size of the ship is usually defined by its purpose and target route. Ports of call, seaways, and canals may impose further limits. Therefore, changes to displacement and principal dimensions for hydrodynamic reasons will be small. The same is usually true for the design speed of merchant vessels. A vessel will become part of a transport chain sustaining a more or less well defined flow of goods. If the vessel is slower than envisioned, it becomes a bottleneck in the transport chain. If it is too fast, it will become idle awaiting the next batch of goods.

Optimization problem Within the design constraints a naval architect composes a hull shape with minimized resistance and develops an optimum propulsor for it. Even better would be to formally optimize the hull–propulsor system for overall high efficiency throughout the operational profile. A formal optimization is unfortunately often skipped for merchant vessels since it requires sophisticated numerical flow simulations and supporting model tests. Steadily increasing and more affordable computational capacities and more robust simulation methods will eventually integrate formal optimization into day-to-day ship design.

Performance prediction

For a successful ship design, it is very important to accurately predict the power necessary to achieve the desired cruising or flank speed. The machinery for a main propulsor is usually the most expensive nonmilitary equipment item. An oversized engine wastes space in the ship which could have been used for more (paying) cargo or a smaller (cheaper) ship. Operating costs tend to increase with engine size as well. If the predicted rating for an engine is larger than necessary, your design will be more expensive and your customer might order your competitor's design instead. The ship might not reach its design speed if the naval architect happens to underpredict the power requirements. In many cases there is no easy fix, and shipbuilding contracts commit the shipbuilder or design agent to paying hefty fines if the contracted speed is not achieved. In the worst case, the customer might refuse to take ownership of the ship, leaving a shipbuilder with unpaid bills and an unwanted asset.

Challenges

The essential challenge in ship hydrodynamics is to get it right the first time. Most ships are one of a kind designs with investment cost so high that a prototype cannot be built and tested before the actual vessel is constructed. Small series exist for naval vessels and smaller ships like pleasure craft and work boats. Merchant vessel series rarely show more than single digit repeats. Numbers that are typical for aircraft or car models are never reached in shipbuilding. This does not imply that engineering of cars and aircraft is easier. It just means that the economic risk of engineering failures per unit is bigger for ships. In small series, results for the lead ship are exploited in the construction of repeat designs. Again, only minor changes will occur. If the lead vessel does not perform as desired, repeats will be unlikely.

The difficulty of our fluid dynamics problem is augmented by the fact that reliable information about hull shape, resistance, and propulsion is needed early in the design. Except for minor details, the hull shape needs to be settled as soon as possible since it affects not only resistance and propulsion, but also stability, structure, and functionality of the vessel. This leaves very little time for extensive computational analyses or model tests for merchant vessels where the early design time is measured in weeks rather than months. Naval vessels tend to have longer lead times, but in their case additional and often changing mission constraints complicate the design problem.

1.3 Available Tools

Task

A moving ship creates a complex flow field in its vicinity. As naval architects we are interested in the physical properties of the flow field like fluid particle velocities, pressure distributions, and their ultimate effects on the performance of a ship. Obtaining quantitative descriptions of the flow processes around a ship hull poses a complex exterior flow problem, characterized by:

- complicated three-dimensional geometries,
- two fluids (air and water), and

- moving boundaries.

Moving boundaries Although the shape of hull and propeller are known in later design stages, their position in the water and the boundary between air and water will change when the vessel is underway. Changing water pressure along the hull causes it to deviate from its position at rest. In computational fluid dynamics the boundary between air and water is often called free surface. Its wavy shape has to be found as part of the solution.

About 250 years of research and advances in engineering sciences have developed a set of tools which naval architects may employ to make accurate predictions of resistance and propulsion properties. However, ship hydrodynamics is still an active field of research, and improvements to existing methods and new tools for old and new design problems will be developed for the foreseeable future.

Tools In order to obtain the desired data for our ship design, we can perform:

Model tests

- model tests, since we have no prototype to test at full scale:
  - Model tests are well established and arguably the most reliable performance prediction method. The trial speed at a certain power output of the engine has to be predicted within ±0.2 kn. Model tests are performed in later stages of the design.
  - We cannot satisfy all required physical scaling laws at the same time. Therefore, model test data have to be extrapolated to full scale. This can introduce significant errors if not done properly. Model testing will be discussed later in the book (Chapters 26–30 and 41–44).
  - Model tests are considered expensive. However, they are cheap compared with the overall cost of a ship. Model tests are usually done just for the final design to confirm earlier predictions. Modifications might be tested if problems arise which invalidate earlier estimates, and the contracted trial speed might not be reached.

Estimates

- estimates:
  - Estimates are only acceptable for early design phases.
  - Available methods are the result of regression analyses of model test data and ship trial data. The methods are quite simple but not always accurate (about ±10%).
  - Despite their limited accuracy, estimates are still the method of choice in early design stages. We will discuss current methods in Chapters 31, 50 and 51.

Numerical analysis

- CFD (computational fluid dynamics) simulations:
  - CFD simulations are time consuming, especially the grid generation, and actual computation may take several hours or even days to complete.
  - CFD is not very reliable yet and a lot of experience is needed to produce a valid grid and to select appropriate boundary conditions and process parameters. An expert user can produce results which are as good as a model test, but the occasional user should apply CFD results with great caution.

- Reliable results can be obtained for partial problems, e.g. wave resistance computations and related shape optimization of the forebody.
- CFD results are useful to compare and rank design alternatives.
- With the steady increase in computational speed and improvements in flow modeling, CFD will play an increasingly important role in ship resistance and propulsion predictions.

We will discuss marine model testing and practical estimates at length. CFD, although of increasing accuracy and importance, is not presented here because it requires additional background in numerical methods which does not fit the format of this book.

Self Study Problems

1. Define the calm water condition.
2. Discuss pros and cons of investigating hull and propulsor separately.
3. Explain in your own words what total resistance means in ship hydrodynamics.
4. Explain what open water condition means and where it is used.
5. State the advantages and disadvantages of the separate analyses of hull and propeller.
6. Which tools are at the naval architect's disposal for resistance estimates? Discuss their advantages and disadvantages.

2

Ship Resistance

Resistance, or drag, is a force acting against the motion of a vessel. This chapter defines resistance in the context of ship hydrodynamics and explores possible subdivisions to better understand how characteristics of the ship and the fluid contribute to the resultant force. All subdivisions of resistance provide useful insight but are also ambiguous because clear boundaries do not exist between the components. The reader should always keep in mind that the total resistance is the only force which can be directly measured.

We will discuss three different subdivisions of the resistance: first, we observe flow phenomena associated with the moving vessel and qualitatively explain how they contribute to the resistance. Next, a more pragmatic subdivision is introduced which forms the basis for meaningful model tests. Finally, we will turn to the physical aspects of how fluids interact with the hull which are exploited in computational tools.

Learning Objectives

At the end of this chapter students will be able to

- discuss the concept of calm water total resistance
- distinguish flow phenomena around a ship hull
- define components of ship resistance
- summarize how water and air interact with a vessel

2.1 Total Resistance

Total resistance

Naval architects link ship resistance or, better, total resistance R_T with three distinct conditions:

- Total resistance is the horizontal force acting on the hull. The effects of a propulsor are deliberately excluded.
- The ship is moving with constant speed on a straight course.

Fundamentals of Ship Hydrodynamics: Fluid Mechanics, Ship Resistance and Propulsion, First Edition.
Lothar Birk.

Companion website: www.wiley.com/go/birk/hydrodynamics

- The water is calm, i.e. without a current and without wind or wind generated waves. In addition, we assume the water is deep enough so that the sea bottom does not affect the waves generated by a moving vessel.

The total resistance R_T of a hull may be measured as a towing force.

Total resistance coefficient

The total resistance for displacement type vessels is usually reported as a dimensionless total resistance coefficient C_T.

$$C_T = \frac{R_T}{\frac{\rho}{2} v_S^2 S} \tag{2.1}$$

Resistance is normalized by the product of fluid density ρ, constant vessel speed v_S squared, and wetted surface S. The wetted surface S is taken for the ship at rest at the selected displacement and trim. We ignore changes in wetted surface caused by ship generated waves while the ship is underway. Typically, total resistance coefficient values range from 0.001 to 0.005.

2.2 Phenomenological Subdivision

Whether you watch a ship sail by or observe a duck paddling across a pond, you will notice similar flow phenomena. Some are obvious, others less so.

Ship waves

- All objects moving on or close to the water surface will create waves. Anyone who has sailed through a storm at sea or who has seen the destruction of coastal installations caused by high surf can attest that waves contain a lot of energy. Strong winds will build up high waves given the opportunity to act over a prolonged period. A ship sailing in calm water involuntarily spends some of its kinetic energy to generate waves. Ship waves appear in a distinct pattern and their creation is associated with the wave resistance. The higher the waves are, the higher is the wave resistance.

Eddies

- All images of a moving vessel show a disturbed flow region along the hull and in its wake. Most visible is a band of white froth next to and behind the vessel which mostly stems from the breaking of the bow wave. Underneath the froth, large and small eddies form in the flow. Like the generation of ship waves, creation of eddies siphons kinetic energy away from the ship. In other words, the hull is performing additional work per unit time. For steady, straight motion the work per unit time or power is equal to the product of speed and force. Since the speed is constant and work increases due to the energy transferred into the eddies, resistance will increase as well. The eddies are caused by friction in the fluid and friction between hull and fluid. The eddy resistance is considered part of the broader category of viscous resistance. The kinetic energy stored in the eddies is eventually converted into other forms of energy and the eddies dissipate.

Boundary layer

- Less obvious are the effects of friction between hull and water. The associated wall shear stress forms the frictional resistance which is typically the biggest contribution to the total resistance. Friction between hull surface and fluid and within the fluid is significant only in the boundary layer. A boundary layer is a thin sheet of flow around material surfaces in which viscosity of the water plays a dominant role. Frictional resistance is typically the largest part of the resistance.

- Like a submerged hull moves through water, everything above the waterline moves through air. As a result, we have additional frictional and eddy resistance components summarized as air resistance. It is, however, much smaller than the resistance of the submerged part because the density of air is 800 times smaller than that of water. Note that the air, like the water, is assumed to be initially at rest.

Additional resistance components

Additional resistance components may be associated with flow phenomena that occur only in special cases. Some examples are:

- Spray resistance: fast boats and ships create a spray pattern at the bow. Water is forced up and to the sides, which results in an additional resistance component.
- Induced resistance: lifting surfaces like stabilizer fins and rudders generate flow patterns with distinct vortices. Vortices create flow patterns with an asymmetric pressure distribution. The main resultant force acts perpendicular to the onflow (lift force). However, lift is always accompanied by a force component which acts against the motion. We call this component induced resistance.

Some components of the resistance are associated with specific parts of the ship hull. Examples are

- transom resistance,
- bow thruster tunnel resistance, and
- appendage resistance.

In subsequent chapters we will study the major flow phenomena more closely and explore how to estimate their contribution to the total resistance.

2.3 Practical Subdivision

The problem of scale

As early as the 17th century, scientists investigated the resistance of ships by towing scale models across ponds (Calero, 2008). In 1794, the Swedish admiral and shipbuilder Fredrik Henrik af Chapman (*1721 – †1808) even built a towing tank and tested systematically varied geometries and model ship hulls (Harris, 1989). However, the results of all these efforts were inconclusive because two important questions had not yet been answered.

(i) When a scale model is towed at a speed v_M, what is the corresponding speed v_S of the full scale vessel?

(ii) How do you convert the measured model resistance R_{TM} into the resistance R_{TS} of the full scale vessel?

William Froude (*1810 – †1879), British engineer and naval architect, was the first to postulate and successfully apply answers to these fundamental questions (Froude, 1868). We will explain his ideas in due course.

Length scale

Models of ship hulls are build at a fixed geometric scale

$$\lambda = \frac{L_S}{L_M} \tag{2.2}$$

The subscript $_S$ indicates a characteristic of the full scale vessel and the subscript $_M$ stands for the model. All linear dimensions L_S of the real vessel are shrunk by the factor λ to construct the model. Consequently, surfaces S and volumes V of full scale vessel and model are connected via the following relationships:

$$\lambda^2 = \frac{S_S}{S_M} \qquad \text{and} \qquad \lambda^3 = \frac{V_S}{V_M}. \tag{2.3}$$

Time scale

A corresponding relationship for velocities requires an additional constant time scale τ.

$$\tau = \frac{T_S}{T_M} \tag{2.4}$$

T_S is the duration of a process at full scale and T_M the equivalent time at model scale. The average velocity is defined by the ratio of the distance traveled L and the time T spent to cover the distance L.

$$\frac{v_S}{v_M} = \frac{\frac{L_S}{T_S}}{\frac{L_M}{T_M}} = \frac{\frac{L_S}{L_M}}{\frac{T_S}{T_M}} = \frac{\lambda}{\tau} \tag{2.5}$$

Given a geometric scale λ, what is the appropriate time scale τ? Can τ be chosen independently of λ? We will answer these questions shortly.

Force scale

In essence, a scale κ is wanted for the forces F acting on ship and model.

$$\kappa = \frac{F_S}{F_M} \tag{2.6}$$

The weight force of a model of geometric scale λ, for example, may be derived by applying Archimedes' principle:[1] a floating body displaces as much water as it weighs. Therefore, the scale κ_w of the weight forces is

$$\kappa_w = \frac{W_S}{W_M} = \frac{\rho_S\, g_S\, V_S}{\rho_M\, g_M\, V_M} = \frac{\rho_S\, g_S}{\rho_M\, g_M}\lambda^3 \tag{2.7}$$

If we conduct model tests on Earth, the gravitational acceleration should be the same for full scale vessel and model: $g_S = g_M = g$. Density of salt water and fresh water differ by approximately 2.6%. Ignoring this for the moment, you can state that the weight forces scale like volumes with the scale cubed.

$$\kappa_w \approx \lambda^3 \tag{2.8}$$

Unfortunately, this simple force scale does not apply to a ship's resistance.

[1] Named after the Greek scientist and engineer Archimedes (*287 BC – †212 BC), who formulated it first.

2.3.1 Froude's hypothesis

Corresponding speeds

Froude answered the first question (i) with the proposal of 'corresponding speeds:'

$$\frac{v_S}{v_M} = \sqrt{\lambda} \tag{2.9}$$

Comparing this with Equation (2.5), we see that this is equivalent to a time scale of $\tau = \sqrt{\lambda}$. This relationship had previously been presented by the French naval engineer Ferdinand Reech (*1805 –†1884). However, it had never been applied to model tests. Today, the relationship of corresponding speeds is called Froude's law of similarity in honor of Froude's extraordinary contributions to ship model testing. In order to make sense of model tests, we enforce that the selected model speed satisfies the requirement of an equal Froude number.

$$Fr = \frac{v}{\sqrt{g\,L_{WL}}} = Fr_S = Fr_M \tag{2.10}$$

L_{WL} is the length of vessel or model at the waterline. Assuming that the gravitational acceleration g is the same for model and ship, we may retrieve Froude's expression for the corresponding speeds.

$$\begin{aligned} Fr_S &= Fr_M \\ \frac{v_S}{\sqrt{g\,L_{WL_S}}} &= \frac{v_M}{\sqrt{g\,L_{WL_M}}} \\ \frac{v_S}{v_M} &= \frac{\sqrt{g\,L_{WL_S}}}{\sqrt{g\,L_{WL_M}}} = \sqrt{\frac{L_{WL_S}}{L_{WL_M}}} = \sqrt{\lambda} \end{aligned} \tag{2.11}$$

A formal justification for this particular time scale is presented in Chapter 28.

With respect to the force scale κ, Froude realized that a single scale is insufficient. The wave resistance part requires a different scale factor than the viscous resistance. This is due to the fact that we cannot attain full dynamic similarity of the fluid flow around the model if we test the model in water. In order to achieve full dynamic similarity, the water would have to be scaled somehow like the model itself. This is impossible and, in fact, a fluid which has an appropriately scaled kinematic viscosity does not exist.

Froude's hypothesis

In lieu of a fluid with the correct, scaled viscosity, Froude proposed to split the total resistance into two parts:

(i) frictional resistance R_F of a flat plate with the same wetted surface S as the ship hull, and

(ii) residuary resistance R_R, which mainly comprises the wave resistance.

The total resistance is the sum of these two parts:

$$R_T(Re, Fr) = R_F(Re) + R_R(Fr) \tag{2.12}$$

This is known as Froude's hypothesis or Froude's method. The hypothesis includes two assumptions:

- The frictional resistance R_F is a function of the Reynolds number Re. The Reynolds number Re for a ship is defined as the dimensionless ratio of ship speed v_S times length over wetted surface L_{OS} divided by the kinematic viscosity of the fluid ν.

$$Re = \frac{v_S L_{OS}}{\nu} \qquad (2.13)$$

It is named in honor of British scientist Osborne Reynolds (*1842 – †1912), who investigated the phenomenon of laminar and turbulent flows.

- The residuary resistance R_R is solely a function of the Froude number Fr. Froude did not call it the Froude number, though. He called it corresponding speed.

Froude was well aware that his hypothesis is not entirely true. The residuary resistance still includes a part of the viscous resistance which is dependent on the Reynolds number. However, Froude's hypothesis proved workable and laid the path to meaningful model testing.

Froude's method

Froude (1872) conducted model tests with flat plates of varying length and surface roughness. From the results he developed formulas to compute the frictional resistance component R_F. The formulas could be cast into a form of

$$R_F = c\, v_S^m\, S \qquad \text{(no longer in use)} \qquad (2.14)$$

The exponent m of the ship velocity v_S changed with the length of the plates but seemed to level out at $m = 1.83$ for long plates.

Knowing the frictional resistance component R_F enabled William Froude to separate the residuary resistance from the measured total resistance of the model.

$$R_R(Fr) = R_{TM}(Re, Fr) - R_{FM}(Re) \qquad \text{for the model} \qquad (2.15)$$

The dimensionless residuary resistance coefficient

$$C_R = \frac{R_R}{\frac{\rho}{2}\, v_M^2\, S_M} = C_{TM} - C_{FM} \qquad (2.16)$$

is applicable to both model and full scale vessel at the Froude number of the model test.

$$R_R(Fr) = \frac{\rho}{2}\, v_S^2\, S_S\, C_R \qquad \text{for the full scale vessel} \qquad (2.17)$$

The total resistance of the full scale vessel is obtained by adding a computed frictional resistance R_F and small corrections which will be discussed in Chapter 30.

2.3.2 ITTC's method

ITTC 1957 model–ship correlation line

Froude's initial procedure has been continuously refined and improved over the past 150 years. The International Towing Tank Conference (ITTC) publishes recommended procedures for model testing and CFD calculations on its website at http://www.ittc.

info. In its current set of recommended procedures, ITTC uses the 1957 model–ship correlation line as friction coefficient C_F.

$$C_F = \frac{0.075}{\left[\log_{10}(Re) - 2\right]^2} \tag{2.18}$$

Although not completely physical, this formula is widely used today. Chapter 13 discusses this topic in more detail. Equation (2.18) applies only to smooth surfaces, like the surface of a ship model which has been sanded with very fine grit sandpaper. For full scale ship hulls, a correction is added which depends on the roughness of the hull surface. The dimensional frictional resistance is computed in the standard way:

$$R_F = \frac{\rho}{2} v_S^2 \, S \, C_F \tag{2.19}$$

Form factor

Hughes (1954) proposed using a form factor k to augment the flat plate frictional resistance. ITTC adopted this approach in an attempt to remove the viscous components from Froude's residuary resistance. The total resistance is now split into the viscous resistance

$$R_V = (1 + k) R_F \tag{2.20}$$

and the remaining wave resistance R_W.

$$R_T = (1 + k) R_F + R_W \tag{2.21}$$

The form factor k of the model is also applied to the full scale vessel. Determining the form factor by experiments is quite challenging because it involves resistance measurements at low speeds for which the forces are small and are subject to considerable uncertainty.

ITTC's residuary resistance coefficient

ITTC held on to the term residuary resistance until recently. Only in its 2017 revision of the recommended procedures was the name changed from residuary resistance to wave resistance. Therefore, you will find many publications which state Equation (2.21) with R_R instead of R_W.

Wave resistant coefficient

In ITTC's recommended procedure, the wave resistance coefficient C_W is obtained from model tests by subtracting the viscous resistance from the total resistance (ITTC, 2017b).

$$C_W = C_{TM} - (1 + k) C_{FM} \tag{2.22}$$

Full scale total resistance coefficient

The full scale total resistance is obtained by adding an augmented frictional resistance and corrections to the wave resistance (ITTC, 2017a). In dimensionless coefficients this reads as

$$C_{TS} = (1 + k) C_{FS} + C_W + \text{corrections} \tag{2.23}$$

Chapters 29 and 30 discuss the ITTC method in detail.

Although wave resistance is the major component of Froude's residuary resistance, it is important to realize that the wave resistance coefficient C_W obtained by ITTC's method (2.22) is substantially different from the residuary resistance coefficient C_R resulting from Froude's method (2.16). Figure 2.1 illustrates the effect of the form factor on the size of the remaining resistance components C_R and C_W, respectively. For the same total resistance, the wave resistance coefficient is considerably smaller than the residuary resistance coefficient. As a consequence, great care should be taken not to mix the two methods.

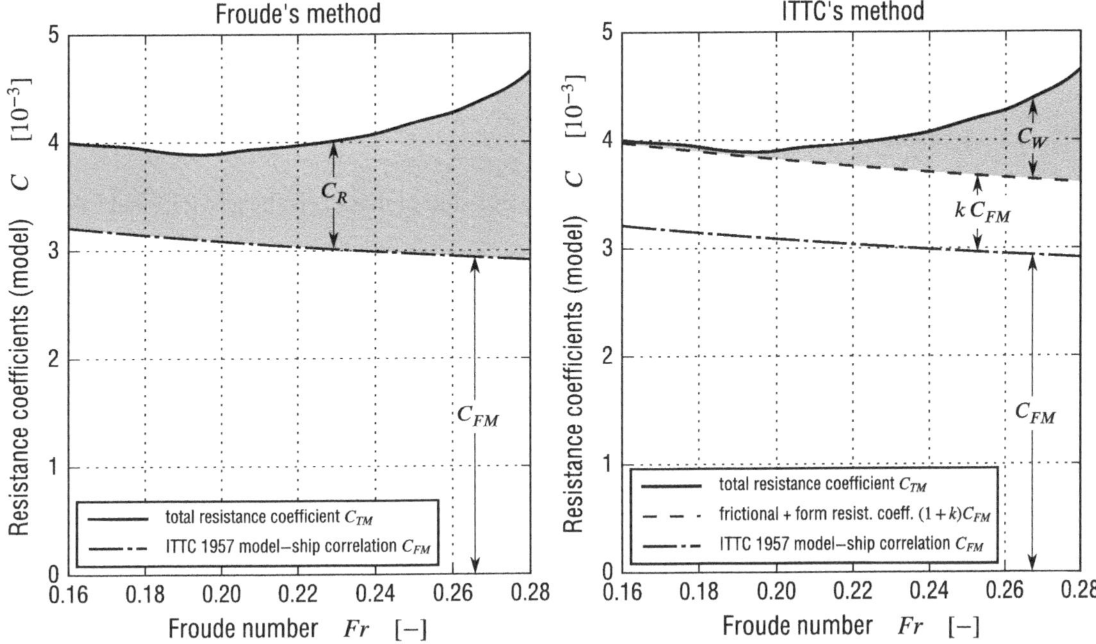

Figure 2.1 Comparison between Froude's and ITTC's current method of derivation for the residuary resistance coefficient C_R and wave resistance coefficient C_W

2.4 Physical Subdivision

Phenomenological and practical subdivision of resistance do not lend themselves to numerical evaluations of the resistance. For this purpose we need a subdivision which closely follows the physics of fluid mechanics and the mathematical models used to capture the kinematic and dynamic relationships of flow properties.

In general, three generic types of forces act on any moving body:

Forces acting on a ship

- inertia forces,
- body forces, and
- surface forces.

Inertia is the property of all bodies to maintain their state of motion. The inertia force acts against the motion and is present when the body accelerates or decelerates. Newton's second law states that the inertia force is in equilibrium with the resultant of all external forces. External forces interact with the vessel as body forces or surface forces. Body forces act on a vessel as a whole and are present in force fields like gravity or magnetism. Surface forces act on the surface of the body, usually via pressure and shear stress.

A body moving with constant speed and without change in direction experiences no acceleration. Consequently, inertia forces will be zero and the resultant of all external forces has to vanish as well. We consider our ship in this type of steady motion, i.e. sailing with constant speed on a straight course.

2.4.1 Body forces

Body forces

The only notable body force acting on our ship of mass Δ is the gravitational force.

$$\underline{W} = \begin{pmatrix} 0 \\ 0 \\ -g\,\Delta \end{pmatrix} \tag{2.24}$$

Our Cartesian coordinate system $< x, y, z >$ has its positive x-axis pointing forward (in the direction of motion), the y-axis pointing to port and the z-axis pointing upwards. The gravitational force or weight (2.24) points in negative z-direction. Therefore, it acts perpendicular to the resistance force (x-direction).

According to Archimedes' principle, the weight of a vessel is equal to the weight of the displaced water at rest.

$$g\,\Delta = \rho\, g\, V \tag{2.25}$$

Δ is the mass of the vessel and V is its volumetric displacement. For displacement type vessels, weight is put into equilibrium by the resultant hydrostatic pressure force. However, the total pressure distribution changes when a vessel is in motion, causing small changes in vertical position, called dynamic sinkage, and trim.

2.4.2 Surface forces

Surface forces

Fluids interact with a ship hull in two ways:

- normal stress σ, acting normal to the surface, and
- shear stress τ, acting tangential to the surface.

The three-dimensional stress in a fluid is described by a stress tensor with three normal stresses and six shear stresses.

$$\text{Stress tensor:}\quad \underline{\underline{\hat{\sigma}}} = \begin{pmatrix} \hat{\sigma}_{xx} & \tau_{xy} & \tau_{xz} \\ \tau_{yx} & \hat{\sigma}_{yy} & \tau_{yz} \\ \tau_{zx} & \tau_{zy} & \hat{\sigma}_{zz} \end{pmatrix} \tag{2.26}$$

The tensor is symmetric, i.e. $\tau_{xy} = \tau_{yx}$, $\tau_{xz} = \tau_{zx}$, and $\tau_{yz} = \tau_{zy}$. In total, we have to find six stresses to completely describe the stress at a point in the fluid.

In fluid mechanics, pressure p represents the mean normal stress:

$$p = \frac{1}{3}\left(\hat{\sigma}_{xx} + \hat{\sigma}_{yy} + \hat{\sigma}_{zz}\right) \tag{2.27}$$

Pressure acts equally in all directions. Separating pressure from the normal stresses allows us to use the same basic equations for a fluid at rest and in motion.

$$\underline{\underline{\hat{\sigma}}} = -p + \underline{\underline{\sigma}}$$

The remaining stress tensor $\underline{\underline{\sigma}}$ is dependent on the viscosity of the fluid.

Planing boats

In contrast to displacement type vessels, planing boat hulls develop an upwards oriented pressure force at higher speeds, which carries part of their weight. The corresponding reduction in displacement and resistance allows them to reach much higher speeds than displacement type vessels.

Ship resistance

The total surface force $\underline{F}_S$ acting on a hull is obtained by the integration of pressure and stresses over the actual wetted surface S_{actual} of the ship when it is underway.

$$\underline{F}_S = \iint_{S_{\text{actual}}} p\,\underline{n}\,\mathrm{d}S + \iint_{S_{\text{actual}}} \underline{\underline{\sigma}}\,\underline{n}\,\mathrm{d}S \tag{2.28}$$

The x-component of $\underline{F}_S$ is the force commonly referred to as resistance.

Wave resistance

The first integral captures pressure changes caused by the formation of a ship wave pattern and is associated with the wave resistance.

D'Alembert's paradox

Wave resistance may be eliminated by deeply submerging a body. However, a submerged moving body will still experience a resistance force because water is a viscous fluid. The viscosity of water is, fortunately, very small. In fact, it is so small that scientists in the 19th century widely assumed that frictional forces must be negligible. The French scientist Jean-Baptiste le Rond d'Alembert (*1717 – †1783) attempted to compute the resistance of a submerged body assuming the fluid is frictionless (inviscid). As it turns out, the resultant pressure force vanishes for inviscid fluids. This clearly does not match up to experimental observations and is therefore known as d'Alembert's paradox. The question of how a fluid of such low viscosity can produce significant frictional forces is partially explained by boundary layer theory. We will discuss viscous and inviscid flow theories in greater detail in later chapters.

Propulsion

Pressure forces are also the main source of propeller thrust, which is necessary to overcome the resistance.

Prandtl's boundary layer theory

The second integral in Equation (2.28) considers the effects of the boundary layer and is associated with the viscous resistance. The boundary layer concept was introduced by German engineer Ludwig Prandtl (*1875 – †1953) in 1904. For now it may suffice that the boundary layer is a region of flow close to material surfaces in which shear stresses within the fluid and between the fluid and the hull surface play an important role.

Figure 2.2 illustrates the flow regions. Boundary layers cover all surfaces and grow in thickness from bow to stern. The thickness is exaggerated in Figure 2.2. Typically, boundary layers at the stern are roughly 1% of the ship length thick.

Friction between fluid and hull

As the enlarged section in Figure 2.2 shows, fluid particles are actually sticking to material surfaces like the hull. The velocity grows rapidly with distance from the hull until it reaches the exterior flow speed. Boundary layer thickness is commonly defined as the distance from the surface to where the local velocity has reached the exterior flow speed. Figure 2.3 shows the result of a paint flow test. A custom made paint is put in stripes on the dry hull of a ship model. The paint stripes roughly follow the ship's stations. While the paint is still wet, the model is put back into the water and towed at the speed of interest. The water rushing over the hull surface smears the paint in the direction of the flow, thus revealing the flow patterns across the hull.

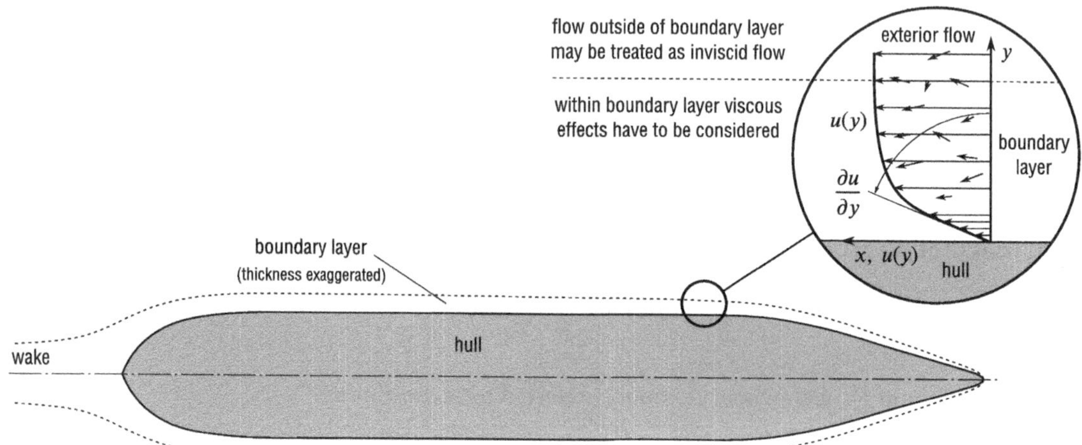

Figure 2.2 Viscosity of the fluid has significant effect on the flow within the boundary layer around a ship hull

Frictional resistance Paint flow tests are very useful for the proper placement of appendages like bilge keels, struts, and gratings covering thruster tunnels. In our discussion of resistance components, the paint flow test is of interest because it confirms what we know from experience: friction occurs between fluid and the hull surface. The associated shear stress drags the paint over the surface. The shear stress results in frictional resistance, which is the largest resistance component for slow ships.

Viscous pressure resistance The boundary layer also has an effect on the pressure distribution around the hull. Compared with inviscid flow patterns, pressure remains lower at the downstream side (stern) of vessels when a boundary layer is present. The combination of high pressure at the bow and lower pressure at the stern creates the viscous pressure resistance.

2.5 Major Resistance Components

Major resistance components Three resistance components stand out from our discussion of possible subdivisions of the ship resistance:

- frictional resistance (R_F, C_F) caused by the shear stress between fluid and hull,
- viscous pressure resistance (R_{VP}, C_{VP}), which reflects the changes in pressure over a hull due to viscous flow effects, and
- wave resistance (R_W, C_W) associated with the generation of the ship wave system.

Superposition The total resistance is obtained by summing up the components:

$$R_T = R_F + R_{VP} + R_W \tag{2.29}$$

The sum of frictional and viscous pressure resistance is the viscous resistance.

$$R_V = R_F + R_{VP} \qquad \text{(viscous resistance, smooth)} \tag{2.30}$$

(a) Entrance

(b) Midbody

(c) Run

Figure 2.3 Results of a paint flow test. Photos courtesy of Dr. Alfred Kracht, Versuchsanstalt für Wasserbau und Schiffbau (VWS), Berlin, Germany

Summations like Equations (2.29) and (2.30) imply that the summands are independent of each other. For resistance components this is generally not the case. We can neither measure the resistance components independently nor compute them all separately from first principles. Equation (2.29) is not practically used because viscous pressure resistance may not, if at all, be determined accurately enough.

Resistance curve

Figure 2.4 shows a typical total resistance curve for a container ship with common subdivisions in dimensionless and dimensional form as a function of Froude number Fr. For low speeds ($Fr < 0.2$), the total resistance coefficient C_{TS} slightly declines with

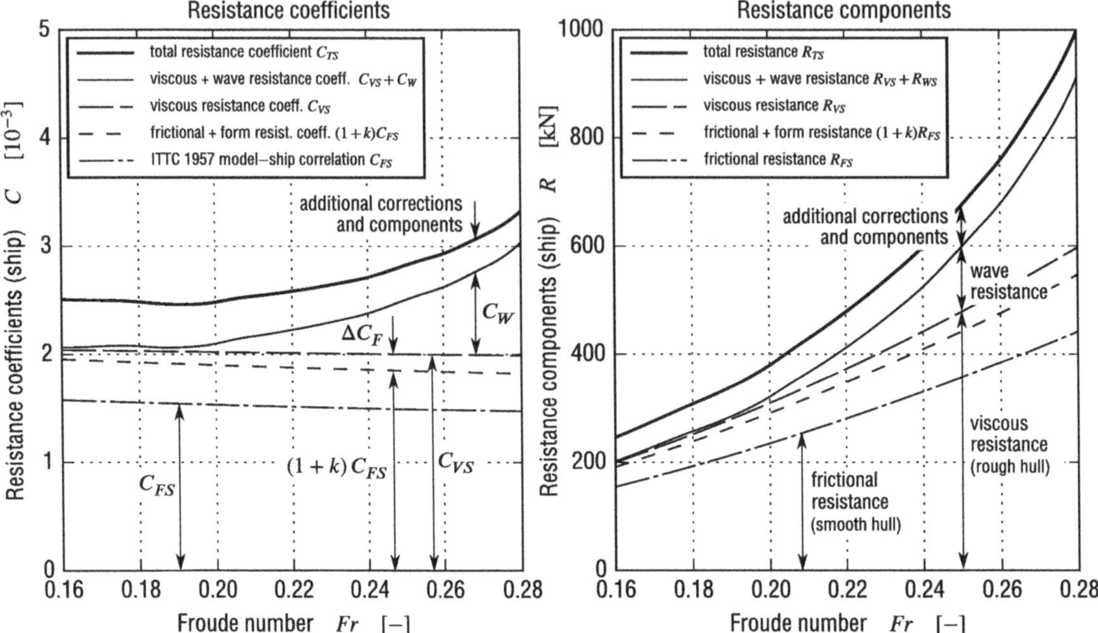

Figure 2.4 Resistance coefficients and resistance for a container ship as functions of the Froude number (velocity)

increasing Froude number. The resistance R_{TS} is still increasing, however, at a rate somewhat lower than velocity squared v_S^2. For higher Froude numbers, the slope of the total resistance coefficient curve grows continuously, indicating that the resistance increases at a rate higher than v_S^2.

Humps and hollows The total resistance curves (C_{TS}, R_{TS}) show minor humps and hollows in contrast to the frictional resistance curves (C_{FS}, R_{FS}) which are smooth. Hollows, with relatively low resistance, are achieved when the waves generated by the ship hull superimpose into a wave pattern with low wave heights due to wave cancellation. The humps represent regions of unfavorable wave superposition with high wave heights. Humps tend to appear around Froude numbers 0.23, 0.3, and 0.48, whereas favorable hollows appear near 0.21, 0.25, and 0.34. The details will depend on the actual hull shape. We will discuss this in greater detail in Chapter 25 on ship wave resistance.

The lines for frictional resistance (C_{FS}, R_{FS}), viscous resistance (C_{VS}, R_{VS}), and wave resistance (C_W, R_W) show that at typical speeds frictional resistance is the largest resistance component. For low Froude numbers ($Fr < 0.12$), total resistance is almost equal to the viscous resistance. Wave resistance is negligible at low speeds, but, with increasing velocity, wave resistance grows faster than viscous resistance. At Froude numbers above 0.35, wave resistance will exceed frictional resistance for most vessels.

Additional resistance components Other resistance components indicated in Figure 2.4 include:

- Roughness allowance ΔC_F: the frictional resistance is increased if the surface is not hydraulically smooth, i.e. the surface is rough. This applies to the full scale ship only (see Chapter 13).

- Correlation allowance C_A: for values derived from model tests, a correlation allowance may be included, a catch-all for scaling effects not considered elsewhere in the performance prediction procedure (see Chapter 30).

- Appendage resistance C_{APP}: additional viscous or induced resistance from appendages (rudder, struts, bossings, etc.) not yet considered in the viscous resistance.

Speed is relative

In Figure 2.4 the speed of the vessel is represented by the Froude number Fr. This choice is deliberate. Although the curves will not change if the actual vessel speed is substituted for the Froude number, it is important to note that vessel speed is a relative quantity. For a short vessel like a trawler, a speed of $v_S = 10$ kn is fast, whereas for an 18 000 TEU[2] container ship of 400 m length, $v_S = 10$ kn is considered slow steaming. The Froude number, rather than the absolute speed, has a major impact on the subdivision of ship resistance.

Relative size of resistance components

Figure 2.5 shows the subdivisions of resistance for three different types of displacement vessels. The data have been taken from the excellent reference by Larsson and Raven (2010). The decompositions of the actual total resistance coefficients on the left (Figure 2.5(a)) reflect the same trends shown in Figure 2.4. Although the fishing vessel is the slowest of the three in absolute velocity, it is sailing at the highest Froude number. Because of the high Froude number, wave resistance has become the largest component of its resistance. The high total resistance coefficient of $C_T = 8.1 \cdot 10^{-3}$ for the fishing vessel indicates that it will need a larger engine relative to vessel size than a container ship or a tanker. However, absolute value of resistance and installed engine power will still be the smallest of the three.

The tanker sails at the lowest Froude number, and, as a consequence, its viscous resistance makes up 92.5% of the total resistance (Figure 2.5(b)). Tankers usually have high block coefficients $C_B \approx 0.8$. Note that the wave breaking portion of the wave resistance is twice as large as the wave pattern resistance. Blunt bows create a high bow wave that dissipates a significant amount of energy by wave breaking. The fine lines of the container vessel create a lower bow wave and most of the wave resistance is associated with the energy contained in the wave pattern trailing the ship. Slender vessels also tend to have less viscous pressure resistance.

Only total resistance can be measured directly

As mentioned, subdivisions of the resistance provide insight into how vessel speed and form contribute to the overall resistance. Naval architects have to be mindful of the contributions to the resistance so that design efforts can be properly prioritized. However, one should also keep in mind that all subdivisions of resistance are based on assumptions and that boundaries among components are blurry. The influence of a resistance component on other components is usually ignored. Again, only the total resistance can be measured directly.

References

Calero, J. (2008). *The genesis of fluid mechanics 1640–1780*. Springer, Dordrecht, The Netherlands.

[2]TEU stands for Twenty-foot Equivalent Unit, i.e. a 20 ft long container

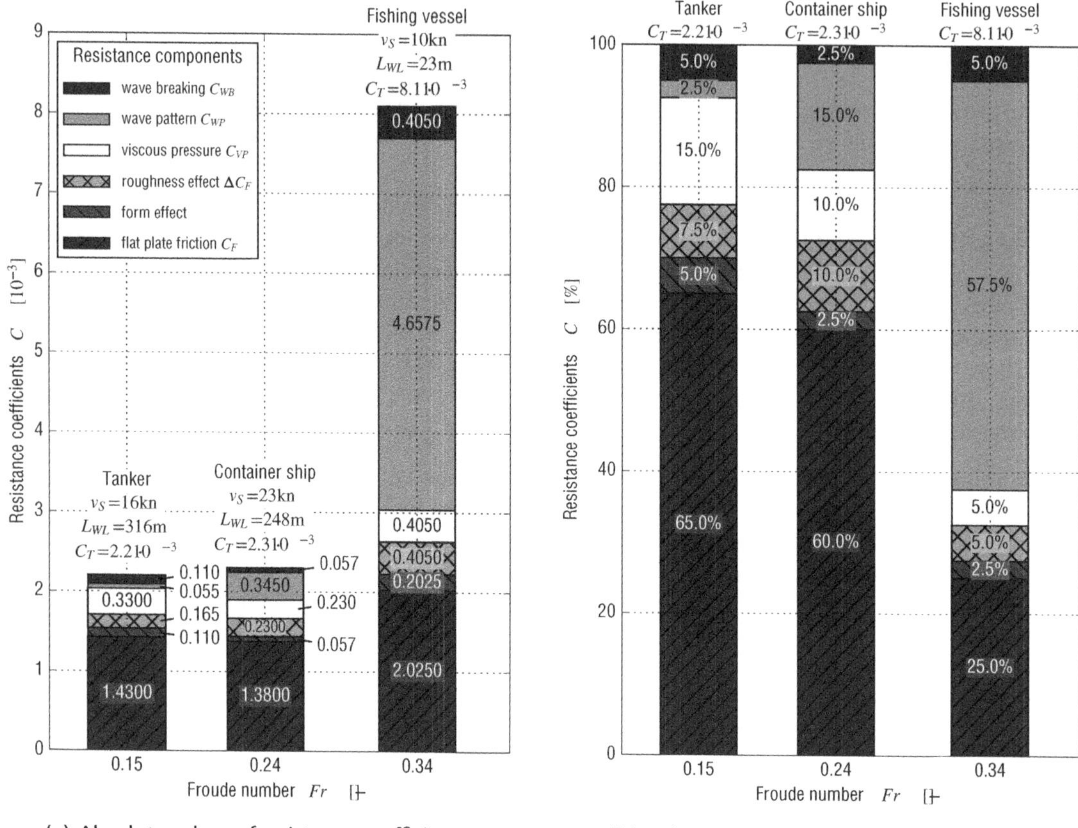

(a) Absolute values of resistance coefficients

(b) Relative values of resistance coefficients

Figure 2.5 Comparison of absolute and relative size of resistance components for three different displacement type vessels at design speed. The data are taken from Larsson and Raven (2010, pages 13,14)

Froude, W. (1868). Observations and suggestions on the subject of determining by experiment the resistance of ships. Correspondence with the British Admiralty. Published in *The Papers of William Froude*, A.D. Duckworth, The Institution of Naval Architects, London, United Kingdom, pp. 120–128, 1955.

Froude, W. (1872). Experiments on surface-friction experienced by a plane moving through water. Read before the British Association for the Advancement of Science at Brighton. Published in *The Papers of William Froude*, A.D. Duckworth, The Institution of Naval Architects, London, United Kingdom, pp. 138–147, 1955.

Harris, D. (1989). *FH Chapman – The first naval architect and his work.* Naval Institute Press, Annapolis, Maryland.

Hughes, G. (1954). Friction and form resistance in turbulent flow, and a proposed formulation for use in model and ship correlation. *RINA Transactions and Annual Report*, 96(4):314–376. doi:10.3940/rina.trans.1954.16.

ITTC (2017a). *1978 ITTC performance prediction method.* International Towing Tank Conference, Recommended Procedures and Guidelines 7.5-02-03-01.4. Revision 4.

ITTC (2017b). *Resistance test.* International Towing Tank Conference, Recommended Procedures and Guidelines 7.5-02-02-01. Revision 4.

Larsson, L. and Raven, H. (2010). *Ship resistance and flow.* The Principles of Naval Architecture Series. Society of Naval Architects and Marine Engineers (SNAME), Jersey City, New Jersey.

Self Study Problems

1. The next time you come near a pond, observe the wave systems created by water fowl moving on the water surface.

2. Throw two pebbles into a pond so that they hit the water a few feet apart. Observe the radiating waves. Watch when the two wave systems intersect and what happens to the wave amplitude as the waves spread.

3. A ship model is built at scale $\lambda = 30$. If the full scale vessel has the following principal dimensions, what are the corresponding values for the model?

full scale vessel		model
dimension	**value**	**value**
length between perpendiculars L_{PP}	179.3 m	
length in water line L_{WL}	182.2 m	
length over wetted surface L_{OS}	186.1 m	
molded beam B	29.2 m	
molded draft T	11.2 m	
wetted surface S	6627.2 m^2	
displacement V	36355.7 m^3	
block coefficient C_B	0.62	

4. If the speed of the vessel in the previous problem is $v_S = 18$ kn, what are its Froude and Reynolds numbers? Assuming the same Froude number, what would be the tow speed of the model at scale $\lambda = 30$?

5. Use the ITTC 1957 model–ship correlation line Equation (2.18) and the vessel particulars from Problem 3 to compute the frictional resistance for the vessel. Assume the vessel sails at Froude number $Fr = 0.231$.

6. Which component(s) of the resistance is (are) captured by the form factor k?

7. What is the difference between Froude's residuary resistance and ITTC's wave resistance?

8. Explain to a friend the three major resistance components.

3

Fluid and Flow Properties

Ship hydrodynamics is concerned with the flow of water and air around hull and superstructure. In this chapter we formulate basic concepts about fluids and summarize those physical properties important to ship hydrodynamics. Fluid mechanics employs different methods of observation and mathematical formulation to find the physical properties in a fluid flow: velocity, pressure, density, and temperature. The difference between Lagrange's and Euler's formulation of fluid motion is explained, and a brief review of hydrostatics reinforces the concepts of pressure and pressure forces.

Learning Objectives

At the end of this chapter students will be able to:

- explain the difference between solids and fluids
- know basic properties of water and air
- distinguish between Lagrangian and Eulerian mechanics
- understand and apply the concept of fluid pressure

3.1 A Word on Notation

Vectors

In this book, equations are represented in vector form wherever possible. Whether a quantity is a scalar or a vector should be obvious from the notation and not just from the context. In many textbooks vectors are represented by symbols in bold print or little arrows stacked on top of the symbol. I find both methods impractical. Bold symbols cannot be reproduced on a blackboard and are sometimes hard to distinguish from normal weighted letters. Arrows simply take too much time. For convenience, I adopted the notation I use in my handwritten notes. Vectors are identified by a single underscore, e.g. a velocity vector in space is

$$\underline{v} = \begin{pmatrix} u \\ v \\ w \end{pmatrix} \tag{3.1}$$

Fundamentals of Ship Hydrodynamics: Fluid Mechanics, Ship Resistance and Propulsion, First Edition.
Lothar Birk.

Companion website: www.wiley.com/go/birk/hydrodynamics

Note that the vector $\underline{v}$ stands for the velocity, which has a magnitude and a direction, whereas the scalar v denotes the transverse velocity component or in some cases the magnitude of velocity. The vector components u, v, and w indicate how much a basis vector contributes to the total. For the Cartesian coordinate system we use the basis vectors

$$\underline{i} = \begin{pmatrix} 1 \\ 0 \\ 0 \end{pmatrix} \qquad \underline{j} = \begin{pmatrix} 0 \\ 1 \\ 0 \end{pmatrix} \qquad \underline{k} = \begin{pmatrix} 0 \\ 0 \\ 1 \end{pmatrix} \tag{3.2}$$

The basis vectors are of unit length and are orthogonal to each other. Obviously, the velocity vector is equal to $\underline{v} = u\underline{i} + v\underline{j} + w\underline{k}$. The magnitude or length of a vector is defined as the square root of the sum of its squared components. For example, the magnitude of a velocity vector is

$$|\underline{v}| = \sqrt{u^2 + v^2 + w^2} \tag{3.3}$$

Matrices

Matrices are marked by double underscores as an extension of the vector notation.

$$\underline{\underline{A}} = \begin{pmatrix} a_{11} & a_{12} & a_{13} \\ a_{21} & a_{22} & a_{23} \\ a_{31} & a_{32} & a_{33} \end{pmatrix} \tag{3.4}$$

Equation (3.4) shows a square matrix $\underline{\underline{A}}$ with dimension $[3 \times 3]$, i.e. with three rows and three columns. A vector is equivalent to a matrix with a single column.

Transposition

The superscript operator ^{T} transposes a column vector into a row vector. Appropriate use of transposition ^{T} allows us to employ the rules of matrix multiplication for the dot product.

Dot product

The dot product of two vectors is also called the inner product. It results in a scalar and is one of the most important vector operations. You will encounter it frequently in this book. A dot product is executed like a matrix multiplication:

1. Two matrices of size $[m \times n]$ and $[n \times l]$ can be multiplied only if the number of columns of the first matrix n is equal to the number of rows n of the second matrix. The resulting matrix has as many rows as the first matrix and as many columns as the second matrix, i.e. $[m \times n] \cdot [n \times l] \rightarrow [m \times l]$. For the dot product of two three-dimensional vectors, we multiply a row matrix of size $[1 \times 3]$ (hence the transposition of the column vector) with a $[3 \times 1]$ matrix. The result is a trivial matrix of size $[1 \times 1]$ which is equivalent to a scalar.

2. A matrix multiplication is executed by multiplying a row of the first matrix with a column of the second matrix element by element and then summing up the products. In a dot product, each operand has only one row or column to multiply. Therefore, the result is a scalar.

Cross product

In contrast to the dot product, which results in a scalar, a cross product of two vectors yields a vector. The cross product is properly defined only in three-dimensional space.

$$\underline{a} \times \underline{b} = \underline{c} \tag{3.5}$$

The resulting vector $\underline{c}$ is perpendicular to the plane defined by the vectors $\underline{a}$ and $\underline{b}$. If $\underline{a}$ and $\underline{b}$ are parallel, their cross product will result in the null vector $\underline{c} = (0, 0, 0)^T$, i.e. it has neither a direction nor a length.

Dot product examples Given are two vectors $\underline{p}^T = (x, y, z)$ and $\underline{q}^T = (\xi, \eta, \zeta)$. Assuming the basis vectors form an orthogonal set, the dot product is:

$$\underline{p}^T \underline{q} = (x, y, z) \begin{pmatrix} \xi \\ \eta \\ \zeta \end{pmatrix} = x\xi + y\eta + z\zeta \tag{3.6}$$

With actual numbers for the vector components, it looks like this

$$\underline{p}^T \underline{q} = (1, 2, 3) \begin{pmatrix} 4 \\ 5 \\ 6 \end{pmatrix} = 1 \cdot 4 + 2 \cdot 5 + 3 \cdot 6 = 32 \tag{3.7}$$

Geometrically the dot product represents the projected length of one vector onto the direction of another.

Dot product properties The dot product is commutative, i.e. the order of operands may be reversed without affecting the result.

$$\underline{p}^T \underline{q} = \underline{q}^T \underline{p} \tag{3.8}$$

The dot product is zero if $\underline{p}$ and $\underline{q}$ are perpendicular to each other. If you take the dot product of a vector $\underline{v}^T = (u, v, w)$ with itself, the result is equal to the squared magnitude of $\underline{v}$.

$$\underline{v}^T \underline{v} = |\underline{v}|^2 = u^2 + v^2 + w^2 \tag{3.9}$$

Cross product examples Using the vectors $\underline{p}$ and $\underline{q}$ from above in a cross product yields

$$\begin{aligned} \underline{p} \times \underline{q} &= \begin{pmatrix} x \\ y \\ z \end{pmatrix} \times \begin{pmatrix} \xi \\ \eta \\ \zeta \end{pmatrix} = \begin{vmatrix} \underline{i} & \underline{j} & \underline{k} \\ x & y & z \\ \xi & \eta & \zeta \end{vmatrix} \\ &= \underline{i}(y\zeta - z\eta) + \underline{j}(z\xi - x\zeta) + \underline{k}(x\eta - y\xi) = \begin{pmatrix} y\zeta - z\eta \\ z\xi - x\zeta \\ x\eta - y\xi \end{pmatrix} \end{aligned} \tag{3.10}$$

With the numbers from above:

$$\begin{aligned} \underline{c} = \underline{p} \times \underline{q} &= \begin{pmatrix} 1 \\ 2 \\ 3 \end{pmatrix} \times \begin{pmatrix} 4 \\ 5 \\ 6 \end{pmatrix} = \begin{vmatrix} \underline{i} & \underline{j} & \underline{k} \\ 1 & 2 & 3 \\ 4 & 5 & 6 \end{vmatrix} \\ &= \underline{i}(2 \cdot 6 - 3 \cdot 5) + \underline{j}(3 \cdot 4 - 1 \cdot 6) + \underline{k}(1 \cdot 5 - 2 \cdot 4) = \begin{pmatrix} -3 \\ 6 \\ -3 \end{pmatrix} \end{aligned} \tag{3.11}$$

Check the result by confirming that $\underline{c}$ is perpendicular to $\underline{a}$ and $\underline{b}$ by showing that the dot products are zero, i.e. $\underline{c}^T \underline{a} = 0$ and $\underline{c}^T \underline{b} = 0$.

3.2 Fluid Properties

What is a fluid?

Ships travel in an environment formed by the fluids water and air. Although we all have a notion about fluids, it is not easily put into words. In contrast to solids, a fluid at rest cannot bear external shear stresses. A solid loaded with external shear stresses will deform until its internal stresses balance the external ones. The deformation of a solid stops once equilibrium is reached. A fluid will continue to move until the external shear stresses have vanished.

Liquid versus gas

Fluids comprise liquids and gases. A mass of gas has neither a unique shape nor a definite volume. It will completely fill any container, due to the small forces of attraction between its molecules. If the container is not closed, gas will escape. Liquids do not have a unique shape either, but they have a distinct volume. Due to moderate forces of attraction between its molecules, a liquid will take the shape of its storage container, but maintain its volume rather than fill the whole container. In the presence of gravity, liquids typically form a clearly defined surface. Some materials, e.g. glass, have fluid properties but flow so slowly that we rather treat them as solids.

Fresh water

The liquid considered in ship hydrodynamics is water. Chemically it is a combination of hydrogen and oxygen (H_2O). Depending on the combination of temperature and pressure, water exists in three phases: solid as ice, liquid as water, and gaseous as vapor. Varying levels of salt, air, and microscopic particles are dissolved in naturally occurring water. To bring order into chaos, the international community organized the International Association for the Properties of Water and Steam (IAPWS, www.iapws.org), which is tasked with the definition of the standard properties for pure water. The ITTC has adopted the IAPWS guidelines and incorporated them into its own set of recommended procedures (IAPWS, 2011; ITTC, 2011).

Seawater

The essential difference between fresh water and seawater is the salt content or short salinity. Salinity varies quite a bit across the Earth's oceans. The Intergovernmental Oceanographic Commission (IOC) has set a standard absolute salinity of 35.16504 g salt per one kilogram of seawater. In ITTC's recommended procedures, seawater properties are taken from IOC et al. (2010) and complemented with correlations from Sharqawy et al. (2010).

3.2.1 Properties of water

Only a few of the water properties defined in the cited documents are relevant for our discussion of ship hydrodynamics:

- density,
- viscosity, and
- vapor pressure.

All three quantities are a function of fluid temperature.

Density

Density measures the compactness of matter and is given as mass unit per volume unit, e.g. kg/m^3. As mentioned before, for any given temperature, density is treated as a constant in most applications of ship hydrodynamics.

Viscosity Viscosity is a measure for the internal friction and expresses how sticky a fluid is. The viscosity of a fluid is either provided as:

- dynamic viscosity μ, measured in units of force times time per area, i.e. Ns/m^2 or kg/(ms), or as
- kinematic viscosity ν, which is defined as the ratio of dynamic viscosity and density.

$$\nu = \frac{\mu}{\rho} \tag{3.12}$$

Kinematic viscosity ν is stated in units of m^2/s.

In ship hydrodynamics, kinematic viscosity ν is used more often because it is more convenient for fluids with constant density.

Vapor pressure Sometimes molecules of a liquid escape and form a vapor (gas) above the liquid. Vapor pressure is a measure for the tendency of molecules to transfer from liquid into vapor stage. The higher the vapor pressure of a liquid, the more likely are transfers. Vapor pressure is low for water at room temperature. Only when you heat water to its boiling point (around 100 °C) does its vapor pressure reach atmospheric pressure levels. Transfers from liquid to vapor phase happen not only at the surface of a liquid but also within a liquid. Just observe the bubbles that appear at the bottom of a pan filled with boiling water next time you prepare pasta. High temperatures are unlikely in ship hydrodynamics. However, sometimes the pressure in a flow falls below the vapor pressure and bubbles filled with vapor form in the fluid. This is called cavitation, which has many detrimental effects. Details on this phenomenon and how to prevent it are presented in Chapters 45 and 46.

Fresh water properties Most model tests are conducted in tanks filled with fresh water to avoid exposing sensitive sensors and electronic equipment to the corrosive force of seawater. Table 3.1 provides fresh water properties for a range of water temperatures typical for experimental facilities. ITTC's recommended procedure should be consulted for values at other temperatures (ITTC, 2011).

Standard seawater properties Water temperature varies considerably in the world's oceans: from a balmy 30 °C in the Caribbean to just below freezing in arctic waters. However, resistance predictions for ship designs are typically conducted for seawater of 15 °C temperature. The relevant quantities from the ITTC Recommended Procedure for fresh water and seawater properties (ITTC, 2011) are:

- density: $\rho_S = 1026.021\ kg/m^3$
- kinematic viscosity: $\nu_S = 1.1892 \cdot 10^{-6}\ m^2/s$ seawater
- vapor pressure: $p_{vs} = 1670.82$ Pa

Figure 3.1 compares density, kinematic viscosity, and vapor pressure of fresh and seawater for the common range of water temperatures. The values in Table 3.1 and Figure 3.1 have been computed for a standard atmospheric pressure of 101 325 Pa. The standards should be consulted if the properties are needed for substantially different pressure levels as may occur in deep sea diving.

Table 3.1 Fresh water properties

t_W [°C]	ρ [kg/m^3]	ν [10^{-6}m^2/s]	p_v [Pa]
15.0	999.103	1.1386	1705.74
15.1	999.088	1.1356	1716.76
15.2	999.072	1.1326	1727.84
15.3	999.057	1.1296	1738.98
15.4	999.041	1.1267	1750.19
15.5	999.026	1.1237	1761.46
15.6	999.010	1.1208	1772.79
15.7	998.994	1.1179	1784.18
15.8	998.978	1.1150	1795.64
15.9	998.962	1.1121	1807.17
16.0	998.946	1.1092	1818.76
16.1	998.930	1.1064	1830.41
16.2	998.913	1.1035	1842.13
16.3	998.897	1.1007	1853.92
16.4	998.880	1.0978	1865.77
16.5	998.863	1.0950	1877.69
16.6	998.847	1.0922	1889.68
16.7	998.830	1.0894	1901.73
16.8	998.813	1.0866	1913.85
16.9	998.795	1.0839	1926.04
17.0	998.778	1.0811	1938.29
17.1	998.761	1.0784	1950.61
17.2	998.743	1.0756	1963.01
17.3	998.725	1.0729	1975.47
17.4	998.708	1.0702	1988.00
17.5	998.690	1.0675	2000.60
17.6	998.672	1.0648	2013.27
17.7	998.654	1.0621	2026.01
17.8	998.635	1.0594	2038.82
17.9	998.617	1.0568	2051.70
18.0	998.599	1.0541	2064.66
18.1	998.580	1.0515	2077.68
18.2	998.561	1.0489	2090.78
18.3	998.543	1.0462	2103.95
18.4	998.524	1.0436	2117.19
18.5	998.505	1.0410	2130.50
18.6	998.486	1.0385	2143.89
18.7	998.467	1.0359	2157.35
18.8	998.447	1.0333	2170.89
18.9	998.428	1.0308	2184.50
19.0	998.408	1.0282	2198.18
19.1	998.389	1.0257	2211.94
19.2	998.369	1.0232	2225.78
19.3	998.349	1.0207	2239.69
19.4	998.329	1.0182	2253.68
19.5	998.309	1.0157	2267.74
19.6	998.289	1.0132	2281.88
19.7	998.269	1.0107	2296.10
19.8	998.248	1.0083	2310.39
19.9	998.228	1.0058	2324.76
20.0	998.207	1.0034	2339.21

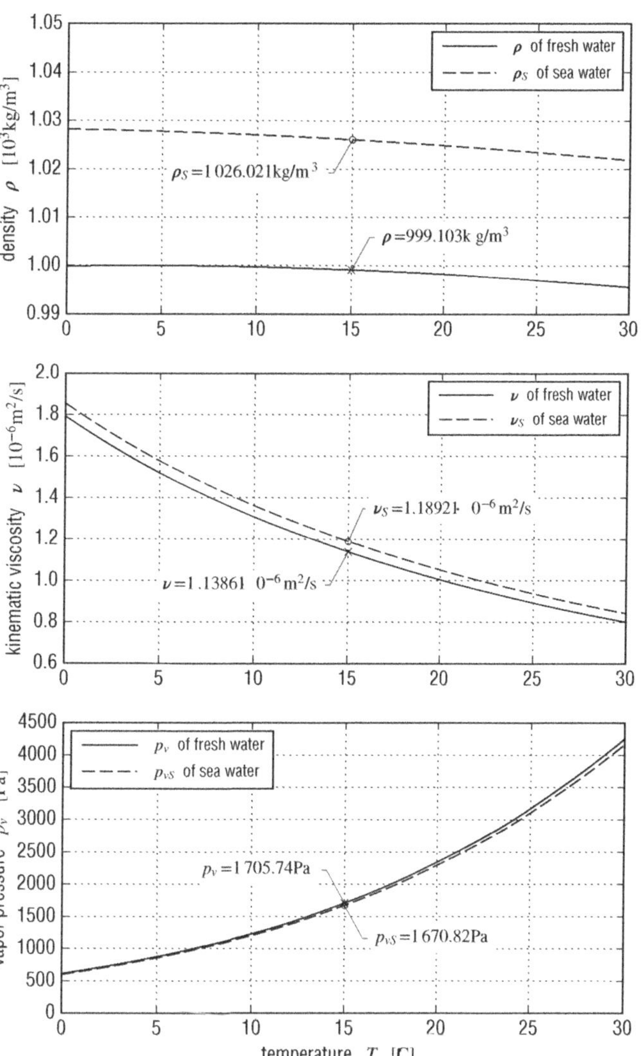

Figure 3.1 Fresh and seawater properties as a function of temperature

3.2.2 Properties of air

Density of air

The properties of air are somewhat neglected in ship hydrodynamics. The density of air is about 800 times smaller than that of water. Hence, the forces on a ship's hull are of greater importance for ship resistance and propulsion in calm water conditions than the forces on its superstructure. We only make use of the air density in the context of resistance predictions for ships. The density of air depends on temperature, height above sea level, and moisture content. For dry air of temperature 15 °C, the International Standard Atmosphere (ISA) model states an air density of $\rho_A = 1.225$ kg/m^3 (ISO, 1975).

Kinematic viscosity of air

Simulations of the flow around a ship superstructure also need a value for the kinematic viscosity. For dry air of temperature 15 °C at sea level, kinematic viscosity is $\nu_A = 1.4657 \cdot 10^{-5}\mathrm{m}^2/\mathrm{s}$. This is an order of magnitude higher than the kinematic viscosity of water, which seems counterintuitive. Air is actually less viscous than water. However, you have to compare the dynamic viscosities to make this obvious. The dynamic viscosity for dry air is $\mu_S = 1.7955 \cdot 10^{-5}$ Ns/m^2 and for seawater it is $\mu_S = 122.0144 \cdot 10^{-5}$ Ns/m^2. Only because of its considerably lower density does the kinematic viscosity of air become larger than that of seawater (see Equation (3.12)).

3.2.3 Acceleration of free fall

Standard gravitational acceleration

The gravitational acceleration g or acceleration of free fall is obviously not particular to any fluid but is nonetheless important for fluid flows. The international community set a standard value for the gravitational acceleration (BIPM, 1901):

$$g = 9.80665 \frac{\mathrm{m}}{\mathrm{s}^2} \qquad \text{standard acceleration of gravity} \tag{3.13}$$

For engineering applications we often employ a rounded value of $g = 9.807$ m/s^2.

Effect of latitude

However, gravitational acceleration varies with geographical latitude and height above sea level. Therefore, the standard value may not be accurate enough for the calibration of model testing equipment. The maximum gravitational acceleration occurs at the poles with $g_p = 9.8321849378$ m/s^2. Due to the elliptical cross section of the globe, locations toward the Equator are farther away from the Earth's center. As a result, they experience lower g values. The gravitational acceleration is $g_e = 9.7803253359$ m/s^2 at the Equator. For a location at latitude θ, the gravitational acceleration may be computed from the following formula:

$$g(\theta) = g_e \frac{\left[1 + k\sin^2(\theta)\right]}{\sqrt{1 - e^2\sin^2(\theta)}} \tag{3.14}$$

with constants $k = 0.00193185265$ and $e^2 = 0.00669437999$

Data and formula have been stated according to the World Geodetic System 1984 (WGS 84, see e.g. NIMA, 2000).

Local gravitational acceleration

Gravitational acceleration is also affected by various mineral deposits and changes in thickness of the Earth's crust. The Physikalisch-Technische Bundesanstalt in Germany maintains a web page at https://www.ptb.de/cartoweb3/SISproject.php, which allows the user to enter longitude, latitude, and height above sea level to retrieve the gravitational acceleration for a specific location.

3.3 Modeling and Visualizing Flow

Kinematics and kinetics

Basic courses in dynamics introduce the kinematics and kinetics of moving bodies. Kinematic describes the movement of matter and kinetic connects the motion with its causes, i.e. the forces acting on the objects. In particular, you studied Newton's second law stating that the inertia force $m\,\underline{a}$ is equal to the resultant external force acting on the body.

$$m\,\underline{a} = \underline{F}_{\text{external}} = \sum_i \underline{F}_i$$

Lagrangian mechanics

Following individual bodies and describing their motions is the Lagrangian formulation of mechanics. It is named after Joseph-Louis Lagrange (*1733 – †1813), an Italian mathematician who succeeded Leonard Euler as the director of mathematics at the Prussian Academy of Sciences in Berlin. You can find details in your textbook on dynamics or, for starters, read the article on Langrangian mechanics on Wikipedia (http://en.wikipedia.org/wiki/Lagrangian_mechanics).[1]

The Lagrangian description states physical quantities (velocity, momentum, etc.) as properties of a piece of matter with mass m. This is necessary to formulate Newton's second law of motion. However, the Lagrangian approach does not lend itself well to describe the motion of fluids because we would have to track an infinite number of fluid particles to entirely describe a flow.

Eulerian mechanics

It is more practical to study the development of physical properties over time at various positions in the fluid domain. Jean-Baptiste le Rond d'Alembert, a French mathematician, physicist, philosopher, and music theorist, introduced this type of mathematical modeling of continua. However, we now call it the Eulerian description. It is named after Swiss mathematician and physicist Leonard Euler (pronounced 'Oiler', *1707 – †1783), who used this method to formulate the basic equations of motions for an inviscid fluid. We will study these later in Chapters 7 and 14. In the Eulerian formulation, physical properties of the flow are described as functions of space $\underline{x}^T = (x, y, z)$ and time t. For example, the pressure function is $p = p(x, y, z, t)$. This is also known as field theory. Chapter 4 reviews the substantial derivative, which converts flow properties from Eulerian into Lagrangian coordinates.

The description of a flow field in Eulerian coordinates has two advantages:

(i) For many flow patterns, a coordinate system exists in which the Eulerian description becomes independent of time. For example, describing the flow around a body moving forward with constant speed in a coordinate system which is fixed to the body (moves with the same speed), makes the flow field steady. When you are a passenger on a ship moving steadily in calm water, the wave pattern you observe behind the vessel will not change over time.

(ii) The Eulerian description concurs with our instruments measuring flow properties. We do not have pressure sensors which can be attached to a fluid particle and track the changes in velocity along the particle's path. However, we stick Pitot-static tubes into the fluid to measure velocity over time at selected locations. Pitot-static tubes are named after French hydraulic engineer Henri Pitot (*1695–†1771). They actually measure dynamic pressure which is proportional to the squared flow velocity.

[1]There are two old but very instructive videos which explain the differences between Lagrangian and Eulerian description of fluid flows. Part 1: http://www.youtube.com/watch?v=Xg6L-dnUZ8c and Part 2: http://www.youtube.com/watch?v=BRj3BptzhnA

Flow properties Since density and temperature are considered constants in ship hydrodynamics, the objective is to find the spatial and time dependent distribution of velocity and pressure.

$$\text{velocity vector} \qquad \underline{v} = \begin{pmatrix} u(x,y,z,t) \\ v(x,y,z,t) \\ w(x,y,z,t) \end{pmatrix} \tag{3.15}$$

$$\text{pressure} \qquad p = p(x,y,z,t) \tag{3.16}$$

Knowledge of these flow properties enables us to compute forces acting on a vessel.

Basic principles Three basic principles of continuum mechanics are available to formulate a mathematical model of the flow:

- conservation of mass,
- conservation of momentum, and
- conservation of energy.

We will concentrate on the first two of these principles. Conservation of mass leads to the continuity equation. Conservation of momentum is equivalent to Newton's second law and results in the Navier-Stokes equations. For applications in ship hydrodynamics, the equation for the conservation of energy is often not directly used because the remaining two are sufficient to describe flows with constant temperature, viscosity, and density. However, conservation of energy is often employed to model turbulence.

Pathline Flow patterns can be visualized in different ways. If you follow an object and record its path, you see its pathline. Just observe the flight of a bird or watch a leaf floating down a stream. Of course, observing the pathline of a fluid particle in a flow is almost impossible, because we cannot easily identify individual particles. For visualization purposes, however, small, highly reflective particles may be used.

Streaklines Flow can also be visualized by inserting a constant stream of colored fluid or smoke into the flow. The smoke enters the flow at a fixed point. The visible lines of smoke emanating from that point are called streaklines. For instance, the smoke rising from a candle which has just been extinguished forms a streakline.

Streamlines Most widely known are streamlines. These are lines which are tangent to the velocity vector at all points they include. Since the flow is tangent to the velocity vector, fluid does not cross streamlines. Therefore, streamlines divide the flow into regions. Streamlines may be computed from the Eulerian formulation. Visualizing streamlines, however, is not quite as simple. You may seed a flow with reflective particles and take a photo with a suitable long exposure time. Particles will appear as short dashes on the picture which may be interpreted as velocity vectors. Drawing curves tangent to the vectors reconstructs the streamlines. In an unsteady flow, streamlines will constantly change. For time independent flows (steady flows), pathlines, streaklines, and streamlines are identical, which makes their visualization a lot easier.

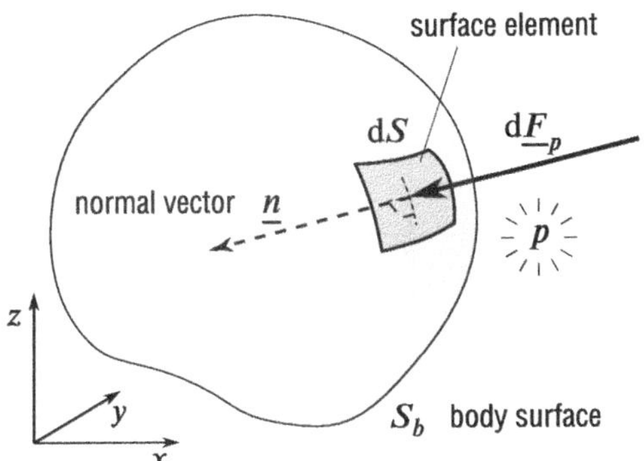

Figure 3.2 The pressure force $d\underline{F}_p$ acting on a small surface element

3.4 Pressure

Pressure and pressure forces

Pressure is the magnitude of a force per unit area exerted by one object onto another. It is comparable to the distributed load in beam theory. Pressure is a scalar. The direction of the resultant force is defined by the normal vector $\underline{n}$ of the surface a pressure acts upon. Thus, a pressure p exerts the force $d\underline{F}_p$ on an infinitesimally small surface area dS with normal vector $\underline{n}$ (Figure 3.2).

$$d\underline{F}_p = p\,\underline{n}\,dS \tag{3.17}$$

Depending on the orientation of the normal vector $\underline{n}$, the equation may feature a preceding minus sign. In this text, the normal vector points into the body, i.e. out of the fluid domain. Thus, a positive pressure is generating a force in the direction of the normal vector.

Integration over the entire body surface S_b yields the total pressure force $\underline{F}_p$.

$$\underline{F}_p = \iint_{S_b} p\,\underline{n}\,dS \tag{3.18}$$

For complex geometries like a ship hull, the integral has to be solved by numerical methods. In ship hydrodynamics, finding the pressure p is the more challenging problem though.

Hydrostatics

A simpler case is a body in fluid at rest. Consider the small cube shown in Figure 3.3 submerged in water. Its sides are aligned with the coordinate system axes and the z-axis points upward. The volume of the cube is $\Delta V = \Delta x \Delta y \Delta z$. It is not moving, hence it is in hydrostatic equilibrium. The external forces acting on the cube are its weight $\underline{F}_g$ and pressure forces on its six faces. Archimedes' principle dictates that the weight of the body is equal to the weight of the displaced water.

$$\underline{F}_g = \rho \begin{pmatrix} 0 \\ 0 \\ -g \end{pmatrix} \Delta V \tag{3.19}$$

For convenience, we assume that the cube is so small that the pressure may be considered constant over each face of the cube. In that case, the pressure force integral (3.18)

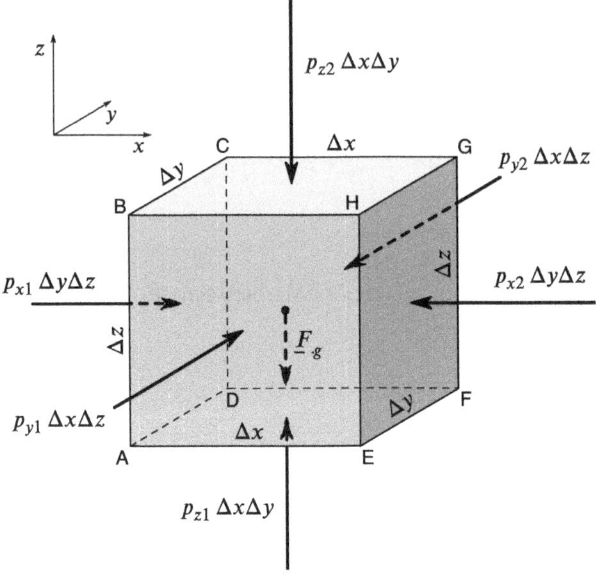

Figure 3.3 Forces on a small cube in hydrostatic equilibrum

simplifies to pressure times area, since the normal vectors are constant on each face as well. The components of the pressure force acting on the cube are:

$$
\begin{aligned}
&x\text{-direction:} \qquad & F_{px} &= -(p_{x2} - p_{x1})\,\Delta y\Delta z \\
&y\text{-direction:} \qquad & F_{py} &= -(p_{y2} - p_{y1})\,\Delta x\Delta z \\
&z\text{-direction:} \qquad & F_{pz} &= -(p_{z2} - p_{z1})\,\Delta x\Delta y
\end{aligned}
\tag{3.20}
$$

Hydrostatic equilibrium

Since the cube is not moving, all forces must be in equilibrium:

$$
-\begin{pmatrix} (p_{x2} - p_{x1})\,\Delta y\Delta z \\ (p_{y2} - p_{y1})\,\Delta x\Delta z \\ (p_{z2} - p_{z1})\,\Delta x\Delta y \end{pmatrix} + \rho \begin{pmatrix} 0 \\ 0 \\ -g \end{pmatrix} \Delta V = 0 \tag{3.21}
$$

We divide the equation by the cube volume $\Delta V = \Delta x\Delta y\Delta z$.

$$
-\begin{pmatrix} \frac{(p_{x2}-p_{x1})}{\Delta x} \\ \frac{(p_{y2}-p_{y1})}{\Delta y} \\ \frac{(p_{z2}-p_{z1})}{\Delta z} \end{pmatrix} + \rho \begin{pmatrix} 0 \\ 0 \\ -g \end{pmatrix} = 0 \tag{3.22}
$$

Infinitesimally small cube

Letting the cube shrink to infinitesimal size, i.e. $\Delta x \rightarrow \mathrm{d}x$, $\Delta y \rightarrow \mathrm{d}y$, and $\Delta z \rightarrow \mathrm{d}z$, converts the first vector into the gradient of pressure grad p.

$$
-\begin{pmatrix} \frac{\partial p}{\partial x} \\ \frac{\partial p}{\partial y} \\ \frac{\partial p}{\partial z} \end{pmatrix} + \rho \begin{pmatrix} 0 \\ 0 \\ -g \end{pmatrix} = 0 \tag{3.23}
$$

Several important conclusions follow from this analysis:

(i) The x- and y-components of the equation state that the hydrostatic pressure is constant in horizontal directions because the derivatives vanish.

$$\frac{\partial p}{\partial x} = \frac{\partial p}{\partial y} = 0 \quad \longrightarrow \quad p \text{ is constant in } x \text{ and } y \text{ when fluid at rest} \tag{3.24}$$

(ii) From the z-component of the equation we conclude:

Basic theorem of hydrostatics

$$-\frac{\partial p}{\partial z} - \rho g = 0$$

Since, according to conclusion (i), the pressure p only depends on z, we may write:

$$\mathrm{d}p + \rho g \,\mathrm{d}z = 0 \tag{3.25}$$

The latter is known as the basic theorem of hydrostatics. Formal integration with respect to z results in:

$$p + \rho g z = \text{const.} \tag{3.26}$$

(iii) Less obvious is a result for the pressure force on the infinitesimal cube: the first vector in Equation (3.23) represents the resultant pressure force on the fluid element $\mathrm{d}V$ per unit volume. Thus, it yields the same result as Equation (3.17):

$$\mathrm{d}\underline{F}_p = p\,\underline{n}\,\mathrm{d}S = -\mathrm{grad}\, p\,\mathrm{d}V \tag{3.27}$$

The gradient of a scalar is a vector of its spatial derivatives

$$\mathrm{grad}\, p = \begin{pmatrix} \frac{\partial p}{\partial x} \\ \frac{\partial p}{\partial y} \\ \frac{\partial p}{\partial z} \end{pmatrix} \tag{3.28}$$

Formal integration of Equation (3.27) yields Gauss' integral theorem:

$$\iint_S p\,\underline{n}\,\mathrm{d}S = -\iiint_V \mathrm{grad}\, p\,\mathrm{d}V \tag{3.29}$$

Hydrostatic pressure

Figure 3.4 depicts a water column in hydrostatic equilibrium. The column reaches from the water surface at z_0 down to the level $z = -h < 0$. An atmospheric pressure p_A is recorded above the water surface. According to conclusion (i), no forces act in the horizontal plane. Therefore, we may limit our static equilibrium condition to the z-direction.

$$0 = -p_A \mathrm{d}S + p(z)\mathrm{d}S - \rho g \mathrm{d}V \tag{3.30}$$

Division by the cross section area $\mathrm{d}S$ and reshuffling the terms yields the well known equation for the distribution of static pressure in a fluid.

$$p(z) = p_A + \rho g\, h \tag{3.31}$$

If we place the origin of the coordinate system at the water surface $z_0 = 0$, the equation simplifies to:

$$p(z) = p_A - \rho g\, z \tag{3.32}$$

Figure 3.4 Hydrostatic pressure in a water column

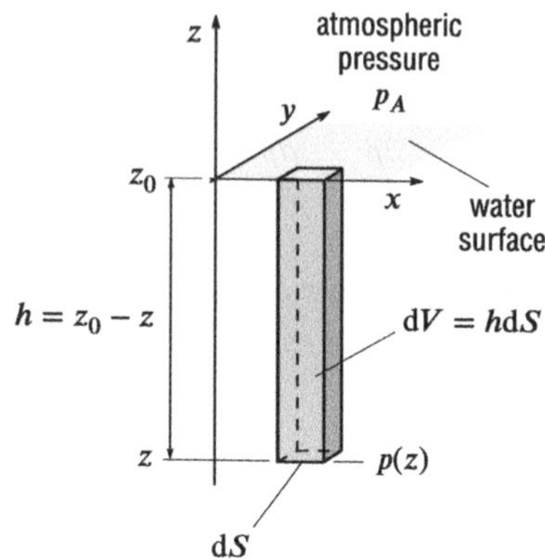

Figure 3.5 Pressure distribution around a ship

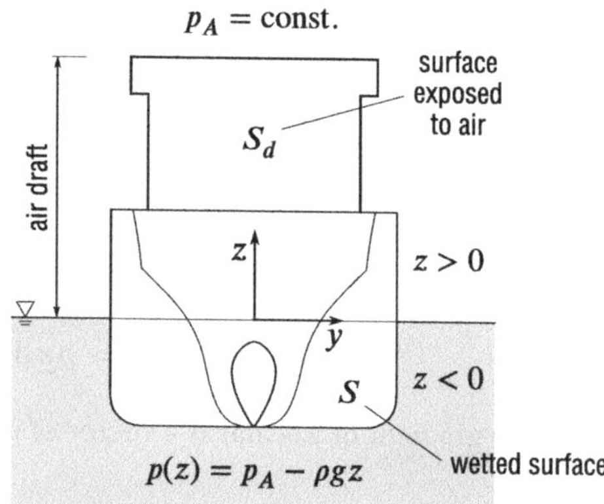

Note that z is negative for locations underwater.

Pressure force on a floating ship The relationship (3.29) can be used to prove Archimedes' principle. We compute the pressure force acting on a floating ship. As indicated in Figure 3.5, a constant atmospheric pressure p_A is acting on all surfaces S_d above the waterline, i.e. we ignore the small changes in atmospheric pressure over the air draft of the vessel. The wetted surface S below the waterline is subject to the pressure $p(z)$ from Equation (3.32), which varies over the draft.

Solving the integral (3.18) for these complex surfaces would be a formidable task. Luckily, Equation (3.29) provides a much simpler solution. It allows us to convert the surface integral into a volume integral.

$$\begin{aligned} \underline{F}_p &= \iint\limits_{S+S_d} p\underline{n}\,\mathrm{d}S \\ &= -\iiint\limits_{V+V_d} \operatorname{grad} p\,\mathrm{d}V \end{aligned} \tag{3.33}$$

V and V_d are the volumes associated with the underwater and above water surfaces S and S_d.

The pressure p is a function in two parts.

$$p = \begin{cases} p_A & \text{for } z > 0 \\ p_A - \rho g z & \text{for } z \leq 0 \end{cases} \tag{3.34}$$

For the first case, $z > 0$, above the water surface, the gradient is a zero vector, because the derivatives of a constant vanish.

$$\operatorname{grad} p = \operatorname{grad} p_A = \begin{pmatrix} 0 \\ 0 \\ 0 \end{pmatrix} \quad \text{for } z > 0 \tag{3.35}$$

Below the water line, the pressure gradient turns into

$$\operatorname{grad} p = \operatorname{grad} \left(p_A - \rho g z\right) = \begin{pmatrix} 0 \\ 0 \\ -\rho g \end{pmatrix} \quad \text{for } z \leq 0 \tag{3.36}$$

As a result, the constant part of the pressure will not yield a resultant force. Since we did not restrict ourselves to a specific shape, this result is general: a constant pressure *acting on all sides* of a body does not yield a resultant force.

Buoyancy force

As expected, the nonzero part of the pressure force points in vertical direction. Since the gradient of p is zero above the waterline, the volume integral can be restricted to the submerged volume V.

$$F_{pz} = -\iiint_V (-\rho g)\,\mathrm{d}V$$

The integrand is constant across the volume and can be extracted from the integral.

$$F_{pz} = -(-\rho g) \iiint_V \mathrm{d}V$$

The two minus signs cancel and the remaining integral yields the volumetric displacement V of the vessel.

$$F_{pz} = \rho g\, V \tag{3.37}$$

The result states that the buoyancy force points upwards and is equal to the weight of the displaced water, which is Archimedes' principle.

References

BIPM (1901). Resolution of the 3rd CGPM. https://www1.bipm.org/en/CGPM/db/3/2/. Online; accessed 10-July-2018.

IAPWS (2011). Revised supplementary release on properties of liquid water at 0.1 MPa. IAPWS SR6-08(2011), The International Association for the Properties of Water and Steam, Plzen, Czech Republic.

IOC, SCOR, and IAPSO (2010). The international thermodynamic equation of seawater – 2010: Calculation and use of thermodynamic properties. Manuals and Guides No. 56, Intergovernmental Oceanographic Commission. UNESCO (English), 196 pp.

ISO (1975). Standard atmosphere. ISO 2533:1975, International Organization for Standardization, Geneva, Switzerland.

ITTC (2011). *Fresh water and seawater properties.* International Towing Tank Conference, Recommended Procedures and Guidelines 7.5-02-01-03. Revision 2.

NIMA (2000). Department of Defense World Geodetic System 1984. Technical Report NIMA TR8350.2, third edition, Amendment 1, National Imagery and Mapping Agency.

Sharqawy, M., Lienhard V, J., and Zubair, S. (2010). Thermophysical properties of seawater: A review of existing correlations and data. *Desalination and Water Treatment,* 16:354–380.

Self Study Problems

1. Compute the dot products of the resulting vector $\underline{c}$ of Equation (3.11) with the arguments $\underline{p}$ and $\underline{q}$ of the cross product. Both should be zero as $\underline{c}$ must be perpendicular to $\underline{p}$ and $\underline{q}$.

2. Download the latest version of the fresh water and seawater properties document from ITTC's website at www.ittc.info (ITTC, 2011).

3. Read the Wikipedia pages about L. Euler and J.-L. Lagrange.

4. A sunken ship is to be raised by a SSCV (semisubmersible crane vessel) with a lifting capacity of 12 000 metric tons. The ship's displacement afloat was $V = 6\,000\ \text{m}^3$.

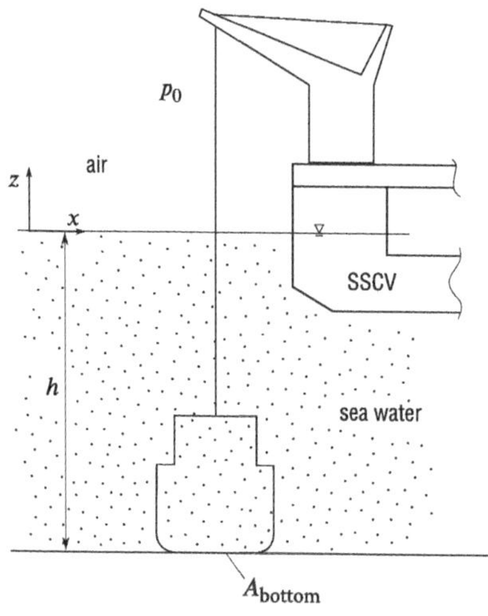

It is now completely filled with water and sitting on the ground in water of depth $h = 45$ m. The contact area with the ground is $A_{\text{bottom}} = 160\ \text{m}^2$. There is no water between ship bottom and sand. The density of seawater is $\rho = 1026.021\ \text{kg/m}^3$ and the atmospheric pressure is $p_A = 101\,325$ Pa. Gravitational acceleration in SI units is $g = 9.807\ \text{m/s}^2$.

Is the crane capable of lifting the sunken ship? Support your answer with a calculation of the required lifting force.

4

Fluid Mechanics and Calculus

Many of the fundamental pieces of calculus are related to fluid mechanics: total derivative, gradient, divergence, and rotation, among others. We will explore this connection and the application of differential operators in fluid mechanics in this chapter.

Learning Objectives

At the end of this chapter students will be able to:

- recapture selected concepts from calculus
- understand the connection of differential operators with fluid mechanics

4.1 Substantial Derivative

Before we derive the important continuity and Navier-Stokes equations, we consider a mathematical tool necessary in the Eulerian description of fluid flows. In the preceding chapter, we noted that in fluid mechanics we are interested in the spatial distribution of velocity and pressure over time. In ship hydrodynamics we assume that the density ρ of the fluid is constant. Such a fluid is called incompressible. For the limited space around a ship, water temperature T and dynamic viscosity μ are typically constant as well. Thus, the remaining unknowns are the velocity field $\underline{v}$ and the pressure distribution p.

Let us consider an infinitesimally small fluid element moving with the flow and track the development of the pressure p (Figure 4.1). At $t = t_1$ the fluid element is at position $\underline{x}_1$ and will be subject to the pressure $p = p_1(\underline{x}_1, t_1)$ at its center. Some time later, at $t = t_2$, the fluid element will be at position $\underline{x}_2$ with pressure $p = p_2(\underline{x}_2, t_2)$. We use a Taylor series[1] expansion to compute the value of p_2 assuming we know p_1 and its derivatives.

A short reminder for those who have long forgotten their calculus lessons: Taylor series allow us to express an initially unknown value of a function $f(x_2)$ at x_2 by the known **Taylor series**

[1]Named after English mathematician Brook Taylor, *1685 – †1731.

Fundamentals of Ship Hydrodynamics: Fluid Mechanics, Ship Resistance and Propulsion, First Edition.
Lothar Birk.

Companion website: www.wiley.com/go/birk/hydrodynamics

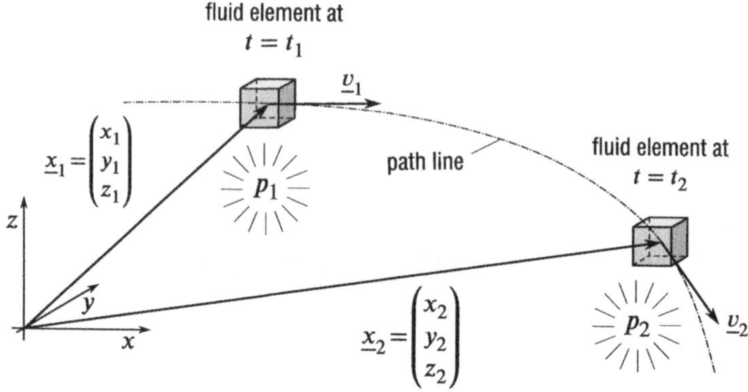

Figure 4.1 Following a fluid particle and the flow properties it encounters along the way

value and derivatives of the function f at a nearby location x_1:

$$\begin{aligned} f(x_2) &= \sum_{n=0}^{\infty} \frac{f^{(n)}(x_1)}{n!}(x_2 - x_1)^n \\ &= f^{(0)}(x_1) + f'(x_1)(x_2 - x_1) + \frac{f''(x_1)}{2}(x_2 - x_1)^2 \\ &\quad + \frac{f'''(x_1)}{6}(x_2 - x_1)^3 + \mathcal{O}(4) \end{aligned} \tag{4.1}$$

Of course, we have to make sure that the function is continuous and the derivatives actually exist. Landau's symbol $\mathcal{O}$ is used to represent terms of higher order. In the equation above, $\mathcal{O}(4)$ stands for the terms with derivatives of order $n \geq 4$.

If there is more than one free variable, things become a bit more complicated. With two variables (x, y), for instance, the Taylor series expansion of f is

$$\begin{aligned} f(x_2, y_2) &= f^{(0)}(x_1, y_1) + \frac{\partial f(x_1, y_1)}{\partial x}(x_2 - x_1) + \frac{\partial f(x_1, y_1)}{\partial y}(y_2 - y_1) \\ &\quad + \frac{1}{2}\frac{\partial^2 f(x_1, y_1)}{\partial x^2}(x_2 - x_1)^2 + \frac{1}{2}\frac{\partial^2 f(x_1, y_1)}{\partial y^2}(y_2 - y_1)^2 \\ &\quad + \frac{\partial^2 f(x_1, y_1)}{\partial x \partial y}(x_2 - x_1)(y_2 - y_1) \\ &\quad + \mathcal{O}(3) \end{aligned} \tag{4.2}$$

Note that there are two first order derivatives and three second order derivatives.

Taylor series example The pressure distribution $p = p(x, y, z, t)$ has four variables. Writing only the zero and first order terms explicitly yields:

$$\begin{aligned} p_2 &= p_1 + \left(\frac{\partial p}{\partial x}\right)_1 (x_2 - x_1) + \left(\frac{\partial p}{\partial y}\right)_1 (y_2 - y_1) + \left(\frac{\partial p}{\partial z}\right)_1 (z_2 - z_1) \\ &\quad + \left(\frac{\partial p}{\partial t}\right)_1 (t_2 - t_1) + \mathcal{O}(2) \end{aligned} \tag{4.3}$$

We subtract p_1 on both sides and divide the equation by the time difference $(t_2 - t_1)$. Resorting zero and first order terms and omitting the higher order terms $\mathcal{O}(2)$ yields the linear approximation of the pressure:

Linear approximation

$$\frac{(p_2 - p_1)}{(t_2 - t_1)} = \left(\frac{\partial p}{\partial t}\right)_1 + \left(\frac{\partial p}{\partial x}\right)_1 \frac{(x_2 - x_1)}{(t_2 - t_1)} + \left(\frac{\partial p}{\partial y}\right)_1 \frac{(y_2 - y_1)}{(t_2 - t_1)} + \left(\frac{\partial p}{\partial z}\right)_1 \frac{(z_2 - z_1)}{(t_2 - t_1)} \tag{4.4}$$

Of course, by omitting terms of second and higher order, the left and right sides of the equation are no longer truly identical. The equal sign is justified, however, if we assume that changes in position and time are small.

The left-hand side of Equation (4.4) can be interpreted as the mean rate of change of pressure during the time interval $t \in [t_1, t_2]$. We write:

$$\frac{(p_2 - p_1)}{(t_2 - t_1)} = \frac{\Delta p}{\Delta t} \tag{4.5}$$

On the right-hand side of Equation (4.4), we obtain similar expressions for the change in position for each of the three coordinate axes.

$$\frac{(x_2 - x_1)}{(t_2 - t_1)} = \frac{\Delta x}{\Delta t} \qquad \frac{(y_2 - y_1)}{(t_2 - t_1)} = \frac{\Delta y}{\Delta t} \qquad \frac{(z_2 - z_1)}{(t_2 - t_1)} = \frac{\Delta z}{\Delta t} \tag{4.6}$$

Replacing the terms in Equation (4.4) with the abbreviations from Equations (4.5) and (4.6) yields:

$$\frac{\Delta p}{\Delta t} = \left(\frac{\partial p}{\partial t}\right)_1 + \frac{\Delta x}{\Delta t}\left(\frac{\partial p}{\partial x}\right)_1 + \frac{\Delta y}{\Delta t}\left(\frac{\partial p}{\partial y}\right)_1 + \frac{\Delta z}{\Delta t}\left(\frac{\partial p}{\partial z}\right)_1 \tag{4.7}$$

Shrinking the length of the time interval Δt to zero results in the total change of pressure per unit time:

$$\lim_{\Delta t \to 0} \frac{\Delta p}{\Delta t} \equiv \frac{\mathrm{D}p}{\mathrm{D}t} \tag{4.8}$$

On the right-hand side, the fractions become the local components u, v, w of the velocity for $\Delta t \to 0$:

$$\lim_{\Delta t \to 0} \frac{\Delta x}{\Delta t} = u \qquad \lim_{\Delta t \to 0} \frac{\Delta y}{\Delta t} = v \qquad \lim_{\Delta t \to 0} \frac{\Delta z}{\Delta t} = w \tag{4.9}$$

Performing the transition $\Delta t \to 0$ in Equations (4.8) and (4.9) and substituting the results into Equation (4.7) yields:

Substantial derivative

$$\frac{\mathrm{D}p}{\mathrm{D}t} = \left(\frac{\partial p}{\partial t}\right) + u\left(\frac{\partial p}{\partial x}\right) + v\left(\frac{\partial p}{\partial y}\right) + w\left(\frac{\partial p}{\partial z}\right) \tag{4.10}$$

The term $\mathrm{D}p/\mathrm{D}t$ is known as the substantial derivative (or material derivative) of the pressure.

The substantial derivative consists of two parts:

Local derivative

- the local derivative $\dfrac{\partial p}{\partial t}$

Convective derivative

- and the convective derivative $u\left(\frac{\partial p}{\partial x}\right) + v\left(\frac{\partial p}{\partial y}\right) + w\left(\frac{\partial p}{\partial z}\right)$

In the Eulerian formulation, the total rate of change of fluid properties is composed of a change over time at a selected location and a change caused by the movement of the fluid element from one place to another. The former we call local derivative, and the latter we call convective derivative.

For instance, if you start the air conditioning in a house, the temperature inside will drop over a period of time – a local derivative. If you step outside, you might experience a higher temperature, although the temperatures inside and outside have not changed during the time span you needed to leave the house – a convective derivative.

Substantial derivative operator

The substantial differential Dp is equivalent to the total differential dp from calculus. The capital D is retained in many references to indicate the physical meaning of the substantial derivative as the total rate of change. We introduce the substantial derivative operator

$$\frac{\mathrm{D}}{\mathrm{D}t} = \left(\frac{\partial}{\partial t}\right) + u\left(\frac{\partial}{\partial x}\right) + v\left(\frac{\partial}{\partial y}\right) + w\left(\frac{\partial}{\partial z}\right) = \left(\frac{\partial}{\partial t}\right) + \underline{v}^T\underline{\nabla} \tag{4.11}$$

It may be applied to all flow properties of interest. $\underline{\nabla}$ is the Nabla operator, which is explained in the next section.

4.2 Nabla Operator and Its Applications

Nabla operator $\underline{\nabla}$

The Nabla operator is a differential operator. It is also known as the del-operator. In three-dimensional Cartesian space it is a vector composed of the three partial spatial derivative operators

$$\underline{\nabla} = \begin{pmatrix} \frac{\partial}{\partial x} \\ \frac{\partial}{\partial y} \\ \frac{\partial}{\partial z} \end{pmatrix} \tag{4.12}$$

The Nabla operator is very useful in fluid mechanics because it significantly simplifies notation and makes the meaning of individual terms more transparent; however, by itself it has no physical meaning.

Throughout this book the Nabla operator will be used on numerous occasions. Hence, it will be advantageous if you make yourself familiar with the following applications of $\underline{\nabla}$.

4.2.1 Gradient

Gradient

First, we apply the Nabla operator to a scalar physical property like the pressure p. In this case, we get a vector with the spatial changes of pressure. This vector is known as the pressure gradient. If s represents a scalar property, the following expressions are equivalent:

$$\underline{\nabla} s = \text{grad}\, s = \begin{pmatrix} \frac{\partial s}{\partial x} \\ \frac{\partial s}{\partial y} \\ \frac{\partial s}{\partial z} \end{pmatrix} \tag{4.13}$$

The gradient of a scalar quantity s has some important qualities:

- The gradient points into the direction of steepest ascent, i.e. if you always follow this direction, you take the shortest path to the 'summit' or largest value.
- A direction normal to the gradient is part of an isoline, i.e. s does not change.
- If the gradient vanishes, i.e. grad $s = \underline{0}$, you have a possible extreme value of the quantity s. It is a necessary but not a sufficient condition for a minimum or maximum. The point could just be a saddle point.

4.2.2 Divergence

Nabla may also operate on vectorial quantities $\underline{v}$. Since the Nabla operator itself is a vector, three specific operations exist:

- a dot product $\underline{\nabla}^T \underline{v}$ (see Section 4.2.2) which results in a scalar,
- a cross product $\underline{\nabla} \times \underline{v}$ (see Section 4.2.3) which results in a vector, or
- a dyad $\underline{\nabla}\, \underline{v}^T$ resulting in a matrix.

The dot and cross products with the Nabla operator are frequently used in fluid mechanics and warrant a closer look.

Moving control volume

The divergence of the velocity vector $\underline{v}$ has a special meaning as the rate of change of volume of a fluid. For illustration, imagine a balloon filled with helium rising into the air (Figure 4.2). If the balloon is properly sealed, the mass contained in the volume is constant as the balloon will always contain the same fluid particles. However, the balloon may change its shape.

The divergence of a vector is defined as a dot product between the Nabla operator and the vector $\underline{v}$:

$$\begin{aligned} \text{div}\, \underline{v} = \underline{\nabla}^T \underline{v} &= \left(\frac{\partial}{\partial x}, \frac{\partial}{\partial y}, \frac{\partial}{\partial z}\right) \begin{pmatrix} u \\ v \\ w \end{pmatrix} \\ &= \frac{\partial}{\partial x} u + \frac{\partial}{\partial y} v + \frac{\partial}{\partial z} w \\ &= \frac{\partial u}{\partial x} + \frac{\partial v}{\partial y} + \frac{\partial w}{\partial z} \end{aligned} \tag{4.14}$$

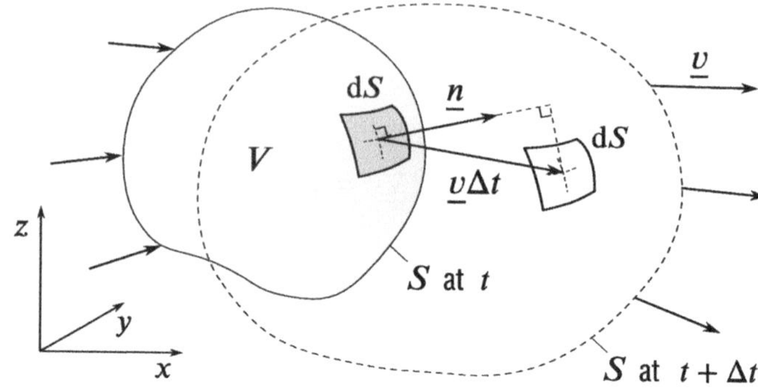

Figure 4.2 A moving, finite control volume V which changes over time

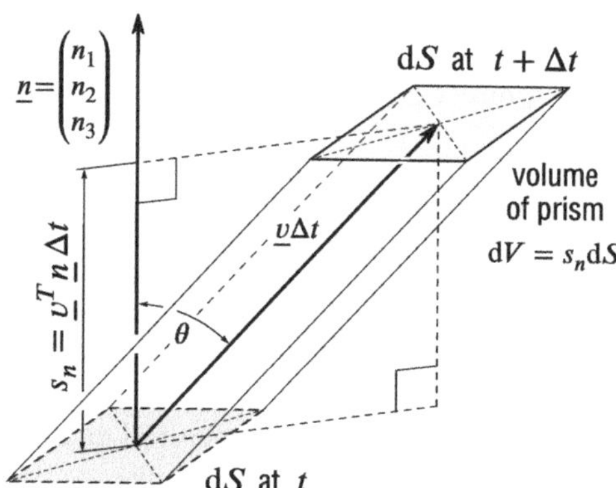

Figure 4.3 The distance s_n traveled by a surface element in normal direction

Differential change in volume

We observe a small area $\mathrm{d}S$ on the balloon's surface. Points on the surface S of the control volume move with the flow velocity $\underline{v}$. A surface element travels a distance $\Delta\underline{s} = \underline{v}\,\Delta t$ during a small time interval Δt. In the direction of the surface normal $\underline{n}$, the surface element will travel the distance s_n (see Figure 4.3).

$$s_n = \left(\underline{v}\,\Delta t\right)^T \underline{n} \tag{4.15}$$

The movement of the surface element $\mathrm{d}S$ causes a small change in the volume V. The volume of a prism is equal to the product of base surface $\mathrm{d}S$ and height s_n normal to the base surface.

$$\mathrm{d}V = s_n\,\mathrm{d}S = \left(\underline{v}\,\Delta t\right)^T \underline{n}\,\mathrm{d}S \tag{4.16}$$

Of course, if the integrity of the control volume is to be maintained, surface elements cannot move independently. Over the time interval Δt, the control volume will change by the integral of all differential volume changes $\mathrm{d}V$.

$$\Delta V = \iint\limits_S \left(\underline{v}\,\Delta t\right)^T \underline{n}\,\mathrm{d}S \tag{4.17}$$

The time interval is independent of the integration variable $\mathrm{d}S$ which allows us to extract it from the integral and to divide the equation by Δt.

$$\frac{\Delta V}{\Delta t} = \iint\limits_S \underline{v}^T \underline{n}\,\mathrm{d}S \tag{4.18}$$

In the limit case $\Delta t \longrightarrow 0$, we obtain the substantial derivative of the control volume V.

$$\frac{\mathrm{D}V}{\mathrm{D}t} = \iint_S \underline{v}^T \underline{n}\,\mathrm{d}S \tag{4.19}$$

Divergence theorem

The divergence theorem from calculus states that for any continuous vector field $\underline{v}$ the following relationship exists:

$$\iint_S \underline{v}^T \underline{n}\,\mathrm{d}S = \iiint_V \left(\underline{\nabla}^T \underline{v}\right)\mathrm{d}V \tag{4.20}$$

Combining Equations (4.19) and (4.20), we can express the total derivative of the volume V as:

$$\frac{\mathrm{D}V}{\mathrm{D}t} = \iiint_V \left(\underline{\nabla}^T \underline{v}\right)\mathrm{d}V \tag{4.21}$$

In a fluid we may subdivide a control volume until we end up with a small volume δV. In fact, we make δV so small that the integrand $\left(\underline{\nabla}^T \underline{v}\right)$, in good approximation, is constant across δV. Then, the volume integral on the right-hand side reduces to a product of the divergence of the velocity with the volume.

$$\frac{\mathrm{D}(\delta V)}{\mathrm{D}t} = \iiint_{(\delta V)} \left(\underline{\nabla}^T \underline{v}\right)\mathrm{d}V \approx \left(\underline{\nabla}^T \underline{v}\right)\delta V \tag{4.22}$$

Divergence of velocity

Dividing this equation by δV yields an expression for the divergence of the velocity which bears a physical meaning.

$$\underline{\nabla}^T \underline{v} = \frac{\left(\frac{\mathrm{D}(\delta V)}{\mathrm{D}t}\right)}{\delta V} = \frac{\text{rate of change of volume}}{\text{volume}} \tag{4.23}$$

The divergence of a velocity field is a measure for the rate of change in volume per unit volume.

4.2.3 Rotation

Rotation = curl

The rotation or curl is defined as the cross product between the Nabla operator and a vector. It is therefore only properly defined in three dimensions.[2] For instance, the

[2] You can also define it for two dimensions; however, the result needs special interpretation as it is not part of the plane of the two original vectors.

rotation of the velocity vector is rot $\underline{v}$ with:

$$\text{rot}\,\underline{v} = \underline{\nabla} \times \underline{v} = \begin{pmatrix} \frac{\partial}{\partial x} \\ \frac{\partial}{\partial y} \\ \frac{\partial}{\partial z} \end{pmatrix} \times \begin{pmatrix} u \\ v \\ w \end{pmatrix} = \begin{vmatrix} \underline{i} & \underline{j} & \underline{k} \\ \frac{\partial}{\partial x} & \frac{\partial}{\partial y} & \frac{\partial}{\partial z} \\ u & v & w \end{vmatrix}$$

$$= \left(\frac{\partial w}{\partial y} - \frac{\partial v}{\partial z}\right)\underline{i} + \left(\frac{\partial u}{\partial z} - \frac{\partial w}{\partial x}\right)\underline{j} + \left(\frac{\partial v}{\partial x} - \frac{\partial u}{\partial y}\right)\underline{k} \tag{4.24}$$

$$= \begin{pmatrix} \frac{\partial w}{\partial y} - \frac{\partial v}{\partial z} \\ \frac{\partial u}{\partial z} - \frac{\partial w}{\partial x} \\ \frac{\partial v}{\partial x} - \frac{\partial u}{\partial y} \end{pmatrix}$$

The rotation of the velocity vector plays an important role in the definition of potential flows and finds application in lifting flows.

4.2.4 Laplace operator

Since the Nabla operator is a vector, we can form a dot product of the Nabla operator with itself. In Cartesian coordinates we get:

$$\underline{\nabla}^T\underline{\nabla} = \left(\frac{\partial}{\partial x}, \frac{\partial}{\partial y}, \frac{\partial}{\partial z}\right)\begin{pmatrix} \frac{\partial}{\partial x} \\ \frac{\partial}{\partial y} \\ \frac{\partial}{\partial z} \end{pmatrix} = \frac{\partial^2}{\partial x^2} + \frac{\partial^2}{\partial y^2} + \frac{\partial^2}{\partial z^2} \tag{4.25}$$

As expected, the result of a dot product is a scalar. This scalar differential operator computes the sum of the second order spatial derivatives. As we will see later, it occurs quite frequently in the mathematical treatment of fluid mechanics and is known as the Laplace operator $\mathbf{\Delta}$, so named after the man who defined it, the French mathematician Pierre-Simon Laplace (*1749–†1827).

$$\mathbf{\Delta} = \underline{\nabla}^T\underline{\nabla} = \frac{\partial^2}{\partial x^2} + \frac{\partial^2}{\partial y^2} + \frac{\partial^2}{\partial z^2} \tag{4.26}$$

The Laplace operator applied to a scalar quantity ϕ yields:

$$\mathbf{\Delta}\phi = (\underline{\nabla}^T\underline{\nabla})\phi = \frac{\partial^2\phi}{\partial x^2} + \frac{\partial^2\phi}{\partial y^2} + \frac{\partial^2\phi}{\partial z^2} \tag{4.27}$$

The condition that the sum of second order spatial derivatives of ϕ vanishes is known as the Laplace equation.

$$\mathbf{\Delta}\phi = 0 \qquad \text{Laplace equation} \tag{4.28}$$

This linear, homogeneous, second order partial differential equation forms the foundation of potential theory. The Laplace equation expresses conservation of mass for potential flow, a specific class of inviscid flows. See Chapter 15 for an introduction and Kellogg (1929) for details. Besides fluid mechanics, potential theory is also used in the field of magneto-electrodynamics.

References

Kellogg, O. (1929). *Foundations of potential theory.* Springer, Berlin, Germany.

Self Study Problems

1. Compute the gradient $\underline{\nabla}\,\phi$ for the function

$$\phi(\underline{x}) = \frac{-1}{4\pi r}$$

with the distance $r = \sqrt{(x-\xi)^2 + (y-\eta)^2 + (z-\zeta)^2}$ between the field point $\underline{x} = (x, y, z)^T$ and the fixed point $\underline{\xi} = (\xi, \eta, \zeta)^T$ (source point). You may assume that $\underline{x} \neq \underline{\xi}$.

2. Show that $\boldsymbol{\Delta}\phi = 0$ for the function

$$\phi(\underline{x}) = \frac{-1}{4\pi r}$$

with the distance $r = \sqrt{(x-\xi)^2 + (y-\eta)^2 + (z-\zeta)^2}$ between the field point $\underline{x} = (x, y, z)^T$ and the fixed point $\underline{\xi} = (\xi, \eta, \zeta)^T$ (source point). You may assume that $\underline{x} \neq \underline{\xi}$.

3. Explain the physical meaning of the divergence of the velocity.

5

Continuity Equation

In Eulerian fluid mechanics, conservation laws may be set up for four different types of control volumes. A control volume may be infinitesimally small or finite in size. Independent of size, a control volume may be fixed in space or moving along with the flow. In this chapter, we discuss the application of all four types of control volumes to the conservation of mass principle. As a result, we obtain four equivalent mathematical models for the conservation of mass principle known as the continuity equation.

Learning Objectives

At the end of this chapter students will be able to:

- define and apply different types of control volumes
- discuss the conservation of mass principle
- derive and apply the continuity equation

5.1 Mathematical Models of Flow

Finite, infinitesimal, fixed or moving control volume

Different mathematical models may be derived for the same conservation principle. The models differ with respect to the size of the control volume (CV) and whether the control volume is fixed or moving. In total, we have four types of control volumes:

(i) infinitesimally small fluid element fixed in space

(ii) finite fluid element fixed in space

(iii) infinitesimally small fluid element moving with the fluid

(iv) finite fluid element moving with the fluid

Figure 5.1 shows which combination of fixed or moving and finite or infinitesimal control volume yields which form of the conservation law. Finite control volumes lead to integral forms, whereas infinitesimally small control volumes result in differential forms of the conservation laws. Moving control volumes always contain the same fluid

Fundamentals of Ship Hydrodynamics: Fluid Mechanics, Ship Resistance and Propulsion, First Edition.
Lothar Birk.

Companion website: www.wiley.com/go/birk/hydrodynamics

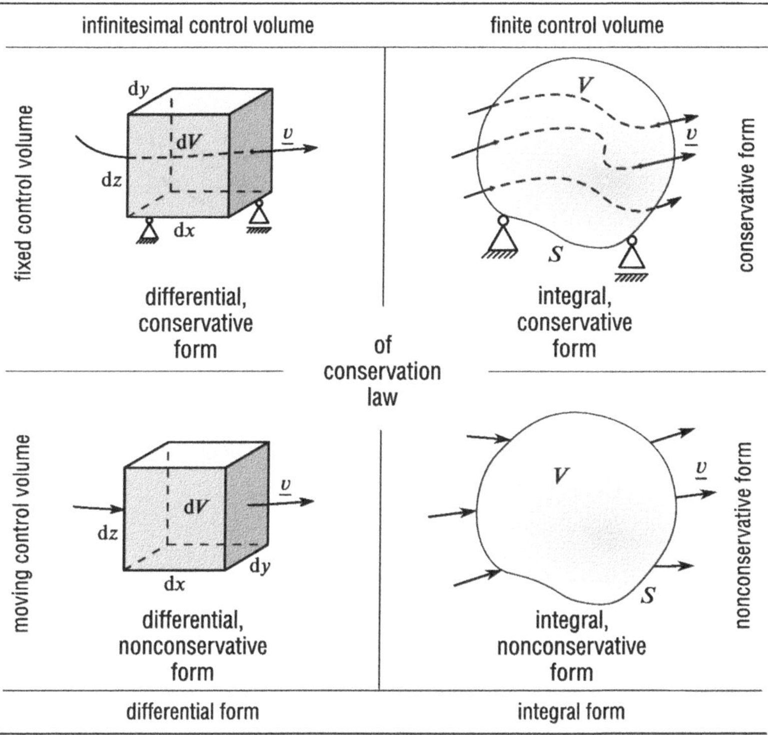

Figure 5.1 Four types of mathematical models for fluid flows and the resulting form of the conservation law

mass but may change their shape. They yield so-called nonconservative forms of the equations. Fixed control volumes cannot change their shape or location. Fluid is passing through the CV and may accumulate or recede in it. The resulting equations are in conservative form.

Conservative versus nonconservative

The distinction conservative versus nonconservative is rather confusing. It refers to properties of the discretized versions of the equations as they appear in numerical solvers. Although the resulting equations may look different, all forms represent the physical principle of conservation of mass, momentum, or energy. For details see Anderson, Jr. (1995, Chapter 2).

As an example, we will derive mathematical models for the conservation of mass principle for all cases (i) through (iv).

5.2 Infinitesimal Fluid Element Fixed in Space

The sketch in Figure 5.2 shows a fixed, infinitesimally small control volume $\mathrm{d}V$. The cube is so small that flow velocities are assumed to be constant across each of its six faces. An arrow on each face indicates the mass flux into and out of the fixed fluid element. The arrows in Figure 5.2 point in a positive coordinate direction if they are

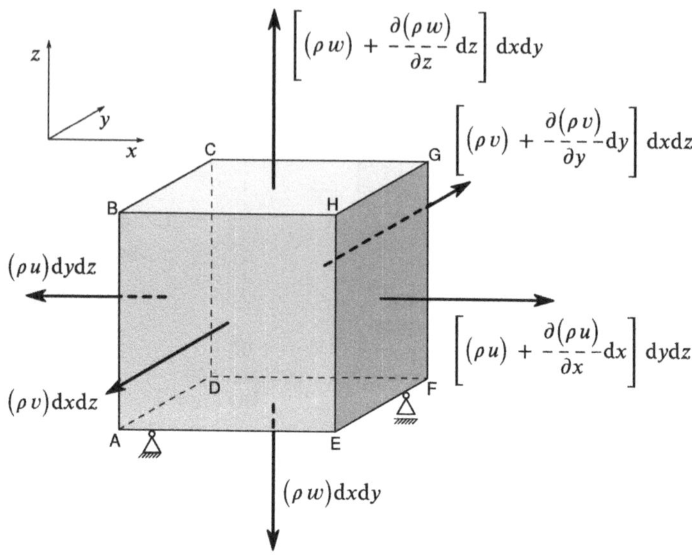

Figure 5.2 Mass flux through the surface of a fluid element

on a positive face of the element and vice versa for negative faces of the element. As a consequence, outflow is considered positive and inflow is considered a negative mass flow rate. Conservation of mass requires that

$$\text{time rate } \textit{decrease} \text{ of mass inside } dV \; = \; \text{mass flow rate } \textit{out} \text{ of } dV \text{ through } dS$$

$$\text{or short} \qquad -C = B \tag{5.1}$$

In other words, the negative rate of change of mass C in dV is equal to the net mass outflow per time unit B over the boundary dS of the control volume dV.

Net mass flux through surface

A mass flow rate or mass flux is measured in kilograms per second. The mass flux across a surface element dS is equal to the product of density, velocity normal to dS, and the surface area. On the face ABCD of the cube, the normal vector is pointing in negative x-direction, i.e. $\underline{n} = -\underline{i}$. We get for the mass flux:

$$\text{mass flux through face ABCD:} \quad -(\rho\, \underline{v}^T \underline{i})\, dS \; = \; -(\rho u)\, dydz \tag{5.2}$$

On the opposite face EFGH with normal vector $\underline{n} = +\underline{i}$, the mass flow rate per area is expressed as a truncated Taylor series of $\rho\underline{v}$. Then, the mass flux through face EFGH is given by:

mass flux through face EFGH:

$$\left[(\rho\underline{v}) + \frac{\partial(\rho\underline{v})}{\partial x} dx\right]^T \underline{i}\, dS \; = \; \left[(\rho u) + \frac{\partial(\rho u)}{\partial x} dx\right] dydz \tag{5.3}$$

Application of the Taylor series avoids introducing additional unknowns.

Summing the mass flow rates in x-direction (5.2) and (5.3) yields:

$$-(\rho u)\,\mathrm{d}y\mathrm{d}z + \left[(\rho u) + \frac{\partial(\rho u)}{\partial x}\mathrm{d}x\right]\mathrm{d}y\mathrm{d}z = \left(-\rho u + \rho u + \frac{\partial(\rho u)}{\partial x}\mathrm{d}x\right)\mathrm{d}y\mathrm{d}z$$

$$= \frac{\partial(\rho u)}{\partial x}\mathrm{d}x\mathrm{d}y\mathrm{d}z$$

$$= \frac{\partial(\rho u)}{\partial x}\mathrm{d}V \qquad (5.4)$$

The resulting mass fluxes for the remaining four faces are:

y-direction, faces AEHB and CGFD: $\frac{\partial(\rho v)}{\partial y}\mathrm{d}V$ (5.5)

z-direction, faces ADFE and BHGC: $\frac{\partial(\rho w)}{\partial z}\mathrm{d}V$ (5.6)

The sum of Equations (5.4), (5.5), and (5.6) is equal to the total mass flow rate over the surface $\mathrm{d}S$:

$$B = \left[\frac{\partial(\rho u)}{\partial x} + \frac{\partial(\rho v)}{\partial y} + \frac{\partial(\rho w)}{\partial z}\right]\mathrm{d}V \qquad (5.7)$$

Rate of change of mass

For the rate of change of mass in $\mathrm{d}V$ on the left-hand side of Equation (5.1), we first determine the total mass contained in $\mathrm{d}V$:

$$\mathrm{d}m = \rho\,\mathrm{d}V = \rho\,\mathrm{d}x\mathrm{d}y\mathrm{d}z \qquad (5.8)$$

We choose a fluid element $\mathrm{d}V$ so small that the density ρ may be assumed constant throughout $\mathrm{d}V$.

The control volume $\mathrm{d}V$ is fixed in space. As a consequence, $\mathrm{d}V$ does not change over time and the rate of change of mass will be

$$C = \frac{\partial m}{\partial t} = \frac{\partial \rho}{\partial t}\mathrm{d}V \qquad (5.9)$$

We now state the conservation of mass by equating $-C$ and B:

$$-\frac{\partial \rho}{\partial t}\mathrm{d}V = \left[\frac{\partial(\rho u)}{\partial x} + \frac{\partial(\rho v)}{\partial y} + \frac{\partial(\rho w)}{\partial z}\right]\mathrm{d}V \qquad (5.10)$$

or after division by the volume $\mathrm{d}V$

$$\frac{\partial \rho}{\partial t} + \frac{\partial(\rho u)}{\partial x} + \frac{\partial(\rho v)}{\partial y} + \frac{\partial(\rho w)}{\partial z} = 0 \qquad (5.11)$$

Continuity equation (differential, conservative)

Introducing the Nabla operator yields

$$\frac{\partial \rho}{\partial t} + \underline{\nabla}^T(\rho\underline{v}) = 0 \qquad (5.12)$$

Equation (5.12) represents the conservative, differential form of the continuity equation.

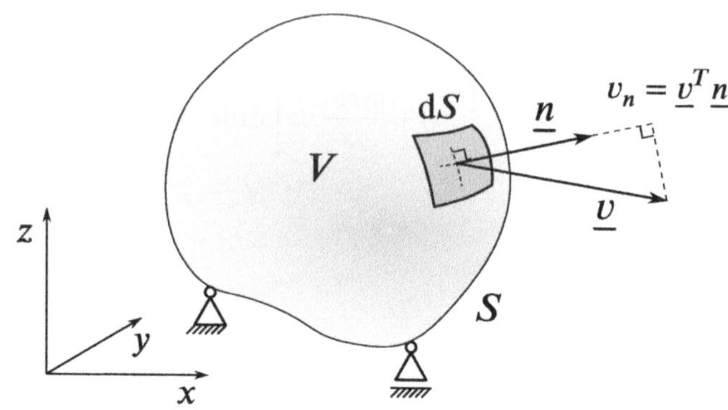

Figure 5.3 Flux through the surface S of a finite volume V fixed in space

5.3 Finite Control Volume Fixed in Space

Next, we investigate a finite control volume V that is fixed in space (Figure 5.3). S denotes the surface enclosing the control volume V. If mass is neither produced nor destroyed within the control volume V, conservation of mass requires that

time rate *decrease* of mass inside V = net mass flow rate *out* of V through S

$$\text{or short} \qquad -C = B \tag{5.13}$$

Net mass flux through surface

We begin again with the mass flux B across the control surface S. The orientation of the surface is defined by its normal vector $\underline{n}$. By definition, the normal vector is of unit length $|\underline{n}| = 1$ and points out of the fluid domain. The mass flow rate through an infinitesimally small part dS of the surface is a function of its size and the fluid velocity normal to the surface element:

$$\rho\, v_n \mathrm{d}S = \rho\, \underline{v}^T \mathrm{d}\underline{S} = \rho\, \underline{v}^T \underline{n}\, \mathrm{d}S \tag{5.14}$$

The dot product $v_n = \underline{v}^T \underline{n}$ yields the component of the velocity vector, which points in the direction of the normal vector.

The net mass flow rate through all of S is obtained by integrating Equation (5.14):

$$B = \iint\limits_S \rho\, \underline{v}^T \underline{n}\, \mathrm{d}S \tag{5.15}$$

If the integrand is positive, mass leaves the control volume V.

Rate of change of mass

The mass contained in a fluid element is $\mathrm{d}m = \rho\, \mathrm{d}V$. The total mass in V will be

$$\int \mathrm{d}m = \iiint\limits_V \rho\, \mathrm{d}V \tag{5.16}$$

The rate of change with respect to time is:

$$C = \frac{\partial}{\partial t} \iiint\limits_V \rho\, \mathrm{d}V = \iiint\limits_V \frac{\partial \rho}{\partial t}\, \mathrm{d}V \tag{5.17}$$

Since the volume V is fixed in space, i.e. its limits do not change over time, interchanging integration and time differentiation is allowed.

In summary, we have $-C = B$ or

Continuity equation (integral, conservative)

$$-\iiint_V \frac{\partial \rho}{\partial t}\,\mathrm{d}V = \iint_S \rho\,\underline{v}^T \underline{n}\,\mathrm{d}S \tag{5.18}$$

This is the integral, conservative form of the continuity equation.

5.4 Infinitesimal Element Moving With the Fluid

Next, we consider an infinitesimally small fluid element $\mathrm{d}V$ moving with the flow. By definition, it always contains the same amount of fluid. However, it may change its shape over time. Since the mass in $\mathrm{d}V$ is constant, its substantial derivative must vanish.

$$\frac{\mathrm{D}(\mathrm{d}m)}{\mathrm{D}t} = 0 \tag{5.19}$$

We expand the substantial derivative of $\mathrm{d}m = \rho\mathrm{d}V$ by applying the product rule:

$$\frac{\mathrm{D}(\rho\,\mathrm{d}V)}{\mathrm{D}t} = \mathrm{d}V\frac{\mathrm{D}\rho}{\mathrm{D}t} + \rho\,\frac{\mathrm{D}(\mathrm{d}V)}{\mathrm{D}t} = 0 \tag{5.20}$$

Division by $\mathrm{d}V$ yields:

$$\frac{\mathrm{D}\rho}{\mathrm{D}t} + \rho\left[\frac{1}{\mathrm{d}V}\frac{\mathrm{D}(\mathrm{d}V)}{\mathrm{D}t}\right] = 0 \tag{5.21}$$

The term in brackets is the divergence of the velocity field (4.23): $\operatorname{div}\underline{v} = \underline{\nabla}^T\underline{v}$

Continuity equation (differential, non-conservative)

$$\frac{\mathrm{D}\rho}{\mathrm{D}t} + \rho\underline{\nabla}^T\underline{v} = 0 \tag{5.22}$$

We obtained the differential, nonconservative form of the continuity equation.

5.5 Finite Control Volume Moving With the Fluid

Continuity equation (integral, non-conservative)

Finally, a very simple case! By definition, the mass in the moving finite control volume V does not change. Think of a weather balloon released into the atmosphere. It contains a fixed amount of helium, which expands as the outside pressure decreases with increasing altitude. At take-off it will be slack and will burst eventually at high altitude when the pressure difference becomes too great.

The total mass in the control volume V is

$$m = \iiint_V \rho\,\mathrm{d}V \tag{5.23}$$

If the mass does not change over time, its total derivative must vanish.

$$\frac{\mathrm{D}m}{\mathrm{D}t} = \frac{\mathrm{D}}{\mathrm{D}t} \iiint_V \rho \,\mathrm{d}V = 0 \tag{5.24}$$

This is the continuity equation in its integral, nonconservative form.

5.6 Summary

Despite the different appearances, all four forms of the continuity equation express the same physical principle: conservation of mass. With rigorous mathematical transformations, each form can be converted into the others.

Conversion For instance, the integral, conservative form of Equation (5.18) may be converted into the differential, conservative form of the continuity Equation (5.12). As a first step, we move the integrals in Equation (5.18) onto the same side.

$$\iiint_V \frac{\partial \rho}{\partial t} \mathrm{d}V + \iint_S \rho \underline{v}^T \underline{n} \,\mathrm{d}S = 0$$

With the help of the divergence theorem (4.20), the surface integral may be converted into a volume integral.

$$\iiint_V \frac{\partial \rho}{\partial t} \mathrm{d}V + \iiint_V \underline{\nabla}^T (\rho \underline{v}) \,\mathrm{d}V = 0$$

Both volume integrals have the same limits. Therefore, they may be combined.

$$\iiint_V \left[\frac{\partial \rho}{\partial t} + \underline{\nabla}^T (\rho \underline{v}) \right] \mathrm{d}V = 0 \tag{5.25}$$

In general, a vanishing integral does not imply that the integrand must be zero as well. Just check the sine function $\sin(x)$ and its definite integral with limits 0 and 2π. However, we know that the conservation of mass principle must hold true in any control volume, i.e. it must hold true not only in V but also in any subdivision of V. As a consequence, the integrand must vanish in any infinitesimally small part of V and the integrand must be zero everywhere:

$$\frac{\partial \rho}{\partial t} + \underline{\nabla}^T (\rho \underline{v}) = 0 \tag{5.12}$$

This is, of course, the differential, conservative form of the continuity equation from page 53. Anderson, Jr. (1995) presents additional conversions.

Application The continuity equation has immediate engineering applications. Study the flow of water through the contraction nozzle depicted in Figure 5.4.

As a simplification, we assume that the flow velocity vectors are constant across inlet ① and outlet ②. The task is to find the outlet velocity as a function of the geometry

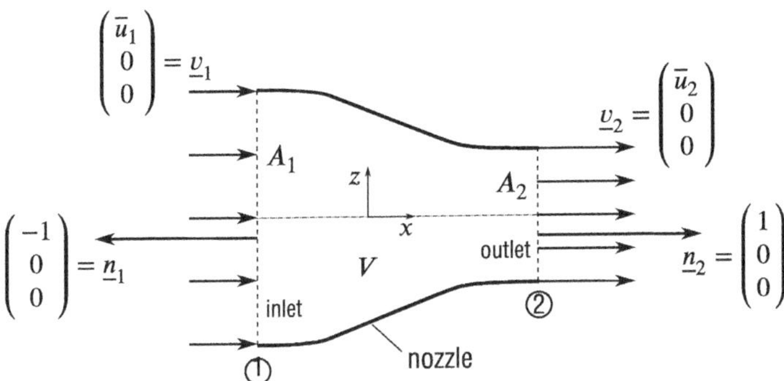

Figure 5.4 Flow through a contraction nozzle

and inlet velocity. Obviously, the nozzle is a finite, fixed control volume. Hence, the continuity equation (5.18) in its conservative integral form applies.

Water is considered incompressible, i.e. if the temperature does not change, the density is constant and its time derivative vanishes.

$$\text{because of} \quad \frac{\partial \rho}{\partial t} = 0 \quad \text{the volume integral vanishes:} \quad \iiint_V \frac{\partial \rho}{\partial t}\, \mathrm{d}V = 0$$

The continuity Equation (5.18) converts into

$$0 = \iint_S \rho\, \underline{v}^T \underline{n}\, \mathrm{d}S$$

This means that the net flow rate across the boundary S of the control volume is zero. In other words, since no fluid may escape through the nozzle wall, the inflow must be equal to the outflow.

$$0 = \iint_{A_1} \rho\, \underline{v}_1^T \underline{n}_1\, \mathrm{d}S + \iint_{A_2} \rho\, \underline{v}_2^T \underline{n}_2\, \mathrm{d}S$$

We substitute the vectors from Figure 5.4 into this equation and execute the dot products.

$$0 = \iint_{A_1} -\rho\, \bar{u}_1\, \mathrm{d}S + \iint_{A_2} \rho\, \bar{u}_2\, \mathrm{d}S$$

As stated above, the velocities $\bar{u}_1$ and $\bar{u}_2$ are assumed constant across inlet and outlet respectively. Consequently, the velocities may be extracted and the surface integrals result in the inlet and outlet cross section areas.

$$0 = -\rho\, \bar{u}_1\, A_1 + \rho\, \bar{u}_2\, A_2$$

Therefore,

$$A_1\, \bar{u}_1 = A_2\, \bar{u}_2$$

and we find that the outlet velocity is proportional to the ratio of inlet to outlet cross section areas.

$$\bar{u}_2 = \frac{A_1}{A_2}\, \bar{u}_1$$

If $A_1 > A_2$, the velocity will increase, i.e. $\bar{u}_2 > \bar{u}_1$. For $A_1/A_2 = 2.5$ and $\bar{u}_1 = 5.0\,\mathrm{m/s}$, we get $\bar{u}_2 = 12.5\,\mathrm{m/s}$.

References

Anderson, Jr., J. (1995). *Computational fluid dynamics – The basics with applications.* McGraw-Hill, Inc., New York, NY.

Self Study Problems

1. A balloon is completely filled with water and thrown into a river. What type of control volume does it represent?
2. Convert the differential, nonconservative form of the continuity equation (5.22) into its differential, conservative form (5.12). Justify the mathematical operation in every step.
3. Convert the integral, nonconservative form of the continuity equation (5.24) into its integral, conservative form (5.18). Justify the mathematical operation in every step. *Hint: make use of the divergence theorem.*

6

Navier-Stokes Equations

The Navier-Stokes equations were independently derived in the first half of the 19th century by the French engineer and physicist Claude Louis Marie Henri Navier (*1785 – †1836) and the Irish born scientist Sir George Gabriel Stokes (*1819 – †1903). Although we have known the equations for almost 200 years, we do not know much about them except that they are incredibly hard to solve. Currently, we can solve only simplified versions of the equations by means of sophisticated numerical algorithms and substantial computer power. These algorithms are discussed in books about CFD.

We will derive the equations here because they are so fundamental to fluid mechanics. Afterwards, we will simplify the equations and study some cases that are relevant to ship hydrodynamics. The Navier-Stokes equations express the conservation of momentum principle and represent Newton's second law (simplified $F = m\,a$) applied to a continuum. Newton's second law states that the momentum of a body will change only if an external force acts on it. If no resultant external force exists, a body will continue to move with constant speed on a straight path.

Learning Objectives

At the end of this chapter students will be able to:

- discuss momentum in the context of fluid mechanics
- set up the equilibrium of forces for a control volume
- follow the derivation of the Navier-Stokes equations
- identify terms contained in the Navier-Stokes equations

6.1 Momentum

Momentum is the product of mass m and velocity vector $\underline{v}$. Consequently, momentum $m\,\underline{v}$ is a vector quantity with magnitude and direction. It follows from Newton's second law that the time rate of change in momentum is equal to the vector sum of all external forces acting on the mass. **Momentum**

$$\frac{\mathrm{d}(m\underline{v})}{\mathrm{d}t} = \sum \underline{F}_{\text{external}} \tag{6.1}$$

Fundamentals of Ship Hydrodynamics: Fluid Mechanics, Ship Resistance and Propulsion, First Edition.
Lothar Birk.

Companion website: www.wiley.com/go/birk/hydrodynamics

In solid mechanics we commonly assume that the mass is constant and independent of time, which leads to the widely known form $m\frac{\mathrm{d}\underline{v}}{\mathrm{d}t} = \sum \underline{F}_{\text{external}}$ of Newton's second law. In addition, we concentrate the mass into a single point. Then $\frac{\mathrm{d}\underline{v}}{\mathrm{d}t}$ is the acceleration of a mass point caused by the external forces. A similar equation can be stated for the rotational motion of a mass point.

In fluid mechanics we need a more general approach because we cannot track every water molecule individually. In our discussion of the continuity equation, we employed four mathematical models based on four different control volumes which resulted in varying forms of the continuity equation. Derivation of the Navier-Stokes equations may be done using either of these models. However, since it is a lengthier process, only the derivation for an infinitesimally small control volume fixed in space is presented here. It is arguably the easiest to follow. Other forms are presented by Anderson, Jr. (1995).

6.2 Conservation of Momentum

Like mass, momentum is a conserved quantity. Changes in momentum may be caused by the flow of particles across the boundaries of a control volume (momentum transport) or by external forces acting on the control volume, which cause the fluid particles to accelerate, decelerate, or change direction. In a closed system, momentum will change only if there is a resultant external force. We will first define the rate of change of momentum before we consider the transport of momentum across the boundary of a control volume and the external forces acting on a control volume.

6.2.1 Time rate of change of momentum

The total amount of mass in a fluid element $\mathrm{d}V$ is $\mathrm{d}m = \rho\mathrm{d}V$ with momentum $\rho\mathrm{d}V\underline{v}$. Since the fluid element is fixed, the change of momentum per unit time is equal to the local time derivative.

$$\text{time rate of change of momentum:} \quad \frac{\partial}{\partial t}\left(\rho \mathrm{d}V \underline{v}\right) \tag{6.2}$$

Time rate of change of momentum

By definition, a fixed control volume $\mathrm{d}V$ may not change. Thus, $\mathrm{d}V$ represents a constant factor for the time derivative.

$$\text{time rate of change of momentum:} \quad \frac{\partial(\rho\underline{v})}{\partial t}\mathrm{d}V \tag{6.3}$$

Like the momentum itself, the time rate of change of momentum (6.3) is a vector too.

6.2.2 Momentum flux over boundary

We now compute the momentum flux across the boundaries of the fluid element. It expresses the change of momentum a mass experiences because it moves in the flow field. Figure 6.1 shows the fluid element with only the x-component of the momentum

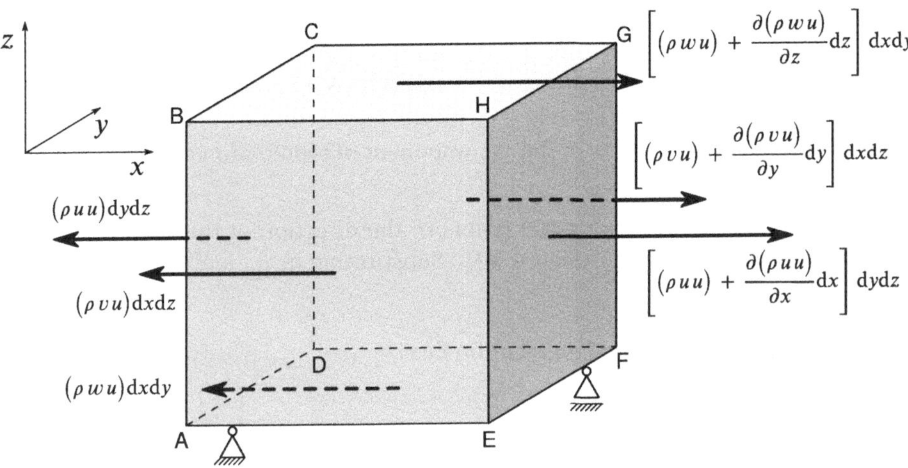

Figure 6.1 Momentum flux in x-direction through the surface of an infinitesimal, fixed fluid element $\mathrm{d}V$

flux across all six faces. For clarity, we omitted the y-component and z-component. Arrows indicate the sign associated with a specific momentum flux.

The mass flow rate (or mass flux) over an infinitesimal surface area $\mathrm{d}S$ is equal to $\rho \underline{v}^T \underline{n}\mathrm{d}S$. The mass flux depends on density, the velocity $\underline{v}^T \underline{n}$ normal to the surface element, and the area $\mathrm{d}S$ of the surface element. Note that the mass flux through $\mathrm{d}S$ is a scalar. As before, we let the normal vector $\underline{n}$ point out of the fluid domain (here the control volume $\mathrm{d}V$). The velocity vector has the components $\underline{v} = (u, v, w)^T$.

Transport of momentum

Momentum is tied to mass. As mass moves, it carries momentum with it. The amount of momentum transported per time unit through a surface element $\mathrm{d}S$ is

$$\text{momentum flux across } \mathrm{d}S \quad = \left(\rho\, \underline{v}^T \underline{n}\right) \underline{v}\, \mathrm{d}S \tag{6.4}$$

The momentum flux has the dimension of a force measured in Newtons:

$$\left[\left(\frac{\mathrm{kg}}{\mathrm{m}^3} \cdot \frac{\mathrm{m}}{\mathrm{s}} \cdot 1\right) \frac{\mathrm{m}}{\mathrm{s}} \cdot \mathrm{m}^2\right] = \left[\frac{\mathrm{kg{\cdot}m}}{\mathrm{s}^2}\right] = [\mathrm{N}]$$

The dot product $(\underline{v}^T \underline{n})$ is a scalar. Therefore, the components of Equation (6.4) are given by

$$x\text{-component of momentum flux across } \mathrm{d}S = \left(\rho \underline{v}^T \underline{n}\right) u\, \mathrm{d}S \tag{6.5}$$

$$y\text{-component of momentum flux across } \mathrm{d}S = \left(\rho \underline{v}^T \underline{n}\right) v\, \mathrm{d}S \tag{6.6}$$

$$z\text{-component of momentum flux across } \mathrm{d}S = \left(\rho \underline{v}^T \underline{n}\right) w\, \mathrm{d}S \tag{6.7}$$

We now apply this to the infinitesimal control volume $\mathrm{d}V$, which we consider fixed in space as indicated by the supports (Figure 6.1). This control volume has eight corners A, B, C, D, E, F, G, and H, as well as six quadrilateral faces which we identify by their corner points:

- with a normal vector parallel to the x-axis faces ABCD and EFGH

- with a normal vector parallel to the y-axis faces AEHB and CGFD
- with a normal vector parallel to the z-axis faces ADFE and BHGC

Flux of the x-component of momentum

We explicitly consider the flux of the x-component of momentum across all faces of the fluid element dV.

Individual faces

The normal vector of the face ABCD points in the direction of the negative x-axis, i.e. $\underline{n}_{\mathsf{ABCD}} = (-1, 0, 0)^T = -\underline{i}$ (Figure 6.1). Substituting $\underline{n}_{\mathsf{ABCD}}$ and $dS = dydz$ into Equation (6.5) yields

$$\begin{aligned} \text{ABCD: } x\text{-component of momentum flux} &= \left(\rho \underline{v}^T \underline{n}_{\mathsf{ABCD}}\right) u \, dydz \\ &= \rho \left(u, v, w\right) \begin{pmatrix} -1 \\ 0 \\ 0 \end{pmatrix} u \, dydz \\ &= -\left(\rho u u\right) dydz \end{aligned} \tag{6.8}$$

In Figure 6.1 the minus sign of the right-hand side term in Equation (6.8) is represented by the arrow which points in negative x-direction.

Over the length dx of a fluid element, the momentum flux per unit surface area (ρuu) may change. As for the conservation of mass, we express the change as a Taylor series expansion that we truncate after the linear term. The normal vector for the face EFGH points in positive x-direction $\underline{n}_{\mathsf{EFGH}} = (1, 0, 0)^T = \underline{i}$.

$$\text{EFGH: } x\text{-component of momentum flux} = \left[\rho u u + \frac{\partial(\rho u u)}{\partial x} dx\right] dydz \tag{6.9}$$

The momentum flux contributions for the remaining four faces of the fluid element are obtained in a similar manner by substituting the respective normal vectors and surface edge lengths into Equation (6.5).

$$\text{AEHB: } x\text{-component of momentum flux} = -\left(\rho v u\right) dxdz \tag{6.10}$$

$$\text{CGFD: } x\text{-component of momentum flux} = \left[\rho v u + \frac{\partial(\rho v u)}{\partial y} dy\right] dxdz \tag{6.11}$$

$$\text{ADFE: } x\text{-component of momentum flux} = -\left(\rho w u\right) dxdy \tag{6.12}$$

$$\text{BHGC: } x\text{-component of momentum flux} = \left[\rho w u + \frac{\partial(\rho w u)}{\partial z} dz\right] dxdy \tag{6.13}$$

What may be confusing at first is the fact that we have a flow of x-momentum over faces which are oriented in y- and z-directions. You have to remind yourself that the mass flux across a surface is ultimately a scalar. Momentum, however, is a vectorial quantity. Therefore, the mass flux over, for example, the top face BHGC, which is oriented in positive z-direction, can carry momentum in x-direction.

Summing up all parts of the x-component of the momentum flux (6.8) through (6.13) yields:

Total x-component of momentum flux

$$\begin{aligned}(\rho \underline{v}^T \underline{n} u)\, \mathrm{d}S &= -(\rho u u)\, \mathrm{d}y\mathrm{d}z + \left[\rho u u + \frac{\partial(\rho u u)}{\partial x}\mathrm{d}x\right] \mathrm{d}y\mathrm{d}z \\ &- (\rho v u)\, \mathrm{d}x\mathrm{d}z + \left[\rho v u + \frac{\partial(\rho v u)}{\partial y}\mathrm{d}y\right] \mathrm{d}x\mathrm{d}z \\ &- (\rho w u)\, \mathrm{d}x\mathrm{d}y + \left[\rho w u + \frac{\partial(\rho w u)}{\partial z}\mathrm{d}z\right] \mathrm{d}x\mathrm{d}y\end{aligned} \tag{6.14}$$

After simplifying Equation (6.14), only the differential terms remain on the right-hand side. They represent the total flux of the x-component of momentum across the boundary of $\mathrm{d}V$. Each term is proportional to the volume $\mathrm{d}V = \mathrm{d}x\mathrm{d}y\mathrm{d}z$ of the fluid element.

$$x\text{-component:} \quad (\rho \underline{v}^T \underline{n} u)\, \mathrm{d}S = \left[\frac{\partial(\rho u u)}{\partial x} + \frac{\partial(\rho v u)}{\partial y} + \frac{\partial(\rho w u)}{\partial z}\right] \mathrm{d}V \tag{6.15}$$

The y- and z-components of the momentum flux follow in similar fashion using Equations (6.6) and (6.7) instead of Equation (6.5):

Fluxes of the y- and z-component of momentum

$$y\text{-component:} \quad (\rho \underline{v}^T \underline{n} v)\, \mathrm{d}S = \left[\frac{\partial(\rho u v)}{\partial x} + \frac{\partial(\rho v v)}{\partial y} + \frac{\partial(\rho w v)}{\partial z}\right] \mathrm{d}V \tag{6.16}$$

$$z\text{-component:} \quad (\rho \underline{v}^T \underline{n} w)\, \mathrm{d}S = \left[\frac{\partial(\rho u w)}{\partial x} + \frac{\partial(\rho v w)}{\partial y} + \frac{\partial(\rho w w)}{\partial z}\right] \mathrm{d}V \tag{6.17}$$

6.2.3 External forces

Two types of external forces may act on a fluid element:

- body forces and
- surface forces

Figure 6.2 shows only the x-components of the external forces for clarity.

Body forces

Body forces are also called volume forces. They act on the mass element as a whole. Gravity and electromagnetic forces are of this type. The only body force we will consider is the gravity force $\underline{F}_g$, which we model as mass $\rho \mathrm{d}V$ times gravitational acceleration vector $\underline{f} = (f_x, f_y, f_z)^T = (0, 0, -g)^T$.

$$\underline{F}_g = \rho \underline{f} \mathrm{d}V \tag{6.18}$$

The x- and y-components of the weight force (6.18), $\rho f_x \mathrm{d}V$, actually vanish in Earth's gravity field. However, we keep them for the sake of completeness.

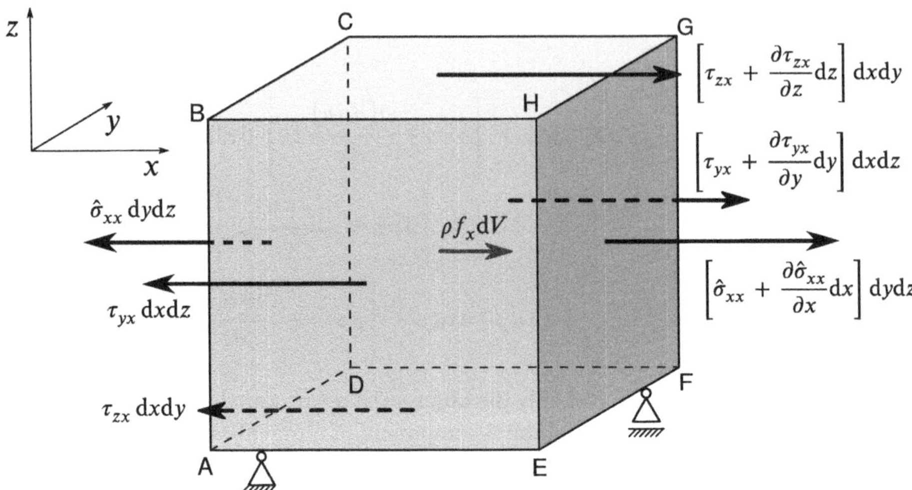

Figure 6.2 x-components of surface and body forces acting on the fixed, infinitesimal fluid element dV

Surface forces Surface forces are a result of normal and shear stresses acting on the fluid element. Collecting the stress related forces in x-direction for all six faces yields:

$$\begin{aligned} x\text{-comp. of surface force:} \quad & -\hat{\sigma}_{xx}\,\mathrm{d}y\mathrm{d}z + \left[\hat{\sigma}_{xx} + \frac{\partial \hat{\sigma}_{xx}}{\partial x}\mathrm{d}x\right]\mathrm{d}y\mathrm{d}z \\ & -\tau_{yx}\mathrm{d}x\mathrm{d}z + \left[\tau_{yx} + \frac{\partial \tau_{yx}}{\partial y}\mathrm{d}y\right]\mathrm{d}x\mathrm{d}z \\ & -\tau_{zx}\,\mathrm{d}x\mathrm{d}y + \left[\tau_{zx} + \frac{\partial \tau_{zx}}{\partial z}\mathrm{d}z\right]\mathrm{d}x\mathrm{d}y \end{aligned} \tag{6.19}$$

Combining the terms and substituting dV = dxdydz results in

$$x\text{-component of surface force:} \quad \left[\frac{\partial \hat{\sigma}_{xx}}{\partial x} + \frac{\partial \tau_{yx}}{\partial y} + \frac{\partial \tau_{zx}}{\partial z}\right]\mathrm{d}V \tag{6.20}$$

Repeating the surface force collection for the other two coordinate directions reveals:

$$y\text{-component of surface force:} \quad \left[\frac{\partial \tau_{xy}}{\partial x} + \frac{\partial \hat{\sigma}_{yy}}{\partial y} + \frac{\partial \tau_{zy}}{\partial z}\right]\mathrm{d}V \tag{6.21}$$

$$z\text{-component of surface force:} \quad \left[\frac{\partial \tau_{xz}}{\partial x} + \frac{\partial \tau_{yz}}{\partial y} + \frac{\partial \hat{\sigma}_{zz}}{\partial z}\right]\mathrm{d}V \tag{6.22}$$

There are nine stresses, but they are symmetric, i.e. $\tau_{xy} = \tau_{yx}$, $\tau_{xz} = \tau_{zx}$, and $\tau_{yz} = \tau_{zy}$. This leaves us with six stresses which have to be determined in addition to the velocity components and the pressure.

6.2.4 Conservation of momentum equations

Conservation of momentum in x-direction

In the three previous subsections 6.2.1, 6.2.2, and 6.2.3 we derived all parts needed to assemble the actual conservation of momentum equations. The x-component of the momentum time rate of change (6.3) and the momentum flux across the fluid element boundary (6.15) form the left-hand side of the conservation of momentum equation. It represents the total rate of change in x-momentum for the fluid element $\mathrm{d}V$. According to Newton's second law, this must be equal to the sum of external forces, which consists of the x-components of body force (6.18) and surface force (6.20). Together, we get the conservation equation for the x-component of the momentum:

$$\frac{\partial(\rho u)}{\partial t}\mathrm{d}V + \left[\frac{\partial(\rho u u)}{\partial x} + \frac{\partial(\rho v u)}{\partial y} + \frac{\partial(\rho w u)}{\partial z}\right]\mathrm{d}V = \rho f_x \mathrm{d}V + \left[\frac{\partial \hat{\sigma}_{xx}}{\partial x} + \frac{\partial \tau_{yx}}{\partial y} + \frac{\partial \tau_{zx}}{\partial z}\right]\mathrm{d}V \quad (6.23)$$

Obviously, all terms are proportional to the small but finite volume $\mathrm{d}V$, which allows us to divide Equation (6.23) by $\mathrm{d}V$:

$$\frac{\partial(\rho u)}{\partial t} + \left[\frac{\partial(\rho u u)}{\partial x} + \frac{\partial(\rho v u)}{\partial y} + \frac{\partial(\rho w u)}{\partial z}\right] = \rho f_x + \left[\frac{\partial \hat{\sigma}_{xx}}{\partial x} + \frac{\partial \tau_{yx}}{\partial y} + \frac{\partial \tau_{zx}}{\partial z}\right] \quad (6.24)$$

Conservation of momentum in y- and z- direction

The corresponding y- and z-components of the conservation of momentum equations are:

$$\frac{\partial(\rho v)}{\partial t} + \left[\frac{\partial(\rho u v)}{\partial x} + \frac{\partial(\rho v v)}{\partial y} + \frac{\partial(\rho w v)}{\partial z}\right] = \rho f_y + \left[\frac{\partial \tau_{xy}}{\partial x} + \frac{\partial \hat{\sigma}_{yy}}{\partial y} + \frac{\partial \tau_{zy}}{\partial z}\right] \quad (6.25)$$

$$\frac{\partial(\rho w)}{\partial t} + \left[\frac{\partial(\rho u w)}{\partial x} + \frac{\partial(\rho v w)}{\partial y} + \frac{\partial(\rho w w)}{\partial z}\right] = \rho f_z + \left[\frac{\partial \tau_{xz}}{\partial x} + \frac{\partial \tau_{yz}}{\partial y} + \frac{\partial \hat{\sigma}_{zz}}{\partial z}\right] \quad (6.26)$$

Closure problem

The three components of the conservation of momentum equations describe the relationships between ten yet unknown flow properties:

1. density ρ
2. three velocity components u, v, w
3. six stresses $\hat{\sigma}_{xx}$ $\hat{\sigma}_{yy}$, $\hat{\sigma}_{zz}$, τ_{xy}, τ_{xz}, and τ_{yz}

Together with the continuity equation, we have four equations to solve for ten unknowns. Obviously, we have to gather more equations to properly model the problem and to ultimately obtain a solution.

6.3 Stokes' Hypothesis

Hooke's Law The same problem occurs in mechanics of materials with the equilibrium of elastic material: more unknowns than equations. The missing equations are derived from observations and material properties. Robert Hooke (*1635 – †1703), a 17th century English natural philosopher and architect, stated that the normal stress σ in elastic bodies is proportional to the change in length as a fraction of total length called strain ε: $\sigma = E\varepsilon$, the material property being the modulus of elasticity E.

Newtonian fluids Newton was the first to show that the shear stress in some fluids is a function of the rate of change of strain, i.e. the shear stress depends on how fast the fluid is deformed.

$$\tau = \mu \frac{\partial u}{\partial y} \tag{6.27}$$

He derived this theoretically, based on flow observations and basic assumptions. However, experiments later confirmed this approach. In honor of his discovery, fluids are called Newtonian fluids if they have a linear relationship between the stresses and the rates of change of strain. The constant of proportionality is the dynamic viscosity μ which is measured in [Pa s] or [(N s)/m^2]. For flow conditions typically encountered in ship hydrodynamics, water and air are considered Newtonian fluids. We further assume that the fluid is isotropic, i.e. the dynamic viscosity does not depend on the direction in which it is measured.

In analogy to the general stress–strain relationship for elastic bodies, a relationship is derived for fluids which links stresses with the rate of change of strain. Note the difference!

- Elastic body: stress is proportional to the deformation (strain).
- Fluid: stress is a function of *how fast* (rate of change of strain) the fluid is deformed.

This exploits physical observations and mathematical properties of tensors.

The Irish mathematician and physicist Sir George Gabriel Stokes (*1819 – †1903) introduced the necessary equations for normal and shear stresses (6.28) during a lecture in 1845 (Stokes, 1849).

$$\begin{aligned}
\hat{\sigma}_{xx} &= -p + \lambda \underline{\nabla}^T \underline{v} + 2\mu \frac{\partial u}{\partial x} & \tau_{xy} &= \tau_{yx} = \mu \left[\frac{\partial v}{\partial x} + \frac{\partial u}{\partial y}\right] \\
\hat{\sigma}_{yy} &= -p + \lambda \underline{\nabla}^T \underline{v} + 2\mu \frac{\partial v}{\partial y} & \tau_{xz} &= \tau_{zx} = \mu \left[\frac{\partial u}{\partial z} + \frac{\partial w}{\partial x}\right] \\
\hat{\sigma}_{zz} &= -p + \lambda \underline{\nabla}^T \underline{v} + 2\mu \frac{\partial w}{\partial z} & \tau_{yz} &= \tau_{zy} = \mu \left[\frac{\partial w}{\partial y} + \frac{\partial v}{\partial z}\right]
\end{aligned} \tag{6.28}$$

For convenience, normal stresses are split into three parts:

1. The pressure p, which is the negative mean of the three normal stresses:

$$p = -\frac{1}{3}\left(\hat{\sigma}_{xx} + \hat{\sigma}_{yy} + \hat{\sigma}_{zz}\right) \tag{6.29}$$

   It is common to all coordinate directions.

2. A term which is proportional to the volume dilation $\underline{\nabla}^T \underline{v}$, i.e. the divergence of velocity.

3. The part which is proportional to the rate of change of strain in the respective coordinate direction, e.g. $\partial u/\partial x$.

By separating the pressure from the remaining normal stress, the limiting case for vanishing velocity lets the equations collapse to the hydrostatic case.

A more elaborate discussion about the analogy of the stress–rate of change of strain relationship for fluids and the stress–strain relationship of elastic bodies can be found in Schlichting and Gersten (2003, Chapter 3).

Two material properties appear in the Equations (6.28) for the stresses:

Stokes' hypothesis

- the dynamic viscosity μ and
- the second viscosity λ.

Determining these properties is a task of the scientific field of rheology.

In principle, the viscosity can depend on time, position, as well as the state of a flow field itself (temperature, stress, velocity, etc.). Luckily for us, air and water belong to the class of Newtonian fluids. For Newtonian fluids, the material properties μ and λ are independent of the actual stress and velocity values.

Dynamic viscosity differs from fluid to fluid. Viscosities of water and air are a function of temperature (see Section 3.2.1). Very little is known about the second viscosity λ. Since it is associated with changes in volume, it cannot be observed with incompressible fluids. Remember that the divergence of the velocity vanishes for fluids with constant density: $\underline{\nabla}^T \underline{v} = 0$.

Stokes solved the difficulty posed by the unknown second viscosity by drawing an analogy to elastic bodies. He postulated the relationship $3\lambda + 2\mu = 0$ between dynamic and second viscosity, which has been named in his honor:

$$\text{Stokes' hypothesis for second viscosity} \quad \lambda = -\frac{2}{3}\mu \qquad (6.30)$$

Our experience shows that the resulting Navier-Stokes equations describe viscous fluid flow correctly over a wide range of applications. However, the validity of Stokes' hypothesis is still debated among experts. See for instance Gad-el Hak (1995).

6.4 Navier-Stokes Equations for a Newtonian Fluid

Navier-Stokes equations

Finally, substituting the stress–rate of change of strain relationships (6.28) together with Stokes' hypothesis (6.30) into the conservation of momentum equations (6.24), (6.25), and (6.26) yields the Navier-Stokes equations (NSE) for an isotropic Newtonian fluid in their conservative, differential form, which describe viscous flows.

$$\frac{\partial(\rho u)}{\partial t}+\left[\frac{\partial(\rho u u)}{\partial x}+\frac{\partial(\rho v u)}{\partial y}+\frac{\partial(\rho w u)}{\partial z}\right]=\rho f_x-\frac{\partial p}{\partial x}$$
$$-\frac{\partial}{\partial x}\left[\frac{2}{3}\mu\underline{\nabla}^T\underline{v}-2\mu\frac{\partial u}{\partial x}\right]+\frac{\partial}{\partial y}\left[\mu\left(\frac{\partial v}{\partial x}+\frac{\partial u}{\partial y}\right)\right]+\frac{\partial}{\partial z}\left[\mu\left(\frac{\partial u}{\partial z}+\frac{\partial w}{\partial x}\right)\right] \quad (6.31)$$

$$\frac{\partial(\rho v)}{\partial t}+\left[\frac{\partial(\rho u v)}{\partial x}+\frac{\partial(\rho v v)}{\partial y}+\frac{\partial(\rho w v)}{\partial z}\right]=\rho f_y-\frac{\partial p}{\partial y}$$
$$+\frac{\partial}{\partial x}\left[\mu\left(\frac{\partial v}{\partial x}+\frac{\partial u}{\partial y}\right)\right]-\frac{\partial}{\partial y}\left[\frac{2}{3}\mu\underline{\nabla}^T\underline{v}-2\mu\frac{\partial v}{\partial y}\right]+\frac{\partial}{\partial z}\left[\mu\left(\frac{\partial w}{\partial y}+\frac{\partial v}{\partial z}\right)\right] \quad (6.32)$$

$$\frac{\partial(\rho w)}{\partial t}+\left[\frac{\partial(\rho u w)}{\partial x}+\frac{\partial(\rho v w)}{\partial y}+\frac{\partial(\rho w w)}{\partial z}\right]=\rho f_z-\frac{\partial p}{\partial z}$$
$$+\frac{\partial}{\partial x}\left[\mu\left(\frac{\partial u}{\partial z}+\frac{\partial w}{\partial x}\right)\right]+\frac{\partial}{\partial y}\left[\mu\left(\frac{\partial w}{\partial y}+\frac{\partial v}{\partial z}\right)\right]-\frac{\partial}{\partial z}\left[\frac{2}{3}\mu\underline{\nabla}^T\underline{v}-2\mu\frac{\partial w}{\partial z}\right] \quad (6.33)$$

All terms in equations (6.31), (6.32), and (6.33) represent forces per unit fluid volume. Figure 6.3 identifies the force components in the NSE. On the left-hand side we have the inertia force and on the right-hand side we see the sum of external forces which comprises body and surface forces.

C.L. Navier, G.G. Stokes

The Navier-Stokes equations are named after Stokes and Navier, who published a form of the NSE earlier than Stokes (Navier, 1822). However, Navier had no physical concept for the material parameter we now know as viscosity (Navier, 1822).

Mathematical classification

The Navier-Stokes equations are a set of coupled, nonlinear partial differential equations. They are nonlinear because of the convection terms in brackets on the left side of (6.31), (6.32), and (6.33), which contain products of the unknown velocity components. If you also consider the density and viscosity distributions as unknown, then more terms are nonlinear. Luckily, we may assume that kinematic viscosity and density are constant for most problems in ship hydrodynamics. We will exploit this in the next chapter to simplify the equations.

Even today we know very little about the existence and uniqueness of solutions to the Navier-Stokes equations. The Clay Mathematical Institute (CMI) in Cambridge, MA made the Navier-Stokes equations one of their famous Millenium Problems. A contribution which represents *substantial progress toward a mathematical theory which will unlock the secrets hidden in the Navier-Stokes equations* can earn you a $1 million award (see http://www.claymath.org/millennium-problems/navier-stokes-equation).

NSE solvers

All modern approaches of CFD in ship hydrodynamics actually do not solve the equations above but a simplified version of them called the Reynolds averaged Navier-Stokes equations or short RANSE (see Chapter 8). The key is that the time dependent flow velocity is subdivided into a time dependent average velocity and a highly fluctuating turbulence. This leads to an additional term in the equations. It has the form of a stress

$$\frac{\partial(\rho u)}{\partial t} + \left[\frac{\partial(\rho u u)}{\partial x} + \frac{\partial(\rho v u)}{\partial y} + \frac{\partial(\rho w u)}{\partial z}\right] = \rho f_x - \frac{\partial p}{\partial x} - \frac{\partial}{\partial x}\left[\frac{2}{3}\mu\underline{\nabla}^T\underline{v} - 2\mu\frac{\partial u}{\partial x}\right] + \frac{\partial}{\partial y}\left[\mu\left(\frac{\partial v}{\partial x} + \frac{\partial u}{\partial y}\right)\right] + \frac{\partial}{\partial z}\left[\mu\left(\frac{\partial u}{\partial z} + \frac{\partial w}{\partial x}\right)\right]$$

$$\frac{\partial(\rho v)}{\partial t} + \left[\frac{\partial(\rho u v)}{\partial x} + \frac{\partial(\rho v v)}{\partial y} + \frac{\partial(\rho w v)}{\partial z}\right] = \rho f_y - \frac{\partial p}{\partial y} - \frac{\partial}{\partial y}\left[\frac{2}{3}\mu\underline{\nabla}^T\underline{v} - 2\mu\frac{\partial v}{\partial y}\right] + \frac{\partial}{\partial x}\left[\mu\left(\frac{\partial v}{\partial x} + \frac{\partial u}{\partial y}\right)\right] + \frac{\partial}{\partial z}\left[\mu\left(\frac{\partial w}{\partial y} + \frac{\partial v}{\partial z}\right)\right]$$

$$\frac{\partial(\rho w)}{\partial t} + \left[\frac{\partial(\rho u w)}{\partial x} + \frac{\partial(\rho v w)}{\partial y} + \frac{\partial(\rho w w)}{\partial z}\right] = \rho f_z - \frac{\partial p}{\partial z} - \frac{\partial}{\partial z}\left[\frac{2}{3}\mu\underline{\nabla}^T\underline{v} - 2\mu\frac{\partial w}{\partial z}\right] + \frac{\partial}{\partial x}\left[\mu\left(\frac{\partial u}{\partial z} + \frac{\partial w}{\partial x}\right)\right] + \frac{\partial}{\partial y}\left[\mu\left(\frac{\partial w}{\partial y} + \frac{\partial v}{\partial z}\right)\right]$$

Figure 6.3 Forces comprising the Navier-Stokes equations for an isotropic Newtonian fluid

tensor and in honor of Osborne Reynolds is named the Reynolds stress tensor. For the newly introduced unknowns additional equations are needed, which are known as turbulence models.

References

Anderson, Jr., J. (1995). *Solutions manual to accompany computational fluid dynamics – The basics with applications.* McGraw-Hill, Inc., New York, NY.

Gad-el Hak, M. (1995). Stokes' hypothesis for a Newtonian, isotropic fluid. *Journal of Fluids Engineering*, 117(1):3–5.

Navier, C. (1822). Mémoire sur les lois du mouvement des fluides. *Mém. Acad. Sci. Inst. France*, 6:389–440.

Schlichting, H. and Gersten, K. (2003). *Boundary-layer theory.* Springer-Verlag, Berlin, Heidelberg, New York, eigth edition. Corrected printing.

Stokes, G. (1849). On the theories of the internal friction of fluids in motion, and of the equilibrium and motion of elastic solids. *Transactions of the Cambridge Philosophical Society*, 8:287–319.

Self Study Problems

1. Draw duplicates of the fluid element dV in Figure 6.1 and sketch in momentum fluxes for y and z directions. Follow the steps of Equations (6.4) through (6.15) and derive the momentum flux in y- and z-direction. You should arrive at the result shown in Equations (6.16) and (6.17).
2. Summarize in your own words Stokes' hypothesis.
3. Why are we not able to determine the second viscosity λ for water?

7

Special Cases of the Navier-Stokes Equations

The Navier-Stokes equations developed in the previous chapter contain the fluid properties density ρ and dynamic viscosity μ. The density is also part of the continuity equation (5.12). We will simplify both the continuity equation and the NSE for the ship resistance and propulsion problem by making reasonable assumptions about density and viscosity. A dimensionless form of the NSE will provide insight into the different scales of the forces driving the flow.

Learning Objectives

At the end of this chapter students will be able to:

- find the effects of constant density on conservation of mass and momentum
- identify the terms in the NSE as forces per unit mass
- formulate a dimensionless version of the NSE
- introduce characteristic, dimensionless parameters
- explain the challenge of high Reynolds numbers

7.1 Incompressible Fluid of Constant Temperature

Constant temperature

Ships sail at the phase boundary of water and air. We discussed important properties of these fluids in Chapter 3. Besides salinity, density and dynamic viscosity of water are mainly a function of fluid temperature. At the speeds achievable by ships, friction is too low to cause a measurable change in water temperature. Therefore, it is opportune to treat fluid temperature as constant.

Constant viscosity and density

Salinity tends to be uniform in a stretch of ocean or waterway. Thus, if the temperature does not change either, we can assume that the dynamic viscosity is constant for a specific flow problem. We also observe that water changes its volume only slightly even under high pressure. With small dependency on pressure and constant temperature we

Fundamentals of Ship Hydrodynamics: Fluid Mechanics, Ship Resistance and Propulsion, First Edition.
Lothar Birk.

Companion website: www.wiley.com/go/birk/hydrodynamics

treat the density of water as a constant as well. In particular, we assume that density is independent of space and time and commonly say water is incompressible.

Air changes its volume a lot more easily under pressure. Just think of the hand pump you use to inflate a ball. Even if you block the nozzle you can move the piston quite a bit, thereby compressing the air within the pump. However, for the modest velocities of ships, variations in pressure are not big enough to cause considerable changes in air density. Therefore, air is treated as incompressible as well.

Continuity equation for incompressible fluid

With density ρ and dynamic viscosity μ fixed, continuity and Navier-Stokes equations may be simplified. We start with the conservative, differential form of the continuity equation (5.12):

$$\frac{\partial \rho}{\partial t} + \underline{\nabla}^T(\rho \underline{v}) = 0 \tag{5.12}$$

Obviously, the time derivative of the density (first term) will vanish, if the density does not change over time. From the second term we factor out the space independent density ρ:

$$\underbrace{\frac{\partial \rho}{\partial t}}_{0} + \rho \underline{\nabla}^T \underline{v} = 0$$

Finally, we divide by the density, which is always nonzero. Otherwise there would be a vacuum and we would not be discussing fluid mechanics.

$$\underline{\nabla}^T \underline{v} = 0 \tag{7.1}$$

The final expression (7.1) is the continuity equation for an incompressible fluid. The left-hand side is the divergence of the velocity which we introduced previously. In Section 4.2.2, we discovered that the divergence of the velocity is related to the change in volume. If the divergence of velocity vanishes, a given mass will always occupy the same volume. Imagine a balloon filled with water. You can change its form but not its volume, i.e. the fluid is incompressible.

NSE for constant density and viscosity

Now let's consider the components of the Navier-Stokes equations (6.31), (6.32), and (6.33). Constant density ρ and dynamic viscosity μ can be extracted from the derivatives on the left-hand and right-hand sides:

$$\rho \frac{\partial u}{\partial t} + \rho \left[\frac{\partial (u\,u)}{\partial x} + \frac{\partial (v\,u)}{\partial y} + \frac{\partial (w\,u)}{\partial z}\right] = \rho f_x - \frac{\partial p}{\partial x}$$
$$- \mu \frac{\partial}{\partial x}\left[\frac{2}{3}\underline{\nabla}^T \underline{v} - 2\frac{\partial u}{\partial x}\right] + \mu \frac{\partial}{\partial y}\left[\frac{\partial v}{\partial x} + \frac{\partial u}{\partial y}\right] + \mu \frac{\partial}{\partial z}\left[\frac{\partial u}{\partial z} + \frac{\partial w}{\partial x}\right]$$

$$\rho \frac{\partial v}{\partial t} + \rho \left[\frac{\partial (u\,v)}{\partial x} + \frac{\partial (v\,v)}{\partial y} + \frac{\partial (w\,v)}{\partial z}\right] = \rho f_y - \frac{\partial p}{\partial y}$$
$$+ \mu \frac{\partial}{\partial x}\left[\frac{\partial v}{\partial x} + \frac{\partial u}{\partial y}\right] - \mu \frac{\partial}{\partial y}\left[\frac{2}{3}\underline{\nabla}^T \underline{v} - 2\frac{\partial v}{\partial y}\right] + \mu \frac{\partial}{\partial z}\left[\frac{\partial w}{\partial y} + \frac{\partial v}{\partial z}\right] \tag{7.2}$$

$$\rho\frac{\partial w}{\partial t}+\rho\left[\frac{\partial(u\,w)}{\partial x}+\frac{\partial(v\,w)}{\partial y}+\frac{\partial(w\,w)}{\partial z}\right]=\rho f_z-\frac{\partial p}{\partial z}$$
$$+\mu\frac{\partial}{\partial x}\left[\frac{\partial u}{\partial z}+\frac{\partial w}{\partial x}\right]+\mu\frac{\partial}{\partial y}\left[\frac{\partial w}{\partial y}+\frac{\partial v}{\partial z}\right]-\mu\frac{\partial}{\partial z}\left[\frac{2}{3}\underline{\nabla}^T\underline{v}-2\frac{\partial w}{\partial z}\right]$$

The viscous forces on the right-hand side contain a term with the divergence of velocity. According to the continuity equation for incompressible fluids, this term vanishes as shown here for the x-component of Equation (7.2).

$$\rho\frac{\partial u}{\partial t}+\rho\left[\frac{\partial(u\,u)}{\partial x}+\frac{\partial(v\,u)}{\partial y}+\frac{\partial(w\,u)}{\partial z}\right]=\rho f_x-\frac{\partial p}{\partial x}$$
$$-\mu\frac{\partial}{\partial x}\left[\frac{2}{3}\underbrace{\underline{\nabla}^T\underline{v}}_{=0}-2\frac{\partial u}{\partial x}\right]+\mu\frac{\partial}{\partial y}\left[\frac{\partial v}{\partial x}+\frac{\partial u}{\partial y}\right]+\mu\frac{\partial}{\partial z}\left[\frac{\partial u}{\partial z}+\frac{\partial w}{\partial x}\right] \quad (7.3)$$

All components are divided by the density ρ, and we introduce the kinematic viscosity $\nu=\mu/\rho$.

$$\frac{\partial u}{\partial t}+\frac{\partial(u\,u)}{\partial x}+\frac{\partial(v\,u)}{\partial y}+\frac{\partial(w\,u)}{\partial z}=f_x-\frac{1}{\rho}\frac{\partial p}{\partial x}$$
$$-\nu\frac{\partial}{\partial x}\left[-2\frac{\partial u}{\partial x}\right]+\nu\frac{\partial}{\partial y}\left[\frac{\partial v}{\partial x}+\frac{\partial u}{\partial y}\right]+\nu\frac{\partial}{\partial z}\left[\frac{\partial u}{\partial z}+\frac{\partial w}{\partial x}\right]$$

$$\frac{\partial v}{\partial t}+\frac{\partial(u\,v)}{\partial x}+\frac{\partial(v\,v)}{\partial y}+\frac{\partial(w\,v)}{\partial z}=f_y-\frac{1}{\rho}\frac{\partial p}{\partial y}$$
$$+\nu\frac{\partial}{\partial x}\left[\frac{\partial v}{\partial x}+\frac{\partial u}{\partial y}\right]-\nu\frac{\partial}{\partial y}\left[-2\frac{\partial v}{\partial y}\right]+\nu\frac{\partial}{\partial z}\left[\frac{\partial w}{\partial y}+\frac{\partial v}{\partial z}\right] \quad (7.4)$$

$$\frac{\partial w}{\partial t}+\frac{\partial(u\,w)}{\partial x}+\frac{\partial(v\,w)}{\partial y}+\frac{\partial(w\,w)}{\partial z}=f_z-\frac{1}{\rho}\frac{\partial p}{\partial z}$$
$$+\nu\frac{\partial}{\partial x}\left[\frac{\partial u}{\partial z}+\frac{\partial w}{\partial x}\right]+\nu\frac{\partial}{\partial y}\left[\frac{\partial w}{\partial y}+\frac{\partial v}{\partial z}\right]-\nu\frac{\partial}{\partial z}\left[-2\frac{\partial w}{\partial z}\right]$$

Next, we further differentiate the velocity derivatives on the right-hand side with respect to the coordinate directions x, y, and z and reorganize the resulting second order derivatives. For brevity, we show this transformation only for the x-component of the NSE.

$$\frac{\partial u}{\partial t}+\frac{\partial(u\,u)}{\partial x}+\frac{\partial(v\,u)}{\partial y}+\frac{\partial(w\,u)}{\partial z}$$
$$=f_x-\frac{1}{\rho}\frac{\partial p}{\partial x}+2\nu\frac{\partial^2 u}{\partial x^2}+\nu\left[\frac{\partial^2 v}{\partial x\partial y}+\frac{\partial^2 u}{\partial y^2}\right]+\nu\left[\frac{\partial^2 u}{\partial z^2}+\frac{\partial^2 w}{\partial x\partial z}\right]$$
$$=f_x-\frac{1}{\rho}\frac{\partial p}{\partial x}+\nu\left(\frac{\partial^2 u}{\partial x^2}+\frac{\partial^2 u}{\partial y^2}+\frac{\partial^2 u}{\partial z^2}\right)+\nu\left(\frac{\partial^2 u}{\partial x^2}+\frac{\partial^2 v}{\partial x\partial y}+\frac{\partial^2 w}{\partial x\partial z}\right) \quad (7.5)$$

The first pair of parentheses on the right of (7.5) contains the sum of second order derivatives of the horizontal velocity component u. We summarize this term using the scalar Laplace operator $\mathbf{\Delta} = \frac{\partial^2}{\partial x^2} + \frac{\partial^2}{\partial y^2} + \frac{\partial^2}{\partial z^2}$

$$\mathbf{\Delta} u = \frac{\partial^2 u}{\partial x^2} + \frac{\partial^2 u}{\partial y^2} + \frac{\partial^2 u}{\partial z^2} \tag{7.6}$$

The second pair of parentheses on the right of Equation (7.5) is the x-derivative of the divergence of velocity which – you guessed correctly – vanishes for incompressible fluids (7.1):

$$\begin{aligned}\left(\frac{\partial^2 u}{\partial x^2} + \frac{\partial^2 v}{\partial x \partial y} + \frac{\partial^2 w}{\partial x \partial z}\right) &= \frac{\partial}{\partial x}\underbrace{\left(\frac{\partial u}{\partial x} + \frac{\partial v}{\partial y} + \frac{\partial w}{\partial z}\right)}_{=\underline{\nabla}^T \underline{v}} \\ &= \frac{\partial}{\partial x}\left(\cancelto{=0}{\underline{\nabla}^T \underline{v}}\right) \\ &= 0\end{aligned} \tag{7.7}$$

The continuity equation allows us to simplify the left-hand side as well. We first apply the product rule of differentiation to the convective derivative:

$$\frac{\partial(u\,u)}{\partial x} + \frac{\partial(v\,u)}{\partial y} + \frac{\partial(w\,u)}{\partial z} = u\frac{\partial u}{\partial x} + u\frac{\partial u}{\partial x} + v\frac{\partial u}{\partial y} + u\frac{\partial v}{\partial y} + w\frac{\partial u}{\partial z} + u\frac{\partial w}{\partial z}$$

Reordering the summands results in

$$\frac{\partial(u\,u)}{\partial x} + \frac{\partial(v\,u)}{\partial y} + \frac{\partial(w\,u)}{\partial z} = u\frac{\partial u}{\partial x} + v\frac{\partial u}{\partial y} + w\frac{\partial u}{\partial z} + u\cancelto{=0}{\left(\frac{\partial u}{\partial x} + \frac{\partial v}{\partial y} + \frac{\partial w}{\partial z}\right)}$$

The last term in parenthesis is once more the divergence of the velocity and vanishes for constant density ρ.

$$\frac{\partial(u\,u)}{\partial x} + \frac{\partial(v\,u)}{\partial y} + \frac{\partial(w\,u)}{\partial z} = u\frac{\partial u}{\partial x} + v\frac{\partial u}{\partial y} + w\frac{\partial u}{\partial z}$$

The right-hand side may now be rewritten in a convenient vector form:

$$\frac{\partial(u\,u)}{\partial x} + \frac{\partial(v\,u)}{\partial y} + \frac{\partial(w\,u)}{\partial z} = \left(\underline{v}^T \underline{\nabla}\right)u \tag{7.8}$$

Note that the expression $\left(\underline{v}^T \underline{\nabla}\right)$ is a scalar differential operator and is *not* the divergence of velocity.

Substituting the results of Equations (7.6), (7.7), and (7.8) back into Equation (7.5) yields the x-component of the NSE for incompressible fluids.

$$\frac{\partial u}{\partial t} + \left(\underline{v}^T \underline{\nabla}\right)u = f_x - \frac{1}{\rho}\frac{\partial p}{\partial x} + \nu \mathbf{\Delta} u \tag{7.9}$$

We perform the same transformations for the y- and z-components of the NSE:

$$\frac{\partial v}{\partial t} + \left(\underline{v}^T \underline{\nabla}\right) v = f_y - \frac{1}{\rho}\frac{\partial p}{\partial y} + \nu \Delta v \tag{7.10}$$

$$\frac{\partial w}{\partial t} + \left(\underline{v}^T \underline{\nabla}\right) w = f_z - \frac{1}{\rho}\frac{\partial p}{\partial z} + \nu \Delta w \tag{7.11}$$

The three components of the NSE may now be assembled into a convenient vector equation:

Vector form of NSE ($\rho, \nu = \text{const.}$)

$$\underbrace{\frac{\partial \underline{v}}{\partial t} + \left(\underline{v}^T \underline{\nabla}\right) \underline{v}}_{①} = \underbrace{\underline{f}}_{②} - \underbrace{\frac{1}{\rho}\underline{\nabla} p}_{③} + \underbrace{\nu \Delta \underline{v}}_{④} \tag{7.12}$$

This looks a lot more compact than the Navier-Stokes equations we developed first in Chapter 6. The individual terms still represent the same forces as in Figure 6.3. This time, however, all forces are given as accelerations, i.e. force per unit mass.

1. Inertia force on the left-hand side.
2. Vector of body forces $\underline{f}$.
3. Pressure force $\frac{1}{\rho}\underline{\nabla} p$.
4. Viscous force $\nu \Delta \underline{v}$.

Equation (7.12) represents the conservation of momentum for an incompressible Newtonian fluid at constant temperature. Together with the continuity equation (7.1), it describes the motion of water around ships. We now have four equations – the continuity equation and the three components of the NSE – which in principle allow us to find the four unknowns: pressure p and the components u, v, and w of the velocity vector.

In practice, most available CFD systems do not attempt to solve Equation (7.12) directly. For high Reynolds numbers this direct numerical simulation (DNS) is still beyond our computational capabilities. Instead, we solve the RANSE, which are derived by splitting the instantaneous time dependent velocity into an average velocity (still time dependent) and the turbulent velocity. This introduces the initially unknown Reynolds stress tensor. The added unknowns require additional equations to represent the turbulence, so-called turbulence models. The latter are still a focus of ongoing research and several turbulence models of varying complexity are in use for ship resistance computations. The interested reader can find a derivation of the RANSE form of the Navier-Stokes equations for incompressible flow in Chapter 8.

7.2 Dimensionless Navier-Stokes Equations

Additional insight into the Navier-Stokes equations can be gained by rewriting them in a dimensionless form. This is general practice for implementation of numerical solutions methods. Therefore, this example might suffice to show this useful technique.

We revisit the NSE (7.12):

$$\frac{\partial \underline{v}}{\partial t} + \left(\underline{v}^T \underline{\nabla}\right)\underline{v} = \underline{f} - \frac{1}{\rho}\underline{\nabla} p + \nu \boldsymbol{\Delta} \underline{v} \tag{7.12}$$

As mentioned, both sides represent accelerations and are measured in the physical unit m/s^2.

Dimensionless lengths and $\underline{\nabla}$

We make quantities dimensionless by dividing them by a suitable reference value. For instance, all space variables are divided by a reference length L:

$$x^* = \frac{x}{L} \qquad y^* = \frac{y}{L} \qquad z^* = \frac{z}{L} \tag{7.13}$$

This seems fairly obvious. Less apparent is that the Nabla operator has a dimension as well. The dimensionless Nabla operator $\underline{\nabla}^*$ is equal to

$$\underline{\nabla}^* = \begin{pmatrix} \dfrac{\partial}{\partial x^*} \\ \dfrac{\partial}{\partial y^*} \\ \dfrac{\partial}{\partial z^*} \end{pmatrix} = \begin{pmatrix} \dfrac{\partial}{\partial\left(\frac{x}{L}\right)} \\ \dfrac{\partial}{\partial\left(\frac{y}{L}\right)} \\ \dfrac{\partial}{\partial\left(\frac{z}{L}\right)} \end{pmatrix} = \begin{pmatrix} L\dfrac{\partial}{\partial x} \\ L\dfrac{\partial}{\partial y} \\ L\dfrac{\partial}{\partial z} \end{pmatrix} = L\underline{\nabla} \tag{7.14}$$

Dimensionless Laplace operator $\boldsymbol{\Delta}$

The dimensionless Laplace operator is derived from the identity $\boldsymbol{\Delta} = \underline{\nabla}^T\underline{\nabla}$:

$$\boldsymbol{\Delta}^* = \left(\underline{\nabla}^*\right)^T \underline{\nabla}^* = \left(L\underline{\nabla}\right)^T L\underline{\nabla} = L^2\boldsymbol{\Delta} \tag{7.15}$$

Any geometric characteristic of the flow problem may serve as reference length L, for example the length or beam of a vessel, the diameter of a propeller, the waterdepth, and others.

Dimensionless time

We also introduce a reference period T. Possible choices are the duration of a process or the period of a recurring event. Our dimensionless time t^* and partial derivative will be

$$t^* = \frac{t}{T} \qquad \frac{\partial}{\partial t^*} = T\frac{\partial}{\partial t} \tag{7.16}$$

Dimensionless velocity

For convenience, we will select a reference velocity of magnitude U_∞, e.g. ship or flow speed. In practice, reference length and reference period can be used to define a reference velocity or length and velocity constitute a reference time. The dimensionless velocity vector becomes

$$\underline{v}^* = \begin{pmatrix} u^* \\ v^* \\ w^* \end{pmatrix} = \begin{pmatrix} \dfrac{u}{U_\infty} \\ \dfrac{v}{U_\infty} \\ \dfrac{w}{U_\infty} \end{pmatrix} = \frac{1}{U_\infty}\begin{pmatrix} u \\ v \\ w \end{pmatrix} = \frac{\underline{v}}{U_\infty} \tag{7.17}$$

Finally, we select a reference acceleration for the vector of body forces (per unit mass) $\underline{f}$. The gravity force is the only body force of note we will consider. Therefore, we use the gravitational acceleration $g = 9.807\,\mathrm{m/s^2}$ as a reference value. For the pressure we select p_∞, which often represents the pressure far away from the body.

Dimensionless gravitation and pressure

$$\underline{f}^* = \frac{\underline{f}}{g} \qquad p^* = \frac{p}{p_\infty} \tag{7.18}$$

The NSE (7.12) become dimensionless by replacing the dimensional quantities with their dimensionless counterparts and reference quantities, i.e. $x = Lx^*$, $\underline{v} = U_\infty \underline{v}^*$, $p = p_\infty p^*$, and so on.

Dimensionless NSE

$$\frac{1}{T}\frac{\partial(U_\infty \underline{v}^*)}{\partial t^*} + \left[(U_\infty \underline{v}^*)^T\left(\frac{\underline{\nabla}^*}{L}\right)\right](U_\infty \underline{v}^*)$$

$$= g\underline{f}^* - \frac{\underline{\nabla}^*}{\rho L}(p_\infty p^*) + \nu\left(\frac{\boldsymbol{\Delta}^*}{L^2}\right)(U_\infty \underline{v}^*) \tag{7.19}$$

We factor out the reference quantities in each term:

$$\frac{U_\infty}{T}\frac{\partial \underline{v}^*}{\partial t^*} + \frac{U_\infty^2}{L}(\underline{v}^{*T}\underline{\nabla}^*)\underline{v}^* = g\underline{f}^* - \frac{p_\infty}{\rho L}(\underline{\nabla}^* p^*) + \frac{\nu U_\infty}{L^2}(\boldsymbol{\Delta}^*\underline{v}^*) \tag{7.20}$$

Finally, this equation is multiplied by L/U_∞^2 and we obtain:

$$\left[\frac{L}{TU_\infty}\right]\frac{\partial \underline{v}^*}{\partial t^*} + (\underline{v}^{*T}\underline{\nabla}^*)\underline{v}^* = \left[\frac{gL}{U_\infty^2}\right]\underline{f}^* - \left[\frac{p_\infty}{\rho U_\infty^2}\right]\underline{\nabla}^* p^* + \left[\frac{\nu}{LU_\infty}\right]\boldsymbol{\Delta}^*\underline{v}^* \tag{7.21}$$

All factors in (7.21) are now dimensionless, including the four terms in brackets which are important numbers characterizing the flow around a ship hull.

Characteristic numbers of the flow

Strouhal number

1. On the left-hand side we have a form of the Strouhal number

   $$St = \frac{L}{TU_\infty} \qquad \text{Strouhal number} \tag{7.22}$$

   Vincenz Strouhal (*1850–†1922) was a Czech physicist who introduced this dimensionless number in 1878 (Strouhal, 1878).

   You probably have heard taught wires (rigging, guides of a cell tower) hum in high winds. The sound is created when the periodic shedding of vortices in the flow causes the wire to vibrate. The Strouhal number St is used to describe the frequency of unsteady processes like vortex shedding. Therefore, it precedes the local time derivative of the velocity $\partial\underline{v}^*/\partial t^*$, which represents the acceleration in the dimensionless Navier-Stokes equations (7.21).

Froude number

2. The first dimensionless number on the right-hand side is connected to the Froude number Fr.

   $$\left[\frac{gL}{U_\infty^2}\right] = \frac{1}{Fr^2} \qquad \text{with}$$

   $$Fr = \frac{U_\infty}{\sqrt{gL}} \qquad \text{Froude number} \tag{7.23}$$

The Froude number Fr is named after William Froude, the father of modern ship model testing. Naval architects use the Froude number as a dimensionless velocity. It is connected to the gravity forces and important for the similarity of wave patterns.

Euler number

3. The next pair of brackets encloses a form of Euler number named after Leonard Euler.

$$Eu = \left[\frac{p_\infty}{\rho U_\infty^2}\right] \quad \text{Euler number of fluid mechanics} \tag{7.24}$$

Euler made many important contributions to mathematics, structural mechanics, and fluid mechanics, among others, and various theories, equations, as well as numbers bear his name. In practice, we are more likely to encounter dimensionless pressure coefficients C_p which compare pressure differences with the dynamic pressure $(1/2)\rho U_\infty^2$:

$$C_p = \frac{p - p_\infty}{\frac{1}{2}\rho U_\infty^2} \quad \text{pressure coefficient} \tag{7.25}$$

Reynolds number

4. The last dimensionless number in brackets is the reciprocal of the Reynolds number Re:

$$\left[\frac{\nu}{LU_\infty}\right] = \frac{1}{Re} \quad \text{with}$$

$$Re = \frac{LU_\infty}{\nu} \quad \text{Reynolds number} \tag{7.26}$$

The Reynolds number is named after Osborne Reynolds, who studied, among many other things, the flow in pipes and under which conditions the flow transitioned from laminar to turbulent. The dimensionless Reynolds number is obviously connected to the viscous forces caused by the viscosity of the water.

Froude number and Reynolds number are arguably the most important qualifiers of flow conditions in ship hydrodynamics. We will encounter these numbers many times in subsequent chapters.

Dimensionless NSE

We now substitute the dimensionless numbers St, Fr, Eu, and Re for the square brackets into the Navier-Stokes equations:

$$St\,\frac{\partial \underline{v}^*}{\partial t^*} + \left(\underline{v}^{*T}\underline{\nabla}^*\right)\underline{v}^* = \frac{1}{Fr^2}\,\underline{f}^* - Eu\,\underline{\nabla}^* p^* + \frac{1}{Re}\,\Delta^*\underline{v}^* \tag{7.27}$$

Example

An example will teach us something about the terms in the Navier-Stokes equations. Consider the following data set:

length of vessel	$L_{WL} =$	120.00	m
ship speed	$v_s =$	20.00	kn
period	$T =$	6.59	s
gravitational acceleration	$g =$	9.81	$\mathrm{m/s^2}$
density of salt water at 15 °C	$\rho_s =$	1026.021	$\mathrm{kg/m^3}$
standard atmospheric pressure	$p_\infty =$	101325.00	Pa
kinematic viscosity of seawater at 15 °C	$\nu =$	$1.1892{\cdot}10^{-6}$	$\mathrm{m^2/s}$

We will employ the numbers to calculate the four dimensionless constants of the Navier-Stokes equations.

1. Strouhal number – Before we substitute the values above into Equation (7.22), we must convert the ship speed into SI units. One knot is equal to one nautical mile per hour, i.e. kn = M/h and

$$1\,\text{kn} = \frac{1852\,\text{m/M}}{3600\,\text{s/h}} = 0.5144\bar{4}\,\text{m h/(s M)} \tag{7.28}$$

   Therefore, the ship speed we want to use as a reference velocity is

$$v_s = 20\,\text{kn} = 10.289\,\text{m/s} = U_\infty$$

   As a reference period, we use the period of the transverse waves generated by the ship. We will discuss this in depth later. With the reference length $L = 120\,\text{m}$, reference period $T = 6.59\,\text{s}$, and reference velocity $U_\infty = 10.289\,\text{m/s}$ we obtain a Strouhal number of

$$St = \frac{L}{TU_\infty} = \frac{120\,\text{m}}{6.59\,\text{s}\;10.289\,\text{m/s}} = 1.76984$$

   It is good practice to check that the result is truly dimensionless.

2. Froude number – With the preparation above, the Froude number poses no challenge:

$$Fr = \frac{U_\infty}{\sqrt{gL}} = \frac{10.289\,\text{m/s}}{\sqrt{9.81\,\text{m/s}^2 \cdot 120\,\text{m}}} = 0.29988$$

   Please note, that the Froude number is usually computed on the basis of length in waterline L_{WL}.

   The coefficient for the Navier-Stokes equations is

$$\frac{1}{Fr^2} = 0.29988^{-2} = 11.12$$

3. Euler number – In this example we get for the Euler number

$$Eu = \frac{p_\infty}{\rho U_\infty^2} = \frac{101325\,\text{kg/(ms}^2)}{1026.021\,\text{kg/m}^3 \cdot 10.289^2\,(\text{m/s})^2} = 0.93288$$

4. Reynolds number – The Reynolds number is:

$$Re = \frac{LU_\infty}{\nu} = \frac{120\,\text{m} \cdot 10.289\,\text{m/s}}{1.1892 \cdot 10^{-6}\text{m}^2/\text{s}} = 1.03823 \cdot 10^9$$

   Note that the Reynolds number is usually based on the length over wetted surface L_{OS}.

   The coefficient for the Navier-Stokes equations is the inverse of the Reynolds number

$$\frac{1}{Re} = \frac{1}{1.03823 \cdot 10^9} = 9.632 \cdot 10^{-10}$$

We rewrite the Navier-Stokes equations using the coefficients we computed for this example:

$$1.76984 \frac{\partial \underline{v}^*}{\partial t^*} + \left(\underline{v}^{*T} \underline{\nabla}^*\right)\underline{v}^* = 11.12\,\underline{f}^* - 0.93288\,\underline{\nabla}^* p^* + 9.632 \cdot 10^{-10}\,\Delta^* \underline{v}^* \quad (7.29)$$

Three of the coefficients are in the range of 1, but the coefficient for the body forces is of magnitude 10, and the coefficient for the viscous forces is very small, roughly 10^{-9}. Thus, the factor for the body forces is 10 000 000 000 times larger than the factor for the viscous forces! This seems to indicate that viscous forces may be negligible.

Indeed, there are flow phenomena which support this notion. We know from detailed observations that waves created by storms in the vicinity of the Antarctic Circle travel halfway around the world to hit the shores of the North American Pacific coast. Their height diminishes somewhat along the way but more due to adverse winds than friction.

On the other hand, it is a well known fact that if you give a boat a push, it will come to rest again after a short while. Even if we avoid the generation of waves by considering a well streamlined, submerged body, it will come to rest fairly soon without a continuous propulsive force.

Boundary layer theory

This apparent discrepancy between theory, in the form of the coefficients in the Navier-Stokes equations, and the reality of observations vexed scientists and engineers at the end of the 19th century. It was Ludwig Prandtl who proposed in 1904 to divide the flow around bodies into two regions: a thin sheet of fluid close to the body surface, called the boundary layer, where viscous effects are present and an exterior flow outside the boundary layer where viscous effects are mostly negligible. Prandtl introduced the concept of fluid molecules sticking to the surface and also offered an explanation for the phenomenon of flow separation. More on this later. Readers are encouraged to read the paper by Anderson, Jr. (2005) on Prandtl's boundary layer theory and its impact on aerodynamics and fluid mechanics.

Prandtl's ideas allow a simplification of the Navier-Stokes equations into the boundary layer equations. Although the boundary layer equations are a special case of the Navier-Stokes equations, we will discuss them in subsequent chapters. They form the basis of skin friction computations and provide important insights into the flow around ship hulls.

References

Anderson, Jr., J. (2005). Ludwig Prandtl's boundary layer. *Physics Today*, 58(12):42–48.

Strouhal, V. (1878). Über eine besondere Art der Tonerregung. NF. Bd. V(10):216–251. Deutsches Textarchiv http://www.deutschestextarchiv.de/strouhal_tonerregung_1878, last visited July 12, 2014.

Self Study Problems

1. Look at the following equation of fluid mechanics:

$$\underbrace{\frac{\partial \underline{v}}{\partial t} + \left(\underline{v}^T \underline{\nabla}\right)\underline{v}}_{①} = \underbrace{\underline{f}}_{②} - \underbrace{\frac{1}{\rho}\underline{\nabla} p}_{③} + \underbrace{\nu \Delta \underline{v}}_{④}$$

   Answer the following questions:

   (a) What type of equation is this mathematically?

   (b) What is the name of the equation?

   (c) State what each of the terms ① through ④ represents.

   (d) What is the dimension of the terms in the equation?

2. Provide a definition for each of the four dimensionless numbers: *Re*, *Fr*, *Eu*, *St*.

3. Consider the following form of the continuity equation.

$$\frac{\partial \rho}{\partial t} + \underline{\nabla}^T(\rho \underline{v}) = 0$$

   (a) State whether the equation is in differential or integral form and whether it is the conservative or nonconservative form.

   (b) Convert the equation into a dimensionless form based on reference quantities L for length, T for time, U_∞ for velocity, and ρ_∞ for density.

   (c) Which of the four dimensionless numbers appears in the resulting dimensionless continuity equation?

8

Reynolds Averaged Navier-Stokes Equations (RANSE)

In practice, most available CFD systems do not attempt to solve Equation (7.12) directly. For high Reynolds numbers direct numerical simulation is still beyond our computational capabilities. Instead, we solve the RANS equations, which are derived by splitting the instantaneous time dependent velocity into an average velocity (still time dependent) and the turbulent velocity. This introduces the initially unknown Reynolds stress tensor. The added unknowns require additional equations to represent the turbulence. Turbulence modeling is still a focus of ongoing research, and several turbulence models of varying complexity are in use for ship resistance computations. We derive the RANSE form of the Navier-Stokes equations for incompressible flow below to illustrate the difference.

Learning Objectives

At the end of this chapter students will be able to:

- transform the NSE into RANSE
- formulate a dimensionless version of the NSE
- introduce characteristic dimensionless parameters
- explain the challenge of high Reynolds numbers

8.1 Mean and Turbulent Velocity

Turbulence Most instruments used to measure velocity and pressure report an average quantity. This commonly encompasses averaging over a small volume and a short time period as the instruments are not sensitive enough to follow extremely rapid fluctuations of velocity and pressure like they occur in turbulence. By definition, we assume steady flow to be time independent. Many of our instruments to measure flow velocity, e.g. a Pitot-static tube, will register a constant velocity when they are immersed in a constant stream. However, the flow may still experience rapid fluctuations u' around a constant mean velocity $\overline{u}$, especially in a flow at high Reynolds number (see Figure 8.1(a)).

Fundamentals of Ship Hydrodynamics: Fluid Mechanics, Ship Resistance and Propulsion, First Edition.
Lothar Birk.

Companion website: www.wiley.com/go/birk/hydrodynamics

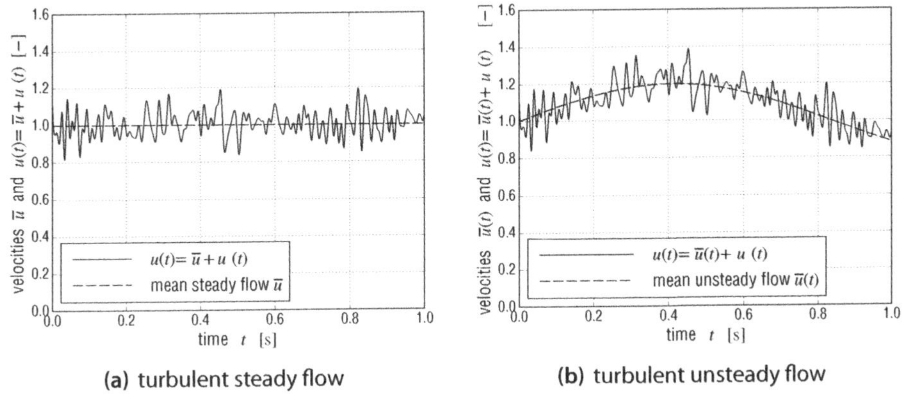

(a) turbulent steady flow

(b) turbulent unsteady flow

Figure 8.1 Mean and actual velocities in steady and unsteady turbulent flow

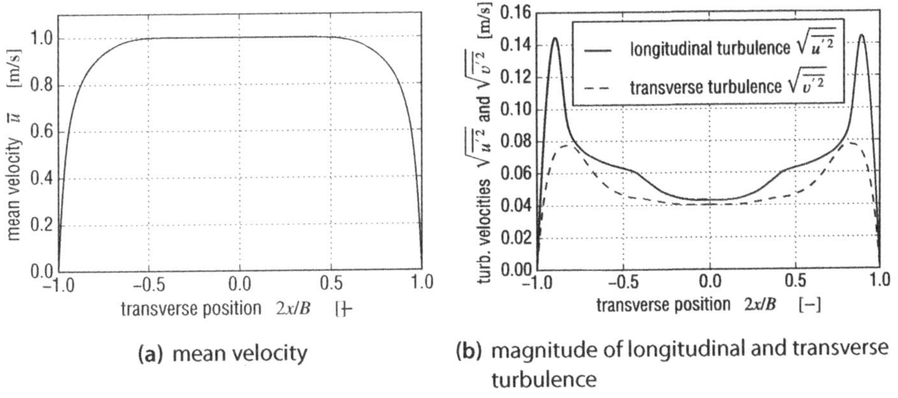

(a) mean velocity

(b) magnitude of longitudinal and transverse turbulence

Figure 8.2 Velocity and turbulence distribution across an air duct

Measuring turbulence

Hot wire anemometers (HWA) are sensitive enough to register rapidly varying turbulent velocities u'. An HWA consists primarily of a heated Tungsten wire exposed to the flow. Passing fluid will cool the wire more or less depending on actual flow velocity. The resistance of Tungsten is a function of temperature and even tiny temperature changes cause measurable voltage changes in the attached circuitry.

Figure 8.2(a) shows a typical mean velocity distribution found in the cross section of an air duct. Figure 8.2(b) presents the magnitude of longitudinal and transverse turbulence. The longitudinal turbulence is parallel to the mean flow. Turbulence occurs in all three coordinate directions and, as the example shows, is not necessarily equal in all coordinate directions. Nevertheless, we often assume that turbulence is isotropic, i.e. equal in all directions. In general, turbulence depends on several factors, including disturbances in the on flow, mean flow velocity, geometry, pressure gradients, etc.

Reynolds' averaging

Our numerical tools and associated discretizations (grids) are not capable of resolving the small scale of turbulence in space and time. Osborne Reynolds proposed to divide all relevant flow properties into a mean value and a turbulent fluctuation.

$$
\begin{aligned}
u(t) &= \overline{u}(t) + u'(t) & \rho(t) &= \overline{\rho}(t) + \rho'(t) \\
v(t) &= \overline{v}(t) + v'(t) & p(t) &= \overline{p}(t) + p'(t) \\
w(t) &= \overline{w}(t) + w'(t) & \underline{f}(t) &= \overline{\underline{f}}(t) + \underline{f}'(t)
\end{aligned}
\tag{8.1}
$$

Mean values of time series are calculated by integrating the series over time and dividing by the length T of the time series.

$$\overline{u} = \frac{1}{T}\int_0^T u(t)\,\mathrm{d}t \tag{8.2}$$

For unsteady flows, T must be small enough to follow the variation of the mean value over time but large enough that the mean value of the turbulent velocity $\overline{u'}$ vanishes.

$$\overline{u'} = \frac{1}{T}\int_0^T u'(t)\,\mathrm{d}t = 0 \tag{8.3}$$

Conversion of the NSE into the RANSE starts with introducing the approach Equation (8.1) into the basic equations. Here we restrict ourselves to incompressible flows and use the continuity equation (7.1) and the NSE in the form of Equation (7.12).

8.2 Time Averaged Continuity Equation

Application of Reynolds' averaging

Starting with the continuity equation, we replace the time dependent velocity by the sum of mean velocity and turbulence.

$$
\begin{aligned}
\underline{\nabla}^T \underline{v}(t) = \underline{\nabla}^T \begin{pmatrix} u(t) \\ v(t) \\ w(t) \end{pmatrix} &= \left(\frac{\partial}{\partial x}, \frac{\partial}{\partial y}, \frac{\partial}{\partial z}\right) \begin{pmatrix} \overline{u}(t) + u'(t) \\ \overline{v}(t) + v'(t) \\ \overline{w}(t) + w'(t) \end{pmatrix} = 0 \\
\frac{\partial}{\partial x}\left(\overline{u}(t) + u'(t)\right) + \frac{\partial}{\partial y}\left(\overline{v}(t) + v'(t)\right) &+ \frac{\partial}{\partial z}\left(\overline{w}(t) + w'(t)\right) = 0
\end{aligned}
\tag{8.4}
$$

Subsequently, we will omit the time argument, which shortens the writing quite a bit.

Repeating this substitution for the NSE (7.12) yields:

$$
\begin{aligned}
&\frac{\partial}{\partial t}\begin{pmatrix} \overline{u} + u' \\ \overline{v} + v' \\ \overline{w} + w' \end{pmatrix} + \left[\left(\overline{u} + u', \overline{v} + v', \overline{w} + w'\right)\underline{\nabla}\right]\begin{pmatrix} \overline{u} + u' \\ \overline{v} + v' \\ \overline{w} + w' \end{pmatrix} \\
&\qquad = \overline{\underline{f}} + \underline{f}' - \frac{1}{\rho}\underline{\nabla}\left(\overline{p} + p'\right) + \nu\Delta\begin{pmatrix} \overline{u} + u' \\ \overline{v} + v' \\ \overline{w} + w' \end{pmatrix}
\end{aligned}
\tag{8.5}
$$

So far, we have not modified the continuity and Navier-Stokes equations, we have just introduced the identities from Equation (8.1).

Next, we average Equations (8.4) and (8.5) over time. That means we integrate both sides of the equations over time and divide by the time period T.

$$\frac{1}{T}\int_0^T \dots \,\mathrm{d}t = \frac{1}{T}\int_0^T \dots \,\mathrm{d}t \tag{8.6}$$

This is a lot of cumbersome writing work. We can take a shortcut by exploiting the rules for forming averages.

Rules for averages

In a nutshell, the following computation rules are applied, which can all be derived from basic rules of integration.

...	equals	...
(a) average of a sum	$\overline{g+h} = \overline{g} + \overline{h}$	sum of averages
(b) average of an average	$\overline{\overline{g}} = \overline{g}$	average
(c) average of a derivative	$\overline{\frac{\partial g}{\partial s}} = \frac{\partial \overline{g}}{\partial s}$	derivative of the average (8.7)
(d) average of an integral	$\overline{\int g\,\mathrm{d}s} = \int \overline{g}\,\mathrm{d}s$	integral of the average

and the average of an average $\overline{g}$ multiplied by another function h

$$(e) \qquad \overline{\overline{g}\cdot h} = \overline{g}\cdot\overline{h}$$

is the product of averages. However, the average of a product is not equal to the average of products:

$$\overline{g\cdot h} \neq \overline{g}\cdot\overline{h} \tag{8.8}$$

Just consider the sine function $g = \sin(t)$. Its average is zero over one period $T = 2\pi$.

$$\begin{aligned}
\overline{g} = \overline{\sin(t)} &= \frac{1}{T}\int_0^T \sin(t)\,dt \\
&= \frac{1}{2\pi}\Big[-\cos(t)\Big]_0^{2\pi} \\
&= \frac{1}{2\pi}\big[-1-(-1)\big] \\
&= 0
\end{aligned}$$

However, the average of $g \cdot h$ with $g = \sin(t)$ and $h = \sin(t)$ is clearly not equal to the product of the averages.

$$\begin{aligned}\overline{g \cdot h} = \overline{\sin(t) \cdot \sin(t)} &= \frac{1}{T}\int_0^T \sin(t)\sin(t)\,dt \\ &= \frac{1}{2\pi}\left[\frac{t}{2} - \frac{\sin(2t)}{4}\right]_0^{2\pi} \\ &= \frac{1}{2\pi}\left[\pi - 0 - 0 + 0\right] \\ &= \frac{1}{2} \neq \overline{\sin(t)} \cdot \overline{\sin(t)}\end{aligned}$$

Equipped with the rules for averages, we return to the continuity equation (8.4) and Navier-Stokes equations (8.5) for incompressible flow.

Reynolds' average of the continuity equation

Taking the average of the continuity equation (8.4) on both sides yields:

$$\overline{\frac{\partial}{\partial x}\left(\overline{u} + u'\right) + \frac{\partial}{\partial y}\left(\overline{v} + v'\right) + \frac{\partial}{\partial z}\left(\overline{w} + w'\right)} = \overline{0}$$

The average of a constant is equal to the constant. Applying rule (a) for averages of a sum results in

$$\overline{\frac{\partial}{\partial x}\left(\overline{u} + u'\right)} + \overline{\frac{\partial}{\partial y}\left(\overline{v} + v'\right)} + \overline{\frac{\partial}{\partial z}\left(\overline{w} + w'\right)} = 0$$

and the average of a derivative is equal to the derivative of the average according to rule (c):

$$\frac{\partial}{\partial x}\left(\overline{\overline{u} + u'}\right) + \frac{\partial}{\partial y}\left(\overline{\overline{v} + v'}\right) + \frac{\partial}{\partial z}\left(\overline{\overline{w} + w'}\right) = 0 \tag{8.9}$$

Again, we apply rule (a):

$$\frac{\partial}{\partial x}\left(\overline{\overline{u}} + \overline{u'}\right) + \frac{\partial}{\partial y}\left(\overline{\overline{v}} + \overline{v'}\right) + \frac{\partial}{\partial z}\left(\overline{\overline{w}} + \overline{w'}\right) = 0$$

Then, forming the derivatives of the sums, we apply rule (b) and resort the terms:

$$\frac{\partial \overline{u}}{\partial x} + \frac{\partial \overline{v}}{\partial y} + \frac{\partial \overline{w}}{\partial z} + \frac{\partial \overline{u'}}{\partial x} + \frac{\partial \overline{v'}}{\partial y} + \frac{\partial \overline{w'}}{\partial z} = 0 \tag{8.10}$$

The averages of the turbulent velocities vanish by definition, i.e.

$$\frac{\partial \overline{u}}{\partial x} + \frac{\partial \overline{v}}{\partial y} + \frac{\partial \overline{w}}{\partial z} + \cancelto{0}{\frac{\partial \overline{u'}}{\partial x}} + \cancelto{0}{\frac{\partial \overline{v'}}{\partial y}} + \cancelto{0}{\frac{\partial \overline{w'}}{\partial z}} = 0$$

What remains is the fact that the mean velocities satisfy the continuity equation (8.11).

$$\frac{\partial \bar{u}}{\partial x} + \frac{\partial \bar{v}}{\partial y} + \frac{\partial \bar{w}}{\partial z} = 0 \tag{8.11}$$

Subtracting Equation (8.11) from Equation (8.10) reveals that the turbulent velocities themselves also satisfy the continuity equation.

$$\overline{\frac{\partial u'}{\partial x}} + \overline{\frac{\partial v'}{\partial y}} + \overline{\frac{\partial w'}{\partial z}} = 0 \tag{8.12}$$

We will use this result to our advantage when we deal with the averaging of the Navier-Stokes equations.

8.3 Time Averaged Navier-Stokes Equations

Reynolds' average of inertia force

For the NSE, we illustrate the process using the x-component of Equation (8.5). Let us start with the inertia forces (per unit mass) on the left-hand side. With the help of rules (a) and (c), the average of the local derivative turns into

$$\overline{\frac{\partial}{\partial t}(\bar{u} + u')} = \frac{\partial \bar{\bar{u}}}{\partial t} + \underbrace{\frac{\partial \overline{u'}}{\partial t}}_{0} = \frac{\partial \bar{u}}{\partial t} \tag{8.13}$$

The first summand of the convective derivative transforms into

$$\begin{aligned}
\overline{(\bar{u} + u')\frac{\partial}{\partial x}(\bar{u} + u')} &= \overline{\bar{u}\frac{\partial \bar{u}}{\partial x} + \bar{u}\frac{\partial u'}{\partial x} + u'\frac{\partial \bar{u}}{\partial x} + u'\frac{\partial u'}{\partial x}} \\
&= \overline{\bar{u}\frac{\partial \bar{u}}{\partial x}} + \overline{\bar{u}\frac{\partial u'}{\partial x}} + \overline{u'\frac{\partial \bar{u}}{\partial x}} + \overline{u'\frac{\partial u'}{\partial x}} \\
&= \bar{u}\frac{\partial \bar{u}}{\partial x} + \bar{u}\frac{\partial \overline{u'}}{\partial x} + \overline{u'}\frac{\partial \bar{u}}{\partial x} + \overline{u'\frac{\partial u'}{\partial x}}
\end{aligned}$$

The second and third terms on the right-hand side vanish because the mean value of the turbulence is zero: $\overline{u'} \equiv 0$. However, the last term cannot be simplified:

$$\overline{(\bar{u} + u')\frac{\partial}{\partial x}(\bar{u} + u')} = \bar{u}\frac{\partial \bar{u}}{\partial x} + \overline{u'\frac{\partial u'}{\partial x}} \tag{8.14}$$

Applying the same operations to the second and third terms of the convective derivative in Equation (8.5) yields

$$\overline{(\bar{v} + v')\frac{\partial}{\partial y}(\bar{u} + u')} = \bar{v}\frac{\partial \bar{u}}{\partial y} + \overline{v'\frac{\partial u'}{\partial y}} \tag{8.15}$$

and

$$\overline{(\bar{w} + w')\frac{\partial}{\partial z}(\bar{u} + u')} = \bar{w}\frac{\partial \bar{u}}{\partial z} + \overline{w'\frac{\partial u'}{\partial z}} \tag{8.16}$$

Reynolds' average of external forces

Next are the external forces per unit mass on the right-hand side of Equation (8.5). Again, the process is shown using the x-component as an example.

1. Body force

$$\overline{\overline{f}_x + f'_x} = \overline{f}_x + \overline{f'_x} = \overline{f}_x \tag{8.17}$$

2. Pressure force

$$\overline{\frac{\partial}{\partial x}\left(\overline{p} + p'\right)} = \frac{\partial \overline{p}}{\partial x} + \frac{\partial \overline{p'}}{\partial x} = \frac{\partial \overline{p}}{\partial x} \tag{8.18}$$

3. Viscous force

$$\begin{aligned}
&\overline{\frac{\partial^2}{\partial x^2}\left(\overline{u} + u'\right) + \frac{\partial^2}{\partial y^2}\left(\overline{u} + u'\right) + \frac{\partial^2}{\partial z^2}\left(\overline{u} + u'\right)} \\
&= \frac{\partial^2 \overline{u}}{\partial x^2} + \frac{\partial^2 \overline{u}}{\partial y^2} + \frac{\partial^2 \overline{u}}{\partial z^2} + \underbrace{\frac{\partial^2 \overline{u'}}{\partial x^2} + \frac{\partial^2 \overline{u'}}{\partial y^2} + \frac{\partial^2 \overline{u'}}{\partial z^2}}_{0} \\
&= \frac{\partial^2 \overline{u}}{\partial x^2} + \frac{\partial^2 \overline{u}}{\partial y^2} + \frac{\partial^2 \overline{u}}{\partial z^2} \\
&= \mathbf{\Delta} \overline{u}
\end{aligned} \tag{8.19}$$

Reynolds' average of NSE x-component

We take stock and collect the time averaging results from Equations (8.13) through (8.19).

$$\frac{\partial \overline{u}}{\partial t} + \overline{u}\frac{\partial \overline{u}}{\partial x} + \overline{u'\frac{\partial u'}{\partial x}} + \overline{v}\frac{\partial \overline{u}}{\partial y} + \overline{v'\frac{\partial u'}{\partial y}} + \overline{w}\frac{\partial \overline{u}}{\partial z} + \overline{w'\frac{\partial u'}{\partial z}} = \overline{f}_x + \frac{\partial \overline{p}}{\partial x} + \mathbf{\Delta}\overline{u} \tag{8.20}$$

The equation looks similar to Equation (8.5) from which we started, except that there are now three additional summands on the left-hand side. We subtract the additional terms to move them over to the right-hand side.

$$\frac{\partial \overline{u}}{\partial t} + \overline{u}\frac{\partial \overline{u}}{\partial x} + \overline{v}\frac{\partial \overline{u}}{\partial y} + \overline{w}\frac{\partial \overline{u}}{\partial z} = \overline{f}_x + \frac{\partial \overline{p}}{\partial x} + \mathbf{\Delta}\overline{u} - \overline{u'\frac{\partial u'}{\partial x}} - \overline{v'\frac{\partial u'}{\partial y}} - \overline{w'\frac{\partial u'}{\partial z}} \tag{8.21}$$

The physics of the new terms will be explained in due course, but first we will modify them slightly. Consider the following expression, which we expand by application of the product rule of differentiation.

$$\begin{aligned}
&\frac{\partial}{\partial x}\left(u'u'\right) + \frac{\partial}{\partial y}\left(v'u'\right) + \frac{\partial}{\partial z}\left(w'u'\right) \\
&= u'\frac{\partial u'}{\partial x} + \frac{\partial u'}{\partial x}u' + v'\frac{\partial u'}{\partial y} + \frac{\partial v'}{\partial y}u' + w'\frac{\partial u'}{\partial z} + \frac{\partial w'}{\partial z}u' \\
&= u'\underbrace{\left(\frac{\partial u'}{\partial x} + \frac{\partial v'}{\partial y} + \frac{\partial w'}{\partial z}\right)}_{0} + u'\frac{\partial u'}{\partial x} + v'\frac{\partial u'}{\partial y} + w'\frac{\partial u'}{\partial z}
\end{aligned} \tag{8.22}$$

In the last line, the terms with factor u' are summarized, and the continuity equation for turbulent velocities (8.12) allows us to eliminate the term. The result of this exercise is the following identity:

$$\frac{\partial}{\partial x}\left(u'u'\right) + \frac{\partial}{\partial y}\left(v'u'\right) + \frac{\partial}{\partial z}\left(w'u'\right) = u'\frac{\partial u'}{\partial x} + v'\frac{\partial u'}{\partial y} + w'\frac{\partial u'}{\partial z} \tag{8.23}$$

We average both sides of the equation over time.

$$\frac{\partial}{\partial x}\overline{(u'u')}+\frac{\partial}{\partial y}\overline{(v'u')}+\frac{\partial}{\partial z}\overline{(w'u')} = \overline{u'\frac{\partial u'}{\partial x}}+\overline{v'\frac{\partial u'}{\partial y}}+\overline{w'\frac{\partial u'}{\partial z}} \tag{8.24}$$

Obviously, the right-hand side is identical to the three new terms in our time averaged x-component of the NSE (8.21). We will use the left-hand side of Equation (8.24) to replace the three new terms.

RANSE

It is left to the reader to repeat the time averaging process with the y- and z-components of the NSE. As a result, we obtain the RANS equations.

$$\begin{aligned}
\frac{\partial \overline{u}}{\partial t}+(\underline{\overline{v}}^T\underline{\nabla})\overline{u} &= \overline{f}_x-\frac{1}{\rho}\frac{\partial \overline{p}}{\partial x}+\nu\boldsymbol{\Delta}\overline{u}-\left[\frac{\partial}{\partial x}\overline{(u'u')}+\frac{\partial}{\partial y}\overline{(v'u')}+\frac{\partial}{\partial z}\overline{(w'u')}\right]\\
\frac{\partial \overline{v}}{\partial t}+(\underline{\overline{v}}^T\underline{\nabla})\overline{v} &= \overline{f}_y-\frac{1}{\rho}\frac{\partial \overline{p}}{\partial y}+\nu\boldsymbol{\Delta}\overline{v}-\left[\frac{\partial}{\partial x}\overline{(u'v')}+\frac{\partial}{\partial y}\overline{(v'v')}+\frac{\partial}{\partial z}\overline{(w'v')}\right]\\
\frac{\partial \overline{w}}{\partial t}+(\underline{\overline{v}}^T\underline{\nabla})\overline{w} &= \overline{f}_z-\frac{1}{\rho}\frac{\partial \overline{p}}{\partial z}+\nu\boldsymbol{\Delta}\overline{w}-\left[\frac{\partial}{\partial x}\overline{(u'w')}+\frac{\partial}{\partial y}\overline{(v'w')}+\frac{\partial}{\partial z}\overline{(w'w')}\right]
\end{aligned} \tag{8.25}$$

At this point, the bars on $\overline{u}, \overline{v}, \overline{w}, \overline{p}$ are omitted with the understanding that u, v, w, and p now represent mean flow quantities without turbulence.

8.4 Reynolds Stresses and Turbulence Modeling

Reynolds stresses

If you study the RANSE (8.25), you will find that the new terms look similar to the stress terms in the conservation of momentum Equations (6.24), (6.25), and (6.26). The averaging process has introduced six, yet unknown, averages of the turbulent velocity products $\overline{(u'u')}$, $\overline{(u'v')}$, $\overline{(u'w')}$, $\overline{(v'v')}$, $\overline{(v'w')}$, and $\overline{(w'w')}$. There is a total of nine terms, but six appear in pairs.

$$\overline{(u'v')}=\overline{(v'u')} \qquad \overline{(u'w')}=\overline{(w'u')} \qquad \overline{(v'w')}=\overline{(w'v')} \tag{8.26}$$

The new unknowns are often referred to as 'Reynolds stresses', which is not quite correct. In order to obtain a stress unit [N/m^2], the products of turbulent velocity [m^2/s^2] have to be multiplied by the density of the fluid first.

Reynolds stress tensor

The unknown terms are collected into the Reynolds stress tensor $\underline{\underline{\tau}}$.

$$\underline{\underline{\tau}} = \begin{pmatrix} \tau'_{xx} & \tau'_{yx} & \tau'_{zx}\\ \tau'_{xy} & \tau'_{yy} & \tau'_{zy}\\ \tau'_{xz} & \tau'_{yz} & \tau'_{zz}\end{pmatrix} = -\rho\begin{pmatrix} \overline{u'u'} & \overline{v'u'} & \overline{w'u'}\\ \overline{u'v'} & \overline{v'v'} & \overline{w'v'}\\ \overline{u'w'} & \overline{v'w'} & \overline{w'w'}\end{pmatrix} \tag{8.27}$$

Note that the Reynolds stresses are only apparent stresses. Their appearance is a consequence of separating mean and turbulent velocities and averaging the equation over time.

Vector form of RANSE

The RANSE (8.25) can be cast into a vector form by introducing the Reynolds stress tensor $\underline{\underline{\tau}}$.

$$\frac{\partial \underline{v}}{\partial t} + \left(\underline{v}\underline{\nabla}\right)\underline{v} = \underline{f} - \frac{1}{\rho}\underline{\nabla}\, p + \nu \boldsymbol{\Delta} \underline{v} + \frac{1}{\rho}\left(\underline{\nabla}^T \underline{\underline{\tau}}\right)^T \tag{8.28}$$

Turbulence models

Similar to Stokes' approach, which connected the unknown stresses in the conservation of momentum equations to rates of change in the velocity field, we need to find additional equations for the unknown Reynolds stresses. This is where turbulence models come into play. Turbulence modeling is a very active field of research and we will not dive into details here. The reader may start with the classic book on boundary layer theory by Schlichting and Gersten (2003) and then study turbulence models in depth with Wilcox (2006). There are also excellent websites, like NASA's Turbulence Modeling Resource (Rumsey et al., 2018).

Boussinesq's eddy viscosity hypothesis

Some of the turbulence models use only one or two additional equations, which requires that the Reynolds stress tensor is expressed as a function of one or two parameters. In the following, we outline the classic approach introduced by Joseph Valentin Boussinesq (*1842 – †1929), a French mathematician and physicist. Boussinesq (1877) proposed to link the apparent Reynolds stresses to the space derivatives of the mean velocities. This is known as Boussinesq's eddy viscosity hypothesis. Details can also be found in Schmitt (2007).

The hypothesis is based on three basic assumptions.

Apparent viscosity

1. Reynolds stresses are proportional to the eddy viscosity μ_t, also called apparent viscosity. In contrast to the dynamic viscosity μ for Newtonian fluids, eddy viscosity is not a material constant. μ_t still depends on the velocity distribution.

Isotropic turbulence

2. Turbulence is treated as isotropic, which means the turbulent velocity components u', v', and w' are all of identical magnitude. We require specifically that $\overline{u'u'} = \overline{v'v'} = \overline{w'w'}$. This is definitely a stretch and cannot be true close to walls. As we approach a wall, turbulence normal to the surface must diminish while other components are not affected as much.

Mean kinetic energy of turbulence

3. The mean kinetic turbulent energy is

$$k = \frac{1}{2}\left(\overline{u'u'} + \overline{v'v'} + \overline{w'w'}\right) \tag{8.29}$$

The assumptions lead to the following expression for the Reynolds stress tensor.

$$\underline{\underline{\tau}} = \begin{pmatrix} 2\mu_t \frac{\partial \overline{u}}{\partial x} - \frac{2}{3}\rho\, k & \mu_t\left(\frac{\partial \overline{u}}{\partial y} + \frac{\partial \overline{v}}{\partial x}\right) & \mu_t\left(\frac{\partial \overline{u}}{\partial z} + \frac{\partial \overline{w}}{\partial x}\right) \\ \mu_t\left(\frac{\partial \overline{u}}{\partial y} + \frac{\partial \overline{v}}{\partial x}\right) & 2\mu_t \frac{\partial \overline{v}}{\partial y} - \frac{2}{3}\rho\, k & \mu_t\left(\frac{\partial \overline{w}}{\partial y} + \frac{\partial \overline{v}}{\partial z}\right) \\ \mu_t\left(\frac{\partial \overline{u}}{\partial z} + \frac{\partial \overline{w}}{\partial x}\right) & \mu_t\left(\frac{\partial \overline{w}}{\partial y} + \frac{\partial \overline{v}}{\partial z}\right) & 2\mu_t \frac{\partial \overline{w}}{\partial z} - \frac{2}{3}\rho\, k \end{pmatrix} \tag{8.30}$$

This leaves us with the mean kinetic energy k and the apparent viscosity μ_t as remaining unknowns.

We employ the first row of the new Reynolds stress tensor (8.30) and substitute it for the Reynolds stresses in the x-component of the RANSE (8.25), i.e.

Substituting Reynolds stresses

$$
\begin{aligned}
&-\left[\frac{\partial}{\partial x}\overline{(u'u')}+\frac{\partial}{\partial y}\overline{(v'u')}+\frac{\partial}{\partial z}\overline{(w'u')}\right] \\
&=\frac{1}{\rho}\left\{\frac{\partial}{\partial x}\left[2\mu_t\frac{\partial\overline{u}}{\partial x}-\frac{2}{3}\rho k\right]+\frac{\partial}{\partial y}\left[\mu_t\left(\frac{\partial\overline{u}}{\partial y}+\frac{\partial\overline{v}}{\partial x}\right)\right]+\frac{\partial}{\partial z}\left[\mu_t\left(\frac{\partial\overline{u}}{\partial z}+\frac{\partial\overline{w}}{\partial x}\right)\right]\right\} \\
&=\frac{\partial}{\partial x}\left[2\nu_t\frac{\partial\overline{u}}{\partial x}-\frac{2}{3}\rho k\right]+\frac{\partial}{\partial y}\left[\nu_t\left(\frac{\partial\overline{u}}{\partial y}+\frac{\partial\overline{v}}{\partial x}\right)\right]+\frac{\partial}{\partial z}\left[\nu_t\left(\frac{\partial\overline{u}}{\partial z}+\frac{\partial\overline{w}}{\partial x}\right)\right]
\end{aligned}
\tag{8.31}
$$

For incompressible fluids, the apparent viscosity μ_t is replaced with the eddy kinematic viscosity $\nu_t = \mu_t/\rho$.

The derivatives on the right-hand side of Equation (8.31) must be taken with the product rule because the kinematic eddy viscosity ν_t is a function of spatial variables x, y, and z. We expand the right-hand side accordingly and rearrange its terms.

$$
\begin{aligned}
&-\left[\frac{\partial}{\partial x}\overline{(u'u')}+\frac{\partial}{\partial y}\overline{(v'u')}+\frac{\partial}{\partial z}\overline{(w'u')}\right] \\
&\quad=2\nu_t\frac{\partial^2\overline{u}}{\partial x^2}+2\frac{\partial\nu_t}{\partial x}\frac{\partial\overline{u}}{\partial x}-\frac{2}{3}\frac{\partial k}{\partial x} \\
&\qquad+\frac{\partial\nu_t}{\partial y}\left(\frac{\partial\overline{u}}{\partial y}+\frac{\partial\overline{v}}{\partial x}\right)+\nu_t\left(\frac{\partial^2\overline{u}}{\partial y^2}+\frac{\partial^2\overline{v}}{\partial x\partial y}\right) \\
&\qquad+\frac{\partial\nu_t}{\partial z}\left(\frac{\partial\overline{u}}{\partial z}+\frac{\partial\overline{w}}{\partial x}\right)+\nu_t\left(\frac{\partial^2\overline{u}}{\partial z^2}+\frac{\partial\overline{w}}{\partial x\partial z}\right)
\end{aligned}
\tag{8.32}
$$

Rearranging the terms on the right-hand side results in

$$
\begin{aligned}
&-\left[\frac{\partial}{\partial x}\overline{(u'u')}+\frac{\partial}{\partial y}\overline{(v'u')}+\frac{\partial}{\partial z}\overline{(w'u')}\right] \\
&=\nu_t\left(\frac{\partial^2\overline{u}}{\partial x^2}+\frac{\partial^2\overline{u}}{\partial y^2}+\frac{\partial^2\overline{u}}{\partial z^2}\right)+\nu_t\left(\frac{\partial^2\overline{u}}{\partial x^2}+\frac{\partial^2\overline{v}}{\partial x\partial y}+\frac{\partial^2\overline{w}}{\partial x\partial z}\right) \\
&\quad-\frac{2}{3}\frac{\partial k}{\partial x}+\frac{\partial\nu_t}{\partial x}\left(\frac{\partial\overline{u}}{\partial x}+\frac{\partial\overline{u}}{\partial x}\right)+\frac{\partial\nu_t}{\partial y}\left(\frac{\partial\overline{u}}{\partial y}+\frac{\partial\overline{v}}{\partial x}\right)+\frac{\partial\nu_t}{\partial z}\left(\frac{\partial\overline{u}}{\partial z}+\frac{\partial\overline{w}}{\partial x}\right)
\end{aligned}
\tag{8.33}
$$

The first pair of parentheses on the right involves the Laplace operator

$$
\nu_t\left(\frac{\partial^2\overline{u}}{\partial x^2}+\frac{\partial^2\overline{u}}{\partial y^2}+\frac{\partial^2\overline{u}}{\partial z^2}\right)=\mathbf{\Delta}\overline{u}
\tag{8.34}
$$

The second parenthesis vanishes because it contains the divergence of the velocity, i.e. the continuity equation for incompressible fluids.

$$
\nu_t\left(\frac{\partial^2\overline{u}}{\partial x^2}+\frac{\partial^2\overline{v}}{\partial x\partial y}+\frac{\partial^2\overline{w}}{\partial x\partial z}\right)=\nu_t\frac{\partial}{\partial x}\underbrace{\left(\frac{\partial\overline{u}}{\partial x}+\frac{\partial\overline{v}}{\partial y}+\frac{\partial\overline{w}}{\partial z}\right)}_{\text{conti. eq. incompress. }\underline{\nabla}^T\underline{v}=0}=0
\tag{8.35}
$$

The last three summands can be summarized in a convenient vector form.

$$\frac{\partial \nu_t}{\partial x}\left(\frac{\partial \overline{u}}{\partial x}+\frac{\partial \overline{u}}{\partial x}\right)+\frac{\partial \nu_t}{\partial y}\left(\frac{\partial \overline{u}}{\partial y}+\frac{\partial \overline{v}}{\partial x}\right)+\frac{\partial \nu_t}{\partial z}\left(\frac{\partial \overline{u}}{\partial z}+\frac{\partial \overline{w}}{\partial x}\right) = \left(\underline{\nabla}\,\nu_t\right)^T\left(\underline{\nabla}\,\overline{u}+\frac{\partial \underline{\overline{v}}}{\partial x}\right) \quad (8.36)$$

RANSE based on Boussinesq's hypothesis

Finally, by substituting Equations (8.34), (8.35), and (8.36) into Equation (8.33), we obtain for the three Reynolds stress terms on the x-component of RANSE the following expressions:

$$-\left[\frac{\partial}{\partial x}\overline{(u'u')}+\frac{\partial}{\partial y}\overline{(v'u')}+\frac{\partial}{\partial z}\overline{(w'u')}\right] = \nu_t\boldsymbol{\Delta}\overline{u}-\frac{2}{3}\frac{\partial k}{\partial x}+\left(\underline{\nabla}\,\nu_t\right)^T\left(\underline{\nabla}\,\overline{u}+\frac{\partial \underline{\overline{v}}}{\partial x}\right) \quad (8.37)$$

Application of the same procedure to the Reynolds stress terms of the y- and z-components yields a vector form of the RANS equations based on Boussinesq's eddy viscosity hypothesis.

$$\frac{\partial \underline{v}}{\partial t}+\left(\underline{v}\underline{\nabla}\right)\underline{v} = \underline{f}-\frac{1}{\rho}\underline{\nabla}\,p + \left(\nu+\nu_t\right)\boldsymbol{\Delta}\underline{v}-\frac{2}{3}\underline{\nabla}\,k+\left[\left(\underline{\nabla}\,\nu_t\right)^T\underline{\nabla}\right]\underline{v}+\left[\underline{\nabla}\,\underline{v}^T\right]\left(\underline{\nabla}\,\nu_t\right) \quad (8.38)$$

Note that the dyadic product $\underline{\nabla}\,\underline{v}^T$ results in a [3×3] matrix. One can imagine how much more effort turbulent flow computations require by comparing the RANSE (8.38) with the equivalent incompressible NSE (7.12) applied to laminar flow, especially if we consider that Equation (8.38) is already a simplified version of RANSE.

Need for turbulence models

Together with the continuity equation, we now have four equations for six unknowns.

$$\overline{u}=\overline{u}(x,y,z,t) \qquad \overline{v}=\overline{v}(x,y,z,t) \qquad \overline{w}=\overline{w}(x,y,z,t) \quad (8.39)$$
$$\overline{p}=\overline{p}(x,y,z,t) \qquad \overline{\nu_t}=\overline{\nu_t}(x,y,z,t) \qquad \overline{k}=\overline{k}(x,y,z,t)$$

Consequently, two additional equations for the eddy viscosity ν_t and the mean turbulent energy k have to be found.

This is accomplished with two equations turbulence models like the k-ω or the k-ε models among others. In the latter, eddy viscosity is expressed as a function of the mean turbulent kinetic energy k and a dissipation rate ε.

$$\nu_t = C_\mu\frac{k^2}{\varepsilon}$$

C_μ is an adjustable constant. For k and ε, two transport equations are provided which are similar to the NSE themselves.

The resulting system of six coupled, nonlinear partial differential equations has to be discretized and solved considering boundary conditions on the faces of the computational domain. All in all, a difficult and time consuming process. An introduction into CFD can be found in Anderson, Jr. (1995) and Versteeg and Malalasekera (2007).

References

Anderson, Jr., J. (1995). *Computational fluid dynamics – The basics with applications.* McGraw-Hill, Inc., New York, NY.

Boussinesq, J. (1877). Essai sur la théorie des eaux courantes. *Mémoires présentés par divers savants à l'Académie des Sciences*, 23(1):1–680.

Rumsey, C., Smith, B., and Huang, G. (2018). Turbulence modeling resource. http://turbmodels.larc.nasa.gov. Last visited December 12, 2018.

Schlichting, H. and Gersten, K. (2003). *Boundary-layer theory.* Springer-Verlag, Berlin, Heidelberg, New York, eigth edition. Corrected printing.

Schmitt, F. (2007). About Boussinesq's turbulent viscosity hypothesis: historical remarks and a direct evaluation of its validity. *Comptes Rendus Mécanique*, 335(9–10):617–627.

Versteeg, H. and Malalasekera, W. (2007). *An introduction to computational fluid dynamics: the finite volume method.* Pearson, Harlow, England, second edition.

Wilcox, D. (2006). *Turbulence modeling for CFD.* DCW Industries, La Cañada, California, third edition.

Self Study Problems

1. Explain turbulence in your own words.
2. What is the time average of the turbulent velocity components u', v', and w'?
3. Derive the Reynolds averaged y- and z- component of the external forces in the Navier-Stokes equation following the example in Equations (8.17) to (8.19).
4. Why is turbulence model required to solve the Reynolds Averaged Navier Stokes Equations (RANSE)?

9

Application of the Conservation Principles

The conservation of mass and momentum equations – also known as continuity equation and Navier-Stokes equations – form a system of coupled, nonlinear, partial differential equations. They are without doubt complicated and no generally applicable analytical solution is known. We will explain the basic concepts of using the conservation of mass and conservation of momentum principles with two examples:

- calculating the drag of an object in a wind tunnel, and
- computing the drag of a submerged vessel in an unbounded fluid.

Admittedly, the examples are constructed and simplified to allow an analytical solution. However, they show the principal procedures employed in a wake analysis, and the use of such analysis in determining the resistance of a body in a flow. Along the way we will use appropriate boundary conditions. This discussion will also provide an introduction to an integral form of the conservation of momentum equations.

Learning Objectives

At the end of this chapter students will be able to:

- apply conservation of mass and momentum principles
- determine the drag of a body with known wake field

9.1 Body in a Wind Tunnel

Problem Consider a streamlined body of revolution in a wind tunnel with a circular cross section (diameter $\varnothing D = 2r_0 = 2$ m) (see Figure 9.1). You are tasked with the computation of the drag force $\underline{F}_D$ acting on the body. The installed measurement equipment allows us to read pressure and velocity distribution at the inlet ① upstream of the body and outlet ② downstream of the body.

Fundamentals of Ship Hydrodynamics: Fluid Mechanics, Ship Resistance and Propulsion, First Edition.
Lothar Birk.

Companion website: www.wiley.com/go/birk/hydrodynamics

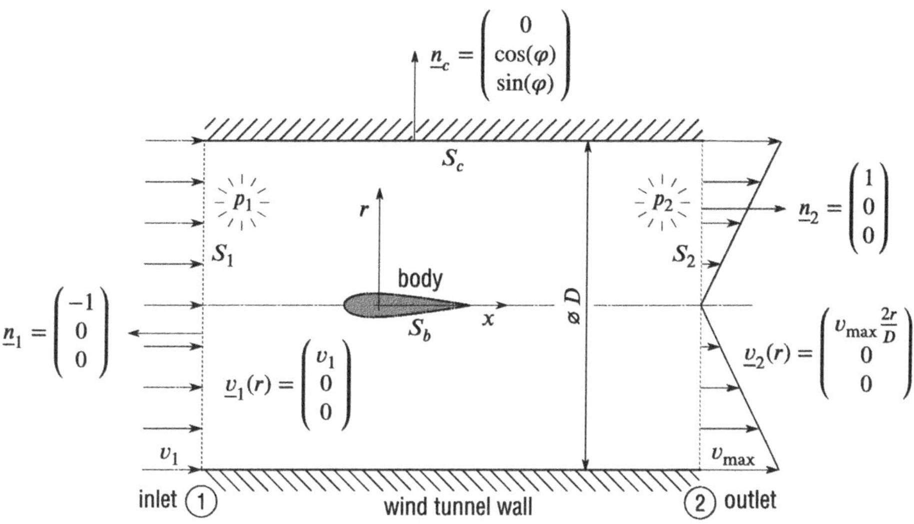

Figure 9.1 Body of revolution in a wind tunnel (simplified)

Input data

Measured pressure at the inlet is $p_1 = 1005\,\mathrm{hPa}$ ($= 100\,500\,\mathrm{N/m^2}$) and at the outlet $p_2 = 1000\,\mathrm{hPa}$. The pressure is constant across the tunnel cross section. We neglect the boundary layer at the tunnel wall and assume the inflow velocity is constant with a steady x-component of $v_1 = 10\,\mathrm{m/s}$. The measured local velocity at the outlet was simplified to a radial distribution (constant on circles concentric to the x-axis) $\underline{v}_2(r) = \left(v_{max} r / r_0, 0, 0\right)^T$. The air density is $\rho = 1.225\,\mathrm{kg/m^3}$.

Solution

The reaction of the drag force is an external force (surface force) acting on the fluid between inlet and outlet. We can employ the conservation of mass and momentum principles to solve this problem. Since the equations are applicable to any flow problem, we need additional equations to make the solution specific to the given problem. These additional equations are derived from boundary conditions. You may remember from mechanics of materials that the deflection of a beam under load depends on how the beam is supported. We will specify appropriate boundary conditions below.

Conservation of momentum states that the rate of change of momentum is equal to the sum of external forces. The integral form of the conservation of momentum principle reads as

$$\frac{\partial}{\partial t} \iiint_V \rho\, \underline{v}\, \mathrm{d}V + \iint_S \rho\, \underline{v} \left(\underline{v}^T \underline{n}\right) \mathrm{d}S = \sum (\text{external forces}) \tag{9.1}$$

The fixed control volume V is the wind tunnel content between inlet and outlet. Its surface S consists of inlet S_1, outlet S_2, the cylinder S_c, and indeed the body surface S_b.

Rate of change of momentum

The left-hand side consists of two terms. The volume integral describes the change of momentum over time. Since the flow is steady and the control volume is fixed, the time derivative of the volume integral must vanish.

$$\frac{\partial}{\partial t} \iiint_V \rho \underline{v}\, \mathrm{d}V = 0 \tag{9.2}$$

Momentum flux for wall and body We split the surface integral for the momentum flux into four parts: inlet, outlet, cylinder and body.

$$\iint_S \rho \underline{v}(\underline{v}^T \underline{n})\, \mathrm{d}S = \iint_{S_1} \ldots \mathrm{d}S + \iint_{S_2} \ldots \mathrm{d}S + \iint_{S_c} \ldots \mathrm{d}S + \iint_{S_b} \ldots \mathrm{d}S \tag{9.3}$$

The tunnel wall will not allow any air to escape, and the dot product of velocity and normal vector vanishes everywhere on the cylindrical tunnel wall

$$\underline{v}^T \underline{n}_c = 0 \qquad \text{on tunnel wall } S_c$$

In addition, we consider the test body S_b impermeable, i.e. no fluid flows through it, i.e. $\underline{v}^T \underline{n}_b = 0$ on S_b as well. Consequently, the integrals over S_c and S_b vanish.

Momentum flux through inlet Velocity distribution $\underline{v}$ and normal vector $\underline{n}_1$ are constant at the inlet. Therefore, the surface integral over the inlet reduces to

$$\iint_{S_1} \rho\, \underline{v}_1 (\underline{v}_1^T \underline{n}_1)\, \mathrm{d}S = -\rho \pi r_0^2 \begin{pmatrix} v_1^2 \\ 0 \\ 0 \end{pmatrix} \tag{9.4}$$

Momentum flux through outlet For this exercise, we assume a heavily simplified, analytical radial velocity distribution. In a real world application this would be replaced by a detailed wake measurement. Using polar coordinates for the cross section, the surface element becomes $\mathrm{d}S = r\mathrm{d}\varphi\mathrm{d}r$ (with $r \in [0, r_0]$ and $\varphi \in [0, 2\pi]$). We expand the integrand for the integral over S_2:

$$\iint_{S_2} \rho \underline{v}_2 (\underline{v}_2^T \underline{n}_2)\, \mathrm{d}S = \rho \int_0^{r_0} \int_0^{2\pi} \begin{pmatrix} v_{\max} \frac{r}{r_0} \\ 0 \\ 0 \end{pmatrix} \left[\left(v_{\max} \frac{r}{r_0}, 0, 0 \right) \begin{pmatrix} 1 \\ 0 \\ 0 \end{pmatrix} \right] r\, \mathrm{d}\varphi \mathrm{d}r \tag{9.5}$$

The integrand on the right-hand side does not depend on the polar coordinate φ, and the inner integral simplifies to multiplication with the difference between its upper and lower boundary.

$$\begin{aligned} \iint_{S_2} \rho \underline{v}_2 (\underline{v}_2^T \underline{n}_2)\, \mathrm{d}S &= 2\pi\rho \begin{pmatrix} v_{\max}^2 \\ 0 \\ 0 \end{pmatrix} \int_0^{r_0} \frac{r^3}{r_0^2}\, \mathrm{d}r = 2\pi\rho \begin{pmatrix} v_{\max}^2 \\ 0 \\ 0 \end{pmatrix} \left[\frac{r^4}{4r_0^2} \right]_0^{r_0} \\ &= \frac{1}{2} \rho \pi r_0^2 \begin{pmatrix} v_{\max}^2 \\ 0 \\ 0 \end{pmatrix} \end{aligned} \tag{9.6}$$

In summary, the rate of change of momentum on the left-hand side of (9.1) is equal to

$$\frac{\partial}{\partial t} \iiint_V \rho \underline{v}\, \mathrm{d}V + \iint_S \rho \underline{v}(\underline{v}^T \underline{n})\, \mathrm{d}S = \rho \pi r_0^2 \left[\frac{1}{2} \begin{pmatrix} v_{\max}^2 \\ 0 \\ 0 \end{pmatrix} - \begin{pmatrix} v_1^2 \\ 0 \\ 0 \end{pmatrix} \right] \tag{9.7}$$

Conservation of mass

The only unknown term in this expression is the maximum velocity v_{max} in the outlet. If it cannot be measured properly, we can compute it from the conservation of mass principle. At a relatively low speed[1] in the wind tunnel, we can assume that the flow is incompressible (constant density).

For incompressible, steady flow the continuity equation for a fixed control volume states

$$\iint_S \rho\left(\underline{v}^T \underline{n}\right) \mathrm{d}S = 0 \tag{9.8}$$

The vanishing mass flux integral indicates that as much fluid is flowing out of the control volume as is flowing into it. Again, we split the surface S of the control volume into four parts. Zero flow boundary conditions are imposed on the solid walls of S_c and S_b. The mass flux through inlet and outlet must be equal but of opposite sign:

$$0 = \iint_{S_1} \rho\left(\underline{v}^T \underline{n}\right) \mathrm{d}S + \iint_{S_2} \rho\left(\underline{v}^T \underline{n}\right) \mathrm{d}S \tag{9.9}$$

$$= \rho \int_0^{r_0}\int_0^{2\pi} \left(v_1, 0, 0\right)\begin{pmatrix}-1\\0\\0\end{pmatrix} r\,\mathrm{d}\varphi \mathrm{d}r + \rho \int_0^{r_0}\int_0^{2\pi} \left(v_{max}\frac{r}{r_0}, 0, 0\right)\begin{pmatrix}1\\0\\0\end{pmatrix} r\,\mathrm{d}\varphi \mathrm{d}r \tag{9.10}$$

$$= \rho\pi r_0^2 \left[\frac{2}{3}\, v_{max} - v_1\right] \tag{9.11}$$

or

$$v_{max} = \frac{3}{2}\, v_1 \tag{9.12}$$

This simple relationship should not be mistaken for a general solution. It is a consequence of our simplified velocity distribution at the outlet.

Total momentum flux

Substituting this result into the momentum flux (9.7) yields:

$$\frac{\partial}{\partial t} \iiint_V \rho\, \underline{v}\, \mathrm{d}V + \iint_S \rho\, \underline{v}\left(\underline{v}^T \underline{n}\right) \mathrm{d}S = \frac{1}{8}\rho\pi r_0^2 \begin{pmatrix} v_1^2 \\ 0 \\ 0 \end{pmatrix} \tag{9.13}$$

Sum of external forces

We now turn to the sum of external forces making up the right-hand side of our conservation of momentum equation (9.1).

There are body and surface forces acting on the fluid in our control volume. Gravity will act as a body force on the fluid. However, the gravity force is acting vertically to the flow and is balanced by the aero-static pressure distribution in the fluid.[2] The influence of gravity on our wind tunnel flow can safely be neglected.

Only the surface forces remain. There will be pressure forces on inlet $\underline{F}_{p_1}$ and outlet $\underline{F}_{p_2}$. Of course, there is also the force exerted by the body onto the fluid $-\underline{F}_D$. The

[1]More exact: low Mach number $Ma = \frac{v}{a}$ with a being the speed of sound in the respective medium. For $Ma < 0.2$ air flows may be treated as incompressible.

[2]Our assumption that the pressure is constant across inlet and outlet is not 100% true. There is a small variation of static pressure depending on the tunnel diameter (about 24 Pa or 0.024% of p_2 in the example).

minus sign expresses the fact that the reaction of the drag force (fluid acting on body) is acting against the motion of the fluid. It is consistent to ignore the friction force on the wind tunnel wall S_c, since we neglected the wall influence on the velocity distribution near the wall as well. A more accurate assessment would measure complete velocity profiles without and with body to distinguish between the change of momentum caused by the tank wall and that caused by the body.

In our simplified case the sum of external forces reduces to

$$\sum(\text{external forces}) = \underline{F}_{p_1} + \underline{F}_{p_2} - \underline{F}_D \tag{9.14}$$

Inlet pressure force Integration of the pressure over the surface yields the pressure forces. For the inlet this results in

$$\underline{F}_{p_1} = -\iint_{S_1} p_1\, \underline{n}_1 \,\mathrm{d}S \tag{9.15}$$

The minus sign in front of the integral is a consequence of our choice for the normal vector direction: it is pointing *out* of the control volume. Therefore, a positive pressure will cause a force acting in the negative normal direction. Since the inlet is a flat disc with constant normal vector and the pressure p_1 is assumed constant, the integration simplifies to a multiplication with the tunnel cross section area.

$$\underline{F}_{p_1} = -\pi r_0^2 p_1 \begin{pmatrix} -1 \\ 0 \\ 0 \end{pmatrix} = \pi r_0^2 p_1 \begin{pmatrix} 1 \\ 0 \\ 0 \end{pmatrix} \tag{9.16}$$

As expected, the resulting force on S_1 points in the positive x-direction.

Outlet pressure force Analysis of the pressure force on the outlet yields

$$\underline{F}_{p_2} = -\iint_{S_2} p_2\, \underline{n}_2 \,\mathrm{d}S = -\pi r_0^2 p_2 \begin{pmatrix} 1 \\ 0 \\ 0 \end{pmatrix} \tag{9.17}$$

This force points in the negative x-direction.

Body drag force Our test object is a body of revolution, which allows the conclusion that only the x-component of the drag force will be nonzero. Therefore, the sum of external forces is equal to

$$\sum(\text{external forces}) = \pi r_0^2 p_1 \begin{pmatrix} 1 \\ 0 \\ 0 \end{pmatrix} - \pi r_0^2 p_2 \begin{pmatrix} 1 \\ 0 \\ 0 \end{pmatrix} - \begin{pmatrix} F_D \\ 0 \\ 0 \end{pmatrix} \tag{9.18}$$

Drag force Equating the x-components of the rate of change in momentum (9.13) and the external force (9.18) results in

$$\frac{1}{8}\rho \pi r_0^2 v_1^2 = \pi r_0^2 (p_1 - p_2) - F_D \tag{9.19}$$

We solve for the unknown drag force F_D:

$$F_D = \pi r_0^2 \left[(p_1 - p_2) - \frac{1}{8}\rho v_1^2 \right] \tag{9.20}$$

Finally, substituting values for the known quantities on the right-hand side results in:

$$F_D = \pi 1^2\,\mathrm{m}^2 \left[\left(100500\ \mathrm{Pa} - 100000\ \mathrm{Pa}\right) - \frac{1}{8}\,1.22\,\frac{\mathrm{kg}}{\mathrm{m}^3}\,10^2\,\frac{\mathrm{m}^2}{\mathrm{s}^2} \right]$$

$$= 1523\,\mathrm{N} \tag{9.21}$$

Do not forget to check the units: pressure $[\mathrm{Pa} = \mathrm{N/m^2}]$ multiplied by surface area $[\mathrm{m}^2]$ results in a force [N]. The units in the last term reduce to

$$\left[\mathrm{m}^2 \left(\frac{\mathrm{kg}}{\mathrm{m}^3}\right)\left(\frac{\mathrm{m}^2}{\mathrm{s}^2}\right)\right] = \left[\frac{\mathrm{kg\,m}}{\mathrm{s}^2}\right]$$

which is also equal to Newton [N].

9.2 Submerged Vessel in an Unbounded Fluid

Our second example considers a body moving with constant velocity u_0 in positive x-axis direction. The principal difference from the first example is the fact that there is no tunnel wall. Instead, the body is moving in an unbounded fluid. The resistance of the body is to be determined using conservation of mass and momentum principles:

$$\frac{\partial}{\partial t}\iiint_V \rho\,\mathrm{d}V + \iint_S \rho\left(\underline{v}^T\underline{n}\right)\mathrm{d}S = 0 \tag{9.22}$$

$$\frac{\partial}{\partial t}\iiint_V \rho\underline{v}\,\mathrm{d}V + \iint_S \rho\underline{v}\left(\underline{v}^T\underline{n}\right)\mathrm{d}S = \sum(\text{external forces}) \tag{9.1}$$

Fixed finite but open control volume

We select a $< x, y, z >$-coordinate system that is fixed to the moving body (Figure 9.2). The control volume V is a cylinder with mantle S_c. Its length and its radius R_c are chosen so large that the disturbance due to the body has declined. Specifically, we require that the pressure has returned to its reference value p_0 everywhere outside of the control volume. The horizontal component of the velocity vector is again u_0. The zero flow boundary condition of the wind tunnel example is replaced by a constant pressure boundary condition. As a consequence, a small mass flow across the mantle S_c must exist for the submerged body in an unbounded fluid.

Steady flow assumption

The flow is steady and the control volume is fixed with respect to the chosen coordinate system. Consequently, the terms with time derivatives vanish in Equations (9.22) and (9.1), and they simplify to

$$\iint_S \rho\left(\underline{v}^T\underline{n}\right)\mathrm{d}S = 0 \qquad \text{conservation of mass, steady flow} \tag{9.23}$$

$$\iint_S \rho\,\underline{v}\left(\underline{v}^T\underline{n}\right)\mathrm{d}S = \sum(\text{external forces}) \qquad \text{conservation of momentum, steady flow} \tag{9.24}$$

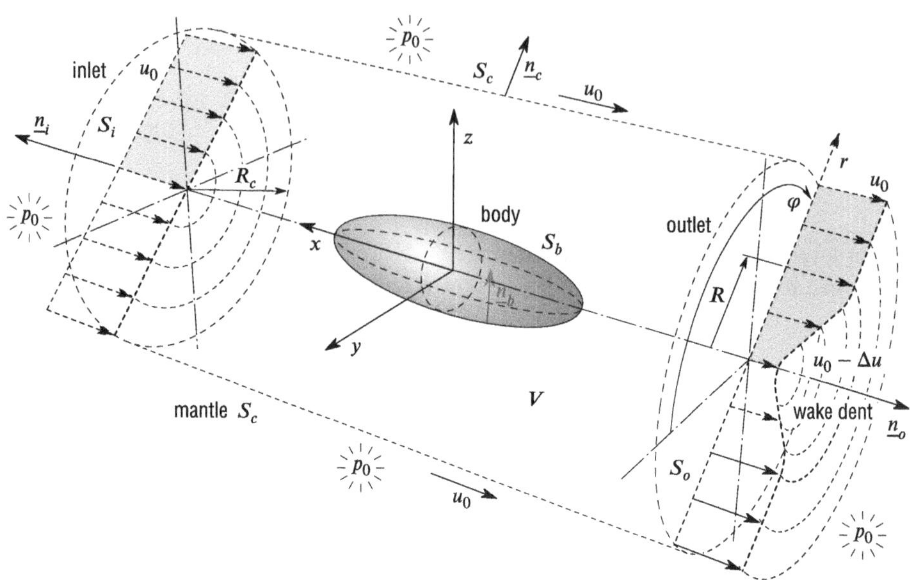

Figure 9.2 Ellipsoid moving in an unbounded fluid

Velocity field At the inlet S_i, fluid enters the control volume V with constant velocity u_0. Again, for the outside observer the fluid is at rest, and the body moves with speed u_0. However, for an observer on the body, fluid is streaming toward the body. Due to friction between water and body, a boundary layer develops and the momentum loss will cause a drag force F_D acting on the body. The momentum loss becomes apparent in the 'wake dent' behind the body (Figure 9.2). At the outlet S_o of the control volume fluid velocity is reduced by Δu from its initial value u_0.

In this example we assume the velocity loss Δu is radially symmetric to the x-axis and given by

$$\frac{\Delta u(r)}{u_0} = \begin{cases} \frac{1}{16R^4}(R+r)^2(R-r)^2 & \text{for } |r| < R \\ 0 & \text{elsewhere} \end{cases} \tag{9.25}$$

The velocity is reduced by a variable Δu in a region of the outlet centered around the x-axis. Obviously, we must choose the control volume radius R_c to be larger than the radius R of the region with reduced velocity. In an actual application, careful measurement of the wake field behind the body would provide Δu values, which then would be integrated numerically.

9.2.1 Conservation of mass

Mass flux The boundary surface S of the control volume consists of four parts: inlet S_i, outlet S_o, mantle S_c, and body S_b. The interior of the body is not part of the control volume. We split the surface integral in Equation (9.23) into four parts.

$$0 = \iint_{S_i} \ldots \, dS + \iint_{S_o} \ldots \, dS + \iint_{S_c} \ldots \, dS + \underbrace{\iint_{S_b} \ldots \, dS}_{=0} \tag{9.26}$$

Nothing flows through the body surface S_b and the integrand $\underline{v}^T \underline{n}$ vanishes.

Mass flux over mantle

In contrast to the wind tunnel example, we may not claim that the mass flow integral vanishes over the mantle surface S_c. To the contrary, if the flow through the outlet S_o is smaller than the inflow through S_i, a mass flow $\dot{m}_c$ over the mantle exists. From Equation (9.26) follows

$$\dot{m}_c = \iint_{S_c} \rho\left(\underline{v}^T \underline{n}\right) dS = -\iint_{S_i} \rho\left(\underline{v}^T \underline{n}\right) dS - \iint_{S_o} \rho\left(\underline{v}^T \underline{n}\right) dS \tag{9.27}$$

At the inlet, velocity vector and normal vector are

$$\underline{v} = \begin{pmatrix} -u_0 \\ 0 \\ 0 \end{pmatrix} \qquad \underline{n}_i = \begin{pmatrix} 1 \\ 0 \\ 0 \end{pmatrix} \tag{9.28}$$

The corresponding vectors for the outlet are

$$\underline{v} = \begin{pmatrix} -(u_0 - \Delta u) \\ 0 \\ 0 \end{pmatrix} \qquad \underline{n}_o = \begin{pmatrix} -1 \\ 0 \\ 0 \end{pmatrix} \tag{9.29}$$

Substituting the vectors into Equation 9.27 for the mass flow across the mantle yields:

$$\begin{aligned} \dot{m}_c &= -\iint_{S_i} \rho\left(-u_0, 0, 0\right) \begin{pmatrix} 1 \\ 0 \\ 0 \end{pmatrix} dS - \iint_{S_o} \rho\left[-(u_0 - \Delta u), 0, 0\right] \begin{pmatrix} -1 \\ 0 \\ 0 \end{pmatrix} dS \\ &= +\iint_{S_i} \rho\, u_0 \, dS - \iint_{S_o} \rho\left(u_0 - \Delta u\right) dS \end{aligned}$$

After extraction of all constant terms in the integrals over S_i and S_o, a single integral remains to be solved:

$$\dot{m}_c = \rho\, \pi R_c^2\, u_0 - \rho\, \pi R_c^2\, u_0 + \iint_{S_o} \rho\, \Delta u \, dS = +\iint_{S_o} \rho\, \Delta u \, dS \tag{9.30}$$

The integration is straightforward in polar coordinates $< r, \varphi >$. The radial distance r from the x-axis takes values from 0 to R_c, or more precisely $r \in [0, R < R_c]$, because

$\Delta u = 0$ for $r > R$. The angle φ describes a full circle $\varphi \in [0, 2\pi]$. Substituting expression (9.25) for the velocity difference Δu gives us

$$\begin{aligned} \dot{m}_c &= \iint\limits_{S_o} \rho\, \Delta u\, \mathrm{d}S = \int\limits_0^R \int\limits_0^{2\pi} \rho\, u_0 \frac{1}{16R^4} (R+r)^2 (R-r)^2\, r\, \mathrm{d}\varphi \mathrm{d}r \\ &= 2\pi\, \rho\, u_0 \frac{1}{16R^4} \int\limits_0^R (R+r)^2 (R-r)^2\, r\, \mathrm{d}r \end{aligned} \tag{9.31}$$

Expanding all terms leaves a simple polynomial integral to solve.

$$\begin{aligned} \dot{m}_c &= \frac{2\pi\, \rho\, u_0}{16R^4} \int\limits_0^R \left(R^4 - 2R^2r^2 + r^4\right) r\, \mathrm{d}r = \frac{2\pi\, \rho\, u_0}{16R^4} \left[\frac{R^4 r^2}{2} - \frac{R^2 r^4}{2} + \frac{r^6}{6} \right]_0^R \\ &= \frac{2\pi\, \rho\, u_0}{16R^4} \left[\frac{R^6}{2} - \frac{R^6}{2} + \frac{R^6}{6} \right] = \frac{\rho}{48}\, \pi\, R^2\, u_0 \end{aligned} \tag{9.32}$$

$\dot{m}_c$ represents how much mass is 'escaping' over the mantle surface S_c per time unit. We will employ this result shortly to estimate the amount of longitudinal momentum that passes through S_c.

9.2.2 Conservation of momentum

Conservation of momentum states that the rate of change in momentum is equal to the sum of external forces. We are most interested in the longitudinal component of the equation. So, flow will be mostly axially symmetric if the body is axially symmetric. Vortex shedding might cause a time varying transverse force, but we assume here that the flow is steady.

Sum of external forces

Pressure force vanishes External forces consist of body and surface forces. The only body force is the gravity force. It is acting perpendicular to the flow and will not affect the resistance of the body. At the outer surfaces of the control volume S_i, S_o, and S_c the pressure acts with a constant value p_0 and the resultant pressure force vanishes. The velocity has no gradient perpendicular to the surfaces S_i, S_o, and S_c. Therefore, no shear stresses occur.

Drag force The only boundary with a resultant pressure and shear stress related force is the body. The resultant force in x-direction is the resistance F_D we are looking for. The resistance will act against the body's motion. For the sum of external forces we need the resistance's reaction which acts on the fluid in positive x-direction.

$$\sum(\text{external forces}) = \begin{pmatrix} F_D \\ 0 \\ 0 \end{pmatrix} \tag{9.33}$$

Rate of change of momentum

Momentum flux

The surface integral (9.1) for the momentum flux is broken into four parts.

$$\iint_S \rho\, \underline{v}\left(\underline{v}^T \underline{n}\right) \mathrm{d}S = \iint_{S_i} \ldots \mathrm{d}S + \iint_{S_o} \ldots \mathrm{d}S + \iint_{S_c} \ldots \mathrm{d}S + \underbrace{\iint_{S_b} \ldots \mathrm{d}S}_{=0} \tag{9.34}$$

The body S_b is a stream surface and consequently no flow crosses into the body ($\underline{v}^T \underline{n}_c = 0$) and the corresponding integral vanishes.

Momentum flux through mantle

In the case of the wind tunnel, the third integral vanishes as well. However, in the current case S_c is not a solid boundary, as we discovered applying the conservation of mass principle. Instead, we made the radius R_c of the control volume V so big that the velocity vector on the mantle S_c is in good approximation constant and equal to

$$\underline{v} = \begin{pmatrix} -u_0 \\ 0 \\ 0 \end{pmatrix} \quad \text{for } r = R_c \tag{9.35}$$

Since there is a small mass flow across surface S_c, the velocity cannot be exactly parallel to the x-axis everywhere. However, we may always increase the radius R to justify that $v, w \ll u_0$. As a consequence, the momentum flux across the mantle S_c can be expressed as a function of the mass flux $\dot{m}_c$ (9.27).

$$\begin{aligned} \iint_{S_c} \rho\, \underline{v}\left(\underline{v}^T \underline{n}_c\right) \mathrm{d}S &\approx \iint_{S_c} \rho \begin{pmatrix} -u_0 \\ 0 \\ 0 \end{pmatrix} \left(\underline{v}^T \underline{n}_c\right) \mathrm{d}S = \begin{pmatrix} -u_0 \\ 0 \\ 0 \end{pmatrix} \iint_{S_c} \rho \left(\underline{v}^T \underline{n}_c\right) \mathrm{d}S \\ &\approx \begin{pmatrix} -u_0 \\ 0 \\ 0 \end{pmatrix} \dot{m}_c \end{aligned} \tag{9.36}$$

Momentum flux through inlet

The momentum flux over the inlet S_i is

$$\iint_{S_i} \rho\, \underline{v}\left(\underline{v}^T \underline{n}_i\right) \mathrm{d}S = \iint_{S_i} \rho \begin{pmatrix} -u_0 \\ 0 \\ 0 \end{pmatrix} \left[\left(-u_0, 0, 0\right) \begin{pmatrix} 1 \\ 0 \\ 0 \end{pmatrix}\right] \mathrm{d}S = \rho \pi R_c^2 \begin{pmatrix} u_0^2 \\ 0 \\ 0 \end{pmatrix} \tag{9.37}$$

Momentum flux through outlet

For the momentum flux integral over the outlet S_o, we employ polar coordinates. The surface element becomes $\mathrm{d}S = r\mathrm{d}\varphi\mathrm{d}r$.

$$\begin{aligned} \iint_{S_o} \rho\, \underline{v}\left(\underline{v}^T \underline{n}_o\right) \mathrm{d}S &= \iint_{S_o} \rho \begin{pmatrix} -(u_0 - \Delta u) \\ 0 \\ 0 \end{pmatrix} \left[\left(-(u_0 - \Delta u), 0, 0\right) \begin{pmatrix} -1 \\ 0 \\ 0 \end{pmatrix}\right] \mathrm{d}S \\ &= \rho \int_0^{R_c} \int_0^{2\pi} \begin{pmatrix} -u_0^2 + 2u_0 \Delta u - (\Delta u)^2 \\ 0 \\ 0 \end{pmatrix} r\, \mathrm{d}\varphi \mathrm{d}r \end{aligned}$$

Splitting the integral into parts and solving the first integral yields:

$$\iint_{S_o} \rho\,\underline{v}(\underline{v}^T\underline{n}_o)\,\mathrm{d}S = \rho\pi R_c^2\begin{pmatrix}-u_0^2\\0\\0\end{pmatrix} + \rho\int_0^R\int_0^{2\pi}\begin{pmatrix}2u_0\Delta u\\0\\0\end{pmatrix} r\,\mathrm{d}\varphi\mathrm{d}r \qquad (9.38)$$

$$+\,\rho\int_0^R\int_0^{2\pi}\begin{pmatrix}-(\Delta u)^2\\0\\0\end{pmatrix} r\,\mathrm{d}\varphi\mathrm{d}r$$

The first term in the momentum flux over the outlet (9.38) will cancel with the contribution from the inlet (9.37).

The two remaining integrals involve the change in velocity Δu and because $\Delta u = 0$ for $r > R$, the upper limit of the integrals has been adjusted to R (which is smaller than R_c). Substitution of Equation (9.25) into the first remaining integral results in an integral similar to the one already solved for the mass flux $\dot{m}_c$.

$$\rho\int_0^R\int_0^{2\pi}\begin{pmatrix}2u_0\Delta u\\0\\0\end{pmatrix} r\,\mathrm{d}\varphi\mathrm{d}r = \frac{2\rho\,u_0^2}{16R^4}\int_0^R\int_0^{2\pi}\begin{pmatrix}(R+r)^2\,(R-r)^2\\0\\0\end{pmatrix} r\,\mathrm{d}\varphi\mathrm{d}r$$

$$= \frac{4\pi\,\rho\,u_0^2}{16R^4}\begin{pmatrix}\frac{R^6}{6}\\0\\0\end{pmatrix}$$

$$= \rho\pi R^2 u_0^2\begin{pmatrix}\frac{1}{24}\\0\\0\end{pmatrix} \qquad (9.39)$$

The second integral over $(\Delta u)^2$ is often neglected if the changes in velocity at the outlet are small. We have an analytical expression, therefore solving the integral poses no further difficulty.

$$\rho\int_0^R\int_0^{2\pi}\begin{pmatrix}-(\Delta u)^2\\0\\0\end{pmatrix} r\,\mathrm{d}\varphi\mathrm{d}r = \frac{2\pi\,\rho\,u_0^2}{256\,R^8}\int_0^R\begin{pmatrix}-\left[(R+r)^2\,(R-r)^2\right]^2\\0\\0\end{pmatrix} r\,\mathrm{d}r$$

$$= \frac{2\pi\,\rho\,u_0^2}{256\,R^8}\int_0^R\begin{pmatrix}-R^8+4R^6r^2-6R^4r^4+4R^2r^6-r^8\\0\\0\end{pmatrix} r\,\mathrm{d}r$$

$$= \frac{2\pi\,\rho\,u_0^2}{256\,R^8}\left[\begin{pmatrix}-\frac{R^8r^2}{2}+R^6r^4-R^4r^6+\frac{R^2r^8}{2}-\frac{r^{10}}{10}\\0\\0\end{pmatrix}\right]_0^R$$

$$\rho \int_0^R \int_0^{2\pi} \begin{pmatrix} -(\Delta u)^2 \\ 0 \\ 0 \end{pmatrix} r\,\mathrm{d}\varphi \mathrm{d}r = \frac{2\pi\,\rho\,u_0^2}{256\,R^8} \begin{pmatrix} -\frac{R^{10}}{2} + R^{10} - R^{10} + \frac{R^{10}}{2} - \frac{R^{10}}{10} \\ 0 \\ 0 \end{pmatrix}$$

$$= \rho\,\pi\,R^2\,u_0^2 \begin{pmatrix} -\frac{1}{1280} \\ 0 \\ 0 \end{pmatrix} \tag{9.40}$$

Finally, we can collect the results for the momentum flux over the boundary S of the control volume. With results from Equations (9.32), (9.36), (9.37), (9.38), (9.39), and (9.40 we get

Momentum flux result

$$\iint_S \rho\underline{v}\left(\underline{v}^T\underline{n}\right)\mathrm{d}S = \rho\,\pi\,R^2\,u_0^2 \left[\begin{pmatrix} \frac{1}{24} \\ 0 \\ 0 \end{pmatrix} + \begin{pmatrix} -\frac{1}{1280} \\ 0 \\ 0 \end{pmatrix} + \begin{pmatrix} -\frac{1}{48} \\ 0 \\ 0 \end{pmatrix}\right]$$

$$= \rho\,\pi\,R^2\,u_0^2 \begin{pmatrix} \frac{77}{3840} \\ 0 \\ 0 \end{pmatrix} \tag{9.41}$$

Equating the result for the momentum flux with the sum of external forces provides us with an equation for the drag force F_D acting on the body.

Drag force

$$\rho\,\pi\,R^2\,u_0^2 \begin{pmatrix} \frac{77}{3840} \\ 0 \\ 0 \end{pmatrix} = \begin{pmatrix} F_D \\ 0 \\ 0 \end{pmatrix} \tag{9.42}$$

$$F_D \approx 0.020052\,\rho\,\pi\,R^2\,u_0^2 \tag{9.43}$$

As it should be, the drag force will not depend on the cross section πR_c^2 of the control volume, if it is chosen large enough. It solely depends on magnitude and extent of the wake dent.

If we assume a body velocity of $u_0 = 2.5$ m/s and a radius of the wake $R = 1.0$ m, the resistance of the body in salt water of density $\rho = 1026.021$ kg/m^3 will be

$$\begin{aligned} F_D &= 0.020052 \cdot 1026.021\ \mathrm{kg/m^3}\,\pi\,1.0^2\ \mathrm{m^2}\,2.5^2\ (\mathrm{m/s})^2 \\ &= 403.96\ \frac{\mathrm{kg\ m}}{\mathrm{s^2}} \\ &= 403.96\ \mathrm{N} \end{aligned}$$

The units work out as expected and the result is a force in Newton.

As you hopefully noted, this type of analysis does not require any knowledge about the geometry of the body. It could have any shape. The only requirement is the accurate measurement of magnitude and extent of the wake field. If the body causes noticeable asymmetries in the flow, careful assessment of the three-dimensional wake will also deliver the force components in transverse and vertical direction.

10

Boundary Layer Theory

This chapter introduces the basic equations of boundary layer theory, which was originally proposed by Ludwig Prandtl in 1904 (see Prandtl et al., 1989). Together with the following three chapters, this is necessarily a fairly short treatise on this complex subject, which is still the topic of ongoing research. The most comprehensive text on boundary layer theory is the book by Schlichting and Gersten (2003). Hermann Schlichting (*1907 – †1982) was a student of Ludwig Prandtl. Another good reference is the book by Young (1989).

We will first discuss some phenomena associated with the boundary layer and then return to the Navier-Stokes equations. By making certain assumptions on flow conditions in the boundary layer, we will derive the actual boundary layer equations.

Learning Objectives

At the end of this chapter students will be able to:

- discuss the concept of boundary layers
- describe the differences between laminar and turbulent boundary layer flow
- explain flow separation
- work with dimensionless equations
- derive the boundary layer equations

10.1 Boundary Layer

Conundrum of viscous forces

In Section 7.2, we derived the dimensionless factors in the NSE (7.27) and computed them for a 120 m long ship. The factor for the viscous forces equals the reciprocal value $1/Re$ of the Reynolds number. In the example, this viscous force factor is about 10^{-10} times smaller than the factors associated with other forces, which supports the notion that viscous forces possibly could be ignored in ship hydrodynamics. However, simply omitting the viscous forces leads to obviously erroneous resistance predictions.

Boundary layer

Ludwig Prandtl solved this apparent dilemma by splitting the flow into two regions:

Fundamentals of Ship Hydrodynamics: Fluid Mechanics, Ship Resistance and Propulsion, First Edition.
Lothar Birk.

Companion website: www.wiley.com/go/birk/hydrodynamics

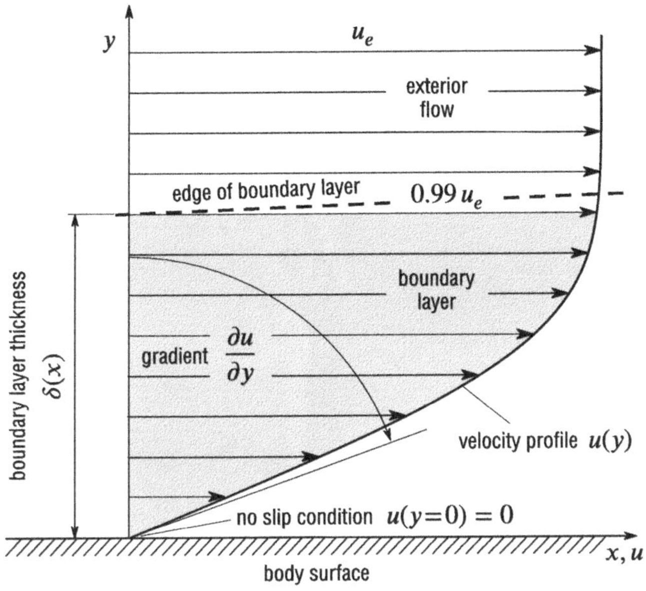

Figure 10.1 Basic properties of the velocity distribution in the boundary layer

- A region close to the body which we call the boundary layer. Boundary layers cover solid surfaces and boundaries between fluids that do not completely mix (like air and water). Within this thin sheet of flow, viscous forces cannot be neglected, that is, $(1/Re)\,\mathbf{\Delta}^*\underline{v}^*$ represents a noticeable force despite its small leading factor.
- An exterior flow outside of the boundary layer where viscous forces can be neglected in most practical cases.

Basic properties of the boundary layer are discussed below.

10.1.1 Boundary layer thickness

Definition of boundary layer thickness

Figure 10.1 shows the velocity distribution near a body surface. For convenience, we assume that the curvature of the surface is small and that the x-axis is parallel to the wall. With increasing distance y from the wall, velocity grows continuously from zero to the exterior flow velocity u_e. Because there is no obvious break or knuckle in the velocity distribution, it is hard to determine where the boundary layer ends and the exterior flow begins. In experiments, we usually mark the outer edge of the boundary layer at a distance from the wall where the velocity u reaches 99% of the exterior flow velocity (Figure 10.1). The distance from the surface to the outer edge is called boundary layer thickness δ.

No slip condition

At the body surface, fluid molecules stick to the body. We call this the no slip condition of boundary layer theory. This assumption has been verified in experiments by observation and measurements.

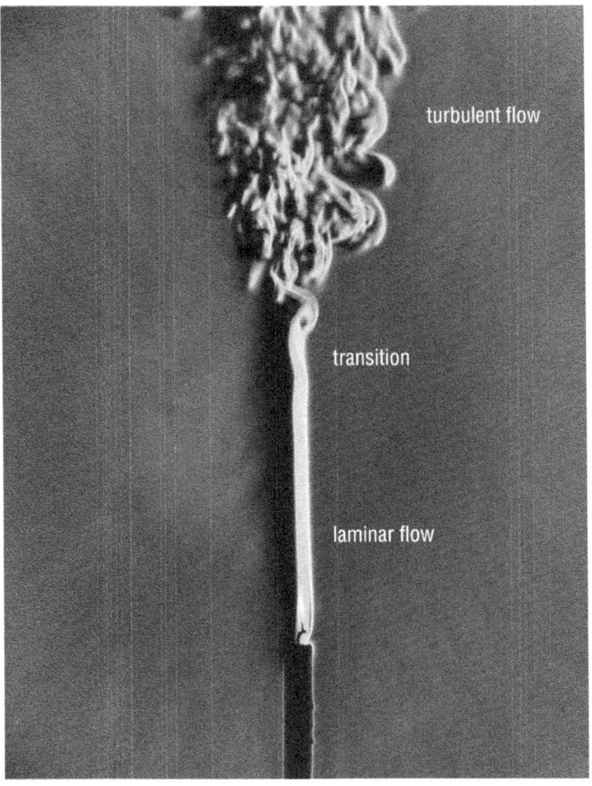

Figure 10.2 Transition from laminar to turbulent flow of the air rising from a burning candle. Reproduced with kind permission by Dr. Gary S. Settles, Floviz, Inc.

10.1.2 Laminar and turbulent flow

Laminar and turbulent flow

You have probably watched the rising smoke of a cigarette or from a candle which just got extinguished. The smoke forms a smooth band for some distance above the tip but soon starts meandering and then seems to break up in chaotic swirls and eddies. Figure 10.2 visualizes the air flow above a burning candle. We clearly see the straight flow right above the flame. Apparently, the flow is stable and does not seem to be time dependent. We call this flow laminar, derived from the Latin word 'lamina', which means 'thin sheet.' After a short distance the flow starts swerving left and right before it becomes fully turbulent. In turbulent flow, fluid particles deviate from the general flow direction in seemingly random moves, sometimes even moving against the general flow for a short while. Therefore, turbulent flow is constantly changing over time. However, more often than not, we treat turbulent flow as steady by splitting the time dependent velocity $\underline{v}(t)$ into a mean velocity $\underline{\overline{v}}$ and a turbulent fluctuation $\underline{v}'(t)$: $\underline{v}(t) = \underline{\overline{v}} + \underline{v}'(t)$. For details see Chapter 8.

Laminar flow

Boundary layer flows are laminar only for small local Reynolds numbers $Re_x < 500\,000$ (Fox and McDonald, 1998). The local Reynolds number Re_x is based on the current position x rather than on a fixed reference length L.

$$Re_x = \frac{u_e\, x}{\nu} \quad \text{local Reynolds number} \tag{10.1}$$

Re_x changes along the body surface!

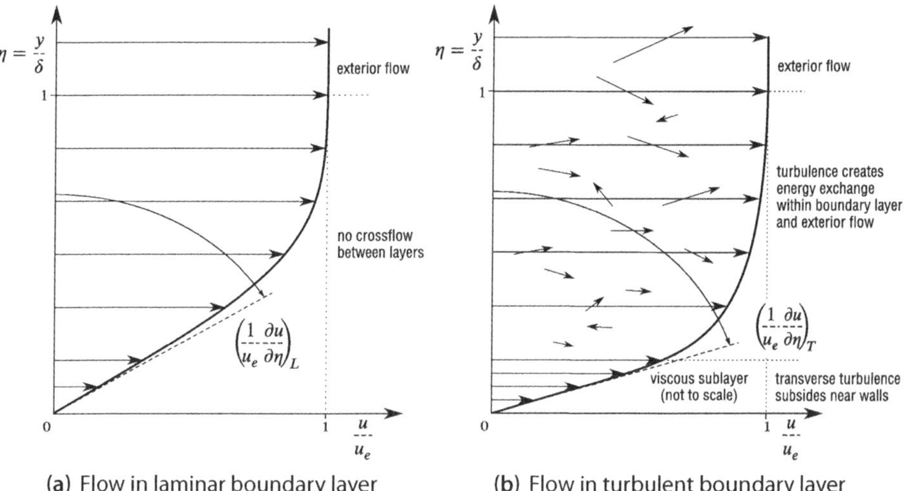

(a) Flow in laminar boundary layer

(b) Flow in turbulent boundary layer

Figure 10.3 Flow characteristics of laminar and turbulent boundary layers

In laminar boundary layers, the flow appears to be neatly stacked in tiers (see Figure 10.3(a)). There is neither momentum exchange with the exterior flow nor between the tiers.

Laminar-turbulent transition

Under ideal conditions, a boundary layer can stay laminar beyond $Re_x > 500\,000$. However, every flow experiences a varying level of small disturbances, called preturbulence. Preturbulence, surface roughness, or obstacles on the surface may trigger the boundary layer to transition from laminar to turbulent flow at lower Reynolds numbers. With increasing local Reynolds number, the flow transitions from laminar to turbulent somewhere in the range $200\,000 < Re_x < 1\,000\,000$. The exact position depends on the amount of preturbulence, surface roughness, and other factors.

Turbulent flow

For Reynolds numbers $Re_x > 500\,000$ the boundary layer is usually treated as turbulent (see Figure 10.3(b)). With turbulence, fluctuations in the transverse velocity v – though small overall – become large enough to cause momentum exchange within the boundary layer and with the exterior flow. Transverse velocity must diminish close to the wall. Therefore, turbulent boundary layers usually have a laminar sublayer. In modern literature it is called viscous sublayer because it is governed by the viscosity of the fluid.

Boundary layer thickness

Figure 10.4 illustrates the development of a boundary layer over a flat plate. Its thickness is zero at the beginning of a plate and continuously grows toward the end of the plate. Note that we exaggerate the thickness in essentially all schematic drawings of the boundary layer in order to show what is happening in it. The boundary layer thickness at the stern of a ship is in the order of 1% of the vessel's length.

Laminar versus turbulent

Initially, the flow is laminar in the boundary layer. The formula given in Figure 10.4 for the laminar boundary layer thickness δ_L reflects Blasius' solution of the boundary layer equations (Blasius, 1907). δ_L is a function of distance from the leading edge x and the local Reynolds number Re_x. However, with increasing local Reynolds number Re_x, the boundary layer becomes unstable and transitions into a turbulent flow. The boundary layer thickens more quickly in a turbulent boundary layer. The equation

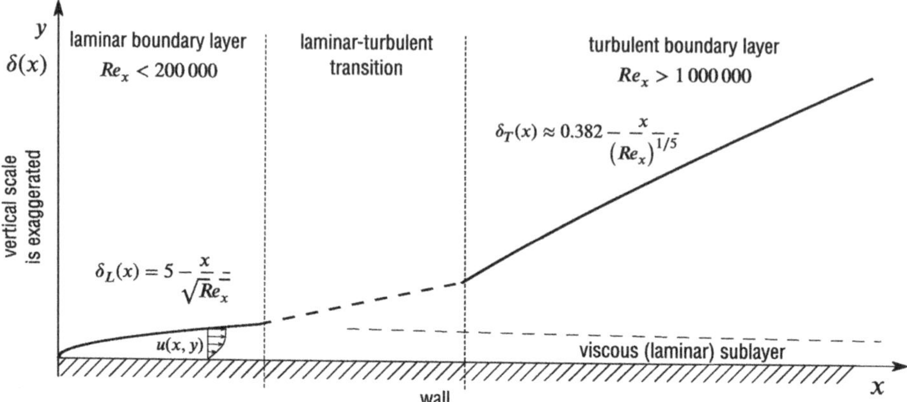

Figure 10.4 Development of the boundary layer along a flat surface. Note that the outer limit of the boundary layer is not a streamline

stated in Figure 10.4 for the turbulent boundary layer thickness δ_T is an approximate solution for moderate Reynolds numbers ($\approx 10^7$). It ignores the initial laminar part of the boundary layer.

The exact line of transition is hard to predict because it depends not only on the local Reynolds number but also on preturbulence, surface roughness, and even the shape of the leading edge of the plate. Low preturbulence, smooth surface, and round leading edges tend to delay the transition.

The distinction between laminar and turbulent boundary layers is essential in ship model testing. Boundary layers of full scale ships are turbulent due to the high Reynolds numbers ($Re > 10^8$) for ship flows. Naval architects must ensure that the boundary layer on a ship model is also turbulent. Otherwise, flow patterns observed on a model may not be comparable to those of the full scale vessel, rendering the model test results pretty much useless.

10.1.3 Flow separation

Separation On curved surfaces we may encounter an additional flow feature called separation. It commonly occurs with blunt bodies like spheres. A well-designed ship hull should not suffer measurably from flow separation, especially with a working propeller at the stern.

Water sticks to the hull surface. Therefore, the velocity tangential to the wall vanishes there: $u(y = 0) = 0$. Due to the viscosity, fluid sticking to the wall slows down fluid particles close to the wall. Those, in turn, will slow down neighboring particles. With increasing position x, the zone of slowed down particles stretches farther out into the flow. Consequently, the boundary layer thickens. The initially positive transverse derivative $\partial u/\partial y$ diminishes (Figure 10.5). Especially in regions where the exterior flow slows down ($\partial u_e/\partial x < 0$), the flow near but not quite at the wall stops as well. Eventually, the flow reaches a separation point with $\partial u/\partial y = 0$. Further downstream a region of separated flow develops in which the direction of flow is reversed.

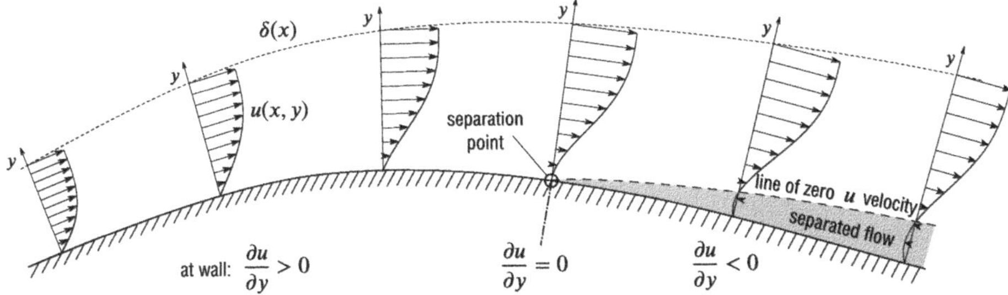

Figure 10.5 Development of velocity profile in the boundary layer along a curved surface with flow separation

Effect on resistance

Shear stress in laminar boundary layers is smaller than shear stress in turbulent boundary layers. As a consequence, frictional resistance is lower for laminar boundary layers. However, the latter are more prone to flow separation. Within a region of separated flow, pressure is not rebuilt to the levels at the bow of a vessel, which causes a significant increase in viscous pressure resistance. Turbulent boundary layers withstand separation better because the momentum exchange with the exterior flow replenishes energy lost to friction. With delayed separation, pressure is rebuilt over a larger portion of the hull surface. As a consequence, the total viscous resistance for turbulent boundary layers with less or no separation in many cases is smaller than the total viscous resistance of laminar boundary layers.

10.2 Simplifying Assumptions

Two-dimensional flow

Exterior flow and boundary layer flow must satisfy the principles of conservation of mass and momentum. Therefore, continuity and Navier-Stokes equations apply to both boundary layer and exterior flow. For our discussion of the boundary layer, we restrict ourselves to two-dimensional, incompressible flow in the x-y–plane. Two dimensional flow simply means that the same flow field is observed in any plane with constant z coordinate. Consequently, derivatives in z-direction vanish and the velocity vector becomes $\underline{v} = (u, v)^T$.

Incompressible, steady 2D-flow

We assume here that the fluid is incompressible and hence the density ρ is constant. Consequently, the continuity equation (5.12) reduces to the requirement that the divergence of the velocity vanishes:

$$\underline{\nabla}^T \underline{v} = 0 \tag{7.1}$$

The Navier-Stokes equations for incompressible flow of a Newtonian fluid read as

$$\frac{\partial \underline{v}}{\partial t} + \left(\underline{v}^T \underline{\nabla}\right) \underline{v} = \underline{f} - \frac{1}{\rho} \underline{\nabla} p + \nu \underline{\Delta v} \tag{7.12}$$

We further assume that the flow is steady, which eliminates the first term in Equation (7.12):

$$\frac{\partial \underline{v}}{\partial t} = 0 \qquad \text{for steady flow} \tag{10.2}$$

Restating continuity and Navier-Stokes equations in components for two-dimensional, incompressible, steady flow yields:

$$\begin{aligned}
\frac{\partial u}{\partial x} + \frac{\partial v}{\partial y} &= 0 && \text{2D continuity equation} \\
u\frac{\partial u}{\partial x} + v\frac{\partial u}{\partial y} &= f_x - \frac{1}{\rho}\frac{\partial p}{\partial x} + \nu\Delta u && x\text{-component of 2D NSE} \\
u\frac{\partial v}{\partial x} + v\frac{\partial v}{\partial y} &= f_y - \frac{1}{\rho}\frac{\partial p}{\partial y} + \nu\Delta v && y\text{-component of 2D NSE}
\end{aligned} \tag{10.3}$$

Unfortunately, this system of nonlinear, partial differential equations is very difficult to solve.

2D conservation of mass and momentum equations

Following Prandtl's ideas, we will eliminate some terms in Equations (10.3). For details see Schlichting and Gersten (2003, Chapter 7). In order to decide which terms are of lesser importance, we have to determine their relative magnitude. This is best done by studying the dimensionless equivalent of Equations (10.3). External flow speed u_e, length of body L, gravitational acceleration g, and twice the dynamic pressure ρu_e^2 serve as reference quantities. Following the process illustrated in Section 7.2, we get the two-dimensional conservation of mass and momentum equations of steady, incompressible flow in dimensionless form:

$$\begin{aligned}
\frac{\partial u^*}{\partial x^*} + \frac{\partial v^*}{\partial y^*} &= 0 \\
u^*\frac{\partial u^*}{\partial x^*} + v^*\frac{\partial u^*}{\partial y^*} &= \frac{1}{Fr^2}f_x^* - Eu\frac{\partial p^*}{\partial x^*} + \frac{1}{Re}\left(\frac{\partial^2 u^*}{\partial x^{*2}} + \frac{\partial^2 u^*}{\partial y^{*2}}\right) \\
u^*\frac{\partial v^*}{\partial x^*} + v^*\frac{\partial v^*}{\partial y^*} &= \frac{1}{Fr^2}f_y^* - Eu\frac{\partial p^*}{\partial y^*} + \frac{1}{Re}\left(\frac{\partial^2 v^*}{\partial x^{*2}} + \frac{\partial^2 v^*}{\partial y^{*2}}\right)
\end{aligned} \tag{10.4}$$

Geometric relationships

The longitudinal coordinate $x^* = x/L$ is pointing along the body contour, whereas the transverse coordinate $y^* = y/L$ is pointing in normal direction to the body surface. Although we use Cartesian coordinates in the equation, x^* and y^* may also be read as more general curvilinear coordinates. The longitudinal coordinate x takes values between zero and the reference length L. Consequently, the dimensionless coordinate x^* ranges from zero to one. We say the magnitude of a dimensionless coordinate x^* is of order one.

$$x^* \sim 1 \tag{10.5}$$

The transverse coordinate $y^* = y/L$ grows from zero up to the boundary layer thickness $\delta^* = \delta/L$. We conclude that y^* is of the same order of magnitude as the boundary layer thickness δ^*.

$$y^* \sim \delta^* \tag{10.6}$$

Longitudinal velocity and derivatives

The longitudinal velocity u reaches the value u_e at the outer edge of the boundary layer. Consequently, the dimensionless longitudinal velocity $u^* = u/u_e$ is also of order one.

$$u^* \sim 1 \tag{10.7}$$

We further assume that the magnitude of u^* remains of the same order along the body.

This implies that the derivative $\partial u^*/\partial x^*$ must be of order one as well:

$$\begin{array}{lccc} & u^* & \approx & x^* \dfrac{\partial u^*}{\partial x^*} \\ \text{order of magnitude} & \sim 1 & \approx & \sim 1 \cdot 1 \end{array} \tag{10.8}$$

Otherwise the velocity would grow beyond magnitude one. In practice this means that the body surface changes only gradually and has no sharp corners. With both $\partial u^*/\partial x^*$ and x^* of magnitude one, the second order derivative $\partial^2 u^*/\partial x^{*2}$ will be of magnitude one as well.

The transverse derivative of the horizontal velocity $\partial u^*/\partial y^*$ must be of magnitude $1/\delta^*$ in order for the velocity u to grow from zero at the wall (no-slip condition) to its full value at the outer edge of the boundary layer:

$$\begin{array}{lccc} & u^* & \approx & y^* \dfrac{\partial u^*}{\partial y^*} \\ \text{order of magnitude} & \sim 1 & \approx & \sim \delta^* \dfrac{1}{\delta^*} = 1 \end{array} \tag{10.9}$$

Then, since the transverse distance y^* is of magnitude δ^*, the second order transverse derivative $\partial^2 u^*/\partial y^{*2}$ must be of magnitude $1/\delta^{*2}$:

$$\begin{array}{lccc} & \dfrac{\partial u^*}{\partial y^*} & \approx & y^* \dfrac{\partial^2 u^*}{\partial y^{*2}} \\ \text{order of magnitude} & \sim \dfrac{1}{\delta^*} & \approx & \sim \delta^* \dfrac{1}{\delta^{*2}} = \dfrac{1}{\delta^*} \end{array} \tag{10.10}$$

Transverse velocity and derivatives

The transverse velocity v^* vanishes at the wall (no flow through wall). From the dimensionless continuity equation follows that $\partial u^*/\partial x^*$ and $\partial v^*/\partial y^*$ must be of the same order of magnitude (one).

$$\begin{array}{ccc} \dfrac{\partial u^*}{\partial x^*} + \dfrac{\partial v^*}{\partial y^*} & = & 0 \\ \sim 1 - \sim 1 & = & 0 \end{array} \tag{10.11}$$

With the derivative in y^*-direction being of order one, the transverse velocity v^* will be of the same order as dimensionless boundary layer thickness $\delta^* = \delta/L$.

$$\begin{array}{lccc} & v^* & \approx & y^* \dfrac{\partial v^*}{\partial y^*} \\ \text{order of magnitude} & \sim \delta^* & \approx & \sim \delta^* \, 1 = \delta^* \end{array} \tag{10.12}$$

With v^* of magnitude δ^* and x^* of magnitude one, its derivative $\partial v^*/\partial x^*$ will be of magnitude δ^*.

$$\begin{array}{lccc} & v^* & \approx & x^* \dfrac{\partial v^*}{\partial x^*} \\ \text{order of magnitude} & \sim \delta^* & \approx & \sim 1 \quad \delta^* = \delta^* \end{array} \tag{10.13}$$

Its second order derivative $\partial^2 v^*/\partial x^{*2}$ will also be of magnitude δ^*:

$$\begin{array}{lccc} & \dfrac{\partial v^*}{\partial x^*} & \approx & x^* \dfrac{\partial^2 v^*}{\partial x^{*2}} \\ \text{order of magnitude} & \sim \delta^* & \approx & \sim 1 \quad \delta^* = \delta^* \end{array} \tag{10.14}$$

Relative magnitude of terms

Now that you are thoroughly confused, let's rewrite first the longitudinal conservation of momentum Equation (10.4) and note underneath each term the order of magnitude we found in the paragraphs above.

$$\begin{array}{ccccc} u^* \dfrac{\partial u^*}{\partial x^*} + v^* \dfrac{\partial u^*}{\partial y^*} & = & \dfrac{1}{Fr^2} f_x^* & - \; Eu \dfrac{\partial p^*}{\partial x^*} & + \dfrac{1}{Re} \left(\dfrac{\partial^2 u^*}{\partial x^{*2}} + \dfrac{\partial^2 u^*}{\partial y^{*2}} \right) \\ \sim 1 \cdot 1 + \delta^* \dfrac{1}{\delta^*} & \approx & \sim 0 & - \quad 1 & + \; \delta^{*2} \left(1 \quad + \dfrac{1}{\delta^{*2}} \right) \end{array} \tag{10.15}$$

Obviously, the left-hand side representing the x-component of the inertia force is of magnitude one. The right-hand side of Equation (10.15) is equivalent to the sum of external forces in x-direction.

We know from observation that the boundary layer thickness is small compared with the length of a body, i.e. $\delta \ll L$ or $\delta^* \ll 1$. Accordingly, the second term dominates the first term in the parentheses on the right-hand side:

$$\frac{1}{\delta^{*2}} \gg 1$$

However, the components of an external force may not be larger than one in magnitude. Otherwise, the equality cannot be maintained in Equation (10.15). As a consequence, the inverse of the Reynolds number must be of small magnitude to reduce the viscous forces to magnitude one.

$$\frac{1}{Re} \sim \delta^{*2} \qquad \text{or} \qquad \delta^* \sim \frac{1}{\sqrt{Re}} \tag{10.16}$$

In turn, the dimensionless boundary layer thickness is proportional to the square root of the inverse Reynolds number. We will come back to this important conclusion. Note that for the limit case $Re \to \infty$, the boundary layer vanishes as $\lim_{Re \to \infty} \delta = 0$. This applies to a hypothetical, inviscid fluid.

Before we simplify this equation, we repeat the process for the y-component of the conservation of momentum equation, assuming the same relationship between Reynolds number and boundary layer thickness.

$$\begin{array}{ccccc} u^* \dfrac{\partial v^*}{\partial x^*} + v^* \dfrac{\partial v^*}{\partial y^*} & = & \dfrac{1}{Fr^2} f_y^* & - \; Eu \dfrac{\partial p^*}{\partial y^*} & + \dfrac{1}{Re} \left(\dfrac{\partial^2 v^*}{\partial x^{*2}} + \dfrac{\partial^2 v^*}{\partial y^{*2}} \right) \\ \sim 1 \; \delta^* + \delta^* \; 1 & \approx & \sim 0 & - \quad 1 & + \; \delta^{*2} \left(\delta^* \quad + \dfrac{1}{\delta^*} \right) \end{array} \tag{10.17}$$

Again, inertia force and viscous force are of the same magnitude. However, the y-components are of magnitude $\delta^* \ll 1$, which renders them considerably smaller than their x-direction counterparts.

Assumptions of boundary layer theory

Based on these observations, Prandtl formulated his basic assumptions of boundary layer theory:

1. The boundary layer is thin, i.e.

Boundary layer must be thin

$$\delta \ll x \leq L \tag{10.18}$$

2. Fluid sticks to the wall, i.e.

No-slip condition

$$u = 0 \qquad \text{for } y = 0, \text{ no-slip condition} \tag{10.19}$$

3. Flow speed changes rapidly across the boundary layer from 0 at the wall (no-slip condition (10.19)) to the exterior flow speed at the boundary layer thickness δ. This means the derivative of the flow speed in direction normal to the wall is large.

$$\frac{\partial u}{\partial y} \sim \frac{u_e}{\delta} \qquad \text{is large} \tag{10.20}$$

4. Flow speed changes slowly along x so that

Slope and curvature of body are small

$$\frac{\partial u}{\partial x} \sim \frac{u_e}{L} \qquad \text{is moderate} \tag{10.21}$$

5. The second order derivative of u with respect to x is negligible compared with its second order derivative with respect to y:

$$\frac{\partial^2 u}{\partial x^2} \ll \frac{\partial^2 u}{\partial y^2} \tag{10.22}$$

6. Transverse velocity v and its x-derivatives are small within the boundary layer

$$v \sim u_e \frac{\delta}{L} \qquad \frac{\partial v}{\partial x} \sim u_e \frac{\delta}{L^2} \qquad \frac{\partial^2 v}{\partial x^2} \sim u_e \frac{\delta}{L^3} \tag{10.23}$$

7. The y-derivatives of the transverse velocity are moderate to large:

$$\frac{\partial v}{\partial y} \sim \frac{u_e}{L} \qquad \text{and} \qquad \frac{\partial^2 v}{\partial y^2} \sim \frac{u_e}{L\delta} \tag{10.24}$$

8. In addition, we will ignore the effect of gravity on the boundary layer flow. In a control volume completely filled with incompressible fluid, gravity forces and hydrostatic forces are in equilibrium. When we omit the gravity forces in Equation (10.3), the pressure p is understood to be the dynamic pressure only.

Body forces will be ignored

In the next section, the assumptions of boundary layer theory are employed to derive the simplified boundary layer equations.

10.3 Boundary Layer Equations

We start again with the two-dimensional continuity equation and Navier-Stokes equations for steady, incompressible flow (10.3). The body forces f_x and f_y are omitted based on the facts that the gravity force is the only body force of interest and that it

Two-dimensional flow

may be neglected according to assumption No. 8. Some terms are annotated to indicate whether they are small or very small based on the assumptions of boundary layer theory above.

$$\frac{\partial u}{\partial x} + \frac{\partial v}{\partial y} = 0 \qquad \text{2D continuity equation}$$

$$u\frac{\partial u}{\partial x} + v\frac{\partial u}{\partial y} = -\frac{1}{\rho}\frac{\partial p}{\partial x} + \overset{\text{very small}}{\cancel{\nu\frac{\partial^2 u}{\partial x^2}}} + \nu\frac{\partial^2 u}{\partial y^2} \tag{10.25}$$

$$\overset{\text{small}}{\cancel{u\frac{\partial v}{\partial x}}} + \overset{\text{small}}{\cancel{v\frac{\partial v}{\partial y}}} = -\frac{1}{\rho}\frac{\partial p}{\partial y} + \overset{\text{very small}}{\cancel{\nu\frac{\partial^2 v}{\partial x^2}}} + \overset{\text{small}}{\cancel{\nu\frac{\partial^2 v}{\partial y^2}}}$$

2D boundary layer equations

Finally, omitting all terms that are labeled 'small' or 'very small,' we obtain the two-dimensional boundary layer equations

$$\frac{\partial u}{\partial x} + \frac{\partial v}{\partial y} = 0 \qquad \text{2D continuity equation} \tag{10.26}$$

$$u\frac{\partial u}{\partial x} + v\frac{\partial u}{\partial y} = -\frac{1}{\rho}\frac{\partial p}{\partial x} + \nu\frac{\partial^2 u}{\partial y^2} \tag{10.27}$$

$$0 = -\frac{1}{\rho}\frac{\partial p}{\partial y} \tag{10.28}$$

This set of coupled partial differential equations is much less complex than the original Navier-Stokes equations (10.3). However, the set is still nonlinear due to the convection terms on the left-hand side of the conservation of momentum equation (10.27). An analytic solution is not known for general cases, but this set of equations can be integrated numerically. Schlichting and Gersten (2003, Chapter 23) provides an overview and further details may be found in Krause (1967).

Blasius (1907) solved the equations by transformation and series expansion for the case of a flat plate. An English version of his text can be found in Blasius (1950). We will study the case of a flat plate in Chapter 12.

Pressure is constant across thin boundary layers

However, without actually solving Equations (10.26) through (10.28), we may already draw an important conclusion. Take a closer look at the y-component of the conservation of momentum equation (10.28)! It states that the derivative of pressure vanishes across the boundary layer:

$$0 = \frac{\partial p}{\partial y} \quad \text{or} \quad p(y) = \text{const.} \quad \text{across boundary layer} \tag{10.29}$$

In other words, the pressure does not change across the boundary layer. As a consequence, the pressure of the exterior flow will also act on the body surface.

This conclusion is important because it allows us to derive pressure forces from exterior flow solutions, which we may compute by neglecting viscous effects. This is still far from trivial and we will discuss some basics later on. Of course, you have to remember that this will yield useful results only as long as the boundary layer is thin.

Next, we will apply the mass and momentum conservation principles to a piece of the boundary layer to study the relationship between flow velocities governed by Equations (10.26)–(10.28) and the shear stress acting on the wall. Knowing the shear stress will enable us to compute the frictional resistance.

References

Blasius, H. (1907). *Grenzschichten in Flüssigkeiten mit kleiner Reibung.* PhD thesis, Universität Göttingen.

Blasius, H. (1950). The boundary layers in fluids with little friction. Technical Report TM-1256, National Advisory Committee for Aeronautics (NACA). Translation of Blasius (1908). Grenzschichten in Flüssigkeiten mit kleiner Reibung. *Zeitschrift für Mathematik und Physik,* 56(2):1–37.

Fox, R. and McDonald, A. (1998). *Introduction to Fluid Mechanics.* John Wiley & Sons, New York, NY.

Krause, E. (1967). Numerical solution of the boundary-layer equations. *AIAA Journal,* 5(7):1231–1237. doi: 10.2514/3.4176.

Prandtl, L., Oswatitsch, K., and Wieghardt, K. (1989). *Führer durch die Strömungslehre.* Friedrich Vieweg & Sohn, Braunschweig/Wiesbaden, ninth edition. Verbesserte und erweiterte Auflage.

Schlichting, H. and Gersten, K. (2003). *Boundary-layer theory.* Springer-Verlag, Berlin, Heidelberg, New York, eigth edition. Corrected printing.

Young, A. (1989). *Boundary layers.* AIAA Education Series. American Institute of Aeronautics and Astronautics (AIAA), Washington, DC.

Self Study Problems

1. How is the outer limit of the boundary layer determined in experiments?
2. Discuss the differences of laminar and turbulent boundary layers and their effect on resistance.
3. What characterizes the separation point?
4. Summarize the assumptions of boundary layer theory.
5. Explain under which conditions and why we may assume that the pressure is constant across the boundary layer.

11

Wall Shear Stress in the Boundary Layer

In this chapter we apply our knowledge of boundary layers to discover a relationship between flow properties and the resultant wall shear stress which is at the root of frictional resistance. The discussion makes use of the conservation of mass and momentum principles, and you will see that it is very similar to the examples from Chapter 9.

Learning Objectives

At the end of this chapter students will be able to:

- apply conservation principles for mass and momentum to boundary layer flow problems
- further understand the implications of the assumptions made in boundary layer theory
- define boundary layer thickness, displacement thickness, and momentum thickness

11.1 Control Volume Selection

As a control volume V, we cut a thin sliver with length $\mathrm{d}x$ out of the boundary layer (see Figure 11.1). For convenience, we assume that the x-axis points along the wall and the y-axis normal to the wall. In addition, we restrict ourselves to two-dimensional, incompressible, and steady flow. In order to check the physical units in the equations, we assume that the control volume is of width b in the direction of the third, invisible z-axis. Note that the normal vector $\underline{n}$ points out of the control volume everywhere on S and that the control volume is fixed in space.

The boundary S of the control volume is split into four parts. Starting on the left and going clockwise we have:

1. An inlet surface S_{AB}, marked by the letters AB. Fluid is flowing into the control volume through the inlet.

Fundamentals of Ship Hydrodynamics: Fluid Mechanics, Ship Resistance and Propulsion, First Edition.
Lothar Birk.

Companion website: www.wiley.com/go/birk/hydrodynamics

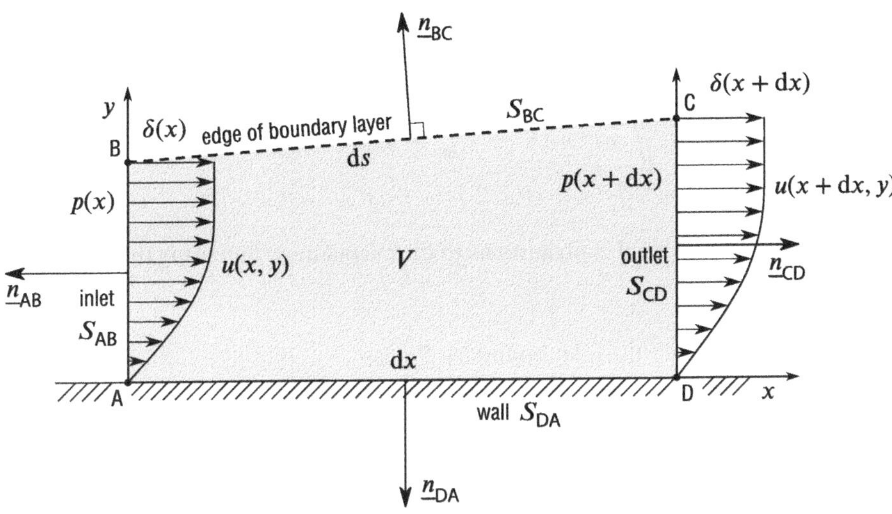

Figure 11.1 Cross section through a finite, fixed control volume V in the boundary layer

2. The outer edge of the boundary layer S_{BC} stretches between B and C. Note that this boundary does not represent a streamline. There will be a mass flow across this boundary (see Equation (11.10)).
3. On the right side, we have the outlet S_{CD} marked by letters CD.
4. Finally, a piece of the wall forms the lower boundary S_{DA} marked with letters DA.

11.2 Conservation of Mass in the Boundary Layer

Conservation of mass states that in a control volume V the rate of change of mass must be equal to the resultant mass flow across the boundary S of the control volume. We express this physical principle with the continuity equation. Its integral, conservative form was derived in Chapter 5.

Integral form of continuity equation

$$\underbrace{\iiint\limits_V \frac{\partial \rho}{\partial t}\,\mathrm{d}V}_{\substack{\text{rate of change}\\ \text{of mass in } V}} = \underbrace{-\iint\limits_S \rho\,\underline{v}^T\underline{n}\,\mathrm{d}S}_{\substack{\text{resultant mass flow}\\ \text{across boundary } S}} \tag{5.18}$$

Since the density ρ is constant and the control volume is fixed in space, the volume integral on the left-hand side vanishes for incompressible, steady flow.

Steady flow in a fixed control volume

$$\iiint\limits_V \frac{\partial \rho}{\partial t}\,\mathrm{d}V = 0 \tag{11.1}$$

Consequently, at any given moment, as much fluid is flowing out as is flowing into the control volume.

Mass flux across boundaries

We split the surface integral into four integrals, one for each part of the boundary. Each integral represents the mass flux $\dot{m}_{ij}$ across the boundary S_{ij}.

$$0 = \iint_S \rho \underline{v}^T \underline{n} \, \mathrm{d}S = \dot{m}_{\mathrm{AB}} + \dot{m}_{\mathrm{BC}} + \dot{m}_{\mathrm{CD}} + \dot{m}_{\mathrm{DA}} \tag{11.2}$$

Let us discuss the individual contributions to the overall mass flux using the definitions from Figure 11.1:

Mass flux through inlet

1. The mass flux over the inlet boundary S_{AB} is

$$\dot{m}_{\mathrm{AB}} = \iint_{S_{\mathrm{AB}}} \rho \underline{v}^T \underline{n} \, \mathrm{d}S \quad \text{with} \tag{11.3}$$

$$\text{velocity } \underline{v}(x,y) = \begin{pmatrix} u(x,y) \\ v(x,y) \end{pmatrix} \text{ and normal } \underline{n}_{\mathrm{AB}}(x,y) = \begin{pmatrix} -1 \\ 0 \end{pmatrix} \tag{11.4}$$

The surface element $\mathrm{d}S$ is simply the product of depth b (not visible in Figure 11.1) and a differential step $\mathrm{d}y$ in y-direction:

$$\mathrm{d}S = b \, \mathrm{d}y \tag{11.5}$$

The double integral reduces to a single integral over the boundary layer thickness $\delta(x)$. We get for the mass flux

$$\dot{m}_{\mathrm{AB}} = \int_0^{\delta(x)} \rho (u,v)^T \begin{pmatrix} -1 \\ 0 \end{pmatrix} b \, \mathrm{d}y = -b \int_0^{\delta(x)} \rho \, u(x,y) \, \mathrm{d}y \tag{11.6}$$

Without any knowledge about the velocity distribution, we cannot proceed any further at the moment.

Mass flux through outlet

2. We keep the boundary layer limit S_{BC} for last and tackle the outlet S_{CD} next. Its normal vector points in the positive x-axis direction.

$$\underline{n}_{\mathrm{CD}}(x + dx, y) = \begin{pmatrix} 1 \\ 0 \end{pmatrix} = -\underline{n}_{\mathrm{AB}}$$

Over the short distance $\mathrm{d}x$, boundary layer thickness δ and horizontal velocity u have changed in a yet unknown manner. To avoid introducing even more unknowns, we develop the mass outflow as a Taylor series expansion of the inflow. Since the normal vector of S_{CD} points in the opposite direction, we take the negative of the Taylor series. It may be truncated after its linear term because $\mathrm{d}x$ is small.

$$\dot{m}_{\mathrm{CD}} = -\left(\dot{m}_{\mathrm{AB}} + \frac{\partial \dot{m}_{\mathrm{AB}}}{\partial x} \mathrm{d}x \right) \tag{11.7}$$

Substituting the result for $\dot{m}_{\text{AB}}$ from Equation (11.6) yields:

$$\begin{aligned}\dot{m}_{\text{CD}} &= -b\left[-\int_0^{\delta(x)} \rho\,u(x,y)\,\mathrm{d}y - \frac{\partial}{\partial x}\left(\int_0^{\delta(x)} \rho\,u(x,y)\,\mathrm{d}y\right)\mathrm{d}x\right] \\ &= b\left[\int_0^{\delta(x)} \rho\,u(x,y)\,\mathrm{d}y + \frac{\partial}{\partial x}\left(\int_0^{\delta(x)} \rho\,u(x,y)\,\mathrm{d}y\right)\mathrm{d}x\right] \qquad (11.8)\end{aligned}$$

3. Determining the mass flow across the wall S_{DA} is easy for a change: it is a wall, so nothing can flow through it.

No mass flux through wall

$$\dot{m}_{\text{DA}} = 0 \qquad (11.9)$$

4. Finally, we have to address the mass flow across the boundary layer edge S_{BC}. Considering the facts that nothing flows through the wall and that the mass fluxes for inlet and outlet are different, something must flow across the outer boundary to balance inflow and outflow.

Mass flux through boundary layer limit

We employ Equation (11.2) to determine the respective mass flow.

$$\dot{m}_{\text{BC}} = -\dot{m}_{\text{AB}} - \dot{m}_{\text{CD}} - \dot{m}_{\text{DA}}$$

With $\dot{m}_{\text{DA}} = 0$ we have

$$\begin{aligned}\dot{m}_{\text{BC}} &= -\dot{m}_{\text{AB}} - \dot{m}_{\text{CD}} \\ &= -b\,\frac{\partial}{\partial x}\left(\int_0^{\delta(x)} \rho\,u(x,y)\,\mathrm{d}y\right)\mathrm{d}x \qquad (11.10)\end{aligned}$$

The mass flow $\dot{m}_{\text{BC}}$ is the difference between inflow and outflow which makes sense if nothing flows through the wall part of the boundary. Let us keep this result (11.10) in mind, since we will need it shortly.

11.3 Conservation of Momentum in the Boundary Layer

Conservation of momentum states that the rate of change of momentum in V is equal to the sum of external forces acting on the control volume according to Newton's second law. We are interested in the force and shear stress acting parallel to the wall, i.e. in this case the x-direction.

For our finite, fixed control volume, we apply an integral form of the conservation of momentum equations (see Chapter 9), but use only its x-component. The y- and z-components are obtained by replacing $\rho\,u$ with $\rho\,v$ or $\rho\,w$ and the corresponding external force component.

Integral form of conservation of momentum principle

$$\frac{\partial}{\partial t}\iiint_V \rho u\,\mathrm{d}V + \iint_S \rho\,u\left(\underline{v}^T\underline{n}\right)\mathrm{d}S = \sum(\text{external forces in } x\text{-direction}) \qquad (9.1)$$

The volume integral vanishes for steady, incompressible flow in a fixed control volume.

Body and surface forces contribute to the sum of external forces. Recalling assumption 8 of boundary layer theory from page 115, we neglect the body forces. Only the sum of surface forces F_{S_x} remains.

$$\iint_S \rho u \left(\underline{v}^T \underline{n}\right) \mathrm{d}S = F_{S_x} \tag{11.11}$$

11.3.1 Momentum flux over boundary of control volume

As for the conservation of mass, we split the surface integral on the left-hand side representing the x-momentum flux through S into four parts.

$$\iint_S \rho u \left(\underline{v}^T \underline{n}\right) \mathrm{d}S = \dot{M}_{\mathrm{AB}} + \dot{M}_{\mathrm{BC}} + \dot{M}_{\mathrm{CD}} + \dot{M}_{\mathrm{DA}} \tag{11.12}$$

Momentum flux through inlet

1. We start with the inlet boundary S_{AB}. Velocity vector $\underline{v}$, normal vector $\underline{n}$, and surface element $\mathrm{d}S$ remain the same as in Equations (11.4) and (11.5). Thus, we get for the momentum flux in x

$$\begin{aligned} \dot{M}_{\mathrm{AB}} &= \iint_{S_{\mathrm{AB}}} \rho u \left(\underline{v}^T \underline{n}\right) \mathrm{d}S = \int_0^{\delta(x)} (\rho u)(u, v)^T \begin{pmatrix} -1 \\ 0 \end{pmatrix} b \,\mathrm{d}y \\ &= -b \int_0^{\delta(x)} (\rho u) u \,\mathrm{d}y \end{aligned} \tag{11.13}$$

Momentum flux through outlet

2. The strategy used for the mass transport through the outlet is also employed for the momentum flux: we develop the flux through the outlet as a Taylor series expansion of the flux through the inlet. Again, the change in orientation of the normal vector from inlet to outlet triggers a change in sign for the Taylor series:

$$\dot{M}_{\mathrm{CD}} = -\left(\dot{M}_{\mathrm{AB}} + \frac{\partial \dot{M}_{\mathrm{AB}}}{\partial x} \mathrm{d}x\right) \tag{11.14}$$

Substituting the result for $\dot{M}_{\mathrm{AB}}$ from Equation (11.13) yields:

$$\begin{aligned} \dot{M}_{\mathrm{CD}} &= -b\left[-\int_0^{\delta(x)} (\rho u) u \,\mathrm{d}y - \frac{\partial}{\partial x}\left(\int_0^{\delta(x)} (\rho u) u \,\mathrm{d}y\right) \mathrm{d}x\right] \\ &= b\left[\int_0^{\delta(x)} (\rho u) u \,\mathrm{d}y + \frac{\partial}{\partial x}\left(\int_0^{\delta(x)} (\rho u) u \,\mathrm{d}y\right) \mathrm{d}x\right] \end{aligned} \tag{11.15}$$

3. Like the mass transport, momentum flux through the wall S_{DA} vanishes because the velocity and normal vectors are perpendicular to each other and, consequently, their dot product vanishes, i.e. $\left(\underline{v}^T \underline{n}\right) = 0$.

No momentum flux through wall

$$\dot{M}_{DA} = 0 \tag{11.16}$$

4. Finally, the momentum flux $\dot{M}_{BC}$ through the outer boundary S_{BC} needs some careful consideration.

Momentum flux through boundary layer limit

At the edge of the boundary layer, the velocity is equal to the exterior flow speed u_e. In the general case of a curved body, the exterior flow velocity u_e will change over the length dx.

$$u_e(x + dx) = u_e(x) + \frac{\partial u_e(x)}{\partial x} dx \tag{11.17}$$

We cannot say how much specifically without any knowledge about the body shape. However, the derivative $\partial u_e/\partial x$ in Equation (11.17) is considered constant if dx is sufficiently small. That allows us to accurately estimate the mean horizontal velocity $\bar{u}_e$ for the boundary BC.

Mean velocity

$$\bar{u}_e = \frac{1}{2}\left(u_e + u_e + \frac{\partial u_e}{\partial x} dx\right) = u_e + \frac{1}{2}\frac{\partial u_e}{\partial x} dx \tag{11.18}$$

The momentum (in x-direction) flowing through the boundary BC is approximately equal to the product of mass flow times the mean velocity $\bar{u}_e$.

$$\begin{aligned} \dot{M}_{BC} &= \iint_{S_{BC}} \rho u \left(\underline{v}^T \underline{n}\right) dS \\ &\approx \bar{u}_e \iint_{S_{BC}} \rho \left(\underline{v}^T \underline{n}\right) dS \\ &= \bar{u}_e \, \dot{m}_{BC} \end{aligned} \tag{11.19}$$

Now it is time to substitute the result for the mass flux over BC (11.10) into the Equation (11.19) for the momentum flux. Using Equation (11.18) to replace the mean velocity $\bar{u}_e$ yields:

$$\begin{aligned} \dot{M}_{BC} &= \left(u_e + \frac{1}{2}\frac{\partial u_e}{\partial x} dx\right)\left[-b\frac{\partial}{\partial x}\left(\int_0^{\delta(x)} \rho u \, dy\right) dx\right] \\ &= -b u_e \frac{\partial}{\partial x}\left(\int_0^{\delta(x)} \rho u \, dy\right) dx - \underbrace{b \frac{1}{2}\frac{\partial u_e}{\partial x}\frac{\partial}{\partial x}\left(\int_0^{\delta(x)} \rho u \, dy\right)(dx)^2}_{\approx 0} \end{aligned} \tag{11.20}$$

The last term is proportional to $(dx)^2$ and therefore of second order small. Like before in the Taylor series, we neglect second order and higher order terms.

$$\dot{M}_{BC} = -b u_e \frac{\partial}{\partial x}\left(\int_0^{\delta(x)} \rho u \, dy\right) dx \tag{11.21}$$

Resultant momentum flux All parts of the momentum flux have been computed and are ready to be assembled into the resultant momentum flux equation. We substitute the results for $\dot{M}_{\mathrm{AB}}$ (11.13), $\dot{M}_{\mathrm{CD}}$ (11.15), $\dot{M}_{\mathrm{DA}}$ (11.16), and $\dot{M}_{\mathrm{BC}}$ (11.21) back into Equation (11.12).

$$\begin{aligned}\iint_S \rho u(\underline{v}^T \underline{n})\,\mathrm{d}S &= \dot{M}_{\mathrm{AB}} + \dot{M}_{\mathrm{CD}} + \dot{M}_{\mathrm{DA}} + \dot{M}_{\mathrm{BC}} \\ &= -b\int_0^{\delta(x)} (\rho u) u\,\mathrm{d}y + b\left[\int_0^{\delta(x)} (\rho u) u\,\mathrm{d}y + \frac{\partial}{\partial x}\left(\int_0^{\delta(x)} (\rho u) u\,\mathrm{d}y\right)\mathrm{d}x\right] \\ &\quad + 0 - b u_e \frac{\partial}{\partial x}\left(\int_0^{\delta(x)} \rho u\,\mathrm{d}y\right)\mathrm{d}x \end{aligned} \tag{11.22}$$

Obviously, the first two integrals in Equation (11.22) cancel each other out. We obtain the following expression for the total x-momentum flux across the boundary S of our fluid element V.

$$\iint_S \rho u(\underline{v}^T \underline{n})\,\mathrm{d}S = b\frac{\partial}{\partial x}\left(\int_0^{\delta(x)} (\rho u) u\,\mathrm{d}y\right)\mathrm{d}x - b u_e \frac{\partial}{\partial x}\left(\int_0^{\delta(x)} \rho u\,\mathrm{d}y\right)\mathrm{d}x \tag{11.23}$$

11.3.2 Surface forces acting on control volume

We now turn to the right-hand side of the conservation of momentum Equation (11.11), the sum of external forces. As mentioned above, only surface forces F_{S_x} in the x-direction need to be considered.

Surface forces to consider Before we dive into the mathematics of it, let us take a look at physics, geometry, and our assumptions. Surface forces result from pressure acting normal to the surface or from shear stress acting tangential to the surface. Since the normal vectors of inlet and outlet boundaries AB and CD are parallel to the x-axis, only pressure forces will have an effect. Shear forces on AB and CD act perpendicular to the x-axis. On the outer boundary BC, we have reached the exterior flow and, as per boundary layer theory assumptions, we may neglect viscous forces there. Consequently, only a pressure force is acting on boundary BC. On the wall DA, shear forces and pressure forces will be present. However, the wall is parallel to the x-axis and its normal vector $\underline{n}_{\mathrm{DA}}$ points in y-direction. The resulting pressure force will have no component in x-direction. Finally, remember that, according to our conclusions from the boundary layer equations, pressure $p(x)$ is constant across the boundary layer as long as the boundary layer is sufficiently thin.

We move on to the forces acting on the individual boundaries.

Pressure force on inlet

1. The exterior flow pressure $p(x)$ acts on the inlet surface S_{AB}. As usual, a pressure force is computed by integration of pressure over the surface. Directions of differential forces are determined by the direction of the normal vector.

$$\underline{F}_{S_{\mathrm{AB}}} = -\iint_{S_{\mathrm{AB}}} p\,\underline{n}_{\mathrm{AB}}\,\mathrm{d}S \tag{11.24}$$

For the inlet surface, this simplifies considerably as both pressure and normal vector are constant across the inlet surface. As a consequence, the resultant pressure force in x-direction reduces to the product of pressure and area. The rectangular inlet area is equal to the product of boundary layer thickness $\delta(x)$ (height) and depth b.

$$F_{S_{\mathrm{AB}x}} = p(x)\,\delta(x)\,b \tag{11.25}$$

2. The same is true for the outlet surface S_{CD}. Again, pressure and normal vector are constant. However, they both have changed by a differential amount. In addition, we have to consider that the pressure force will act against the positive x-direction.

Pressure force on outlet

$$F_{S_{\mathrm{CD}x}} = -\left(p(x) + \frac{\partial p(x)}{\partial x}\mathrm{d}x\right)\left(\delta(x) + \frac{\partial \delta(x)}{\partial x}\mathrm{d}x\right) b$$

Expanding the products results in:

$$F_{S_{\mathrm{CD}x}} = -p(x)\,\delta(x)\,b - \delta(x)\frac{\partial p(x)}{\partial x}\mathrm{d}x\,b - p(x)\frac{\partial \delta(x)}{\partial x}\mathrm{d}x\,b - \underbrace{\frac{\partial p(x)}{\partial x}\frac{\partial \delta(x)}{\partial x}(\mathrm{d}x)^2\,b}_{\approx 0}$$

The last term is proportional to $(\mathrm{d}x)^2$ and thus of second order small. Neglecting the second order term, we obtain for the pressure force on the outlet:

$$F_{S_{\mathrm{CD}x}} = -p(x)\,\delta(x)\,b - \delta(x)\frac{\partial p(x)}{\partial x}\mathrm{d}x\,b - p(x)\frac{\partial \delta(x)}{\partial x}\mathrm{d}x\,b \tag{11.26}$$

3. As discussed above, the pressure force has no component in x-direction. The force acting on the fluid in y-direction is the reaction force which keeps the wall in place. The only x-component for the exterior force results from the yet unknown shear stress τ_w.

Friction force on wall

$$\underline{F}_{S_{\mathrm{DA}}} = -\iint_{S_{\mathrm{DA}}} \tau_w\,\underline{i}\,\mathrm{d}S \tag{11.27}$$

The vector $\underline{i} = (1,0)^T$ is the unit vector in positive x-direction and represents the tangent vector to the surface S_{DA}.

As with all other flow quantities, we develop the value at point D as a Taylor series based on the value $\tau_w(x)$ at point A (Figure 11.2(a)). Like the mean exterior flow velocity (11.18), we estimate the mean shear stress $\bar{\tau}_w(x)$ acting over the short distance $\mathrm{d}x$ to be:

Mean wall shear stress

$$\bar{\tau}_w = \frac{1}{2}\left(\tau_w + \tau_w + \frac{\partial \tau_w}{\partial x}\mathrm{d}x\right) = \tau_w + \frac{1}{2}\frac{\partial \tau_w}{\partial x}\mathrm{d}x \tag{11.28}$$

Note that although the force on the wall points in positive x-direction, i.e. the fluid attempts to drag the plate with it, we have to consider the forces acting on the fluid. The wall exerts a force on the control volume against the motion of the fluid.

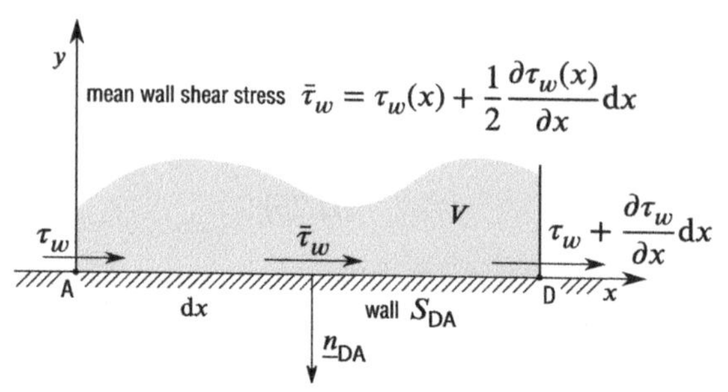

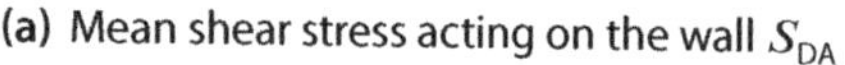
(a) Mean shear stress acting on the wall S_{DA}

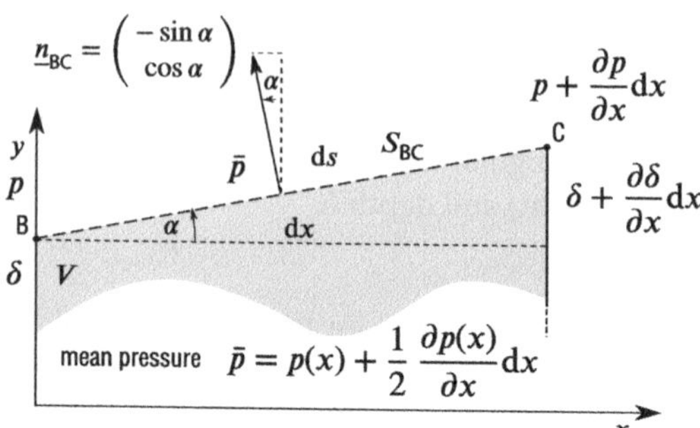

(b) Mean pressure acting on the outer limit of the boundary layer S_{BC}

Figure 11.2 Surface forces acting on the control volume

With the mean shear stress being constant ($\bar{\tau}_w$ = const.) over the small area $S_{\mathrm{DA}} = \mathrm{d}x\, b$, the integral in Equation (11.27) reduces to the product of mean shear stress and area.

$$F_{S_{\mathrm{DA}x}} = -\bar{\tau}_w\, \mathrm{d}x\, b \tag{11.29}$$

Pressure force on boundary layer limit

4. Finally, a pressure force is acting on the outer edge of the boundary layer S_{BC}. The transverse velocity derivative vanishes, $\partial u/\partial y = 0$ for $y = \delta(x)$. Therefore, no shear stress acts on the outer limit of the boundary layer. This is in accord with our assumption that the exterior flow may be treated as inviscid flow.

In analogy to the mean shear stress on the wall, we compute a mean pressure $\bar{p}$ acting on the surface $S_{\mathrm{BC}} = \mathrm{d}s\, b$ (see Figure 11.2(b)).

$$\bar{p} = \frac{1}{2}\left(p + p + \frac{\partial p}{\partial x}\mathrm{d}x\right) = p + \frac{1}{2}\frac{\partial p}{\partial x}\mathrm{d}x \tag{11.30}$$

The resultant pressure force will be:

$$\underline{F}_{S_{\mathrm{BC}}} = -\iint_{S_{\mathrm{BC}}} p\, \underline{n}_{\mathrm{BC}}\, \mathrm{d}S \tag{11.31}$$

Assuming that pressure p and normal vector $\underline{n}_{\mathrm{BC}}$ are constant with

$$p = \bar{p} \qquad \underline{n}_{\mathrm{BC}} = \begin{pmatrix} -\sin(\alpha) \\ \cos(\alpha) \end{pmatrix}$$

the vectorial pressure force is

$$\underline{F}_{S_{\mathrm{BC}}} = -\bar{p}\, \underline{n}_{\mathrm{BC}}\, \mathrm{d}s\, b \tag{11.32}$$

The length $\mathrm{d}s$ of the boundary layer edge will be slightly larger than the length of the wall $\mathrm{d}x$ and the angle α is derived from the geometry of the cross section through the control volume V. The sine of α is equal to the ratio of $\mathrm{d}\delta$ and $\mathrm{d}s$

$$\sin(\alpha) = \frac{\mathrm{d}\delta}{\mathrm{d}s}$$

Using this geometric relationship, the x-component of the pressure force becomes

$$\begin{aligned} F_{S_{BC\,x}} &= \bar{p}\,\sin(\alpha)\,\mathrm{d}s\,b \\ &= \bar{p}\,\frac{\mathrm{d}\delta}{\mathrm{d}s}\,\mathrm{d}s\,b \\ &= \bar{p}\,\mathrm{d}\delta\,b \end{aligned} \tag{11.33}$$

The total differential $\mathrm{d}\delta$ of the boundary layer thickness is equal to

$$\mathrm{d}\delta = \frac{\partial \delta}{\partial x}\,\mathrm{d}x \tag{11.34}$$

Therefore, using Equation (11.34) and the equation for the mean pressure (11.30), the pressure force (11.33) becomes:

$$\begin{aligned} F_{S_{BC\,x}} &= \left(p + \frac{1}{2}\frac{\partial p}{\partial x}\mathrm{d}x\right)\left(\frac{\partial \delta}{\partial x}\mathrm{d}x\right) b \\ &= p\,\frac{\partial \delta}{\partial x}\mathrm{d}x\,b + \underbrace{\frac{1}{2}\frac{\partial p}{\partial x}\frac{\partial \delta}{\partial x}(\mathrm{d}x)^2\,b}_{\approx 0} \end{aligned} \tag{11.35}$$

Yet again, we will ignore the second order term proportional to the small quantity $(\mathrm{d}x)^2$.

Sum of external forces

Summarizing the results (11.25), (11.26), (11.29), and (11.35), we compute the total force F_{S_x} on the surface S acting in x-direction. For convenience we omit the position argument, knowing that pressure $p = p(x)$, boundary layer thickness $\delta = \delta(x)$, and wall shear stress $\tau_w = \tau_w(x)$ are functions of x.

$$\begin{aligned} F_{S_x} &= \cancel{p\delta b} - \cancel{p\delta b} - \delta\,\frac{\partial p}{\partial x}\mathrm{d}x\,b - \cancel{p\,\frac{\partial \delta}{\partial x}\mathrm{d}x\,b} + \cancel{p\,\frac{\partial \delta}{\partial x}\mathrm{d}x\,b} - \tau_w\,\mathrm{d}x\,b \\ &= -\tau_w\,\mathrm{d}x\,b - \delta\,\frac{\partial p}{\partial x}\,\mathrm{d}x\,b \end{aligned} \tag{11.36}$$

Conservation of momentum equation

Equipped with the resultant momentum flux (11.23) and the resultant exterior force (11.36), the conservation of momentum equation is assembled.

$$\begin{aligned} &\cancel{b}\,\frac{\partial}{\partial x}\left(\int_0^{\delta(x)} (\rho u)\,u\,\mathrm{d}y\right)\cancel{\mathrm{d}x} - \cancel{b}\,u_e\frac{\partial}{\partial x}\left(\int_0^{\delta(x)} \rho\,u(x,y)\,\mathrm{d}y\right)\cancel{\mathrm{d}x} \\ &\qquad = -\tau_w\,\cancel{\mathrm{d}x\,b} - \frac{\partial p}{\partial x}\delta(x)\,\cancel{\mathrm{d}x\,b} \end{aligned} \tag{11.37}$$

Dividing by the wall surface $(\mathrm{d}x b)$ and solving for the unknown wall shear stress τ_w yields:

$$\tau_w(x) = -\frac{\partial}{\partial x}\left(\int_0^{\delta(x)} (\rho u)\,u\,\mathrm{d}y\right) + u_e\frac{\partial}{\partial x}\left(\int_0^{\delta(x)} \rho\,u\,\mathrm{d}y\right) - \delta(x)\frac{\partial p}{\partial x} \tag{11.38}$$

Equation (11.38) relates the wall shear stress to the flow properties in the control volume. In order to compute the shear stress and the resultant resistance, we would have to know the exterior flow velocity $u_e(x)$, the boundary layer thickness $\delta(x)$, the velocity distribution within the boundary layer $u(x, y)$, and the pressure distribution $p(x)$. Right now, we do not have enough equations to resolve this.

Express pressure as function of velocity

We can eliminate the pressure from this equation by employing the Bernoulli equation (14.8) which is derived from the Euler equations (see later in Chapter 14). The Bernoulli equation relates pressure and flow velocity. As stated before, we consider the flow to be steady and neglect the influence of body forces. The exterior flow is inviscid as per boundary layer assumptions. If we add the condition that the flow is irrotational, i.e. rot $\underline{v} = \underline{0}$, the exterior flow may be treated as potential flow (see Chapter 15). Under these conditions the Bernoulli equation states that the sum of pressure p and dynamic pressure $1/2\,\rho\,u_e^2$ remains constant at all points in the flow.

$$p + \frac{1}{2}\rho u_e^2 = \text{const.} \tag{11.39}$$

At point B we have pressure p and exterior flow velocity u_e. At point C both may have changed by small amounts, which we approximate by Taylor series expansion. Thus, setting up Bernoulli's equation between points B and C yields:

$$p + \frac{1}{2}\rho u_e^2 = p + \frac{\partial p}{\partial x}\mathrm{d}x + \frac{1}{2}\rho\left(u_e + \frac{\partial u_e}{\partial x}\mathrm{d}x\right)^2 = \text{const.} \tag{11.40}$$

We expand the terms and ignore the second order term at the end:

$$\cancel{p} + \cancel{\frac{1}{2}\rho u_e^2} = \cancel{p} + \frac{\partial p}{\partial x}\mathrm{d}x + \cancel{\frac{1}{2}\rho u_e^2} + \rho\, u_e \frac{\partial u_e}{\partial x}\mathrm{d}x + \cancel{\frac{1}{2}\rho\left(\frac{\partial u_e}{\partial x}\right)^2(\mathrm{d}x)^2}^{\approx 0}$$

The result is an expression for the pressure gradient as a function of the exterior flow velocity and its derivative.

$$\frac{\partial p}{\partial x} = -\rho\, u_e \frac{\partial u_e}{\partial x} \tag{11.41}$$

We use Equation (11.41) to replace the pressure term in our Equation (11.38) for the shear stress:

$$\tau_w(x) = -\frac{\partial}{\partial x}\left(\int_0^{\delta(x)} (\rho u)\, u\, \mathrm{d}y\right) + u_e \frac{\partial}{\partial x}\left(\int_0^{\delta(x)} \rho\, u\, \mathrm{d}y\right) + \delta(x)\,\rho\, u_e \frac{\partial u_e}{\partial x} \tag{11.42}$$

This is still not the final form of the equation we need.

Simplifying the equation for wall shear stress

Study the following expression where the right-hand side is derived by applying the product rule: $u_e(x)$ is representing the first factor and the integral the second x-dependent factor.

$$\frac{\partial}{\partial x}\left(u_e \int_0^{\delta(x)} (\rho u)\, \mathrm{d}y\right) = \frac{\partial u_e}{\partial x}\int_0^{\delta(x)} (\rho u)\, \mathrm{d}y + u_e \frac{\partial}{\partial x}\left(\int_0^{\delta(x)} \rho\, u\, \mathrm{d}y\right) \tag{11.43}$$

The last integral also appears in Equation (11.42) for the shear stress. We rearrange Equation (11.43) to solve for this integral:

$$u_e \frac{\partial}{\partial x}\left(\int_0^{\delta(x)} \rho u \, dy\right) = \frac{\partial}{\partial x}\left(u_e \int_0^{\delta(x)} (\rho u)\, dy\right) - \frac{\partial u_e}{\partial x}\int_0^{\delta(x)} (\rho u)\, dy \tag{11.44}$$

We substitute the right-hand side for the integral in Equation (11.42) which – at first glance – seems to complicate the equation. We are almost there!

$$\begin{aligned}\tau_w(x) &= -\frac{\partial}{\partial x}\left(\int_0^{\delta(x)} (\rho u) u \, dy\right) + \frac{\partial}{\partial x}\left(u_e \int_0^{\delta(x)} (\rho u)\, dy\right) \\ &\quad - \frac{\partial u_e}{\partial x}\int_0^{\delta(x)} (\rho u)\, dy + \delta(x)\, \rho\, u_e \frac{\partial u_e}{\partial x}\end{aligned} \tag{11.45}$$

Remember that the exterior flow velocity $u_e(x)$ is a function of x only. Consequently, we can move the factor u_e in front of the second integral underneath the integral sign. From the last two terms, we factor out the derivative $\partial u_e/\partial x$:

$$\begin{aligned}\tau_w(x) &= -\frac{\partial}{\partial x}\left(\int_0^{\delta(x)} (\rho u)\, u \, dy\right) + \frac{\partial}{\partial x}\left(\int_0^{\delta(x)} (\rho u)\, u_e \, dy\right) \\ &\quad + \frac{\partial u_e}{\partial x}\left[-\int_0^{\delta(x)} (\rho u)\, dy + \delta(x)\, \rho\, u_e\right]\end{aligned} \tag{11.46}$$

The first two integrals can be combined because a sum of derivatives is equal to the derivative of a sum.

$$\tau_w(x) = \frac{\partial}{\partial x}\left(\int_0^{\delta(x)} (\rho u)\left(u_e - u\right) dy\right) + \frac{\partial u_e}{\partial x}\left[-\int_0^{\delta(x)} (\rho u)\, dy + \delta(x)\, \rho\, u_e\right] \tag{11.47}$$

In order to combine the terms in square brackets, we rewrite the last term:

$$\delta(x)\, \rho\, u_e = \int_0^{\delta(x)} 1 \, dy\, \rho\, u_e = \int_0^{\delta(x)} \rho\, u_e \, dy \tag{11.48}$$

Again, we utilized that density ρ and exterior flow velocity u_e are not functions of y within the boundary layer.

$$\tau_w(x) = \frac{\partial}{\partial x}\left(\int_0^{\delta(x)} \rho\, u\left(u_e - u\right) dy\right) + \frac{\partial u_e}{\partial x}\int_0^{\delta(x)} \rho\left(u_e - u\right) dy \tag{11.49}$$

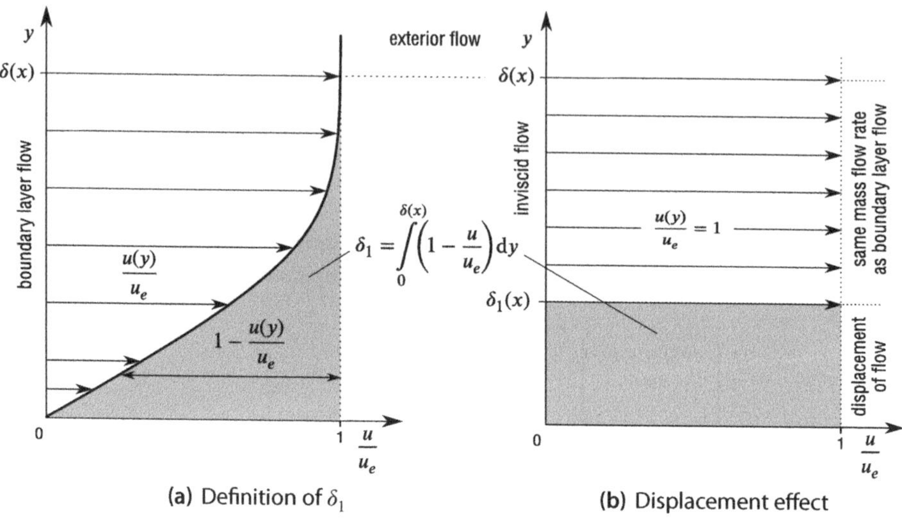

(a) Definition of δ_1 (b) Displacement effect

Figure 11.3 Definition of displacement thickness δ_1 and displacement effect on exterior flow

Finally, we normalize the velocities in the integrands by multiplying them with u_e^2/u_e^2 and u_e/u_e respectively and extract from the integrals all factors which are independent of the integration variable y.

$$\tau_w(x) = \frac{\partial}{\partial x}\left[\rho u_e^2 \int_0^{\delta(x)} \frac{u}{u_e}\left(1 - \frac{u}{u_e}\right) \mathrm{d}y\right] + \frac{\partial u_e}{\partial x}(\rho u_e) \int_0^{\delta(x)} \left(1 - \frac{u}{u_e}\right) \mathrm{d}y \tag{11.50}$$

The remaining integrals in Equation (11.50) have special meaning in boundary layer theory and will be explained in more detail in the following two subsections.

11.3.3 Displacement thickness

The last integral in Equation (11.50) is the displacement thickness.

$$\delta_1 = \int_0^{\delta(x)} \left(1 - \frac{u}{u_e}\right) \mathrm{d}y \tag{11.51}$$

The displacement thickness δ_1 can be interpreted as a 'thickening' of the body. Mathematically, the integral in Equation 11.51 is equal to the gray shaded area in Figure 11.3(a). It is a measure of how much less viscous fluid flows through the boundary layer compared with an inviscid fluid. The displacement thickness can be used to include the displacement effects of the boundary layer in inviscid flow calculations. The body is enlarged by δ_1 which usually improves the prediction of wave patterns and streamlines (see Figure 11.3(b)).

11.3.4 Momentum thickness

The first integral in Equation (11.50) is the momentum thickness.

$$\Theta = \int_0^{\delta(x)} \frac{u}{u_e}\left(1 - \frac{u}{u_e}\right) dy \tag{11.52}$$

Momentum thickness Θ is a measure of how much momentum is lost in the boundary layer. As we will see, Θ plays an important role in the computation of the wall shear stress and the frictional resistance.

11.4 Wall Shear Stress

Introducing displacement thickness δ_1 (11.51) and momentum thickness Θ (11.52) into Equation (11.50) for the wall shear stress yields:

$$\tau_w(x) = \frac{\partial}{\partial x}\left(\rho u_e^2 \Theta\right) + \rho u_e \frac{\partial u_e}{\partial x}\delta_1 \tag{11.53}$$

This is a general result of boundary layer theory. The partial differential equation (11.53) for the shear stress involves three unknown functions and their derivatives:

1. exterior flow velocity u_e,
2. displacement thickness δ_1, and
3. momentum thickness Θ.

Displacement and momentum thickness, in turn, depend on the boundary layer thickness δ, which is used as the upper limit for the defining integrals. We will discuss an approximate solution of the shear stress equation (11.53) in the next chapter. It will get us a step closer to an answer on how to determine the frictional resistance of a ship.

Self Study Problems

1. Annotate the text in this chapter by indicating which assumption of boundary layer theory is used in each step.
2. Why is the outer limit of the boundary layer not a streamline?
3. Consider the following velocity profile for a laminar boundary layer:

   $$\frac{u}{u_e} = \sin\left(\frac{\pi y}{2\delta}\right)$$

   Compute displacement and momentum thickness (as functions of boundary layer thickness δ).
4. How does Equation (11.53) change when a flow with vanishing pressure gradient $\partial p/\partial x = 0$ is considered? *Hint: Go back to Equation (11.38) and perform the steps used in simplifying the equation for shear stress.*

12

Boundary Layer of a Flat Plate

In previous chapters we derived the simplified conservation of mass and momentum equations which describe the flow in a thin boundary layer. Here, these equations are solved for the special case of a flat plate. The approximate solution provides estimates for boundary layer thickness, displacement thickness, momentum thickness, and the important wall shear stress. We derive flat plate frictional coefficients for laminar flow. A solution for turbulent flow will be presented in the following chapter.

Learning Objectives

At the end of this chapter students will be able to:

- solve boundary layer flow problems
- interpret results from boundary layer theory
- understand the origin of flat plate friction coefficients

12.1 Boundary Layer Equations for a Flat Plate

Flat plate

Consider the flow over a flat plate of length L and beam b (Figure 12.1), with the latter measured in the invisible z-direction. The onflow is uniform in positive x-direction with constant velocity u_0 (parallel flow). As usual, thickness of the boundary layer is greatly exaggerated in Figure 12.1.

Constant exterior velocity

A flat plate does not force the fluid to change course. Therefore, the exterior flow speed $u_e = u_0$ will be maintained everywhere outside the boundary layer.

$$u_e = u_0 = \text{const.} \qquad \text{or} \qquad \frac{\partial u_e}{\partial x} = 0 \qquad \text{for } y > \delta(x) \tag{12.1}$$

This assumption is not completely true. In the previous chapter we learned that the boundary layer has a displacement effect. This will also occur on the plate and, consequently, the exterior flow experiences a 'body' with finite thickness which will increase exterior flow speed. However, since the boundary layer is thin and much smaller than the length of the plate, this increase in exterior velocity will be very small and the assumption of $u_0 = \text{const.}$ is justified.

Fundamentals of Ship Hydrodynamics: Fluid Mechanics, Ship Resistance and Propulsion, First Edition.
Lothar Birk.

Companion website: www.wiley.com/go/birk/hydrodynamics

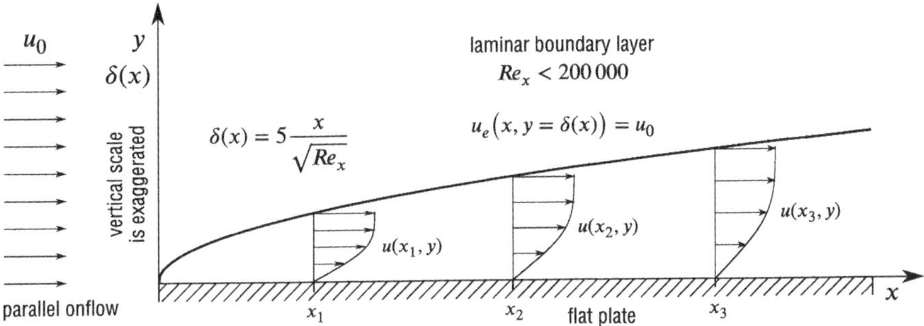

Figure 12.1 Laminar boundary layer along a flat plate

Constant pressure along plate

With the vanishing x-derivative of the external flow velocity, the pressure derivative must vanish as a consequence of the Bernoulli equation (11.41).

$$\frac{\partial p}{\partial x} = -\rho\, u_e \frac{\partial u_e}{\partial x} = 0 \qquad \text{for a flat plate}$$

Thus, for a flat plate, pressure is constant along and across the boundary layer ($\partial p/\partial x = 0 \;\wedge\; \partial p/\partial y = 0$), the latter being one of the conclusions from the boundary layer equations (10.28).

Wall shear stress equation

Utilizing this simplification enables us to eliminate the last term in Equation (11.53) for the wall shear stress. With $u_e(x) = u_0 = \text{const.}$ and $\partial u_0/\partial x = 0$ we get:

$$\tau_w(x) = \frac{\partial}{\partial x}\left(\rho\, u_0^2 \Theta\right) + \rho\, u_0 \underbrace{\frac{\partial u_0}{\partial x}}_{=0} \delta_1$$

Extracting the constant density ρ and the squared flow velocity u_0 from the first term on the right-hand side yields a relationship between wall shear stress τ_w and the momentum thickness Θ.

$$\tau_w(x) = \rho\, u_0^2 \frac{\partial \Theta}{\partial x} \tag{12.2}$$

We replace Θ with its defining integral.

$$\tau_w(x) = \rho\, u_0^2 \frac{\partial}{\partial x} \int_0^{\delta(x)} \frac{u}{u_0}\left(1 - \frac{u}{u_0}\right) \mathrm{d}y \tag{12.3}$$

The solution of Equation (12.3) is defined by a dimensionless velocity distribution u/u_0 over the thickness of the boundary layer. In fact, we may define the velocity distribution first and employ it to find an expression for the unknown boundary layer thickness δ, which marks the upper limit of the integral. Suitable velocity distributions may be found from measurements.

12.2 Dimensionless Velocity Profiles

Shape of velocity profile assumed to stay similar

Based on his observations, Prandtl suggested that the basic shape of the velocity distribution remains approximately the same for increasing x-position along the boundary layer. It is simply stretched in y-direction as indicated in Figure 12.1. If a simple equation for the velocity distribution can be derived from measurements, Equation (12.3) can be solved for the wall shear stress. An example will demonstrate this.

Parabolic velocity profile

Let us assume the velocity distribution is reasonably well described by a parabola. It is one of several usable shapes.

$$\frac{u}{u_0} = \bar{a} + \bar{b}\,y + \bar{c}\,y^2 \tag{12.4}$$

Please note that simple formulas like Equation (12.4) apply only to boundary layers with laminar flow. The velocity distribution in turbulent boundary layers is more complicated (see next chapter).

Dimensionless coordinates

Next, proper values need to be found for the coefficients $\bar{a}$, $\bar{b}$, and $\bar{c}$ in Equation (12.4). To simplify this task, we introduce the dimensionless y-coordinate η.

$$\eta = \frac{y}{\delta(x)} \qquad \text{or} \qquad y = \eta\,\delta(x) \tag{12.5}$$

η grows from zero at the wall to the value one at the edge of the boundary layer (Figure 12.1). We replace the y-coordinate in Equation (12.4) with η.

$$\begin{aligned} \frac{u}{u_0} &= \bar{a} + \bar{b}\,y + \bar{c}\,y^2 \\ &= a + \bar{b}\,\eta\,\delta + \bar{c}\,\eta^2\,\delta^2 \\ &= a + b\,\eta + c\,\eta^2 \end{aligned} \tag{12.6}$$

The new coefficients a, b, and c are

$$a = \bar{a} \qquad\qquad b = \bar{b}\,\delta \qquad\qquad c = \bar{c}\,\delta^2$$

Matching profile to boundary conditions

With the coefficients a, b, and c, the assumed velocity profile may be adjusted to the specific boundary conditions of the flow problem at hand. Three conditions are required to determine the coefficients:

(i) Based on the basic boundary layer assumptions (see page 114), fluid sticks to the plate surface.

(ii) Also, the flow velocity u increases from zero to the external flow speed across the boundary layer.

(iii) At the boundary layer edge, the transverse derivative of u vanishes, i.e. the velocity remains constant outside of the boundary layer.

Mathematically, this results in the set of equations below. The no-slip boundary condition at the wall requires:

$$\text{for } y = 0: \quad u = 0 \qquad \text{or for } \eta = 0: \quad \frac{u}{u_0} = 0 \tag{12.7}$$

At the outer edge of the boundary layer we impose:

$$\text{for } y = \delta: \quad u = u_0 \qquad \text{or for } \eta = 1: \quad \frac{u}{u_0} = 1 \tag{12.8}$$

$$\text{also for } y = \delta: \quad \frac{\partial u}{\partial y} = 0 \qquad \text{or for } \eta = 1: \quad \frac{\partial}{\partial \eta}\left(\frac{u}{u_0}\right) = 0 \tag{12.9}$$

Coefficient *a*

From Equation (12.6) and the first boundary condition (12.7) follows the value for coefficient a:

$$\text{from } \left.\frac{u}{u_0}\right|_{\eta=0} = 0 \quad \text{follows} \quad a + b \cdot 0 + c \cdot 0^2 = 0 \quad \text{and} \quad a = 0 \tag{12.10}$$

Obviously, the y-axis part of the velocity distribution must vanish or otherwise the no-slip condition would be violated.

Coefficients *b* and *c*

The second boundary condition Equation (12.8), provides a relationship between coefficients b and c.

$$\text{from } \left.\frac{u}{u_0}\right|_{\eta=1} = 1 \quad \text{follows} \quad \not{a} + b \cdot 1 + c \cdot 1^2 = 1 \quad \text{and} \quad b + c = 1 \tag{12.11}$$

Of course, we exploit that the coefficient a is equal to zero.

From the third boundary condition, Equation (12.9), we derive a second expression connecting b and c. The derivative of the velocity profile with respect to η is

$$\frac{\partial}{\partial \eta}\left(\frac{u}{u_0}\right) = \frac{\partial}{\partial \eta}\left(a + b\eta + c\eta^2\right) = b + 2c\,\eta \tag{12.12}$$

Equating it to zero at the edge of the boundary layer yields:

$$\left.\frac{\partial}{\partial \eta}\left(\frac{u}{u_0}\right)\right|_{\eta=1} = 0 \qquad \text{or} \qquad b + 2c \cdot 1 = 0 \tag{12.13}$$

We find that

$$b = -2c \tag{12.14}$$

Substituting the result (12.14) into condition (12.11) results in

$$c = -1 \qquad \text{and} \qquad b = 2$$

Dimensionless velocity profile

Thus, the dimensionless velocity profile within the boundary layer is defined as

$$\frac{u}{u_0} = 2\eta - \eta^2 \tag{12.15}$$

It is important to note that although Equation (12.15) satisfies the boundary conditions of boundary layer theory, we only assume that it represents a realistic approximation of the actual velocity distribution. Other useful approximations are left as self study problems. Having an expression for the velocity distribution across the boundary layer enables us to compute the boundary layer thickness.

12.3 Boundary Layer Thickness

Momentum thickness integral

First, we apply the velocity profile to the definition of the momentum thickness (11.52).

$$\Theta = \int_0^{\delta(x)} \frac{u}{u_0}\left(1 - \frac{u}{u_0}\right) \mathrm{d}y \qquad (12.16)$$

The integration variable and limits of integration are replaced with their dimensionless counterparts.

$$y = \eta\,\delta \qquad \mathrm{d}y = \delta\,\mathrm{d}\eta \qquad y=0 \longrightarrow \eta=0 \qquad y=\delta \longrightarrow \eta=1$$

Considering that the boundary layer thickness δ can be extracted from the integral because it is only a function of x and not of y, we obtain:

$$\int_0^{\delta(x)} \frac{u}{u_0}\left(1 - \frac{u}{u_0}\right) \mathrm{d}y = \delta \int_0^1 \frac{u}{u_0}\left(1 - \frac{u}{u_0}\right) \mathrm{d}\eta \qquad (12.17)$$

Next, the assumed velocity profile $2\eta - \eta^2$ is substituted for u/u_0.

$$\Theta(x) = \delta \int_0^1 \left(2\eta - \eta^2\right)\left(1 - 2\eta + \eta^2\right)\mathrm{d}\eta = \delta \int_0^1 \left(2\eta - 5\eta^2 + 4\eta^3 - \eta^4\right)\mathrm{d}\eta$$

$$= \delta \left[\eta^2 - \frac{5}{3}\eta^3 + \eta^4 - \frac{1}{5}\eta^5\right]_0^1 = \frac{2}{15}\,\delta(x) \qquad (12.18)$$

For our assumed velocity profile $u/u_0 = 2\eta - \eta^2$, the momentum thickness is 2/15 of the geometric boundary layer thickness. Note that both boundary layer thickness $\delta(x)$ and momentum thickness $\Theta(x)$ are functions of position x along the plate. Only the ratio $\Theta(x)/\delta(x)$ is constant, which is a consequence of assuming that the shape of the velocity profile u/u_0 is independent of x.

Wall shear stress (1st equation)

We dump the result for the momentum thickness (12.18) into the equation for the wall shear stress (12.3)

$$\tau_w = \rho\,u_0^2 \frac{\partial}{\partial x}\left(\frac{2}{15}\,\delta\right) = \frac{2}{15}\,\rho\,u_0^2 \frac{\partial \delta}{\partial x} \qquad \text{for } u/u_0 = 2\eta - \eta^2 \qquad (12.19)$$

This equation has two unknowns: the boundary layer thickness $\delta(x)$ and the wall shear stress $\tau_w(x)$. Obviously, we need to consult a second equation to solve this.

Netwon's shear stress (2nd equation)

The observant reader may remember that we came across a similar problem when we derived the Navier-Stokes equations. There the unknown stresses in the fluid were related to the spatial rate of change of velocities. In that context, we stated the simplest form of this relationship as it was derived by Newton (see Equation (6.27) on page 66). Newton derived this relationship for simple shear flow between two plates. In our case, the second plate is very far away but the flow conditions are similar (2D over a flat surface). If we employ Newton's relationship for shear stress and use the derivative of the velocity at the wall ($y = 0$), we get

$$\tau_w = \mu \left.\frac{\partial u}{\partial y}\right|_{y=0} \tag{6.27}$$

Again, we introduce the dimensionless velocity profile u/u_0 and the coordinate $\eta = y/\delta$ by multiplying the numerator with $1 = u_0/u_0$ and the denominator with $1 = \delta/\delta$, respectively.

$$\tau_w = \mu \left.\frac{\partial u \left(\frac{u_0}{u_0}\right)}{\partial y \left(\frac{\delta}{\delta}\right)}\right|_{y=0} = \mu \frac{u_0}{\delta} \left.\frac{\partial \left(\frac{u}{u_0}\right)}{\partial \left(\frac{y}{\delta}\right)}\right|_{y=0} = \mu \frac{u_0}{\delta} \left[\frac{\partial}{\partial \eta}\left(\frac{u}{u_0}\right)\right]_{\eta=0} \tag{12.20}$$

Extracting external flow speed u_0 and boundary layer thickness δ from the differential operators is possible because they do not depend on the variables y or η.

The additional, required equation for wall shear stress and boundary layer thickness is obtained by introducing the derivative of the assumed velocity profile (12.15) into Equation (12.20).

$$\tau_w = \mu \frac{u_0}{\delta} \left[2 - 2\eta\right]_{\eta=0} = 2\mu \frac{u_0}{\delta} \qquad \text{for } u/u_0 = 2\eta - \eta^2 \tag{12.21}$$

ODE for boundary layer thickness

Next, we equate the right-hand sides of Equations (12.19) and (12.21) for the wall shear stress.

$$\frac{2}{15} \rho u_0^2 \frac{\partial \delta}{\partial x} = 2\mu \frac{u_0}{\delta}$$

Division by $2/15$ and ρu_0^2 and reintroducing the kinematic viscosity $\nu = \mu/\rho$ results in

$$\frac{\partial \delta}{\partial x} - \frac{15\nu}{u_0} \frac{1}{\delta} = 0 \tag{12.22}$$

Finally, we have derived an equation which only bears the boundary layer thickness as an unknown quantity.

In fact, Equation (12.22) is an ordinary differential equation (ODE) for the boundary layer thickness. δ is only a function of the x-coordinate. The equation is of type $\delta' - \alpha\delta^{-1} = 0$. This means we are looking for a function whose derivative is equal to the inverse of the function itself. You know a function with this property! It is the square root function. We state the boundary layer thickness as a generic square root function:

$$\delta = C_1 \sqrt{x} + C_2 \qquad \text{and its derivative:} \quad \frac{\partial \delta}{\partial x} = \frac{C_1}{2} x^{-\frac{1}{2}} \tag{12.23}$$

Integration constants

The constant $C_2 = 0$ in Equation (12.23) must vanish because the boundary layer thickness is zero at the leading edge of the plate $\delta(x = 0) = 0$, as stated in one of our basic assumptions of boundary layer theory.

Substituting the function and its derivative from Equation (12.23) into the ordinary differential equation (12.22) yields:

$$\frac{C_1}{2}x^{-\frac{1}{2}} - \frac{15\nu}{u_0}\frac{1}{C_1 x^{\frac{1}{2}}} = 0$$

or

$$x^{-\frac{1}{2}}\left(\frac{C_1}{2} - \frac{15\nu}{u_0}\frac{1}{C_1}\right) = 0$$

This must hold true for all x, therefore the term in parentheses must vanish:

$$C_1 - \frac{30\nu}{u_0}\frac{1}{C_1} = 0$$

The last equation yields for the constant C_1:

$$C_1 = \sqrt{\frac{30\nu}{u_0}} \qquad \text{for } u/u_0 = 2\eta - \eta^2 \tag{12.24}$$

Laminar boundary layer thickness approximation

With the constant C_1 our expression is complete for the boundary layer thickness of a flat plate.

$$\delta(x) = \sqrt{\frac{30\nu}{u_0}}\sqrt{x} \tag{12.25}$$

The boundary layer grows with the square root of the position x along the plate. If you study sketches and pictures of boundary layers, you will recognize the basic shape of the square root function.

Finally, we introduce the local Reynolds number Re_x with

$$Re_x = \frac{u_0\, x}{\nu} \tag{10.1}$$

Note that Re_x differs from our usual definition of the Reynolds number. Re_x uses a variable position x as reference length. As a consequence, the local Reynolds number changes from $Re_x = 0$ at the beginning of the plate ($x = 0$) to $Re_x = u_0 L/\nu$ at the end of the plate ($x = L$). Introducing Re_x into Equation (12.25) yields the following equation for the boundary layer thickness based on the parabolic velocity profile:

$$\delta(x) = \sqrt{30}\frac{x}{\sqrt{Re_x}} \approx 5.48\frac{x}{\sqrt{Re_x}} \qquad \text{for } u/u_0 = 2\eta - \eta^2 \tag{12.26}$$

As expected, the boundary layer thickness is proportional to the inverse square root of the local Reynolds number as we discussed in Section 10.2.

It is important to remember that we achieved this result under the assumption that the velocity profile in the boundary layer is well represented by the parabolic function of Equation (12.15): $u/u_0 = 2\eta - \eta^2$. It turns out that this is a reasonable, but inaccurate, assumption for laminar boundary layers. Boundary layers on smooth, flat plates without excessive disturbances in the onflow are laminar up to a position x for which $Re_x = 200\,000$ (Fox and McDonald, 1998).

Blasius' laminar boundary layer thickness solution

Blasius used a mathematically more rigorous approach. He first converted the boundary layer Equations (10.26)–(10.28) into an ODE of third order by introducing a stream function. Then he solved the third order, nonlinear ODE for the stream function by a series expansion which allows control of the desired accuracy (Blasius, 1907).

Blasius' result for the boundary layer thickness for laminar flow over a flat plate is (Schlichting and Gersten, 2003)

$$\delta(x) \approx 5.0 \frac{x}{\sqrt{Re_x}} \qquad \text{Blasius' result} \qquad (12.27)$$

According to Blasius' more accurate result, our approximation (12.26) overpredicts the thickness by about 10%.

Boundary layer thickness example

For a flat plate of length $L = 0.2$ m in a parallel flow of velocity $u_0 = 1.0$ m/s in fresh water at 15 °C, we can expect the following boundary layer thickness:

1. The kinematic viscosity of fresh water (15 °C) is $\nu = 1.1386 \cdot 10^{-6}$ m²/s, according to the ITTC water properties tables (ITTC, 2011).
2. The resulting local Reynolds number at the end of the plate will be

$$Re_x = \left. \frac{u_0\, x}{\nu} \right|_{x=L} = \frac{1.0\,\text{m/s} \cdot 0.2\,\text{m}}{1.1386\, 10^{-6}\,\text{m}^2/\text{s}} = 175654.3 < 200000$$

   The result is indeed dimensionless as it should be and the value of $Re_x = 175654.3$ indicates that the assumption of laminar flow is acceptable.

3. Finally, we compute the maximum boundary layer thickness for the assumed parabolic velocity profile (12.15):

$$\delta(x = L) = 5.48 \left. \frac{x}{\sqrt{Re_x}} \right|_{x=0.2\,\text{m}} = 5.48 \frac{0.2\,\text{m}}{\sqrt{175654.3}} = 0.0026\,\text{m}$$

   With 2.6 mm the boundary layer thickness has reached 1.25% of the plate length. Although this is fairly thin, we must expect that any predictions based on the boundary layer equations will deviate more and more from the actual flow as the boundary layer becomes thicker.

Laminar boundary layer results do not apply to ships

Luckily, boundary layers of ships are thin as well. Reynolds numbers for ships are in the range of 10^9 which results in low δ/L ratios as the boundary layer thickness is proportional to the inverse of $\sqrt{Re_x}$. However, ship boundary layers are turbulent and the formulas presented in this chapter do not apply. Thickness of turbulent boundary layers grows faster along the plate than the thickness of laminar boundary layers. We will discuss more relevant solutions for turbulent boundary layers in Chapter 13.

Having finally found an expression for the thickness of a laminar boundary layer, we can compute the remaining boundary layer characteristics and compare them with Blasius' more accurate results.

12.4 Wall Shear Stress

Shear stress approximation

We have two equations to solve for the wall shear stress: Equations (12.19) and (12.21). Of course, both yield the same result. For convenience, we use Equation (12.21) and substitute the result for the boundary layer thickness (12.26) into it:

$$\tau_w = 2\mu\frac{u_0}{\delta} = 2\mu\frac{u_0}{\sqrt{30}\frac{x}{\sqrt{Re_x}}} \qquad \text{for } u/u_0 = 2\eta - \eta^2 \tag{12.28}$$

Rearranging the last term and expanding it by the factor $1 = \rho u_0/\rho u_0$ yields:

$$\tau_w = 2\frac{\sqrt{Re_x}}{\sqrt{30}}\frac{\mu\, u_0}{x}\cdot\frac{\rho\, u_0}{\rho\, u_0} = 2\frac{\sqrt{Re_x}}{\sqrt{30}}\rho\, u_0^2\left(\frac{\mu}{\rho\, u_0\, x}\right) \tag{12.29}$$

The last factor in parentheses is the inverse of the local Reynolds number Re_x.

$$\tau_w = \frac{2}{\sqrt{30}}\rho\, u_0^2\frac{1}{\sqrt{Re_x}} \qquad \text{for } u/u_0 = 2\eta - \eta^2 \tag{12.30}$$

Skin friction coefficient approximation

In order to compare this result with measurements and other approximation, it is common practice to present the wall shear stress as a dimensionless skin friction coefficient C_{f_x}.

$$C_{f_x}(x) = \frac{\tau_w}{\frac{1}{2}\rho\, u_0^2} \tag{12.31}$$

For the case of the quadratic velocity profile with wall shear stress from Equation (12.30), the local skin friction coefficient is equal to

$$C_{f_x}(x) = \frac{4}{\sqrt{30}}\frac{1}{\sqrt{Re_x}} \approx 0.730\frac{1}{\sqrt{Re_x}} \qquad \text{for } u/u_0 = 2\eta - \eta^2 \tag{12.32}$$

Blasius' more accurate solution

With Blasius' solution for the boundary layer thickness, a local skin friction coefficient of

$$C_{f_x}(x) = 0.665\frac{1}{\sqrt{Re_x}} \qquad \text{Blasius' result} \tag{12.33}$$

is obtained.

Figure 12.2 compares our approximation with Blasius' result. It is interesting to note that both solutions predict an infinitely high shear stress at the beginning of the plate at $x = 0$. This is a singularity of the mathematical boundary layer theory model rather than a real phenomenon. Studying the ODE (12.22), we see that we cannot get a valid solution for $\delta(x = 0) = 0$ because of the $1/\delta$-term.

We noted in Figure 12.1 that the velocity profiles are geometrically similar along the plate but stretched in y-direction. The stretching results in a visible reduction of the transverse derivative of the velocity

$$\left.\frac{\partial u}{\partial y}\right|_{x=x_1} > \left.\frac{\partial u}{\partial y}\right|_{x=x_2} > \left.\frac{\partial u}{\partial y}\right|_{x=x_3} \qquad \text{for } x_1 > x_2 > x_3 \tag{12.34}$$

This decline in the slope of the velocity profile is also reflected in a decreasing shear stress along the plate.

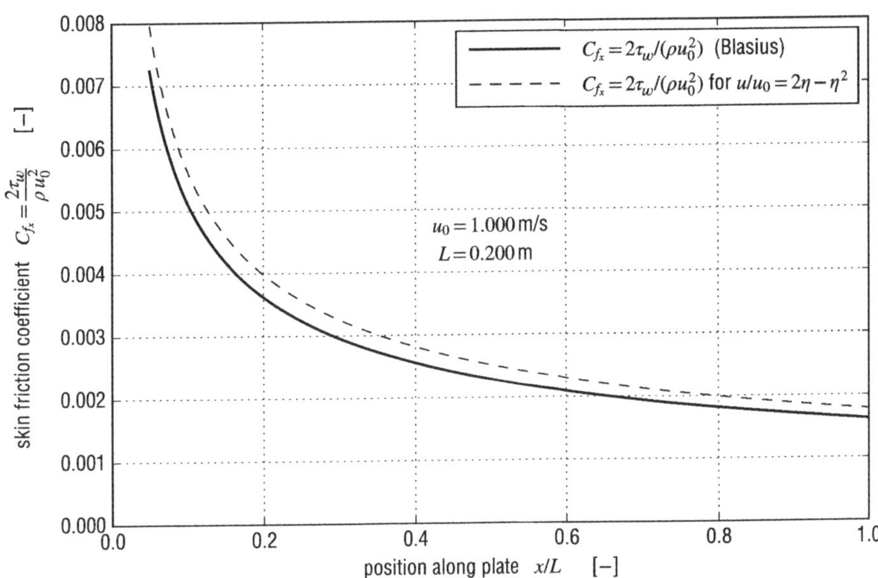

Figure 12.2 Boundary layer shear stress for laminar flow over a flat plate as dimensionless position dependent skin friction coefficient C_{f_x}

12.5 Displacement Thickness

The displacement thickness is defined as the integral

Displacement thickness approximation

$$\delta_1 = \int_0^{\delta(x)} \left(1 - \frac{u}{u_0}\right) \mathrm{d}y \tag{11.51}$$

Its computation is straight forward for our assumed velocity profile from Equation (12.15) $u/u_0 = 2\eta - \eta^2$:

$$\begin{aligned} \delta_1 &= \delta \int_0^1 \left(1 - 2\eta + \eta^2\right) \mathrm{d}\eta = \delta \left[\eta - \eta^2 + \frac{1}{3}\eta^3\right]_0^1 = \frac{1}{3}\delta \\ &= \frac{\sqrt{30}}{3} \frac{x}{\sqrt{Re_x}} \approx 1.826 \frac{x}{\sqrt{Re_x}} \qquad \text{for } u/u_0 = 2\eta - \eta^2 \end{aligned} \tag{12.35}$$

Based on Blasius' results for a laminar boundary layer, the displacement thickness should be

Blasius' more accurate result

$$\delta_1(x) = 1.72 \frac{x}{\sqrt{Re_x}} \qquad \text{Blasius' result} \tag{12.36}$$

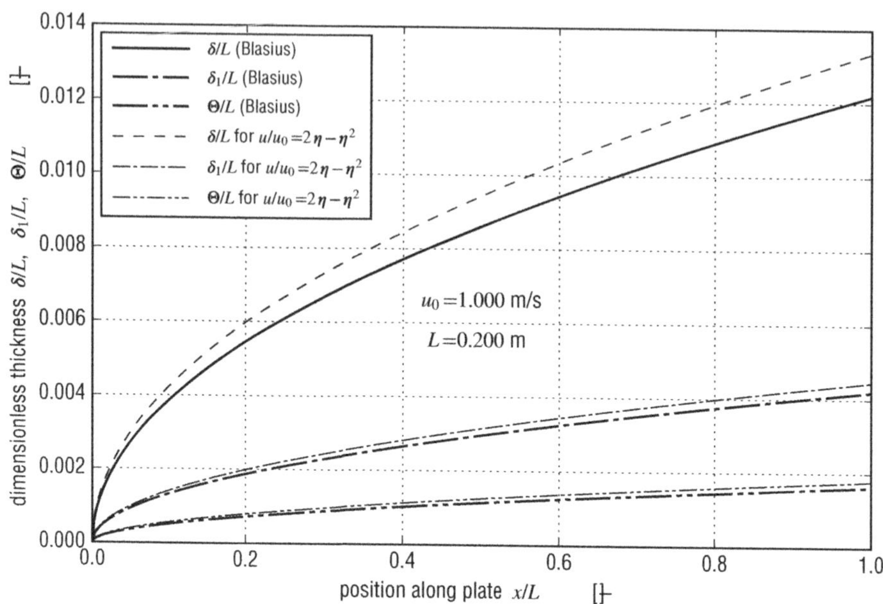

Figure 12.3 Boundary layer thickness δ, displacement thickness δ_1, and momentum thickness Θ for laminar flow over a flat plate

12.6 Momentum Thickness

Momentum thickness approximation

The integral for the momentum thickness of a flat plate boundary layer has been solved as part of our quest to find the boundary layer thickness (see page 136).

$$\Theta = \frac{2}{15}\delta = \frac{2\sqrt{30}}{15}\frac{x}{\sqrt{Re_x}} \approx 0.730\frac{x}{\sqrt{Re_x}} \qquad \text{for } u/u_0 = 2\eta - \eta^2 \tag{12.37}$$

Blasius' more accurate result

Blasius' solution for the momentum thickness of a laminar boundary layer is

$$\Theta(x) = 0.665\frac{x}{\sqrt{Re_x}} \qquad \text{Blasius' result} \tag{12.38}$$

Figure 12.3 compares the results for boundary layer thickness, displacement thickness, and momentum thickness of Blasius' with our approximation based on the velocity profile $u/u_0 = 2\eta - \eta^2$. The approximate solution somewhat overpredicts the more accurate values of Blasius. In the self study problems readers can explore other approximations. In any case, the following relationship holds true between the three characteristic boundary layer thicknesses:

$$\Theta < \delta_1 < \delta \tag{12.39}$$

With Blasius' results for the laminar boundary layer, the following ratios are found:

$$\frac{\delta_1}{\delta} = \frac{1.72}{5.0} \approx 0.344 \qquad \frac{\Theta}{\delta} = \frac{0.665}{5.0} \approx 0.133 \tag{12.40}$$

12.7 Friction Force and Coefficients

Friction force

Integration of the shear stress distribution (12.2) across the plate results in the total friction force F_{F_x} on one side of the plate.

$$F_{F_x} = \iint \tau_w \mathrm{d}S = b \int_0^L \tau_w \,\mathrm{d}x = \rho u_0^2 b \int_0^L \frac{\partial \Theta}{\partial x} \mathrm{d}x = \rho u_0^2 b \int_0^L \mathrm{d}\Theta$$
$$= \rho u_0^2 b \Big[\Theta \Big]_0^L = \rho u_0^2 b \Theta_L \tag{12.41}$$

Θ_L represents the value of the momentum thickness at the end of the plate at $x = L$. As expected, the total friction force is equal to the momentum lost over the length of the plate.

For the parabolic velocity profile (12.15), momentum thickness (12.37) is $\Theta(x) = 0.730x/\sqrt{Re_x}$. The momentum thickness at the end of the plate is

$$\Theta(x = L) = 0.730 \frac{L}{\sqrt{Re_L}} \qquad \text{for } u/u_0 = 2\eta - \eta^2 \tag{12.42}$$

Substituting this result into the formula for the friction force and using the standard Reynolds number $Re = Re_L = u_0 L/\nu$ yields:

$$F_{F_x} = 0.730 \rho u_0^2 b L \frac{1}{\sqrt{Re}} \qquad \text{for } u/u_0 = 2\eta - \eta^2 \tag{12.43}$$

The product of beam b and plate length L represents the wetted surface of the plate.

Based on the data from the example on page 139 ($L = 0.2$ m, $u_0 = 1.1$ m/s, $\nu = 1.1386 \cdot 10^{-6}$ m^2/s, and density $\rho = 999.1026$ kg/m^3), the friction force per unit plate width will be

$$\frac{F_{F_x}}{b} = 0.730 \cdot 999.1026\,\text{kg/m}^3 \left(1.1\,\text{m/s}\right)^2 0.2\,\text{m} \frac{1}{\sqrt{193168.8}}$$
$$= 0.402\,\text{N/m}$$

This is a fairly small force and hints at how difficult it is to conduct measurements which are accurate enough to validate assumptions and theories.

Laminar flow friction coefficient

Analog to the coefficients introduced for the components of ship resistance, we introduce a dimensionless friction coefficient for the flat plate in laminar flow.

$$C_F = \frac{F_{F_x}}{\frac{1}{2}\rho u_0^2 b L} = \frac{1.46}{\sqrt{Re}} \qquad \text{for } u/u_0 = 2\eta - \eta^2 \tag{12.44}$$

With Blasius' result for the momentum thickness the friction coefficient becomes

$$C_F = \frac{1.33}{\sqrt{Re}} \qquad \text{Blasius' result, valid for } Re < 200\,000 \tag{12.45}$$

Unlike the local skin friction coefficient C_{f_x}, flat plate friction coefficients are independent of the position. C_F is valid for plates as a whole.

Unfortunately, the friction coefficient for laminar boundary layers (12.45) is not applicable to ships. Reynolds numbers for ships are considerably higher than the upper limit for laminar flow. Appropriate friction coefficients for turbulent flows will be discussed in the following chapter.

References

Blasius, H. (1907). *Grenzschichten in Flüssigkeiten mit kleiner Reibung.* PhD thesis, Universität Göttingen.

Fox, R. and McDonald, A. (1998). *Introduction to Fluid Mechanics.* John Wiley & Sons, New York, NY.

ITTC (2011). *Fresh water and seawater properties.* International Towing Tank Conference, Recommended Procedures and Guidelines 7.5-02-01-03. Revision 2.

Schlichting, H. and Gersten, K. (2003). *Boundary-layer theory.* Springer-Verlag, Berlin, Heidelberg, New York, eigth edition. Corrected printing.

Self Study Problems

1. Order by increasing magnitude: boundary layer thickness δ, displacement thickness δ_1, and momentum thickness Θ.

2. Show that the integral for the ratio of momentum and boundary layer thickness

$$\frac{\Theta}{\delta} = \int_0^1 \frac{u}{u_0}\left(1 - \frac{u}{u_0}\right) \mathrm{d}\eta$$

is equal to

$$\int_0^1 \frac{u}{u_0}\left(1 - \frac{u}{u_0}\right) \mathrm{d}\eta = \frac{4-\pi}{2\pi}$$

for the following assumed velocity profile

$$\frac{u}{u_0} = \sin\left(\frac{\pi}{2}\eta\right) \qquad \text{for} \quad 0 \leq \eta \leq 1$$

3. Following the procedures explained in Section 12.3, compute boundary layer thickness δ, displacement thickness δ_1, momentum thickness Θ, and wall shear stress τ_w assuming the velocity profile is given by

$$\frac{u}{u_0} = \sin\left(\frac{\pi}{2}\frac{y}{\delta}\right) \qquad \text{for} \quad 0 \leq y \leq \delta$$

4. Following the procedures explained in Section 12.3 compute boundary layer thickness δ, displacement thickness δ_1, momentum thickness Θ, and wall shear stress τ_w assuming the velocity profile is given by

$$\frac{u}{u_0} = \frac{3}{2}\eta - \frac{1}{2}\eta^3 \qquad \text{for} \quad 0 \le \eta \le 1$$

Remember that $\eta = y/\delta$.

5. Why is the following equation *not* a suitable velocity profile for the boundary layer flow?

$$\frac{u}{u_0} = \frac{2}{3}\eta - \frac{1}{2}\eta^3 \qquad \text{for} \quad 0 \le \eta \le 1$$

13

Frictional Resistance

Major resistance component

For most ships, frictional resistance is the major component of their calm water resistance. Wave resistance exceeds the frictional component only at Froude numbers above 0.3. Obviously, the accurate prediction of frictional resistance R_F is essential in a ship's performance assessment. Notwithstanding more than a century of research and great advances in fluid mechanics, we still lack a unified theory of turbulent flow and the resulting surface friction and momentum loss. As a consequence, our frictional resistance predictions rely on approximations extracted from experimental observations combined with physical insight drawn from conservation of mass, momentum, and energy principles.

Ongoing research

The frictional resistance of ships results from their turbulent boundary layers, which are the topic of ongoing research efforts all over the world. Progress is regularly published through a number of dedicated conferences and journals. This chapter provides only a short summary of turbulent boundary layer features. For further study, the textbooks by Young (1989) and White (2006) may be of help. An in-depth discussion of the topic is given by Schlichting and Gersten (2003).

Learning Objectives

At the end of this chapter students will be able to:

- discuss fundamental properties of turbulent boundary layers
- select flat plate friction coefficients
- apply the ITTC 1957 model–ship correlation coefficient
- estimate effects of surface roughness

13.1 Turbulent Boundary Layers

Ships have turbulent boundary layers

In preceding chapters, we discussed the basics of boundary layer theory and derived flat plate friction coefficients applicable to laminar flow. Reynolds numbers for boats and ships are well beyond the limit for laminar flow. Consequently, ship boundary layers will be turbulent except for a negligible fraction of their length at the bow. So

Fundamentals of Ship Hydrodynamics: Fluid Mechanics, Ship Resistance and Propulsion, First Edition.
Lothar Birk.

Companion website: www.wiley.com/go/birk/hydrodynamics

why did we discuss laminar boundary layers at all? Simply put, we derived important relationships which also apply to turbulent boundary layers.

Drag force

The frictional resistance or drag of a flat plate of length x follows from the total momentum loss along the plate as expressed by Equation (12.41). The momentum lost up to the position x is quantified by the momentum thickness $\Theta(x)$.

$$F_{F_x} = \iint \tau_w \mathrm{d}S = b \int_0^x \tau_w(\xi)\,\mathrm{d}\xi = \rho u_0^2\, b\, \Theta(x) \tag{13.1}$$

Accordingly, the frictional resistance coefficient C_F for a flat plate of width b and length x is equal to

$$C_F = \frac{F_{F_x}}{\frac{1}{2}\rho u_0^2 b\, x} = 2\frac{\Theta(x)}{x} \tag{13.2}$$

Shear stress

Flat plate boundary layers are characterized by a vanishing pressure gradient in the exterior flow. With constant pressure along the boundary layer, shear stress is proportional to the lengthwise change in momentum thickness. See Equation (12.2). Therefore, the local skin friction coefficient is

$$C_{f_x}(x) = \frac{\tau_w}{\frac{1}{2}\rho u_0^2} = 2\frac{\mathrm{d}\Theta(x)}{\mathrm{d}x} \tag{13.3}$$

Equations (13.2) and (13.3) are derived with flat plate boundary layer theory assumptions, i.e. the boundary layer is assumed to be thin and the streamwise pressure gradient $\partial p/\partial x$ vanishes. However, they are valid for laminar as well as turbulent flow. Critical in both expressions is the momentum thickness Θ as a function of the lengthwise position x. The momentum thickness, in turn, is a function of the velocity profile u/u_0 in the boundary layer (Equation (11.52)). A substantial amount of research is dedicated to the development of suitable mathematical models for the velocity distribution in turbulent boundary layers.

Turbulence

The basic nature of turbulence has been explained in Section 8.1. Essentially, turbulence results in a rapid, irregular fluctuation of velocity which may be registered with fast reacting instruments like hot wire anemometers. As introduced by Reynolds (1895), it is convenient to split the velocity within a turbulent boundary layer into a time independent mean velocity $\overline{u}$ and a time dependent turbulence u'.

$$u(t) = \overline{u} + u'(t) \qquad \text{Reynolds decomposition} \tag{13.4}$$

Boundary layer thickness

Similar to laminar flat plate boundary layers, the thickness of turbulent boundary layers is defined by the point where the flow velocity u approaches the exterior flow speed u_0. Due to the unsteady turbulence, the distance of this point from the wall is fluctuating over time at every position x along the plate. A snapshot for a selected time $t = t_0$ is shown on the upper right in Figure 13.1. Most researchers define the boundary layer thickness δ as the distance where the mean velocity $\overline{u}$ reaches the exterior flow speed. This is indicated by the dashed line in the snapshot and used as the definition of δ in this chapter. However, some authors define the boundary layer thickness as the envelope of all turbulent fluctuations. For example, see Grigson (1993).

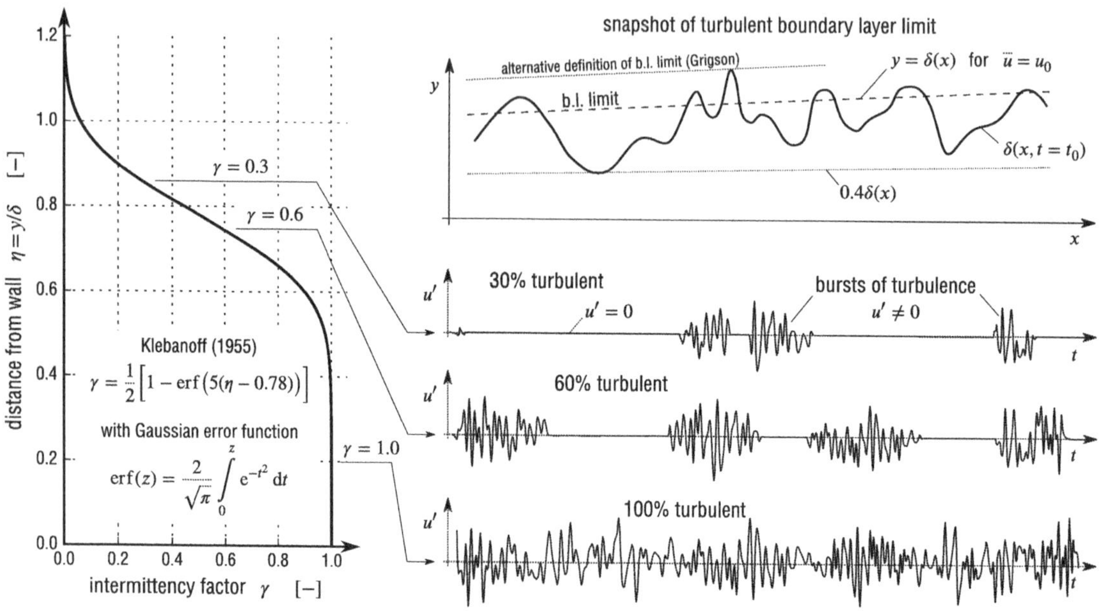

Figure 13.1 Features of a turbulent boundary layer over a flat plate (zero pressure gradient)

Intermittency The flow outside of the boundary layer is free of turbulence and usually treated as inviscid. Approaching the boundary layer limit from the exterior flow, short bursts of turbulence occur where the boundary layer temporarily overshoots the limit $y = \delta(x)$. Klebanoff (1955) found that these bursts may be detected at a distance of up to 120% of δ from the wall. The duration of turbulence bursts rapidly increases within the boundary layer. The ratio of periods with turbulence to total time observed is expressed in the intermittency factor γ (Klebanoff, 1955). An intermittency factor of zero ($\gamma = 0$) indicates that the flow is free of turbulence. As shown in Figure 13.1, the intermittency factor grows rapidly toward the wall and reaches 100% at approximately 40% of the boundary layer thickness.

Outer scaling When velocity profiles for turbulent boundary layers are plotted in the usual manner as dimensionless ratio u/u_0 over a dimensionless wall distance $\eta = y/\delta$, very little can be deduced about the flow close to the wall. This is known as outer scaling because it uses the exterior flow speed u_0 and the boundary layer thickness δ as reference scales. An example is shown on the left side of Figure 13.2.

Wall friction velocity and wall units Prandtl (1961b) suggested that it is beneficial to use the wall friction velocity u_τ and wall units y^+ as reference quantities instead. Both parameters are extensively used in the theory of turbulent boundary layers.

$$u_\tau = \sqrt{\frac{\tau_w}{\rho}} \qquad y^+ = \frac{y u_\tau}{\nu} \tag{13.5}$$

The wall friction velocity u_τ scales the velocity profile by the amount of shear stress it creates. Observant readers may notice that the wall units y^+ are a form of local Reynolds number, which uses the distance from the wall y as reference length and the friction velocity u_τ as reference speed.

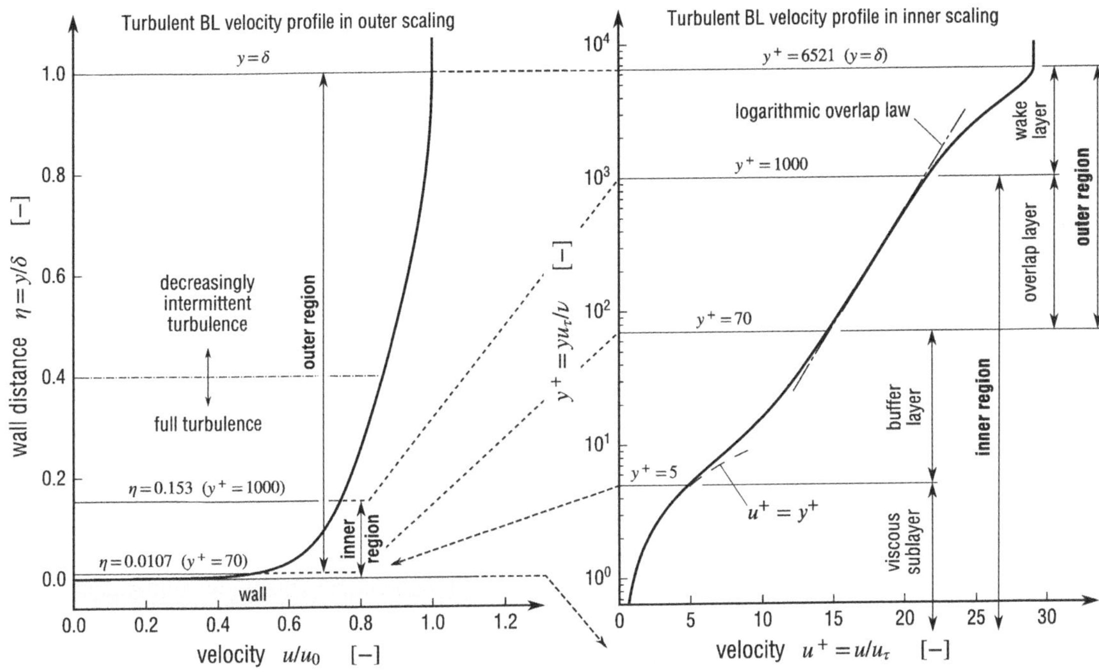

Figure 13.2 A typical turbulent boundary layer velocity profile depicted in outer and inner scaling

Inner scaling

Velocity profiles for turbulent boundary layers are usually presented in inner scaling. The dimensionless velocity $u^+ = u/u_\tau$ is plotted with a logarithmic axis of wall units y^+ as shown on the right-hand side of Figure 13.2. The logarithmic scaling of the wall unit axis stretches the distances close to the wall so that the inner region covers most of the graph.

Turbulent boundary layer structure

Turbulent boundary layers have a more complex structure than laminar boundary layers. Broadly speaking, the turbulent boundary layer is subdivided into an inner and an outer region. The inner region makes up about 15% of the boundary layer thickness in the example shown in Figure 13.2. Inner and outer region share an overlap layer.

Outer region

For the outer region, von Kármán (1931) stated that the dimensionless velocity profile should only be a function of the exterior flow velocity u_0 and the distance from the wall η. This is formulated as the velocity defect law.

$$\frac{u_0 - \overline{u}}{u_\tau} = f(\eta) \qquad \text{velocity defect law} \tag{13.6}$$

The difference $u_0 - \overline{u}$ expresses how much the fluid has slowed down within the boundary layer due to friction. In flows with nonzero pressure gradient ($\partial p/\partial x \neq 0$), the velocity distribution in the outer layer will also depend on $\partial p/\partial x$.

Inner region

Prandtl (1961a) argued that the mean velocity distribution near the wall should be governed by the wall friction, fluid properties, and the distance from the wall

$$\overline{u} = G(\tau_w, \rho, \mu, y) \tag{13.7}$$

Dimensional analysis, which we will discuss in a later chapter, reveals that the dimensionless velocity profile in the inner layer must be of the form

$$u^+ = \frac{\overline{u}}{u_\tau} = g(y^+) \qquad \text{inner law} \tag{13.8}$$

The inner region is subdivided into three parts: a viscous sublayer, the buffer layer, and the overlap layer, which may also be considered a part of the outer region (Figure 13.2).

Viscous sublayer Directly adjacent to the wall lies the viscous sublayer. The presence of the wall dampens the transverse turbulence and the flow attains laminar characteristics. Hence, some older publications name this layer the laminar sublayer. The viscous sublayer stretches only up to about $y^+ = 5$. Schlichting and Gersten (2003, Chapter 17, p. 523) report a universal law of the wall which applies to flat plates. As shown in Figure 13.2, $u^+ = y^+$ may be used as a first approximation.

Buffer layer The connection between viscous sublayer and the overlap layer is formed by the buffer layer. In Figure 13.2, the velocity profile is completed with Spalding's approximation of the velocity profile for the inner region (White, 2006, Chapter 6). The approximation is given in an implicit form.

$$y^+ = u^+ + \mathrm{e}^{-\kappa B}\left[\mathrm{e}^{\kappa u^+} - 1 - \kappa u^+ - \frac{\left(\kappa u^+\right)^2}{2} - \frac{\left(\kappa u^+\right)^3}{6}\right] \qquad \text{inner region} \tag{13.9}$$

Spalding (1961) used a value of 0.4 for the von Kármán constant κ and the value $B = 4.1$. Spalding's approximation fits velocity profiles measured at high Reynolds numbers reasonably well in the inner region (see Figure 13.3).

Wall layer The combination of viscous sublayer and buffer layer is sometimes referred to as wall layer. The wall layer is very thin and takes up roughly 1% of the boundary layer thickness, as shown in the velocity profile in outer scaling of Figure 13.2.

Overlap layer The velocity profile must be continuous throughout the boundary layer. Therefore, inner and outer region must share an overlap layer in which both velocity defect law and inner law must be valid. Setting Equations (13.6) and (13.8) equal yields:

$$\frac{\overline{u}}{u_\tau} = g(y^+) = \frac{u_0}{u_\tau} - f(\eta) \tag{13.10}$$

The general form of the velocity profile in the overlap layer is given by the logarithmic overlap law.

$$u^+ = \frac{1}{\kappa}\ln(y^+) + B \qquad \text{logarithmic overlap law} \tag{13.11}$$

Some publications call this the 'logarithmic law of the wall,' which is somewhat misleading. It is not valid at the wall. The logarithmic overlap law fits experimental data fairly well in the range of $70 < y^+ < 1000$ for high Reynolds number flows.

The constants κ and B have to be determined from experiments. κ is known as the von Kármán constant and usually assigned the value $\kappa = 0.41$. However, reported values range from 0.38 to 0.461. There is indication from recent experiments that the von Kármán constant has some dependency on the Reynolds number (Österlund et al.,

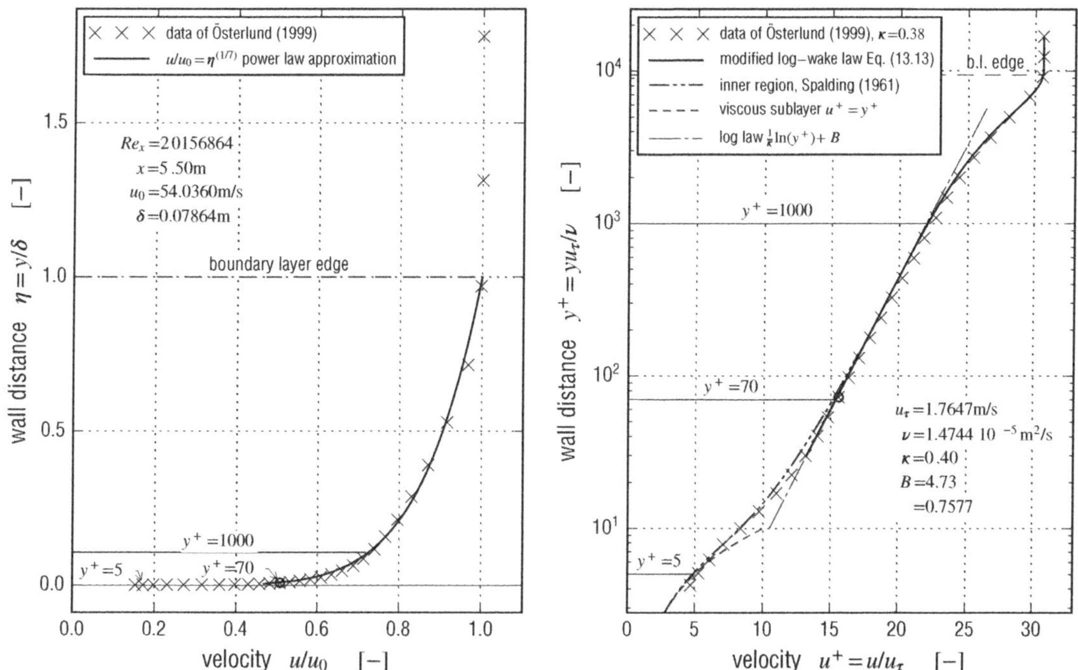

Figure 13.3 Comparing the modified log–wake law with experimental data from Österlund (1999) (Profile SW981113F)

1999). Reported values for the constant B vary from 4.0 to 7.15, but most frequently a value of $B = 5.0$ is used (Grigson, 2000).

Log-wake law for the outer region

Close to the boundary layer limit of flat plates, measured velocities exceed the values predicted by the logarithmic overlap law. Coles (1956) added the wake function $W(\eta)$ to the logarithmic law to extend its use to the outer edge of the boundary layer. So far, the wake function lacks a concise physical theory. Expressions are derived by data fitting. A commonly used sine-squared form of the wake function is attributed to Hinze (1975). Together with the logarithmic law, we get the log–wake law.

$$u^+ = \frac{1}{\kappa}\ln(y^+) + B + \frac{2\Pi}{\kappa}\sin^2\left(\frac{\pi\eta}{2}\right) \qquad \text{log–wake law} \qquad (13.12)$$

The constant Π is Coles' wake strength parameter. Its value depends on the pressure gradient along the body. For high Reynolds number flows on flat plates, a value of $\Pi = 0.7577$ is used (Guo et al., 2005). A drawback of this equation is that it violates the boundary condition of vanishing slope $\partial u/\partial y = 0$ at the boundary layer limit $y = \delta$. Guo (1998) added a correction term and introduced the modified log–wake law which has a vanishing derivative $\partial u/\partial y = 0$ at $y = \delta$.

$$u^+ = \frac{1}{\kappa}\ln(y^+) + B + \frac{2\Pi}{\kappa}\sin^2\left(\frac{\pi\eta}{2}\right) - \frac{\eta^3}{3\kappa} \qquad \text{modified log–wake law} \qquad (13.13)$$

The modified log–wake law fits experimental data very well over the complete outer region of the boundary layer. Figure 13.3 compares the modified log–wake law (13.13) with the experimental data provided by Österlund (1999). As of May 2018, Österlund's data sets are available at http://www.mech.kth.se/~jens/zpg/.

Power law The left-hand side plot in Figure 13.3 with the velocity profile in outer scaling also contains an approximation of the velocity profile which was introduced by Prandtl (1961a).

$$\frac{\overline{u}}{u_0} = \eta^{1/7} \qquad \text{power law} \tag{13.14}$$

It fits the data astonishingly well, but it neither satisfies the zero slope condition at the outer edge of the boundary layer nor provides a finite slope at the wall. For higher Reynolds number flows, the exponent may be lowered to 1/9.

Ongoing and future work There are still many unanswered questions which hamper our ability to predict the behavior of turbulent boundary layers. Unfortunately, we have not yet been able to solve the Navier-Stokes equations for high Reynolds number flows even for the simple looking case of a flat plate. Available direct numerical solutions are so far restricted to low Reynolds numbers. Experimental investigations are also scarce for high Reynolds numbers. There are several research groups actively working on the subject, though (see for instance Monkewitz et al., 2007; Marusic et al., 2009). Accurate measurements of wall shear stress and fluid velocity are very difficult in unsteady turbulent boundary layers. Even the smallest hot wire anemometers have a finite diameter and locally disturb the flow. Pressure taps have to be accurately drilled and the edges carefully deburred to avoid affecting the flow. Any wind tunnel or circulating water tunnel is limited in the achievable Reynolds number by its dimensions and attainable flow speeds. An outdoor laboratory on a flat salt lake near Salt Lake City, Utah, is an attempt to remove the dimensional restrictions. However, weather and geographical influences render control of test conditions more difficult (Metzger et al., 2007).

13.2 Shear Stress in Turbulent Flow

Wall shear stress In contrast to viscous laminar flow, turbulent boundary layers feature two mechanisms for shear stress generation. They share the viscosity driven shear stress with laminar boundary layers, but also generate shear stress by dissipation of energy through turbulent eddies. Using Reynolds' decomposition of the flow speed (13.4), wall shear stress may be expressed as a sum of viscous and turbulent shear stress.

$$\tau_w = \tau_v + \tau_t \tag{13.15}$$

Viscous shear stress Close to the wall, viscosity of the fluid and its associated viscous stress will dominate. Viscous stress is computed as in the case of a laminar boundary layer:

$$\tau_v = \mu \frac{\partial \overline{u}}{\partial y} = \rho \nu \frac{\partial \overline{u}}{\partial y} \qquad \text{viscous shear stress} \tag{13.16}$$

Turbulent shear stress In the outer region of the boundary layer flow, turbulent stress will be the dominant component. Energy is dissipated in small and large eddies which form in the boundary layer. Turbulent stress can be related to averages of products of turbulence velocities.

For a flat plate the dominant component is

$$\tau_t = \rho \overline{u'v'} = \rho \nu_t \frac{\partial \overline{u}}{\partial y} \quad \text{turbulent shear stress} \tag{13.17}$$

The eddy viscosity ν_t is introduced to unify the notation. Turbulence models are used to calculate the eddy viscosity (see Section 8.4).

Combining Equations (13.16) and (13.17) for viscous and turbulent shear stresses and dividing Equation (13.15) by the density yields

$$\frac{\tau_w}{\rho} = (\nu + \nu_t)\frac{\partial \overline{u}}{\partial y} \tag{13.18}$$

This equation is not very helpful in determining shear stress because neither the eddy viscosity nor the derivative of the mean velocity is easily determined. It is more practical to rely on the relationship between momentum thickness and shear stress expressed in Equations (13.2) and (13.3) to derive friction coefficient formulas.

13.3 Friction Coefficients for Turbulent Flow

Prandtl's semi-empirical formula

Some of the first theoretical formulas for friction coefficients were developed by Ludwig Prandtl and his research group in Göttingen, Germany. A number of these formulas are based on the power law velocity profile (13.14), and we outline the procedure below.

Momentum thickness

Initially, we follow the procedure from Section 12.3 and establish a relationship between boundary layer thickness δ and momentum thickness Θ. To that effect, the power law $u/u_0 = \eta^{1/7}$ is substituted into the definition of the momentum thickness (12.16)

$$\begin{aligned}\Theta &= \delta \int_0^1 \frac{u}{u_0}\left(1 - \frac{u}{u_0}\right) \mathrm{d}\eta = \delta \int_0^1 \eta^{1/7}\left[1 - \eta^{1/7}\right] \mathrm{d}\eta \\ &= \delta \int_0^1 \left[\eta^{1/7} - \eta^{2/7}\right] \mathrm{d}\eta = \delta \left[\frac{7}{8}\eta^{8/7} - \frac{7}{9}\eta^{9/7}\right]_0^1 = \frac{7}{72}\delta\end{aligned} \tag{13.19}$$

Thus, the local skin friction coefficient from Equation (13.3) becomes:

$$C_{f_x} = \frac{7}{36}\frac{\partial \delta}{\partial x} \tag{13.20}$$

With the chosen velocity profile, we cannot exploit Newton's shear stress formula, Equation (6.27), because the y-derivative of $\eta^{1/7}$ does not exist at the wall ($y = \eta = 0$). Prandtl (1961a) circumvented this problem by utilizing a shear stress formula that Blasius (1913) had previously derived for turbulent flow in pipes.

$$C_{f_x} = \frac{0.045}{Re_\delta^{1/4}} \tag{13.21}$$

Combining Equations (13.20) and (13.21) and expanding the Reynolds number based on boundary layer thickness $Re_\delta = u_0\delta/\nu$ results again in an ordinary differential equation for the boundary layer thickness:

$$0.045\left(\frac{\nu}{u_0\delta}\right)^{1/4} - \frac{7}{36}\frac{\partial\delta}{\partial x} = 0 \tag{13.22}$$

With the approach $\delta = Cx^{4/5}$, a constant $C = 0.3707$ is derived, and turbulent boundary layer thickness and momentum thickness are given by

$$\delta = 0.3707\frac{x}{Re_x^{1/5}} \quad \text{and} \quad \Theta = 0.036\frac{x}{Re_x^{1/5}} \tag{13.23}$$

Note that turbulent boundary layers grow much faster than laminar boundary layers: $x^{0.8}$ compared with $x^{0.5}$.

The desired flat plate friction coefficient is derived from Equation (13.2) with $x = L$:

$$C_F = \frac{0.0721}{Re^{1/5}} \quad \text{(do not use for ships)} \tag{13.24}$$

This formula is not applicable to ships because the underlying Equation (13.21) for shear stress in pipes is not accurate enough for high Reynolds numbers.

Prandtl–Schlichting formula

In a later paper, Prandtl (1961b) reported a flat plate friction formula which is based on the logarithmic overlap law velocity profile (13.11). The momentum thickness was computed numerically for a set of Reynolds numbers and converted into a convenient regression formula.

$$C_F = \frac{0.455}{\left[\log_{10}(Re)\right]^{2.58}} \quad \text{Prandtl–Schlichting, } Re \leq 10^9 \tag{13.25}$$

Equation (13.25) is known as the Prandtl–Schlichting formula and widely used in fluid mechanics for cases with $Re \leq 10^9$.

Friction coefficient based on inner variables

White (2006) recommends the following flat plate friction formula for general use:

$$C_F = \frac{0.523}{\left[\ln(0.06Re)\right]^2} \quad \text{flat plate, turbulent flow} \tag{13.26}$$

It is based on an analysis of Spalding's formula for the inner region (13.9) by Kestin and Persen (1962).

Summary

Dozens of these friction coefficient formulas have been developed over the past 100 years. A selection of friction coefficient curves is compared in Figure 13.4. They all lie within the band of available experimental data, and none of them may really claim it is better than the others. It is the engineer's responsibility to select a formula that fits the posed problem and range of Reynolds numbers.

None of the formulas stated in this section is used for the evaluation of model test results. Figure 13.4 also includes model–ship correlation lines, which are discussed in the following section.

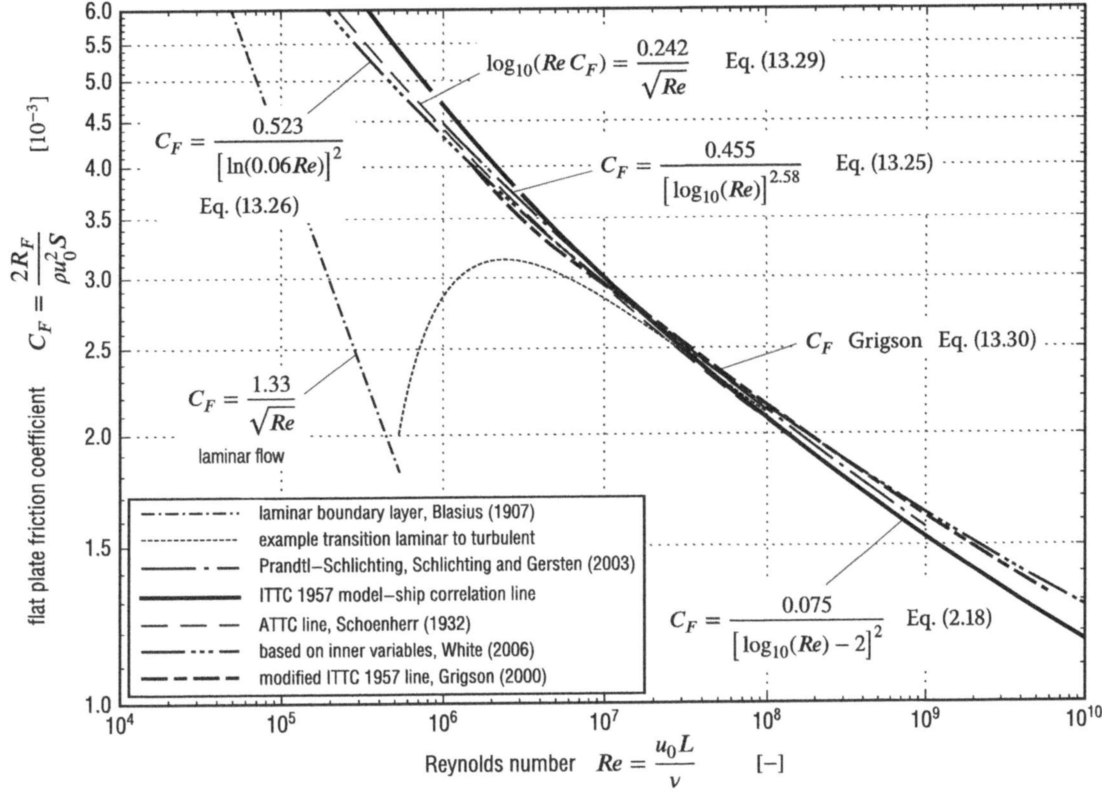

Figure 13.4 Flat plate friction coefficients for smooth surfaces

13.4 Model–Ship Correlation Lines

Importance of the friction coefficient

William Froude's successful method of predicting full scale resistance from model tests requires an accurate method to calculate the frictional resistance for model and full scale vessel (see Section 2.3.1). The frictional resistance is subtracted from the measured total resistance to determine the residual resistance coefficient C_R.

$$C_R = C_{TM} - C_{FM} \qquad \text{(Froude)} \tag{13.27}$$

In the current recommended procedures of the ITTC, frictional resistance is augmented by a form factor k to capture effects of the three-dimensional hull form. This method was originally proposed by Hughes (1954). Subtracting the augmented flat plate friction from the total resistance results in the wave resistance C_W.

$$C_W = C_{TM} - (1 + k)\, C_{FM} \qquad \text{(ITTC)} \tag{13.28}$$

ITTC used to call the result residuary resistance, until its most recent 2017 edition of the recommended procedures. This latest convention mirrors physics more closely, because an accurate form factor will capture all viscous effects beyond the flat plate resistance.

ITTC 1978 performance prediction method

The wave resistance coefficient C_W is considered a function of the Froude number alone. Consequently, it is identical for model and full scale vessel. The full scale resistance is then reassembled using the wave resistance coefficient C_W, an augmented frictional resistance coefficient $(1 + k)C_{FS}$ for the full scale vessel, plus some smaller corrections. This is part of the ITTC 1978 performance prediction method (ITTC, 2017). Details are presented in Chapters 30 and 44.

Model basins

In order to make full scale resistance predictions from model tests, Froude obtained a grant from the British Royal Navy to build a towing tank, which is a long water basin for testing ship models (Brown, 2006). He towed flat planks of up to 15.24 m length and plotted their resistance as a function of speed (Froude, 1872). Since flat plates have essentially no displacement, the measured resistance was taken as the frictional resistance. Froude's measurements were later cast into formula with the basic form of $R_F = cv_S^m S$. S is the wetted surface and c is a length and surface roughness dependent constant. The exponent m approached a value of 1.83 for higher Reynolds numbers (Schoenherr, 1932).

Schoenherr, ATTC line

Froude's work initiated the advent of model basins all over the world, and numerous researchers conducted similar tests in the late 19th and early 20th centuries. Schoenherr (1932) collected the results and added his own experiments. Subsequently, Schoenherr created a curve fit following an implicit function model suggested originally by von Kárman.

$$0 = \frac{0.242}{\sqrt{Re}} - \log_{10}\left(Re\, C_{F_{ATTC}}\right) \qquad \text{Schoenherr, 1932} \qquad (13.29)$$

In 1947, this friction line was adopted by the American Towing Tank Conference (ATTC) for use with Froude's method in the evaluation of ship model tests and henceforth called the ATTC line. Although the underlying data covers only Reynolds numbers up to about $Re < 5 \cdot 10^8$, it was applied to model and full scale vessels alike.

ITTC 1957 model–ship correlation line

Equation (13.29) is awkward to use because it is implicit and nonlinear. Comparisons of results with geometrically similar models of different scales indicated that a steeper slope in the range of model Reynolds numbers would be desirable. After a long and sometimes contentious discussion of various proposed friction lines, the ITTC adopted the following model–ship correlation line (ITTC, 1957).

$$C_F = C_{F_{ITTC}} = \frac{0.075}{\left[\log_{10}(Re) - 2\right]^2} \qquad \text{ITTC, 1957} \qquad (2.18)$$

The ITTC 1957 line is a compromise for the specific purpose of making full scale predictions from model test results, and its physical foundation is weak. To cite from the ITTC's decision (Saunders et al., 1957):

> In view of the above, the Conference decides that the line given by the formula [Equation (2.18)] is adopted as the 'ITTC 1957 model–ship correlation line,' it being clearly understood that this is regarded only as an interim solution to the problem for practical engineering purposes.

Use in model tests

Chapters 29 and 30 explain in detail how the ITTC 1957 line is used in the evaluation of model tests and the prediction of full scale resistance.

Possible improvements

Equation (2.18) is an astonishingly simple solution to a very complicated problem. The 'interim solution' has now been in use for more than 60 years. Research on turbulent boundary layers has progressed considerably, as summarized in Section 13.1. Several proposals for an improved model–ship correlation line have been made but none has been universally accepted yet. Grigson (2000) developed an algorithm for the flat plate friction coefficient considering a wake–log form of the velocity distribution in the boundary layer. The algorithm is fairly complex, but Grigson provides a simple modification to the ITTC 1957 line which represents the results of his algorithm quite nicely.

$$C_F = \begin{cases} G_1(x_1)\, C_{F_{ITTC}} & \text{for } 1.5 \cdot 10^6 < Re < 2 \cdot 10^7 \\ G_2(x_2)\, C_{F_{ITTC}} & \text{for } 2 \cdot 10^7 \leq Re < 6 \cdot 10^9 \end{cases} \tag{13.30}$$

The factors $G_1(x_1)$ and $G_2(x_2)$ are given as

$$G_1(x_1) = 0.9335 + 0.147\, x_1^2 - 0.071\, x_1^3 \quad \text{with } x_1 = \log_{10}(Re) - 6.3 \tag{13.31}$$

$$G_2(x_2) = 1.0096 + 0.0456\, x_2 - 0.013944\, x_2^2 + 0.0019444\, x_2^3 \quad \text{with } x_2 = \log_{10}(Re) - 7.3 \tag{13.32}$$

Grigson's friction line is also shown in Figure 13.4. The line is very similar to Equation (13.26) by White (2006), which is reassuring, since both are based on formulations of the boundary layer velocity profile in inner variables. So far however, the ITTC 1957 model–ship correlation line Equation (2.18) is still the recommended formula for the evaluation of model test results.

13.5 Effect of Surface Roughness

Effect of surface roughness

The ITTC 1957 model–ship correlation line and the other C_F-formulas presented above are designed for hydraulically smooth plates. Hydraulically smooth means that surface roughness has no influence on the flow in the boundary layer. This is the case as long as surface imperfections do not penetrate through the viscous sublayer as illustrated in Figure 13.5(a). The dimensionless height of the bumps is smaller than $y^+ = yu_\tau/\nu = 5$. Surface imperfections which penetrate through the viscous sublayer into the buffer layer, i.e. bumps higher than $y^+ > 5$, the friction coefficient increases beyond the value for the smooth surface. This is known as the transitional regime depicted in Figure 13.5(b). In cases where the surface imperfections reach across the wall layer into the overlap layer, the frictional coefficient becomes independent of the Reynolds number *Re*. In this fully rough regime, the frictional coefficient is only a function of the relative height of the surface roughness (Figure 13.5(c)).

Equivalent sand roughness

Defining the roughness of a surface is actually quite complicated. Surface imperfections may vary in size, form, and distribution. Different statistical parameters can be used to define the technical roughness k_{tech}. Often the measured surface profile is divided into smaller samples and a mean value is taken of the maximum surface imperfection heights in each sample. The concept of equivalent sand roughness k_S is introduced in order to measure the effects of surface roughness independently of its shape and

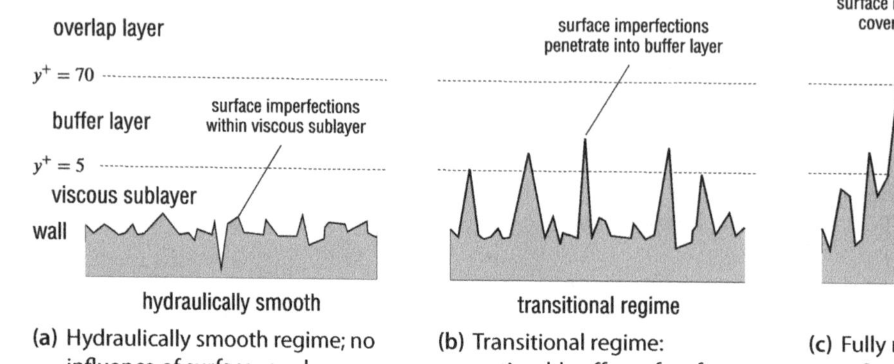

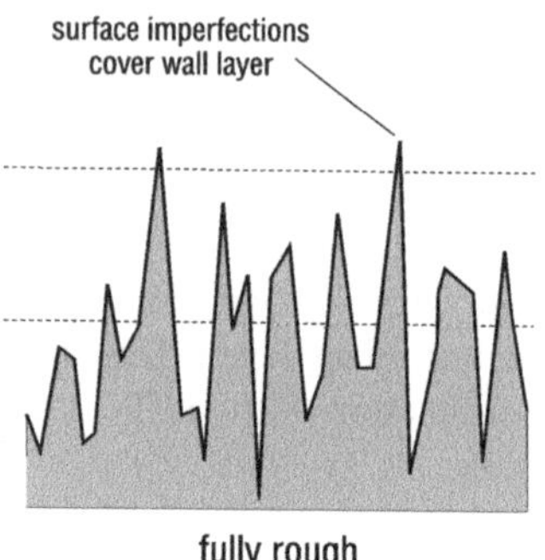

(a) Hydraulically smooth regime; no influence of surface roughness on shear stress and C_F

(b) Transitional regime: noticeable effect of surface roughness on shear stress and C_F

(c) Fully rough regime: vanishing influence of viscous shear stress. C_F becomes independent of Re

Figure 13.5 Types of technical surface roughness and their effect on friction

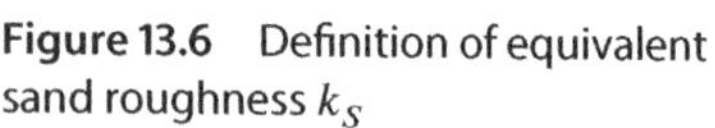

Figure 13.6 Definition of equivalent sand roughness k_S

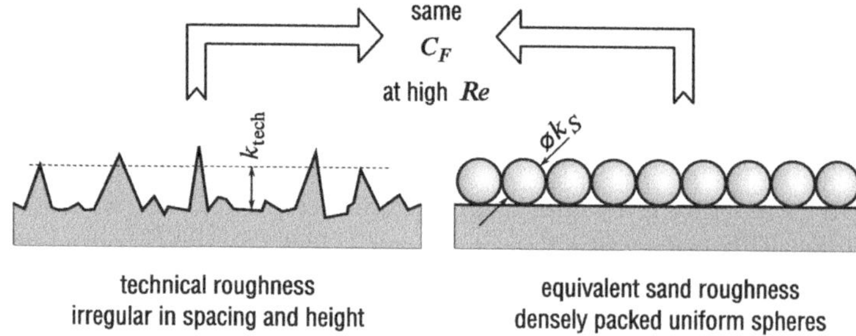

distribution. Imagine a surface completely covered with spheres of diameter k_S, as shown on the right in Figure 13.6. This can be achieved by covering a surface with a uniform grain, like you find on sandpaper.

For each technical roughness an equivalent sand roughness is found by comparing the frictional coefficients at high Reynolds numbers. To be more precise, the value of k_S^+ must be larger than 70 so that we are in the fully rough regime.

$$k_S^+ = \frac{k_S u_\tau}{\nu} > 70 \qquad \text{fully rough regime} \tag{13.33}$$

Again, the friction velocity u_τ of Equation (13.5) is employed to define the Reynolds number k_S^+ based on the equivalent sand roughness as the reference length.

The influence of surface roughness was extensively studied by Prandtl's research group in Göttingen. Nikuradse (1950) studied the flow in pipes with varying sand roughness. Based on his results, Prandtl and Schlichting (1961) developed a theoretical model for the flow over smooth and rough plates. Its details are beyond the scope of this book, but the interested reader may study Chapters 17 and 18 of Schlichting and Gersten (2003) for an up-to-date version. The theoretical model results in a set of implicit equations for the frictional coefficient C_F of a plate as a function of the Reynolds number Re and the relative surface roughness k_S/L.

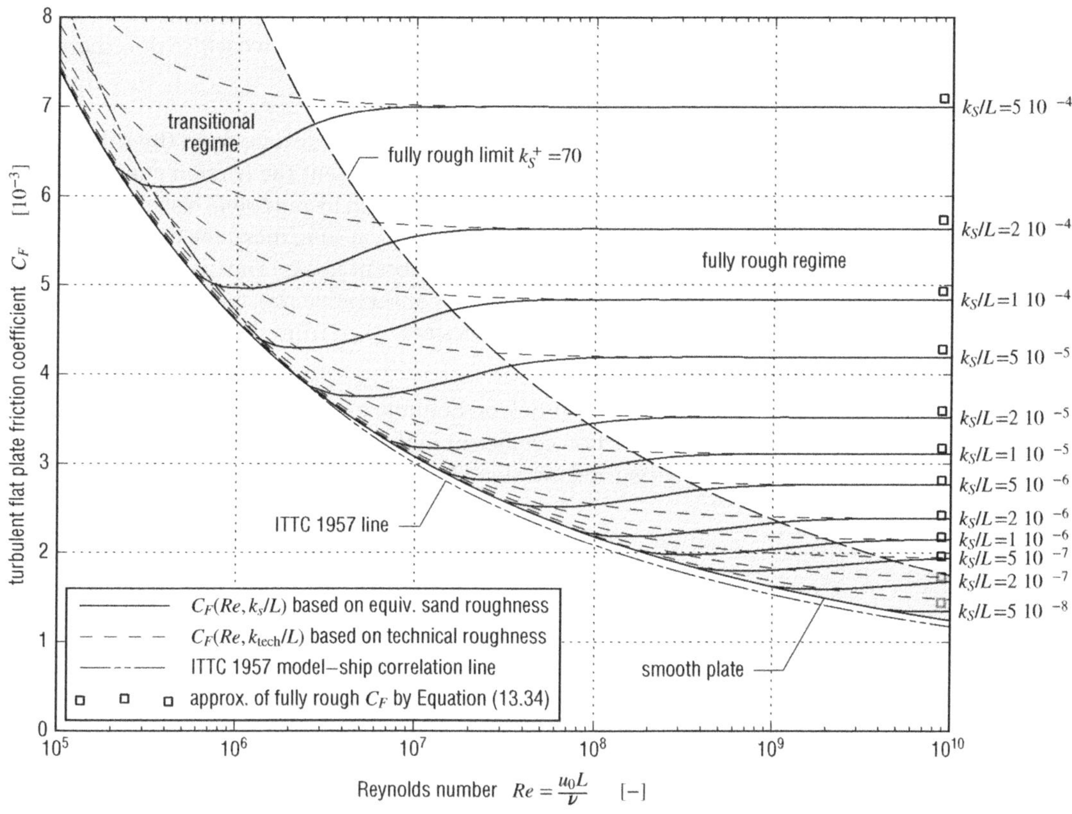

Figure 13.7 Flat plate friction coefficient for turbulent flow and its dependency on Reynolds number and relative surface roughness k_S/L

Friction coefficient for rough plates

Figure 13.7 depicts the resulting friction coefficient curves. Some may recognize the similarity with the Moody chart which is used in the design of piping systems. For resistance predictions, ITTC recommends the use of an equivalent sand roughness for welded steel ship hulls of $k_S = 150 \cdot 10^{-6}$ m. Thus, for a ship of length $L = 300$ m, relative surface roughness would be $k_S/L = 2 \cdot 10^{-6}$. If we pick up the C_F-curve for $k_S/L = 2 \cdot 10^{-6}$ at the right side of the chart and follow it to lower Reynolds numbers (to the left), we see that the frictional coefficient is initially constant with $C_F = 2.3858 \cdot 10^{-3}$. Around $Re = 2 \cdot 10^9$ the C_F-curve starts to dip, and we enter the transitional regime where the surface roughness affects the viscous flow in the wall layer. Close to $Re = 10^8$ it merges with the curve for smooth plates.

Two important conclusions can be drawn from the C_F-curves in Figure 13.7:

- In the fully rough regime, the frictional resistance coefficient becomes independent of the Reynolds number. As a consequence, frictional resistance grows quadratic with speed for high Reynolds numbers and the higher the relative surface roughness k_S/L the higher the final value of C_F.
- A surface that is hydraulically smooth at low Reynolds numbers may show effects of surface roughness at high Reynolds numbers. A plate with relative roughness

$k_S/L = 2 \cdot 10^{-5}$ shows almost no roughness effect at Reynolds number $Re = 10^7$ (model scale) whereas at $Re = 10^9$ (full scale) its friction coefficient is more than twice that of a smooth plate.

Technical versus equivalent sand roughness

Figure 13.7 also features a set of dashed curves which deviate from the solid curves in the transitional regime. The dashed curves represent the friction coefficients for plates with technical roughness, i.e. plates roughened by randomly distributed surface imperfections. In contrast to the plates with uniform roughness, the dip in C_F-curves through the transitional regime is noticeably absent in experiments with plates with technical roughness k_{tech}. Therefore, more conservative resistance estimates are obtained for technical systems if the C_F-vales are taken from the dashed curves in the transitional regime.

Approximation of fully rough C_F

In older editions of Schlichting's book on boundary layer theory, an approximation is provided for the flat plate friction coefficient in the fully rough regime. It seems to have been omitted in Schlichting and Gersten (2003) but is restated in White (2006).

$$C_{F\,\text{rough}} = \left[1.89 + 1.62 \log_{10}\left(\frac{L}{k_S}\right)\right]^{-2.5} \qquad (13.34)$$

The values are shown as squares on the right-hand side of Figure 13.7. The estimates are fairly accurate for the relative roughness of ship hulls $k_S/L \approx 10^{-6}$.

For comparison, the ITTC 1957 model–ship correlation line (2.18) has been added to Figure 13.7. It emphasizes that the ITTC 1957 line predicts lower frictional coefficients at full scale than modern theory and experiments suggest.

Roughness allowance

Towing tank models are built to be hydraulically smooth, but actual ship hulls do not satisfy this criterion. Therefore, we will augment the resistance curve for the full scale vessel by a roughness allowance ΔC_F (see Chapter 30, page 358).

Keep the bow/leading edge smooth

Revisit the distribution of wall shear stress over the length of a plate in Figure 12.2. Although the specific graph is for laminar flow, the fact that wall shear stress is highest at the beginning of a plate or body and then declines also applies to turbulent boundary layers. High wall shear stress τ_w, in turn, implies high friction velocity $u_\tau = \sqrt{\tau_w/\rho}$. With higher friction velocity the Reynolds number $k_S^+ = k_S u_\tau / nu$ is more likely to exceed the limit for fully rough regime ($k_S^+ = 70$) at the beginning than at the end of a plate, i.e. the initial part of a plate may fall into the fully rough regime although the plate overall may not. Therefore, it is more important to smooth and clean the bow of a vessel or the leading edge of foil than their downstream parts.

13.6 Effect of Form

Thin plate versus ship

Evidently, ship hulls are not flat plates. An infinitely thin plate would not produce waves and would not have a wave resistance. However, an infinitely flat plate has no volume and, as a result, no buoyancy. Consequently, our imaginary thin ship would have a low resistance but could carry neither cargo nor its own weight.

Varying external pressure

For flat plates we assumed that the exterior flow pressure is constant along the plate. Then, the exterior flow velocity u_e is constant according to the Bernoulli equation. In the literature this is called a zero pressure gradient flow. The assumption of $p =$ const. allows us to eliminate the second term in Equation (11.53) for the wall shear stress. Pressure varies over a three-dimensional, voluminous body, and the second term in Equation (11.53) has to be retained. As a consequence, computation of shear stress and boundary layer thickness becomes significantly more complicated. The changes in exterior flow pressure affect the growth of the boundary layer thickness and the actual shear stress distribution. In addition, the boundary layer pushes the exterior flow away from the body: the displacement effect. We will revisit this problem in Chapter 18, after we have learned more about the pressure distribution in exterior flows.

13.7 Estimating Frictional Resistance

Frictional resistance

Frictional resistance R_F is the integral of the shear stress vectors over the wetted surface of the ship hull. Especially for ships at lower speeds, R_F is by far the largest resistance component. Although the ITTC 1957 model–ship correlation line (2.18) is technically not an accurate friction line, it is widely used in estimates of the frictional resistance of ships.

Reynolds number

First, the Reynolds number Re needs to be computed:

$$Re = \frac{v_S \, L_{OS}}{\nu} \tag{13.35}$$

As input, we require ship speed v_S, length over wetted surface L_{OS}, and the kinematic viscosity ν. The kinematic viscosity ν is the ratio of dynamic viscosity and density. For example, let us compute the frictional resistance for a ship of 212 m length over wetted surface sailing at 20 knots. It is standard practice to make resistance predictions for salt water at temperature 15 °C.

Unit conversion

Obviously, the speed has to be converted into suitable SI-units first. 1 knot is equivalent to the speed of 1 international nautical mile per hour. A 'knot' is not an official SI-unit but its usage is allowed (BIPM, 2006). Conversion into meter per second is obtained with the following factor:

$$1 \text{ kn} = \frac{1 \text{ international nautical mile}}{1 \text{ hour}} = \frac{1852.0 \text{ m}}{3600.0 \text{ s}} = 0.5144\bar{4} \, \frac{\text{m}}{\text{s}} \tag{13.36}$$

Using an abbreviated value of 0.51444 is sufficiently accurate. Therefore, our ship speed of 20 kn is equivalent to

$$20 \text{ kn} = 20 \cdot 0.51444 \text{ m/s} = 10.289 \text{ m/s}$$

Using the kinematic viscosity value of $1.1892 \cdot 10^{-6}$ m^2/s for seawater at temperature 15 degree Celsius (ITTC, 2011), the Reynolds number in our example is

$$Re = \frac{10.289 \text{ m/s} \cdot 212 \text{ m}}{1.1892 \cdot 10^{-6} \text{ m}^2/\text{s}} = 1\,834\,231\,416.1$$

Reynolds numbers for full scale ships are very high, typically in the range of 10^8 to 10^{10}. When you compute the frictional resistance coefficient according to the 1957 model–ship correlation line (2.18), make sure you use the common logarithm function with base 10 and *not* the natural logarithm.[1]

$$C_F = \frac{0.075}{\left(\log_{10}(1834231416.1) - 2\right)^2} = 1.4216 \cdot 10^{-3}$$

This value is valid only for a hydraulically smooth surface.

The frictional resistance itself will be

$$\begin{aligned} R_F &= \frac{\rho}{2} v_S^2 \, S \, C_F \\ &= \frac{1026.021 \text{ kg m}^{-3}}{2} \left(10.289 \text{ m s}^{-1}\right)^2 5000 \text{ m}^2 \cdot 1.4216 \cdot 10^{-3} \\ &= 386\,026 \,\frac{\text{kg m}}{\text{s}^2} = 386\,026 \text{ N} = 386.03 \text{ kN} \end{aligned}$$

The frictional resistance component is the only part of the resistance for which we have such a simple procedure. Other components, like the viscous pressure resistance and the wave resistance, require model test results or the numerical solution of the fundamental equations. We will come back to these resistance components in Chapters 18 and 25, respectively.

References

BIPM (2006). *The international system of units (SI).* The International Bureau of Weights and Measures (BIPM), Sèvres, France, eigth edition.

Blasius, H. (1913). *Das Ähnlichkeitsgesetz bei Reibungsvorgängen in Flüssigkeiten: über den Gültigkeitsbereich der beiden Ähnlichkeitsgesetze in der Hydraulik,* volume 131 of *Mitteilungen über Forschungsarbeiten auf dem Gebiete des Ingenieurwesens, insbesondere der Technischen Hochschulen.* VDI-Verlag Berlin.

Brown, D. (2006). *The way of a ship in the midst of the sea – The life and work of William Froude.* Periscope Publishing Ltd.

Coles, D. (1956). The law of the wake in turbulent boundary layer. *Journal of Fluid Mechanics,* 1(2):191–226.

Froude, W. (1872). Experiments on surface-friction experienced by a plane moving through water. Read before the British Association for the Advancement of Science at Brighton. Published in *The Papers of William Froude,* A.D. Duckworth, The Institution of Naval Architects, London, United Kingdom, pp. 138–147, 1955.

Grigson, C. (1993). An accurate smooth friction line for use in performance prediction. In *Transactions of The Royal Institution of Naval Architects (RINA),* volume 135, pages 149–162. The Royal Institution of Naval Architects.

Grigson, C. (2000). A planar friction algorithm and its use in analysing hull resistance. In *Transactions of The Royal Institution of Naval Architects (RINA),* volume 142, pages 76–115. The Royal Institution of Naval Architects.

[1] Unfortunately, the natural logarithm is named log() in many programming languages. Logarithm functions with base 10 are often accessible as log10().

Guo, J. (1998). *Turbulent velocity profiles in clear water and sediment-laden flows.* PhD thesis, Colorado State University, Department of Civil Engineering, Fort Collins, CO.

Guo, J., Julien, P., and Meroney, R. (2005). Modified logâŁ"wake law for zero-pressure-gradient turbulent boundary layers. *Journal of Hydraulic Research*, 43(4):421–430.

Hinze, J. (1975). *Turbulence.* McGraw-Hill, New York, NY.

Hughes, G. (1954). Friction and form resistance in turbulent flow, and a proposed formulation for use in model and ship correlation. *RINA Transactions and Annual Report*, 96(4):314–376. doi:10.3940/rina.trans.1954.16.

ITTC (1957). 8th *International Towing Tank Conference*, Madrid Spain. ITTC. Available at www.ittc.info.

ITTC (2011). *Fresh water and seawater properties.* International Towing Tank Conference, Recommended Procedures and Guidelines 7.5-02-01-03. Revision 2.

ITTC (2017). *1978 ITTC performance prediction method.* International Towing Tank Conference, Recommended Procedures and Guidelines 7.5-02-03-01.4. Revision 4.

Kestin, J. and Persen, L. (1962). Application of Schmidt's, method to the calculation of Spalding's function and of the skin-friction coefficient in turbulent flow. *Int. Journal of Heat and Mass Transfer*, 5(3–4):143–152.

Klebanoff, P. (1955). Characteristics of turbulence in a boundary layer with zero pressure gradient. NACA Technical Report 1247, National Advisory Committee for Aeronautics.

Marusic, I., Hutchins, N., and Mathis, R. (2009). High Reynolds number effects in wall turbulence. In *Sixth International Symposium on Turbulence and Shear Flow Phenomena*, pages 993–1002, Seoul, Korea.

Metzger, M., McKeon, J., and Holmes, H. (2007). The near-neutral atmospheric surface layer: turbulence and non-stationarity. *Philosophical Transactions of the Royal Society of London A*, 365:859–âŁ"876.

Monkewitz, P., Chauhan, K., and Nagib, H. (2007). Self-consistent high-Reynolds-number asymptotics for zero-pressure-gradient turbulent boundary layers. *Physics of Fluids*, 19(115101).

Nikuradse, J. (1950). Laws of flow in rough pipes. Technical Memorandum TM 1292, National Advisory Committee for Aeronautics (NACA). Translation of Strömungsgesetze in rauhen Rohren, published in 1933 as VDI Forschungsheft 361, Beilage zu Forschung auf dem Gebiet des Ingenieurwesens, Ausgabe B, Band 4, Juli/August.

Österlund, J. (1999). *Experimental studies of zero pressure-gradient turbulent boundary layer flow.* PhD thesis, Royal Institute of Technology, Department of Mechanics, Stockholm, Sweden. ISRN KTH/MEK/TR–99/16–SE.

Österlund, J., Johansson, A., Nagib, H., and Hites, M. (1999). Wall shear stress measurements in high Reynolds number boundary layers from two facilities. In 30th *AIAA Fluid Dynamics Conference*, Norfolk, VA. American Institute of Aeronautics and Astronautics. AIAA 99-3814.

Prandtl, L. (1961a). Über den Reibungswiderstand strömender Luft. In Tollmien, W., Schlichting, H., and Görtler, H., editors, *Gesammelte Abhandlungen zur angewandten Hydro- und Aerodynamik*, volume 2, pages 620–626. Springer Verlag, Berlin, Heidelberg. Original published in 1927 as Ergebnisse der Aerodynamischen Versuchsanstalt zu Göttingen, Oldenbourg München-Berlin, III. Lief., S.1–5.

Prandtl, L. (1961b). Zur turbulenten Strömung in Rohren und längs Platten. In Tollmien, W., Schlichting, H., and Görtler, H., editors, *Gesammelte Abhandlungen zur angewandten Hydro- und Aerodynamik*, volume 2, pages 632–648. Springer Verlag, Berlin, Heidelberg. Original published in 1932 as Ergebnisse der Aerodynamischen Versuchsanstalt zu Göttingen, Oldenbourg München-Berlin, IV. Lief., S.18–29.

Prandtl, L. and Schlichting, H. (1961). Das Widerstandsgesetz rauher Platten. In Tollmien, W., Schlichting, H., and Görtler, H., editors, *Gesammelte Abhandlungen zur angewandten Hydro- und Aerodynamik*, volume 2, pages 649–662. Springer Verlag, Berlin, Heidelberg. Original published 1934 in Werft, Reederei, Hafen 15. Jg., S.1–4.

Reynolds, O. (1895). On the dynamical theory of incompressible, viscous fluids and the determination of the criterion. *Philosophical Transactions of the Royal Society London. A*, 186:123–164.

Saunders, H., Walker, W., and Mazarredo, I. (1957). Concluding technical session. In *8th International Towing Tank Conference*, pages 323–332, Madrid Spain. ITTC.

Schlichting, H. and Gersten, K. (2003). *Boundary-layer theory*. Springer-Verlag, Berlin, Heidelberg, New York, eigth edition. Corrected printing.

Schoenherr, K. (1932). Resistance of flat surfaces moving through a fluid. *SNAME Transactions*, 40:279–313.

Spalding, D. (1961). A single formula for the law of the wall. *Journal of Applied Mechanics*, 28(3):455–458. doi:10.1115/1.3641728.

von Kármán, T. (1931). Mechanical similitude and turbulence. Technical Memorandum 611, National Advisory Committee for Aeronautics.

White, F. (2006). *Viscous fluid flow*. McGraw-Hill, New York, NY, third edition.

Young, A. (1989). *Boundary layers*. AIAA Education Series. American Institute of Aeronautics and Astronautics (AIAA), Washington, DC.

Self Study Problems

1. State and describe the regions and layers in a turbulent boundary layer.
2. Download a velocity profile from Österlund's data sets, available at http://www.mech.kth.se/~jens/zpg/. Plot the data in outer and inner scaling (see Figure 13.2). Add the power law profile Equation (13.14) and compare!
3. Explain why surface roughness at the bow is more detrimental to its drag than roughness at the stern.
4. Repeat the estimate of frictional resistance from Section 13.7 for the speeds 10 kn and 15 kn.
5. Repeat the estimate of the frictional resistance from Section 13.7 for the speeds 10 kn, 15 kn, and 20 kn with the friction formula (13.26). Compare the results with those from the text and previous problem.

14

Inviscid Flow

As stated in Section 7.2, the coefficient for the shear stress forces is very small for water. We will discuss a number of flow conditions, where we treat the fluid as inviscid, i.e. we set its kinematic viscosity to zero:

$$\nu = 0 \qquad \text{for inviscid fluid} \tag{14.1}$$

This assumption leads to many practical solutions in fluid mechanics. However, we have to be careful in interpreting the results because an actual inviscid fluid does not exist. The hypothetical inviscid fluid is also known as ideal fluid.

After defining the Euler equations, we derive the important Bernoulli equation for a streamline. It connects pressure and flow velocity and is widely used in fluid mechanics. Later, we will derive more specific forms of the Bernoulli equation for specific flow types. This chapter also introduces the concepts of rotation, vorticity, and circulation.

Learning Objectives

At the end of this chapter students will be able to:

- identify the Euler equations
- discuss and apply the Bernoulli equation
- define vorticity and circulation

14.1 Euler Equations for Incompressible Flow

Euler equations

The conservation of momentum equations for inviscid fluids were formulated by Euler (1757) roughly 75 years before Navier included viscous forces in his version of the conservation of momentum equations. We can obtain the Euler equations for the conservation of momentum for an incompressible (ρ = const.) and inviscid fluid by setting the kinematic viscosity to zero ($\nu = 0$) and, consequently, omitting the last term in the Navier-Stokes equations (7.12).

$$\frac{\partial \underline{v}}{\partial t} + \left(\underline{v}^T \underline{\nabla}\right)\underline{v} = \underline{f} - \frac{1}{\rho}\underline{\nabla}\, p \qquad \text{Euler equations} \tag{14.2}$$

Fundamentals of Ship Hydrodynamics: Fluid Mechanics, Ship Resistance and Propulsion, First Edition.
Lothar Birk.

Companion website: www.wiley.com/go/birk/hydrodynamics

This is the differential, conservative form of the Euler equations in Cartesian coordinates.

Careful application required

Solving the Euler equations will lead to incorrect results for the total resistance of a body. They ignore the viscous effects which occur in the boundary layer. Although less demanding than the Navier-Stokes equations, the Euler equations (14.2) still represent a formidable challenge for numerical solutions. The fundamental tasks of discretizing a three-dimensional domain and solving large systems of linear equations remain. Nonetheless, the Euler equations still provide the basis for practical solutions of important problems in ship hydrodynamics like wave theory, wave resistance, lifting surfaces, and propeller theory to name a few. This requires some extra restrictions in addition to the underlying assumptions in Equations (14.2) that the fluid is incompressible and inviscid.

Alternative form of Euler equations

A slightly different form of the Euler equations (14.2) is obtained by modifying the left-hand side with the following vector identity.

$$(\underline{v}^T \underline{\nabla})\underline{v} = \frac{1}{2}\underline{\nabla}(\underline{v}^T \underline{v}) - \underline{v} \times (\underline{\nabla} \times \underline{v}) \tag{14.3}$$

This vector identity is confirmed by expanding the terms on both sides.

Substituting the identity (14.3) into the Euler equations (14.2) above yields

$$\frac{\partial \underline{v}}{\partial t} + \frac{1}{2}\underline{\nabla}(\underline{v}^T \underline{v}) - \underline{v} \times (\underline{\nabla} \times \underline{v}) = \underline{f} - \frac{1}{\rho}\underline{\nabla} p \tag{14.4}$$

We will later use this second form of the Euler equations to derive the important Bernoulli equation for *potential flow*. In potential flows, the last term on the left-hand side vanishes because the curl of the fluid velocity vanishes, i.e. $\nabla \times \underline{v} = \underline{0}$. We call these flows *irrotational*. More on this in Chapter 15.

14.2 Bernoulli Equation

The terms in the Euler equations (14.2) represent forces per unit mass. An integration of the forces along a streamline yields the work done by the forces. Of course, for unsteady flow this is just a snapshot as the streamlines may change over time. Before we perform this integration, we take a look at the geometry and kinematics of the problem (Figure 14.1).

Tangent vector

The tangent $\underline{s}$ to the streamline is a vector of unit length $|\underline{s}| = 1$.

$$\underline{s} = (s_x, s_y, s_z)^T$$

Its coordinates s_x, s_y, and s_z represent the cosines of the angles a streamline forms with the coordinate axes at that point.

A line element $\mathrm{d}\underline{s}$ of the streamline is defined as a vector stretching an infinitesimally small distance $\mathrm{d}s$ in tangent direction.

$$\mathrm{d}\underline{s} = \begin{pmatrix} \mathrm{d}x \\ \mathrm{d}y \\ \mathrm{d}z \end{pmatrix} \qquad \text{with length} \quad \mathrm{d}s = \sqrt{\mathrm{d}x^2 + \mathrm{d}y^2 + \mathrm{d}z^2} \tag{14.5}$$

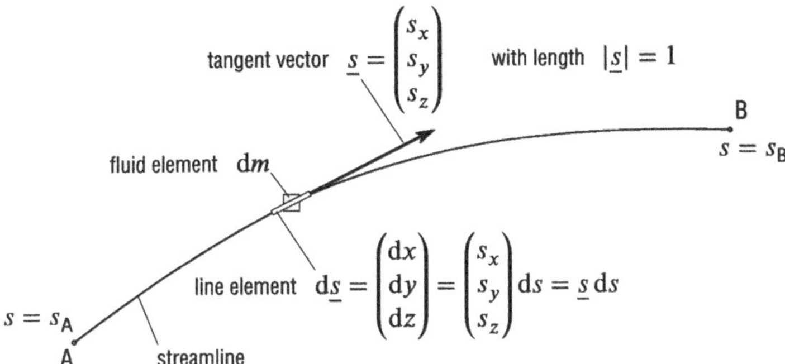

Figure 14.1 A fluid element dm moves from point A to point B along a streamline

Velocity vector

Per definition, the velocity vector $\underline{v}$ is tangent to the streamline in every instant.

$$\underline{v} = \begin{pmatrix} u \\ v \\ w \end{pmatrix} \qquad \text{with speed} \quad |\underline{v}| = \sqrt{u^2 + v^2 + w^2}$$

Consequently, both line element d$\underline{s}$ and velocity $\underline{v}$ can be rewritten as a product of the tangent vector $\underline{s}$ and their respective length.

$$d\underline{s} = \underline{s}\, ds \qquad \underline{v} = \underline{s}\, |\underline{v}| \tag{14.6}$$

Work in a vector field

Work is often defined as force times distance traveled. In 3D space, however, force vector and motion are not always parallel. Only the portion of the forces pointing in the direction of motion will contribute to the work done on a fluid element. The dot product allows us to compute the component of a force vector $\underline{F}$ pointing in the direction of the line element d$\underline{s}$. The work done by the force $\underline{F}$ along the distance ds is equal to

$$dW = d\underline{s}^T \underline{F} = \underline{s}^T \underline{F}\, ds \tag{14.7}$$

Of course, we assumed that ds is so small that the force is considered constant over the distance ds.

Line integral

Integration of the differential work element dW along the streamline results in the total work done between the points A and B. We dot-multiply Equation (14.2) with the unit length tangent vector $\underline{s}$ and obtain the forces per unit mass that point in the direction of motion. Subsequently, we integrate the forces per unit mass along the streamline from point A to point B.

$$\int_{s_A}^{s_B} \underline{s}^T \frac{\partial \underline{v}}{\partial t}\, ds + \int_{s_A}^{s_B} \underline{s}^T \left(\underline{v}^T \underline{\nabla}\right) \underline{v}\, ds = \int_{s_A}^{s_B} \underline{s}^T \underline{f}\, ds - \int_{s_A}^{s_B} \underline{s}^T \frac{1}{\rho} \underline{\nabla} p\, ds \tag{14.8}$$

Study of second term

We will study the integrals individually and start with the second integral on the left-hand side of Equation (14.8). Its integrand contains the dot product of velocity vector and Nabla operator. We substitute the velocity vector with the product of tangent vector times speed, i.e. $\underline{v} = \underline{s}|\underline{v}|$.

$$\left(\underline{v}^T \underline{\nabla}\right) = \left(\underline{s}|\underline{v}|\right)^T \underline{\nabla} = |\underline{v}|\underline{s}^T \underline{\nabla} \tag{14.9}$$

The speed $|\underline{v}|$ is a scalar and can be extracted from the dot product. The dot product of a unit vector and the Nabla operator is called a *directional derivative.*

$$\underline{s}^T \underline{\nabla} = \frac{\partial}{\partial s} \tag{14.10}$$

It specifies the component of the gradient that points in direction of the unit vector (here $\underline{s}$). Together we have:

$$\left(\underline{v}^T \underline{\nabla}\right) = |\underline{v}| \frac{\partial}{\partial s} \tag{14.11}$$

We substitute this into the second integral of Equation (14.8).

$$\int_{s_A}^{s_B} \underline{s}^T \left(\underline{v}^T \underline{\nabla}\right) \underline{v} \, \mathrm{d}s = \int_{s_A}^{s_B} \underline{s}^T \left(|\underline{v}| \frac{\partial}{\partial s}\right) \underline{v} \, \mathrm{d}s = \int_{s_A}^{s_B} \left(\underline{s} \, |\underline{v}|\right)^T \frac{\partial \underline{v}}{\partial s} \, \mathrm{d}s = \int_{s_A}^{s_B} \underline{v}^T \frac{\partial \underline{v}}{\partial s} \, \mathrm{d}s \tag{14.12}$$

Again, we utilized the facts that a scalar factor in a dot product can be moved from one operand to the other and that $\underline{v} = \underline{s} \, |\underline{v}|$.

We apply the product rule to the following expression:

$$\frac{\partial}{\partial s} \left(\underline{v}^T \underline{v}\right) = \underline{v}^T \frac{\partial \underline{v}}{\partial s} + \left(\frac{\partial \underline{v}}{\partial s}\right)^T \underline{v} = 2 \, \underline{v}^T \frac{\partial \underline{v}}{\partial s} \tag{14.13}$$

Multiplication with 1/2 yields an alternative expression for the integrand of the second integral.

$$\underline{v}^T \frac{\partial \underline{v}}{\partial s} = \frac{1}{2} \frac{\partial}{\partial s} \left(\underline{v}^T \underline{v}\right) \tag{14.14}$$

$$\int_{s_A}^{s_B} \underline{s}^T \left(\underline{v}^T \underline{\nabla}\right) \underline{v} \, \mathrm{d}s = \int_{s_A}^{s_B} \underline{v}^T \frac{\partial \underline{v}}{\partial s} \, \mathrm{d}s = \frac{1}{2} \int_{s_A}^{s_B} \frac{\partial}{\partial s} \left(\underline{v}^T \underline{v}\right) \mathrm{d}s \tag{14.15}$$

The integral of a derivative of a function is the function itself. Therefore, the second integral in Equation (14.8) results in

$$\int_{s_A}^{s_B} \underline{s}^T \left(\underline{v}^T \underline{\nabla}\right) \underline{v} \, \mathrm{d}s = \frac{1}{2} \int_{s_A}^{s_B} \frac{\partial}{\partial s} \left(\underline{v}^T \underline{v}\right) \mathrm{d}s = \frac{1}{2} \left[\left(\underline{v}^T \underline{v}\right)\right]_{s_A}^{s_B} \tag{14.16}$$

Since the dot product of a vector with itself is equal to its length squared, the work per unit mass associated with the convective acceleration is

$$\int_{s_A}^{s_B} \underline{s}^T \left(\underline{v}^T \underline{\nabla}\right) \underline{v} \, \mathrm{d}s = \frac{1}{2} \left[|\underline{v}|^2\right]_{s_A}^{s_B} = \frac{1}{2} |\underline{v}_\mathrm{B}|^2 - \frac{1}{2} |\underline{v}_\mathrm{A}|^2 \tag{14.17}$$

$\underline{v}_\mathrm{A}$ being the velocity vector at point A and $\underline{v}_\mathrm{B}$ being the velocity vector at point B, respectively. The expression on the right-hand side is equivalent to the increase in kinetic energy per unit mass.

Next, we take a closer look at the integral of the body forces per unit mass. It is the first integral on the right-hand side of Equation (14.8). The gravity force is the only body force of note that we consider here. Thus, the vector of body forces $\underline{f}$ per unit mass is pointing in negative z-direction (z is positive upwards) with strength g (gravitational acceleration).

Study of third term

$$\int_{s_A}^{s_B} \underline{s}^T \underline{f}\, ds = \int_{s_A}^{s_B} (s_x, s_y, s_z) \begin{pmatrix} 0 \\ 0 \\ -g \end{pmatrix} ds = \int_{s_A}^{s_B} (-g\, s_z)\, ds \tag{14.18}$$

The product of the cosine s_z of the angle between path element and vertical axis and the length ds of the path elements is equal to the differential $dz = s_z\, ds$. The work per unit mass due to the gravity force then depends only on the vertical distance $(z_B - z_A)$ between points A and B, a fact you hopefully remember from your physics classes.

$$\int_{s_A}^{s_B} \underline{s}^T \underline{f}\, ds = -\int_{z_A}^{z_B} g\, dz = -g\,(z_B - z_A) \tag{14.19}$$

For the integral of the work done by pressure forces (surface force), we employ findings from previous integrals.

$$\begin{aligned} -\int_{s_A}^{s_B} \underline{s}^T \frac{1}{\rho} \underline{\nabla}\, p\, ds &= -\frac{1}{\rho} \int_{s_A}^{s_B} \underline{s}^T \underline{\nabla}\, p\, ds = -\frac{1}{\rho} \int_{s_A}^{s_B} \frac{\partial p}{\partial s}\, ds = -\frac{1}{\rho} \Big[p \Big]_{s_A}^{s_B} \\ &= -\frac{1}{\rho} (p_B - p_A) \end{aligned} \tag{14.20}$$

Finally, we substitute the results for the integrals (14.17), (14.19), and (14.20) into Equation (14.8).

Bernoulli equation for unsteady flow

$$\int_{s_A}^{s_B} \underline{s}^T \frac{\partial \underline{v}}{\partial t}\, ds + \frac{1}{2} |\underline{v}_B|^2 - \frac{1}{2} |\underline{v}_A|^2 = -g\,(z_B - z_A) - \frac{1}{\rho}(p_B - p_A) \tag{14.21}$$

We multiply the equation by the density, move terms depending on the initial position A to the right-hand side, and move terms depending on the final point B to the left. The result is known as the Bernoulli equation for unsteady flow.

$$\rho \int_{s_A}^{s_B} \underline{s}^T \frac{\partial \underline{v}}{\partial t}\, ds + \frac{1}{2} \rho\, |\underline{v}_B|^2 + \rho\, g\, z_B + p_B = \frac{1}{2} \rho\, |\underline{v}_A|^2 + \rho\, g\, z_A + p_A \tag{14.22}$$

This equation may be used, for instance, to compute the water hammer pressure which occurs in a flow line when a valve is rapidly closed.

Note that points A and B lie on the same streamline. The right-hand side of the unsteady Bernoulli equation (14.22) may conveniently be expressed as a constant.

$$\rho \int_{s_A}^{s_B} \underline{s}^T \frac{\partial \underline{v}}{\partial t}\, ds + \frac{1}{2} \rho\, |\underline{v}_B|^2 + \rho\, g\, z_B + p_B = C_A \tag{14.23}$$

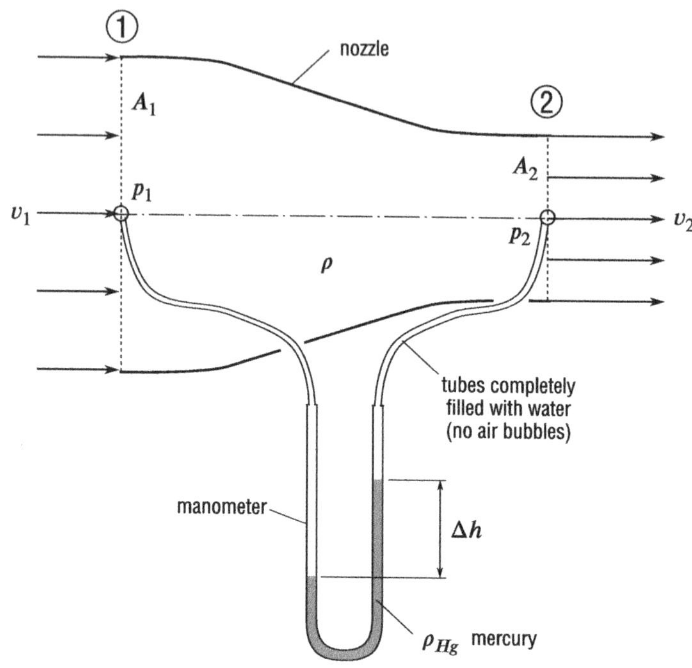

Figure 14.2 Determining the flow speed by measuring pressure difference in a contraction nozzle

The constant C_A usually changes from streamline to streamline.

Bernoulli equation for steady flow

The remaining path integral on the left-hand side vanishes for steady flow because $\partial \underline{v}/\partial t = 0$.

$$\frac{1}{2}\rho\,|\underline{v}_B|^2 + \rho\,g\,z_B + p_B = \frac{1}{2}\rho\,|\underline{v}_A|^2 + \rho\,g\,z_A + p_A \tag{14.24}$$

Equation (14.24) emphasizes the important result that the sum of dynamic, hydrostatic, and ambient pressure is constant along a streamline (inviscid flow).

$$\frac{1}{2}\rho\,|\underline{v}|^2 + \rho\,g\,z + p = \text{const.} \tag{14.25}$$

We will derive more specialized versions of the Bernoulli equation later.

Application of Bernoulli equation

The Bernoulli equation is an important tool in the evaluation of flow properties. Consider the flow through the contraction nozzle of a cavitation tunnel (Figure 14.2). We intend to determine the flow velocity from the difference in pressure between inlet and outlet. The circulation pump is moving water through the nozzle from left to right. We assume the flow is steady and incompressible. The tunnel is equipped with pressure taps before and after the nozzle. The taps are perpendicular to the flow direction. A u-tube manometer, filled with mercury or more environmentally friendly fluids, measures the pressure difference between inlet and outlet.

Conservation of mass

The conservation of mass principle dictates that mass fluxes are equal at inlet ① and ②. For steady, incompressible flow, the continuity equation for a fixed control volume requires that the net flow across the surface of the control volume vanishes.

$$\iint_S \rho\,(\underline{v}^T\underline{n})\,\mathrm{d}S = -v_1\,A_1 + v_2\,A_2 = 0 \qquad \text{or} \qquad v_1 = v_2\,\frac{A_2}{A_1} \tag{14.26}$$

Conservation of momentum

The Bernoulli equation is our representation of the conservation of momentum principle. We restate Equation (14.24) for the steady flow along a streamline between inlet and outlet.

$$\frac{1}{2}\rho\,|\underline{v}_1|^2 + \rho\,g\,z_1 + p_1 = \frac{1}{2}\rho\,|\underline{v}_2|^2 + \rho\,g\,z_2 + p_2 \qquad (14.27)$$

Along the center line of the tunnel hydrostatic pressure remains unchanged ($z_1 = z_2$). This eliminates the hydrostatic pressure term $\rho g z$ on both sides. We solve the remaining equation for the velocity v_2.

$$v_2 = \sqrt{\frac{2}{\rho}\,\frac{p_1 - p_2}{\left[1 - \left(\frac{A_2}{A_1}\right)^2\right]}} \qquad (14.28)$$

Pressure difference

The u-tube manometer measures the pressure difference $p_1 - p_2$. Hydrostatic equilibrium between left and right arm of the manometer requires that

$$p_1 = p_2 + \rho_{Hg}\,g\,\Delta h \quad \text{or} \quad p_1 - p_2 = \rho_{Hg}\,g\,\Delta h \qquad (14.29)$$

We substitute this result into Equation (14.28):

$$v_2 = \sqrt{\frac{\rho_{Hg}}{\rho}\,\frac{2\,g\,\Delta h}{\left[1 - \left(\frac{A_2}{A_1}\right)^2\right]}} \qquad (14.30)$$

The density of the fresh water in the tank is $\rho = 998.2072\ \text{kg/m}^3$ (for a temperature of 20 °C) and the density of mercury is $\rho_{Hg} = 13579.04\ \text{kg/m}^3$. Given a cavitation tunnel with a contraction ratio of $A_2/A_1 = 0.4$ and a reading of $\Delta h = 85.2$ mm mercury on the manometer, the velocity at the outlet ② will be

$$v_2 = \sqrt{\frac{13579.04}{998.2072}\cdot\frac{2\cdot 9.807\ \text{m/s}^2\ 0.0852\ \text{m}}{[1 - 0.4^2]}} \approx 5.202\ \text{m/s} \qquad (14.31)$$

High density ratios of gauge fluid and water (here ρ_{Hg}/ρ) enable the measurement of large pressure differences. Mercury is about 13.5 times heavier than water. However, mercury is a potent poison and an environmental hazard. Therefore, manometers filled with mercury have to be handled with great care. Colored oil might be used as an alternative gauge fluid.

14.3 Rotation, Vorticity, and Circulation

Fluid element movement

We study a differential fluid element moving with the flow. In general, a fluid element will change position and shape in four different ways:

1. translation

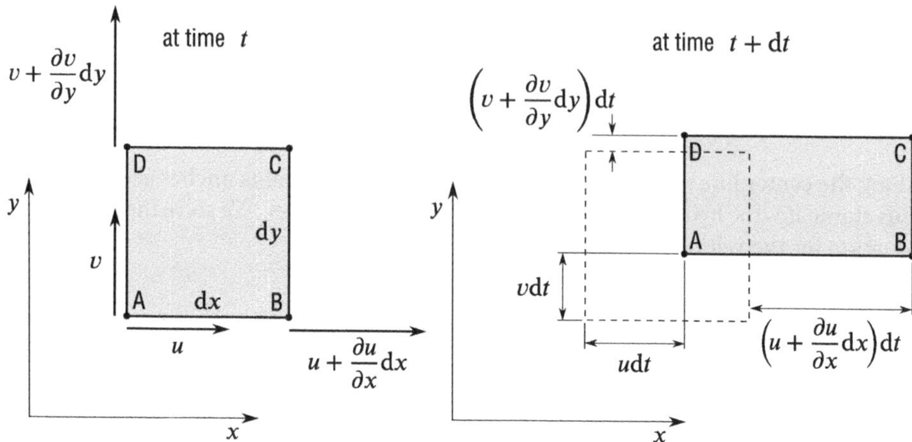

Figure 14.3 Translation and linear deformation of a fluid element

2. stretching (linear deformation)
3. rotation
4. shearing (angular deformation)

Translation and stretching

For better visualization, we momentarily restrict ourselves to planar (two-dimensional) flow. As the fluid element changes position in the x-y-plane (translation) it may also lengthen and shorten (stretching). Assume that the fluid element has the side lengths $\mathrm{d}x$ and $\mathrm{d}y$ and an area $\mathrm{d}S = \mathrm{d}x\mathrm{d}y$ at time t (Figure 14.3). Horizontal velocity at corner A will be u and $u + (\partial u/\partial x)\,\mathrm{d}x$ at corner B. The latter is again obtained by a Taylor series expansion which is truncated after the linear term.

The distances $u\mathrm{d}t$ and $v\mathrm{d}t$ express the translation of the fluid element as a whole. At time t+$\mathrm{d}t$, the length of side $\overline{\mathsf{AB}}$ will have changed from $\mathrm{d}x$ to the new length $\mathrm{d}x$+$(\partial u/\partial x)\,\mathrm{d}x\mathrm{d}t$ due to the difference in velocities at points A and B. In the vertical direction, the length of side $\overline{\mathsf{AD}}$ will change from $\mathrm{d}y$ to $\mathrm{d}y + (\partial v/\partial y)\,\mathrm{d}y\mathrm{d}t$ (Figure 14.3). For a moving fluid element filled with incompressible fluid (ρ = const.), conservation of mass requires that the area (planar flow, or volume for 3D-flow) of the element remains constant. If one side lengthens, the other side must shrink.

$$
\begin{aligned}
\text{area at time } t &= \text{area at time } t + \mathrm{d}t \\
\mathrm{d}x\mathrm{d}y &= \overline{\mathsf{AB}} \cdot \overline{\mathsf{AD}} \\
&= \left(\mathrm{d}x + \frac{\partial u}{\partial x}\mathrm{d}x\mathrm{d}t\right)\left(\mathrm{d}y + \frac{\partial v}{\partial y}\mathrm{d}y\mathrm{d}t\right)
\end{aligned}
$$

We expand the product on the right-hand side and divide by the area $\mathrm{d}S = \mathrm{d}x\mathrm{d}y$ on both sides. The remaining change of the surface area must vanish:

$$
0 = \frac{\partial u}{\partial x}\mathrm{d}x\mathrm{d}y\mathrm{d}t + \frac{\partial v}{\partial y}\mathrm{d}x\mathrm{d}y\mathrm{d}t + \frac{\partial u}{\partial x}\frac{\partial v}{\partial y}\mathrm{d}x\mathrm{d}y\mathrm{d}t^2
$$

The last term is proportional to $\mathrm{d}t^2$, i.e. it is of second order small. Since we already cut off higher order terms in the Taylor expansions for the velocities, it is only consequent

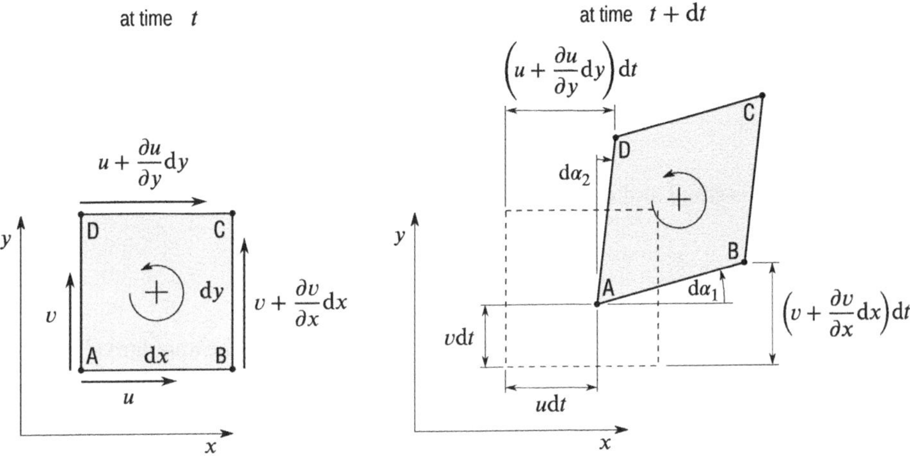

Figure 14.4 Rotation and angular deformation of a fluid element

to ignore this second order term as well. The remaining equation is divided by the area $dS = dxdy$, and we obtain the two-dimensional continuity equation for incompressible fluids.

$$\frac{\partial u}{\partial x} + \frac{\partial v}{\partial y} = 0 \tag{14.32}$$

This form of the continuity equation enforces that elongation in one direction is accompanied by shortening in another direction. As a result, area or volume stay constant. This is not necessarily the case for compressible fluids like air of course.

Rotation and shearing

Figure 14.4 explains the effects of changing velocities on rotation and angular deformation of the fluid element. Like the horizontal velocity u, the transverse velocity v may also change along the side $\overline{\mathsf{AB}}$. Over a short time period dt point B will rise above point A by the distance

$$\left(v + \frac{\partial v}{\partial x}dx\right)dt - v\,dt = \frac{\partial v}{\partial x}dx\,dt$$

As a result, side $\overline{\mathsf{AB}}$ rotates by the angle $d\alpha_1$ around point A with

$$\tan(d\alpha_1) = \frac{\frac{\partial v}{\partial x}dx\,dt}{dx} = \frac{\partial v}{\partial x}dt \approx d\alpha_1$$

For small angles $\sin(d\alpha_1) \approx d\alpha_1$ and $\cos(d\alpha_1) \approx 1$. Hence, $\tan(d\alpha_1) \approx d\alpha_1$. The angular velocity ω_1 of side $\overline{\mathsf{AB}}$ becomes

$$\omega_1 = \frac{d\alpha_1}{dt} = \frac{\partial v}{\partial x} \tag{14.33}$$

Side $\overline{\mathsf{AD}}$ of the fluid element is moving in x direction. If point D is moving faster than point A, side $\overline{\mathsf{AD}}$ will turn clockwise around point A (Figure 14.4). Since we used the positive sign for the anti-clockwise rotation of side $\overline{\mathsf{AB}}$, clockwise rotation is negative.

Side $\overline{\mathsf{AD}}$ rotates by the angle $\mathrm{d}\alpha_2$ with

$$\tan(\mathrm{d}\alpha_2) = -\frac{\frac{\partial u}{\partial y}\mathrm{d}y\,\mathrm{d}t}{\mathrm{d}y} = -\frac{\partial u}{\partial y}\mathrm{d}t$$

We assume $\mathrm{d}\alpha_2$ is small and obtain the angular velocity of side $\overline{\mathsf{AD}}$

$$\omega_2 = \frac{\mathrm{d}\alpha_2}{\mathrm{d}t} = -\frac{\partial u}{\partial y} \tag{14.34}$$

Rotation We define the mean value of the angular velocities ω_1 and ω_2 as the angular velocity of the element. The axis of rotation is perpendicular to the x-y-plane and parallel to the z-axis.

$$\omega_z = \frac{1}{2}(\omega_1 + \omega_2) = \frac{1}{2}\left(\frac{\partial v}{\partial x} - \frac{\partial u}{\partial y}\right) \tag{14.35}$$

Considering the rotation of fluid elements in the y-z-plane and z-x-plane respectively, leads to angular velocities about the x- and y-axes.

$$\omega_x = \frac{1}{2}\left(\frac{\partial w}{\partial y} - \frac{\partial v}{\partial z}\right) \tag{14.36}$$

$$\omega_y = \frac{1}{2}\left(\frac{\partial u}{\partial z} - \frac{\partial w}{\partial x}\right) \tag{14.37}$$

Summarizing Equations (14.35) through (14.37) yields the general vector of angular velocity $\underline{\omega}$ for a three-dimensional fluid element $\mathrm{d}V$.

$$\underline{\omega} = \frac{1}{2}\begin{pmatrix}\omega_x\\ \omega_y\\ \omega_z\end{pmatrix} = \frac{1}{2}\begin{pmatrix}\frac{\partial w}{\partial y} - \frac{\partial v}{\partial z}\\ \frac{\partial u}{\partial z} - \frac{\partial w}{\partial x}\\ \frac{\partial v}{\partial x} - \frac{\partial u}{\partial y}\end{pmatrix} = \frac{1}{2}(\underline{\nabla} \times \underline{v}) = \frac{1}{2}\,\mathrm{rot}\,\underline{v} \tag{14.38}$$

The rotation of the velocity rot $\underline{v}$ is also known as curl $\underline{v}$.

Vorticity The vector of vorticity $\underline{\gamma}$ is defined as twice the angular velocity.

$$\underline{\gamma} = 2\underline{\omega} = \underline{\nabla} \times \underline{v} = \mathrm{rot}\,\underline{v} \tag{14.39}$$

According to Helmholtz's theorems of fluid mechanics, vorticity cannot change in an inviscid fluid (Helmholtz, 1978). More information on Helmholtz's theorems is provided in Section 40.1. Generation and dissipation of vorticity may only be explained by viscous flow theory.

Circulation This leads us to the concept of circulation. Circulation Γ is defined as the line integral of the projection of the velocity $\underline{v}$ onto a closed path C (Figure 14.5). The projection is computed as a dot product and, consequently, a scalar quantity.

$$\Gamma = \oint_C \underline{v}^T \underline{s}\,\mathrm{d}s \tag{14.40}$$

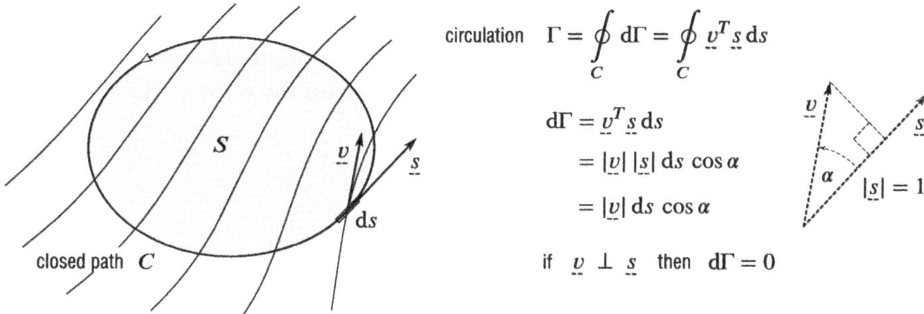

Figure 14.5 Definition of circulation Γ

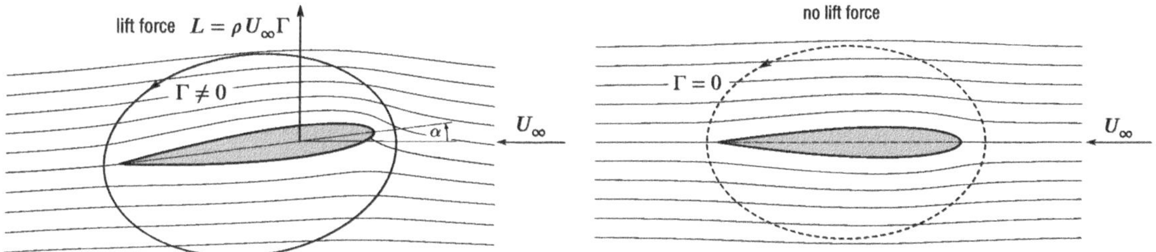

Figure 14.6 Symmetric foil with lifting flow ($\Gamma \neq 0$) and nonlifting flow ($\Gamma = 0$)

The integral is similar to the work integral which we discussed at the beginning of the chapter.

Stokes' integral theorem reveals a connection between circulation and vorticity:

$$\Gamma = \oint_C \underline{v}^T \underline{s}\, ds = \iint_S (\underline{\nabla} \times \underline{v})^T \underline{n}\, dS = \iint_S \underline{\gamma}^T \underline{n}\, dS \tag{14.41}$$

Circulation is equal to the vorticity integrated over the surface S enclosed by the path C.

At first glance, circulation Γ appears to be a rather abstract, mathematical concept. However, circulation is important in potential theory and for the description of lifting flows as they occur on foils, e.g. rudders and propeller blades (Figure 14.6). An asymmetric foil with camber, or a symmetric foil at a nonzero angle of attack, produces a lifting force perpendicular to the onflow. According to Kutta-Joukowsky's lift theorem, the lift force L is proportional to density ρ of the fluid, onflow velocity U_∞, and the circulation Γ. A symmetric foil at zero angle of attack will not develop a lift force, and the circulation is zero. We will come back to this when we discuss foil theory and propellers.

References

Euler, L. (1757). Principes généraux du mouvement des fluides. In *Mémoires de l'Académie des Sciences de Berlin*, volume 11, pages 274–315.

Helmholtz, H. (1978). On the integrals of the hydrodynamic equations that correspond to vortex motions. Translation by Uwe Parpart. *Int. Journal of Fusion Energy*, 1(3–4):41–68. Original published in German in Journal für die reine und angewandte Mathematik, 56:25–55, 1858.

Self Study Problems

1. Show that the vector identity of Equation (14.3) is true.

$$\left(\underline{v}^T \underline{\nabla}\right)\underline{v} = \frac{1}{2}\underline{\nabla}\left(\underline{v}^T \underline{v}\right) - \underline{v} \times \left(\underline{\nabla} \times \underline{v}\right) \tag{14.3}$$

2. Explain the difference between the Euler and the Navier-Stokes equations.
3. Which important function does the Bernoulli equation serve?
4. State the conditions under which the Bernoulli equation (14.24) may be applied.
5. Define and explain circulation.
6. Consider the following velocity field:

$$\underline{v} = \frac{\gamma}{2\pi}\begin{pmatrix} \dfrac{-y}{\left[x^2 + y^2\right]} \\ \dfrac{x}{\left[x^2 + y^2\right]} \\ 0 \end{pmatrix}$$

   Compute the vector of vorticity $\underline{\gamma}$.

15

Potential Flow

In this section we discuss potential flow. Although viscous effects are ignored, it has many practical applications. Potential flow is used to determine basic properties of hydrofoils and lifting wings. That includes keels, rudders, and propellers. It is also employed in the prediction of cavitation inception and in seakeeping. We start this section by introducing the concept of the velocity potential. The velocity potential is then used to derive the important Laplace and Bernoulli equations from the principles of conservation of mass and momentum.

Learning Objectives

At the end of this chapter students will be able to:

- explain the concept of a velocity potential
- apply the principles of conservation of mass and conservation of momentum to potential flow
- predict velocities and pressure for potential flow
- identify limitations of potential flow

15.1 Velocity Potential

In Chapter 14, we introduced the circulation as the integral of the velocity vector along a closed path through the flow field. **Circulation**

$$\Gamma = \oint_C \underline{v}^T \underline{s}\, \mathrm{d}s \tag{14.41}$$

The dot product between velocity vector and path element $\underline{v}^T \underline{s}\mathrm{d}s$ looks similar to the dot product of force vector and path element when work is computed.

The work to move an object from point A to point B is determined by integration of the **Work**

Fundamentals of Ship Hydrodynamics: Fluid Mechanics, Ship Resistance and Propulsion, First Edition.
Lothar Birk.

Companion website: www.wiley.com/go/birk/hydrodynamics

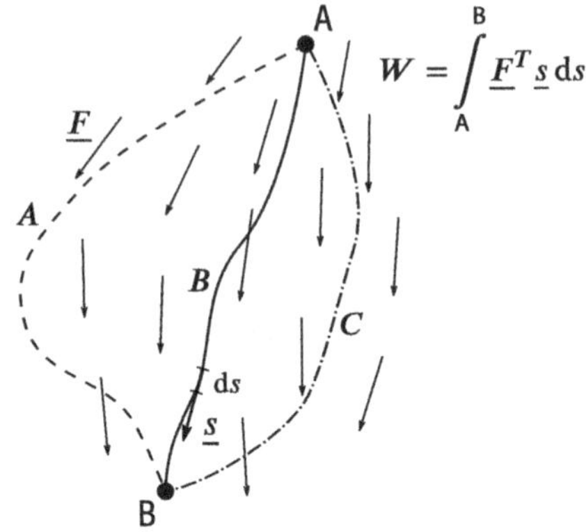

Figure 15.1 The work spent on moving an object from point A to point B

external force $\underline{F}$ acting on the object along the path from A to B.

$$W = \int_{\mathrm{A}}^{\mathrm{B}} \underline{F}^T \underline{s}\, \mathrm{d}s \tag{15.1}$$

Again, vector $\underline{s}$ represents the unit length tangent vector to the path. A major difference to the definition of circulation is that the path for the work does not have to be closed. As the dot product between force $\underline{F}$ and tangent vector indicates, only the component of the force vector pointing in the tangent direction is contributing to the work. No work is done when the force is perpendicular to the path.

Conservative force field

Usually, we prefer to move objects along the shortest path from A to B to conserve energy. Sometimes a longer path is used because the required force might be too large otherwise. Mountain roads have switchbacks because your car engine would not be strong enough to push the car directly up the hill. Does that mean you spend more energy following the switchbacks? Not really, because the car works in a *conservative* force field. In conservative force fields, like Earth's gravity, the work done in moving an object from A to B becomes independent of the path. Assuming the force field $\underline{F}$ depicted in Figure 15.1 is conservative, the work integrals for paths A, B, and C will be equal.

$$W = \int_{\mathrm{A}}^{\mathrm{B}} \underline{F}^T \underline{s}\, \mathrm{d}s = W_A = W_B = W_C \tag{15.2}$$

As a consequence, the work W is only a function of the location of the end points A and B. To be fair, in our mountain road analogy we probably spend some more energy on the switchback road due to the longer exposure to roll and wind resistance. However, the work related to just lifting the car is the same for the steep hill climb and the switchbacks.

Potential

The capability of a conservative force field to perform work is expressed by its potential. Like work, potential is a scalar function. Work is equal to the difference in potential between point B and point A.

$$W = E_{\mathrm{B}} - E_{\mathrm{A}} \tag{15.3}$$

In Earth's gravity field, the potential for an object of mass m is

$$E = -g\, z\, m \tag{15.4}$$

The minus sign stems from the chosen positive z-direction (positive upwards). The work related to lifting the car will be

$$W = E_\mathrm{B} - E_\mathrm{A} = -g\, z_\mathrm{B}\, m - (-g\, z_\mathrm{A}\, m) = -g\, m\left(z_\mathrm{B} - z_\mathrm{A}\right) \tag{15.5}$$

The value is negative if $z_\mathrm{B} > z_\mathrm{A}$ because we have to work against the force field to lift the car higher. If it runs downhill, the work is done by the force field and is positive. Note that the amount of work follows from the change in potential energy between A and B. The absolute size of the potential is irrelevant, there is no difference in work if you lift the car from 2 m to 4 m or from 1002 m to 1004 m.

The rate of change in potential is equal to the vector field. If we take the gradient of the potential (15.4), we obtain the gravity force acting on mass m.

$$\underline{\nabla}\, E = \begin{pmatrix} \dfrac{\partial E}{\partial x} \\ \dfrac{\partial E}{\partial y} \\ \dfrac{\partial E}{\partial z} \end{pmatrix} = \begin{pmatrix} \dfrac{\partial(-g\, m\, z)}{\partial x} \\ \dfrac{\partial(-g\, m\, z)}{\partial y} \\ \dfrac{\partial(-g\, m\, z)}{\partial z} \end{pmatrix} = \begin{pmatrix} 0 \\ 0 \\ -g\, m \end{pmatrix} \tag{15.6}$$

We will come back to this important relationship between a vector field and its potential.

Sufficient condition

Not all force fields are conservative. So you may ask, when is a force field $\underline{F}$ conservative and when does a potential exist? Mathematically speaking, a force field is conservative and a potential exists, if the rotation[1] of the vector field vanishes in a simply connected region (Kellogg, 1929).

$$\mathrm{rot}(\underline{F}) = \underline{\nabla} \times \underline{F} = \underline{0} \tag{15.7}$$

Such a vector field is called irrotational. Equation (15.7) is a sufficient condition, i.e. if it holds true, the vector field $\underline{F}$ is conservative. If $\underline{\nabla} \times \underline{F} \neq \underline{0}$, we cannot draw a conclusion. Additional checks will be necessary to determine whether a potential exists or not.

Simply and multiply connected regions

A simply connected region is a space where all closed curves can be contracted to a point without leaving the region. Study Figure 15.2: In a simply connected region a sling can be pulled tight to any point within the region without any part of the sling leaving the gray shaded area. In Figure 15.2(a) no part of the loop leaves the gray shaded area during the tightening of the sling. If the region contains an object like a foil, any sling which encompasses the object can no longer be reduced to a point. This situation is depicted in Figure 15.2(b). The closest you get is the dotted line. Such a region is called multiply connected. Any region that does not contain the object is still simply connected.

Let us switch back to fluid mechanics. We started this section by pointing out that the definition of the circulation (14.41) looks similar to the work integral (15.1). However, the path C is always closed and the force field $\underline{F}$ is replaced with the velocity field $\underline{v}$.

Velocity potential

We apply the concept of a potential to the velocity field. A flow field is called irrotational if it satisfies equation (15.8) in all simply connected regions.

$$\mathrm{rot}(\underline{v}) = \underline{\nabla} \times \underline{v} = \underline{0} \tag{15.8}$$

[1] See Section 4.2.3 for an introduction to rotation, also called curl

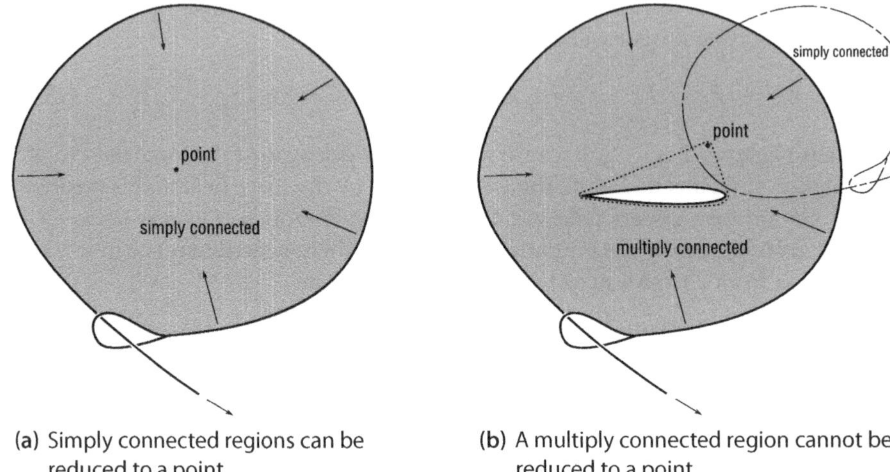

(a) Simply connected regions can be reduced to a point

(b) A multiply connected region cannot be reduced to a point

Figure 15.2 Definition of simply and multiply connected regions

A velocity potential ϕ will exist under this condition.

Potential flow A flow field represented by a velocity potential is a potential flow. The only formal condition is that the flow is irrotational. In practice, this usually implies that the fluid is also inviscid. Shear stresses in the fluid will cause rotation.

Analog to the relationship (15.6) between weight force and the potential of the gravity field, the velocity vector is equal to the gradient of the velocity potential

$$\underline{v} = \underline{\nabla}\,\phi \tag{15.9}$$

This relationship is extremely useful. For a potential flow, we may replace the velocity vector which has three unknown components with the yet unknown scalar potential leading to a reduction in equations and unknowns to solve for.

Two basic examples may illustrate this reduction.

Parallel flow (i) Consider the following velocity potential:

$$\phi(x, y, z) = -U x \tag{15.10}$$

The velocity potential has the physical units m^2/s. Taking the gradient yields the velocity vector.

$$\underline{v}(\underline{x}) = \underline{v}(x, y, z) = \underline{\nabla}\,\phi = \begin{pmatrix} \frac{\partial}{\partial x}\left(-Ux\right) \\ \frac{\partial}{\partial y}\left(-Ux\right) \\ \frac{\partial}{\partial z}\left(-Ux\right) \end{pmatrix} = \begin{pmatrix} -U \\ 0 \\ 0 \end{pmatrix} \tag{15.11}$$

At every point $\underline{x} = (x, y, z)^T$ in the flow field, the velocity in x-direction is equal to $-U$ and zero in all other directions. This represents parallel flow in negative

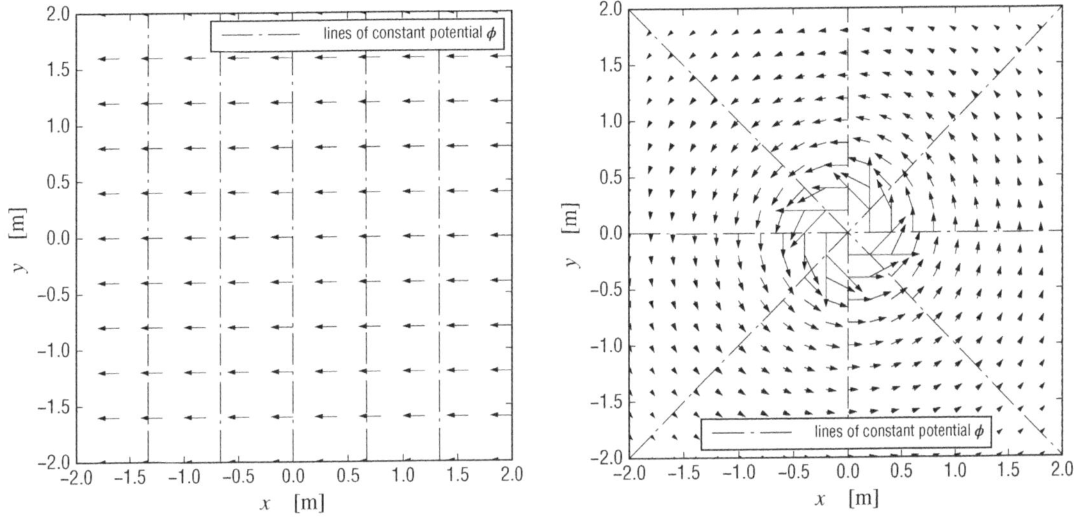

(a) Parallel flow $\phi = -Ux$ in the x-y-plane

(b) Vortex flow $\phi = \gamma/(2\pi)\arctan(y/x)$ in the x-y-plane

Figure 15.3 Examples of basic potential flows

x-direction. Figure 15.3(a) shows the velocity field and selected isolines of the potential ($\phi =$ const.).

(ii) The next example is a bit more complex. **Vortex flow**

$$\phi(x, y, z) = \frac{\gamma}{2\pi} \arctan\left(\frac{y}{x}\right) \tag{15.12}$$

The flow field described by this potential has this velocity distribution:

$$\begin{aligned} \underline{v}(x, y, z) &= \underline{\nabla}\,\phi \\ &= \frac{\gamma}{2\pi}\begin{pmatrix} \frac{\partial}{\partial x}\left[\arctan\left(\frac{y}{x}\right)\right] \\ \frac{\partial}{\partial y}\left[\arctan\left(\frac{y}{x}\right)\right] \\ \frac{\partial}{\partial z}\left[\arctan\left(\frac{y}{x}\right)\right] \end{pmatrix} = \frac{\gamma}{2\pi}\begin{pmatrix} \frac{-y}{x^2+y^2} \\ \frac{x}{x^2+y^2} \\ 0 \end{pmatrix} \end{aligned} \tag{15.13}$$

The strength of the vortex γ is given in units of m^2/s.

At points $\underline{x}_1 = \left(1, 0, 0\right)$ m, $\underline{x}_2 = \left(0, 1, 0\right)$ m, and $\underline{x}_3 = \left(1, 1, 0\right)$ m the velocity vectors are

$$\underline{v}_1(\underline{x}_1) = \frac{\gamma}{2\pi}\begin{pmatrix} 0\,\frac{1}{\mathrm{m}} \\ 1\,\frac{1}{\mathrm{m}} \\ 0\,\frac{1}{\mathrm{m}} \end{pmatrix} \qquad \underline{v}_2(\underline{x}_2) = \frac{\gamma}{2\pi}\begin{pmatrix} -1\,\frac{1}{\mathrm{m}} \\ 0\,\frac{1}{\mathrm{m}} \\ 0\,\frac{1}{\mathrm{m}} \end{pmatrix}$$

$$\underline{v}_3(\underline{x}_3) = \frac{\gamma}{2\pi} \begin{pmatrix} -\frac{1}{2}\frac{1}{\mathrm{m}} \\ \frac{1}{2}\frac{1}{\mathrm{m}} \\ 0\frac{1}{\mathrm{m}} \end{pmatrix}$$

When we evaluate the velocity vectors at many points, we see that the fluid moves in concentric circles around the z-axis (Figure 15.3(b)). The magnitude of the velocity declines rapidly as we move further away from the origin.

At the origin $(x, y)^T = (0, 0)$ m, we cannot evaluate the velocity. The derivatives $\partial\phi/\partial x$ and $\partial\phi/\partial y$ of the potential do not exist, and the velocity components become infinite:

$$u, v \longrightarrow (-\infty, \infty) \qquad \text{for } (x, y)^T \longrightarrow (0, 0)$$

A point at which a function approaches infinity is called a singular point or, succinctly, a singularity. Therefore, we call the vortex potential a singularity. We will discuss more of these special velocity potentials in the following chapter.

Velocity is perpendicular to isolines of potential

The gradient of a function always points in the direction of its steepest ascent and is always perpendicular to isolines of the function. Therefore, the velocity – the gradient of a potential – is always perpendicular to the lines of constant potential (see Figure 15.3).

15.2 Circulation and Velocity Potential

Before we study the equations that allow us to actually compute a potential, we explore the somewhat confusing connection between circulation Γ and velocity potential ϕ.

No circulation – no lift force

We started this chapter recapping the definition of circulation Γ:

$$\Gamma = \oint_C \underline{v}^T \underline{s}\, \mathrm{d}s = \oint_C \mathrm{d}\Gamma \tag{14.41}$$

We also mentioned previously that the lift force L generated by hydrofoils, sails, etc. is proportional to the circulation Γ. This is known as Kutta-Joukowsky's lift theorem (Moran, 1984).

$$L = \rho\, U\, \Gamma \qquad \text{(2D, force per unit length)} \tag{15.14}$$

Thus, vanishing circulation $\Gamma = 0$ means there is no lift force L either.

With Stokes' integral theorem (see Equation (14.41) on page 175), we also found a relationship between circulation and rotation (curl) of the velocity vector.

$$\Gamma = \oint_C \underline{v}^T \underline{s}\, \mathrm{d}s = \iint_S (\underline{\nabla} \times \underline{v})^T \underline{n}\, \mathrm{d}S = \iint_S (\mathrm{rot}\, \underline{v})^T \underline{n}\, \mathrm{d}S \tag{14.41}$$

The last integral of Equation (14.41) above will vanish in potential flow fields because $\mathrm{rot}\, \underline{v} = \underline{0}$ in simply connected regions. Therefore, the circulation is zero for all paths C enclosing a simply connected region S.

$$\Gamma = \oint_C \mathrm{d}\Gamma = 0 \qquad \text{for potential flow (simply connected)} \tag{15.15}$$

So, how can we use potential flow to describe lifting flows around hydrofoils?

The answer to that question lies in the difference between simply and multiply connected regions. Consider the differential line element in the path integral of Equation (14.41):

$$\underline{s}\,\mathrm{d}s = (\mathrm{d}x, \mathrm{d}y, \mathrm{d}z)^T \tag{15.16}$$

A comparison of the integrands in (14.41) above enables us to state the total differential $\mathrm{d}\Gamma$ of the circulation:

$$\mathrm{d}\Gamma = \underline{v}^T\underline{s}\,\mathrm{d}s = u\mathrm{d}x + v\mathrm{d}y + w\mathrm{d}z \tag{15.17}$$

Total differentials $\mathrm{d}\Gamma$ and $\mathrm{d}\phi$

For potential flow, the components u, v, and w of the velocity vector may be replaced with the spatial derivatives of the velocity potential.

$$\mathrm{d}\Gamma = \frac{\partial \phi}{\partial x}\mathrm{d}x + \frac{\partial \phi}{\partial y}\mathrm{d}y + \frac{\partial \phi}{\partial z}\mathrm{d}z \tag{15.18}$$

The term on the right-hand side also represents the definition of the total differential for the potential ϕ.

$$\mathrm{d}\phi = \frac{\partial \phi}{\partial x}\mathrm{d}x + \frac{\partial \phi}{\partial y}\mathrm{d}y + \frac{\partial \phi}{\partial z}\mathrm{d}z = \mathrm{d}\Gamma \quad \text{(potential flow)} \tag{15.19}$$

Thus, the total differentials of circulation and potential are identical for potential flow: $\mathrm{d}\phi = \mathrm{d}\Gamma$.

Combining the findings of Equations (15.19) and (15.15) with the definition of circulation (14.41) yields:

$$\Gamma = \oint_C \mathrm{d}\Gamma = \oint_C \mathrm{d}\phi = 0 \qquad \text{for potential flow (simply connected)} \tag{15.20}$$

A definite integral is equal to the difference of antiderivative values at the limits.

$$\int_{\mathrm{A}}^{\mathrm{B}} \mathrm{d}\phi = \phi_{\mathrm{B}} - \phi_{\mathrm{A}} \tag{15.21}$$

Start and end point will be equal for a closed path A=B. Employing this in Equation (15.20) yields

$$\Gamma = \oint_{\mathrm{A}}^{\mathrm{B=A}} \mathrm{d}\phi = \phi_{\mathrm{B=A}} - \phi_{\mathrm{A}} = 0 \tag{15.22}$$

Therefore, the potential must be single valued in a simply connected region:

$$\phi_{\mathrm{A}} = \phi_{\mathrm{B=A}} \qquad \text{for a simply connected region} \tag{15.23}$$

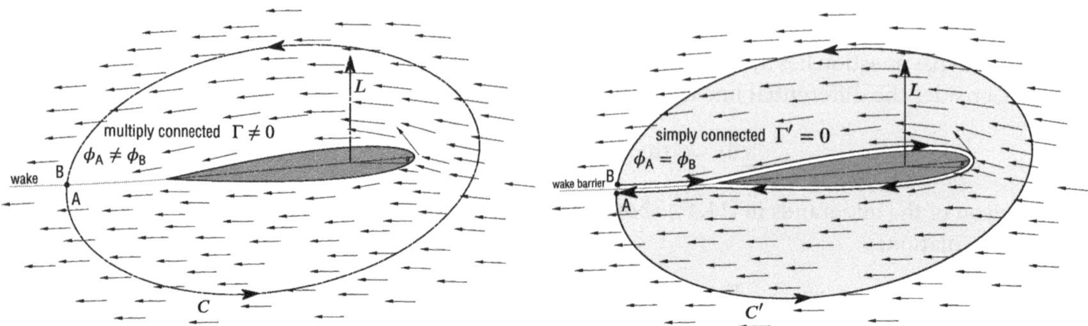

(a) Path C encloses the foil. The region is multiply connected and the circulation does not vanish

(b) Path C' excludes the foil. The region is simply connected and the circulation vanishes

Figure 15.4 Flow field around a symmetric foil at angle of attack α

Lifting foil Let's apply this to the lifting flow around a foil. Figure 15.4(a) shows a path C enclosing a foil. The symmetric foil is operating at a small angle of attack α. We know from experience and observation that the foil will generate a lift force L perpendicular to the onflow. The presence of a lift force $L \neq 0$ requires that $\Gamma \neq 0$. Consequently, the potential at points A=B cannot be single valued.

$$\Gamma = \phi_B - \phi_A \neq 0 \qquad \text{for a multiply connected region} \tag{15.24}$$

This is not a contradiction of potential flow theory because the region enclosed by the path C is multiply connected with the foil in it.

A simply connected region is constructed by extending the path to exclude the foil (Figure 15.4(b)). The extended path C' follows the wake streamline from point B to the trailing edge of the foil, then wraps around the foil and returns along the wake streamline to point A. Since fluid does not cross the wake streamline this artificial barrier will not disrupt the flow. In contrast to the figure, there is no distance between the path elements above and below the wake barrier. Integration of $d\Gamma$ along path C' will find $\Gamma' = 0$ because the region enclosed by C' is simply connected.

15.3 Laplace Equation

Velocity potentials are a powerful tool for the description of a large class of flows, so-called potential flows. The application of the conservation of mass and conservation of momentum principles for Newtonian fluids led us to the continuity equation and Navier-Stokes equations (see Chapters 5 and 6). We restricted ourselves to fluids with constant density $\rho = \text{const.}$, which we call incompressible. The differential conservative forms of continuity equation and Navier-Stokes equations are (see Chapter 7).

Viscous, incompressible flow

$$\underline{\nabla}^T \underline{v} = 0 \qquad \text{continuity equation, } \rho = \text{const.} \tag{5.12}$$

$$\frac{\partial \underline{v}}{\partial t} + \left(\underline{v}^T \underline{\nabla}\right) \underline{v} = \underline{f} - \frac{1}{\rho} \underline{\nabla} p + \nu \boldsymbol{\Delta} \underline{v} \qquad \text{NSE, } \rho = \text{const.} \tag{7.12}$$

Outside the boundary layer, viscous effects are small and may often be neglected. This simplification leads to the Euler equation (see Chapter 14). The continuity equation does not change of course.

Inviscid, incompressible flow

$$\underline{\nabla}^T \underline{v} = 0 \qquad \text{continuity equation, } \rho = \text{const.} \tag{5.12}$$

$$\frac{\partial \underline{v}}{\partial t} + \left(\underline{v}^T \underline{\nabla}\right)\underline{v} = \underline{f} - \frac{1}{\rho}\underline{\nabla}\, p \qquad \text{Euler eq., } \nu = 0,\ \rho = \text{const.} \tag{14.2}$$

For potential flow, we add the restriction that the flow must be irrotational in simply connected regions: rot $\underline{v} = \underline{0}$. In that case, the velocity vector may be replaced by the gradient of a velocity potential. We substitute $\underline{v} = \underline{\nabla}\,\phi$ into the continuity equation (5.12).

Inviscid, incompressible, irrotational flow

$$\begin{aligned} \underline{\nabla}^T \underline{v} &= 0 \\ \underline{\nabla}^T \underline{\nabla}\,\phi &= 0 \\ \boldsymbol{\Delta}\phi &= 0 \end{aligned} \tag{15.25}$$

The dot product of the Nabla operator with itself results in the Laplace operator, which has been introduced in Section 4.2.4. Rewriting the last equation in Cartesian coordinates yields the important Laplace equation.

$$\frac{\partial^2 \phi}{\partial x^2} + \frac{\partial^2 \phi}{\partial y^2} + \frac{\partial^2 \phi}{\partial z^2} = 0 \qquad \text{Laplace equation} \tag{15.26}$$

Laplace equation

The Laplace equation forms the basis of potential theory. Besides in fluid mechanics, potential theory is used in other fields of physics, like gravity, electrodynamics, and magnetism. In mathematics, potential theory and the Laplace equation are closely linked to harmonic functions. We will get back to this when we discuss wave theories.

Properties of Laplace equation

Its favorable properties make the Laplace equation so important. The Laplace equation is

(i) a partial differential equation (PDE)

(ii) of second order (derivatives of second order),

(iii) homogeneous (because the right-hand side is zero), and

(iv) linear.

The linearity (iv) of the Laplace equation is its most consequential property. Do not get distracted by the 'exponents' of 2. They indicate second order derivatives and not that the terms are multiplied by themselves. Note that

$$\frac{\partial^2 \phi}{\partial x^2} \neq \left(\frac{\partial \phi}{\partial x}\right)^2$$

Laplace equation is linear

Any linear combination of two or more solutions of a linear equation is again a solution of the equation. Specifically, if ϕ_1 and ϕ_2 are solutions of the Laplace equation (15.26), i.e.

$$\boldsymbol{\Delta}\phi_1 = 0 \qquad \text{and} \qquad \boldsymbol{\Delta}\phi_2 = 0$$

then the function ϕ formed by a linear combination of ϕ_1 and ϕ_2

$$\phi = a_1\phi_1 + a_2\phi_2$$

will also satisfy the Laplace equation. The finite valued constants a_1 and a_2 have to be independent of the spatial coordinates x, y, and z.

$$\boldsymbol{\Delta}\phi = \boldsymbol{\Delta}\left(a_1\phi_1 + a_2\phi_2\right) = a_1\boldsymbol{\Delta}\phi_1 + a_2\boldsymbol{\Delta}\phi_2 = 0 \tag{15.27}$$

Laplace equation is homogeneous

The Laplace equation is also homogeneous (iii). A homogeneous equation has an infinite number of solutions. In order to find the velocity potential for a particular flow problem, we will need boundary conditions in addition to the Laplace equation. Examples will be presented in Chapters 17 and 20.

Polar coordinates

In many practical applications the Laplace equation is stated in polar, cylindrical, or spherical coordinates. Planar flows make use of polar coordinates with the transformations

$$(x, y) \longrightarrow (r, \varphi) \quad \text{with} \quad x = r\cos(\varphi),\ y = r\sin(\varphi) \text{ and } \sqrt{x^2 + y^2} = r \tag{15.28}$$

Laplace equation in polar coordinates

The variable r measures distance from the origin, and φ measures the angle between the vector $(x, y)^T$ and the positive x-axis. Performing the variable substitution in the Laplace equation results in

$$\frac{\partial^2\phi}{\partial r^2} + \frac{1}{r}\frac{\partial\phi}{\partial r} + \frac{1}{r^2}\frac{\partial^2\phi}{\partial\varphi^2} = 0 \tag{15.29}$$

The velocity vector in polar coordinates is derived from the gradient in polar coordinates.

$$\underline{v}(r, \varphi) = \begin{pmatrix} \dfrac{\partial\phi}{\partial r} \\ \dfrac{1}{r}\dfrac{\partial\phi}{\partial\varphi} \end{pmatrix} \tag{15.30}$$

Note that the components of $\underline{v}(r, \varphi)$ are not pointing in x- and y-direction. The first component points radially away from the origin, and the second component is normal to the radial.

For axially symmetric three-dimensional flows, or flows that rotate around a common axis, cylindrical coordinates might be helpful. The transformation of Cartesian coordinates into cylindrical coordinates is accomplished by adding the longitudinal z-axis to the planar polar coordinates:

$$\begin{aligned} &(x, y, z) \longrightarrow (r, \varphi, z) \\ &\text{with} \quad x = r\cos(\varphi),\ y = r\sin(\varphi) \quad \text{and} \quad z = z \end{aligned} \tag{15.31}$$

Laplace equation in cylindrical coordinates

The Laplace equation becomes

$$\frac{\partial^2\phi}{\partial r^2} + \frac{1}{r}\frac{\partial\phi}{\partial r} + \frac{1}{r^2}\frac{\partial^2\phi}{\partial\varphi^2} + \frac{\partial^2\phi}{\partial z^2} = 0 \tag{15.32}$$

The corresponding velocity vector in cylindrical coordinates is

$$\underline{v}(r,\varphi,z) = \begin{pmatrix} \dfrac{\partial\phi}{\partial r} \\ \dfrac{1}{r}\dfrac{\partial\phi}{\partial\varphi} \\ \dfrac{\partial\phi}{\partial z} \end{pmatrix} \tag{15.33}$$

For point symmetric spatial flows, spherical coordinates may be used.

$$\begin{aligned} &(x,y,z) \longrightarrow (r,\varphi,\theta) \\ &\quad\text{with}\quad x = r\sin(\theta)\cos(\varphi),\ \ y = r\sin(\theta)\cos(\varphi)\quad\text{and}\quad z = r\cos(\theta) \end{aligned} \tag{15.34}$$

The first angle φ (azimuth) describes the longitude with $\varphi \in [0, 2\pi]$ and is measured with respect to the positive x-axis. The second angle defines the latitude and is measured from the positive z-axis: $\theta \in [0, \pi]$ with $\theta = 0$ at the North pole, $\theta = \pi/2$ at the equator, and $\theta = \pi$ at the South pole of the sphere.

Laplace equation in spherical coordinates

The Laplace equation now reads

$$\frac{\partial^2\phi}{\partial r^2} + \frac{2}{r}\frac{\partial\phi}{\partial r} + \frac{1}{r^2\sin^2(\theta)}\frac{\partial^2\phi}{\partial\varphi^2} + \frac{1}{r^2}\frac{\partial^2\phi}{\partial\theta^2} + \frac{1}{r^2}\frac{\cos(\theta)}{\sin(\theta)}\frac{\partial\phi}{\partial\theta} = 0 \tag{15.35}$$

The corresponding velocity vector in spherical coordinates is

$$\underline{v}(r,\varphi,\theta) = \begin{pmatrix} \dfrac{\partial\phi}{\partial r} \\ \dfrac{1}{r\sin(\theta)}\dfrac{\partial\phi}{\partial\varphi} \\ \dfrac{1}{r}\dfrac{\partial\phi}{\partial\theta} \end{pmatrix} \tag{15.36}$$

Laplace eq. represents conservation of mass

The Laplace equation is derived from the continuity equation for incompressible fluids. For irrotational flow, the velocity vector is replaced by the gradient of the velocity potential. In conclusion, the Laplace equation represents the conservation of mass principle for incompressible, inviscid, and irrotational flow. We did not specifically use the assumption of inviscid fluid ($\nu = 0$) in its derivation. However, a viscous fluid is usually never really irrotational.

If the Laplace equation replaces the continuity equation, what happens to the conservation of momentum equations? We will discuss this next.

15.4 Bernoulli Equation for Potential Flow

In Section 14.2, we integrated the Euler equation along a streamline and obtained the important Bernoulli equation (14.22).

We revisit the alternative form of the Euler equation for incompressible flow.

$$\frac{\partial \underline{v}}{\partial t} + \frac{1}{2}\underline{\nabla}(\underline{v}^T \underline{v}) - \underline{v} \times (\underline{\nabla} \times \underline{v}) = \underline{f} - \frac{1}{\rho}\underline{\nabla} p \tag{14.4}$$

For potential flow the last term on the left-hand side of (14.4) will vanish:

$$\text{with} \quad \text{rot}\, \underline{v} = \underline{\nabla} \times \underline{v} = \underline{0} \qquad \text{follows} \qquad \underline{v} \times (\underline{\nabla} \times \underline{v}) = \underline{0}$$

Only conservative force fields allowed

In addition, we will allow only conservative volume forces. The only volume force we usually consider is the gravity force. We already stated that the weight force is conservative and its potential is given on page 15.4: $E = -g\,z\,m$. The Euler equation is expressed as force per unit mass. Therefore, the vector of weight force per unit mass $\underline{f}$ may be replaced with the gradient of its potential per unit mass.

$$\underline{f} = \underline{\nabla}(-gz) \tag{15.37}$$

We substitute the new expression for the body forces $\underline{f}$ and replace the velocity vector with the gradient of the velocity potential.

$$\frac{\partial \underline{\nabla} \phi}{\partial t} + \frac{1}{2}\underline{\nabla}\left[(\underline{\nabla}\phi)^T \underline{\nabla}\phi\right] = \underline{\nabla}(-gz) - \frac{1}{\rho}\underline{\nabla} p \tag{15.38}$$

Assuming that the velocity potential has continuous second order derivatives, we may interchange the time derivative with the Nabla operator (taking spatial derivatives). This is also known as Schwartz's or Clairaut's theorem (Rogawski, 2008).

$$\underline{\nabla}\frac{\partial \phi}{\partial t} + \frac{1}{2}\underline{\nabla}\left[(\underline{\nabla}\phi)^T \underline{\nabla}\phi\right] = \underline{\nabla}(-gz) - \frac{1}{\rho}\underline{\nabla} p \tag{15.39}$$

We notice that every term in Equation (15.39) involves a gradient. Since a sum of derivatives is equal to the derivative of the sum, we shuffle all terms to the left-hand side and extract the Nabla operator.

$$\underline{\nabla}\left[\frac{\partial \phi}{\partial t} + \frac{1}{2}(\underline{\nabla}\phi)^T \underline{\nabla}\phi + gz + \frac{p}{\rho}\right] = \underline{0} \tag{15.40}$$

Bernoulli equation for potential flow

The term in brackets has to be constant and independent from x, y, and z for the spatial derivatives to be zero everywhere. The constant may be a function of time, though. Finally, we have found the Bernoulli equation for potential flow (incompressible, inviscid, and irrotational fluid).

$$\frac{\partial \phi}{\partial t} + \frac{1}{2}(\underline{\nabla}\phi)^T \underline{\nabla}\phi + gz + \frac{p}{\rho} = C(t) \tag{15.41}$$

The major difference to the Bernoulli equation for general incompressible, inviscid flow (14.23) is that the constant C is now valid for the whole domain and not just for a single streamline. This gives us great freedom to adjust the constant according to practical considerations.

The main application of the Bernoulli equation is to compute pressure after the velocity potential, and hence the fluid velocity vector, are known. If we choose a reference point far away from the body where the potential is steady $\partial\phi_\infty/\partial t = 0$, velocity id equal to $\underline{v})_\infty$, height is z_∞, and the pressure is p_∞, the constant becomes

$$C = \frac{1}{2}\underline{v}_\infty^T\,\underline{v}_\infty + g z_\infty + \frac{p_\infty}{\rho}$$

Remember that the dot product of a vector with itself yields its length (magnitude) squared. Substituting C into Bernoulli's equation (15.41) yields

$$\frac{\partial\phi}{\partial t} + \frac{1}{2}|\underline{\nabla}\,\phi|^2 + g z + \frac{p}{\rho} = \frac{1}{2}|\underline{v}_\infty|^2 + g z_\infty + \frac{p_\infty}{\rho} \tag{15.42}$$

Pressure in potential flow

We can solve this for the change in pressure

$$p - p_\infty = -\rho\frac{\partial\phi}{\partial t} + \frac{1}{2}\rho\left[|\underline{v}_\infty|^2 - |\underline{\nabla}\,\phi|^2\right] + \rho\,g(z_\infty - z) \tag{15.43}$$

Pressure forces are obtained by integration of the pressure difference (15.43) over the body surface.

Pressure in steady potential flow

In steady flow all local time derivatives vanish and Equation (15.43) reduces to

$$p - p_\infty = \frac{1}{2}\rho\left[|\underline{v}_\infty|^2 - |\underline{\nabla}\,\phi|^2\right] + \rho\,g(z_\infty - z) \tag{15.44}$$

Linearized Bernoulli equation for unsteady flow

We often assume in wave theory and seakeeping that the local velocity $\underline{\nabla}\,\Phi$ does not deviate much from the far field velocity $\underline{v}_\infty$. This allows us to neglect the nonlinear term $\frac{1}{2}\rho\left[|\underline{v}_\infty|^2 - |\underline{\nabla}\,\phi|^2\right]$ in Equation (15.43). For convenience, if the reference height is put at the calm water level $z_\infty = 0$, we obtain the linearized Bernoulli equation:

$$p - p_\infty = -\rho\,\frac{\partial\phi}{\partial t} - \rho\,g\,z \tag{15.45}$$

The z-axis is pointing upwards and all points in the fluid have negative z values.

Conservation of mass and momentum

In summary of this section on potential flow, we see that the physical principles of conservation of mass and momentum are cast into the Laplace equation and Bernoulli equation for potential flow.

$$\Delta\phi = 0 \qquad \text{Laplace eq.} \tag{15.46}$$

$$\rho\frac{\partial\phi}{\partial t} + \frac{1}{2}\rho|\underline{\nabla}\,\phi|^2 + \rho\,g z + p = C(t) \qquad \text{Bernoulli eq.} \tag{15.47}$$

The constant $C(t)$ is valid for the whole fluid domain.

References

Kellogg, O. (1929). *Foundations of potential theory.* Springer, Berlin, Germany.

Moran, J. (1984). *An introduction to theoretical and computational aerodynamics.* John Wiley & Sons, New York, NY.

Rogawski, J. (2008). *Multivariable calculus – early transcendentals.* W.H. Freeman and Company, New York, NY.

Self Study Problems

1. Which principle of fluid mechanics is expressed by the Laplace equation, and what are the necessary assumptions for applying the Laplace equation?
2. State the general assumptions necessary to apply potential theory.
3. How do you find the velocity vector once a scalar velocity potential is known?
4. What are advantages of potential flow theory compared with solving the Euler equations directly?
5. Explain which assumptions are made to get from the Navier-Stokes equations to the Bernoulli equation for potential flow.
6. What is the key difference of the two forms of the Bernoulli equation stated in Equations (14.23) and (15.41)?

16

Basic Solutions of the Laplace Equation

The Laplace equation forms the basis of potential theory.

$$\Delta\phi = 0 \tag{16.1}$$

It is a linear, homogeneous, partial differential equation of second order. Because it is homogeneous, it has an infinite number of solutions. Its linearity implies that linear combinations of solutions form another solution to the Laplace equation (16.1). In fluid mechanics, solutions to the Laplace equation represent velocity potentials. This chapter introduces basic potentials which are often used as building blocks for potentials which describe more complicated flow patterns. Besides the parallel flow potential, we discuss singularities like source, vortex, and dipole, as well as combinations of these and their resulting flow fields.

Learning Objectives

At the end of this chapter students will be able to:

- discuss basic building blocks of potential flows
- derive velocity fields from potentials
- superimpose singularities to represent more complex flows

16.1 Uniform Parallel Flow

The most basic flow pattern is a uniform flow field, which features the same velocity vector at every point of the domain. This flow is known as parallel flow. For example, the velocity vector of a flow of constant magnitude U_∞ in negative x-direction is given as

Uniform parallel flow

$$\underline{v}(x, y, z) = \underline{\nabla}\,\phi_\infty = -\underline{\nabla}\,(U_\infty\, x) = -\begin{pmatrix} U_\infty \\ 0 \\ 0 \end{pmatrix} \tag{16.2}$$

The corresponding velocity potential is the scalar function

$$\phi_\infty(x, y, z) = -U_\infty\, x \tag{16.3}$$

Fundamentals of Ship Hydrodynamics: Fluid Mechanics, Ship Resistance and Propulsion, First Edition.
Lothar Birk.

Companion website: www.wiley.com/go/birk/hydrodynamics

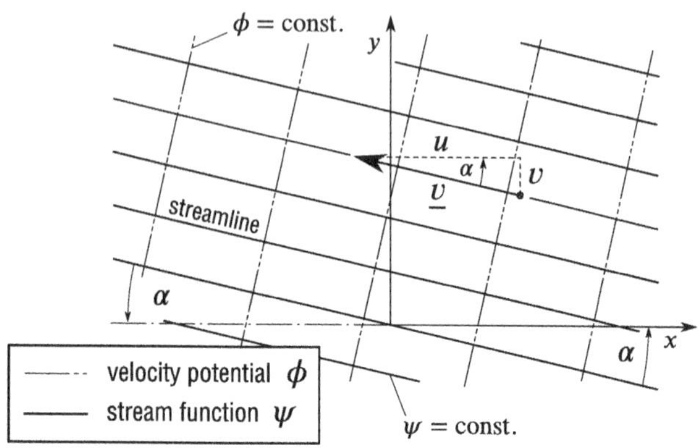

Figure 16.1 Planar uniform flow at angle α

Sometimes we desire uniform flow at a given angle α with respect to the coordinate system. The velocity potential

$$\phi_\infty(x, y, z) = -U_\infty \left[x \cos(\alpha) - y \sin(\alpha)\right] \tag{16.4}$$

describes a planar flow rotated clockwise by α (Figure 16.1). Its gradient yields the following velocity vector:

$$\underline{v}(x, y, z) = \underline{\nabla}\,\phi_\infty = \begin{pmatrix} u \\ v \\ w \end{pmatrix} = \begin{pmatrix} -U_\infty \cos(\alpha) \\ U_\infty \sin(\alpha) \\ 0 \end{pmatrix} \tag{16.5}$$

The speed is as before $|\underline{v}| = U_\infty$.

Obviously, all second order space derivatives of the parallel flow potential ϕ_∞ vanish. Therefore, it satisfies the Laplace equation.

$$\frac{\partial^2 \phi_\infty}{\partial x^2} + \frac{\partial^2 \phi_\infty}{\partial y^2} + \frac{\partial^2 \phi_\infty}{\partial z^2} = 0 \tag{16.6}$$

16.2 Sources and Sinks

Source

A source is a point in space, or on a plane, from which a volume flow emanates. Picture the water flowing out from a natural spring or a garden stone with a hole drilled through fed by a pond pump. Fluid will flow equally in every direction if the surrounding landscape is radially symmetric. The thick, solid lines in Figure 16.2 represent such a flow pattern.

2D source potential

The corresponding velocity potential in two dimensions is

$$\phi_s(\underline{p}, \underline{q}) = \frac{\sigma}{2\pi} \ln(|\underline{p} - \underline{q}|) \tag{16.7}$$

The source strength σ controls how much volume is discharged by the source per unit time. It has the unit cubic meter per second in space and meter squared per second in

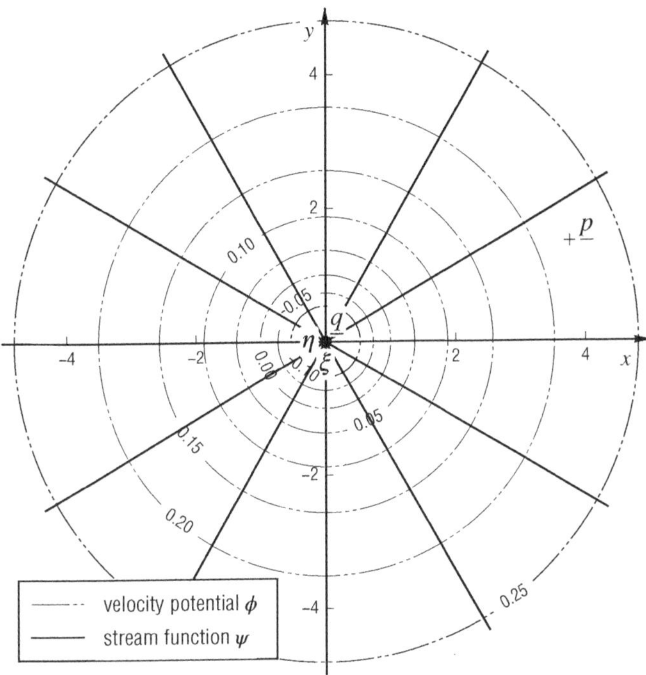

Figure 16.2 Streamlines (ψ = const.) and isolines of velocity potential for a planar source/sink flow. $\underline{q}^T = (\xi, \eta)$ is the location of the source

a planar flow. The source turns into a sink if the strength σ becomes negative which means that a sink removes fluid from the domain. Lines of constant velocity potential form concentric circles around the source (see ——- - -—— lines in Figure 16.2). A proof that the velocity potential in Equation (16.7) satisfies the Laplace equation is left as a self study problem.

Source and field point

The point $\underline{q}$ in Equation (16.7) fixes the location of the source. We call it a source point.

$$\text{2D flows} \quad \underline{q} = \begin{pmatrix} \xi \\ \eta \end{pmatrix} \qquad \text{3D flows} \quad \underline{q} = \begin{pmatrix} \xi \\ \eta \\ \zeta \end{pmatrix} \tag{16.8}$$

In Figure 16.2, the source point is identical to the origin $\underline{q}^T = (0, 0)$. The vector $\underline{p}$ marks the point where we inquire about flow properties. We call it field or collocation point.

$$\text{2D flows} \quad \underline{p} = \begin{pmatrix} x \\ y \end{pmatrix} \qquad \text{3D flows} \quad \underline{p} = \begin{pmatrix} x \\ y \\ z \end{pmatrix} \tag{16.9}$$

We will make use of the distance r_{pq} between source point and field point in a number of cases. For two-dimensional problems r_{pq} is

$$r_{pq} = |\underline{p} - \underline{q}| = \sqrt{(x-\xi)^2 + (y-\eta)^2} \qquad \text{in 2D} \tag{16.10}$$

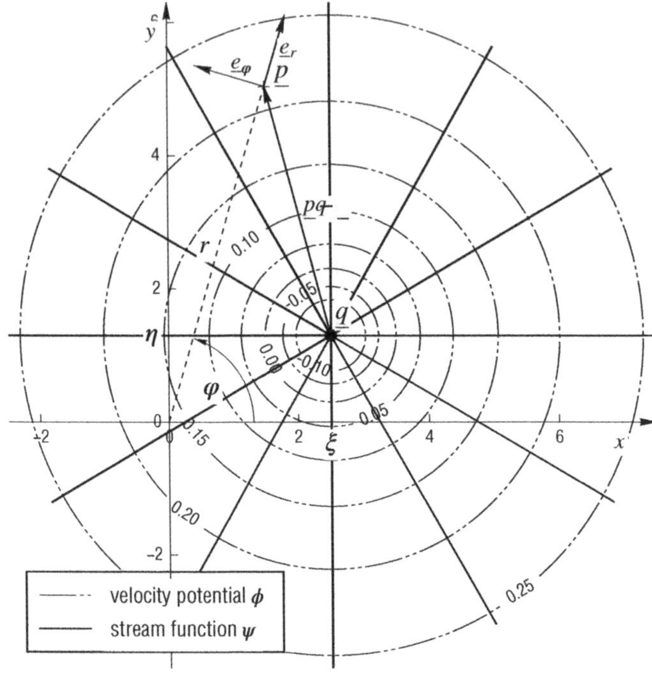

Figure 16.3 Streamlines (ψ = const.) and equipotential lines for a planar source/sink flow; the source/sink is located at $(\xi, \eta) = (2.5, 1.3)^T$

In three-dimensional problems use

$$r_{pq} = \sqrt{(x-\xi)^2 + (y-\eta)^2 + (z-\zeta)^2} \qquad \text{in 3D} \tag{16.11}$$

Velocity field The velocity vector of the two-dimensional source flow in Cartesian coordinates is

$$\underline{v}_s(\underline{p}) = \underline{\nabla}\phi_s = \underline{\nabla}\left[\frac{\sigma}{2\pi}\ln\left(|\underline{p}-\underline{q}|\right)\right] = \frac{\sigma}{2\pi}\begin{pmatrix}\dfrac{(x-\xi)}{r_{pq}^2} \\ \dfrac{(y-\eta)}{r_{pq}^2}\end{pmatrix} \tag{16.12}$$

The velocity vector is tangent to the streamlines. As shown in Figures 16.2 and 16.3, streamlines for source flows are radials emanating from the source in a regular pattern. Fluid is flowing away from the source point if σ is positive and it is flowing toward the source point if $\sigma < 0$. In Figure 16.3, the source point $\underline{q}$ has been moved off the origin of the coordinate system. The resulting formulas for source potential (16.7) and velocity field (16.12) are useful for the assembly of more complicated flow patterns.

Singularity Note that the velocity vector does not exist if field point and source point are identical, i.e. $\underline{p} = \underline{q}$. The flow velocity becomes infinite as we approach the source:

$$|\underline{v}_s| \longrightarrow \infty \quad \text{for} \quad \underline{p} \longrightarrow \underline{q}$$

Because the Function (16.7) is singular for $\underline{p} = \underline{q}$, we call the source/sink a singularity. Later we will discuss other basic solutions of the Laplace equation which classify as singularities. Strictly speaking, a singularity does not satisfy the Laplace equation at the source point. Therefore, equations involving singularities always have to be treated with care during integration and differentiation.

2D source in polar coordinates

Sometimes it is more convenient to describe a two-dimensional flow in polar coordinates. The source potential and its gradient are given by

$$\phi_s(r, \varphi) = \frac{\sigma}{2\pi} \ln\left(\sqrt{\left(r\cos(\varphi) - \xi\right)^2 + \left(r\sin(\varphi) - \eta\right)^2}\right) \tag{16.13}$$

Here, the polar coordinate r is measured from the origin and not from the position of the source (Figure 16.3). It is only equal to the distance between source and field point if the source point is at the origin.

$$\begin{aligned}
\underline{v}_s(r, \varphi) &= \underline{\nabla}\, \phi_s(r, \varphi) \\
&= \begin{pmatrix} \dfrac{\partial}{\partial r}\left[\dfrac{\sigma}{2\pi} \ln\left(\sqrt{\left(r\cos(\varphi) - \xi\right)^2 + \left(r\sin(\varphi) - \eta\right)^2}\right)\right] \\ \dfrac{1}{r}\dfrac{\partial}{\partial \varphi}\left[\dfrac{\sigma}{2\pi} \ln\left(\sqrt{\left(r\cos(\varphi) - \xi\right)^2 + \left(r\sin(\varphi) - \eta\right)^2}\right)\right] \end{pmatrix} \\
&= \frac{\sigma}{2\pi} \begin{pmatrix} \dfrac{\left[r - \xi\cos(\varphi) - \eta\sin(\varphi)\right]}{\left(r\cos(\varphi) - \xi\right)^2 + \left(r\sin(\varphi) - \eta\right)^2} \\ \dfrac{\left[\xi\sin(\varphi) - \eta\cos(\varphi)\right]}{\left(r\cos(\varphi) - \xi\right)^2 + \left(r\sin(\varphi) - \eta\right)^2} \end{pmatrix}
\end{aligned} \tag{16.14}$$

Always remember that the components of the velocity vector in polar coordinates generally do not point in x- and y-direction. They point in radial $\underline{e}_r$ and tangential $\underline{e}_\varphi$ direction (see point $\underline{p}$ in Figure 16.3).

3D source potential

The velocity potential for a source in three dimensions is slightly different. It is also known as a Rankine source, named after the Scottish engineer William J.M. Rankine (*1820–†1872).

$$\phi_s(\underline{p}, \underline{q}) = \frac{-\sigma}{4\,\pi\, r_{pq}} \quad \text{with } r_{pq} = |\underline{p} - \underline{q}| \tag{16.15}$$

The factor $-1/(4\pi)$ ensures that the flow is directed away from the source point $\underline{q}$ for positive source strength σ and the volume flow is equal to the value of σ in cubic meter per second.

3D source velocity field

Its velocity field consists of streamlines that emanate from $\underline{q}$ equally in all directions.

$$\underline{v}_s(\underline{p}) = \underline{\nabla}\left(\frac{-\sigma}{4\,\pi\, r_{pq}}\right) = \begin{pmatrix} \dfrac{\sigma}{4\pi}\dfrac{(x - \xi)}{r_{pq}^3} \\ \dfrac{\sigma}{4\pi}\dfrac{(y - \eta)}{r_{pq}^3} \\ \dfrac{\sigma}{4\pi}\dfrac{(z - \zeta)}{r_{pq}^3} \end{pmatrix} \tag{16.16}$$

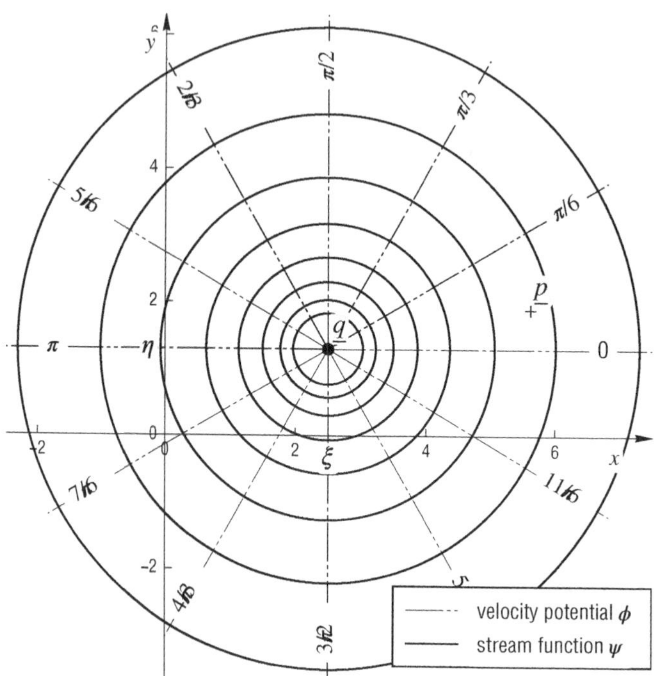

Figure 16.4 Streamlines (ψ = const.) and equipotential lines for a planar vortex flow; the vortex is located at $\underline{q} = (2.5, 1.3)^T$

Distributions of Rankine sources over body surfaces play an important role in modern numerical methods for displacement flow and wave resistance calculations (see Section 25.4).

16.3 Vortex

2D vortex potential Two-dimensional vortex flow was introduced as an example of potential flow in Figure 15.3(b). At first glance, streamlines and equipotential lines of the vortex in Figure 16.4 look like the corresponding illustration of the source flow in Figure 16.3. However, streamlines and equipotential lines are interchanged. For the vortex, streamlines form concentric circles whereas for the source, streamlines are radials. Fluid is rotating around the vortex center. The narrowing of streamlines indicates that the flow velocity is increasing rapidly with decreasing distance from the center $\underline{q}$.

Cartesian coordinates In Cartesian coordinates, the potential of a 2D vortex located at the source point[1] $\underline{q}^T = (\xi, \eta)$ is given by

$$\phi_v(x, y) = \frac{\gamma}{2\pi} \arctan\left(\frac{y - \eta}{x - \xi}\right) \tag{16.17}$$

[1]It is common practice to call the location of any singularity 'source point.'

Its velocity field is provided by the gradient of ϕ_v.

2D vortex velocity field

$$\underline{v}_v = \underline{\nabla}\phi_v = \frac{\gamma}{2\pi}\begin{pmatrix} \dfrac{-(y-\eta)}{\left[(x-\xi)^2+(y-\eta)^2\right]} \\ \dfrac{(x-\xi)}{\left[(x-\xi)^2+(y-\eta)^2\right]} \end{pmatrix} \tag{16.18}$$

Again, we have a singular point at $\underline{q} = \underline{p}$ at which the velocity would be infinite.

The velocity potential of a vortex at the origin is given by the function

Polar coordinates

$$\phi_v(r,\varphi) = \frac{\gamma}{2\pi}\varphi \qquad \text{for } \underline{q} = (0,0)^T \tag{16.19}$$

As expected, the corresponding velocity vector has no component in radial direction and the tangential velocity declines with increasing distance r from the vortex.

$$\underline{v}_v(r,\varphi) = \underline{\nabla}\phi_v(r,\varphi) = \frac{\gamma}{2\pi}\begin{pmatrix} 0 \\ \dfrac{1}{r} \end{pmatrix} \qquad \text{for } \underline{q} = (0,0)^T \tag{16.20}$$

We compute the circulation Γ_v for the vortex flow. We select a streamline located an arbitrary distance $r_{pq} = |\underline{p} - \underline{q}|$ away from the vortex as the integration path C. Since the velocity vector is always tangent to a streamline, the tangent vector of the circular integration path is equal to

2D vortex circulation

$$\underline{s} = \frac{\underline{v}_v}{|\underline{v}_v|} \tag{16.21}$$

Equation (14.41) allows us to compute the circulation of the vortex flow.

$$\Gamma_v = \oint_C \underline{v}_v^T \underline{s}\,\mathrm{d}s = \oint_C \underline{v}_v^T \frac{\underline{v}_v}{|\underline{v}_v|}\,\mathrm{d}s = \oint_C |\underline{v}_v|\,\mathrm{d}s \tag{16.22}$$

The magnitude of velocity (16.18) is constant along each streamline and inversely proportional to the distance r_{pq}.

$$|\underline{v}_v| = \frac{\gamma}{2\pi}\sqrt{\frac{\left[-(x-\xi)\right]^2 + (y-\eta)^2}{\left[(x-\xi)^2+(y-\eta)^2\right]^2}} = \frac{\gamma}{2\pi\, r_{pq}} \tag{16.23}$$

This result is confirmed by Equation (16.20), which shows the velocity vector in polar coordinates for a vortex at the origin. The magnitude of the tangential velocity is equal to (16.23).

Substitution of this result into the integral of Equation (16.22) yields

Circulation and vortex strength

$$\Gamma_v = \frac{\gamma}{2\pi\, r_{pq}}\oint_C \mathrm{d}s = \frac{\gamma}{2\pi\, r_{pq}} 2\pi\, r_{pq} = \gamma \tag{16.24}$$

The remaining path integral is equal to the circumference of a circle with radius r_{pq}. Thus, the vortex strength γ is equal to the circulation of the vortex. In addition, the circulation is constant and independent of the radius r_{pq}.

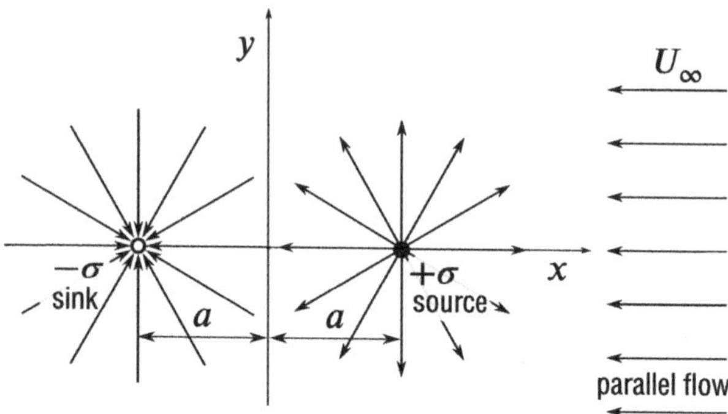

Figure 16.5 Superposition of parallel flow and a source/sink pair

At this point it is worthwhile to point out that the vortex strength and the circulation are nonzero: $\gamma = \Gamma_v \neq 0$. This seems to violate the assumption of irrotational flow which forms the basis of potential flow theory. As explained in Section 15.2, it is sufficient if the curl of the velocity vector vanishes in all simply connected regions. However, any path that includes the vortex is multiply connected. The velocity field is singular for $\underline{p} = \underline{q}$. As a consequence, the vortex location $\underline{q}$ cannot be a part of the domain. In self-study problem #4, you may show that the circulation vanishes for any path that does not include the vortex itself.

16.4 Combinations of Singularities

Before we discuss additional singularities, an example may show how they are used in the description of more complex and more practical flow patterns. Exploiting the linearity of the Laplace equation, new velocity potentials may be formed by linear superposition of singularities.

16.4.1 Rankine oval

Superposition of parallel flow, source, and sink

A classic example is the Rankine oval (Rankine, 1881). Its flow pattern is described by a velocity potential which consists of parallel flow, a source of strength $+\sigma$, and a sink of strength $-\sigma$.

$$\phi = -U_\infty x + \frac{\sigma}{2\pi} \ln \sqrt{(x-a)^2 + y^2} - \frac{\sigma}{2\pi} \ln \sqrt{(x+a)^2 + y^2} \tag{16.25}$$

As illustrated in Figure 16.5, the source at $x = +a$ is facing the parallel flow which points in negative x-axis direction. The sink is located at $x = -a$.

Differentiation of the potential with respect to x (position of the field point) produces the horizontal component of the velocity vector. This requires multiple applications of the chain rule.

$$\begin{aligned} u = \frac{\partial \phi}{\partial x} = -U_\infty &+ \frac{\sigma}{2\pi} \frac{1}{\sqrt{(x-a)^2 + y^2}} \left(\frac{1}{2}\right) \frac{2(x-a)}{\sqrt{(x-a)^2 + y^2}} \\ &- \frac{\sigma}{2\pi} \frac{1}{\sqrt{(x+a)^2 + y^2}} \left(\frac{1}{2}\right) \frac{2(x+a)}{\sqrt{(x+a)^2 + y^2}} \end{aligned}$$

$$u = -U_\infty + \frac{\sigma}{2\pi}\frac{(x-a)}{\left[(x-a)^2+y^2\right]} - \frac{\sigma}{2\pi}\frac{(x+a)}{\left[(x+a)^2+y^2\right]} \tag{16.26}$$

The y-derivative of the potential yields the transverse velocity component:

$$\begin{aligned} v = \frac{\partial\phi}{\partial y} &= \frac{\sigma}{2\pi}\frac{1}{\sqrt{(x-a)^2+y^2}}\left(\frac{1}{2}\right)\frac{(2y)}{\sqrt{(x-a)^2+y^2}} \\ &\quad - \frac{\sigma}{2\pi}\frac{1}{\sqrt{(x+a)^2+y^2}}\left(\frac{1}{2}\right)\frac{(2y)}{\sqrt{(x+a)^2+y^2}} \\ &= \frac{\sigma}{2\pi}\frac{(y)}{\left[(x-a)^2+y^2\right]} - \frac{\sigma}{2\pi}\frac{(y)}{\left[(x+a)^2+y^2\right]} \end{aligned} \tag{16.27}$$

In summary, the velocity field is given by the vector

$$\begin{aligned} \underline{v} = \underline{\nabla}\phi &= \begin{pmatrix} \dfrac{\partial\phi}{\partial x} \\ \dfrac{\partial\phi}{\partial y} \end{pmatrix} \\ &= \begin{pmatrix} -U_\infty + \dfrac{\sigma}{2\pi}\dfrac{(x-a)}{\left[(x-a)^2+y^2\right]} - \dfrac{\sigma}{2\pi}\dfrac{(x+a)}{\left[(x+a)^2+y^2\right]} \\ \dfrac{\sigma}{2\pi}\dfrac{(y)}{\left[(x-a)^2+y^2\right]} - \dfrac{\sigma}{2\pi}\dfrac{(y)}{\left[(x+a)^2+y^2\right]} \end{pmatrix} \end{aligned} \tag{16.28}$$

Rankine oval

Figure 16.6 shows an example of the flow field generated by the superposition of parallel flow in negative x-axis direction, a source of strength σ, and a sink of strength $-\sigma$. Parallel, source, and sink flow are symmetric with respect to the x-axis. It is fair to assume that the flow resulting from superposition is symmetric to the x-axis as well, which is confirmed by the streamlines in Figure 16.6. A striking feature of the flow is the dividing streamline which splits the flow into two independent regions.

- interior flow – Fluid originating from the source is diverted toward the sink which swallows fluid at the same rate as it emanates from the source.
- exterior flow – The parallel flow (from right to left in Figure 16.6 is diverted around the interior flow between source and sink.

The dividing streamline encloses the interior flow. Since fluid particles never cross streamlines, no fluid exchange happens between interior and exterior flow. As a consequence, we may interpret the dividing streamline as a material body contour. The body is known as a Rankine oval or Rankine ovoid.

We investigate the flow field and search for stagnation points, i.e. points at which the fluid velocity vanishes $|\underline{v}| = 0$. In order to find the coordinates $(x_s, y_s)^T$ of the stagnation points we use the conditions $u = 0$ and $v = 0$. Starting with the latter, the second component of Equation (16.28) yields the statement

$$v = \frac{\sigma}{2\pi}\frac{(y_s)}{\left[(x_s-a)^2+y_s^2\right]} - \frac{\sigma}{2\pi}\frac{(y_s)}{\left[(x_s+a)^2+y_s^2\right]} = 0 \tag{16.29}$$

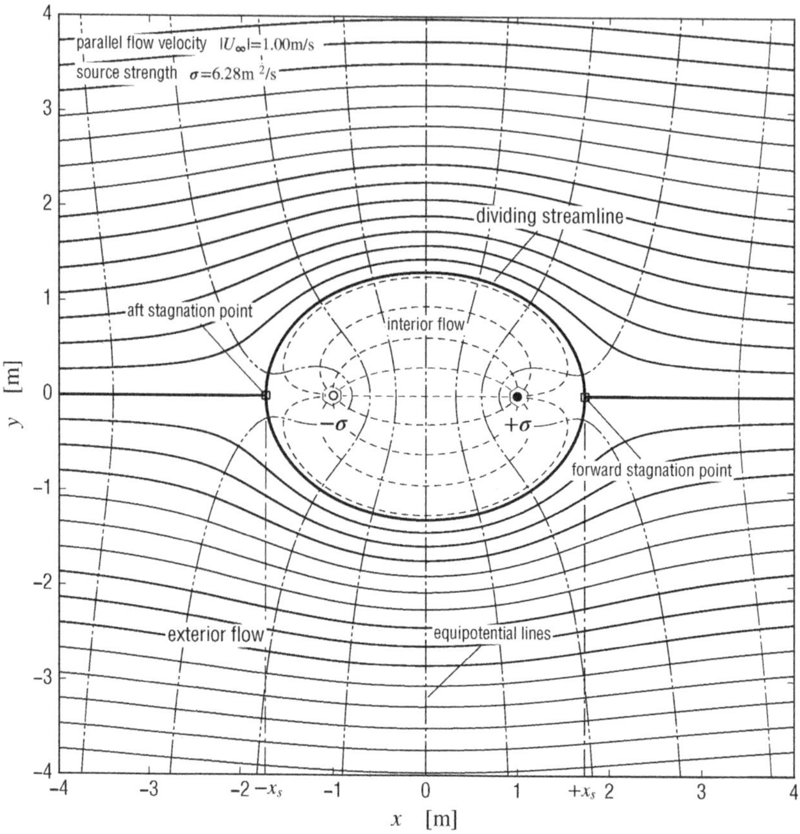

Figure 16.6 Flow field for a Rankine oval, a superposition of parallel flow, source, and sink

The equality is satisfied for $a = 0$ or $y_s = 0$. $a = 0$ would put source and sink together at the origin and they would cancel each other. Only the parallel flow would remain. Therefore, stagnation points must be located on the x-axis with $y_s = 0$. This corresponds to the points where the dividing streamline splits and reunites.

We exploit $y_s = 0$ and study the second condition:

$$u = 0 = -U_\infty + \frac{\sigma}{2\pi}\frac{(x_s - a)}{\left[(x_s - a)^2 + y_s^2\right]} - \frac{\sigma}{2\pi}\frac{(x_s + a)}{\left[(x_s + a)^2 + y_s^2\right]} \tag{16.30}$$

With $y_s = 0$ this simplifies to

$$\begin{aligned} 0 &= -U_\infty + \frac{\sigma}{2\pi}\frac{(x_s - a)}{\left[(x_s - a)^2\right]} - \frac{\sigma}{2\pi}\frac{(x_s + a)}{\left[(x_s + a)^2\right]} \\ &= -U_\infty + \frac{\sigma}{2\pi}\left[\frac{(x_s + a) - (x_s - a)}{(x_s - a)(x_s + a)}\right] \\ &= -U_\infty + \frac{\sigma}{2\pi}\left[\frac{2a}{\left(x_s^2 - a^2\right)}\right] \end{aligned} \tag{16.31}$$

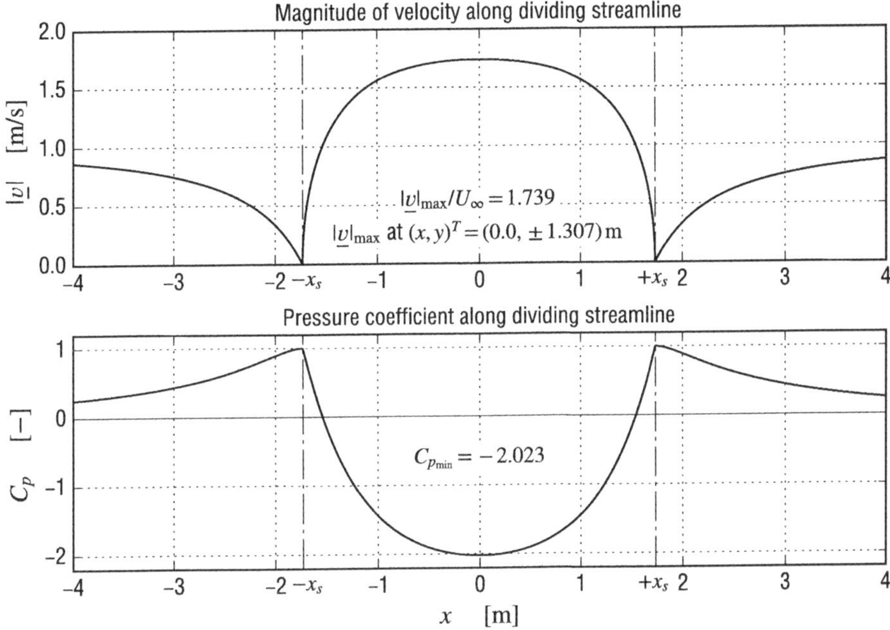

Figure 16.7 Velocity and pressure distribution along the dividing streamline (Rankine oval, $U_\infty = 1.0$ m/s, $\sigma = 2\pi$ m²/s)

We solve this for the x-coordinates of the stagnation points:

$$
\begin{aligned}
U_\infty \left(x_s^2 - a^2\right) &= \frac{\sigma\, a}{\pi} \\
x_s^2 &= \frac{\sigma\, a}{\pi\, U_\infty} + a^2 \\
x_s &= \pm\sqrt{\frac{\sigma\, a}{\pi\, U_\infty} + a^2}
\end{aligned}
\tag{16.32}
$$

For the values $U_\infty = 1.0$ m/s, $\sigma = 2\pi$ m²/s, and $a = 1.0$ m used in Figure 16.6, the half length of the Rankine oval is $x_s = 1.732$ m.

Velocity distribution

Figure 16.7 illustrates velocity and pressure distribution along the dividing streamline. Far in front ($x \longrightarrow \infty$) and far behind ($x \longrightarrow -\infty$) the Rankine oval, the fluid flows in negative x-axis direction with velocity $\underline{v}^T = (-U_\infty, 0)$. Obviously, the velocity vanishes at the forward stagnation point $(+x_s, 0)$. The onflow splits at the stagnation point. While fluid is diverted up or down, it accelerates and reaches its maximum velocity at the largest width of the body at $x = 0$. For $-x_s < x < 0$, fluid decelerates until it reaches the aft stagnation point at $-x_s$. Behind the body fluid accelerates again until the disturbance of the source and sink pair has declined.

Pressure distribution

The dimensionless pressure distribution follows from the Bernoulli equation for steady flow, Equation (15.44).

$$
C_p = \frac{p - p_\infty}{\frac{1}{2}\rho U_\infty^2} = 1 - \frac{|\underline{v}|^2}{U_\infty^2} \tag{16.33}
$$

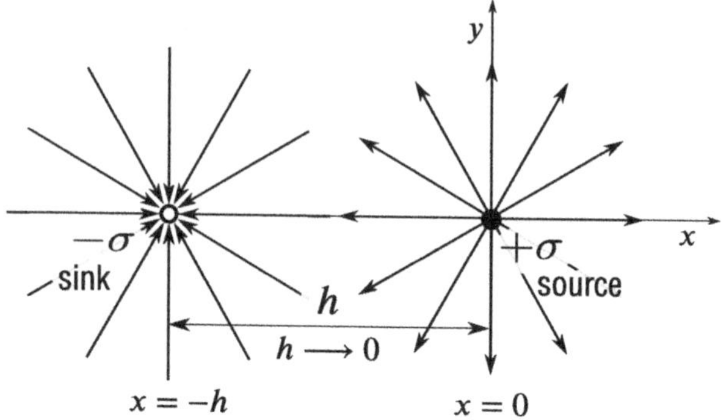

Figure 16.8 Creation of a dipole (doublet) by superposition of source and sink

Obviously, we ignored the hydrostatic pressure in this by assuming it is constant in the plane of the flow.

The actual computation of C_p is quite tedious and a computer program has been employed to create the data in Figure 16.7. First, the dividing streamline was located by tracking a fluid particle over time through integration of the velocity vector. Next, the pressure coefficient has been computed using the stored locations and velocity values.

16.4.2 Dipole

Combination of source and sink

Imagine a source of strength $\sigma > 0$ is placed at the origin $\underline{q}^{(+)} = (0,0)^T$ and a sink of strength $-\sigma$ is placed further to the left on the negative x-axis at $\underline{q}^{(-)} = (-h,0)^T$. Figure 16.8 illustrates the setup.

The combined velocity potential of the source/sink arrangement is

$$\phi_{ss} = \frac{-\sigma}{2\pi} \ln \sqrt{(x+h)^2+y^2} + \frac{\sigma}{2\pi} \ln \sqrt{x^2+y^2} \tag{16.34}$$

If the distance h is reduced to zero, source $+\sigma$ and sink $-\sigma$ will cancel each other, unless we keep the product of source strength and distance constant. We define the new parameter $\mu = \sigma\, h > 0$ and investigate the limit of the potential ϕ_{ss} (16.34) for $h \longrightarrow 0$.

$$\begin{aligned}\lim_{h\to 0}\phi_{ss} &= \lim_{h\to 0}\left[\frac{-\sigma\, h}{2\pi\, h}\left(\ln\sqrt{(x+h)^2+y^2} - \ln\sqrt{x^2+y^2}\right)\right]\\ &= \frac{-\sigma\, h}{2\pi\, h}\lim_{h\to 0}\left[\ln\sqrt{(x+h)^2+y^2} - \ln\sqrt{x^2+y^2}\right]\\ &= \frac{-\mu}{2\pi}\lim_{h\to 0}\left[\frac{\ln\sqrt{(x+h)^2+y^2} - \ln\sqrt{x^2+y^2}}{h}\right]\end{aligned} \tag{16.35}$$

The limit value of the term in square brackets looks like the definition of a first order partial derivative with respect to x.

$$\frac{\partial f}{\partial x} = \lim_{h\to 0}\frac{f(x+h)-f(x)}{h}$$

In our case, the function is $f = \ln\sqrt{x^2+y^2}$. Replacing the limit value in Equation (16.35) with the first order x-derivative results in a new potential which is known as dipole or doublet potential.

$$\phi_d = \frac{-\mu}{2\pi}\frac{\partial}{\partial x}\left(\ln\sqrt{x^2+y^2}\right) \tag{16.36}$$

$$= \frac{-\mu}{2\pi}\frac{(x)}{\left(x^2+y^2\right)} \tag{16.37}$$

Essentially, the dipole potential is the negative derivative of the source potential. Only, the source strength σ is exchanged for the dipole strength μ.

2D dipole in polar coordinates

The dipole potential above may also be expressed in polar coordinates.

$$\phi_d(r,\varphi) = \frac{-\mu}{2\pi}\frac{\cos(\varphi)}{r} \tag{16.38}$$

This follows from Equation (16.37) with the substitutions from Equation (15.28): $x = r\cos(\varphi)$ and $x^2 + y^2 = r^2$.

The velocity field is defined in radial and tangential components as

$$\underline{v}_d(r,\varphi) = \underline{\nabla}\,\phi_d(r,\varphi) = \frac{\mu}{2\pi}\begin{pmatrix}\dfrac{\cos(\varphi)}{r^2}\\[2ex]\dfrac{\sin(\varphi)}{r^2}\end{pmatrix} \quad \text{for } \underline{q} = (0,0)^T \tag{16.39}$$

Cylinder flow

With the specific dipole strength of

$$\mu = 2\pi\,U_\infty\,R^2 \tag{16.40}$$

the velocity potential (16.38) becomes identical to the disturbance potential (17.25), which we will also find as part of the solution of a boundary value problem (see Chapter 17). The dipole potential, with axis in positive x-direction, combined with a parallel flow in negative x-axis direction represents the flow around a long cylinder of radius R. The velocity potential for the cylinder flow is equal to

$$\phi_{\text{cyl}}(r,\varphi) = -U_\infty r\cos(\varphi) - \frac{U_\infty R^2}{r}\cos(\varphi) \tag{16.41}$$

in polar coordinates. In Cartesian coordinates we may state it as

$$\phi_{\text{cyl}}(x,y) = -U_\infty x - U_\infty R^2\frac{x}{\left(x^2+y^2\right)} \tag{16.42}$$

2D dipole potential

If the source point $\underline{q} = (\xi,\eta)^T$ is moved away from the origin, the potential of a dipole in Cartesian coordinates is equal to

$$\phi_d = -\frac{\partial\phi_s}{\partial x} = \frac{-\mu}{2\pi}\frac{(x-\xi)}{\left[(x-\xi)^2+(y-\eta)^2\right]} \tag{16.43}$$

Figure 16.9(a) shows the corresponding pattern of streamlines and equipotential lines. The curves form an orthogonal mesh, with sets of circles that are all tangent to two perpendicular lines through the source point.

2D dipole velocity field

As before, the components of the velocity vector follow from the gradient of the dipole potential.

$$\underline{v}_d = \underline{\nabla}\phi_d = \frac{\mu}{2\pi}\begin{pmatrix} \dfrac{(x-\xi)^2 - (y-\eta)^2}{\left[(x-\xi)^2 + (y-\eta)^2\right]^2} \\ \dfrac{2(x-\xi)(y-\eta)}{\left[(x-\xi)^2 + (y-\eta)^2\right]^2} \end{pmatrix} \tag{16.44}$$

Dipole axis

Dipoles have a distinct axis. The dipole described by the potential in Equation (16.43) is moving fluid through the source point in positive x-axis direction. The axis of the dipole may be oriented in any direction $\underline{s}$. The directional derivative of the source potential in direction $-\underline{s}$ defines the corresponding dipole potential.

$$\phi_d = -\frac{\partial \phi_s}{\partial s} = -\underline{s}^T \underline{\nabla}\phi_s \tag{16.45}$$

This equation is valid in two-dimensional and three-dimensional flow. For a directional vector

$$\underline{s} = \begin{pmatrix} \cos(\alpha) \\ \sin(\alpha) \end{pmatrix}$$

with $|\underline{s}| = 1$ we get

$$\phi_d = \frac{-\mu}{2\pi}\left[\frac{(x-\xi)\cos(\alpha)}{\left[(x-\xi)^2 + (y-\eta)^2\right]} + \frac{(y-\eta)\sin(\alpha)}{\left[(x-\xi)^2 + (y-\eta)^2\right]}\right] \tag{16.46}$$

The axis of the dipole depicted in Figure 16.9(b) is rotated by $\alpha = 65$ degrees with respect to the positive x-axis.

16.5 Singularity Distributions

Superposition of discrete singularities

Cylinder and Rankine oval flow serve as examples of flows around simple geometries modeled by superposition of velocity potentials. More complex flow patterns may be modeled by adding more singularities of varying strength. A closed body contour is usually obtained if the total strength of sources and sinks adds up to zero.

Line and surface distributions

Rather than using discrete singularities, it is more efficient to distribute them along contours or boundary surfaces. Later we will use line distributions of sources and vortices to model the flow past thin lifting foils. Sources and dipoles play an important role in the numerical solution of potential flow problems. It can be shown that any potential flow may be represented by distributions of sources and dipoles over the boundaries of the fluid domain. For details see Katz and Plotkin (2001).

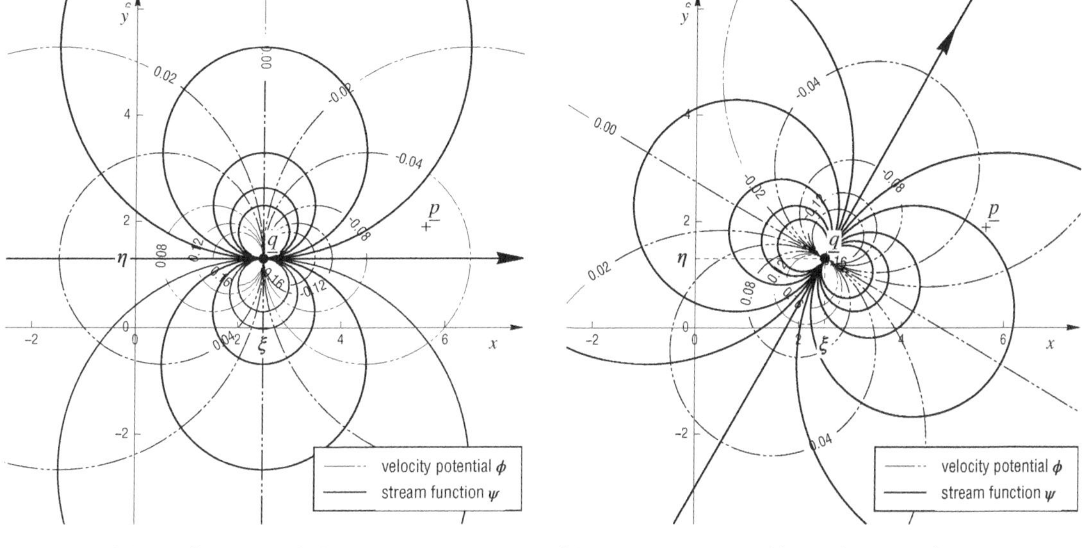

(a) Dipole axis parallel to x-axis

(b) Dipole axis rotated by 65 degrees relative to $+x$-axis

Figure 16.9 Streamlines (ψ = const.) and isolines of velocity potential for planar dipole flows. Dipole is located at $(\xi, \eta) = (2.5, 1.3)^T$

References

Katz, J. and Plotkin, A. (2001). *Low-speed aerodynamics*. Cambridge Aerospace Series. Cambridge University Press, New York, NY, second edition.

Rankine, W. (1881). Elementary demonstrations of principles relating to stream-lines. In Millar, W., editor, *Miscellaneous scientific papers*, chapter XXXI., pages 522–529. Charles Griffin and Company, London. Originally published in *The Engineer*, Otc. 16, 1868.

Self Study Problems

1. Show that the two-dimensional source potential

$$\phi_s(\underline{p}, \underline{q}) = \frac{\sigma}{2\pi} \ln(r) \qquad \text{with } r = \sqrt{(x-\xi)^2 + (y-\eta)^2}$$

satisfies the Laplace equation for $\underline{p} \neq \underline{q}$.

2. Demonstrate that the three-dimensional source potential

$$\phi_s(\underline{p}, \underline{q}) = \frac{-\sigma}{4\pi\, r} \qquad \text{with } r = \sqrt{(x-\xi)^2 + (y-\eta)^2 + (z-\zeta)^2}$$

satisfies the Laplace equation for $\underline{p} \neq \underline{q}$.

3. Employ Equations (16.45) and (16.15) to derive the velocity potential of a three-dimensional dipole whose axis points in positive x-direction.

4. A vortex flow is irrotational in any area which excludes the vortex itself. Show that the circulation vanishes along an arbitrary path C as shown in the figure below.

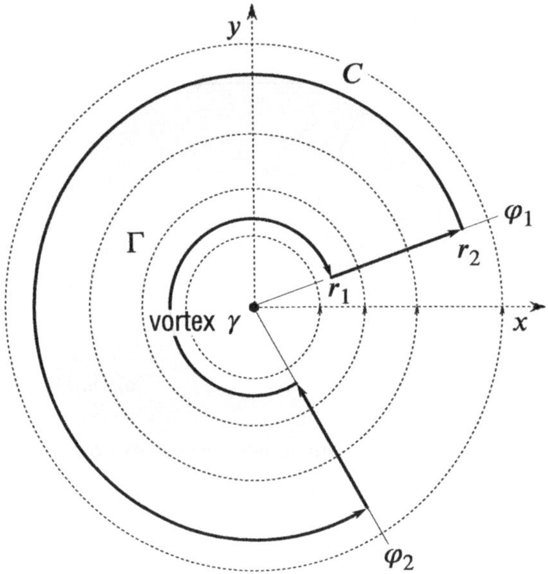

17

Ideal Flow Around A Long Cylinder

As an example of potential flow, we solve the basic equations that describe the fluid flow around an infinitely long cylinder oriented with its axis perpendicular to the flow. Since the flow properties will not change in the direction of the cylinder axis, we may pick any plane transverse to the cylinder axis to set up the necessary equations.

Potential flow assumes that the flow is incompressible, inviscid, and irrotational. With these restrictions, the equations to be solved are the Laplace equation (17.1) and the Bernoulli equation (17.2). From the former follows the velocity potential Φ. Its gradient provides the velocity vector $\underline{v}$. The Bernoulli equation yields the pressure distribution.

$$\Delta\Phi(\underline{x}, t) = 0 \qquad \text{for all } \underline{x} \in V \tag{17.1}$$

$$\rho\frac{\partial\Phi}{\partial t} + \frac{1}{2}\rho|\underline{v}|^2 + p + \rho g z = C(t) \qquad \text{for all } \underline{x} \in V \tag{17.2}$$

We will see that the solution provides useful insight and forms the basis for the solution of other flow problems, despite omitting viscous effects from the get go.

Learning Objectives

At the end of this chapter students will be able to:

- set up and solve a boundary value problem
- compute velocities, pressure, and forces for potential flow around a cylinder
- understand d'Alembert's paradox
- understand the concept of added mass

17.1 Boundary Value Problem

Problem formulation

As stated before, the Laplace equation (17.1) is a linear partial differential equation of second order. Since it is also homogeneous, it will have an infinite number of solutions. Similar to problems in elasticity theory where the deflection of a beam loaded with a force or distributed load depends on the types of supports (clamped, pinned, slide, etc.), we need boundary conditions to make the solution unique to the given problem.

Fundamentals of Ship Hydrodynamics: Fluid Mechanics, Ship Resistance and Propulsion, First Edition.
Lothar Birk.

Companion website: www.wiley.com/go/birk/hydrodynamics

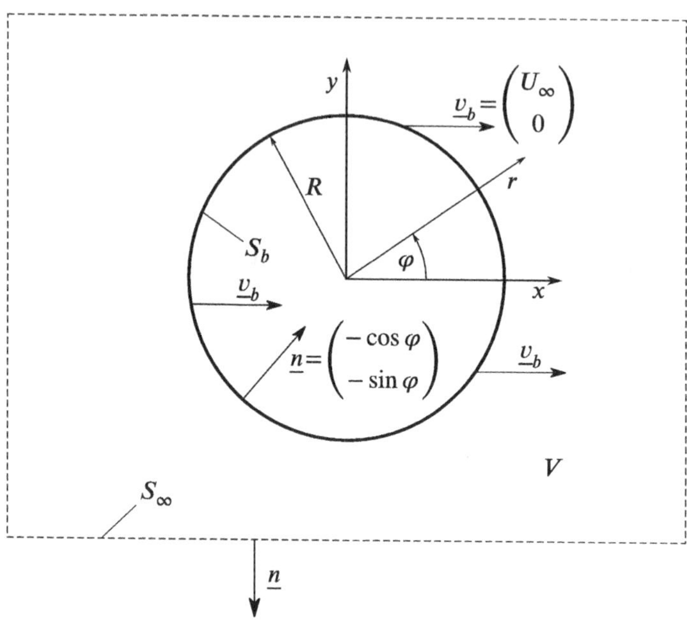

Figure 17.1 An infinitely long cylinder moving with speed U_∞ in positive x-direction in a fluid at rest

Point of view The problem can be formulated in two different ways which ultimately are completely equivalent.

- We can build a mathematical model that reflects the view of an outside observer, who will see an object moving through a fluid. The fluid is initially at rest and moves only to make way for the cylinder.
- Or, we build a model representing the view of an observer moving with the object. For the observer on the cylinder, the object is at rest and only the fluid moves.

We will formulate both below, but show the detailed solution only for the latter case. It is much more convenient because the flow will be steady (invariant with time) in the moving coordinate system. That simplifies the calculation of the velocity and pressure distributions.

17.1.1 Moving cylinder in fluid at rest

Cylinder is moving Figure 17.1 shows a cylinder moving with velocity U_∞ in positive x-direction. The fluid is at rest except for the particles that have to move to make room for the cylinder. The fixed finite control volume V has two boundary surfaces: the body surface S_b and a surface S_∞ in the far field. We assume the surface S_∞ is far enough away from the body that any disturbance due to the moving cylinder has subsided. Note that the normal vector $\underline{n}$ to the surfaces is pointing out of the fluid domain V. As a consequence, the normal vector of the body surface is pointing *into* the cylinder. The interior of the cylinder is not part of the fluid domain.

Types of boundary conditions

In order to properly pose the problem, we must define boundary conditions on all surfaces of the control volume. For *boundary value problems* three types of boundary conditions are considered:

1. Dirichlet boundary conditions impose values on the unknown function itself. In our case the unknown function is the velocity potential.
2. Neumann boundary conditions prescribe the values of derivatives of the unknown function, for instance the velocity components which are the spatial derivatives of the velocity potential.
3. Mixed boundary conditions which involve the unknown functions and their derivatives simultaneously.

Physical boundary conditions

In order to obtain a unique solution, we must define boundary conditions on all surfaces of the control volume. There are two boundary surfaces, and we have to specify a boundary condition for each of them.

- As stated above, the disturbance due to the moving cylinder has subsided in the far field. Therefore, fluid velocity (induced by the moving cylinder) has to vanish in the far field.

  $$\underline{\nabla}\,\Phi \longrightarrow \underline{0} \qquad \text{for } r \longrightarrow \infty$$

  This is a Neumann boundary condition, since we impose a requirement for the derivative of the unknown potential Φ.

- We want the fluid to stay out of the cylinder, i.e. no fluid particle is allowed to flow across the body surface S_b. In a viscous fluid, we expect the molecules to stick to the surface. However, we ignore friction here, which allows molecules to move tangentially along the surface. To prevent fluid from crossing the body surface, we must require that the normal component of the relative velocity between fluid and body vanishes on S_b.

  $$\underline{n}^T\,\underline{v}_{\text{relative}} = 0 \qquad \text{for points on } S_b$$

  The relative velocity is the difference between fluid flow velocity $\underline{v} = \underline{\nabla}\,\Phi$ and the local velocity $\underline{v}_b$ of the body surface.

Boundary value problem for unsteady flow

In summary, at any given time t we have to satisfy the following set of equations which make up a boundary value problem for unsteady flow.

$$\begin{aligned} \boldsymbol{\Delta}\Phi(\underline{x},t) &= 0 && \text{for } \underline{x} \in V \\ \frac{\partial \Phi}{\partial n} - \underline{n}^T\,\underline{v}_b &= 0 && \text{for } \underline{x} \in S_b, \text{ for all } t \\ \lim_{|\underline{x}|\to\infty} \underline{\nabla}\,\Phi(\underline{x},t) &= \underline{0} && \text{for } \underline{x} \in S_\infty \end{aligned} \tag{17.3}$$

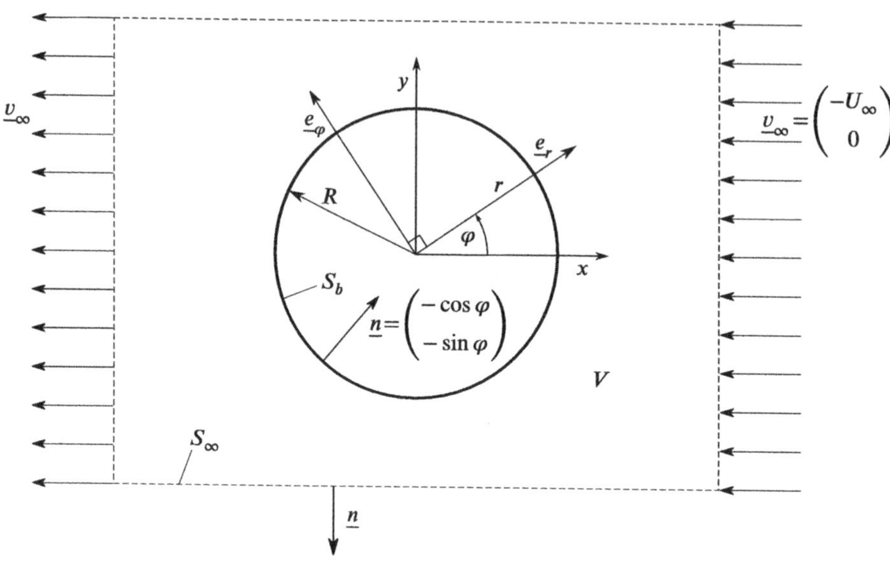

Figure 17.2 An infinitely long cylinder at rest in parallel flow

17.1.2 Cylinder at rest in parallel flow

Fluid is moving The boundary value problem (17.3) formulated above is useful if the cylinder is moving with varying forward speed. However, for a cylinder moving with a constant velocity it is better to formulate the problem in a coordinate system which is moving along with the cylinder. The flow becomes steady, we have to satisfy the following set of equations which make up a boundary value problem.

Figure 17.2 shows the cylinder in a parallel flow which is directed in negative x-axis direction with constant speed U_∞. The parallel flow speed corresponds to the forward speed of the cylinder as seen by an outside observer.

Parallel flow potential The linearity of the Laplace equation allows us to combine solutions of it into new solutions. The velocity potential of parallel flow in negative x-direction is

$$\phi_\infty = -U_\infty x \tag{17.4}$$

Its gradient is

$$\underline{\nabla}\,\phi_\infty = \begin{pmatrix} \dfrac{\partial \phi_\infty}{\partial x} \\ \dfrac{\partial \phi_\infty}{\partial y} \end{pmatrix} = \begin{pmatrix} -U_\infty \\ 0 \end{pmatrix} \tag{17.5}$$

Obviously, all second order derivatives of ϕ_∞ vanish, and the parallel flow velocity potential satisfies the Laplace equation.

In polar coordinates the parallel stream velocity potential is

$$\phi_\infty(r, \varphi) = -U_\infty\, r \cos\varphi \tag{17.6}$$

Introducing the parallel flow potential splits the yet unknown velocity potential ϕ of the flow around a cylinder into two parts: the known parallel flow potential ϕ_∞ and the yet unknown disturbance potential ϕ_d. The latter is so named because it represents the disturbance of parallel flow caused by the cylinder.

Superposition principle

$$\phi = \phi_\infty + \phi_d \tag{17.7}$$

We now reformulate our boundary conditions. There are two boundaries: the far field S_∞ and the body surface S_b.

Boundary conditions

- In the far field, we expect the disturbance caused by the cylinder to subside. A submarine passing by far out at sea will not change flow patterns in the harbor. Mathematically speaking, we request that $\underline{\nabla}\,\phi \to \underline{\nabla}\,\phi_\infty$ for $r \to \infty$, i.e. we have parallel flow conditions in the far field. This is equivalent to

$$\underline{\nabla}\,\phi_d \to \begin{pmatrix} 0 \\ 0 \end{pmatrix} \qquad \text{for } r \to \infty$$

- On the body surface S_b, we now request that the component of the fluid particle velocity normal to the body surface vanishes.

$$\underline{n}^T \underline{\nabla}\,\phi = 0 \qquad \text{for } \underline{x} \in S_b$$

  Remember that the body does not move relative to the chosen coordinate system. Again, this is a Neumann boundary condition. For the dot product of normal vector $\underline{n}$ and gradient of the potential $\underline{\nabla}\,\phi$, we usually write $\partial\phi/\partial n$ (normal derivative).

In summary, the boundary value problem for the cylinder in parallel flow is defined by

Boundary value problem for steady cylinder flow

$$\begin{aligned} \Delta\phi(\underline{x}) &= 0 && \text{for } \underline{x} \in V \\ \frac{\partial\phi}{\partial n} &= 0 && \text{for } \underline{x} \in S_b \\ \lim_{|\underline{x}|\to\infty} \underline{\nabla}\,\phi(\underline{x}) &= \underline{\nabla}\,\phi_\infty(\underline{x}) && \text{for } \underline{x} \in S_\infty \end{aligned} \tag{17.8}$$

17.2 Solution and Velocity Potential

Solution of the boundary value problem (17.8) is easier if it is done in polar coordinates $< r, \varphi >$ rather than Cartesian coordinates $< x, y >$. The Laplace equation in polar coordinates is provided by Equation (15.29) and repeated here for convenience.

Solution by separation of variables

$$\frac{\partial^2\phi}{\partial r^2} + \frac{1}{r}\frac{\partial\phi}{\partial r} + \frac{1}{r^2}\frac{\partial^2\phi}{\partial\varphi^2} = 0 \tag{15.29}$$

The following equations transform coordinates from a polar into a Cartesian coordinate system:

$$x = r\cos(\varphi) \qquad\qquad y = r\sin(\varphi) \tag{17.9}$$

The reverse transformation is given by

$$r = \sqrt{x^2 + y^2} \qquad \varphi = \arctan\left(\frac{y}{x}\right) \tag{17.10}$$

The unknown disturbance potential will be a function of (r, φ). We seek a solution by separation of variables:

$$\phi_d(r, \varphi) = P(r)\,Q(\varphi) \tag{17.11}$$

Instead of $\phi_d(r, \varphi)$, we have to find two functions $P(r)$ and $Q(\varphi)$. P and Q are univariate functions.

We substitute $\phi = \phi_\infty + P\,Q$ into the Laplace equation (15.29). Because of the superposition principle and the fact that ϕ_∞ satisfies the Laplace equation, we need only to investigate the product $P\,Q$. Any derivative with respect to r affects function P only while Q is a constant factor. Derivatives with respect to φ affect$Q$ only, with P being treated as a constant factor.

$$\frac{\partial^2 P}{\partial r^2} Q + \frac{1}{r}\frac{\partial P}{\partial r} Q + \frac{1}{r^2} P \frac{\partial^2 Q}{\partial \varphi^2} = 0 \tag{17.12}$$

Dividing the equation by the product $P\,Q$ yields

$$\frac{\partial^2 P}{\partial r^2}\frac{1}{P} + \frac{1}{r}\frac{\partial P}{\partial r}\frac{1}{P} + \frac{1}{r^2}\frac{\partial^2 Q}{\partial \varphi^2}\frac{1}{Q} = 0 \tag{17.13}$$

Of course, we are excluding the trivial solution $P\,Q \equiv 0$.

Multiplication with r^2 separates the equation into terms which depend on either r or φ but not both.

$$r^2 \frac{\partial^2 P}{\partial r^2}\frac{1}{P} + r\frac{\partial P}{\partial r}\frac{1}{P} + \frac{\partial^2 Q}{\partial \varphi^2}\frac{1}{Q} = 0 \tag{17.14}$$

We move the last term to the right-hand side:

$$r^2 \frac{\partial^2 P}{\partial r^2}\frac{1}{P} + r\frac{\partial P}{\partial r}\frac{1}{P} = -\frac{\partial^2 Q}{\partial \varphi^2}\frac{1}{Q} = C \tag{17.15}$$

For this equation to be true for all possible combinations of r and φ, both sides have to be equal to a constant C, otherwise there would be points where the difference between left and right side would not result in zero. C must be independent of the variables (r, φ).

PDE converted to system of ODEs

The partial differential equation (17.15) is now split into two ordinary differential equations (ODEs). Using subscripts to indicate differentiation we get:

$$r^2 P_{rr} + r P_r - P\,C = 0 \tag{17.16}$$

$$Q_{\varphi\varphi} + Q\,C = 0 \tag{17.17}$$

Note that the constant C must be the same in both equations. It represents the coupling between the ODEs. The differential equations (17.16) and (17.17) are of standard shape. Solutions can be found in any book on differential equations.

We solve for the function P first and assume that it takes the form of a reciprocal function.

Basic solution for $P(r)$

$$
\begin{aligned}
P &= C_1 \frac{1}{r} \\
P_r &= C_1 \frac{\partial P}{\partial r} = -C_1 \frac{1}{r^2} \\
P_{rr} &= C_1 \frac{\partial^2 P}{\partial r^2} = 2C_1 \frac{1}{r^3}
\end{aligned} \tag{17.18}
$$

Substituting this into equation (17.16) yields the desired result.

$$
\begin{aligned}
0 &= r^2\, 2C_1 \frac{1}{r^3} - r\, C_1 \frac{1}{r^2} - C_1 \frac{1}{r} C \\
0 &= \frac{C_1}{r} (1 - C)
\end{aligned} \tag{17.19}
$$

The last equation is satisfied for $C = 1$. The constant C_1 remains to be determined.

With $C = 1$ substituted into (17.17), we need a function Q that is the negative of its second order derivative. The sine and cosine functions are of this type. Because our flow is expected to be symmetric to the y-axis, we select the cosine function.

Basic solution for $Q(\varphi)$

$$
\begin{aligned}
Q &= C_2 \cos \varphi \\
Q_{\varphi\varphi} &= -C_2 \cos \varphi
\end{aligned} \tag{17.20}
$$

Combining the basic solutions $P = C_1/r$ and $Q = C_2 \cos \varphi$, creates a new constant $C_3 = C_1\, C_2$. We obtain for the disturbance potential ϕ_d:

$$
\begin{aligned}
\phi_d(r, \varphi) &= C_1\, C_2 \frac{1}{r} \cos \varphi \\
&= C_3 \frac{1}{r} \cos \varphi
\end{aligned} \tag{17.21}
$$

The boundary conditions are exploited to find the constant C_3. Let us start with the body boundary condition. On the body surface S_b we have to enforce that

Identification of constant

$$
\frac{\partial \phi}{\partial n} = 0 \qquad \text{for} \quad r = R \tag{17.22}
$$

The normal vector on the cylinder surface points always toward the center or, more importantly, in the opposite direction of $\underline{e}_r$. Thus, the normal derivative is equal to the negative radial derivative.

$$
\frac{\partial}{\partial n} = -\frac{\partial}{\partial r} \qquad \text{on } S_b \text{ with } r = R \tag{17.23}
$$

Replacing the normal derivative with $-\partial/\partial r$

$$
-\frac{\partial \phi}{\partial r}\bigg|_{r=R} = 0
$$

and splitting the potential in parallel stream and disturbance part yields:

$$-\left.\frac{\partial \phi_\infty}{\partial r}\right|_{r=R} - \left.\frac{\partial \phi_d}{\partial r}\right|_{r=R} = 0$$
$$-\left.\frac{\partial \phi_d}{\partial r}\right|_{r=R} = \left.\frac{\partial \phi_\infty}{\partial r}\right|_{r=R}$$

Finally, utilizing Equation (17.6) on the right-hand side and Equation (17.21) on the left-hand side yields for the constant C_3:

$$\begin{aligned} -\left.\frac{\partial \phi}{\partial r}\right|_{r=R} &= -U_\infty \cos \varphi \\ C_3 \left.\frac{\cos \varphi}{r^2}\right|_{r=R} &= -U_\infty \cos \varphi \\ C_3 \frac{\cos \varphi}{R^2} &= -U_\infty \cos \varphi \\ C_3 &= -U_\infty R^2 \end{aligned} \tag{17.24}$$

The constant depends on the basic parameters defining the flow problem: the velocity of the parallel stream and the radius R of the cylinder, which makes sense. The unit of the resulting potential is m^2/s, which is also correct.

Potential of disturbance

Substituting the constant C_3 into Equation (17.21) for the disturbance potential gives

$$\phi_d(r, \varphi) = \frac{-U_\infty R^2}{r} \cos \varphi \tag{17.25}$$

This is a dipole or doublet potential, which we already encountered in the previous chapter (see Equation (16.38)).

Velocity potential for the cylinder flow

Finally, the velocity potential for the flow around a cylinder in an unbounded fluid is given by the sum of parallel flow and a dipole with its axis pointing against the parallel flow.

$$\phi(r, \varphi) = -U_\infty r \cos \varphi - \frac{U_\infty R^2}{r} \cos \varphi \tag{17.26}$$

In Cartesian coordinates the velocity potential for the cylinder flow is given by

$$\phi(x, y) = -U_\infty x - U_\infty R^2 \frac{x}{\left(x^2 + y^2\right)} \tag{17.27}$$

Far field boundary condition

Hopefully, you have not forgotten about the boundary condition for the far field surface S_∞. We have not used it yet explicitly but, if you check, you can see that the gradient of the second term of the potential vanishes for $r \longrightarrow \infty$ and the parallel flow part remains. See also the velocity field discussed in the next section.

17.3 Velocity and Pressure Field

Properties of cylinder flow

In the previous section we solved the boundary value problem for the cylinder in parallel flow, and derived the velocity potential (17.26). Now we will study the resulting velocity and pressure distributions.

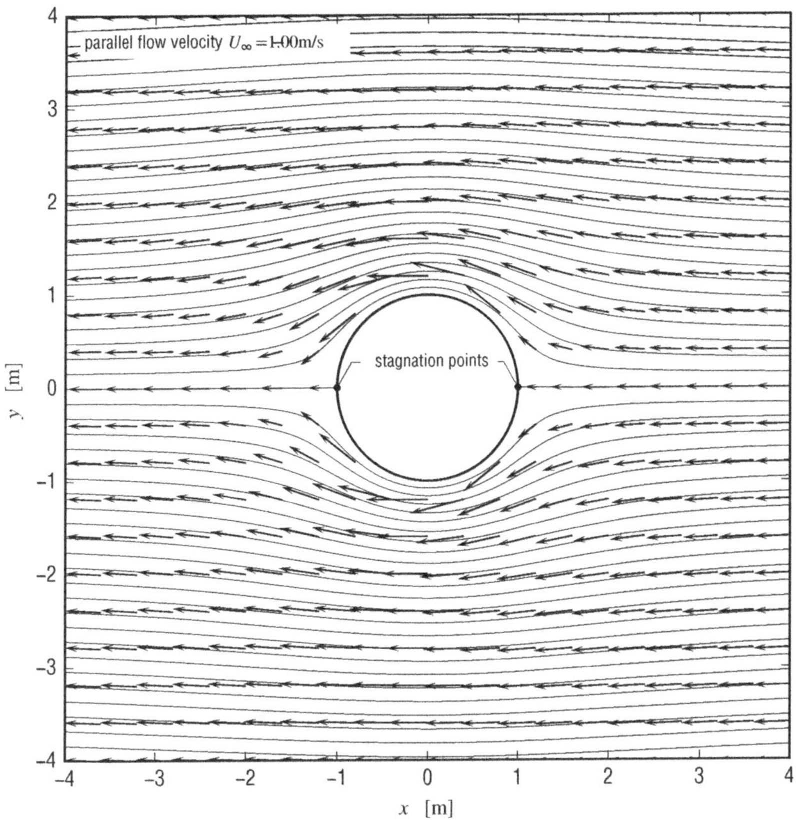

Figure 17.3 Streamlines and velocity field for a cylinder in parallel flow

17.3.1 Velocity field

Velocity vector

The gradient of the potential represents the velocity vector for points (r, φ) in the fluid domain. The gradient is stated with respect to the vector basis formed by the unit vectors $\underline{e}_r$ and $\underline{e}_\varphi$.

$$\underline{\nabla}\, \phi_{<r,\varphi>} = \begin{pmatrix} \dfrac{\partial \phi}{\partial r} \\ \dfrac{1}{r}\dfrac{\partial \phi}{\partial \varphi} \end{pmatrix}_{<r,\varphi>} = \begin{pmatrix} -U_\infty \cos\varphi + \dfrac{U_\infty R^2}{r^2}\cos\varphi \\ U_\infty \sin\varphi + \dfrac{U_\infty R^2}{r^2}\sin\varphi \end{pmatrix}_{<r,\varphi>} \tag{17.28}$$

Figure 17.3 shows the streamlines and velocity arrows for the cylinder in parallel flow. Since fluid cannot move across streamlines, the distance between streamlines may serve as a gage for the magnitude of the velocity. Where streamlines widen, the flow will slow down, and where they narrow, fluid will move faster. The length of the arrows is proportional to the flow speed. Obviously, we have areas of slow flow speeds in front of $(x = +1\text{ m}, y = 0)$ and behind the point $(x = -1\text{ m}, y = 0)$ on the cylinder. High flow speeds occur at the upper and lower side of the cylinder. As expected, the flow is completely symmetric to the x-axis. The minimum velocity is zero, and the maximum velocity is apparently two times the onflow velocity.

Velocity vector Let us investigate the velocity distribution on the body surface itself. On S_b the radius is $r = R$, and the angle takes values from the interval $\varphi \in [0, 2\pi]$. Substituting $r = R$ into Equation (17.28) yields:

$$\underline{v}(r = R) = \begin{pmatrix} v_r \\ v_\varphi \end{pmatrix}_{<r,\varphi>} = \begin{pmatrix} 0 \\ 2U_\infty \sin\varphi \end{pmatrix}_{<r,\varphi>} \tag{17.29}$$

This seems surprising at first but remember that the velocity vector is to the base $< \underline{e}_r, \underline{e}_\varphi >$. The vanishing radial velocity $v_r = 0$ simply reflects the body boundary condition from Equation (17.8).

The component $v_\varphi = 2U_\infty \sin\varphi$ is always tangent to the cylinder surface. Since the sine is limited to the range -1 to $+1$, the maximum velocity of $2U_\infty$ occurs for $\varphi = 90$ degree and $\varphi = 270$ degree which represent the points where the cylinder is widest with respect to the onflow. Stagnation points with $\underline{v} = \underline{0}$ are found for $\varphi = 0$ degree and $\varphi = 180$ degree, i.e. at the front and back of the cylinder.

17.3.2 Pressure field

Pressure distribution The pressure field is derived from the Bernoulli equation. Again, for inviscid, incompressible, and irrotational flow it is

$$\rho \frac{\partial \phi}{\partial t} + \frac{1}{2}\rho \left(\underline{\nabla}\,\phi\right)^T \underline{\nabla}\,\phi + p + \rho g z = C(t) \tag{15.41}$$

The equation has been derived from the conservation of momentum per unit volume equation, essentially by integration in space. Consequently, the terms in the equation represent the energy per volume unit of fluid. We consider the cylinder flow for a steady flow speed. Therefore, the unsteady pressure term $\rho\partial\phi/\partial t$ vanishes. Now the constant C will be independent of location and also be independent of time. We are free to choose a suitable value for C because the potential is only unique up to a constant.

For the cylinder flow a suitable value is

$$C = \frac{1}{2}\rho U_\infty^2 + p_\infty \tag{17.30}$$

Since forces result from pressure differences, we form:

$$p - p_\infty = \underbrace{\frac{1}{2}\rho\left(U_\infty^2 - \left(\underline{\nabla}\,\phi\right)^T \underline{\nabla}\,\phi\right)}_{\text{hydrodynamic pressure}} - \underbrace{\rho g z}_{\text{hydrostatic pressure}} \tag{17.31}$$

Hydrostatic pressure We assume that the cylinder is neutrally buoyant. In that case, the resultant force of the hydrostatic pressure component $\rho g z_0$ compensates the weight of the cylinder.[1]

[1]For a vertical cylinder, we could claim that the static pressure is equal all around the circumference. Hence, it has no resultant force.

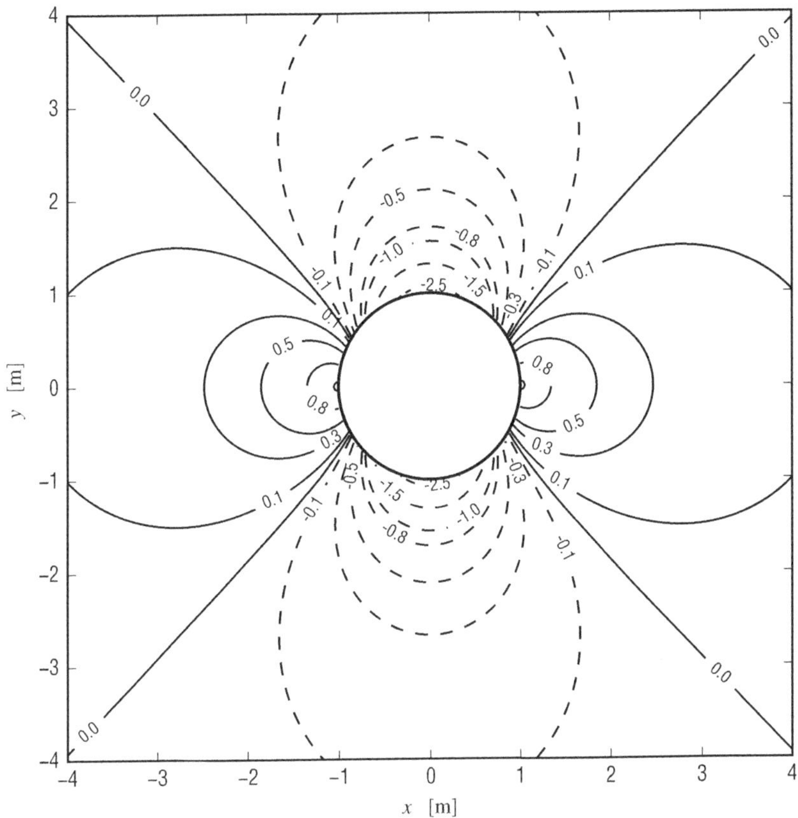

Figure 17.4 Contours of constant pressure coefficient C_p for a cylinder in parallel flow

We divide Equation (17.31) by the dynamic pressure, which yields a dimensionless pressure coefficient C_p.

$$C_p = \frac{p - p_\infty}{\frac{1}{2}\rho U_\infty^2} = 1 - \frac{|\underline{v}|^2}{U_\infty^2} \tag{17.32}$$

The pressure distribution clearly depends on the magnitude of the velocity $|\underline{v}|$. Since the velocity magnitude is symmetric to both axes, the pressure distribution will be symmetric as well (Figure 17.4).

Pressure on cylinder surface

On the cylinder surface, we find with $|\underline{v}|^2 = v_\varphi^2 = 4U_\infty^2 \sin^2\varphi$ that the pressure coefficient is equal to

$$C_p = 1 - 4\sin^2\varphi \tag{17.33}$$

It ranges from $+1$ at the stagnation points ($\varphi = 0$ and 180 degrees) to -3 at the shoulders where the velocity is highest ($\varphi = 90$ and 270 degrees).

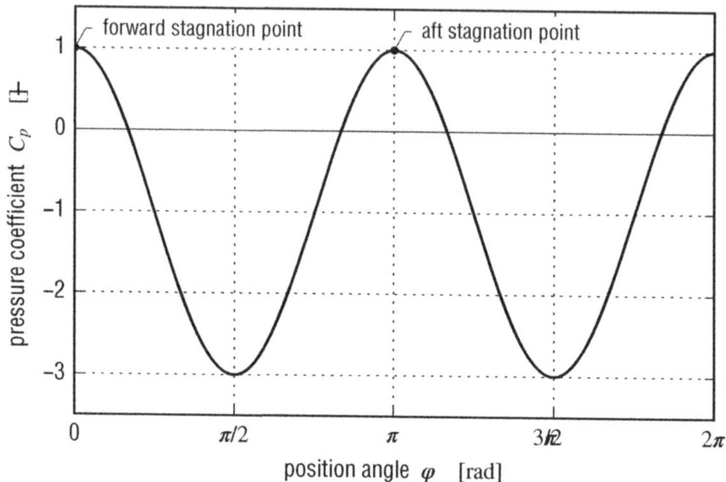

Figure 17.5 Pressure coefficient C_p distribution on the cylinder surface ($r = R$) for a cylinder in parallel flow

17.4 D'Alembert's Paradox

Pressure force We will now use the pressure distribution on the cylinder surface to compute the resultant pressure force. This problem is also known as d'Alembert's paradox because, in contrast to all practical experience, our computation will find a net zero force in all directions.

A pressure force on a body is obtained by integrating the pressure over the body surface:

$$\underline{F}_p = \begin{pmatrix} F_{px} \\ F_{py} \end{pmatrix} = \iint_{S_b} p\underline{n}\,\mathrm{d}S \tag{17.34}$$

This equation will show a minus sign in front of the integral when the normal vector points into the fluid. In our case, a positive pressure (higher than some reference pressure) will cause a force that points in positive normal direction. As mentioned before, we are not interested in the hydrostatic force. We will only integrate the hydrodynamic pressure component of Equation (17.31).

Surface element We assume our cylinder has the length ℓ in z-direction. A surface element of the cylinder surface is then given as

$$\mathrm{d}S = \ell R\,\mathrm{d}\varphi \tag{17.35}$$

$R\mathrm{d}\varphi$ represents a small piece of the circular arc of the cylinder contour. Our double integral over S_b reduces to a single integral over the angle $\varphi \in [0, 2\pi]$ because the flow is not changing in z-direction.

Pressure integration Substituting the hydrodynamic pressure yields for the pressure force

$$\underline{F}_p = \begin{pmatrix} F_{px} \\ F_{py} \end{pmatrix} = \int_0^{2\pi} \frac{1}{2}\rho U_\infty^2 \left(1 - 4\sin^2(\varphi)\right) \begin{pmatrix} -\cos(\varphi) \\ -\sin(\varphi) \end{pmatrix} \ell R \mathrm{d}\varphi \tag{17.36}$$

Extracting all terms independent of the integration variable φ results in the following expression:

$$\underline{F}_p = \frac{1}{2}\rho U_\infty^2 \ell R \int_0^{2\pi} \begin{pmatrix} -\cos(\varphi) + 4\sin^2(\varphi)\cos(\varphi) \\ -\sin(\varphi) + 4\sin^3(\varphi) \end{pmatrix} \mathrm{d}\varphi$$

After splitting the integral at the plus sign, we have

$$\underline{F}_p = \frac{1}{2}\rho U_\infty^2 \ell R \left[\int_0^{2\pi} \begin{pmatrix} -\cos(\varphi) \\ -\sin(\varphi) \end{pmatrix} \mathrm{d}\varphi + 4 \int_0^{2\pi} \begin{pmatrix} \sin^2(\varphi)\cos(\varphi) \\ \sin^3(\varphi) \end{pmatrix} \mathrm{d}\varphi \right]$$

and with the antiderivatives from an integral table follows

$$\underline{F}_p = \frac{1}{2}\rho U_\infty^2 \ell R \left(\left[\begin{pmatrix} -\sin(\varphi) \\ +\cos(\varphi) \end{pmatrix} \right]_0^{2\pi} + 4 \left[\begin{pmatrix} \frac{1}{3}\sin^3(\varphi) \\ \cos(\varphi) - \frac{1}{3}\cos^3(\varphi) \end{pmatrix} \right]_0^{2\pi} \right) \tag{17.37}$$

d'Alembert's paradox

Substituting the integral limits finally gives us

$$\underline{F}_p = \begin{pmatrix} F_{px} \\ F_{py} \end{pmatrix} = \begin{pmatrix} 0 \\ 0 \end{pmatrix} \tag{17.38}$$

Both x- and y-components of the force vanish. Since the body is symmetric to the x-axis, we expected no transverse force F_{py}. However, $F_{px} = 0$ clearly violates any practical observation and experiment. If we place a cylinder in a stream, it will experience a drag force, and we have to secure the cylinder or it will be washed downstream. The theory leads to a contradiction with the experiment, a paradox. The conclusion that something is wrong with the theory was drawn by Jean-Baptiste le Rond d'Alembert in 1752 and is therefore named d'Alembert's paradox. Today we know that neglecting the viscous forces (solving Euler equations instead of Navier-Stokes equations) is the cause for the unrealistic pressure force prediction.

17.5 Added Mass

Of course, pressure forces do not vanish for all potential flows. The picture changes already when we consider, for example, the unsteady potential flow around a cylinder moving through the fluid at rest. Although this thematically belongs to seakeeping and maneuvering, it is shown below that the resultant pressure force gives rise to the so called added mass. It is a hydrodynamic pressure force proportional to the relative acceleration between body and fluid.

Potential of unsteady cylinder flow Solving the boundary value problem (17.3) of the cylinder moving through a fluid at rest yields the velocity potential of a dipole.

$$\phi(r, \varphi, t) = \frac{-U(t)R^2}{r} \cos(\varphi) \tag{17.39}$$

However, here the velocity $U(t)$ of the cylinder (in x-direction) is considered a function of time. The gradient of a potential provides the velocity vector for the fluid. With respect to a polar coordinate vector base we have

$$\underline{v}(r, \varphi, t) = \nabla\phi(r, \varphi, t) = \begin{pmatrix} \dfrac{U(t)R^2}{r^2} \cos(\varphi) \\ \dfrac{U(t)R^2}{r^2} \sin(\varphi) \end{pmatrix} \tag{17.40}$$

The magnitude of the velocity vector is

$$|\underline{v}| = \left| \frac{U(t)R^2}{r^2} \right| \tag{17.41}$$

Unsteady pressure field The corresponding unsteady pressure distribution is obtained from the unsteady Bernoulli equation (15.41).

$$\rho \frac{\partial \phi}{\partial t} + \frac{1}{2}\rho \left(\underline{\nabla}\phi\right)^T \underline{\nabla}\phi + \rho g z + p = C(t) \tag{15.41}$$

Like before, we assume that resultant hydrostatic pressure force and weight are in equilibrium. As a constant $C(t)$, we choose the constant pressure p_∞ in the far field. The remaining dynamic pressure acting on the cylinder surface will be

$$p_{\text{dyn}}(r, \varphi, t) = p - p_\infty = -\rho \frac{\partial \phi}{\partial t} - \frac{1}{2}\rho \left(\underline{\nabla}\phi\right)^T \underline{\nabla}\phi \tag{17.42}$$

Only the body velocity is time dependent in the velocity potential (17.39). Therefore, the time derivative of the potential is

$$\frac{\partial \phi}{\partial t} = \frac{\partial}{\partial t}\left(\frac{-U(t)R^2}{r} \cos(\varphi) \right) = -\frac{\partial U(t)}{\partial t} \frac{R^2}{r} \cos(\varphi) \tag{17.43}$$

Substituting Equations (17.41) and (17.43) into (17.42) and evaluating the expression for $r = R$ results in the pressure on the cylinder surface.

$$p_{\text{dyn}}(r = R, \varphi, t) = \rho R \frac{\partial U(t)}{\partial t} \cos(\varphi) - \frac{1}{2}\rho \left|U(t)\right|^2 \tag{17.44}$$

Unsteady pressure force For the resulting pressure force, we use the definition (17.34) and the surface element dS from Equation (17.35).

$$\underline{F}_p(t) = \begin{pmatrix} F_{px}(t) \\ F_{py}(t) \end{pmatrix} = \int_0^{2\pi} \left(\rho R \frac{\partial U(t)}{\partial t} \cos(\varphi) - \frac{1}{2}\rho \left|U(t)\right|^2 \right) \begin{pmatrix} -\cos(\varphi) \\ -\sin(\varphi) \end{pmatrix} \ell R d\varphi \tag{17.45}$$

As before, we split the integral into terms and extract all factors which do not depend on the integration variable φ.

$$\underline{F}_p(t) = \rho \ell R^2 \frac{\partial U(t)}{\partial t} \int_0^{2\pi} \begin{pmatrix} -\cos^2(\varphi) \\ -\sin(\varphi)\cos(\varphi) \end{pmatrix} \mathrm{d}\varphi - \frac{1}{2} \rho \ell R \, |U(t)|^2 \int_0^{2\pi} \begin{pmatrix} -\cos(\varphi) \\ -\sin(\varphi) \end{pmatrix} \mathrm{d}\varphi \quad (17.46)$$

The second integral in Equation (17.46) already appeared in the steady parallel flow case (17.37) and does not produce a force.

The remaining integral is easily solved.

$$\begin{aligned} \underline{F}_p(t) &= \rho \ell R^2 \frac{\partial U(t)}{\partial t} \int_0^{2\pi} \begin{pmatrix} -\cos^2(\varphi) \\ -\sin(\varphi)\cos(\varphi) \end{pmatrix} \mathrm{d}\varphi \\ &= -\rho \ell R^2 \frac{\partial U(t)}{\partial t} \left[\begin{pmatrix} \frac{1}{2}\varphi + \frac{1}{4}\sin(2\varphi) \\ \frac{1}{2}\sin(2\varphi) \end{pmatrix} \right]_0^{2\pi} \\ &= -\rho \ell R^2 \frac{\partial U(t)}{\partial t} \left[\begin{pmatrix} \pi + 0 - 0 - 0 \\ 0 - 0 \end{pmatrix} \right]_0^{2\pi} \end{aligned} \quad (17.47)$$

Again, due to the symmetry with respect to the x-axis the force in transverse direction vanishes as expected.

Added mass force

However, in contrast to the steady flow case, we now have a resultant force in the x-direction: the added mass force

$$F_{p_x} = -\rho \ell \pi R^2 \frac{\partial U(t)}{\partial t} \quad (17.48)$$

The minus sign indicates that it is acting against the direction of motion. The force is proportional to the acceleration $\partial U/\partial t$ of the cylinder. Obviously, if the cylinder moves at a constant speed and $\partial U/\partial t = 0$, we return to the d'Alembert paradox case. In general, the added mass force is proportional to the relative acceleration between fluid and body.

Added mass

The factor in front of the acceleration must have the dimension of mass. Indeed, $\rho \ell \pi R^2$ is equivalent to the mass of the water displaced by the cylinder and is called added mass. For noncylindrical shapes, the added mass is generally different from the mass of the displaced water. The term 'added mass' is somewhat misleading. Unlike the ballast water pumped into a ship, there is no actual mass added to the cylinder. However, the unsteady pressure force creates an apparent increase in the inertia of the cylinder because it is proportional to the relative acceleration between body and fluid. Added mass is important for the correct assessment of the dynamics of ships in waves and during maneuvering in calm water.

Self Study Problems

1. Derive the velocity vector in Cartesian coordinates for the flow potential (17.27) for a cylinder in parallel flow.

2. The velocity potential for the flow around a rotating cylinder is given by

$$\phi(x, y) = -U_\infty x - U_\infty R^2 \frac{x}{x^2 + y^2} + \frac{\Gamma}{2\pi} \arctan\left(\frac{y}{x}\right) \quad \text{Cartesian coordinates}$$

$$\phi(r, \varphi) = -U_\infty r \cos(\varphi) - \frac{U_\infty R^2}{r} \cos(\varphi) + \frac{\Gamma}{2\pi}\varphi \quad \text{polar coordinates}$$

Γ is the circulation strength, R the radius of the cylinder, and U_∞ the velocity of the parallel flow.

   (a) Derive the equations for the velocity vector in Cartesian and polar coordinates.

   (b) Compute the location of the stagnation points as a function of the position angle φ for circulation strength values $\Gamma \leq 4\pi U_\infty R$.

   (c) What happens to the stagnation point if $\Gamma > 4\pi U_\infty R$?

   (d) Compute the resultant pressure force acting on the cylinder as a function of U_∞, R, and Γ. This is best done in polar coordinates.

18

Viscous Pressure Resistance

The ideal flow assumption used in potential theory is useful in a number of cases. However, as d'Alembert's paradox (Section 17.4) expresses, potential theory is incapable of predicting the total resistance of a body. Even if we include the frictional resistance caused by the shear stresses in the boundary layer (Chapter 13), the result will not match experimental drag data. What is missing is the displacement effect of the boundary layer onto the exterior flow. The displacement effect changes the pressure distribution around a hull, which results in the viscous pressure resistance. In some cases, especially for blunt bodies, flow separation may amplify the effect.

Learning Objectives

At the end of this chapter students will be able to:

- illustrate the effects of a boundary layer on the external flow
- discuss viscous pressure resistance
- recognize flow separation

18.1 Displacement Effect of Boundary Layer

Displacement effect

The viscosity of the fluid results in the frictional resistance. In addition, the resulting boundary layer also affects the exterior flow. Since the fluid in the boundary layer is slowed down by friction, less mass is flowing through the boundary layer per unit time compared to a hypothetical inviscid fluid flow. The displacement thickness δ_1, introduced in Chapter 11, quantifies this effect.

The flow outside of the boundary layer is pushed away from the hull (Figure 18.1). Due to this displacement effect, the body appears to be thicker for the exterior flow than its actual structural dimensions. Since the boundary layer thickness grows from bow to stern, the aft ship is more affected than the bow.

The outer limit of the boundary layer is defined as the position off the hull surface, where local velocity reaches the undisturbed exterior (inviscid) flow speed. As noted before, the curve marking the boundary layer thickness (dashed line in Figure 18.1)

Fundamentals of Ship Hydrodynamics: Fluid Mechanics, Ship Resistance and Propulsion, First Edition.
Lothar Birk.

Companion website: www.wiley.com/go/birk/hydrodynamics

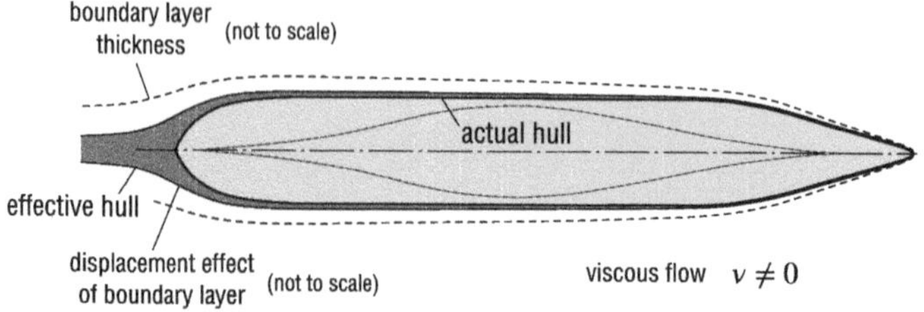

Figure 18.1 The displacement effect of a boundary layer changes the effective hull shape

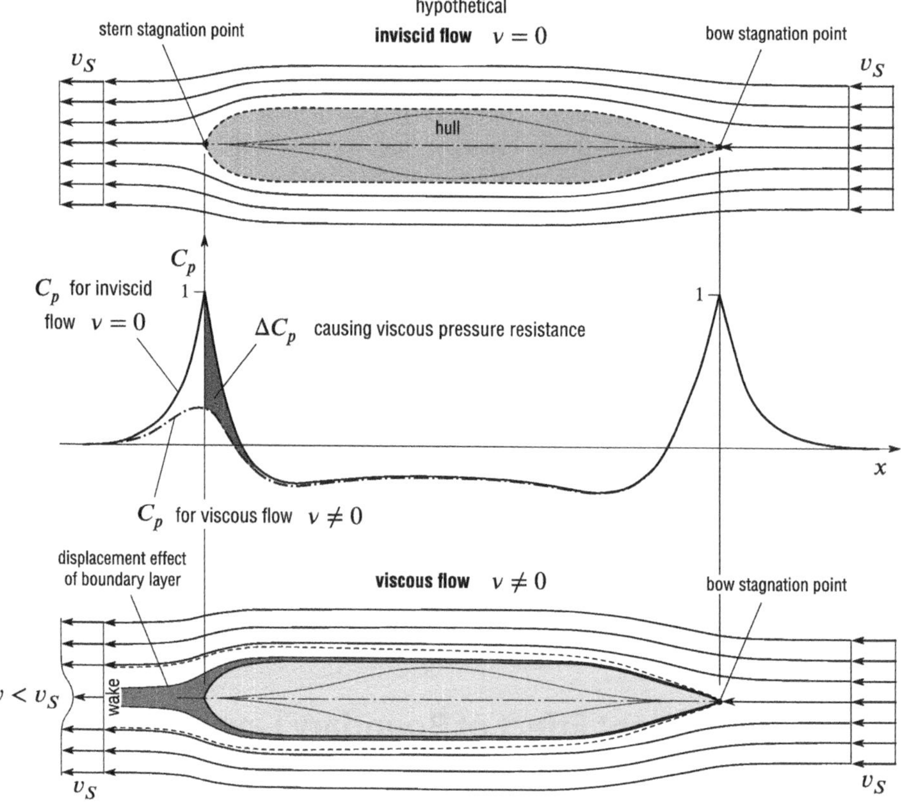

Figure 18.2 The effect of viscous flow on the pressure distribution

is not a streamline of the flow, and neither is the solid line marking the displacement effect. However, the displacement effect boundary may be interpreted as a streamline for a hypothetical, inviscid flow which would have the same mass flow properties as the actual viscous flow. This imaginary streamline extends the hull aft beyond the actual physical surface with significant effects on the pressure distribution around the hull.

Change in pressure distribution

In our studies of potential (inviscid) flow around a cylinder, we found stagnation points at the front and back side of the body. This applies to other geometries submerged in an ideal fluid. The pressure coefficient takes the value $C_p = 1$ at stagnation points. The pressure forces on front and back side are equal but act in opposite directions. As a consequence, the resulting pressure force vanishes in the flow direction if the fluid is considered inviscid (d'Alembert's paradox). This hypothetical case is depicted at the top of Figure 18.2.

Viscous pressure resistance

In a real fluid, only the stagnation point at the bow remains. The aft stagnation point has vanished due to the displacement effect of the boundary layer. Therefore, the pressure remains lower there compared to the ideal flow case. High pressure at the bow (stagnation point) and lower pressure at the stern create a net force which acts against the body's forward motion. We call this force the viscous pressure resistance. Like other resistance components, the viscous pressure resistance cannot be measured directly. In theory, the pressure distribution over the hull may be measured, but we still would be unable to separate the contribution of the displacement effect from other flow features like, for instance, the ship generated waves.

Function of Reynolds number

As the name suggests, viscous pressure resistance R_{VP} is a consequence of the viscosity of fluid. Therefore, it must be dependent on the Reynolds number Re. Despite this fact, William Froude knowingly included R_{VP} in the residuary resistance which we assume to be independent of the Reynolds number and to be solely a function of the Froude number $C_R(Fr)$. Without a practical means of determining R_{VP}, Froude had little choice.

Form factor

The recommended procedures of the International Towing Tank Conference (ITTC, http://www.ittc.info) encapsulate the viscous pressure resistance in the form factor k (Gross, 1975). The major component of viscous resistance for model and full scale vessel is the flat plate equivalent frictional resistance, represented by the ITTC 1957 model–ship correlation coefficient C_F.

$$C_{VM} = (1+k)\,C_{FM} \qquad \text{model} \qquad (18.1)$$

$$C_{VS} = (1+k)\,C_{FS} + \Delta C_F \qquad \text{ship} \qquad (18.2)$$

The portion kC_F captures the viscous pressure resistance and other three-dimensional effects. The same form factor is used for model and ship. ΔC_F is the roughness correction which is applied only to the full scale vessel because models are considered hydrodynamically smooth.

Prohaska (1966) proposed a method to determine the form factor from model tests (see Section 29.3 for details). It is considered the most reliable procedure available. However, the uncertainty of results is still high. Several formulas have been proposed to estimate the form factor k. Unfortunately, neither of them is universal nor does their use significantly improve the resistance prediction from model tests (Gross, 1975). One of the simpler formula has been presented by Watanabe (1973):

$$k = -0.095 + \frac{25.6\,C_B}{\left(\frac{L_{WL}}{B}\right)^2 \sqrt{\frac{B}{T}}} \qquad (18.3)$$

C_B is the block coefficient, B the molded beam, T the molded draft, and L_{WL} the length in the waterline. Granville (1974) derived a formula based on Schoenherr's friction line. The following version of Granville's formula has been corrected for use with the ITTC

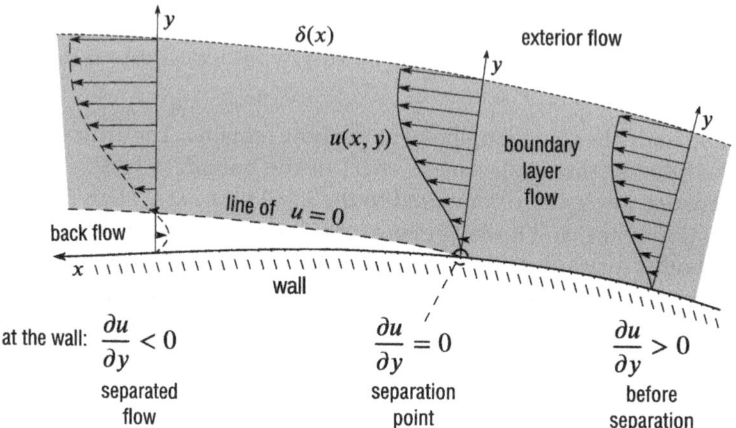

Figure 18.3 Velocity profiles within the boundary layer near a separation point

1957 model–ship correlation line (Gross, 1975).

$$k = -0.03 + 32.8 \frac{C_B^2}{\left(\frac{L_{WL}}{B}\right)^2 \left(\frac{B}{T}\right)} \tag{18.4}$$

Unfortunately, the correlation of form factors derived from Equations (18.3) and (18.4) with results from model tests is poor (Gross, 1975).

Grigson (2000) defines a regression formula specifically for usage with his friction line Equation (13.30).

$$k = 0.028 + 3.30 \left[\frac{S}{(L_{WL})^2} \sqrt{C_B \frac{B}{L_{WL}}}\right] \tag{18.5}$$

use only in conjunction with Equation (13.30)

Additional formulas are available in more recent methods for resistance estimates which use a form factor (see e.g. Chapter 50).

18.2 Flow Separation

Separation point Friction within the fluid, especially at the wall, causes a continuous loss of momentum. The fluid particles slow down in the boundary layer and the boundary layer grows in thickness. Except for turbulent fluctuations, the flow maintains the direction of the external flow and follows the body contour. The velocity profile across the boundary layer has a positive gradient $\partial u/\partial y > 0$ at the wall (Figure 18.3). However, the gradient diminishes due to the continuous loss of momentum. Eventually, the velocity profile has a zero gradient perpendicular to the wall: $\partial u/\partial y = 0$. This condition marks the separation point. Just downstream of the separation point the fluid flows in the opposite direction (back flow). Further downstream we encounter a confused flow field with large and small eddies.

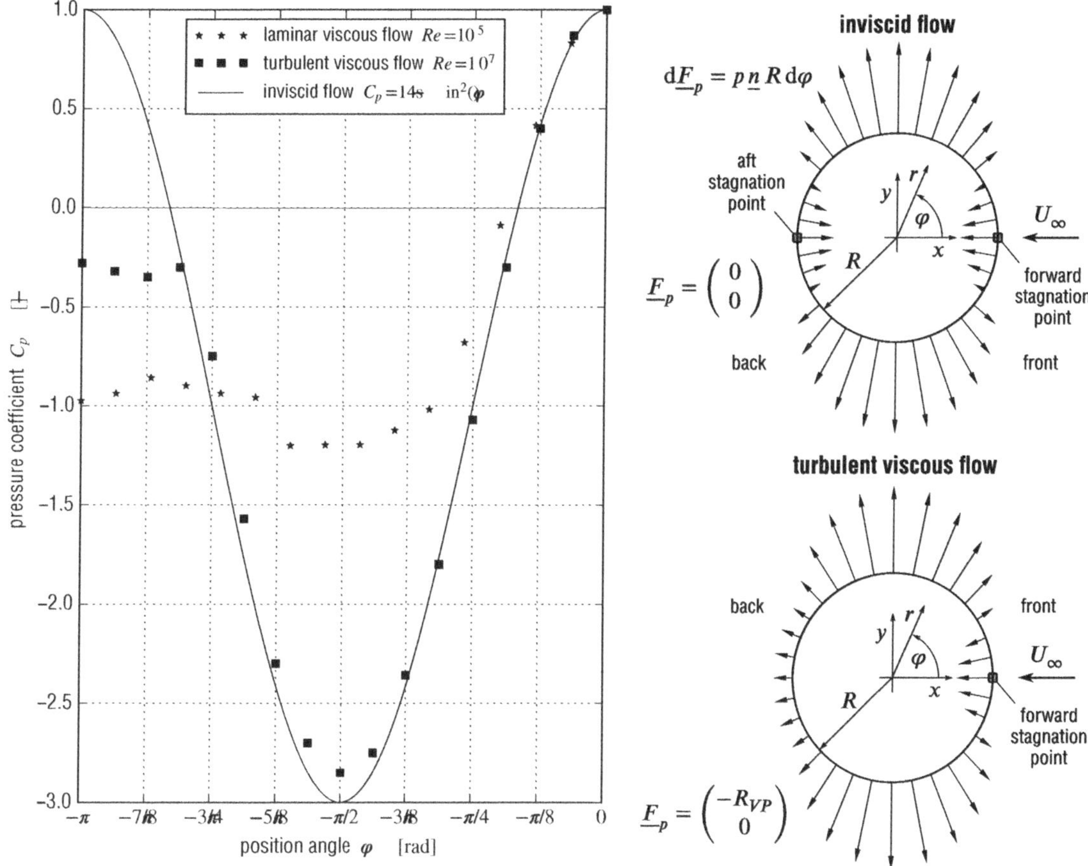

(a) Typical pressure distributions over one side of a cylinder for inviscid flow as well as laminar and turbulent viscous flow

(b) Resulting pressure forces for inviscid flow and turbulent viscous flow

Figure 18.4 Comparison of pressure and forces acting on a cylinder in inviscid and viscous flow

Ludwig Prandtl's boundary layer theory provided the first theory capable of explaining the phenomenon of separation and a method to estimate the location of the separation point. However, the estimates are not necessarily accurate because the boundary layer is no longer thin at the separation point, which violates the basic assumption of Prandtl's boundary layer theory.

It should be emphasized that flow separation does not occur on a well designed ship hull. If at all, ships with high block coefficients, like tankers and bulk carriers, may encounter small areas of flow separation at the stern. Blunt objects like spheres, cylinders, or foils at high angles of attack will suffer from flow separation.

Pressure distribution in ideal and viscous flow

We study the effect of the boundary layer and flow separation on the pressure distribution around a cylinder. Figure 18.4(a) compares the pressure coefficient derived for inviscid flow

$$C_p = 1 - 4\sin^2(\varphi) \qquad (r = R) \tag{17.33}$$

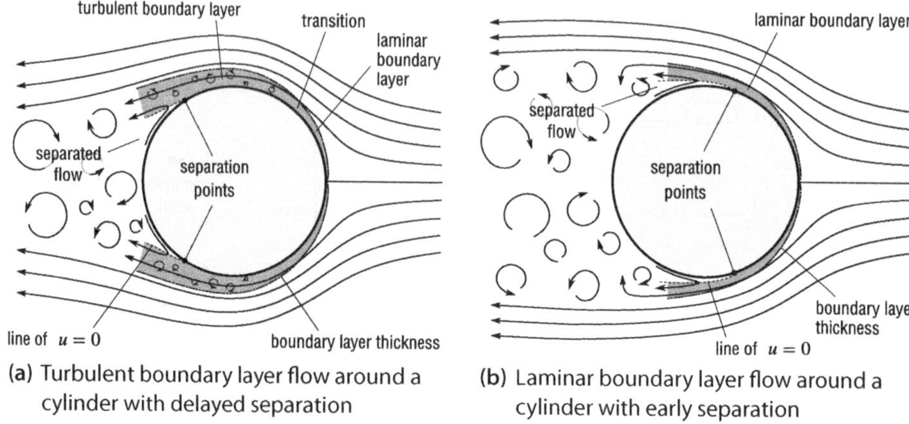

(a) Turbulent boundary layer flow around a cylinder with delayed separation

(b) Laminar boundary layer flow around a cylinder with early separation

Figure 18.5 Comparison of turbulent and laminar boundary layer flow around a cylinder

with typical experimental results for laminar viscous flow and turbulent viscous flow. φ is the position angle with respect to the positive x-axis. The results for turbulent flow ($Re = 10^7$, marked with ■) match the ideal flow results pretty well, except for the backside of the cylinder where the pressure coefficient does not rise back to the level $+1$ of a stagnation point. Instead, it levels off at a roughly constant value. The root cause is the displacement effect of the boundary layer, augmented by flow separation.

Viscous pressure resistance of cylinder

Figure 18.4(b) compares the pressure force distributions on the cylinder surface for ideal flow and turbulent viscous flow. For the latter, the pressure force distribution is no longer symmetric with respect to the y-axis. Both, front and back, feature differential pressure forces $\mathrm{d}\underline{F}_p$ which point in flow direction (against the movement of the cylinder) and result in the viscous pressure resistance R_{VP}.

Turbulent flow separation

The pressure distribution for laminar flow in Figure 18.4(a) ($Re = 10^5$, marked with $\star$) shows little resemblance to the potential flow pressure coefficient (solid line). This is the result of early flow separation and typically results in a higher drag coefficient compared with the turbulent flow case. At first glance, this is surprising because wall friction is higher for turbulent flow. However, momentum in the turbulent boundary layer is replenished through energy exchange with the exterior flow, allowing it to stay attached longer and recover more of the pressure compared with the laminar boundary layer. This is depicted in Figure 18.5(a). The separation points are on the back side of the cylinder and the region of separated flow is comparatively small. The exact locations of the separation points depend on a number of factors: Reynolds number, preturbulence of the flow, surface roughness, and others.

Laminar flow separation

The laminar flow case is shown in Figure 18.5(b). It lacks energy exchange with the exterior flow. Thus, laminar flow reaches the condition $\partial u/\partial y = 0$ for separation earlier. In Figure 18.5(b) separation occurs even before the maximum width of the body. A large area of separated flow is created and, as a consequence, even less of the pressure is recovered, causing a high viscous pressure resistance (Figure 18.4(a)).

As discussed in Chapter 13, boundary layers of ships are turbulent due to the high Reynolds numbers. Ship models use a trip wire, studs, or a strip of sand to trigger the laminar–turbulent transition of the boundary layer. This not only ensures that the boundary layer of the model is mostly turbulent, it also prevents the unwanted effects of possible laminar flow separation.

References

Granville, P. (1974). Progress in the analysis of the viscous resistance of surface ships. Technical Report SPD 581-01, Naval Ship Research and Development Center (NSRDC).

Grigson, C. (2000). A planar friction algorithm and its use in analysing hull resistance. In *Transactions of The Royal Institution of Naval Architects (RINA)*, volume 142, pages 76–115. The Royal Institution of Naval Architects.

Gross, A. (1975). Form factor. In *Proceedings of the 14th ITTC*, pages 576–590, Ottawa, Canada. International Towing Tank Conference. Report of the Performance Committee, Appendix 4.

Prohaska, C. (1966). A simple method for the evaluation of the form factor and the low speed wave resistance. In *Proceedings of the 11th ITTC*, pages 65–66, Tokyo, Japan. International Towing Tank Conference. Resistance Session, Written Contributions.

Watanabe, K. (1973). Note to the Performance Committee. International Towing Tank Conference.

Self Study Problems

1. What is the main cause of the viscous pressure resistance of a ship?
2. Sketch how the pressure distribution around a cylinder in a viscous fluid deviates from the ideal flow case.
3. Golf balls with dimples fly about 30% farther than experimental smooth golf balls. Based on your knowledge of laminar and turbulent flow separation, discuss why a golf ball with dimples flies farther than a perfectly smooth golf ball.
4. How does flow separation increase the resistance of a blunt body?

19

Waves and Ship Wave Patterns

Ships generate a very distinctive wave pattern, also known as the Kelvin wave pattern in honor of William Thomson's (Lord Kelvin, *1824 – †1907) efforts in developing a suitable theory (Thomson , Lord Kelvin). This chapter discusses basic properties of waves and the Kelvin wave pattern utilizing results of linear wave theory. This may serve as a foundation for our more detailed discussion of linear wave theory in subsequent chapters.

Learning Objectives

At the end of this chapter students will be able to:

- discuss basic wave descriptors like wave height, wave length, and wave period.
- understand the phenomenon of wave dispersion
- distinguish phase velocity and group velocity
- describe ship wave patterns

19.1 Wave Length, Period, and Height

Free surface waves Before discussing the ship wave pattern itself, we will review some basic properties of water waves. The surface separating the liquid water from the gaseous air is free of parallel shear stresses when the fluids are in equilibrium. In physics, this is called a free surface. Changes in pressure at the free surface, e.g. due to wind or a moving object, create waves, which subsequently propagate across the water surface.

Wave profile Cutting through the water surface parallel to the direction of propagation reveals the wave elevation ζ, which measures the deviation from the calm water level $z = 0$. The wave profile will resemble the solid curve shown in Figure 19.1.

Wave length and height The horizontal distance between two wave crests marks the wave length L_w. The maximum vertical distance between wave trough and wave crest defines the wave height H. Note that the vertical scale is enlarged in Figure 19.1. Typically waves have steepness ratios of H/L_w less than 0.05. In theory, waves will break once they reach a

Fundamentals of Ship Hydrodynamics: Fluid Mechanics, Ship Resistance and Propulsion, First Edition.
Lothar Birk.

Companion website: www.wiley.com/go/birk/hydrodynamics

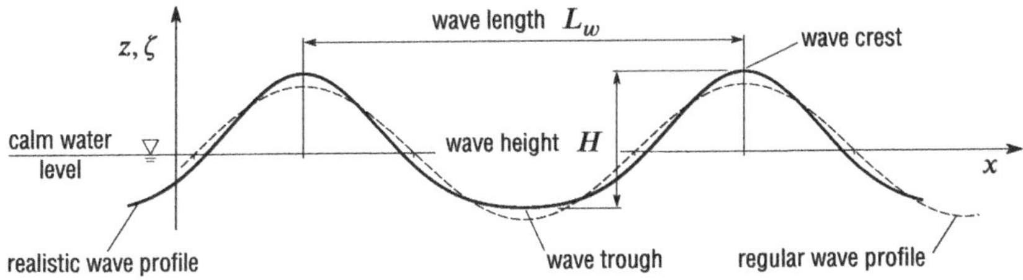

Figure 19.1 Definition of wave length L_w and wave height H; the vertical scale is exaggerated

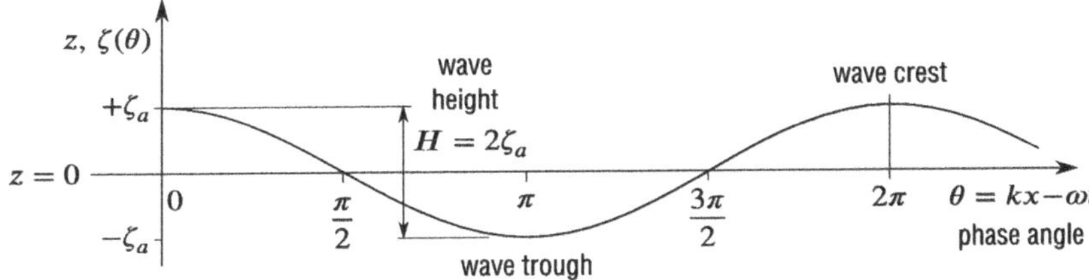

Figure 19.2 Surface elevation of a harmonic, long-crested wave

wave steepness of $H/L_w = 1/7$, but usually break before they reach this limit (Michell, 1893).

Regular, long-crested waves

In simulations, we often repeat the same wave profile over and over again: a regular wave. Natural waves are irregular and have to be described by probabilistic methods. Furthermore, the wave profile is assumed to be the same in every plane parallel to the direction of propagation. The crests and troughs of this long-crested wave stretch to infinity in the transverse direction. In linear wave theory, the actual wave profile is approximated by a cosine function.

$$\zeta(z,t) = \zeta_a \cos(kx - \omega t) \tag{19.1}$$

This simplified, harmonic wave is represented in Figure 19.1 by the dashed curve. Note that the wave crest of the real wave profile is higher than the wave trough is deep. The wave trough is wider than the wave crest.

Wave amplitude

In Equation (19.1), $\theta = kx - \omega t$ defines the phase of a wave. Figure 19.2 shows the wave elevation $\zeta(x,t)$ for one cycle $\theta \in [0, 2\pi]$ from wave crest to wave crest. Since the cosine function is limited to values between ± 1, the wave elevation takes values between $\pm\zeta_a$. Thus, the total range of values for a harmonic wave is two times the wave amplitude ζ_a. The wave amplitude is always positive and defines the maximum deviation of the wave elevation from its mean value $z = 0$.

Wave crests and troughs

Wave crests (maxima of ζ) occur at phases equal to even multiples of π, i.e. 0, 2π, 4π, etc.:

$$\max(\zeta(\theta)) = +\zeta_a \qquad \text{for } \theta = 2k\pi \text{ with k=0,1,2,...}$$

Wave troughs (minima of ζ) occur when the phase θ is equal to odd multiples of π. For harmonic waves, the wave height H is equal to twice the wave amplitude.

$$H = 2\zeta_a \tag{19.2}$$

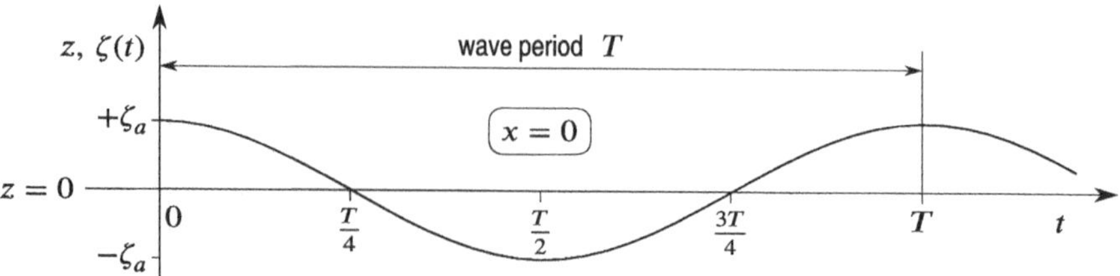

Figure 19.3 Recording of surface elevation of a harmonic, long-crested wave at a fixed position $(x = 0)$

Wave period Let us consider a wave probe which records the free surface elevation $\zeta(t)$ as a function of time at the fixed position $x = 0$ (Figure 19.3). The time that passes between the recording of one wave crest and the next subsequent wave crest is the wave period T.

Wave frequency The completion of a single cycle during one period T requires a change in phase of $\Delta\theta$ equal to two π.

$$\Delta\theta = 2\pi = (kx - \omega t)_{(x=0,t=T)} = -\omega T$$

In other words, the wave frequency ω is reciprocal to the wave period T.

$$\omega = \frac{2\pi}{T} \qquad T = \frac{2\pi}{\omega} \tag{19.3}$$

As an angular frequency, the wave frequency measures how much of a single wave cycle is happening in 2π seconds. Its physical unit is [1/s], however, we often denote it as [rad/s] to distinguish it from a frequency measured in Hz = 1/s. A wave frequency of $\omega = 1$ rad/s means a cycle is completed in $T = 2\pi$ s or approximately 6.283 seconds. With a wave frequency of $\omega = 0.5$ rad/s, only half of the cycle is completed within $2\pi \approx 6.283$ seconds. The period of this wave would be $T = 4\pi$ s = 12.566 seconds. Typical wave frequencies for ocean and ship generated waves are in the range of 0.1 rad/s to 2.0 rad/s.

Wave length If we take a photograph with a very short exposure time, the wave is captured as a function of space x. For convenience, we choose the time $t = 0$. Cutting through the wave in direction of wave progression x reveals the wave length L_w (see Figure 19.4). If the wave length is long, the wave is called long. As we will show later, long waves also have a high wave period (small frequency). Short waves have small wave periods and high wave frequencies.

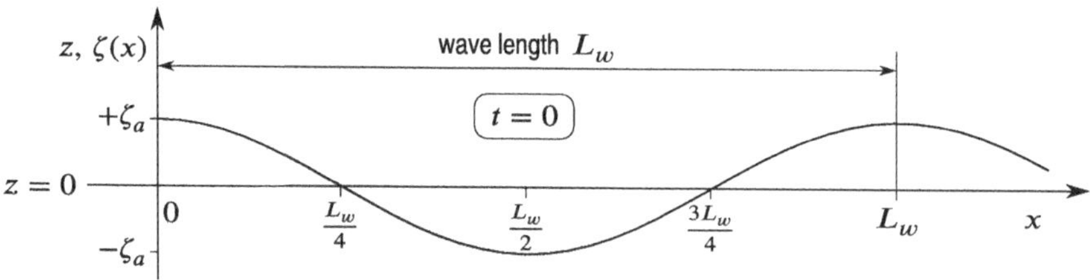

Figure 19.4 Spatial extension of surface elevation of a linear, harmonic, long-crested wave captured at time $(t = 0)$

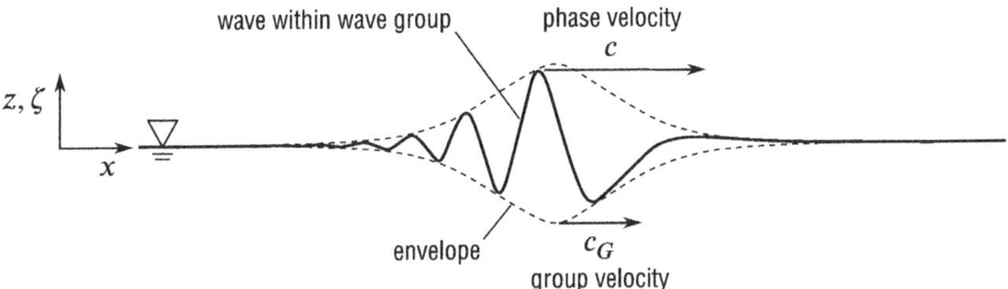

Figure 19.5 A snapshot of the wave elevation in a wave group

Wave number

At any selected point in time t, the change in phase of $\Delta\theta$ from one wave crest to the other is still equal to 2π.

$$\Delta\theta = 2\pi = (kx - \omega t)_{(x=L_w, t=0)} = k\,L_w$$

The wave number k is the spatial equivalent to the wave frequency. Wave number and wave length are reciprocal.

$$k = \frac{2\pi}{L_w} \qquad\qquad L_w = \frac{2\pi}{k} \tag{19.4}$$

The wave number expresses how much of the wave length fits into a distance of 2π meter. Wave numbers are noted in the unit 1/m. Some prefer the unit rad/m to indicate the factor 2π. A wave frequency of $k = 1$ rad/m is equivalent to a wave length of $L_w = 2\pi$ m ≈ 6.283 meter. With a wave number of $k = 0.1$ rad/m, a wave is approximately 62.83 meters long and a tenth of it fits into 2π meters. Since a wave length is always a positive distance, wave numbers are positive as well.

Phase velocity

A wave travels one wave length during one wave period. Therefore, the speed of propagation of a water wave is

$$c = \frac{L_w}{T} = \frac{\omega}{k} \tag{19.5}$$

If you move in lockstep with a wave crest or any other point along the wave, you have to move with velocity c. Since you will have a constant phase relationship to the wave, the velocity of wave propagation is known as phase velocity.

19.2 Fundamental Observations

Wave groups

When you throw a pebble into a calm pond, a group of waves is created. Figure 19.5 shows the vertical displacement of the water surface ζ along the wave group. As the waves propagate and expand in the form of concentric circles, three observations can be made:

(i) Long waves propagate faster than short waves, i.e. if $L_{w1} > L_{w2}$ then $c_1 > c_2$.

(ii) At the front of the wave group, wave amplitude diminishes until the wave vanishes, whereas at the end of the wave group, waves seem to emerge from nothing.

(iii) The waves seem to travel from the back to the front of the group, while the envelope of the wave group moves slower than the individual waves.

Wave dispersion The first observation (i) is known as wave dispersion. Sound waves of varying frequency all travel with a constant speed, i.e. the speed of sound. Electromagnetic waves of different frequency propagate with the speed of light. In contrast, water waves with small wave frequency (long waves) travel faster than short waves with high wave frequency. Very long waves, like the ones created by tsunamis, may cross whole oceans in a few hours. The phase velocity is also influenced by the water depth h. We will derive the exact relationship between wave frequency, water depth, and phase velocity in Section 22.3.

Wave energy Observation (ii) is related to the energy transport in waves. Chapter 24 presents a detailed discussion of wave energy and wave propagation. For now, it may suffice to state that waves contain two types of energy:

- kinetic energy – water particles underneath a wave move in elliptical paths. Since water particles have mass and velocity, they possess kinetic energy.
- potential energy – the up and down of the free surface contains potential energy comparable to the energy of a pendulum. However, potential wave energy propagates with the wave.

The water is initially at rest in front of the wave group. For the water particles to assume the motion characteristic to waves, potential energy is converted into kinetic energy at the front of the wave. As a consequence, the wave amplitude diminishes. At the back of the wave group, kinetic energy is converted into potential energy and waves emerge. Once all kinetic energy has been converted into potential energy and transported away, the water is at rest again.

Viscous effects Viscosity of the water will dissipate some wave energy during this process. This is especially true for waves of very short length (< 0.1 m), so-called capillary waves. They are also affected by surface tension. Ship and ocean waves have longer wave lengths, and the effects of surface tension and viscosity may be neglected for many engineering applications.

Group velocity The observations (ii) and (iii) are connected. Individual waves travel with their respective phase velocity c. As the wave group approaches, water particles are set in motion, but more or less come to rest in their initial position once the wave group has passed. This means that only the potential energy propagates. As we will discuss in Chapter 24, wave energy travels at the group velocity c_G. On deep water it is just half of the phase velocity. The envelope of a wave group reflects the reduced speed of the energy transport. For that reason, you will see waves emerge from the back of the wave group, travel through the group to the front, and finally diminish because their energy is used to set water particles in motion. The relationship between group and phase velocity affects the formation of the Kelvin wave pattern, which we will discuss in the following section.

Wave superposition Next time you pass a quiet pond, throw two pebbles into the water a few yards apart. Watch how the two systems of radiated circular waves spread and interact. For the most part, they will just pass each other and continue on undisturbed. A notable change will occur only if the superposition of waves causes wave breaking. You may notice that

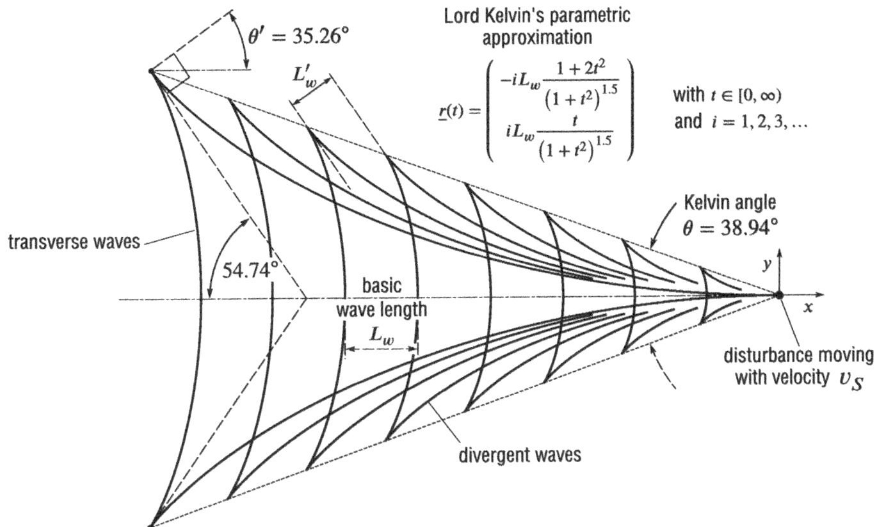

Figure 19.6 Kelvin wave pattern in deep water

the waves appear to be circular after a short distance, no matter what the shapes of the pebbles were.

19.3 Kelvin Wave Pattern

Anything moving at or close enough to the water surface creates waves: You may drag a stick through the water, watch a swan glide across a pond, or observe a koi swimming just underneath the water surface. In all cases, you will observe the wave pattern that is characteristic to all ships traveling the world's oceans, lakes, and rivers.

Thomson (Lord Kelvin) studied the wave pattern created by a traveling point disturbance (Figure 19.6). The solid curves represent lines of constant phase, like the position of wave crests. The parametric equations in Figure 19.6 have been used to create the plot. They represent an approximation of the rather complicated mathematics behind this problem. From Figure 19.6 and observation, the following features may be discerned for wave patterns:

- Waves appear only in a triangular region behind the vessel. Wave elevations rapidly diminish outside of the triangle. In deep water, the interior angle at the tip is 38.94 degrees and is known as the Kelvin angle. **Kelvin angle**

- The Kelvin wave pattern consists of two types of waves: divergent waves and transverse waves. We also call the divergent waves 'diagonal waves' and the transverse waves 'following waves.' Divergent waves appear at the sides of the wave pattern and travel away from the ship's path. Transverse waves follow the ship with their crests and troughs aligned perpendicular to the ship's course. **Kelvin wave pattern**

- Observed from a moving boat, the wave pattern appears to be stationary, which means that the waves themselves propagate with ship speed in the direction of the ship's forward motion. **Waves follow the ship**

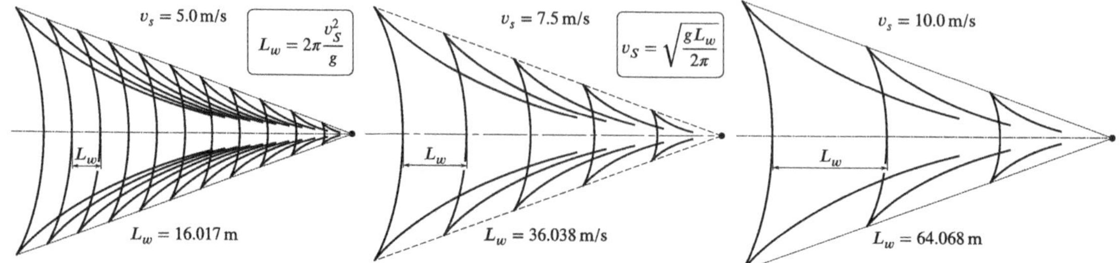

Figure 19.7 Change of Kelvin wave pattern with increasing velocity on deep water

Ship speed determines wave length

- The wave length L_w depends on the ship's speed and, in some cases, also on the water depth (Figure 19.7). The faster a vessel sails, the longer its waves become.

Wave breaking

- Especially at higher speeds or with blunt bows, waves near the bow break, like the waves rolling onto a beach.

Linear wave theory provides a relationship for wave length L_w and wave phase velocity c.

$$L_w = 2\pi \frac{c^2}{g \tanh(kh)} \tag{19.6}$$

On deep water, the product of wave number k and water depth h is larger than π, i.e. $kh > \pi$ and $\tanh(k\,h) \approx 1$. The transverse waves progress with the same speed as the disturbance, i.e. $c = v_S$ or

$$c^2 = v_S^2 = Fr^2\, g\, L_{WL} \tag{19.7}$$

We substitute Equation (19.7) into (19.6) and obtain a formula which provides the length of transverse waves as a function of Froude number for deep water.

$$L_w = 2\pi\, Fr^2\, L_{WL} \qquad \text{for } kh > \pi \tag{19.8}$$

The photo in Figure 19.8 may exemplify that Lord Kelvin's theoretical model is a pretty good match to nature. The satellite photo shows a Kelvin wave pattern visualized by clouds in the wake of Amsterdam Island in the Indian Ocean. The volcano on the island creates a Kelvin pattern at the boundary of two layers of air with different temperatures. As warm, moist air is pushed up into cooler air, water condenses and clouds form (wave crests). When the air sinks, it warms up and the clouds dissolve again (wave troughs).

Ship wave pattern

A ship disturbs the water surface over its whole length rather than just at a single point. Figure 19.9 shows the simulation of the wave pattern behind a ship hull. The specific shape of the vessel influences the height of the waves and the details of the wave pattern. The basic pattern shape, however, does not change as long as the vessel is sailing on deep water and is carried by its buoyancy. The nature of the wave pattern changes if the vessel maneuvers from displacement to planing mode. A planing vessel no longer produces a Kelvin wave pattern.

Wave propagation

Once waves have been generated by a moving vessel, they become independent of the vessel. When the vessel changes its course, newly generated waves will be oriented

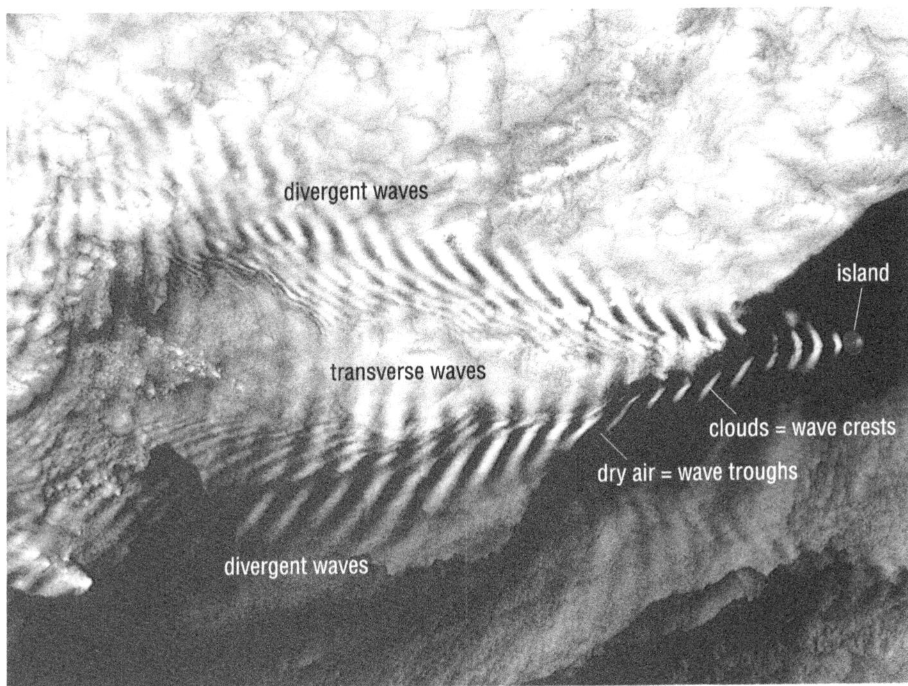

Figure 19.8 Kelvin wave pattern like cloud formation in the slipstream of Amsterdam Island in the southern Indian Ocean. Photo courtesy of NASA Earth Observatory

according to the new course. The waves generated before the course change will continue on their original path. If the vessel stops, wave generation ceases, but the already existing waves continue to spread.

Wave resistance

Waves contain and transport energy. Therefore, the creation of waves requires energy. Wave energy grows quadratic with wave amplitude. Provided two waves have the same wave length, a wave with twice the wave height contains four times the energy. In calm water, the moving ship is the only possible source of energy for wave making. Kinetic energy of the ship is constantly converted into wave energy. The loss of energy for the ship results in its wave resistance. As naval architects we aim to minimize the wave height in the wave pattern to minimize the loss of energy (see Chapter 25).

Shallow water effect

If the water depth is less than half the wave length, the ocean bottom or river bed will influence the behavior of the waves. A first visible sign is a widening of the Kelvin angle. Unfortunately, shallow water effects also result in increased wave resistance (Schlichting, 1934).

References

Michell, J. (1893). The hightest wave in water. *Philosophical Magazine Series 5*, 36:430–437.

Schlichting, O. (1934). Schiffswiderstand auf beschränkter Wassertiefe; Widerstand von

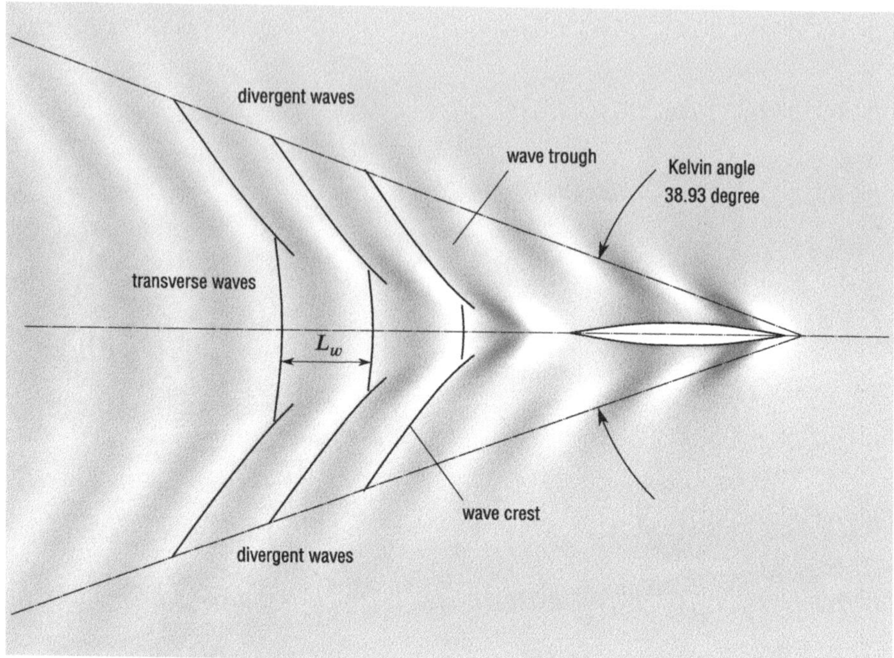

Figure 19.9 Wave pattern of a ship at $Fr = 0.26$

Seeschiffen auf flachem Wasser. In *Jahrbuch der Schiffbautechnischen Gesellschaft (STG)*, volume 35, pages 127–148. Springer Verlag, Berlin.

Thomson (Lord Kelvin), W. (2011). *On ship waves*, volume 3 of *Cambridge Library Collection – Physical Sciences*, pages 450–504. Cambridge University Press, Cambridge, United Kingdom. First presented in 1887.

Self Study Problems

1. Define wave height and wave length in your own words.
2. How does a real wave profile deviate from a harmonic signal (cosine function)?
3. Which geometric properties of a Kelvin wave pattern change when the ship moves into less deep water while maintaining its speed?
4. Compute the wave length for a ship sailing with velocity $v_S = 24\,\text{kn}$?
5. In deep water, a ship creates a wave pattern with a basic wave length of $L_w = 50.0$ m. How fast is the ship? Compute the wave length and phase velocity of the divergent waves!

20

Wave Theory

Waves are fascinating objects. At times, they demurely lap on a beach and at other times they cause havoc and destruction. In this chapter we develop a mathematical model for waves that will enable us to study many of their properties. Because waves independently propagate from their source, we deliberately ignore how they are created. It is helpful to first study the behavior of waves alone.

Learning Objectives

At the end of this chapter students will be able to:

- formulate wave flow as a boundary value problem
- derive free surface boundary conditions
- interpret the physical meaning of boundary conditions in wave theory

20.1 Overview

History

Attempts to capture the physics of waves in mathematical models go way back in time. Craik (2004) provides a discussion of the development of water wave theory over the past three centuries. Many famous scientists and mathematicians contributed to the field: Newton, Lagrange, Laplace, Green, Cauchy, and Poisson to name a few.

Stokes and Airy wave theories

The mathematics of the problem are quite difficult because we deal with nonlinear partial differential equations. Different theories have been developed for general, as well as specialized, applications. The most commonly used were developed by Stokes (1847) who provided approximations of first, second, and higher order for the original nonlinear problem. The approximations are known as Stokes' wave theories of first, second, third, and fifth order. We encountered Sir George Gabriel Stokes already in the context of the Navier-Stokes equations. The first order, or linear solution, is also known as Airy wave theory because Airy published his work a few years before Stokes did (Airy, 1841). Sir George Biddell Airy (*1801–†1892), an English mathematician and astronomer, became Astronomer Royal and established the prime meridian in Greenwich, UK, which is still in use as the reference line for longitude.

Fundamentals of Ship Hydrodynamics: Fluid Mechanics, Ship Resistance and Propulsion, First Edition.
Lothar Birk.

Companion website: www.wiley.com/go/birk/hydrodynamics

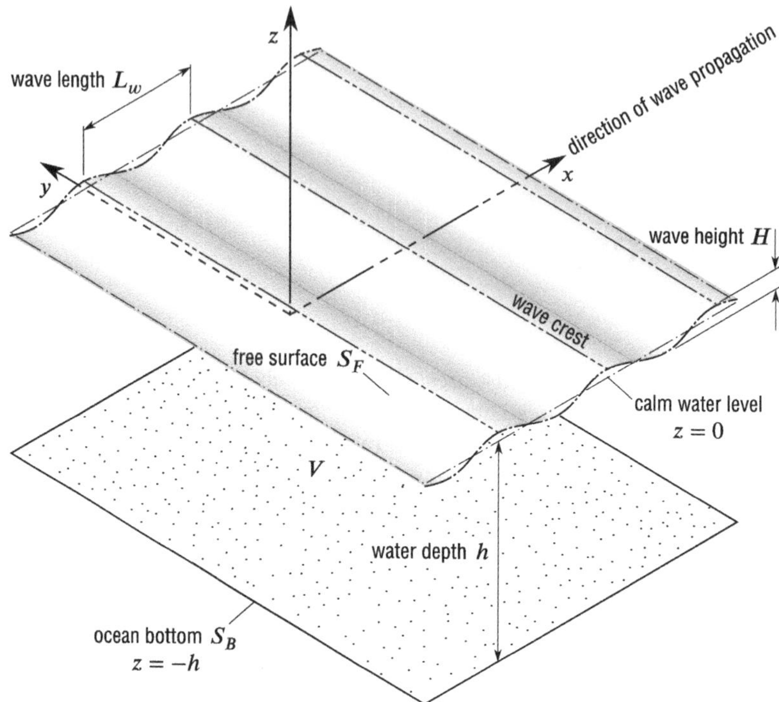

Figure 20.1 Definition of coordinate system and domain boundaries for wave theory of long-crested waves

Linear wave theory Linear wave theory is one of the most important tools in seakeeping analyis and probabilistic prediction of the short and long term behavior of marine vessels in waves. Although it is limited to waves with a small wave height to wave length ratio $H/L_w \ll 1$, linear wave theory provides important insight into the behavior of waves and the wave making of a ship. Therefore, we will discuss linear wave theory in some detail.

Wave resistance The Australian mathematician Michell (1898) developed a model for wave resistance based on linear theory. It can be used to estimate wave resistance of fast, slender vessels, especially multihull vessels. More on this in Section 25.3.

20.2 Mathematical Model for Long-crested Waves

Long-crested waves In order to develop a two-dimensional mathematical model for waves, we assume that their crests and troughs stretch in y-direction from negative infinity to positive infinity (Figure 20.1). These are commonly known as long-crested waves. This simplification renders all flow properties independent of the y-coordinate and restricts wave flow modeling to the x-z-plane. The x-axis points in the direction of wave propagation and the z-axis is pointing upward. We place the origin at the calm water level ($z = 0$). As a consequence, points below the calm water level, i.e. in the water, will have negative z-coordinates ($z < 0$).

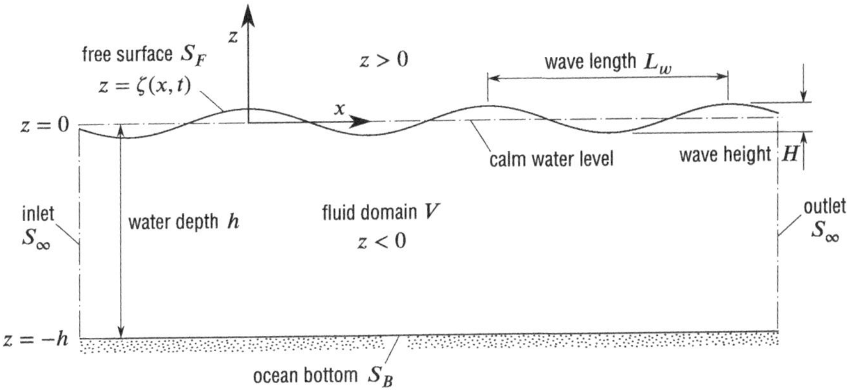

Figure 20.2 Simplified two-dimensional fluid domain for long-crested waves

Figure 20.2 shows the simplified two-dimensional flow domain. The fluid domain V is bordered at the top by the free surface S_F. The ocean bottom S_B forms the lower boundary of the domain. We assume that the ocean bottom is parallel to the calm water surface, i.e. the water depth h is constant throughout the domain. Note that the water depth h is defined as a positive length.

Ignoring viscous effects

As noted before, viscous effects are small for water waves of lengths typically created by a moving ship. However, a notable amount of energy is dissipated in the case of breaking waves. If we exclude the phenomenon of wave breaking, the fluid flow in waves may be treated as inviscid for practical applications. Again, water is considered incompressible (ρ = const.). In addition, we assume that the flow is irrotational (rot $\underline{v} = \underline{0}$), which enables us to employ potential theory.

Conservation of mass and momentum

The Laplace equation and the Bernoulli equation represent the conservation of mass and conservation of momentum principles of fluid mechanics for potential flow.

$$\Delta\phi = 0 \qquad \text{Laplace eq.} \tag{20.1}$$

$$\rho\frac{\partial\phi}{\partial t} + \frac{\rho}{2}|\underline{\nabla}\,\phi|^2 + \rho\, g\, z + p = C(t) \qquad \text{Bernoulli eq. for pot. flow} \tag{20.2}$$

The wave flow is represented by its velocity potential ϕ, which in general is a function of space $\underline{x}$ and time t. In order to define a unique solution of the Laplace equation, boundary conditions are imposed at the borders of the flow region V.

Domain boundaries

The domain V is bordered by the ocean bottom S_B and the free surface S_F (see Figure 20.2). From our discussion of the two-dimensional flow around a cylinder in Chapter 17, we know that an additional condition for the far field boundary S_∞ at the inlet and outlet of the flow is also required.

20.2.1 Ocean bottom boundary condition

No flow through ocean bottom

Let us start with a simple boundary condition. Like the cylinder surface, the ocean bottom S_B is considered a fixed, impenetrable surface. Since we ignore friction, fluid motion parallel to the bottom is allowed. However, the normal velocity has to vanish over S_B.

$$\frac{\partial\phi}{\partial n} = 0 \quad \text{on } S_B \tag{20.3}$$

Since the ocean floor is parallel to the water surface, its outward oriented normal vector points in the negative z-direction, i.e. $\underline{n}^T = (0, 0, -1)$. Consequently, the normal derivative is equal to the derivative in the negative z-direction.

$$\frac{\partial\phi}{\partial n} = \underline{n}^T \underline{\nabla}\phi = (0,\,0,\,-1)\begin{pmatrix}\phi_x\\ \phi_y\\ \phi_z\end{pmatrix} = -\frac{\partial\phi}{\partial z}$$

As a means to shorten the notation, we represent the partial derivatives with subscripts

$$\phi_x = \frac{\partial\phi}{\partial x} \qquad \phi_y = \frac{\partial\phi}{\partial y} \qquad \phi_z = \frac{\partial\phi}{\partial z} \tag{20.4}$$

The minus sign for the z derivative is of no consequence, since we require the vertical velocity component to vanish. Therefore, our final ocean bottom boundary condition reads

$$\phi_z = \frac{\partial\phi}{\partial z} = 0 \qquad \text{for } \underline{x} \text{ on } S_B,\ z = -h \tag{20.5}$$

Physically, it is a kinematic boundary condition because it restricts how the water may move. Mathematically it is classified as a Neumann boundary condition because it applies to a derivative of the unknown potential ϕ.

20.2.2 Free surface boundary conditions

The ocean bottom S_B is easily described mathematically: S_B consists of all points with $z = -h$. Similarly, the calm water surface is formed by points $\underline{x} = (x, y, 0)$. However, the position of the water surface is a priori unknown when waves are present. The shape of the free surface S_F has to be computed as part of the solution of the flow problem. As a consequence the boundary conditions for the free surface have to be satisfied at a yet unknown surface S_F.

Mathematical model for free surface

As a first step, a mathematical model is introduced for the shape of the free surface. We employ an implicit function as you may have seen in the definition of a circle of radius R: $x^2 + y^2 - R^2 = 0$. The contour is formed by all points whose distance from the origin is equal to R. The implicit function for the free surface captures all points whose vertical distance from the calm water level ($z = 0$) is equal to the free surface elevation ζ.

$$F(x, z, t) = z - \zeta(x, t) = 0 \tag{20.6}$$

We call ζ the wave elevation. Because we restricted the problem to long-crested waves, the implicit function F and the wave elevation ζ do not depend on the transverse coordinate y.

Limitations of free surface model

The formulation (20.6) implies that F and ζ are analytic functions. In that case, a unique ζ value must exist for all combinations (x, t) which automatically excludes overturning or breaking waves. As indicated in Figure 20.3, in cases of wave breaking more than one

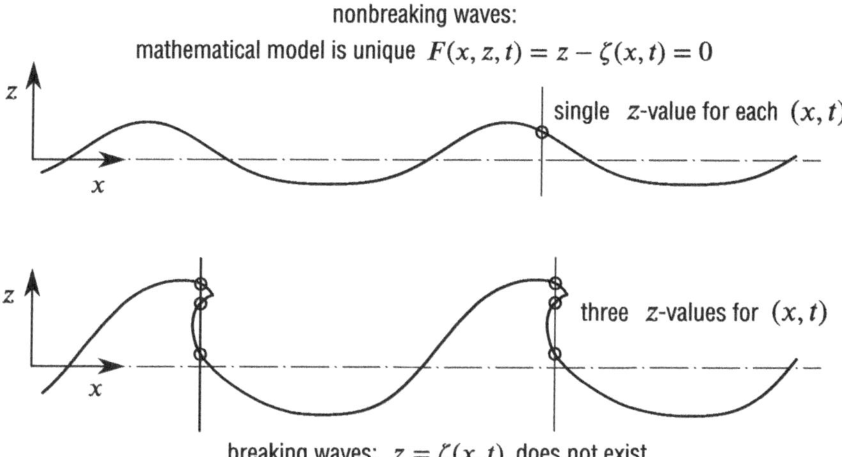

Figure 20.3 The mathematical free surface model is valid for nonbreaking waves only

ζ value might occur for a pair (x, t). As a consequence the function $z = \zeta(x, t)$ is not unique and does not exist. For the discussion of breaking waves, other mathematical models have to be employed.

Two boundary conditions needed

At the ocean bottom, we specify a single boundary condition because its geometry is known. The only unknown is the velocity potential. For the free surface we need two boundary conditions because we have two unknowns: the potential ϕ and the wave elevation ζ. The two free surface boundary conditions will be discussed in the following subsections.

Kinematic free surface boundary condition

The free surface must be a stream surface at all times t. Consequently, no fluid particle may cross the stream surface S_F. Like in the case of the body boundary condition for the cylinder flow, we restrict the movement of the fluid, therefore this will be a kinematic free surface boundary condition. The explicit expression for the kinematic boundary condition at the free surface may be derived in two different ways: (a) using a physical argument or (b) a more mathematical argument.

Physical argument

Let us start with the longer but probably more comprehensible physical argument. If S_F is a stream surface, fluid particles at the surface may move tangential to the surface but not normal to it. Therefore, the relative normal velocity between surface and water particles must vanish. The velocity of a water particle is $\underline{v} = \underline{\nabla}\,\phi$. The surface moves up and down with local velocity

$$\underline{v}_{S_F} = \frac{\partial \zeta}{\partial t}\underline{k}$$

where $\underline{k} = (0, 0, 1)^T$ is the unit vector in vertical direction. Thus, the relative velocity is

$$\underline{v} - \underline{v}_{S_F} = \underline{\nabla}\,\phi - \frac{\partial \zeta}{\partial t}\underline{k} \tag{20.7}$$

According to our stream surface requirement, the component of the relative velocity in normal direction must vanish. The latter is obtained by forming the dot product of normal vector $\underline{n}$ and relative velocity:

$$\underline{n}^T \left(\underline{\nabla} \phi - \frac{\partial \zeta}{\partial t} \underline{k} \right) = 0 \tag{20.8}$$

Normal vector

For implicit surfaces, the normal vector $\underline{n}$ is identical to the gradient normalized to unit length.

$$\underline{n} = \frac{\underline{\nabla} F}{|\underline{\nabla} F|} \tag{20.9}$$

This relationship follows from the characteristic of gradients to always point into the direction of steepest ascent, which in turn is perpendicular to the lines of constant function value.

We substitute the expression (20.9) into Equation (20.8) and multiply it with the length $|\underline{\nabla} F|$ of the gradient:

$$(\underline{\nabla} F)^T \left(\underline{\nabla} \phi - \frac{\partial \zeta}{\partial t} \underline{k} \right) = 0$$

Normal relative velocity must vanish

Expanding the dot products yields

$$\left(\frac{\partial F}{\partial x}, \frac{\partial F}{\partial y}, \frac{\partial F}{\partial z} \right) \begin{pmatrix} \frac{\partial \phi}{\partial x} \\ \frac{\partial \phi}{\partial y} \\ \frac{\partial \phi}{\partial z} \end{pmatrix} - \frac{\partial \zeta}{\partial t} \frac{\partial F}{\partial z} = 0$$

and we get

$$\frac{\partial F}{\partial x} \frac{\partial \phi}{\partial x} + \frac{\partial F}{\partial y} \frac{\partial \phi}{\partial y} + \frac{\partial F}{\partial z} \frac{\partial \phi}{\partial z} - \frac{\partial \zeta}{\partial t} \frac{\partial F}{\partial z} = 0 \tag{20.10}$$

The partial derivatives of the implicit surface function $F = z - \zeta(x, t)$ are

$$\begin{aligned} \frac{\partial F}{\partial x} &= -\frac{\partial \zeta}{\partial x} & \frac{\partial F}{\partial y} &= -\frac{\partial \zeta}{\partial y} \\ \frac{\partial F}{\partial z} &= +1 & \frac{\partial F}{\partial t} &= -\frac{\partial \zeta}{\partial t} \end{aligned} \tag{20.11}$$

Finally, we substitute the derivatives (20.11) into the condition (20.10) that the normal component of the relative velocity between water particles and free surface vanishes.

3D kinematic free surface condition

For a three-dimensional wave field $F(x, y, z, t) = z - \zeta(x, y, t) = 0$, the following is obtained for the kinematic free surface condition:

$$-\frac{\partial \zeta}{\partial t} - \frac{\partial \zeta}{\partial x} \frac{\partial \phi}{\partial x} - \frac{\partial \zeta}{\partial y} \frac{\partial \phi}{\partial y} + \frac{\partial \phi}{\partial z} = 0 \quad \text{on } S_F \text{ with } z = \zeta(x, y, t) \tag{20.12}$$

2D kinematic free surface condition

In the case of long-crested waves, the transverse derivatives $\partial/\partial y$ vanish, and the kinematic boundary condition for long-crested waves requires that the following condition is satisfied at the free surface.

$$-\frac{\partial \zeta}{\partial t} - \frac{\partial \zeta}{\partial x}\frac{\partial \phi}{\partial x} + \frac{\partial \phi}{\partial z} = 0 \quad \text{on } S_F \text{ with } z = \zeta(x,t) \tag{20.13}$$

Mathematical argument

The equivalent mathematical argument will of course lead to the same kinematic boundary condition: The condition $F = z - \zeta(x,t) = 0$ must be valid at all times in order for F to accurately describe the free surface S_F. Consequently, its value $F = 0$ cannot change over time, and its total derivative with respect to time must vanish. Applying the substantial derivative, fluid mechanics' equivalent of the total derivative, to F yields in vector form:

$$\frac{\mathrm{D}F}{\mathrm{D}t} = \frac{\partial F}{\partial t} + \underline{v}^T \underline{\nabla} F = 0$$

and in components

$$\frac{\partial F}{\partial t} + u\frac{\partial F}{\partial x} + v\frac{\partial F}{\partial y} + w\frac{\partial F}{\partial z} = 0 \tag{20.14}$$

We rewrite the latter equation by replacing the partial derivatives of F with the results from Equation (20.11) and by substituting partial derivatives of the potential ϕ for the components u, v, and w of the velocity vector. As expected, the result is the same as Equation (20.12)

$$-\frac{\partial \zeta}{\partial t} \underbrace{- \frac{\partial \phi}{\partial x}\frac{\partial \zeta}{\partial x} - \frac{\partial \phi}{\partial y}\frac{\partial \zeta}{\partial y}}_{\text{nonlinear terms}} + \frac{\partial \phi}{\partial z}1 = 0 \tag{20.12}$$

In long-crested waves, the transverse velocity $v = 0$ vanishes and we obtain again Equation (20.13) as the kinematic boundary condition for the free surface.

Nonlinear and implicit

Unfortunately, the resulting kinematic free surface conditions (20.12) and (20.13) are nonlinear because they contain products of the derivatives of the unknown surface elevation ζ and the velocity potential ϕ. As an added difficulty, the condition has to be satisfied at a yet unknown location $z = \zeta(x,t)$.

Dynamic free surface boundary condition

As mentioned above, a second boundary condition is needed because we do not know the actual location of the free surface. This second condition will be employed to eliminate the unknown surface elevation ζ from the kinematic boundary condition.[1]

Physical argument

Besides kinematic boundary conditions that impose restrictions on how the fluid can move, we can also prescribe dynamic conditions which define how forces interact or do not interact with the fluid. We are interested in the behavior of free waves, i.e. waves that have been created by wind or a ship but are now progressing on their own. That means, we will explicitly ignore the interaction of waves with the atmosphere. Pressure forces on the surface vanish if pressure differences at the water surface vanish.

Mathematical model

To cast this into a mathematical model we recall Bernoulli's equation for potential flow:

[1] Admittedly, this is a simplification which works for the linearized equations only. The exact boundary conditions form a system of coupled equations that has to be solved simultaneously.

$$\frac{\partial \phi}{\partial t} + \frac{1}{2}(\underline{\nabla}\,\phi)^T \underline{\nabla}\,\phi + gz + \frac{p}{\rho} = C(t) \tag{15.41}$$

The fluid is at rest far away from the progressing waves, and the free surface is exposed to atmospheric pressure p_0. This forms our constant for the right-hand side $C(t) = p_0/\rho$.

$$\frac{\partial \phi}{\partial t} + \frac{1}{2}(\underline{\nabla}\,\phi)^T \underline{\nabla}\,\phi + gz + \frac{p}{\rho} = \frac{p_0}{\rho} \qquad \text{at } S_F \tag{20.15}$$

The dynamic boundary condition requests that the pressure difference $p_0 - p$ vanishes at the free surface:

$$\frac{\partial \phi}{\partial t} + \frac{1}{2}(\underline{\nabla}\,\phi)^T \underline{\nabla}\,\phi + gz = 0 \qquad \text{at } S_F \text{ with } z = \zeta(x,y,t) \tag{20.16}$$

Note that all terms in Equation (20.16) have to be evaluated at $z = \zeta(x,y,t)$. Thus, solving Equation (20.16) for the unknown wave elevation does not provide a means to immediately calculate ζ:

$$\zeta(x,y,t) = -\frac{1}{g}\frac{\partial \phi}{\partial t} - \frac{1}{2g}(\underline{\nabla}\,\phi)^T \underline{\nabla}\,\phi \quad \text{at } S_F \text{ with } z = \zeta(x,y,t) \tag{20.17}$$

The wave elevation is also part of the right-hand side in (20.17) as argument of potential ϕ.

Nonlinear dynamic free surface condition

Expansion of the dot product of the gradients of the potential reveals that the dynamic free surface boundary condition is nonlinear, like the kinematic free surface boundary condition.

$$\frac{\partial \phi}{\partial t} + \underbrace{\frac{1}{2}\left[\left(\frac{\partial \phi}{\partial x}\right)^2 + \left(\frac{\partial \phi}{\partial y}\right)^2 + \left(\frac{\partial \phi}{\partial z}\right)^2\right]}_{\text{nonlinear terms}} + gz = 0 \quad \text{at } S_F \text{ with } z = \zeta(x,y,t) \tag{20.18}$$

The second nonlinear term vanishes for the two-dimensional problem of long-crested waves and the dynamic free surface boundary condition reads:

$$\frac{\partial \phi}{\partial t} + \underbrace{\frac{1}{2}\left[\left(\frac{\partial \phi}{\partial x}\right)^2 + \left(\frac{\partial \phi}{\partial z}\right)^2\right]}_{\text{nonlinear terms}} + gz = 0 \quad \text{at } S_F \text{ with } z = \zeta(x,t) \tag{20.19}$$

20.2.3 Far field condition

Radiation condition

We have considered boundary conditions for the ocean bottom S_B and the free surface S_F. What remains to be specified are conditions for the inlet and outlet surfaces S_∞ in the far field. In the case of waves, the far field condition is often referred to as the radiation condition.

Throw a stone into a pond

A radiation condition commonly involves statements about the direction in which waves are traveling and how their amplitude develops. Observe the waves created after you throw a stone into a quiet pond: a set of circular waves emerges which travel away from the spot where the stone hit the water surface. The amplitudes of the waves diminish as the rings expand until they finally fade away. In modeling this wave system, the radiation condition would have to enforce the direction of wave propagation (away from the stone) and that the amplitudes of waves vanish far away from their origin.

Radiation condition for long-crested waves

Friction forces do not exist in an inviscid fluid. Therefore, energy contained in the wave flow cannot be dissipated into heat or eddies. In long-crested waves, no forces act in the y-direction and thus energy cannot be distributed along the wave crests (parallel to the y-axis). As a consequence, the wave height cannot vanish in the far field of our two-dimensional domain. The only far field condition we impose is the requirement that the waves progress with constant velocity c in positive x-direction.

$$c = \text{const.} \geq 0 \qquad \text{at } S_\infty \text{ for } |x| \to \infty \tag{20.20}$$

20.2.4 Nonlinear boundary value problem

3D boundary value problem

Analogous to the cylinder flow problem, the Laplace equation together with the boundary conditions form a boundary value problem whose solution contains the velocity potential ϕ and the wave elevation ζ.

$$\begin{aligned}
\Delta\phi &= 0 && \underline{x} = (x, y, z)^T \in V \\
-\frac{\partial\zeta}{\partial t} - \frac{\partial\zeta}{\partial x}\frac{\partial\phi}{\partial x} - \frac{\partial\zeta}{\partial y}\frac{\partial\phi}{\partial y} + \frac{\partial\phi}{\partial z} &= 0 && \underline{x} \text{ on } S_F,\ z = \zeta(x, y, t) \\
\frac{\partial\phi}{\partial t} + \frac{1}{2}\left|\underline{\nabla}\phi\right|^2 + gz &= 0 && \underline{x} \text{ on } S_F,\ z = \zeta(x, y, t) \\
\frac{\partial\phi}{\partial z} &= 0 && \underline{x} \text{ on } S_B,\ z = -h \\
\text{propagation, energy conservation} & && \text{on } S_\infty
\end{aligned} \tag{20.21}$$

2D boundary value problem

This reduces to the following set of equations for long-crested waves in the two-dimensional case.

$$\begin{aligned}
\Delta\phi &= 0 && \underline{x} = (x, z)^T \in V \\
-\frac{\partial\zeta}{\partial t} - \frac{\partial\zeta}{\partial x}\frac{\partial\phi}{\partial x} + \frac{\partial\phi}{\partial z} &= 0 && \underline{x} \text{ on } S_F,\ z = \zeta(x, t) \\
\frac{\partial\phi}{\partial t} + \frac{1}{2}\left|\underline{\nabla}\phi\right|^2 + gz &= 0 && \underline{x} \text{ on } S_F,\ z = \zeta(x, t) \\
\frac{\partial\phi}{\partial z} &= 0 && \underline{x} \text{ on } S_B,\ z = -h \\
c &= \text{const.} \geq 0 && \text{at } S_\infty \text{ for } |x| \to \infty
\end{aligned} \tag{20.22}$$

A general analytic solution is not known for the exact two- and three-dimensional boundary value problems because of the nonlinear free surface boundary conditions. However, approximate solutions can be derived by using a Fourier series expansion or similar methods. If required, applications use the Stokes' wave theories with approximations from first to fifth order. Most important for practical applications is the first order wave theory, which solves the linearized problem as outlined below and in subsequent chapters.

20.3 Linearized Boundary Value Problem

Linearized boundary value problem

The nonlinear free surface boundary condition can be linearized if the analysis is restricted to waves with very small steepness $H/L \ll 1$. The linearization process is based on a perturbation series expansion of the exact solution. A detailed description of the process is provided in the following chapter. For now we simply study the result of the linearization and point out the major differences to the exact boundary value problem.

Linearized boundary value problem

Obviously, the Laplace equation (20.23) and boundary conditions at the sea bottom S_B (20.26) and the far field S_∞ (20.27) are unchanged. As expected, the nonlinear terms have vanished on the left-hand side of the kinematic and dynamic free surface conditions (20.24) and (20.25). In addition, however, the two free surface conditions are now imposed on the calm water surface $z = 0$ instead of on the actual free surface S_F. This is a major simplification enabling an analytic solution which will be derived in Chapter 22 on linear wave theory.

$$\Delta\phi = 0 \qquad \underline{x} = (x, z)^T \in V \tag{20.23}$$

$$-\frac{\partial \zeta}{\partial t} + \frac{\partial \phi}{\partial z} = 0 \qquad \underline{x} \text{ with } z = 0 \tag{20.24}$$

$$\frac{\partial \phi}{\partial t} + g\zeta = 0 \qquad \underline{x} \text{ with } z = 0 \tag{20.25}$$

$$\frac{\partial \phi}{\partial z} = 0 \qquad \underline{x} \text{ on } S_B,\ z = -h \tag{20.26}$$

$$c = \text{const.} \geq 0 \qquad \text{at } S_\infty \text{ for } |x| \rightarrow \infty \tag{20.27}$$

A less obvious change in the linearized boundary value problem is that potential ϕ and wave elevation ζ are now interpreted as generally small disturbances of the fluid at rest.

Combined linear free surface condition

Now that the free surface conditions are enforced at the calm water level, they are no longer implicit. This allows us to solve the dynamic boundary condition (20.25) for the wave elevation ζ.

$$\zeta(x, t) = -\frac{1}{g}\frac{\partial \phi}{\partial t} \qquad \underline{x} \text{ with } z = 0 \tag{20.28}$$

We take the time derivative of Equation (20.28) and substitute the result for the time derivative of the wave elevation into the linearized kinematic boundary condition (20.24):

$$\frac{\partial^2 \phi}{\partial t^2} + g\frac{\partial \phi}{\partial z} = 0 \qquad \underline{x} \text{ with } z = 0 \tag{20.29}$$

With the combined free surface condition for the free surface, the linearized boundary value problem can be stated with the velocity potential ϕ as the only unknown:

$$\begin{aligned} \Delta\phi &= 0 && \underline{x} = (x, z)^T \in V \\ \frac{\partial^2 \phi}{\partial t^2} + g\frac{\partial \phi}{\partial z} &= 0 && \underline{x} \text{ with } z = 0 \\ \frac{\partial \phi}{\partial z} &= 0 && \underline{x} \text{ on } S_B,\ z = -h \\ c &= \text{const.} \geq 0 && \text{at } S_\infty \text{ for } |x| \rightarrow \infty \end{aligned} \tag{20.30}$$

Once we know the potential, the flow field in the fluid domain is given by the gradient of the potential $\underline{v} = \underline{\nabla}\,\phi$. The pressure follows from the linearized Bernoulli equation (15.45):

$$p - p_\infty = -\rho\,\frac{\partial \phi}{\partial t} - \rho\,g\,z \qquad (15.45)$$

and the wave elevation from Equation (20.28).

References

Airy, G. (1841). Tides and waves. In Smedley, E., Rose, H. J., and Rose, H. J., editors, *Encyclopaedia Metropolitana*, volume V of *Mixed Sciences, Vol. 3*, pages 241–396. B. Fellowes et al., London.

Craik, A. (2004). The origins of water wave theory. *Annual Review of Fluid Mechanics*, 36:1–28.

Michell, J. (1898). The wave resistance of a ship. *Philosophical Magazine Series 5*, 45(272):106–123.

Stokes, G. (1847). On the theory of oscillatory waves. *Transactions of the Cambridge Philosophical Society*, 8:441–455.

Self Study Problems

1. Why may we ignore viscosity for typical wind or ship generated free surface waves?
2. Why do we need two boundary conditions for the free surface but only one for the ocean bottom?
3. Derive the exact kinematic and dynamic free surface conditions for the three-dimensional case, i.e. with $F(x, y, z, t) = z - \zeta(x, y, t) = 0$.
4. Name and explain in words the four boundary conditions of the exact boundary value problem for long-crested waves:

$$\begin{aligned}
-\frac{\partial \zeta}{\partial t} - \frac{\partial \zeta}{\partial x}\frac{\partial \phi}{\partial x} + \frac{\partial \phi}{\partial z} &= 0 && \underline{x} \text{ on } S_F,\ z = \zeta(x,t)\\
\frac{\partial \phi}{\partial t} + \frac{1}{2}\left|\underline{\nabla}\,\phi\right|^2 + gz &= 0 && \underline{x} \text{ on } S_F,\ z = \zeta(x,t)\\
\frac{\partial \phi}{\partial z} &= 0 && \underline{x} \text{ on } S_B,\ z = -d\\
c &= \text{const.} > 0 && \text{at } S_\infty \text{ for } |x| \to \infty
\end{aligned}$$

5. Why is no analytic solution of the exact two-dimensional boundary value problem (20.22) known?

21

Linearization of Free Surface Boundary Conditions

In this chapter, we formally derive the linearized versions of the kinematic and dynamic free surface conditions used in wave theory. The procedure is known as a perturbation approach and is applicable to the linearization of nonlinear equations in general. As results of the linearization process we obtain the kinematic and dynamic free surface boundary conditions which are employed in linear wave theory.

Learning Objectives

At the end of this chapter students will be able to:

- apply the perturbation approach
- perform linearizations of nonlinear equations
- understand the origin of linearized free surface conditions
- know the limitations of linear wave theory

21.1 Perturbation Approach

Nonlinear, implicit free surface conditions

For our potential wave flow problem, we derived two free surface boundary conditions. Two boundary conditions are required because we have two unknown functions: the velocity potential $\phi(x, y, z, t)$ and the actual position $\zeta(x, y, t)$ of the free surface relative to the calm water level $z = 0$. Unfortunately these boundary conditions are:

- *nonlinear* because they contain products of the unknown functions ϕ and ζ or their derivatives. As a consequence the equations cannot be resolved to eliminate one unknown function from either of the two equations,
- *implicit* because all terms have to be evaluated at the unknown free surface S_F with $z = \zeta(x, y, t)$.

Fundamentals of Ship Hydrodynamics: Fluid Mechanics, Ship Resistance and Propulsion, First Edition.
Lothar Birk.

Companion website: www.wiley.com/go/birk/hydrodynamics

A general analytical solution of the nonlinear problem does not exist. Therefore, we will linearize the free surface boundary conditions based on a perturbation approach. Of course, this comes at a price: the solution of the linearized problem will be valid only within certain limits, which we will discuss.

Series expansion

The perturbation approach follows the general concept of mathematical series like the Taylor series or the Fourier series. The idea is that an approximate, basic solution for the unknown function is improved with additional terms that become smaller and smaller. The basic solution is also called the 'zero order' solution and is assumed to be known. The additional terms are of first order, second order, third order, and so on.

The unknown functions, in our case the velocity potential ϕ and the wave elevation ζ, are expressed as a power series expansion with respect to a small parameter ε. The perturbation series for our two unknowns are:

$$\phi = \phi^{(0)} + \varepsilon\phi^{(1)} + \varepsilon^2\phi^{(2)} + \varepsilon^3\phi^{(3)} + \dots \tag{21.1}$$

$$\zeta = \zeta^{(0)} + \varepsilon\zeta^{(1)} + \varepsilon^2\zeta^{(2)} + \varepsilon^3\zeta^{(3)} + \dots \tag{21.2}$$

The superscripts (0), (1), (2), etc. express the order. $\phi^{(\ell)}$ denotes a velocity potential of ℓth-order and $\zeta^{(\ell)}$ an ℓth-order wave elevation. The perturbation series may more concisely be written as the sums:

$$\begin{aligned} \phi(x, y, z, t, \varepsilon) &= \sum_{\ell=0}^{\infty} \varepsilon^{\ell} \phi^{(\ell)}(x, y, z, t), \\ \zeta(x, y, t, \varepsilon) &= \sum_{\ell=0}^{\infty} \varepsilon^{\ell} \zeta^{(\ell)}(x, y, t) \end{aligned} \tag{21.3}$$

The small parameter $\varepsilon \ll 1$ (also called the perturbation parameter) can be interpreted as a dimensionless wave steepness $\varepsilon = k\zeta_a = \frac{2\pi}{L_W}\frac{H}{2} \sim \frac{H}{L_W}$. If this is too mathematical for you, just think of replacing the exact solutions by a sum of potentials or wave elevations that have fast decreasing contributions

$$\begin{aligned} \phi(x, y, z, t) &= \phi^{(0)} + \phi^{(1)} + \phi^{(2)} + \dots \quad \text{with } \phi^{(0)} > \phi^{(1)} > \phi^{(2)} > \dots \\ \zeta(x, y, t) &= \zeta^{(0)} + \zeta^{(1)} + \zeta^{(2)} + \dots \quad \text{with } \zeta^{(0)} > \zeta^{(1)} > \zeta^{(2)} > \dots \end{aligned} \tag{21.4}$$

Different levels of approximation (of the exact solution) are derived by truncating the series expansion after a selected order. Series expansions with respect to ε lead to solutions known as Stokes' wave theories. Stokes' wave theories up to fifth order are used in offshore hydrodynamics.

Linearization

In order to linearize the free surface boundary conditions we truncate the perturbation series after the linear term

$$\phi = \phi^{(0)} + \varepsilon\phi^{(1)} + \mathcal{O}(2) \tag{21.5}$$

$$\zeta = \zeta^{(0)} + \varepsilon\zeta^{(1)} + \mathcal{O}(2) \tag{21.6}$$

Furthermore, we assume the terms of second and higher order represented by $\mathcal{O}(2)$ are negligibly small, and we will omit them from here on.

21.2 Kinematic Free Surface Condition

Kinematic free surface boundary condition

Let us apply the perturbation approach to the kinematic free surface boundary condition (20.12), which expresses the requirement that no particle leave the free surface.

$$-\frac{\partial \zeta}{\partial t} - \frac{\partial \zeta}{\partial x}\frac{\partial \phi}{\partial x} - \frac{\partial \zeta}{\partial y}\frac{\partial \phi}{\partial y} + \frac{\partial \phi}{\partial z} = 0 \quad \text{on } S_F \text{ with } z = \zeta(x, y, t) \qquad (20.12)$$

Another interpretation is that the velocity of a particle normal to the surface has to be equal to the velocity of the surface in that direction.

Simplify notation

Before we proceed, we rewrite the boundary condition using subscripts to indicate partial differentiation.

$$-\zeta_t - \zeta_x\phi_x - \zeta_y\phi_y + \phi_z = 0 \quad \text{on } S_F \text{ with } z = \zeta(x, y, t) \qquad (21.7)$$

Substitute perturbations into boundary condition

Now we replace the unknown functions ϕ and ζ with their linear perturbation approximations.

$$\begin{aligned} -\left(\zeta^{(0)} + \varepsilon\zeta^{(1)}\right)_t - \left(\zeta^{(0)} + \varepsilon\zeta^{(1)}\right)_x\left(\phi^{(0)} + \varepsilon\phi^{(1)}\right)_x & \\ -\left(\zeta^{(0)} + \varepsilon\zeta^{(1)}\right)_y\left(\phi^{(0)} + \varepsilon\phi^{(1)}\right)_y + \left(\phi^{(0)} + \varepsilon\phi^{(1)}\right)_z &= 0 \\ \text{on } S_F \text{ with } z = \zeta^{(0)} + \varepsilon\zeta^{(1)} & \qquad (21.8) \end{aligned}$$

Note that we apply the perturbation also to the position where we satisfy the boundary condition.

The differential operators are applicable to the individual terms, for instance:

$$\left(\zeta^{(0)} + \varepsilon\zeta^{(1)}\right)_t = \zeta_t^{(0)} + \varepsilon\zeta_t^{(1)}$$

Likewise for the other derivatives. The perturbation parameter ε is treated as a constant factor. We first perform the differentiations within the parentheses and then expand the products. Equation (21.8) transforms into

$$\begin{aligned} -\zeta_t^{(0)} - \varepsilon\zeta_t^{(1)} - \zeta_x^{(0)}\phi_x^{(0)} - \zeta_x^{(0)}\varepsilon\phi_x^{(1)} - \varepsilon\zeta_x^{(1)}\phi_x^{(0)} - \underbrace{\varepsilon^2\zeta_x^{(1)}\phi_x^{(1)}}_{\approx 0} & \\ -\zeta_y^{(0)}\phi_y^{(0)} - \zeta_y^{(0)}\varepsilon\phi_y^{(1)} - \varepsilon\zeta_y^{(1)}\phi_y^{(0)} - \underbrace{\varepsilon^2\zeta_y^{(1)}\phi_y^{(1)}}_{\approx 0} & \\ + \phi_z^{(0)} + \varepsilon\phi_z^{(1)} &= 0 \\ \text{on } S_F \text{ with } z = \zeta^{(0)} + \varepsilon\zeta^{(1)} & \qquad (21.9) \end{aligned}$$

Neglect higher order contributions

The products of derivatives of the first order potential $\phi^{(1)}$ and the first order wave elevation $\zeta^{(1)}$ result in second order terms as indicated by the factor ε^2. Since we neglected second order contribution already in our perturbation approximation, it is only logical to neglect the newly created second order terms as well.

Equation (21.9) is now linear if we consider the zero order functions and their derivatives known. There are no longer products of the unknown first order functions $\phi^{(1)}$ and $\zeta^{(1)}$ or their derivatives with each other. This is also reflected in the fact that there are only terms left that have a factor ε of one.

Not done yet

We are not done yet, however. The boundary condition still has to be satisfied at the unknown position $z = \zeta^{(0)} + \varepsilon\zeta^{(1)}$. This affects the derivatives of both zero and first order potentials. In order to remedy this problem, the potentials are developed in a Taylor series with respect to the zero order wave elevation. The x-derivative of the zero order potential may serve as an example.

$$\begin{aligned} \phi_x^{(0)}(x,y,z,t) = & \left.\frac{\partial\phi^{(0)}}{\partial x}\right|_{z=\zeta^{(0)}} + \frac{1}{1!}\left(z-\zeta^{(0)}\right)\left.\frac{\partial^2\phi^{(0)}}{\partial x\partial z}\right|_{z=\zeta^{(0)}} \\ & + \frac{1}{2!}\left(z-\zeta^{(0)}\right)^2\left.\frac{\partial^3\phi^{(0)}}{\partial x\partial z^2}\right|_{z=\zeta^{(0)}} + \mathcal{O}(3) \end{aligned} \quad (21.10)$$

Of course, all terms of order two and higher will be neglected.

$$\phi_x^{(0)}(x,y,z,t) = \left.\frac{\partial\phi^{(0)}}{\partial x}\right|_{z=\zeta^{(0)}} + \left(z-\zeta^{(0)}\right)\left.\frac{\partial^2\phi^{(0)}}{\partial x\partial z}\right|_{z=\zeta^{(0)}} \quad (21.11)$$

At the free surface S_F we have $z = \zeta$. Together with the linear perturbation we obtain

$$z = \zeta = \zeta^{(0)} + \varepsilon\zeta^{(1)} \quad \text{or} \quad \left(z-\zeta^{(0)}\right) = \varepsilon\zeta^{(1)} \quad (21.12)$$

Taylor series approximation

Applying this result to the Taylor series expansion of the zero order potential yields

$$\phi_x^{(0)}(x,y,\zeta,t) = \left.\frac{\partial\phi^{(0)}}{\partial x}\right|_{z=\zeta^{(0)}} + \varepsilon\zeta^{(1)}\left.\frac{\partial^2\phi^{(0)}}{\partial x\partial z}\right|_{z=\zeta^{(0)}} \quad (21.13)$$

The same Taylor series expansion is applied to all other potentials and derivatives in Equation (21.9):

$$\phi_y^{(0)}(x,y,\zeta,t) = \left.\frac{\partial\phi^{(0)}}{\partial y}\right|_{z=\zeta^{(0)}} + \varepsilon\zeta^{(1)}\left.\frac{\partial^2\phi^{(0)}}{\partial y\partial z}\right|_{z=\zeta^{(0)}} \quad (21.14)$$

$$\phi_z^{(0)}(x,y,\zeta,t) = \left.\frac{\partial\phi^{(0)}}{\partial z}\right|_{z=\zeta^{(0)}} + \varepsilon\zeta^{(1)}\left.\frac{\partial^2\phi^{(0)}}{\partial z^2}\right|_{z=\zeta^{(0)}} \quad (21.15)$$

$$\phi_x^{(1)}(x,y,\zeta,t) = \left.\frac{\partial\phi^{(1)}}{\partial x}\right|_{z=\zeta^{(0)}} + \varepsilon\zeta^{(1)}\left.\frac{\partial^2\phi^{(1)}}{\partial x\partial z}\right|_{z=\zeta^{(0)}} \quad (21.16)$$

$$\phi_y^{(1)}(x,y,\zeta,t) = \left.\frac{\partial\phi^{(1)}}{\partial y}\right|_{z=\zeta^{(0)}} + \varepsilon\zeta^{(1)}\left.\frac{\partial^2\phi^{(1)}}{\partial y\partial z}\right|_{z=\zeta^{(0)}} \quad (21.17)$$

$$\phi_z^{(1)}(x,y,\zeta,t) = \left.\frac{\partial\phi^{(1)}}{\partial z}\right|_{z=\zeta^{(0)}} + \varepsilon\zeta^{(1)}\left.\frac{\partial^2\phi^{(1)}}{\partial z^2}\right|_{z=\zeta^{(0)}} \quad (21.18)$$

The first and second order derivatives of the potentials on the right-hand side are now evaluated at the known zero order wave elevation $\zeta^{(0)}$.

The right-hand sides of Equations (21.13) through (21.18) replace the corresponding terms in Equation (21.9). We again revert to subscripts for derivatives and omit the function arguments, but note that all potential derivatives are evaluated at $z = \zeta^{(0)}$ from here on.

$$
\begin{aligned}
-\zeta_t^{(0)} - \varepsilon\zeta_t^{(1)} &- \zeta_x^{(0)}\left(\phi_x^{(0)} + \varepsilon\zeta^{(1)}\phi_{xz}^{(0)}\right) \\
&- \zeta_x^{(0)}\varepsilon\left(\phi_x^{(1)} + \varepsilon\zeta^{(1)}\phi_{xz}^{(1)}\right) - \varepsilon\zeta_x^{(1)}\left(\phi_x^{(0)} + \varepsilon\zeta^{(1)}\phi_{xz}^{(0)}\right) \\
&- \zeta_y^{(0)}\left(\phi_y^{(0)} + \varepsilon\zeta^{(1)}\phi_{yz}^{(0)}\right) - \zeta_y^{(0)}\varepsilon\left(\phi_y^{(1)} + \varepsilon\zeta^{(1)}\phi_{yz}^{(1)}\right) \\
&- \varepsilon\zeta_y^{(1)}\left(\phi_y^{(0)} + \varepsilon\zeta^{(1)}\phi_{yz}^{(0)}\right) + \left(\phi_z^{(0)} + \varepsilon\zeta^{(1)}\phi_{zz}^{(0)}\right) \\
&+ \varepsilon\left(\phi_z^{(1)} + \varepsilon\zeta^{(1)}\phi_{zz}^{(1)}\right) = 0 \qquad \text{on } S_F \text{ with } z = \zeta^{(0)} \qquad (21.19)
\end{aligned}
$$

As a result, the kinematic boundary condition is now enforced at the known zero order wave elevation instead of the unknown exact wave elevation.

Linearized kinematic free surface boundary condition

Expanding the products results in five additional second order terms which we consequently neglect. Finally, we have morphed the nonlinear kinematic boundary condition into its linearized form.

$$
\begin{aligned}
-\zeta_t^{(0)} - \varepsilon\zeta_t^{(1)} &- \zeta_x^{(0)}\phi_x^{(0)} + \zeta_x^{(0)}\varepsilon\zeta^{(1)}\phi_{xz}^{(0)} - \zeta_x^{(0)}\varepsilon\phi_x^{(1)} - \varepsilon\zeta_x^{(1)}\phi_x^{(0)} \\
&- \zeta_y^{(0)}\phi_y^{(0)} + \zeta_y^{(0)}\varepsilon\zeta^{(1)}\phi_{yz}^{(0)} - \zeta_y^{(0)}\varepsilon\phi_y^{(1)} - \varepsilon\zeta_y^{(1)}\phi_y^{(0)} \\
&+ \phi_z^{(0)} + \varepsilon\zeta^{(1)}\phi_{zz}^{(0)} + \varepsilon\phi_z^{(1)} = 0 \qquad \text{on } S_F \text{ with } z = \zeta^{(0)} \qquad (21.20)
\end{aligned}
$$

Only the first order wave elevation $\zeta^{(1)}$ and the first order velocity potential $\phi^{(1)}$ are unknown in Equation (21.20). The equation is linear because it no longer contains products or powers of the unknown functions or their derivatives. At most a first order function is multiplied with known zero order functions.

Zero order solution

What is left is the selection of an appropriate zero order solution! However, before we do so, the same linearization process is applied to the dynamic free surface boundary condition. The zero order solution is introduced in Section 21.4.

21.3 Dynamic Free Surface Condition

The dynamic free surface condition followed from the unsteady Bernoulli equation and expresses the requirement that the pressure is to be constant over the free surface S_F.

$$
\frac{\partial\phi}{\partial t} + \frac{1}{2}|\underline{\nabla}\phi|^2 + gz = 0 \qquad \text{at } S_F \text{ with } z = \zeta(x, y, t) \qquad (20.18)
$$

The same equation reads as follows if again subscripts are used to represent partial differentiation.

$$
\phi_t + \frac{1}{2}\left[\phi_x^2 + \phi_y^2 + \phi_z^2\right] + gz = 0 \qquad \text{at } S_F \text{ with } z = \zeta(x, y, t) \qquad (21.21)
$$

Subsequently, the same process is followed as was used for the kinematic free surface boundary condition. Only major way points are provided here and the step by step execution is left as a self study problem (see page 258).

1. Introduce the linear perturbations

Linear perturbation

$$\phi = \phi^{(0)} + \varepsilon\phi^{(1)} \tag{21.22}$$

$$\zeta = \zeta^{(0)} + \varepsilon\zeta^{(1)} \tag{21.23}$$

into Equation (21.21).

2. Take derivatives and expand all products.

3. Delete all terms proportional to ε^2 (second order, negligibly small)! As a result, we obtain for the dynamic free surface condition the following expression:

Neglect higher order terms

$$\begin{aligned}\phi_t^{(0)} + \varepsilon\phi_t^{(1)} + \frac{1}{2}\left[\left(\phi_x^{(0)}\right)^2 + \left(\phi_y^{(0)}\right)^2 + \left(\phi_z^{(0)}\right)^2\right]& \\ + \varepsilon\phi_x^{(0)}\phi_x^{(1)} + \varepsilon\phi_y^{(0)}\phi_y^{(1)} + \varepsilon\phi_z^{(0)}\phi_z^{(1)} + g\zeta^{(0)} + g\varepsilon\zeta^{(1)} &= 0 \\ \text{at } S_F \text{ with } z = \zeta^{(0)} + \varepsilon\zeta^{(1)}&\end{aligned} \tag{21.24}$$

4. The derivatives of the potential still have to be computed at the unknown free surface location $z = \zeta^{(0)} + \varepsilon\zeta^{(1)}$. The derivatives are, once again, approximated by the truncated Taylor series approximations (21.13) through (21.18) and corresponding Taylor series for the time derivatives $\phi_t^{(0)}$ and $\phi_t^{(1)}$ of the zero and first order potential respectively. The dynamic free surface condition is now satisfied at the zero order wave elevation $z = \zeta^{(0)}$.

Taylor series expansion

5. After renewed expansion of products and the omission of second order terms we have as linearized dynamic free surface boundary condition:

Linearized dynamic free surface condition

$$\begin{aligned}\phi_t^{(0)} + \varepsilon\zeta^{(1)}\phi_{tz}^{(0)} + \varepsilon\phi_t^{(1)} + \frac{1}{2}\left[\left(\phi_x^{(0)}\right)^2 + \left(\phi_y^{(0)}\right)^2 + \left(\phi_z^{(0)}\right)^2\right]& \\ + \phi_x^{(0)}\varepsilon\zeta^{(1)}\phi_{xz}^{(0)} + \phi_y^{(0)}\varepsilon\zeta^{(1)}\phi_{yz}^{(0)} + \phi_z^{(0)}\varepsilon\zeta^{(1)}\phi_{zz}^{(0)}& \\ + \varepsilon\phi_x^{(0)}\phi_x^{(1)} + \varepsilon\phi_x^{(0)}\phi_y^{(1)} + \varepsilon\phi_x^{(0)}\phi_x^{(1)} + g\zeta^{(0)} + g\varepsilon\zeta^{(1)} &= 0 \\ \text{at } S_F \text{ with } z = \zeta^{(0)}&\end{aligned} \tag{21.25}$$

Admittedly, the linearized three-dimensional kinematic and dynamic boundary conditions for the free surface, (21.20) and (21.25) respectively, look even more complicated than their nonlinear equivalents. However, the linearized forms have two distinct advantages:

- The unknown first order wave elevation and potential are multiplied with known constants or zero order functions only.
- The conditions are applied at the known location of the zero order wave elevation instead of the exact position $\zeta(x, y, t)$.

Finally, a suitable pair of zero order wave elevation and velocity has to be selected to formulate the linearized boundary value problem already introduced in Section 20.3.

21.4 Linearized Free Surface Conditions for Waves

Zero order solution must be physical

The zero order potential $\phi^{(0)}$ and zero order wave elevation $\zeta^{(0)}$ were introduced as basic solutions to the problem which we assume to know. In fact, we will select a basic solution. A valid basic solution has to be physical, i.e. it does not violate the underlying conditions of the problem.

Simplest zero order solution

It was mentioned earlier that linear wave theory is only valid for waves with small steepness

$$\frac{\text{wave height}}{\text{wave length}} = \frac{H}{L_w} \ll 1 \tag{21.26}$$

For diminishing wave steepness the disturbance of the free surface eventually becomes invisible. If the surface is not moving, the fluid underneath will not move either (we ignore currents). Thus, a suitable set of zero order solutions is the trivial case of no wave elevation and no fluid velocity, or in other words calm water. We select for the zero order potential:

$$\phi^{(0)}(x, y, z, t) \equiv 0 \tag{21.27}$$

and for the zero order wave elevation:

$$\zeta^{(0)}(x, y, t) \equiv 0 \tag{21.28}$$

Obviously, derivatives of the zero order solutions vanish as well.

Simplifying the kinematic boundary condition

First, the zero order solutions are substituted into the linearized kinematic free surface condition.

$$\begin{aligned}
&- \overset{=0}{\cancel{\zeta^{(0)}_t}} - \varepsilon\zeta^{(1)}_t - \overset{=0}{\cancel{\zeta^{(0)}_x \phi^{(0)}_x}} + \overset{=0}{\cancel{\zeta^{(0)}_x \varepsilon\zeta^{(1)} \phi^{(0)}_{xz}}} - \overset{=0}{\cancel{\zeta^{(0)}_x \varepsilon\phi^{(1)}_x}} - \overset{=0}{\cancel{\varepsilon\zeta^{(1)}_x \phi^{(0)}_x}} \\
&- \overset{=0}{\cancel{\zeta^{(0)}_y \phi^{(0)}_y}} + \overset{=0}{\cancel{\zeta^{(0)}_y \varepsilon\zeta^{(1)} \phi^{(0)}_{yz}}} - \overset{=0}{\cancel{\zeta^{(0)}_y \varepsilon\phi^{(1)}_y}} - \overset{=0}{\cancel{\varepsilon\zeta^{(1)}_y \phi^{(0)}_y}} \\
&+ \overset{=0}{\cancel{\phi^{(0)}_z}} + \overset{=0}{\cancel{\varepsilon\zeta^{(1)} \phi^{(0)}_{zz}}} + \varepsilon\phi^{(1)}_z = 0 \qquad \text{on } S_F \text{ with } z = \zeta^{(0)}
\end{aligned} \tag{21.29}$$

Astonishingly, only two terms remain:

$$- \varepsilon\zeta^{(1)}_t + \varepsilon\phi^{(1)}_z = 0 \qquad \text{on } S_F \text{ with } z = \zeta^{(0)} = 0 \tag{21.30}$$

Kinematic free surface condition for linear wave theory

At this point we may drop the perturbation parameter and omit the superscript for the first order classification. The first order wave elevation and the first order velocity potential are the only nonvanishing components of the solution left.

$$- \frac{\partial \zeta}{\partial t} + \frac{\partial \phi}{\partial z} = 0 \qquad \text{on } S_F \text{ with } z = 0 \tag{21.31}$$

Physical interpretation

Let us take a step back after this long and tedious mathematical derivation. The derivative of the potential with respect to z is the vertical component w of the fluid velocity $\underline{v}$. The time derivative of the wave elevation $\partial\zeta/\partial t$ is the velocity of the free surface in

z-direction. As originally claimed, the kinematic free surface condition enforces that the relative velocity between free surface and fluid particles vanishes. In the case of the linear kinematic free surface condition, the slope of the free surface is ignored. In fact, it is treated as if it is flat. Therefore, the linear kinematic condition is valid only for waves with small steepness H/L_w.

Simplifying the dynamic boundary condition

Applying the selected zero order solution and its derivatives to the dynamic free surface condition (21.25) yields:

$$\underbrace{\phi_t^{(0)}}_{=0} + \varepsilon\zeta^{(1)}\underbrace{\phi_{tz}^{(0)}}_{=0} + \varepsilon\phi_t^{(1)} + \frac{1}{2}\underbrace{\left[\underbrace{\left(\phi_x^{(0)}\right)^2}_{=0} + \underbrace{\left(\phi_y^{(0)}\right)^2}_{=0} + \underbrace{\left(\phi_z^{(0)}\right)^2}_{=0}\right]}_{=0}$$
$$+ \underbrace{\phi_x^{(0)}\varepsilon\zeta^{(1)}\phi_{xz}^{(0)}}_{=0} + \underbrace{\phi_y^{(0)}\varepsilon\zeta^{(1)}\phi_{yz}^{(0)}}_{=0} + \underbrace{\phi_z^{(0)}\varepsilon\zeta^{(1)}\phi_{zz}^{(0)}}_{=0}$$
$$+ \underbrace{\varepsilon\phi_x^{(0)}\phi_x^{(1)}}_{=0} + \underbrace{\varepsilon\phi_x^{(0)}\phi_y^{(1)}}_{=0} + \underbrace{\varepsilon\phi_x^{(0)}\phi_x^{(1)}}_{=0} + g\varepsilon\zeta^{(1)} = 0$$
$$\text{at } S_F \text{ with } z = \zeta^{(0)} = 0 \qquad (21.32)$$

What remains is

$$\varepsilon\phi_t^{(1)} + g\varepsilon\zeta^{(1)} = 0 \qquad \text{at } S_F \text{ with } z = 0 \qquad (21.33)$$

Dynamic free surface condition for linear wave theory

As for the linear kinematic free surface condition, the perturbation parameter and the superscript for first order may be dropped knowing that the first order wave elevation and potential are the only substantial parts of the solution.

$$\frac{\partial\phi}{\partial t} + g\zeta = 0 \qquad \text{at } S_F \text{ with } z = 0 \qquad (21.34)$$

Note that in linear wave theory the free surface conditions are satisfied at the calm water level $z = 0$.

Physical interpretation

In Section (15.4), we introduced the linearized Bernoulli equation (15.45)

$$p - p_\infty = -\rho\frac{\partial\phi}{\partial t} - \rho g z \qquad (15.45)$$

The dynamic boundary condition requires that the pressure is equal everywhere at the free surface with $z = \zeta$. Therefore, the pressure difference must vanish $p - p_\infty = 0$ for $z = \zeta$.

$$0 = -\rho\frac{\partial\phi}{\partial t} - \rho g\zeta \qquad (21.35)$$

Dividing the linear Bernoulli equation applied to the free surface by the negative fluid density $-\rho$ results in the linear dynamic free surface condition (21.34). Here we neglect the influence of the nonlinear dynamic pressure term $1/2\rho|\underline{v}|^2$, which we assume is small.

First order wave elevation

The first order wave elevation is derived from the dynamic free surface boundary condition (21.34).

$$\zeta(x, y, t) = -g\left.\frac{\partial\phi}{\partial t}\right|_{z=0} \qquad (21.36)$$

In long-crested waves, the wave elevation is only a function of x and time t.

Self Study Problems

1. Linearize the following differential equation, assuming the zero order basis solution is linear $f^{(0)} = a \cdot z.$ with a being a constant factor.

$$\frac{\partial f}{\partial x}\frac{\partial f}{\partial z} + \frac{\partial^2 f}{\partial z^2}\phi = 0$$

   Hint: in this case the Taylor series expansion around a specific z value is not needed.

2. Linearize the dynamic free surface condition by executing in full detail the steps 1 through 5 on page 254.

3. Assume steady flow (all time derivatives vanish) and simplify the linearized kinematic and dynamic free surface boundary conditions (21.20) and (21.25) with respect to the following zero order solutions (Neumann-Kelvin linearization):

$$\phi^{(0)} = -U\,x \qquad \text{parallel flow in negative } x\text{-direction}$$

$$\zeta^{(0)} = 0 \qquad \text{calm water}$$

22

Linear Wave Theory

After thoroughly explaining the mathematical model for long-crested waves, it is time to actually solve the boundary value problem. In this chapter the linearized boundary value problem is solved for the two-dimensional case of long-crested waves progressing in positive x-direction. Again, the separation of variables is applied to derive the now time dependent velocity potential. Equipped with the potential, we will study the behavior and properties of waves.

Learning Objectives

At the end of this chapter students will be able to:

- apply potential theory to water waves
- solve potential flow problems
- compute the wave number
- distinguish exact formulation from deep water approximation

22.1 Solution of Linear Boundary Value Problem

Mathematical model recaptured

The linearized boundary value problem for long-crested waves has been introduced in Section 20.3. Chapter 21 provides a detailed account of the linearization process. A solution to the boundary value problem (20.30) (repeated below) will yield a velocity potential that describes the unsteady flow in waves with small wave steepness.

Laplace equation (conservation of mass)

$$\Delta\phi(\underline{x}, t) = \frac{\partial^2\phi}{\partial x^2} + \frac{\partial^2\phi}{\partial z^2} = 0 \qquad \underline{x} = (x, z)^T \in V \tag{22.1}$$

generalized linear free surface boundary condition

$$\frac{\partial^2\phi}{\partial t^2} + g\frac{\partial\phi}{\partial z} = 0 \qquad \underline{x} \text{ with } z = 0 \tag{22.2}$$

Fundamentals of Ship Hydrodynamics: Fluid Mechanics, Ship Resistance and Propulsion, First Edition.
Lothar Birk.

Companion website: www.wiley.com/go/birk/hydrodynamics

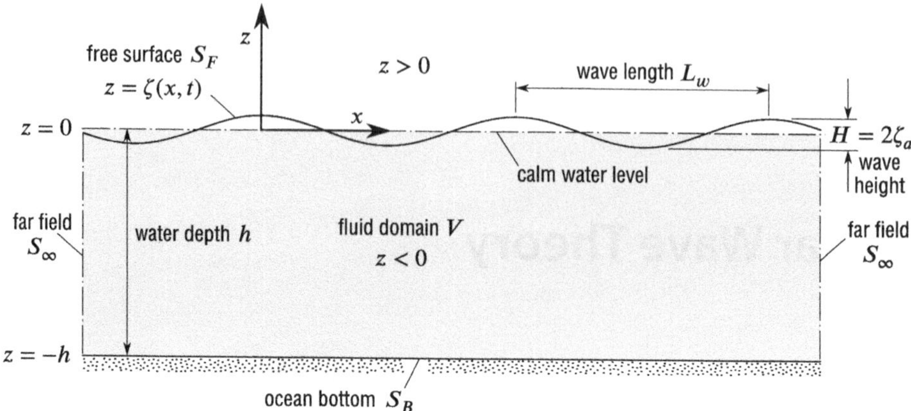

Figure 22.1 Simplified two-dimensional fluid domain for long-crested regular waves

ocean bottom boundary condition

$$\frac{\partial \phi}{\partial z} = 0 \qquad \underline{x} \text{ on } S_B,\ z = -h \tag{22.3}$$

far field boundary condition

$$c = \text{const.} \geq 0 \qquad \text{at } S_\infty \text{ for } |x| \to \infty \tag{22.4}$$

The coordinate system, fluid domain, and its boundaries are depicted in Figure 22.1. The origin is at the calm water level $z = 0$. With the z-axis pointing upwards, points in the fluid domain V have negative z-coordinates $z \leq 0$. Note that the linear free surface boundary condition (22.2) is satisfied at the calm water level $z = 0$ and not at the yet unknown displaced free surface S_F with $z = \zeta(x, t)$.

Separation of variables The velocity potential is a function of x, z, and time t. Therefore, we seek a solution which represents the potential as a product of three functions $X(x)$, $Z(z)$, and $T(t)$. Each of these is a function of just one independent variable.

$$\phi(x, z, t) = X(x)\, Z(z)\, T(t) \tag{22.5}$$

This approach is called separation of variables because it will convert the partial differential Laplace equation into a set of three simultaneous, ordinary differential equations.

The second order spatial derivatives of the potential (22.5) are needed for the Laplace equation.

$$\begin{aligned} \frac{\partial^2 \phi}{\partial x} &= \frac{\partial^2}{\partial x^2}\Big(X\, Z\, T\Big) = \frac{\partial^2 X}{\partial x^2}\, Z\, T = X_{xx}\, Z\, T \\ \frac{\partial^2 \phi}{\partial z} &= \frac{\partial^2}{\partial z^2}\Big(X\, Z\, T\Big) = X\, \frac{\partial^2 Z}{\partial z^2}\, T = X\, Z_{zz}\, T \end{aligned} \tag{22.6}$$

With substitution of the derivatives (22.6) into (22.1) and using subscripts to denote differentiation, the Laplace equation becomes:

$$X_{xx}\, Z\, T + X\, Z_{zz}\, T = 0 \qquad \text{for } \underline{x} = (x, z)^T \text{ in } V \tag{22.7}$$

Remember that the zero order potential $\phi^{(0)} \equiv 0$ describes the fluid at rest. Therefore, it is safe to assume that $X\,Z \neq 0$, and we may divide Equation (22.7) by XZ:

$$\left[\frac{X_{xx}}{X} + \frac{Z_{zz}}{Z}\right] T = 0 \tag{22.8}$$

For this to be true at all times, the term in brackets must vanish, i.e.

$$\frac{X_{xx}}{X} + \frac{Z_{zz}}{Z} = 0 \tag{22.9}$$

The first fraction is a function of x alone. The second fraction only depends on z. As a consequence, the fractions will be equal for all (x, z) only if they are constant in (x, z) and one is the negative of the other.

$$\frac{X_{xx}}{X} = -\frac{Z_{zz}}{Z} = \text{const.} = C \tag{22.10}$$

Ordinary differential equations

Equation (22.10) may be split into two ordinary differential equations that are connected by the same constant C

$$\begin{aligned} \text{for } X(x)\text{:} \quad & \frac{X_{xx}}{X} = C \\ \text{for } Z(z)\text{:} \quad & -\frac{Z_{zz}}{Z} = C \end{aligned} \tag{22.11}$$

We select the constant to be $C = -k^2$ with k being the wave number. Equations (22.11) are multiplied by X and Z respectively, and all terms are moved to the left-hand side.

$$X_{xx} + k^2 X = 0 \tag{22.12}$$

$$Z_{zz} - k^2 Z = 0 \tag{22.13}$$

Solution for X

The complimentary solution of the ordinary differential equation in x (22.12) is a function whose second order derivative is equal to the negative of the function itself, except for a constant factor. The sine and cosine functions are of this type. However, we do not know which of the two. Since combinations of sine and cosine are also plausible, we define

$$X = C_1 \mathrm{e}^{ikx} \tag{22.14}$$

The exponential function combines sine and cosine with Euler's formula for complex numbers $\mathrm{e}^{\mathrm{i}x} = \cos x + \mathrm{i} \sin x$. The complex notation is a useful mathematical tool which simplifies the treatment of our equations significantly. However, the velocity field is real and only the real part of the resulting velocity potential will bear physical meaning.

Differentiating Equation (22.14) twice and substituting it back shows that this is indeed a solution of ODE (22.12). Boundary conditions will be used later to solve for the new constant C_1 which generally differs from constant C above.

Solution for Z

The second ODE (22.13) is satisfied by functions whose second order derivative is equal to the function itself. This time hyperbolic sine and cosine functions will suffice. Selecting the hyperbolic cosine, a general solution to (22.13) is given by

$$Z = C_2 \cosh(kz + \alpha_2) \tag{22.15}$$

Solution for T

Finally, a solution is needed for the time dependent function T. Equation (22.7) is of no use because it would only yield the trivial solution $T \equiv 0$. Instead, the product (22.5) $\phi = X\,Y\,Z$ is introduced into the generalized linear free surface boundary condition (22.2).

$$\phi_{tt} + g\,\phi_z \;=\; X\,Z\,T_{tt} + g\,X\,Z_z\,T \;=\; 0 \qquad \text{for } z = 0 \tag{22.16}$$

Division by the product $X\,Z$ yields

$$T_{tt} + g\,\left.\frac{Z_z}{Z}\right|_{z=0} T \;=\; 0 \tag{22.17}$$

The expression Z_Z/Z has to be evaluated at $z = 0$ because Equation (22.17) is derived from the free surface boundary condition. z is the only variable in the expression Z_Z/Z. Consequently, the term $gZ_z/Z|_{z=0}$ is constant. We select

$$g\,\left.\frac{Z_z}{Z}\right|_{z=0} \;=\; \omega^2 \tag{22.18}$$

ω is the wave frequency.

The ordinary differential equation for T is of the same type as the ODE (22.14) for X.

$$T_{tt} + \omega^2 T \;=\; 0 \tag{22.19}$$

An appropriate complimentary solution is

$$T \;=\; C_3 \mathrm{e}^{-\mathrm{i}\omega t} \tag{22.20}$$

The minus sign is deliberate in the argument of the exponential function. We come back to this when we discuss the far field condition.

So far, the solution for the velocity potential is using the results from Equations (22.14), (22.15), and (22.20):

$$\phi(x,z,t) \;=\; X(x)\,Z(z)\,T(t) \;=\; C_1 C_2 C_3 \cosh(kz + \alpha_2)\,\mathrm{e}^{\mathrm{i}kx}\mathrm{e}^{-\mathrm{i}\omega t} \tag{22.21}$$

The product of constants is yet another constant. We substitute $C_4 = C_1 C_2 C_3$ and get

$$\phi(x,z,t) \;=\; C_4 \cosh(kz + \alpha_2)\,\mathrm{e}^{\mathrm{i}kx}\mathrm{e}^{-\mathrm{i}\omega t} \tag{22.22}$$

Constants

The remaining constants C_4 and α_2 in (22.22) are selected so that the solution $\phi = X\,Y\,Z$ satisfies the remaining boundary conditions of the problem.

The ocean bottom boundary condition requires that the normal velocity vanishes at the ocean floor.

$$\frac{\partial \phi}{\partial z} = 0 \qquad \text{for } z = -h \tag{22.23}$$

Introducing the partial solution (22.22) yields:

$$\frac{\partial}{\partial z}\left[C_4 \cosh(kz + \alpha_2)\,\mathrm{e}^{\mathrm{i}kx}\mathrm{e}^{-\mathrm{i}\omega t}\right] \;=\; 0 \qquad \text{for } z = -h \tag{22.24}$$

Since only the hyperbolic cosine is a function of z and $C_4\,\mathrm{e}^{\mathrm{i}kx}\mathrm{e}^{-\mathrm{i}\omega t} \neq 0$ this simplifies to

$$\frac{\partial}{\partial z}\left[\cosh(kz + \alpha_2)\right] \;=\; 0 \qquad \text{for } z = -h \tag{22.25}$$

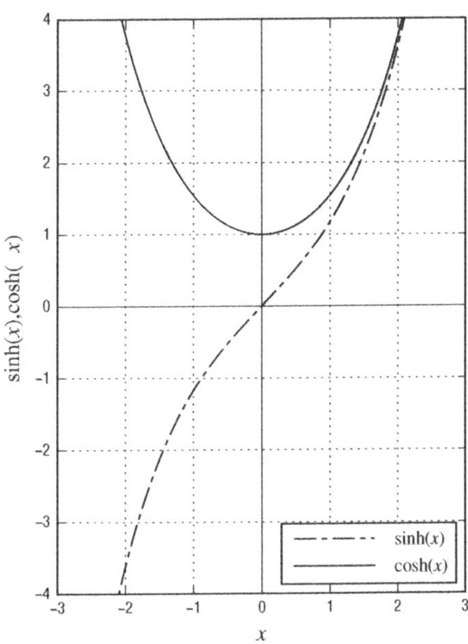

Figure 22.2 The hyperbolic sine and cosine functions

Differentiation results in

$$k \sinh(kz + \alpha_2) = 0 \qquad \text{for } z = -h \tag{22.26}$$

The hyperbolic sine only vanishes if its argument is zero (see Figure 22.2). Substituting the negative water depth for z, we get

$$k(-h) + \alpha_2 = 0 \tag{22.27}$$

Therefore, the constant α_2 is equal to

$$\alpha_2 = kh \tag{22.28}$$

This leaves the constant C_4 to be determined. Except for the far field boundary condition, we have explicitly used all boundary conditions listed in the linearized boundary value problem (22.1) through (22.4). The far field condition is of no use to determine C_4. One has to remember that the generalized free surface condition (22.2) actually consists of two boundary conditions, i.e. the kinematic and the dynamic free surface condition. The linearized dynamic free surface condition (21.34) requires that the pressure at the free surface remains constant.

$$\frac{\partial \phi}{\partial t} + g\zeta = 0 \qquad \text{at } S_F \text{ with } z = 0 \tag{21.34}$$

Solving (21.34) for the unknown wave elevation ζ and substituting the partial solution for the potential (22.22) and the new found constant α_2 from Equation (22.28) yields:

$$\begin{aligned}\zeta(x,t) &= -\frac{1}{g}\left.\frac{\partial \phi}{\partial t}\right|_{z=0} = -\frac{1}{g} C_4 \cosh(k \cdot 0 + kh)\, \mathrm{e}^{\mathrm{i}kx}(-\mathrm{i}\omega)\mathrm{e}^{-\mathrm{i}\omega t}\\ &= \frac{\mathrm{i}\,\omega}{g} C_4 \cosh(kh)\, \mathrm{e}^{\mathrm{i}(kx-\omega t)}\end{aligned} \tag{22.29}$$

Wave amplitude The exponential function represents a general harmonic oscillation of unit amplitude.

$$e^{i(kx-\omega t)} = \cos(kx-\omega t) + i\sin(kx-\omega t)$$

Consequently, the factor preceding the harmonic function may be interpreted as an amplitude ζ_a, which must be real valued since the surface elevation is real.

$$\zeta_a = \frac{i\omega}{g} C_4 \cosh(kh) \tag{22.30}$$

Accordingly, the constant C_4 must be complex and is given by

$$C_4 = -\frac{i\,g}{\omega\cosh(kh)}\zeta_a \tag{22.31}$$

Solution for wave elevation This completes the solution of the linearized boundary value problem. We found that the first order wave elevation is a cosine function with amplitude ζ_a. In complex notation the wave profile is

$$\bar{\zeta}(x,t) = \zeta_a\, e^{i(kx-\omega t)} \tag{22.32}$$

However, only its real part $\Re(\bar{\zeta})$ is of interest to us:

$$\zeta(x,t) = \Re\big(\bar{\zeta}(x,t)\big) = \zeta_a\cos(kx-\omega t) \tag{22.33}$$

Solution for velocity potential The second part of the solution is the first order velocity potential for a long-crested wave in two dimensions. Equation (22.22) with the results for constants α_2 (22.28) and C_4 (22.31) becomes

$$\bar{\phi}(x,z,t) = -\zeta_a \frac{i\,g}{\omega}\frac{\cosh(kz+kh)}{\cosh(kh)} e^{ikx}e^{-i\omega t} \tag{22.34}$$

In order to compute the velocity field, we need the real valued potential ϕ.

$$\begin{aligned}
\phi(x,z,t) &= \Re\big(\bar{\phi}(x,z,t)\big) \\
&= -\frac{\zeta_a g}{\omega}\frac{\cosh(kz+kh)}{\cosh(kh)}\Re\left(i\,e^{ikx}e^{-i\omega t}\right) \\
&= -\frac{\zeta_a g}{\omega}\frac{\cosh(kz+kh)}{\cosh(kh)}\Re\left(i\cos(kx-\omega t) + i^2\sin(kx-\omega t)\right)
\end{aligned}$$

which results in

$$\phi(x,z,t) = \frac{\zeta_a g}{\omega}\frac{\cosh(kz+kh)}{\cosh(kh)}\sin(kx-\omega t) \tag{22.35}$$

Note that the velocity potential (22.35) is valid for all water depths h as long as the general assumptions for linear wave theory are satisfied.

Wave parameters A wave on constant water depth h is defined by two parameters:

- wave frequency ω, and
- wave amplitude ζ_a

Wave frequency and wave number are connected via the dispersion relation (see below). Therefore, the wave number k may be used instead of the wave frequency ω to define the wave.

22.2 Far Field Condition Revisited

Direction of propagation

In the solution of the boundary value problem we hinted that the choice of $e^{-i\omega t}$ (see page 262) as a basis for the time dependent part of the potential is deliberate to satisfy the far field condition.

$$c = \text{const.} \geq 0 \qquad \text{at } S_\infty \text{ for } |x| \to \infty \tag{22.4}$$

The far field condition enforces that the wave is moving with constant velocity in positive x-direction.

Constant phase

The velocity of a wave is determined by how fast the crest-trough-crest contour travels. An observer moving as fast as the wave crest experiences a stationary wave field as we have observed for our ship wave system. Traveling with a specific point of the wave contour, like for instance a crest, requires a constant phase $\theta = kx - \omega t = \text{const.}$ Therefore, the velocity with which a wave progresses is called phase velocity. If the phase is to be constant in the far field, it cannot change over time and its derivative must vanish.

$$\frac{d\theta}{dt} = \frac{d}{dt}\left(kx - \omega t\right) = 0 \tag{22.36}$$

Phase velocity

Wave frequency ω and wave number k are constants for the wave. Therefore, the phase velocity of a regular wave is

$$c = \frac{dx}{dt} = \frac{\omega}{k} = \frac{L_w}{T} \geq 0 \tag{22.37}$$

Since both wave length L_w and wave period T are positive and constant, the phase velocity c is positive and constant as well. The far field condition is satisfied.

If we had opted for $e^{+i\omega t}$ instead of $e^{-i\omega t}$ as the basic time function (see Equation (22.20)), the condition of constant phase (22.36) would have resulted in $c = -L_w/T$ and the wave would move in negative x-direction.

22.3 Dispersion Relation

Link between frequency and wave number

Back on page 234, we stated that wave number and wave frequency are connected. The linearized kinematic free surface condition (21.31) provides this connection.

$$-\frac{\partial \zeta}{\partial t} + \frac{\partial \phi}{\partial z} = 0 \qquad \text{on } S_F \text{ with } z = 0 \tag{21.31}$$

Based on our linear solution for wave elevation (22.33) and potential (22.35), we form the derivatives:

$$\frac{\partial \zeta}{\partial t} = \omega \zeta_a \sin(kx - \omega t)$$

$$\frac{\partial \phi}{\partial z} = k \frac{\zeta_a g}{\omega} \frac{\sinh(kz + kh)}{\cosh(kh)} \sin(kx - \omega t)$$

Both derivatives are evaluated at $z = 0$ and substituted into the kinematic free surface condition above.

$$-\omega \zeta_a \sin(kx - \omega t) + k \frac{\zeta_a g}{\omega} \frac{\sinh(kh)}{\cosh(kh)} \sin(kx - \omega t) = 0$$

The ratio of hyperbolic sine and cosine is equal to the hyperbolic tangent.

$$\zeta_a \sin(kx - \omega t) \left[-\omega + k \frac{g}{\omega} \tanh(kh) \right] = 0 \tag{22.38}$$

The latter equation is satisfied when the sine vanishes for $\theta = n\pi$ with $n = 0, 1, 2, \dots,$ or, much more generally, if the term in brackets vanishes.

$$\left[-\omega + k \frac{g}{\omega} \tanh(kh) \right] = 0 \tag{22.39}$$

Dispersion relation This may be transformed into

$$\omega^2 = k\, g \tanh(kh) \tag{22.40}$$

which is known as the dispersion relation. Since wave frequencies are considered positive, one may also write

$$\omega = \sqrt{k\, g \tanh(kh)} \tag{22.41}$$

The relationship between wave frequency and wave number is a nonlinear, transcendental equation. It is influenced by the gravitational acceleration g and the water depth h. Thus, waves will behave differently depending on the water depth. If one would consider wave experiments on the Moon, with one sixth of Earth's gravity, quite different wave behavior could be expected. Obviously, the wave frequency can be readily determined for any combination of water depth and wave number. In the next chapter, you will learn how to compute the wave number k for a given water depth and wave frequency using the dispersion relation.

Phase velocity As stated above, the surface elevation moves with phase velocity $c = \omega/k = L_w/T$. Replacing the wave frequency ω with the dispersion relation (22.41) yields the phase velocity as a function of wave length. The water depth h serves as a parameter.

$$c = \frac{\omega}{k} = \frac{\sqrt{k\, g \tanh(kh)}}{k} = \sqrt{\frac{g \tanh(kh)}{k}} \tag{22.42}$$

Figure 22.3 visualizes this relationship. The primary horizontal axis is the wave number k. The lower axis shows the corresponding wave length L_w. Both axes use a logarithmic scale. Wave number is increasing from left to right. The reciprocal wave length grows from right to left, and the vertical axis represents the phase velocity c. Note that the vertical axis is logarithmic in the left part of Figure 22.3 but linear in the right subfigure which shows values appropriate for model scale.

Dispersion Several important observations can be made:

- The phase velocity of waves changes with wave length (or wave frequency). As mentioned, this effect is called dispersion. Water waves are therefore quite different from sound and electromagnetic waves, which have constant phase velocities that are independent of wave length (speed of sound, speed of light).
- Waves become faster with increasing wave length (decreasing wave number). Long waves progress astonishingly fast. A wave of 1000 m length reaches a velocity of almost 40 m/s. A wave of 10000 m length travels in deep water as fast as a commercial air liner.

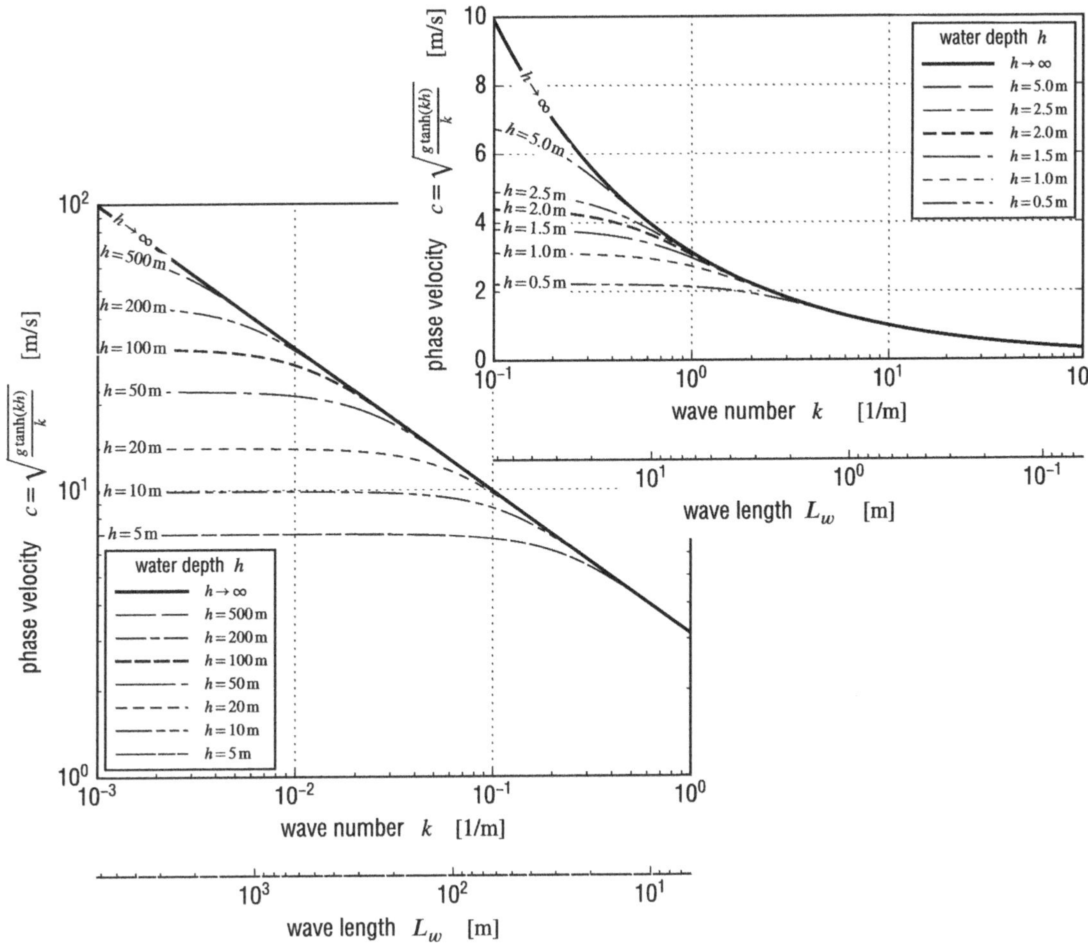

Figure 22.3 Wave phase velocity as function of wave number and water depth based on linear wave theory

- Depending on the water depth h, waves reach a limiting phase velocity. The lower the water depth, the lower the maximum attainable phase velocity. In very shallow water, dispersion vanishes and waves of different wave length have the same phase velocity.

22.4 Deep Water Approximation

Ship trial conditions usually include the requirement of 'deep water.' In this context, water is deep if it does not affect wave motion at the free surface and other parts of the flow around a ship hull. We still have to clarify which water depth is considered deep enough.

Figure 22.4 shows the hyperbolic tangent function as it appears in the dispersion relation. For large arguments $kh \longrightarrow \infty$, the hyperbolic tangent asymptotically approaches the

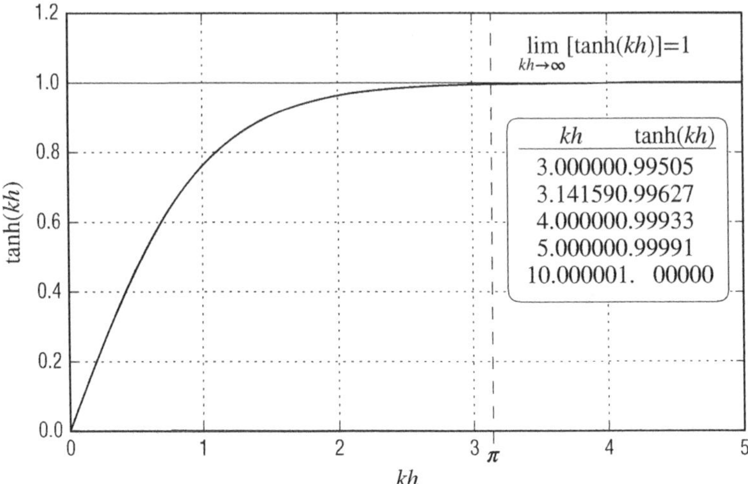

Figure 22.4 The positive arm of the hyperbolic tangent function

value one. For a value of $kh = \pi$, the deviation is only 0.373% from the asymptotic limit 1. For $kh = 5$, the difference is less than 0.01% and for $kh = 10$, the tangent is smaller than one by less than $1.0 \cdot 10^{-8}$.

Deep water condition For practical purposes, water is considered deep if:

$$kh \geq \pi$$

Under this premise, we obtain the following relationship between wave length and water depth:

$$kh = \frac{2\pi}{L_w} h \geq \pi$$

which is equivalent to

$$h \geq \frac{L_w}{2} \quad \text{or} \quad L_w \leq 2\,h \tag{22.43}$$

Thus, if the water depth is larger than half the wave length, the influence of the ocean floor on a wave becomes negligible.

Deep water wave number If kh is large enough, the hyperbolic tangent will be close to one ($\tanh(kh) \approx 1$) and the dispersion relation (22.41) becomes

$$\omega = \sqrt{k_0\,g} \quad \text{or} \quad k_0 = \frac{\omega^2}{g} \tag{22.44}$$

k_0 is the deep water wave number.

Deep water phase velocity With the deep water wave number, phase velocity (22.42) simplifies to

$$c_0 = \sqrt{\frac{g}{k_0}} \tag{22.45}$$

Therefore, the relationships between wave length and phase velocity are

$$c_0 = \sqrt{\frac{g\,L_w}{2\pi}} \quad \text{or} \quad L_w = \frac{2\pi}{g}\,c_0^2 \qquad \text{on deep water } (kh \geq \pi) \tag{22.46}$$

Deep water velocity potential

The assumption of deep water simplifies the velocity potential as well. The only term in the velocity potential (22.35) (or (22.34) for the complex form) that depends on the water depth is the ratio of hyperbolic cosines.

$$\frac{\cosh\big(k(z+h)\big)}{\cosh(kh)}$$

The hyperbolic functions may be replaced by a special combination of exponential functions (Abramowitz and Stegun, 1972, p. 83).

$$\sinh(x) = \frac{1}{2}\,(\mathrm{e}^{x} - \mathrm{e}^{-x}) \tag{22.47}$$

$$\cosh(x) = \frac{1}{2}\,(\mathrm{e}^{x} + \mathrm{e}^{-x}) \tag{22.48}$$

Using the latter for our ratio of hyperbolic cosines yields:

$$\frac{\cosh\big(k(z+h)\big)}{\cosh(kh)} = \frac{\mathrm{e}^{kz+kh} + \mathrm{e}^{-(kz+kh)}}{\mathrm{e}^{kh} + \mathrm{e}^{-kh}} = \frac{\mathrm{e}^{kz}\mathrm{e}^{kh} + \mathrm{e}^{-kz}\mathrm{e}^{-kh}}{\mathrm{e}^{kh} + \mathrm{e}^{-kh}} \tag{22.49}$$

If the product of wave number and water depth becomes very large, i.e. $kh \to \infty$, then the exponential of $(-kh)$ will vanish $\mathrm{e}^{-kh} \to 0$. Therefore, the limit of the ratio of hyperbolic cosines for deep water is equal to the exponential function with argument kz.

$$\lim_{kh\to\infty} \frac{\cosh\big(k(z+h)\big)}{\cosh(kh)} = \lim_{kh\to\infty} \frac{\mathrm{e}^{kz}\mathrm{e}^{kh} + \mathrm{e}^{-kz}\cancelto{0}{\mathrm{e}^{-kh}}}{\mathrm{e}^{kh} + \cancelto{0}{\mathrm{e}^{-kh}}} = \lim_{kh\to\infty} \frac{\mathrm{e}^{kz}\mathrm{e}^{kh}}{\mathrm{e}^{kh}} = \mathrm{e}^{k_0 z} \tag{22.50}$$

Finally, the linear velocity potential for waves in deep water becomes

$$\bar{\phi}(x,z,t) = -\zeta_a\,\frac{\mathrm{i}\,g}{\omega}\,\mathrm{e}^{k_0 z}\,\mathrm{e}^{\mathrm{i}k_0 x}\mathrm{e}^{-\mathrm{i}\omega t} \qquad \text{(complex)} \tag{22.51}$$

$$\phi(x,z,t) = \frac{\zeta_a\,g}{\omega}\,\mathrm{e}^{k_0 z}\,\sin(k_0 x - \omega t) \qquad \text{(real)} \tag{22.52}$$

This simplification was very important in the pre-computer age. With modern desktop computers, it is better to work with the exact solution for the velocity potential (22.35), which is valid for all water depths.

References

Abramowitz, M. and Stegun, I., editors (1972). *Handbook of mathematical functions with formulas, graphs, and mathematical tables.* Number 55 in Applied Mathematics Series. National Bureau of Standards, Washington, DC. Tenth printing, with corrections.

Self Study Problems

1. Summarize the assumptions that have been made in deriving the linearized boundary value problem (22.1) through (22.4).

2. Show that the first order velocity potential for a harmonic wave

$$\phi(x, z, t) = \frac{\zeta_a g}{\omega} \frac{\cosh(kz + kh)}{\cosh(kh)} \sin(kx - \omega t)$$

satisfies the Laplace equation (22.1) and boundary conditions (22.2), (22.3), and (22.4).

3. At which water depth h is a wave of length $L_w = 80$ m considered a deep water wave?

4. Compute wave number and wave frequency for the wave of length $L_w = 80$ m in deep water.

5. You have two waves of different wave length: $L_{w1} = 100$ m and $L_{w2} = 120$ m. Which wave has the greater phase velocity (assume deep water)? Is this still true for a water depth of 20 m? Use Figure 22.3 to justify your answer.

6. A ship is producing a transverse wave of wave length $L_w = 120$ m. Assuming deep water, what is the ship's speed? Use Figure 22.3 to check if your answer is reasonable.

23

Wave Properties

Based on the linear wave theory solution, basic kinematics and dynamics of water waves are investigated in this and the following chapter. First, a method is introduced to solve the nonlinear dispersion relation. With the wave number known, wave particle velocities, accelerations, motions, and unsteady pressure are discussed. A more comprehensive analysis of water wave properties may be found in offshore engineering books by Dean and Dalrymple (1991) and by Clauss et al. (1992).

Learning Objectives

At the end of this chapter students will be able to:

- understand wave kinematics and dynamics
- solve the dispersion relation
- apply potential flow theory results
- compute wave properties

23.1 Linear Wave Theory Results

This section briefly summarizes the most important results of Chapter 22 for the reader who skipped the finer details of linear wave theory. The results form a basis for the discussion of wave properties.

Regular waves

The solution, which satisfies the Laplace equation and the linearized boundary conditions for long-crested progressing waves, predicts a sinusoidal wave elevation with amplitude ζ_a. Waves with a sinusoidal profile are commonly referred to as regular waves of wave frequency ω.

$$\begin{aligned} \bar{\zeta}(x,t) &= \zeta_a\, \mathrm{e}^{\mathrm{i}(kx-\omega t)} \qquad &\text{(complex)} \\ \zeta(x,t) &= \zeta_a \cos(kx-\omega t) \qquad &\text{(real)} \end{aligned} \tag{22.33}$$

Although the discussion here uses almost exclusively the real valued equations, it should be pointed out that the complex notation has many advantages in computational tools.

Fundamentals of Ship Hydrodynamics: Fluid Mechanics, Ship Resistance and Propulsion, First Edition.
Lothar Birk.

Companion website: www.wiley.com/go/birk/hydrodynamics

For example, the superposition of regular waves of different frequencies, amplitudes, and phases into more realistic looking wave patterns is much easier in the complex than the real domain.

Velocity potential The basis for all analysis of kinematics and kinetics of regular waves is the velocity potential:

$$\begin{aligned}
\bar{\phi}(x,z,t) &= -\mathrm{i}\,\frac{\zeta_a\,g}{\omega}\,\frac{\cosh(kz+kh)}{\cosh(kh)}\,\mathrm{e}^{\mathrm{i}kx}\mathrm{e}^{-\mathrm{i}\omega t} && \text{(complex)}\\
\phi(x,z,t) &= \frac{\zeta_a\,g}{\omega}\,\frac{\cosh(kz+kh)}{\cosh(kh)}\,\sin(kx-\omega t) && \text{(real)}
\end{aligned} \qquad (22.35)$$

Equation (22.35) represents the exact solution of the linearized boundary value problem and is valid for all water depths h.

Dispersion relation From the kinematic boundary condition follows the dispersion relation, which expresses the relationship between wave frequency ω, wave number k, and water depth h.

$$\omega = \sqrt{k\,g\,\tanh(kh)} \qquad (22.41)$$

Deep water approximation Influence of the sea bottom on the wave motion becomes negligibly small for water depths larger than half the wave length, i.e. $h > L_w/2$. As shown in Section 22.4, the velocity potential simplifies to the following expression under the assumption that $kh > \pi$.

$$\begin{aligned}
\bar{\phi}(x,z,t) &= -\zeta_a\,\frac{\mathrm{i}\,g}{\omega}\,\mathrm{e}^{k_0 z}\,\mathrm{e}^{\mathrm{i}k_0 x}\mathrm{e}^{-\mathrm{i}\omega t} && \text{(complex)}\\
\phi(x,z,t) &= \frac{\zeta_a\,g}{\omega}\,\mathrm{e}^{k_0 z}\,\sin(k_0 x-\omega t) && \text{(real)}
\end{aligned} \qquad (22.52)$$

The corresponding deep water dispersion relation states

$$\omega^2 = k_0\,g \qquad (23.1)$$

Based on the assumptions made during the solution process, the equations above are valid for fluids with negligible viscosity and constant density as long as the wave steepness $H/L_w \ll 1$ is small.

23.2 Wave Number

Waves are often specified by providing values for wave period T and wave amplitude ζ_a. From the wave period follows the wave frequency $\omega = 2\pi/T$. For deep water ($kh > \pi$), the wave number k_0 follows immediately from Equation (23.1).

$$k_0 = \frac{\omega^2}{g} \qquad (23.2)$$

A wave of period $T = 8$ s has a frequency of

$$\omega = \frac{2\pi}{T} \approx \frac{6.2832}{8\ \mathrm{s}} = 0.7854\ \mathrm{s}^{-1}$$

Therefore, its deep water wave number is

$$k_0 = \frac{\omega^2}{g} \approx \frac{0.7854^2 \text{ s}^{-2}}{9.81 \text{ m s}^{-2}} = 0.06288 \text{ m}^{-1}$$

Its associated wave length is $L_w = 2\pi/k_0 = 99.923$ m.

In the case of restricted water depth h, the nonlinear dispersion relation (22.41) must be solved. Prior to the age of programmable calculators and desktop computers this was a cumbersome task, but with today's computer support it is no longer a hurdle.

Newton-Raphson method

The method of choice to solve Equation (22.41) for the wave number is the Newton-Raphson method (see for example Akai, 1994). Like all methods for nonlinear problems, it is iterative, i.e. it starts with an initial, approximate solution and improves the solution up to the desired accuracy. The Newton-Raphson method follows from a Taylor series expansion around a known solution k_i. The series is truncated after the linear term and higher order terms are ignored.

$$f(k_{i+1}) = f(k_i) + \left(k_{i+1} - k_i\right) \left.\frac{\mathrm{d}f}{\mathrm{d}k}\right|_{k=k_i} \tag{23.3}$$

The function f in Equation (23.3) represents the nonlinear equation to be solved written in an homogeneous form. We square Equation (22.41) and shift all terms to the left side.

$$0 = \frac{\omega^2}{g} - k \tanh(kh)$$

Thus, the function $f(k)$ and its derivative are

$$f(k) = \frac{\omega^2}{g} - k \tanh(kh) \tag{23.4}$$

$$\frac{\mathrm{d}f}{\mathrm{d}k} = -k\frac{h}{\cosh^2(kh)} - \tanh(kh) \tag{23.5}$$

The solution k we seek satisfies the condition $f(k) = 0$. Consequently, the left-hand side of the truncated Taylor series expansion vanishes if we assume that $k = k_{i+1}$ is the desired solution.

$$f(k_{i+1}) = 0 = f(k_i) + \left(k_{i+1} - k_i\right) \left.\frac{\mathrm{d}f}{\mathrm{d}k}\right|_{k=k_i} \tag{23.6}$$

Iterative procedure

The remainder of Equation (23.6) is solved for k_{i+1}.

$$k_{i+1} = k_i - \frac{f(k_i)}{\left.\frac{\mathrm{d}f}{\mathrm{d}k}\right|_{k=k_i}} \tag{23.7}$$

Equation (23.7) represents an iterative procedure. Repeated application will converge toward the real solution in a few steps as long as the derivative in the denominator does not vanish, i.e. $\mathrm{d}f/\mathrm{d}k \neq 0$. The deep water wave number k_0 is readily available as a starting value. The iterations are stopped when the difference between two subsequent approximations is smaller than a tiny threshold value ε.

$$|k_{i+1} - k_i| \leq \varepsilon \tag{23.8}$$

Application example

The following example may serve as an illustration of the procedure. We want to know the wave number k and wave length L_w of a wave with period $T = 8$ s when it travels in water of depth $h = 20$ m. Consequently, the wave frequency is $\omega = 0.7854$ s^{-1} and the deep water wave number is $k_0 = 0.0628797$ m^{-1}. Equipped with k_0 as the starting value, we compute $f(k_0) = 0.0094068$ m^{-1} and the derivative at k_0 as $\mathrm{d}f/dk = -1.1985276$. These values are substituted into Equation (23.7) to derive $k_1 = 0.0707284$ m^{-1}. For the next iteration $i = 1$, the values of $f(k_1)$ and its derivative are computed. Application of Equation (23.7) results in k_2 and so on. After $i = 2$ no changes in the first seven digits can be seen, and $k = 0.0707624$ m^{-1} is considered the result for the wave number with sufficient precision for all practical purposes. Values used in the iteration are summarized in the table below:

i	k_i	$f(k_i)$	$\mathrm{d}f/\mathrm{d}k\,\|_{k_i}$
[–]	[m^{-1}]	[m^{-1}]	[–]
0	0.0628797	0.0094068	−1.1985276
1	0.0707284	0.0000404	−1.1864226
2	0.0707624	0.0000000	−1.1863489
3	0.0707624	−0.0000000	−1.1863489
4	0.0707624	0.0000000	−1.1863489
...	...	...	...

As one can see, the iteration converges very fast. Usually less than a handful of iterations will yield a wave number with high accuracy. The wave length L_w derived from the new wave number is $L_w = 2\pi/k = 88.793$ m which is considerably less than the wave length on deep water ($L_w = 99.923$ m). Note that the length of a wave decreases with decreasing water depth for constant wave period.

Solutions of nonlinear equations should be thoroughly checked.

(i) Does the solution satisfy the equation? In our case the solution clearly satisfies $f(k) = f(0.0707624) = 0$ within numerical precision.

(ii) You may also perform a graphical check by plotting the function $f(k)$ in the vicinity of a solution. Figure 23.1 shows the function $f(k)$ using the input data provided in the example. The intersection of $f(k)$ with the k-axis falls on the value $k = 0.0707624$ m^{-1} as expected.

Therefore, the solution $k = 0.0707624$ m^{-1} makes sense.

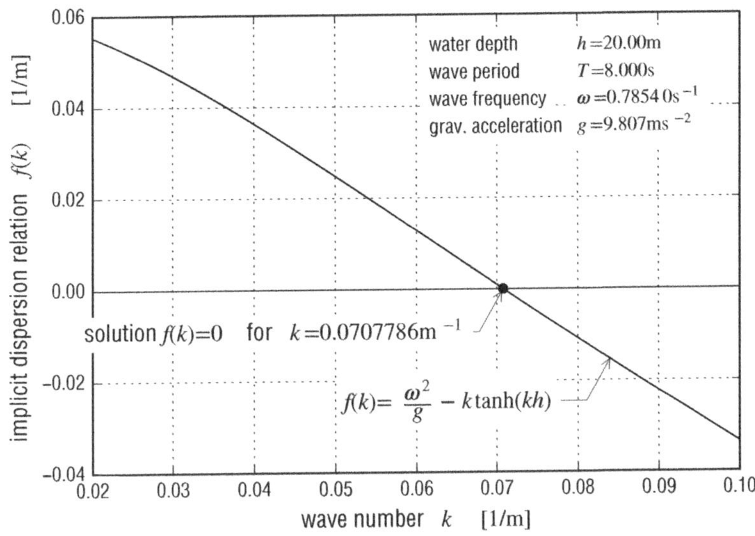

Figure 23.1 Graphical verification of a solution of the nonlinear dispersion relation

23.3 Water Particle Velocity and Acceleration

Equipped with

- water depth h,
- wave frequency ω,
- wave number k,
- and wave amplitude ζ_a

the real-valued velocity potential ϕ from Equation (22.35) provides the tool to investigate water wave kinematics.

Water particle velocity

As we explored in our discussion of potential flow and specifically the flow around a cylinder, flow velocities follow from the spatial derivatives of the velocity potential. The velocity vector $\underline{v} = (u, w)^T$ in the two-dimensional wave flow problem has components in x- and z-directions.

$$
\begin{aligned}
u &= \frac{\partial \phi}{\partial x} \qquad \text{horizontal particle velocity} \\
&= \frac{\partial}{\partial x}\left[\frac{\zeta_a g}{\omega}\frac{\cosh(kz+kh)}{\cosh(kh)}\sin(kx-\omega t)\right] \\
&= k\,\frac{\zeta_a g}{\omega}\frac{\cosh(kz+kh)}{\cosh(kh)}\cos(kx-\omega t) \qquad (23.9)
\end{aligned}
$$

$$\begin{aligned} w &= \frac{\partial \phi}{\partial z} \qquad \text{vertical particle velocity} \\ &= \frac{\partial}{\partial z}\left[\frac{\zeta_a g}{\omega}\frac{\cosh(kz+kh)}{\cosh(kh)}\sin(kx-\omega t)\right] \\ &= k\frac{\zeta_a g}{\omega}\frac{\sinh(kz+kh)}{\cosh(kh)}\sin(kx-\omega t) \end{aligned} \tag{23.10}$$

If the dispersion relation (22.41) is brought into the form (23.11)

$$k g = \frac{\omega^2}{\tanh(kh)} \tag{23.11}$$

and substituted into the equation for the horizontal velocity u, one gets:

$$\begin{aligned} u &= \frac{\omega^2}{\tanh(kh)}\frac{\zeta_a}{\omega}\frac{\cosh(kz+kh)}{\cosh(kh)}\cos(kx-\omega t) \\ &= \zeta_a\omega\frac{\cosh(kz+kh)}{\tanh(kh)\cosh(kh)}\cos(kx-\omega t) \end{aligned}$$

Exploiting that $\tanh(kh) = \sinh(kh)/\cosh(kh)$ yields:

$$u = \zeta_a\omega\frac{\cosh(kz+kh)}{\sinh(kh)}\cos(kx-\omega t) \tag{23.12}$$

The analog result for the vertical particle velocity w is

$$w = \zeta_a\omega\frac{\sinh(kz+kh)}{\sinh(kh)}\sin(kx-\omega t) \tag{23.13}$$

Both velocity components u and w are harmonic functions of time and x-coordinate. The factor preceding the sine and cosine functions may be interpreted as velocity amplitudes u_a and w_a which are a function of submergence z.

Water particle acceleration The water particle acceleration is of great importance in the computation of wave forces and vessel motions. Equations for the horizontal and vertical accelerations are obtained by taking the time derivatives of Equations (23.12) and (23.13). Observe that the time derivative of $\theta = kx - \omega t$ introduces an additional factor $-\omega$.

$$\dot{u} = \frac{\partial u}{\partial t} = \zeta_a\omega^2\frac{\cosh(kz+kh)}{\sinh(kh)}\sin(kx-\omega t) \tag{23.14}$$

$$\dot{w} = \frac{\partial w}{\partial t} = -\zeta_a\omega^2\frac{\sinh(kz+kh)}{\sinh(kh)}\cos(kx-\omega t) \tag{23.15}$$

Phase relationships Figure 23.2 shows a snapshot of several wave properties at time ($t = 0$). From top to bottom, we see wave elevation ζ, particle velocities u and w, and particle accelerations $\dot{u}$ and $\dot{w}$. The graph at the bottom of Figure 23.2 shows the dynamic pressure, which we will discuss in the next section. Each curve represents the spatial distribution of the respective property over 1.5 wave lengths, calculated at the calm water level ($z = 0$).

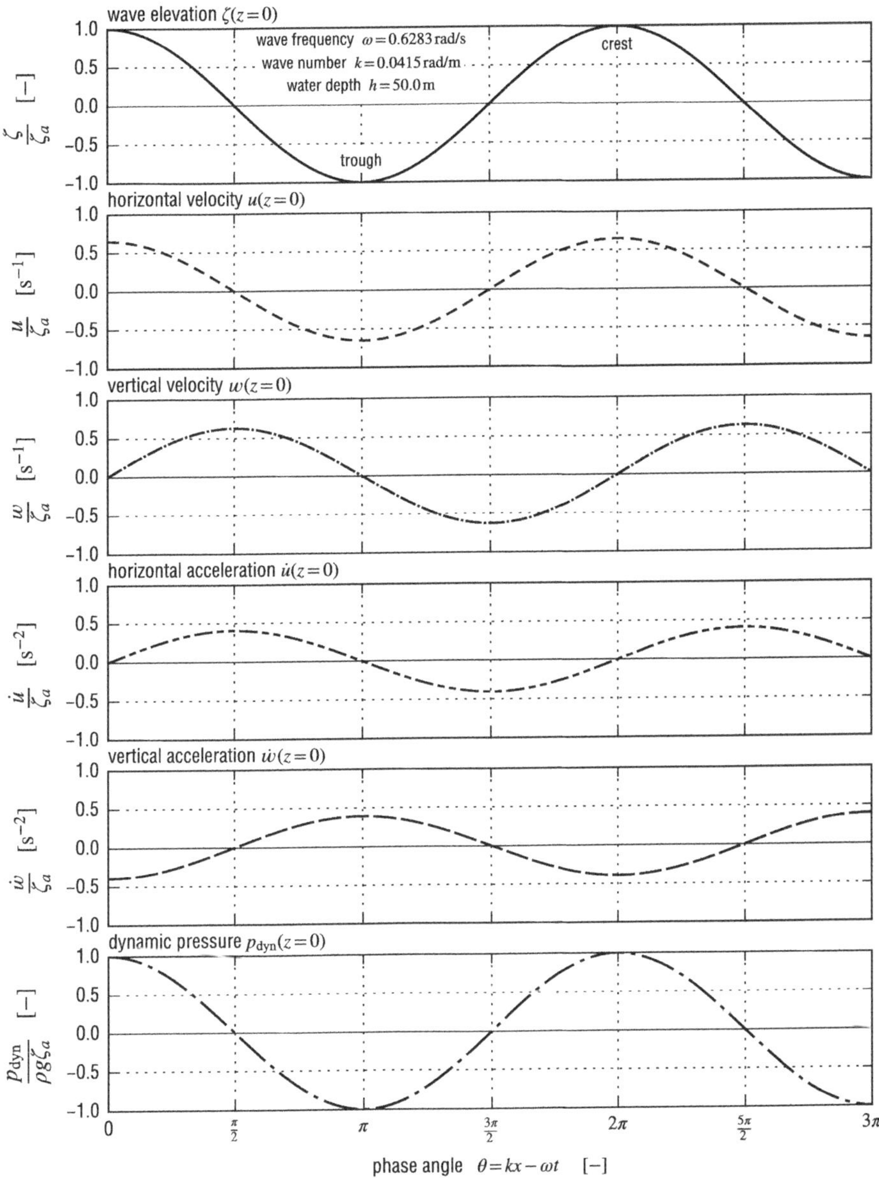

Figure 23.2 Distribution of wave properties over 1.5 wavelength at the calm water level ($z = 0$) for a wave with wave period $T = 10$ s

Velocity amplitudes

Wave elevation ζ and horizontal particle velocity u are in phase, i.e. wave elevation has its positive maximum at the same time as the horizontal particle velocity. Under a wave crest, particles move with maximum velocity in direction of wave propagation, whereas under a wave trough, the particles move with maximum velocity against the direction of wave propagation. The amplitude of the wave elevation is obviously equal to ζ_a. At

$z = 0$ we obtain an amplitude u_a of the velocity of

$$u_a = \zeta_a\omega \left.\frac{\cosh(kz+kh)}{\sinh(kh)}\right|_{z=0} = \frac{\zeta_a\omega}{\tanh(kh)} \quad \text{for } z=0$$

For a deep water wave, the numerator becomes equal to one and the maximum horizontal particle velocity is $u_a = \zeta_a\omega$.

The vertical particle velocity w is phase shifted by a quarter cycle ($\Delta\theta = -\pi/2$). It is zero when the wave elevation is at its maximum or minimum (Figure 23.2). The maximum value w_a at $z = 0$ is equal to

$$w_a = \zeta_a\omega = \frac{2\pi\zeta_a}{T} \quad \text{for } z=0 \tag{23.16}$$

There is no difference in vertical velocity amplitude w_a whether the wave is in deep water or restricted water depths. An expansion of wave amplitude multiplied by wave frequency on the right-hand side reveals that the maximum vertical particle velocity is equal to the ratio of the circumference of a circle with radius ζ_a and the wave period T.

Acceleration amplitudes

The horizontal particle acceleration $\dot{u}$ is phase shifted by $\Delta\theta = -\pi/2$ like the vertical particle velocity (Figure 23.2, fourth graph). Its maximum value is for $z = 0$:

$$\dot{u}_a = \frac{\zeta_a\omega^2}{\tanh(kh)} \quad \text{for } z=0 \tag{23.17}$$

For the amplitude of the vertical particle acceleration we get at the calm water level ($z = 0$):

$$\dot{w}_a = \zeta_a\omega^2 \quad \text{for } z=0 \tag{23.18}$$

$\dot{w}$ is in opposite phase to the wave elevation. Particles experience the largest accelerations downward under a crest and upward under a wave trough.

Vertical distribution of velocities

Figure 23.3 emphasizes the change of particle velocities with increasing submergence. We learned from Figure 23.2 that the horizontal particle velocity has its positive maximum (in the direction of wave propagation) underneath the wave crest. The vertical particle velocity has its positive (upward) maximum where the wave elevation vanishes in front of the approaching wave crest.

The horizontal velocity vanishes where the vertical velocity has an extreme value and vice versa. Velocities are largest at the calm water level and rapidly decline with increasing depth. The decline is governed by the expressions

$$\frac{\cosh(kz+kh)}{\sinh(kh)} \quad \text{and} \quad \frac{\sinh(kz+kh)}{\sinh(kh)}$$

for horizontal and vertical particle velocities respectively. On deep water both velocities decline with e^{kz}.

The vertical particle velocity vanishes at the sea bottom ($z = -h$) as requested by the bottom boundary condition (22.3). However, in restricted water depth the horizontal particle velocity does not vanish at the sea bottom. Every passing wave moves water

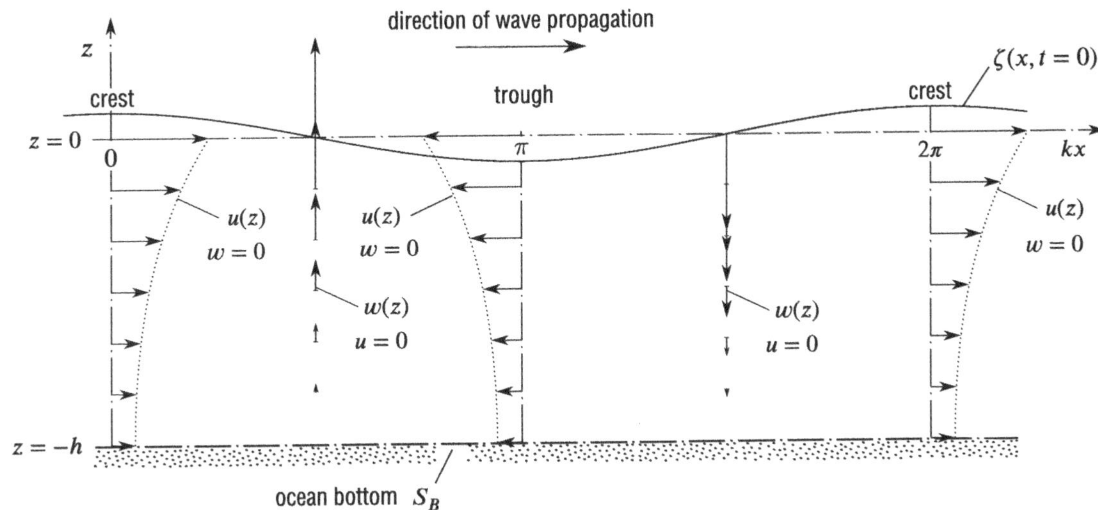

Figure 23.3 Snapshot ($t = 0$) of the velocity field for a wave in restricted water depth

back and forth parallel to the bottom. Just watch the sand rolling back and forth on the sea bottom next time you visit a beach.

Mind the vertical axis

Observe that the origin of the z-axis has been laid into the calm water level. Since the z-axis is positive upward, values of z are negative for all points in the water. The attentive reader will notice that the velocity distributions $u(z)$ and $w(z)$ in Figure 23.3 have been drawn up to the calm water level irrespective of whether the position is under a wave crest or a wave trough. It shall serve as a reminder that in linear wave theory wave amplitudes are assumed to be very small compared with the wave length.

23.4 Dynamic Pressure

The pressure distribution in a wave is obtained from the Bernoulli equation (see Sections 15.4). Since we linearized the boundary value problem, it is opportune to use the linearized form of the Bernoulli equation.

$$p - p_\infty = -\rho \frac{\partial \phi}{\partial t} - \rho g z \tag{15.45}$$

The dynamic pressure in the wave field is equal to the unsteady portion of the pressure:

$$\begin{aligned} p_{\text{dyn}} &= -\rho \frac{\partial \phi}{\partial t} \\ &= -\rho \frac{\partial}{\partial t} \left[\frac{\zeta_a g}{\omega} \frac{\cosh(kz + kh)}{\cosh(kh)} \sin(kx - \omega t) \right] \\ &= \rho \omega \frac{\zeta_a g}{\omega} \frac{\cosh(kz + kh)}{\cosh(kh)} \cos(kx - \omega t) \\ &= \rho g \zeta_a \frac{\cosh(kz + kh)}{\cosh(kh)} \cos(kx - \omega t) \end{aligned} \tag{23.19}$$

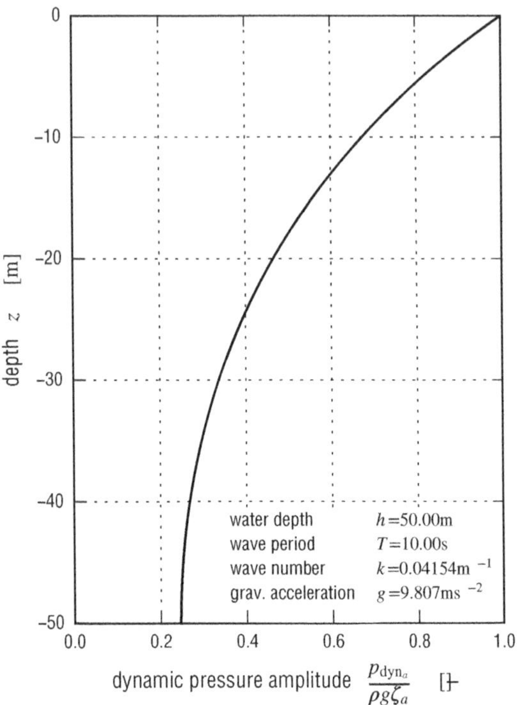

Figure 23.4 Amplitude of dynamic pressure over depth z

As the bottom graph in Figure 23.2 shows, dynamic pressure is in phase with the wave elevation. The maximum value is equivalent to the hydrostatic pressure underneath a water column of height ζ_a. The positive maximum is found at the calm water level ($z = 0$) under wave crests ($\theta = n\pi$, with $n = 0, 2, 4, \dots$).

$$p_{\text{dyn}_a} = \rho\, g\, \zeta_a \qquad \text{for } z = 0 \tag{23.20}$$

The dynamic pressure p_{dyn} at the sea bottom $z = -h$ fluctuates between the values

$$-\frac{\rho\, g\, \zeta_a}{\cosh(kh)} < p_{\text{dyn}} < +\frac{\rho\, g\, \zeta_a}{\cosh(kh)} \qquad \text{for } z = -h.$$

Like wave particle velocities and accelerations, dynamic pressure amplitude declines rapidly with depth z (see Figure 23.4). Dynamic pressure is positive underneath wave crests and negative underneath wave troughs.

23.5 Water Particle Motions

Particle paths Integration of the wave particle velocities Equations (23.12) and (23.13) (or their deep water equivalents) over time t results in the paths of a water particle with initial position (x, z). The photo in Figure 23.5 shows the unsteady movement of small, neutrally buoyant particles underneath a wave. The picture was taken in a small wave flume at the

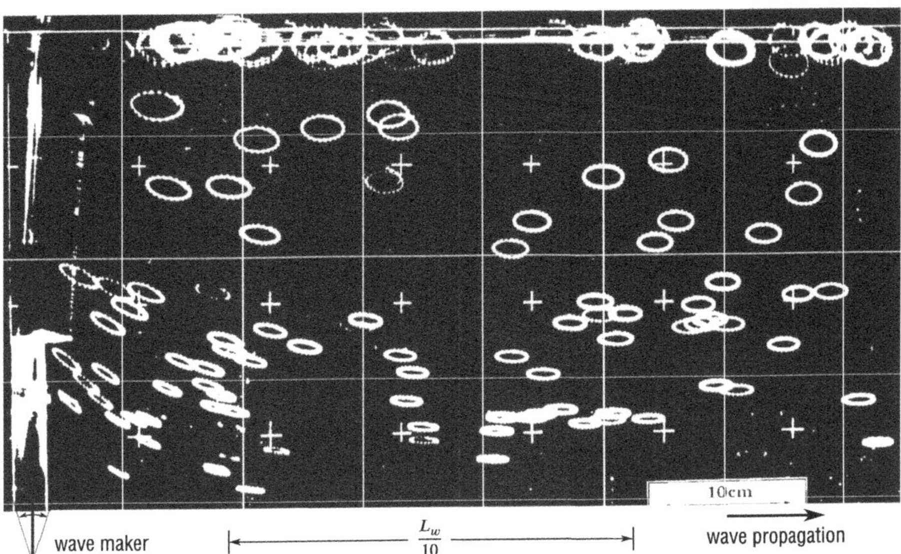

Figure 23.5 Photo of water particle trajectories. Photo courtesy of Dr. Walter L. Kuehnlein, sea2ice Ltd. & Co. KG, www.sea2ice.com

Technical University of Berlin. Its walls are made out of acrylic. The photo's exposure time was chosen to be equal to one wave period. Wave steepness H/L_w is small and, as linear wave theory predicts, paths of particles are closed. Therefore, water particles are not transported along with the wave elevation. On average, they stay at their original location (x, z). The vertical axis of the elliptical paths declines visibly with depths. The horizontal extensions decline very little because the wave length is about 8 times the water depth. As a consequence, shallow water effects are strong.

In steep waves with high H/L_w ratio, particle paths are no longer closed. They look like a spiral and during each wave cycle particles are moved a small distance further in the direction of wave propagation. This mass transport contributes considerably to coastal storm surge in heavy storms like hurricanes.

Deep water trajectories

Figure 23.6 depicts water particle trajectories for deep water (left side) and restricted water depth (on the right). The × mark the position of water particles when they are at rest. For deep water the trajectories are circles. The diameter is largest at the calm water level and equal to twice the wave amplitude. With increasing submergence the diameters of trajectories decline rapidly and vanish almost completely at a depth of half the wave length.

Trajectories on restricted water depth

On the right side of Figure 23.6, trajectories are plotted for the same wave period and amplitude but consider a finite water depth of $h = 50$ m. Due to the influence of the ocean floor, wave length has shrunk by 9%. The ratio of water depth to wave length is about $h/L_w \approx 0.2$. The trajectories have become ellipses. The eccentricity is still small close to the water surface but increases with increasing submergence. At the ocean floor, the vertical axis of all trajectories vanishes and particles simply move back and forth parallel to the bottom (see also Figure 23.5). This is of course a consequence of the bottom boundary condition which prohibits flow through the ocean floor.

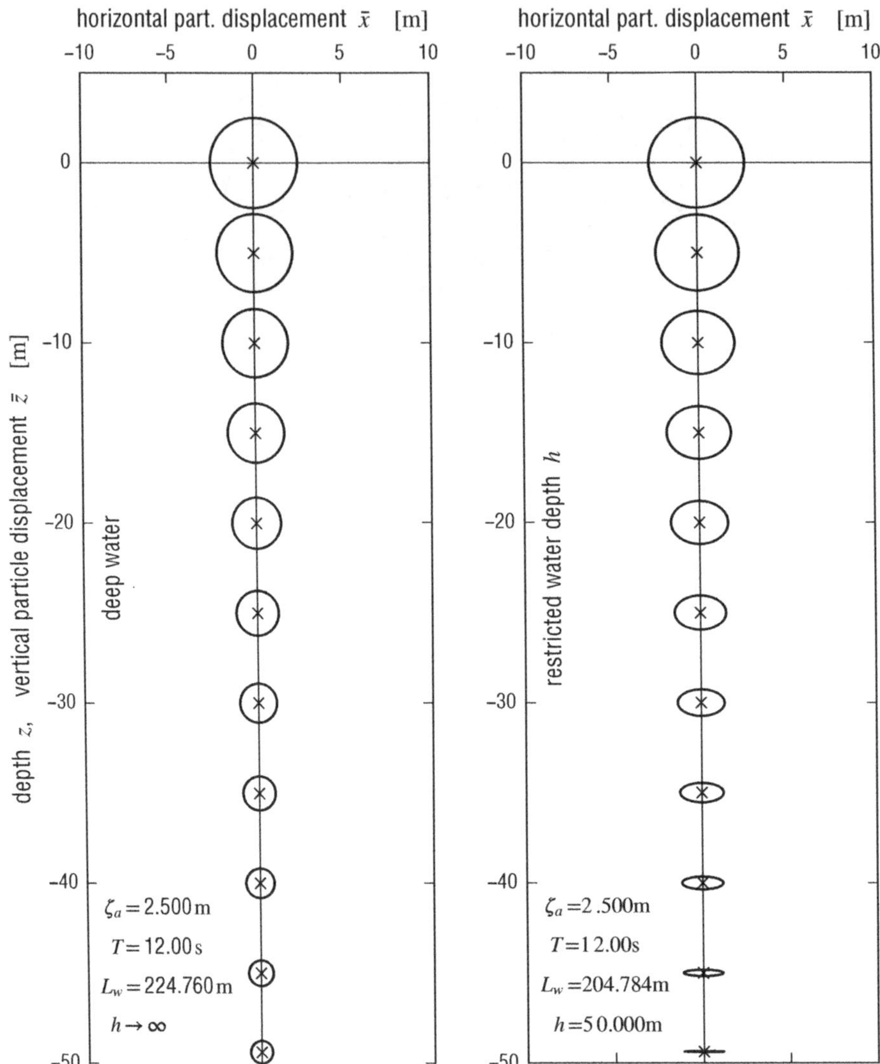

Figure 23.6 Water particle trajectories over one wave period T for deep water (left) and restricted water (right)

References

Akai, T. (1994). *Applied numerical methods for engineers.* John Wiley & Sons, Inc., New York, NY.

Clauss, G., Lehmann, E., and Östergaard, C. (1992). *Offshore structures, volume 1: Conceptual design and hydrodynamics.* Springer Verlag, London.

Dean, R. and Dalrymple, R. (1991). *Water wave mechanics for engineers and scientists,* volume 2 of *Advanced Series on Ocean Engineering.* World Scientific, Hackensack, NJ.

Self Study Problems

1. Compute the wave number k and wave length L_w for a wave with period $T = 6$ s in $h = 12$ m water depth.

2. Will the wave length become shorter or longer when the wave propagates from deep into shallow water?

3. Compute the wave length for the wave with frequency $\omega = 0.95$ s^{-1} in water of depth $h = 25$ m.

4. Find the wave length L_w of a wave of frequency $\omega = 0.95$ s^{-1} in deep water.

5. Use the real-valued deep water potential

$$\phi(x, z, t) = \frac{\zeta_a g}{\omega} \mathrm{e}^{k_0 z} \sin(k_0 x - \omega t) \tag{22.52}$$

   to derive equations for the water particle velocities, accelerations, and dynamic pressure for the case $kh > \pi$.

6. Use the equations from Problem 5 to plot maxima of velocities $u_a(z)$, $w_a(z)$, accelerations $\dot{u}_a(z)$, $\dot{w}_a(z)$, and dynamic pressure $p_{\text{dyn}}(z)$ from $z = -L_w/2$ to $z = 0$ for a wave with period $T = 5$ s and amplitude $\zeta_a = 1$ m.

7. Derive the equations for the complex water particle velocities and accelerations from the complex linear wave theory velocity potential (22.35).

24

Wave Energy and Wave Propagation

The preceding chapter discusses wave kinematics and the distribution of wave particle velocities, accelerations, and dynamic pressure throughout the water column underneath a long crested wave. In this chapter, we will use these results to define the energy contained in a regular wave and how the wave and its energy propagate. The results will form the basis for a brief discussion of ship wave resistance in the following chapter.

Learning Objectives

At the end of this chapter students will be able to:

- distinguish between the kinematic energy and potential energy of waves
- describe and discuss how wave energy propagates
- compute group velocity and energy density of waves

24.1 Wave Propagation

Wave elevation Over the time of a single wave period T a wave travels the distance of one wave length L_w. We used this to define the phase velocity $c = \omega/k$ in Equation (19.5). See also Section 22.2. Figure 24.1 depicts a wave in 13 time instances during one wave period ($\omega t \in [0, 2\pi]$). The wave crest, marked with a small triangle $\triangledown$, travels exactly one wave length from $kx = 0$ to $kx = 2\pi$.

Water particles It is important to understand that the wave elevation is propagating but not the water particles. In the center of Figure 24.1 at $kx = 2\pi$, a particle (small dot) is tracked over one wave period. On average, it stays in the same position. Its horizontal and vertical excursions are equal to the wave amplitude $\pm\zeta_a$. If you look closely, you will notice that the dot marking the water particle is not always exactly on the plotted wave contour. This is most evident at the locations of maximum horizontal excursion. The apparent discrepancy is a consequence of the linearization of the free surface boundary conditions. The harmonic wave elevation $\zeta = \zeta_a \cos(kx - \omega t)$ is only truly valid for vanishing wave amplitude or more precisely if $2\zeta_a/L_w \ll 1$.

Fundamentals of Ship Hydrodynamics: Fluid Mechanics, Ship Resistance and Propulsion, First Edition.
Lothar Birk.

Companion website: www.wiley.com/go/birk/hydrodynamics

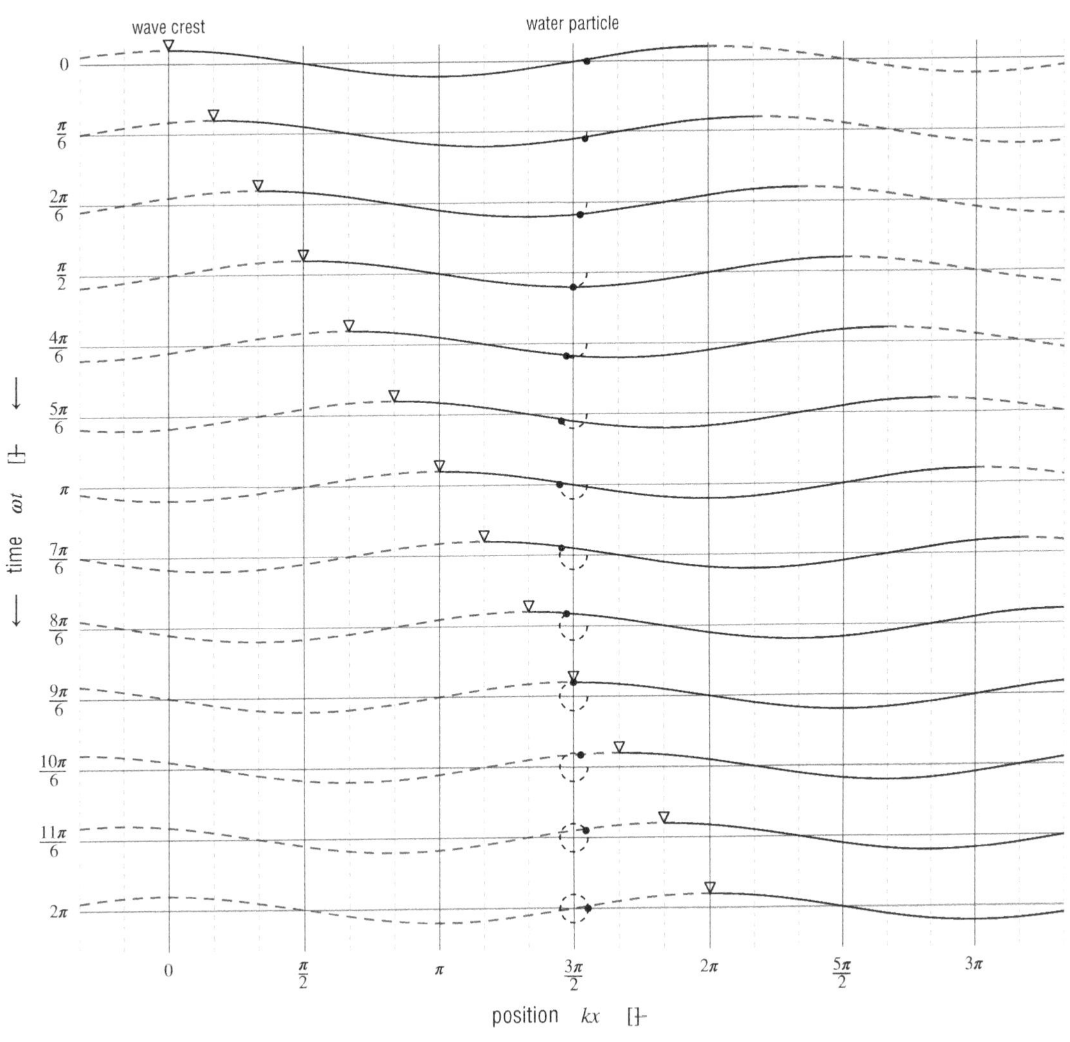

Figure 24.1 Propagation of wave profile and the movement of a water particle over one wave period

Mass transport in steep waves

The larger the wave steepness H/L_w, the less accurate is the sinusoidal wave profile. In fact, as mentioned previously, in steep waves the particle paths are no longer closed loops, and a slow but steady mass transport occurs in the direction of wave propagation.

Effects of finite water depths

Once the water depth h is less than half the wave length, the sea bottom begins to affect the way a wave propagates. Did you ever wonder why waves always roll onto a beach with their crests and troughs parallel to the shoreline? As a wave moves from deeper into shallower water, its wave length becomes shorter. Figure 24.2 shows the wave length L_w as a function of water depth h for a selection of wave periods T. The wave length stays constant in deep water where $2h > L_w$.

A decreasing wave length causes the wave number k to increase. As a consequence, the wave phase velocity $c = \omega/k$ must decrease, i.e. a wave slows down as it moves into

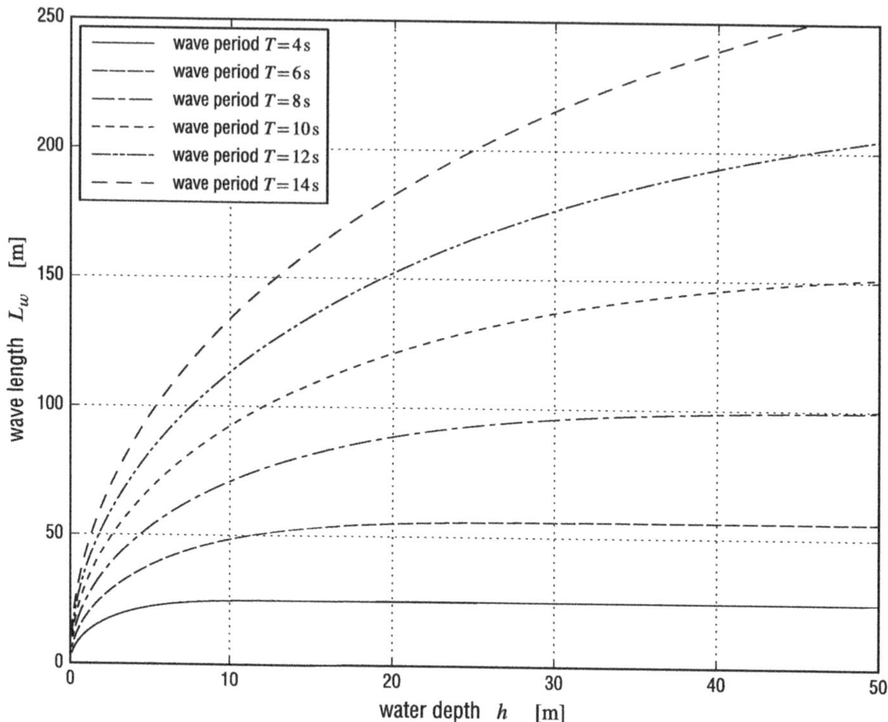

Figure 24.2 Wave length L_w as a function of water depth h for constant wave period T

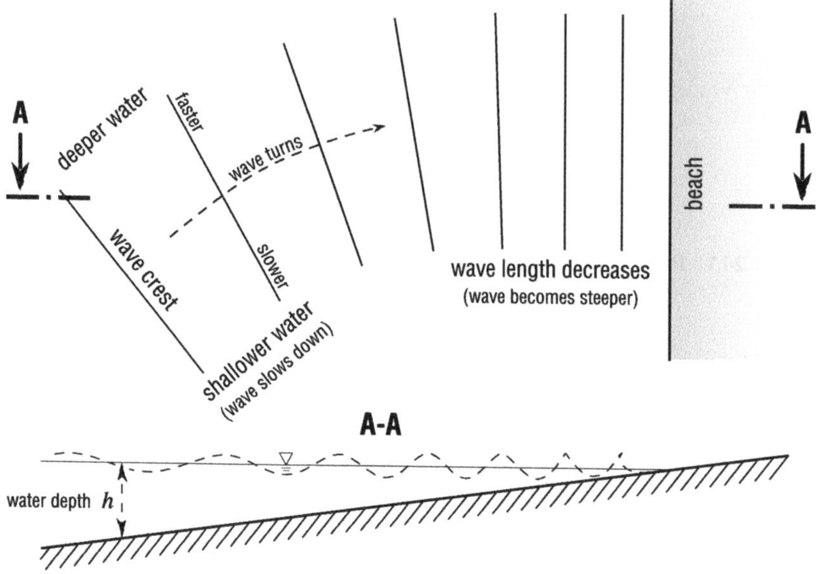

Figure 24.3 Effect of gradually decreasing water depth h on wave propagation and direction

shallower water. Figure 24.3 shows waves approaching a beach. Further offshore, in deeper water, waves move in the direction of prevailing winds. In the figure, waves travel at an oblique angle with respect to the beach. The parts of waves closer to shore will be affected first by the decreasing water depth. Those parts will slow down while the parts further offshore are still traveling at the original phase velocity. The difference in phase velocity causes the wave to turn and eventually its crest becomes parallel to the shoreline. As it further approaches the beach, the front of a wave continues to slow down. The wave becomes higher and steeper, and it will eventually break if the steepness exceeds $H/L_w > 1/7$ (Michell, 1893).

24.2 Wave Energy

Whoever has sailed through a storm at sea or watched waves crashing into a sea wall has experienced the destructive force of waves first hand. In fact, waves carry so much energy that it is a potential source of renewable power. The Bureau of Ocean Energy Management states on its web pages (https://www.boem.gov/Ocean-Wave-Energy/) that in the United States alone the recoverable wave power along the U.S. shelf edge would amount to more than 1000 TWh/yr (i.e. tera-watt hours per year), which is enough to power millions of homes.

Contributions to wave energy

Waves contain two forms of energy:

(i) Kinetic energy: In the previous chapter we discussed the motion of water particles in waves. Obviously, if a mass of fluid moves, it possesses kinetic energy.

(ii) Potential energy: The water surface at a selected point moves up and down in a way similar to the movement of a mass hanging from a spring. In fact, when a wave crest passes, fluid elements have to be lifted above their equilibrium condition. The necessary work done by the flow is converted into potential energy in the gravity field.

Below, both forms of energy are computed and, as before, it is assumed that the wave is regular and its wave steepness is small enough to apply linear wave theory.

24.2.1 Kinetic wave energy

Kinetic energy

Figure 24.4 shows a slice of ocean V which is one wave length L_w long, has a width of b in y-direction, and stretches from the ocean floor at $z = -h$ to the calm water level $z = 0$. The wave propagates in x-direction and the wave crests are parallel to the y-axis (long crested wave). The volume is cut into small volumes $\mathrm{d}V$ of size

$$\mathrm{d}V = \mathrm{d}x\,\mathrm{d}z\,b \tag{24.1}$$

The kinetic energy contained in the volume $\mathrm{d}V$ is

$$\mathrm{d}E_{\mathrm{kin}} = \frac{1}{2}\rho\,|\underline{v}|^2\,\mathrm{d}V \tag{24.2}$$

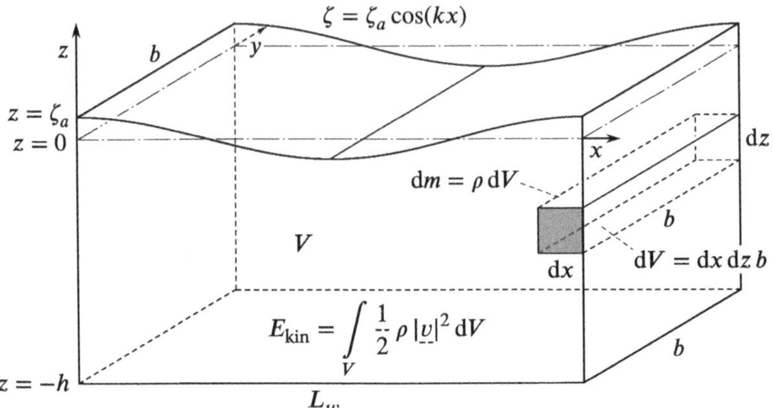

Figure 24.4 Kinetic energy E_{kin} in the control volume V spanning one wave length L_w in x-direction

Obviously, the total kinetic energy in the fluid space V is found by integrating dE_{kin} across the volume.

$$E_{kin} = \int_V \frac{1}{2} \rho |\underline{v}|^2 \, dV \tag{24.3}$$

Since the flow is planar in the x-z-plane, the volume integral reduces to a 2D integral over wave length and water depth. The components of the velocity vector $\underline{v}^T = (u, w)$ are given by Equations (23.12) and (23.13).

$$\begin{aligned} E_{kin} &= \frac{1}{2} \rho b \int_{-h}^{0} \int_{0}^{L_w} (u^2 + w^2) \, dx dz \\ &= \frac{1}{2} \rho b \left(\frac{\zeta_a \omega}{\sinh(kh)} \right)^2 \int_{-h}^{0} \int_{0}^{L_w} \left[\cosh^2(kz + kh) \cos^2(\theta) \right. \\ &\qquad \left. + \sinh^2(kz + kh) \sin^2(\theta) \right] dx dz \end{aligned} \tag{24.4}$$

Simplifying the integrand As a first step, the integrand is simplified using the following identities for the hyperbolic cosine and sine functions.

$$\cosh(\alpha) = \frac{1}{2}\left(e^{\alpha} + e^{-\alpha}\right) \qquad \text{and} \qquad \sinh(\alpha) = \frac{1}{2}\left(e^{\alpha} - e^{-\alpha}\right)$$

from which follow the relationships

$$\cosh^2(\alpha) = \frac{1}{4}\left(e^{2\alpha} + e^{-2\alpha} + 2\right) \qquad \text{and} \qquad \sinh^2(\alpha) = \frac{1}{4}\left(e^{2\alpha} + e^{-2\alpha} - 2\right)$$

Setting $\alpha = kz + kh$, the integrand in Equation (24.4) becomes

$$\begin{aligned}
&\cosh^2(\alpha)\cos^2(\theta) + \sinh^2(\alpha)\sin^2(\theta) \\
&\quad = \frac{1}{4}\left[\left(e^{2\alpha} + e^{-2\alpha} + 2\right)\cos^2(\theta) + \left(e^{2\alpha} + e^{-2\alpha} - 2\right)\sin^2(\theta)\right] \\
&\quad = \frac{1}{4}\left[\left(e^{2\alpha} + e^{-2\alpha}\right)\left(\cos^2(\theta) + \sin^2(\theta)\right) + 2\left(\cos^2(\theta) - \sin^2(\theta)\right)\right] \\
&\quad = \frac{1}{4}\left[\left(e^{2\alpha} + e^{-2\alpha}\right) + 2\left(\cos^2(\theta) - \sin^2(\theta)\right)\right]
\end{aligned}$$

For the second part, we used the identity $1 = \cos^2(\theta) + \sin^2(\theta)$.

Further, with $\cos(2\theta) = \cos^2(\theta) - \sin^2(\theta)$ we finally get a much simpler form of the integrand:

$$\begin{aligned}
&\cosh^2(\alpha)\cos^2(\theta) + \sinh^2(\alpha)\sin^2(\theta) \\
&\quad = \frac{1}{2}\left[\cosh(2kz + 2kh) + \cos(2kx - 2\omega t)\right]
\end{aligned} \tag{24.5}$$

This expression captures the magnitude of velocity $|\underline{v}|$ as a function of x and z.

Equation (24.4) is rewritten based on the result (24.5): **Integration**

$$\begin{aligned}
E_{\text{kin}} = \rho b \left(\frac{\zeta_a \omega}{\sinh(kh)}\right)^2 &\left\{ \int_{-h}^{0}\int_{0}^{L_w} \cosh(2kz + 2kh)\,\mathrm{d}x\mathrm{d}z \right. \\
&\left. + \int_{-h}^{0}\int_{0}^{L_w} \cos(2kx - 2\omega t)\,\mathrm{d}x\mathrm{d}z \right\}
\end{aligned} \tag{24.6}$$

The remaining two integrals are of standard form.

The integrand of the first integral in Equation (24.6) is independent of x. The x-integration simply yields a factor of one wave length and the integral becomes:

$$\begin{aligned}
\int_{-h}^{0}\int_{0}^{L_w} \cosh(2kz + 2kh)\,\mathrm{d}x\mathrm{d}z &= L_w \int_{-h}^{0} \cosh(2kz + 2kh)\,\mathrm{d}z \\
&= L_w \left[\frac{1}{2k}\sinh(2kz + 2kh)\right]_{-h}^{0} \\
&= \frac{L_w}{2k}\sinh(2kh)
\end{aligned} \tag{24.7}$$

The second integrand in Equation (24.6) is only a function of x and time t. Integration over the water depth yields the factor h. However, the remaining integral over the wave length vanishes.

$$\int_{-h}^{0}\int_{0}^{L_w} \cos(2kx - 2\omega t)\,\mathrm{d}x\mathrm{d}z = h\int_{0}^{L_w} \cos(2kx - 2\omega t)\,\mathrm{d}x = h\left[\frac{\sin(2kx - 2\omega t)}{2k}\right]_0^{L_w}$$
$$= h\left[\frac{\sin(2kx)\cos(2\omega t) - \cos(2kx)\sin(2\omega t)}{2k}\right]_0^{L_w} \qquad (24.8)$$
$$= 0$$

Kinetic energy in control volume V With the results from Equations (24.7) and (24.8), we obtain from Equation (24.4) the kinetic wave energy in the control volume V:

$$E_{\text{kin}} = \rho\, b\, L_w \left(\frac{\zeta_a \omega}{2\sinh(kh)}\right)^2 \frac{1}{2k}\sinh(2kh)$$

Utilizing that $\sinh(2kh) = 2\sinh(kh)\cosh(kh)$ results in

$$E_{\text{kin}} = \frac{1}{4}\rho\, b\, L_w\, \zeta_a^2 \left(\frac{\omega^2\cosh(kh)}{k\sinh(kh)}\right) \qquad (24.9)$$

The ratio of hyperbolic cosine and sine is equivalent to the reciprocal hyperbolic tangent:

$$\frac{\cosh(kh)}{\sinh(kh)} = \frac{1}{\tanh(kh)} \qquad (24.10)$$

The resulting expression may be simplified based on the dispersion relation (22.41).

$$\frac{\omega^2}{k\tanh(kh)} = g \qquad (24.11)$$

Thus, the kinetic energy contained in our control volume V is

$$E_{\text{kin}} = \frac{1}{4}\rho\, g\, \zeta_a^2\, b\, L_w \qquad (24.12)$$

Kinetic wave energy density The kinetic energy per unit ocean surface is known as kinetic wave energy density. We divide Equation (24.12) by the area of the free surface $b\,L_w$.

$$\overline{E}_{\text{kin}} = \frac{E_{\text{kin}}}{b\,L_w} = \frac{1}{4}\rho\, g\, \zeta_a^2 \qquad (24.13)$$

Note that the energy density is independent of water depth and wave frequency. It is solely a function of the squared wave amplitude.

24.2.2 Potential wave energy

Potential energy Potential energy in a gravity field is determined by the gravitational acceleration, elevation above a datum level, and mass of the object. When calculating the potential wave energy, we consider the work done when a volume of water is lifted above the calm

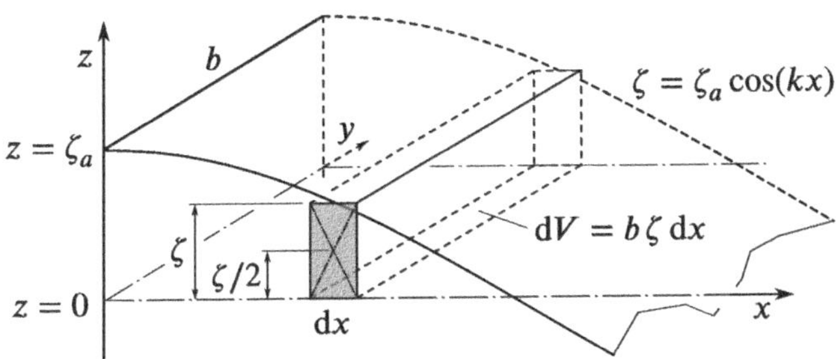

Figure 24.5 Change in potential energy E_{pot} when a fluid element is lifted above the calm water level

water level. Figure 24.5 shows a piece of the free surface. We consider a small volume $\mathrm{d}V$ of length $\mathrm{d}x$, width b, and height equal to the local wave elevation ζ. Its mass will be

$$\mathrm{d}m = \rho\,\mathrm{d}V = \rho\, b\, \zeta\, \mathrm{d}x \tag{24.14}$$

The center of a mass element $\mathrm{d}m$ is lifted above the calm water surface by half of its height. The work necessary for the lift is equal to the change in potential energy.

$$\mathrm{d}E_{pot} = g\,\frac{\zeta}{2}\,\mathrm{d}m = \frac{1}{2}\,\rho\, g\, b\, \zeta^2 \mathrm{d}x \tag{24.15}$$

The same work has to be done to depress the free surface. Outside the range of a progressing wave elevation, overall mass distribution remains unchanged and water fills the control volume. Therefore, changes in potential energy occur at the free surface only.

Potential wave energy

We integrate Equation (24.15) over one wave length $x \in [0, L_w]$.

$$\begin{aligned} E_{pot} &= \frac{1}{2}\,\rho\, g\, b \int_0^{2\pi} \zeta^2\, \mathrm{d}x \\ &= \frac{1}{2}\,\rho\, g\, b \int_0^{2\pi} \zeta_a^2\, \cos^2(kx - \omega t)\, \mathrm{d}x \\ &= \frac{1}{2}\,\rho\, g\, b\, \zeta_a^2 \left[\frac{x}{2} - \frac{1}{4k}\,\sin(2kx - 2\omega t)\right]_0^{L_w} \\ &= \frac{1}{2}\,\rho\, g\, b\, \zeta_a^2 \left(\frac{L_w}{2} - 0\right) \end{aligned} \tag{24.16}$$

Thus, the potential energy distributed over one wave length is equal to the kinetic energy.

$$E_{pot} = \frac{1}{4}\,\rho\, g\, \zeta_a^2\, b\, L_w \tag{24.17}$$

Potential wave energy density

The potential wave energy density is

$$\overline{E}_{pot} = \frac{E_{pot}}{b\, L_w} = \frac{1}{4}\,\rho\, g\, \zeta_a^2 \tag{24.18}$$

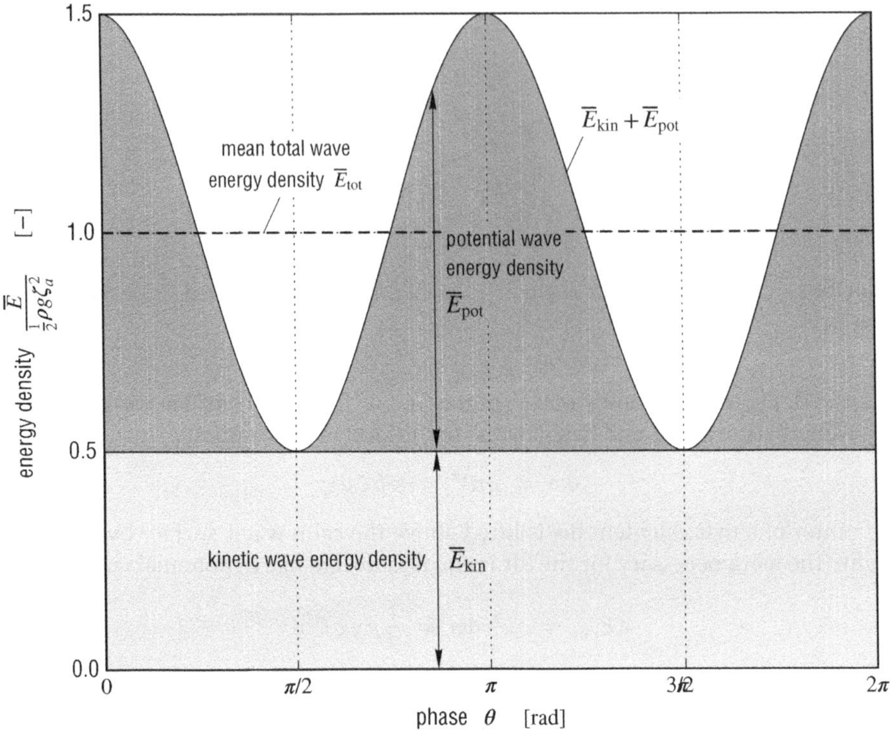

Figure 24.6 Wave energy density distribution of a regular wave over one wave cycle according to linear wave theory

24.2.3 Total wave energy density

The total energy density is equal to the sum of kinetic and potential wave energy density. According to linear wave theory, the energy density of a regular wave amounts to

$$\overline{E} = \overline{E}_{pot} + \overline{E}_{kin} = \frac{1}{2} \rho g \zeta_a^2 \tag{24.19}$$

This result is significant as it relates to the statistical energy density of regular waves $S(\omega)\mathrm{d}\omega = \zeta_a^2/2$. $S(\omega)$ is the value of the spectral energy density spectrum also known as wave spectrum (Dean and Dalrymple, 1991). The latter plays an important role in the statistical treatment of irregular sea states which are modeled as a superposition of many regular components with varying frequency and phase. More on this may be found in offshore engineering references like Barltrop and Adams (1991).

For our discussion of wave resistance, we may draw the conclusion that doubling the wave amplitude in the ship's wave pattern will quadruple its energy density. Low wave resistance is achieved if wave amplitudes in the wave pattern stay as small as possible.

Distribution of wave energy density

Figure 24.6 shows the distribution of wave energy density for a wave over one period. Kinetic energy density is constant across a wave cycle because the amplitude of the wave particle velocity is only varying with submergence. Equation (24.16) reveals that the potential wave energy density follows a $\cos^2$ distribution over one cycle. The potential wave energy density is at a maximum for wave crest and trough. It vanishes at the zero crossings of the wave elevation. However, as we calculated in (24.18), its mean value is equal to the kinetic energy density distribution.

24.3 Energy Transport and Group Velocity

Waves transport energy in the direction of wave propagation. However, since mass transport does not exist in linear waves due to the closed particle paths, only the potential wave energy portion is transported by the wave. In this section we determine how fast wave energy propagates.

First, we study the idealized simulation depicted in Figure 24.7. It follows an idea taken from Clauss et al. (1992). Imagine a wave maker which transfers 4096 energy density units into the water during each wave cycle. The horizontal axis in Figure 24.7 marks the distance from the wave maker which acts at location $x = 0$. The vertical scale depicts time measured from top to bottom in wave periods.

Simulation from cycle 0 to 1

Beginning at the top of Figure 24.7, we track the energy density distribution over 12 wave maker cycles. The wave maker begins by introducing 4096 energy density units, one half of which is converted into kinetic energy; the other half becomes potential energy. During the first cycle, the potential energy propagates one wave length forward along with the wave elevation. The kinetic wave energy remains at the wave maker which adds another 4096 energy density units. At the end of cycle 1 a total of 6144 energy density units are at the wave maker while 2048 units have been propagated to location $x = L_w$.

Simulation from cycle 1 to 2

At station $x = L_w$, half of the 2048 units must be used to get the water particles moving (kinetic energy) and subsequently remain at $x = L_w$. The other half (1024 units) is propagated to station $x = 2L_w$. At the same time, half of the 3072 energy density units from $x = 0$ have moved to station $x = L_w$. The remaining 3072 units at $x = 0$ are augmented by another 4096 units from the wave maker. Thus, the energy distribution after two wave maker cycles is

position x/L_w	0	1	2
energy density units	7168	4096	1024

The lowest row in Figure 24.7 shows the energy density distribution after 12 wave maker cycles. Note that we are slowly approaching equilibrium at the wave maker, in which as much energy is propagated as is added. With increasing distance from the wave maker, energy density declines rapidly. Since the energy density is tied to the wave amplitude, the latter must diminish as well with increasing distance from the wave maker.

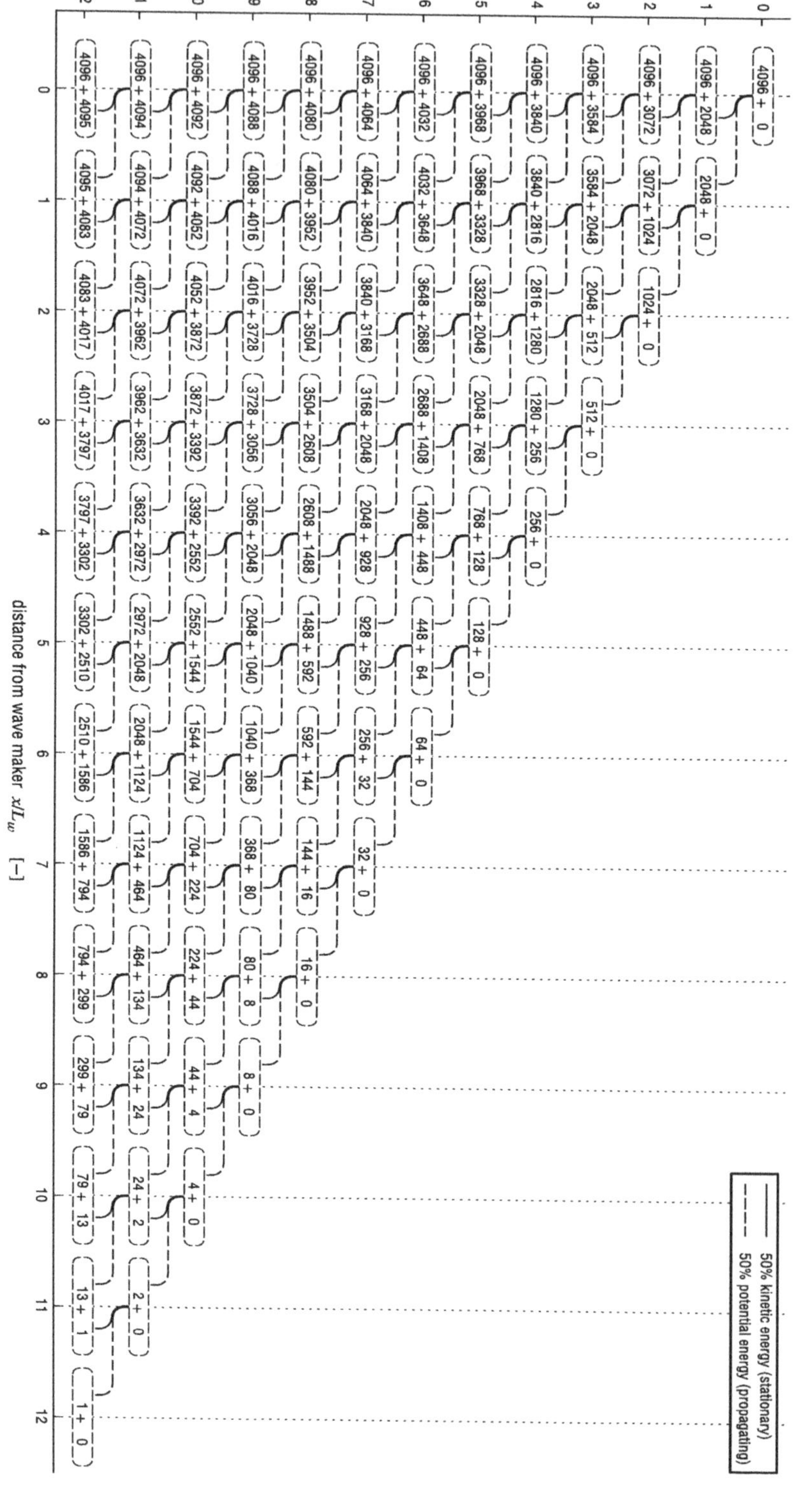

Figure 24.7 Schematic propagation of wave energy for a deep water wave based on linear wave theory

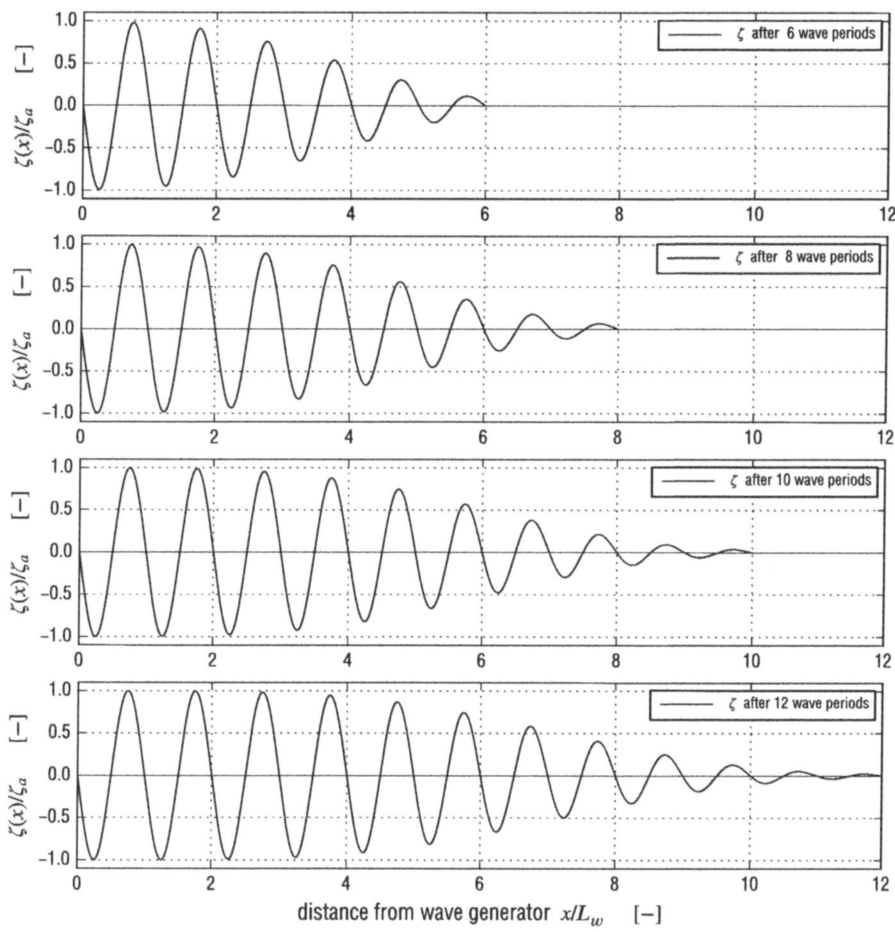

Figure 24.8 Wave elevation profiles after a few selected cycles of wave making (deep water)

Wave elevation

Although the wave maker adds an equal amount of energy in each cycle, it takes six cycles to raise the energy density to within 1% of its asymptotic value. Figure 24.8 depicts the resulting wave elevations for wave maker cycles 6, 8, 10, and 12. The unavoidable build-up of wave amplitude in the tank makes model testing in regular waves difficult as the model will experience the change in amplitude as a wave of different frequency.

Wave energy density distribution

Figure 24.9 shows the wave energy density distributions corresponding to the wave elevations in Figure 24.8. A small amount of the wave energy has progressed as many wave lengths as there are wave cycles. However, the mean value of the wave energy density has traveled only half that distance. This implies that the energy propagates with half the phase velocity.

$$c_G = \frac{c}{2} \qquad \text{for deep water waves} \tag{24.20}$$

The propagation speed of wave energy was first derived by Sir Gabriel Stokes and based on a physical argument (Darrigol, 2003). Later Strutt, Baron Rayleigh (1877) provided

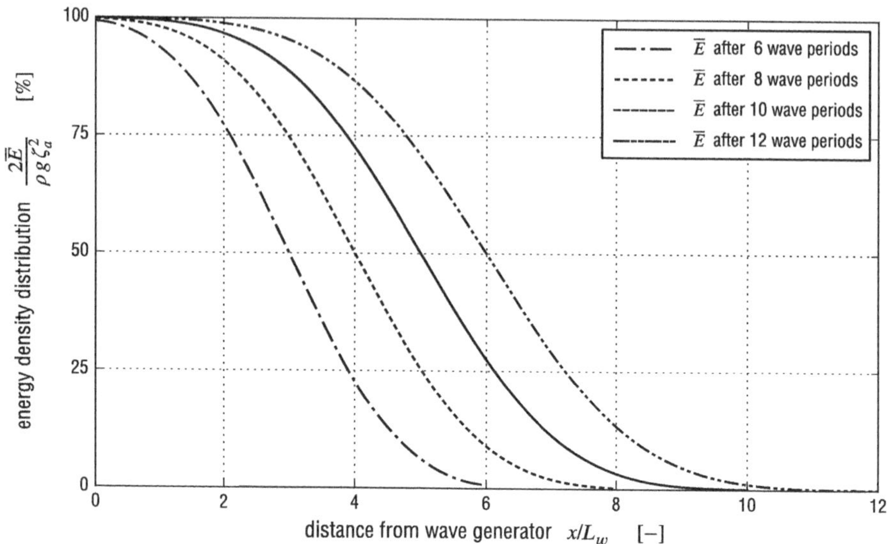

Figure 24.9 Distribution of energy density after a few selected cycles of wave making (deep water)

a formal proof by computing the work done by the dynamic pressure in a cross section of the fluid. In general, the energy propagation velocity of water waves depends also on the water depth.

$$c_G = \frac{c}{2}\left[1 + \frac{2kh}{\sinh(2kh)}\right] \tag{24.21}$$

Group velocity The velocity at which wave energy is transported corresponds to the propagation speed of wave groups. In continuation of our wave making simulation from Figure 24.7, a wave maker is operated for 16 cycles and shut down. Figure 24.10 depicts the resulting group of regular waves after 40, 50, and 60 wave periods. After 60 cycles, the actual front of the wave group has traveled 60 wave lengths, but the amplitudes at the front are very small and cut off on the plots. The dashed lines at the top and bottom of the groups represent group envelopes, which interpolate wave crests and troughs, respectively, A gray triangle marks the maximum peak of the group envelop. During the 10 wave periods between individual plots, the peak of the wave group has traveled exactly 5 wave lengths because the energy transport velocity (24.20) is just half the phase velocity on deep water. Therefore, the energy transport velocity is also known as group velocity c_G. The plots also track a wave labeled #33. As expected, it progresses with phase velocity 10 wave lengths during 10 wave periods.

References

Barltrop, N. and Adams, A. (1991). *Dynamics of fixed marine structures*. Butterworth-Heinemann Ltd, Oxford, United Kingdom, third edition.

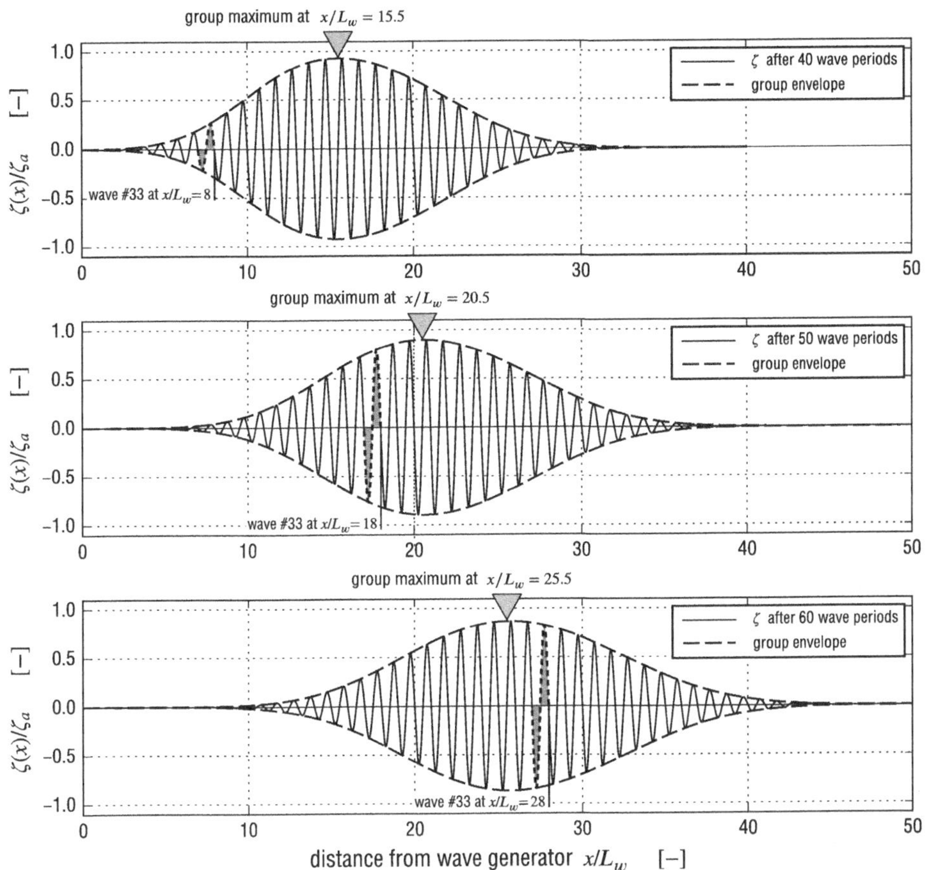

Figure 24.10 Propagation of a group of regular waves over 20 wave periods

Clauss, G., Lehmann, E., and Östergaard, C. (1992). *Offshore structures, volume 1: Conceptual design and hydrodynamics.* Springer Verlag, London.

Darrigol, O. (2003). The spirited horse, the engineer, and the mathematician: water waves in nineteenth-century hydrodynamics. *Archive for History of Exact Sciences,* 58(1):21–95. doi 10.1007/s00407-003-0070-5.

Dean, R. and Dalrymple, R. (1991). *Water wave mechanics for engineers and scientists,* volume 2 of *Advanced Series on Ocean Engineering.* World Scientific, Hackensack, NJ.

Michell, J. (1893). The hightest wave in water. *Philosophical Magazine Series 5,* 36:430–437.

Strutt, Baron Rayleigh, J. (1877). On progressive waves. In *Proceedings of the London Mathematical Society,* volume IX, pages 21–26.

Self Study Problems

1. Compute the wave energy density for a wave of wave height $H = 2$ m and wave length $L_w = 40$ m in $h = 50$ m water depth.

2. Explain to a friend why wave crests always align themselves parallel to a beach with a gentle slope.

3. Throw a stone into a calm pond. Watch the developing ripples closely. Look out for the waves that emerge at the trailing edge of the wave group and observe how they move to the front, decline, and finally vanish at the leading edge of the wave group.

4. Complete 10 cycles of a table like the one shown is Figure 24.7 for a wave maker which adds 3000 energy units during each cycle. Round numbers to one decimal.

25

Ship Wave Resistance

Every object traveling at or close to a free fluid interface will create a wave pattern: a ship sailing across the ocean, a koi close to the water surface, or a sea plane about to land. In this chapter we apply our knowledge about waves to the problem of wave resistance. The actual computation of wave resistance is beyond the scope of this book, but we will discuss basic concepts and introduce Michell's integral and panel methods, which may be employed to assess wave pattern resistance.

Learning Objectives

At the end of this chapter students will be able to:

- understand the origin and mechanism of wave resistance
- discuss favorable and unfavorable wave superposition
- know about methods for wave resistance computation

25.1 Physics of Wave Resistance

Effect of free surface

A deeply submerged body in a frictionless fluid will not experience a drag force. D'Alembert investigated this hypothetical case and found that the resulting pressure force vanishes for a body deeply submerged in an inviscid fluid regardless of the body shape. This result is known as d'Alembert's paradox (see Section 17.4). However, vessels moving at or close to the free surface create the typical ship wave pattern and experience an additional resistance component that we call wave resistance R_W. It exists even in a hypothetical inviscid fluid. Thus, wave resistance is linked to the presence of a free surface. The dimensionless wave resistance coefficient is defined in the usual manner as

$$C_W = \frac{2R_W}{\rho v^2 S} \tag{25.1}$$

Two models may be used to explain how the generation of a wave system results in wave resistance. We may argue either that

(i) continuously generating waves extracts kinetic energy from the ship, or that

Fundamentals of Ship Hydrodynamics: Fluid Mechanics, Ship Resistance and Propulsion, First Edition.
Lothar Birk.

Companion website: www.wiley.com/go/birk/hydrodynamics

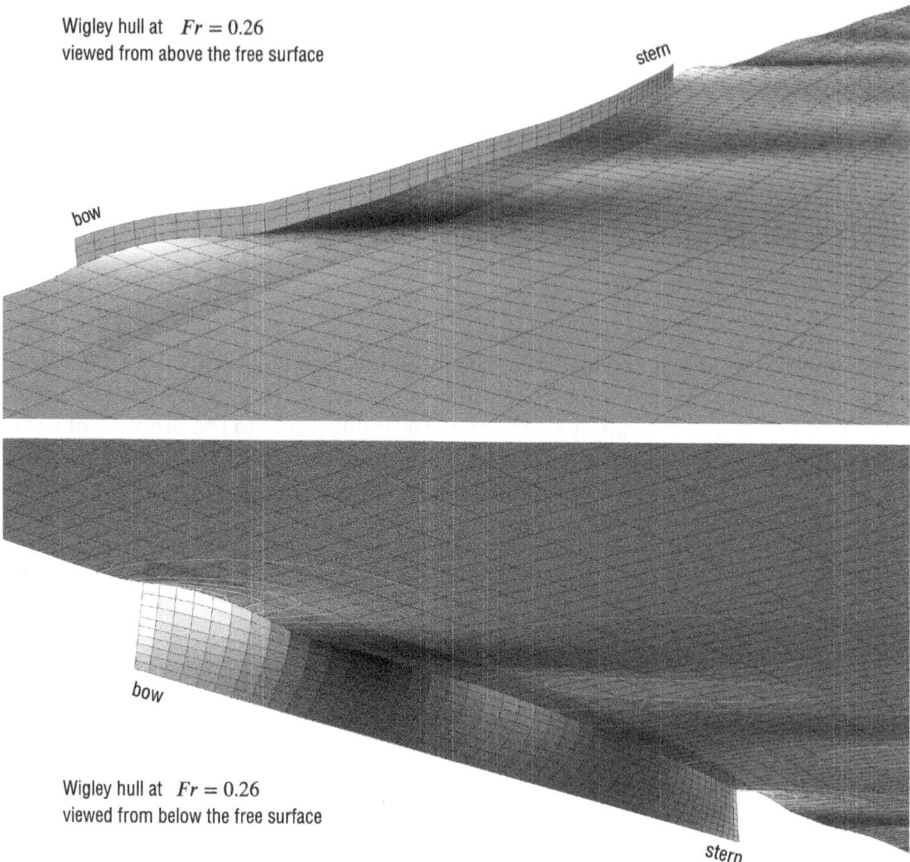

Figure 25.1 Wigley hull at Froude number $Fr = 0.26$ showing the connection between fluid and hull surface pressure and the resulting wave elevation. Light colors indicate high pressure and high wave elevation. Dark colors indicate low pressure and low values of wave elevation

(ii) the generated waves change the pressure distribution around the hull, which results in an additional pressure force.

Both are true, of course, and explain the same phenomenon. Let us discuss the models in more detail.

Loss of energy In the preceding chapter we learned that waves contain and transport energy. As a consequence, creating and sustaining a wave pattern requires energy. The higher the waves are, the more energy is used. In a fluid otherwise at rest, only the moving ship can provide the necessary energy. Apparently, a portion of the ship's kinetic energy is continuously converted into wave energy. The extracted kinetic energy must be replaced with work done by the propulsion system, otherwise the vessel would slow down. The work done per unit time is equivalent to the wave resistance multiplied by ship speed (in steady conditions).

Wind creates waves by locally changing pressure above the water surface. Just blow into your coffee mug and you will create small ripples.[1] A body traveling at or just beneath the free surface also changes the pressure distribution. Figure 25.1 shows the pressure distribution over a Wigley hull and the associated water surface elevations. Low pressure (dark gray to black) on the hull pulls the water surface down and creates a wave trough. High pressure (light gray to white) pushes the water surface upwards and generates wave crests. Any deformation of the free surface, in turn, changes the pressure distribution over the hull. Wave resistance is the x-component of the net pressure force resulting from the changes in pressure.

Pressure force

Wave resistance R_W is commonly subdivided into two parts:

Wave pattern and wave breaking resistance

$$R_W = R_{WP} + R_{WB} \tag{25.2}$$

- wave pattern resistance R_{WP}, associated with the work done to generate the ship wave system, and
- wave breaking resistance R_{WB}, associated with energy dissipation that occurs when waves break. Ships with high block coefficients C_B sail slowly. Wave heights are generally small and wave pattern resistance makes up only a small portion of the overall resistance. Their blunt bows, however, create steep bow waves which immediately break. The swirling motion caused by wave breaking dissipates energy. Again, kinetic energy is lost to the ship, which results in an additional resistance component.

We are unable to measure wave resistance directly. It may, however, be computed from measured wave profiles or deduced from total resistance by subtracting estimates or predictions of other resistance components. Wave resistance is typically the largest portion of the residual resistance.

25.2 Wave Superposition

According to linear wave theory, energy contained in a wave is proportional to half its squared amplitude $\bar{E} \sim \zeta_a^2/2$. Thus, wave resistance is high if the waves in a ship wave pattern are high, which means a lot of energy is spent in creating and sustaining the waves. Ship waves tend to become higher with increasing ship speed. However, wave amplitudes do not grow monotonically with velocity. At certain speeds, waves in the wave pattern are less high than at other speeds. As a result, the wave resistance curve shows distinct humps and hollows. Figure 25.2 shows the wave resistance coefficient C_W of a Wigley hull form as a function of Froude number Fr. At humps, the wave resistance is high compared to the values at adjacent speeds. In contrast, hollows in the C_W-curve seem to indicate favorable conditions with diminished wave resistance.

Humps and hollows

Humps and hollows are the effect of favorable and unfavorable wave superposition. Waves can be superimposed without significant effects onto each other. Just throw two

Wave superposition

[1] In case you wonder why the waves in your mug quickly dissipate once you stop blowing, whereas ship and ocean waves persist for fairly long time periods, here is an explanation: The short waves in your mug are heavily affected by the surface tension of the fluid. In ocean waves with much greater wave lengths, effects of surface tension are negligible.

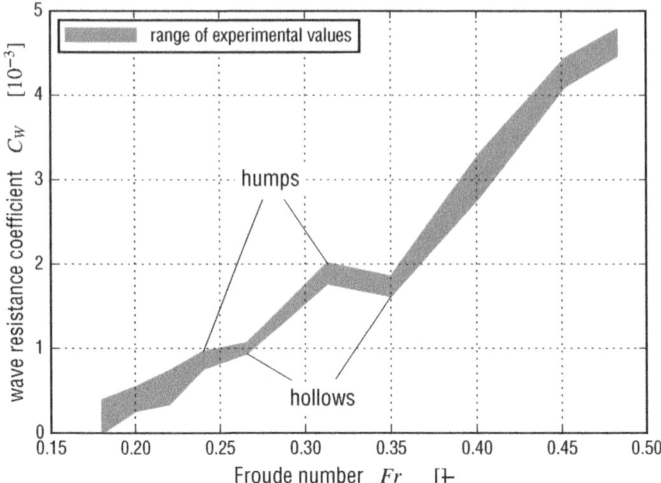

Figure 25.2 Pronounced humps and hollows in a wave resistance curve. Data from model tests with Wigley hulls (Bai and McCarthy, 1979)

stones into a pond some distance apart! The circular waves will cross but continue on their individual paths essentially unimpeded. Significant changes will occur only if the superimposed waves become too steep and break.

Wave resistance of a submerged sphere

In general, analytical solutions of the wave resistance problem are not known for ship shaped bodies. Therefore, we study the phenomenon of wave superposition and its effect on wave resistance with the wave systems created by submerged spheres. Linear wave theory was successfully employed by Havelock (1919) to compute the wave resistance R_W of a sphere traveling at constant speed v with its center submerged at a depth s below the calm water level. Using a Froude depth number

$$Fr_s = \frac{v}{\sqrt{g\,s}} \quad \text{and the factor} \quad \alpha = \frac{1}{Fr_s^2} \tag{25.3}$$

the wave resistance of a submerged sphere with diameter D is given by

$$R_{W_{\text{sphere}}} = \frac{\pi \rho g^4}{64} \frac{D^6}{v^6} \mathrm{e}^{-\alpha} \left[K_0(\alpha) + \left(1 + \frac{1}{2\alpha}\right) K_1(\alpha) \right] \tag{25.4}$$

K_0 and K_1 are modified Bessel functions of the second kind (Abramowitz and Stegun, 1972).

Figure 25.3 shows the resultant wave resistance coefficient as a function of the Froude number based on a reference length $L = 10D$. $S_{\text{sphere}} = \pi D^2$ is the wetted surface of the sphere. Wave resistance coefficient curves are shown for three submergence–diameter ratios $s/D = 1.0, 1.25$, and 1.5. The absolute values of the wave resistance coefficient are of no further use to us. However, the curves allow three general conclusions regarding wave resistance:

- The wave resistance is negligible for small Froude numbers below $Fr < 0.1$.

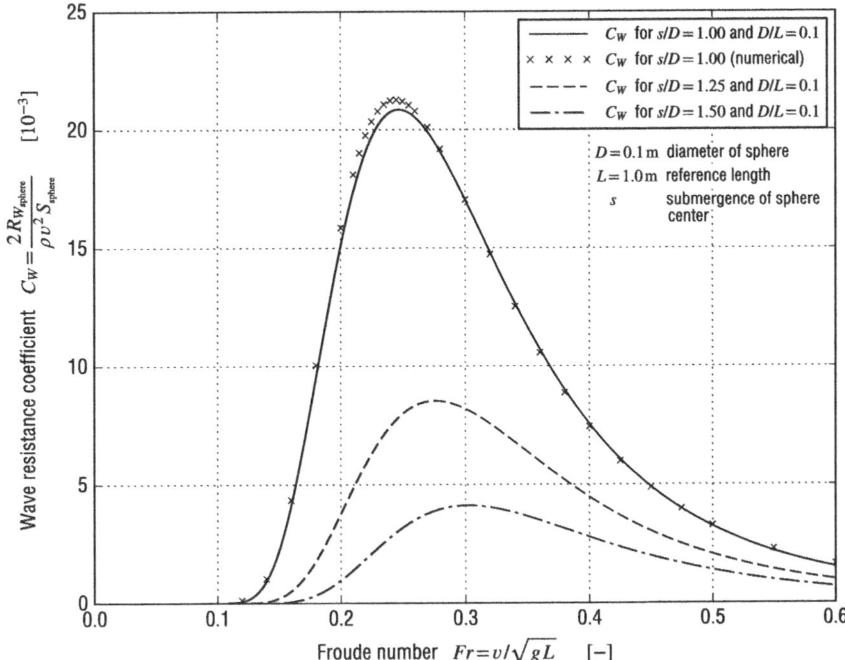

Figure 25.3 Wave resistance coefficient of a single submerged sphere; see Equations (25.4) and (25.1)

- Submergence has a significant influence. Moving the sphere just one radius further away from the free surface reduces the maximum wave resistance of the sphere by 80%. This concurs with our experience that slender ships with deep draft tend to have less wave resistance than beamy vessels with shallow draft where more displacement is concentrated close to the water surface.
- The wave resistance coefficient curve is not monotonic. In the case of the sphere, it has a distinct maximum. Its speed and value strongly depend on submergence. A declining coefficient means that the wave resistance force will grow slower and slower with increasing speed once the maximum has been passed.

Besides the analytical solutions, Figure 25.3 also shows some numerical results for the submergence–diameter ratio $s/D = 1$. They have been obtained with a panel method (see later). Although it overpredicts the maximum by about 2%, the results are generally in good agreement with the exact potential theory result from Equation (25.4).

Wave pattern of a submerged sphere

Figure 25.4 shows the wave pattern created by a sphere of diameter $D/L = 0.1$ positioned forward at $x/L = +0.5$. The sphere center is located one diameter below the calm water surface, i.e. its submergence is $s = D$. A moving reference frame is traveling with the sphere at constant Froude number $Fr = v/\sqrt{g\,L} = 0.252$. Imagine that the position $x/L = +0.5$ marks the bow if the reference length L is interpreted as the ship's length. The line plot at the bottom of Figure 25.4 depicts the wave profile through the sphere's center plane $y/L = 0$. The larger plot above it shows the port side half of its

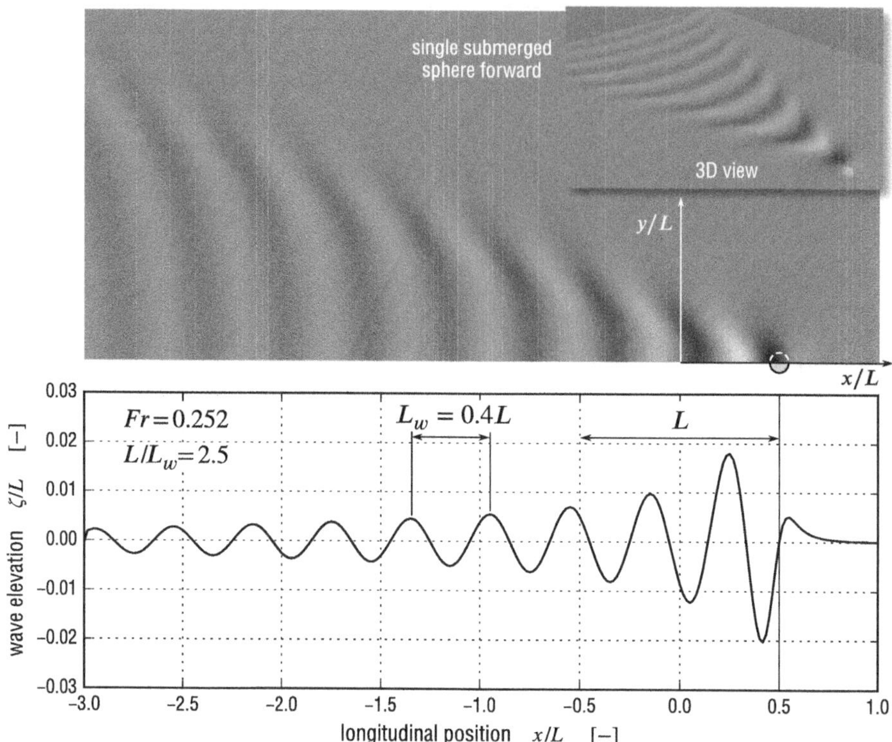

Figure 25.4 Wave pattern and wave profile created by a single submerged sphere at position $x/L = +0.5$ (forward). The sphere's dimensionless speed is the Froude number $Fr = 0.252$, which is based on $L = 10D$

symmetric wave pattern. A 3D view of sphere and wave pattern is presented in the small insert at the top right corner.

According to Equation (19.8), the wave length of the transverse waves is on deep water equal to

$$L_w = 2\pi \, Fr^2 \, L \qquad \text{for } k\,h > \pi \tag{25.5}$$

Then the ratio of wave length and reference length becomes

$$\frac{L_w}{L} = 2\pi \, Fr^2 \tag{25.6}$$

With Froude number $Fr = 0.252$ in Figure 25.4, we get for the sphere:

$$\frac{L_w}{L} = 0.4 \qquad \text{or} \qquad \frac{L}{L_w} = 2.5 \tag{25.7}$$

Wave pattern of second submerged sphere

After this preparatory work, we study the wave pattern of a second sphere. It has the same speed $Fr = 0.252$, diameter $D/L = 0.1$, and submergence but is located aft at $x/L = -0.5$ (Figure 25.5). Comparing the wave profiles of forward sphere (Figure 25.4) and aft sphere (Figure 25.5), we notice that both produce the same wave

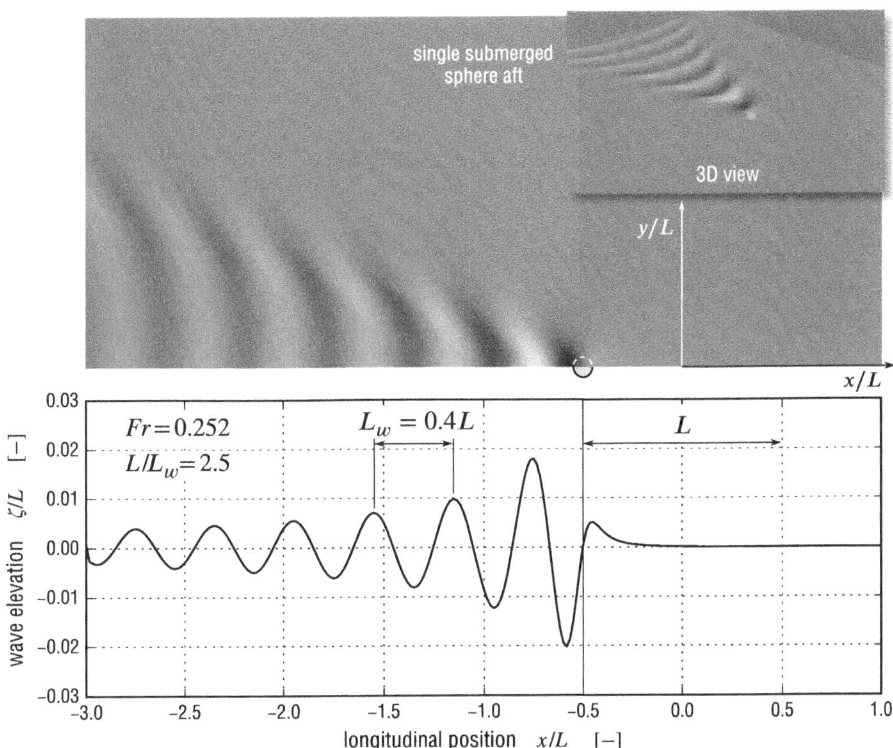

Figure 25.5 Wave pattern and wave profile created by a single submerged sphere at position $x/L = -0.5$ (aft). The sphere's dimensionless speed is the Froude number $Fr = 0.252$

length $L_w = 0.4L$. This is of course expected for identical Froude numbers. The waves following immediately behind the aft sphere also have the same heights as the waves following the forward sphere.

Superposition of wave patterns

Figure 25.6 shows the wave pattern created by both spheres simultaneously. The combined waves trailing the aft sphere are now much lower. The transverse waves in the pattern have almost vanished.

Favorable wave superposition

Figure 25.7 explains this effect. The wave profiles of forward and aft sphere are exactly out of phase, i.e. the bow sphere produces a wave crest where the stern sphere creates a wave trough. As a consequence, waves created by the stern sphere are almost canceled by the waves of the bow sphere. The low wave heights in the wave pattern are an indication for small wave resistance and a hollow in the wave resistance curve.

Unfavorable wave superposition

Next, we study the superposition of waves for the same pair of spheres at the slightly higher Froude number $Fr = 0.282$. The ratio of wave length to reference length is according to Equation (25.6).

$$\frac{L_w}{L} = 0.5 \qquad \text{or} \qquad \frac{L}{L_w} = 2.0 \tag{25.8}$$

The resultant wave pattern is shown in Figure 25.8. It is apparent that the transverse waves are much higher compared with the waves at the smaller Froude number in Figure 25.6.

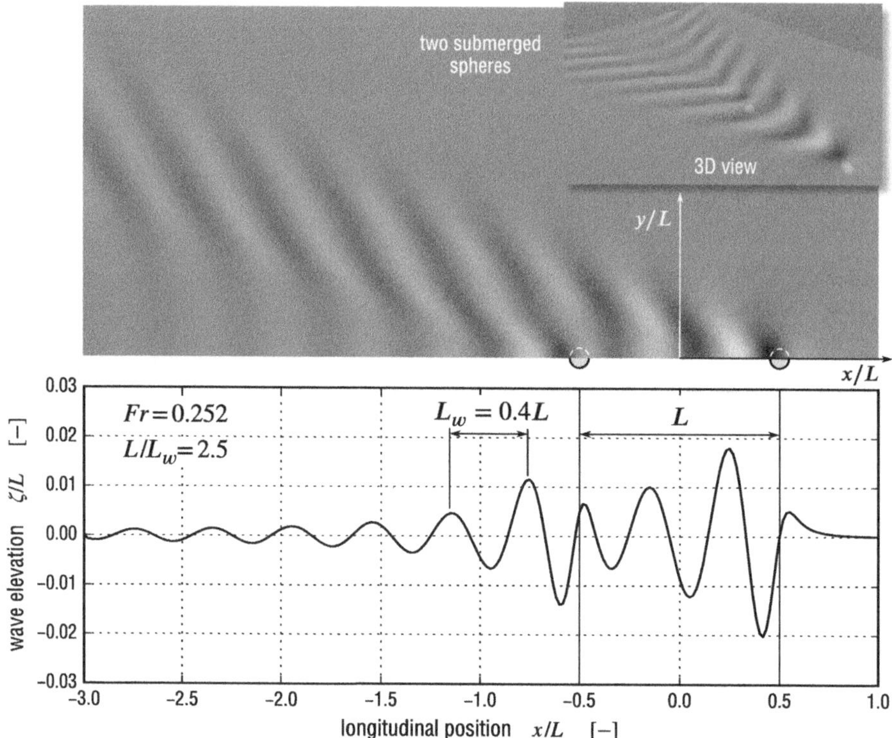

Figure 25.6 Combined wave pattern and profile of two submerged spheres. Froude number $Fr = 0.252$; favorable superposition of waves resulting in low wave heights

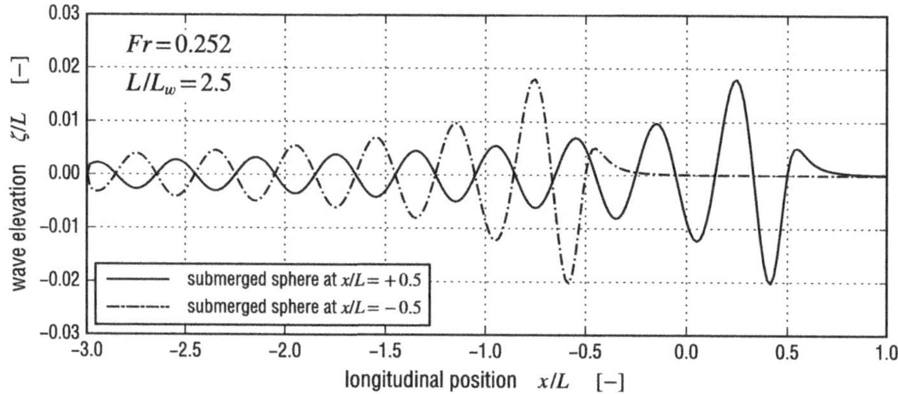

Figure 25.7 Comparison of wave profiles created by submerged spheres at positions $x/L = \pm 0.5$. The spheres' dimensionless speed is the Froude number $Fr = 0.252$

In contrast to the wave profiles at $Fr = 0.252$, the waves of bow and stern sphere now superimpose crest on crest (in phase, see Figure 25.9). This results in wave amplification rather than cancellation. This unfavorable superposition of waves will result in high wave resistance and a hump in the wave resistance curve.

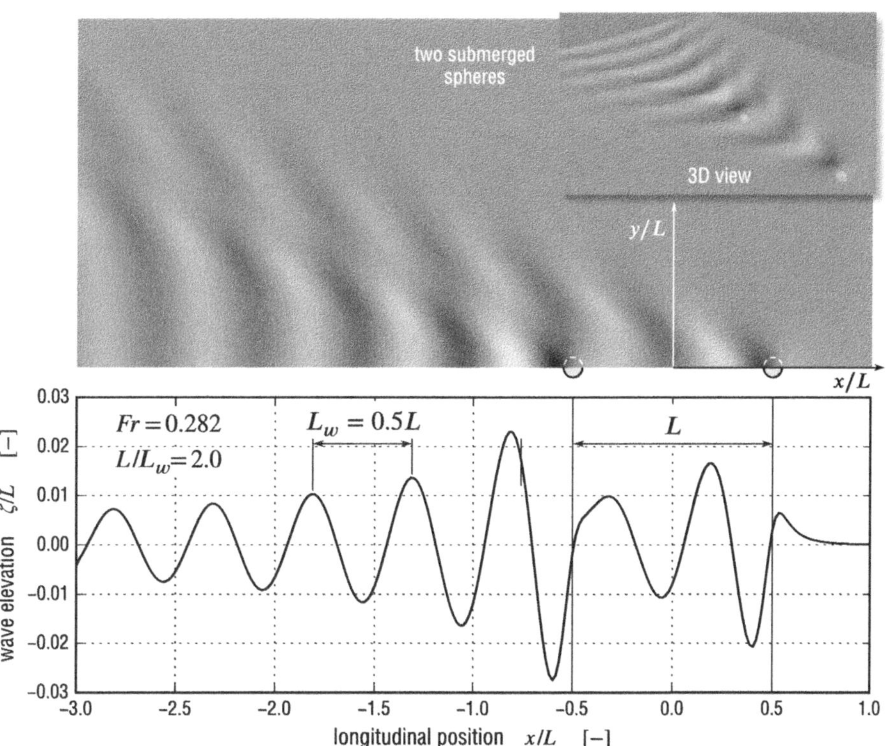

Figure 25.8 Wave pattern and wave profile of two submerged spheres. Froude number $Fr = 0.282$; unfavorable superposition of waves, resulting in high wave heights

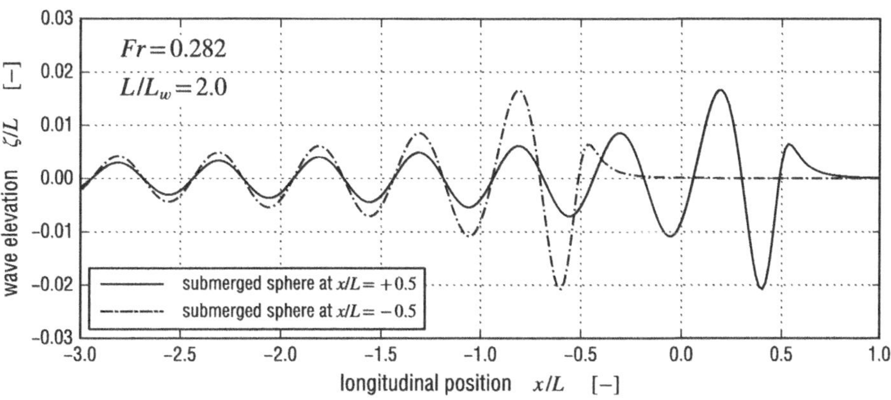

Figure 25.9 Comparison of wave profiles created by submerged spheres at positions $x/L = \pm 0.5$. The spheres' dimensionless speed is Froude number $Fr = 0.282$

Effect on wave resistance

In our study case, bow and stern spheres are separated by ten sphere diameters $L = 10D$. The flow around the bow sphere should have only a minimal impact on the flow around the stern sphere. One might initially guess that the wave resistance of the two sphere system is approximately twice the wave resistance of a single sphere. However, favorable and unfavorable wave superpositions create a much more complicated picture.

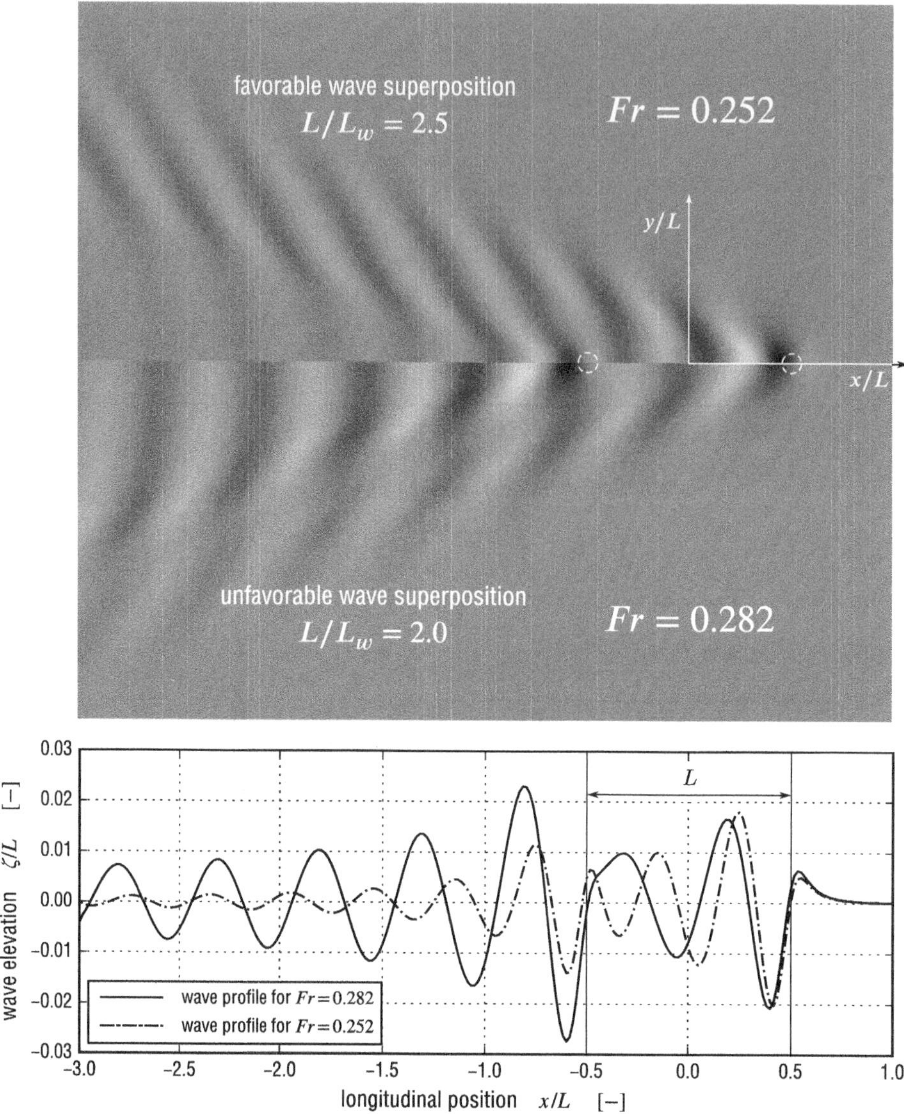

Figure 25.10 Comparison of wave patterns and wave profiles created by two submerged spheres for Froude numbers $Fr = 0.252$ (upper half, favorable superposition) and $Fr = 0.282$ (lower half, unfavorable superposition)

Figure 25.10 compares the wave patterns and profiles created by the spheres moving with different velocities. It is obvious that the waves generated by the moving spheres are much higher for the Froude number with unfavorable wave superposition and, consequently, this results in a much higher wave resistance. $Fr = 0.282$ is not the only case of unfavorable wave superposition. Figure 25.11 compares the wave resistance curves of a single sphere with the curve for the two sphere system. The latter shows large fluctuations around the single sphere wave resistance. In the case of $Fr = 0.252$, the wave resistance coefficient of two spheres is much smaller than the C_W-value of

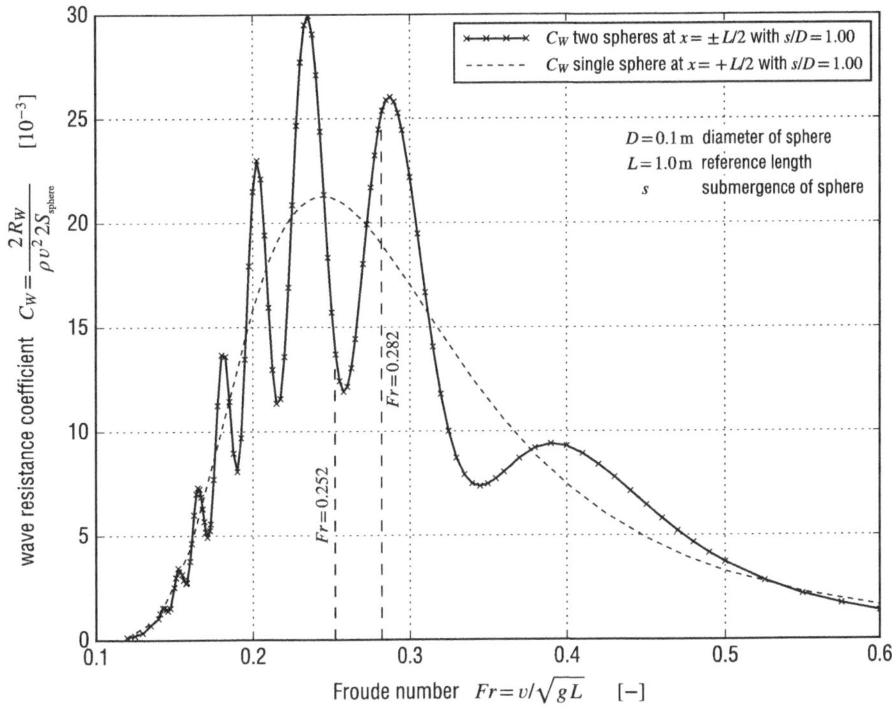

Figure 25.11 Wave resistance coefficient for a system of two submerged spheres; distance between centers is $L = 10D$, submergence is $s = D$

a single sphere. As expected, for $Fr = 0.282$, we see a much higher wave resistance coefficient.

Back to ships

Ships represent continuous pressure disturbances. Therefore, contributions to the ship wave system come from every part of the hull and not just bow and stern. The humps and hollows in the wave resistance curves of ships are much less pronounced than for the two sphere system. In fact, sometimes they are almost invisible. Details of the pressure distribution along the vessel depend on hull shape and speed. Only a detailed assessment with model tests or numerical analysis will provide a clear picture of the wave generation for a specific hull.

The example of the two spheres makes clear that wave resistance may not be effectively minimized for all speeds simultaneously. If the hull is optimized to produce a minimum wave resistance at a specific speed, it may increase wave resistance at other speeds. Therefore, optimization should be applied with care to avoid extreme measures which decrease the overall performance of the ship.

Bulbous bow

One of the most common design measures to lower wave resistance is the bulbous bow. It lengthens and smooths the pressure disturbance created by the bow. The bow wave system becomes lower, which decreases the wave pattern resistance. Even for slow ships with already small wave pattern resistance, a bulbous bow might lower wave breaking resistance by reducing the height of the bow wave.

25.3 Michell's Integral

John Henry Michell

The first reasonably successful attempt to compute the wave resistance of a ship is attributed to the Australian mathematician John H. Michell (*1863–†1940). Michell (1898) derived an equation for the wave resistance that resulted in a triple integral. The derivation of the equation is not for the mathematically fainthearted and is beyond the scope of this book. In his derivation, Michell made the following basic assumptions:

Assumptions

(i) potential flow, i.e. the flow is inviscid, incompressible, and irrotational

(ii) linearized free surface boundary condition, i.e. wave amplitudes are small compared to wave lengths

(iii) the body boundary condition is linearized and satisfied at the center plane, i.e. the beam is small compared with the length of the vessel, and the angle of hull surface elements to the center plane is small.

The limitations of the third and second assumptions render the wave resistance values from Michell's theory close to useless for modern merchant vessel hull forms and speeds. Practical applications are limited to extreme narrow hulls as they are used in catamarans and other multihull vessels and to high Froude numbers. That having been said, one has to acknowledge that Michell's solution of the problem is mathematically sound and proofed to be a big step forward in our understanding of ship waves and the resultant resistance. Michell's solution resulted in a multidimensional integral which can only be solved numerically for realistic hull forms. Michell (1898) included a numerical solution of his wave resistance integral for a ship and he reportedly has solved it for many more cases. At a time where even slide rules were scarce and computers did not exist even in science fiction, this was a formidable accomplishment by itself. Even with today's computer power special algorithms are necessary to accurately solve Michell's wave resistance integral. Readers are encouraged to read the summary on Michell's accomplishment by Tuck (1989).

Michell's integral

The following discussion of Michell's integral uses the notation by Tuck (1987) and Tuck et al. (1999). A coordinate system is fixed to the bow and its x-axis points aft. The vertical z-axis is positive upward and starts at the waterline. In Michell's integral the wave resistance is obtained from an analysis of the wave energy contained in the ship's wave pattern.

$$R_w = \frac{\pi}{2}\,\rho\,v^2 \int_{-\pi/2}^{+\pi/2} |A(\theta)|^2 \cos^3(\theta)\,\mathrm{d}\theta \tag{25.9}$$

The complex function $A(\theta)$ is known as a free wave spectrum or Kochin function, and is a measure for the energy of ship waves traveling in direction θ (Tuck et al., 1999). θ measures the angle between the ship's path and the direction of wave propagation. For $\theta = 0$, waves follow the ship (transverse waves) and for $\theta = \pm\pi/2$ waves have crests parallel to the ship's path.

Free wave spectrum

The free wave spectrum is a function of ship geometry and the propagation angle θ.

$$A(\theta) = \frac{2\mathrm{i}}{\pi}\,\frac{k_0^2}{\cos^4(\theta)}\,\left[P(\theta) + \mathrm{i}Q(\theta)\right] \tag{25.10}$$

The fundamental (transverse) wave number k_0 on deep water is equal to

$$k_0 = \frac{g}{v^2} = \frac{1}{Fr^2\,L} \tag{25.11}$$

The functions $P(\theta)$ and $Q(\theta)$ incorporate the geometry of the hull. They are computed as integrals along the length of a vessel from bow to stern.

$$P(\theta) = \int_0^L F(x,\theta)\cos\left(\frac{k_0 x}{\cos(\theta)}\right)\mathrm{d}x \tag{25.12}$$

$$Q(\theta) = \int_0^L F(x,\theta)\sin\left(\frac{k_0 x}{\cos(\theta)}\right)\mathrm{d}x \tag{25.13}$$

The function $F(x,\theta)$ is an integral over the submerged center plane of the hull from $z = -T$ to the waterline at $z = 0$. *z*-**integral**

$$F(x,\theta) = \int_{-T}^{0} Y(x,z)\,\mathrm{e}^{\left(\frac{k_0 z}{\cos^2(\theta)}\right)}\mathrm{d}z \tag{25.14}$$

Michell's original integral employed the longitudinal derivatives of the hull offsets $\partial Y(x,z)/\partial x$ whereas Tuck's solution works directly with the half breadths $Y(x,z)$ (Tuck, 1987). The equations above assume that the hull offsets at bow and stern vanish, i.e. $Y(0,z) = Y(L,z) = 0$. Practical algorithms for numerical integration of the integrals in Equations (25.14), (25.12), (25.13), and (25.9) are described in Tuck (1987) and Tuck et al. (1999). Read (2009) also contains a description and provides some specifics on the treatment of transoms where $Y(L,z) \neq 0$.

Figure 25.12 shows the wave resistance coefficient for a standard Wigley hull with offsets **Wigley hull**

$$Y(x,z) = \frac{B}{2}\left[1-\left(\frac{2x-L}{L}\right)^2\right]\left[1-\left(\frac{z}{T}\right)^2\right] \tag{25.15}$$

with $x \in [0, L]$ and $z = [-T, 0]$. L, B, and T are length, beam, and draft of the vessel, respectively.

As for the case of two spheres, the theoretical wave resistance curve for the Wigley hull computed with Michell's integral (25.9) features numerous humps and hollows. With decreasing Froude number, fluctuations between humps and hollows become more frequent but also decline in amplitude. Figure 25.12 also contains the range of experimental results for the wave resistance already featured in Figure 25.2 (Bai and McCarthy, 1979). It is pretty obvious that discrepancies between experimental and numerical results are significant in the relevant range of Froude numbers for merchant vessels $Fr \in [0.15, 0.35]$. Even the humps and hollows are not completely aligned. However, one has to bear in mind, that wave resistance cannot be measured directly. Its derivation from experimental results includes estimates of other resistance components which have their own limitations.

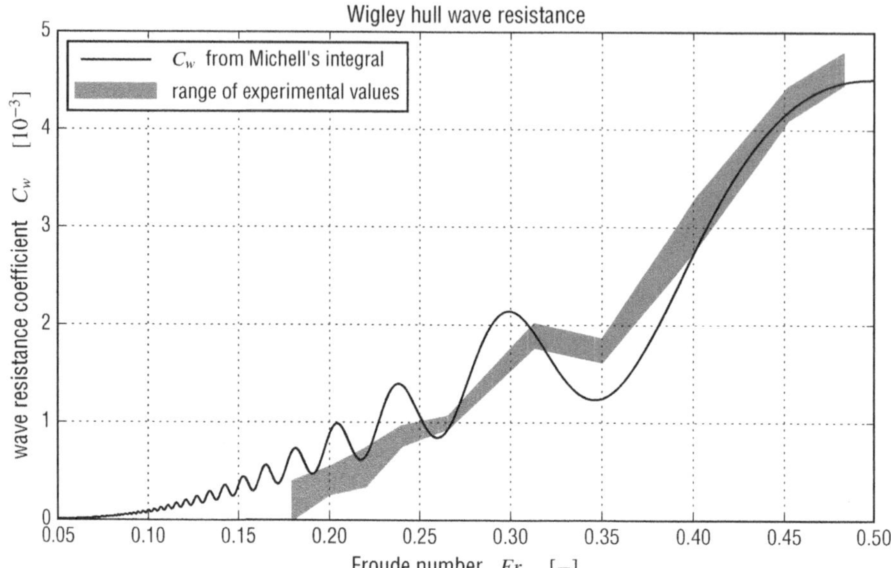

Figure 25.12 Wave resistance coefficient for a Wigley hull with length–beam ratio of $L/B = 10$ and beam–draft ratio of $B/T = 1.6$

Michell's integral predicts wave resistance quite accurately for high Froude numbers $Fr > 0.4$ and for very slender vessels with large L/B and small B/T ratios. However, linear wave theory is of limited use in predicting wave resistance for common merchant and naval vessel hull forms at the low and moderate Froude numbers they usually sail at. Reliable prediction of wave resistance requires the solution of a nonlinear problem using advanced numerical tools which go beyond the scope of this book. A brief introduction and further references are provided in the following section.

25.4 Panel Methods

Panel methods Today, panel methods are the most reliable way of computing the wave resistance of ships. These numerical methods derive their name from the small elements which are used to discretize the underlying mathematical equations. Figure 25.13 shows the bow of a container ship discretized into small quadrilateral and triangular panels. Quadrilateral panels are usually preferred because they reduce the total number of panels. However, quadrilaterals may not be flat. If quadrilateral panels become too twisted, like at the intersection of bulbous bow and stem in Figure 25.13, it is better to split quadrilaterals into triangles.

Panel methods belong to the larger class of boundary element methods (BEM), which are also employed in structural analysis. Boundary element methods solve three-dimensional problems by reducing them to equations which have to be satisfied on the domain boundaries.[2] This effectively reduces the 3D problem to a two-dimensional

[2] Hence the name boundary element method.

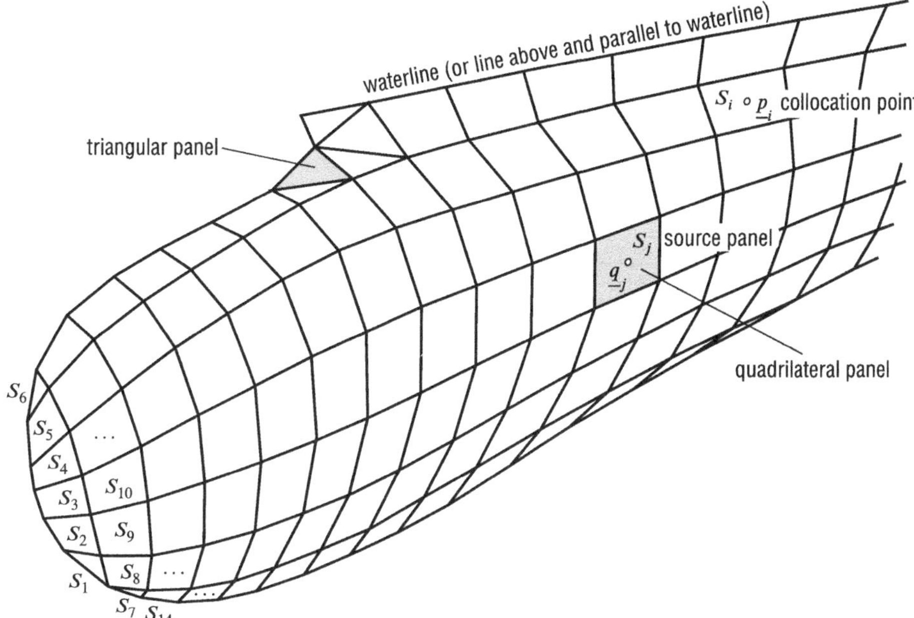

Figure 25.13 Discretization of vessel bow into small panels for wave resistance computation

problem. Instead of subdividing the domain into several hundred thousand or millions of small finite volumes, the boundaries of the domain are subdivided into a couple of hundred or thousand panels $S_0, S_1, \ldots, S_N$ thus substantially reducing the computational effort.

3D panel methods have been successfully used first in the design of aircraft wings and fuselages (Hess and Smith, 1967). A general derivation of the underlying theory and equations can be found in Katz and Plotkin (2001). Two principal methods are in use:

(i) The direct method: Green's integral theorem is used to derive a Fredholm integral equation of the second kind for the velocity potential. It utilizes the Laplace equation and the boundary conditions of the problem to reduce the portion of the boundary which affects the solution. Typically, only the ship hull itself has to be discretized. If a nonlinear free surface boundary condition is employed, the free surface has to be discretized as well (Newman, 1977).

(ii) The indirect method: the potential is assumed to be the sum of singularity distributions over all panels. Different types of singularities or combinations of singularities are in use. Distribution of sources are most common. Typically, the source strength σ_j is assumed to be constant for each panel. The body boundary condition and, if necessary, the free surface boundary condition are employed to determine the source strength values. The boundary conditions will only be truly satisfied at the panel centers, the so-called collocation points.

Important developments and modern panel methods for ship wave resistance computations are described by Jensen (1988), Nakos (1990), Raven (1996), and Janson (1997).

Although viscous effects are ignored in panel methods, they are well suited for wave resistance computations and the shape optimization of bulbous bows and the entrance of ship hulls. They may also serve as a preprocessor for more complex 3D boundary layer and general viscous flow computations, especially if the latter involve a free surface.

References

Abramowitz, M. and Stegun, I., editors (1972). *Handbook of mathematical functions with formulas, graphs, and mathematical tables.* Number 55 in Applied Mathematics Series. National Bureau of Standards, Washington, DC. Tenth printing, with corrections.

Bai, K. and McCarthy, J. (1979). Proceedings of the workshop on ship wave resistance computations. Technical report, David W. Taylor Ship Research and Development Center, Bethesda, Maryland 20084.

Havelock, T. (1919). Wave resistance: some cases of three-dimensional fluid motion. *Proceedings of the Royal Society of London. Series A, Containing Papers of a Mathematical and Physical Character,* 95(670):354–365.

Hess, J. and Smith, A. (1967). Calculation of potential flow about arbitrary bodies. *Progress in Aeronautical Sciences,* 8:1–138.

Janson, C.-E. (1997). *Potential flow panel methods for the calculation of free surface flows with lift.* PhD thesis, Department of Naval Architecture and Ocean Engineering, Chalmers University of Technology, Gothenburg, Sweden.

Jensen, G. (1988). Berechung der stationären Potentialströmung um ein Schiff unter Berücksichtigung der nichtlinearen Randbedingung an der Wasseroberfläche. Technical Report 484, Technische Universität Hamburg-Harburg.

Katz, J. and Plotkin, A. (2001). *Low-speed aerodynamics.* Cambridge Aerospace Series. Cambridge University Press, New York, NY, second edition.

Michell, J. (1898). The wave resistance of a ship. *Philosophical Magazine Series 5,* 45(272):106–123.

Nakos, D. (1990). *Ship wave patterns and motions by a three-dimensional Rankine panel method.* PhD thesis, Massachusetts Institute of Technology, Cambridge, MA, USA.

Newman, J. (1977). *Marine hydrodynamics.* The MIT Press, Cambridge, MA.

Raven, H. (1996). *A solution method for the nonlinear ship wave resistance problem.* PhD thesis, Technical University Delft, Delft, The Netherlands.

Read, D. (2009). *A drag estimate for concept-stage ship design optimization.* PhD thesis, The Graduate School, The University of Maine, Orono, ME.

Tuck, E. (1987). Wave resistance of thin ships and catamarans. Technical Report T8701, Applied Mathematics Department, The University of Adelaide, Australia. Unaltered October 1997 reprint of internal report T8701.

Tuck, E. (1989). The wave resistance formula of J.H. Michell (1898) and its significance to recent research in ship hydrodynamics. *Journal of Australian Math. Soc. Series B,* 30(4):365–377.

Tuck, E., Lazauskas, L., and Scullen, D. (1999). Sea wave pattern evaluation – Part 1 report: Primary code and test results (surface vessels). Technical report, Applied Mathematics Department, The University of Adelaide, Australia.

Self Study Problems

1. A ship of length $L_{WL} = 172$ m creates pronounced bow and stern wave systems. Similar to the system of two spheres you may assume that other parts of the wave making are negligible for this exercise. Compute the length of the transverse waves and reason whether or not the design speed of $Fr = 0.23$ is a good choice regarding low wave resistance.
2. If you want to lower the wave resistance of a ship, would it be better to make it wider and shallower or narrower (smaller beam) and deeper (higher draft)? Give reasons.
3. Read the paper by Tuck (1989) on Michell's wave resistance theory.

26

Ship Model Testing

Resistance and powering requirements have to be determined as early as possible in the design process because size and layout of engine and propulsor heavily influence general arrangements of the ship. Despite enormous advances made in computational methods, model testing is still the most reliable way of predicting resistance and powering requirements for a future ship. In this chapter we discuss the physical assets used in model tests: ship model, propeller model, and the towing tank or cavitation tunnel which represent the waterway. The chapter closes with a selection of websites for model basins around the world.

Learning Objectives

At the end of this chapter students will be able to:

- understand layout and function of scale models and testing facilities
- discuss advantages and disadvantages of different testing facilities
- know about modern model testing facilities

26.1 Testing Facilities

Model basins Facilities for ship model testing are found in research institutions and universities. Ship models are tested either in long water basins called towing tanks or in circulating water tunnels. Basins and circulating water tunnels stand in for oceans, lakes, and waterways and provide a controlled testing environment. Research institutions with a towing tank are often referred to as model basins due to importance and size of the towing tank. Figure 26.1 shows a view of the towing tank of the Hamburg Ship Model Basin (HSVA) in Hamburg, Germany. It is 300 m long, 18 m wide, and 6 m deep and of fairly typical size for a model basin that offers testing and consulting services for marine industry and government clients.

Most universities have smaller tanks for teaching and research purposes. Figure 26.2 shows the towing tank at the University of New Orleans as an example (length 30.8 m, width 4.6 m, water depth 1.8 m). Towing tank buildings often block sun light to slow algae growth which reduces the need for chemical additives and filtering of the water.

Fundamentals of Ship Hydrodynamics: Fluid Mechanics, Ship Resistance and Propulsion, First Edition.
Lothar Birk.

Companion website: www.wiley.com/go/birk/hydrodynamics

Figure 26.1 Towing tank of the Hamburg Ship Model Basin, Photo courtesy of Hamburgische Schiffbau-Versuchsanstalt GmbH (HSVA), www.hsva.de

26.1.1 Towing tank

Towing tank

Towing tanks are long basins filled with fresh water. They range in length from 30 m to more than 1300 m, with widths between 4 m and 18 m, and with water depths between 1.5 m and 8 m. Figure 26.3 shows the principal setup of a towing tank. One end typically features a narrow appendix called the trim tank, which provides easy access to a floating model. Before the actual testing, a model is lowered into the trim tank for ballasting. Afterwards the model is hooked up to sensors and the data acquisition system on the towing carriage. Smaller tanks are equipped with underwater viewing windows which are often considered impractical for basins with great water depth.

Rails

A pair of rails for the towing carriage is mounted on top of the side walls. The rails have to be precisely parallel to each other and parallel to the water surface. The latter requires that the rails follow the Earth's curvature, i.e. the rails form an arc when viewed from the side. This is especially critical for tests where the model is rigidly attached to a force balance to measure maneuvering or wave forces. In the case of a typical resistance test, sinkage would be underestimated in the middle of a towing tank if the rails did not follow the Earth's curvature.

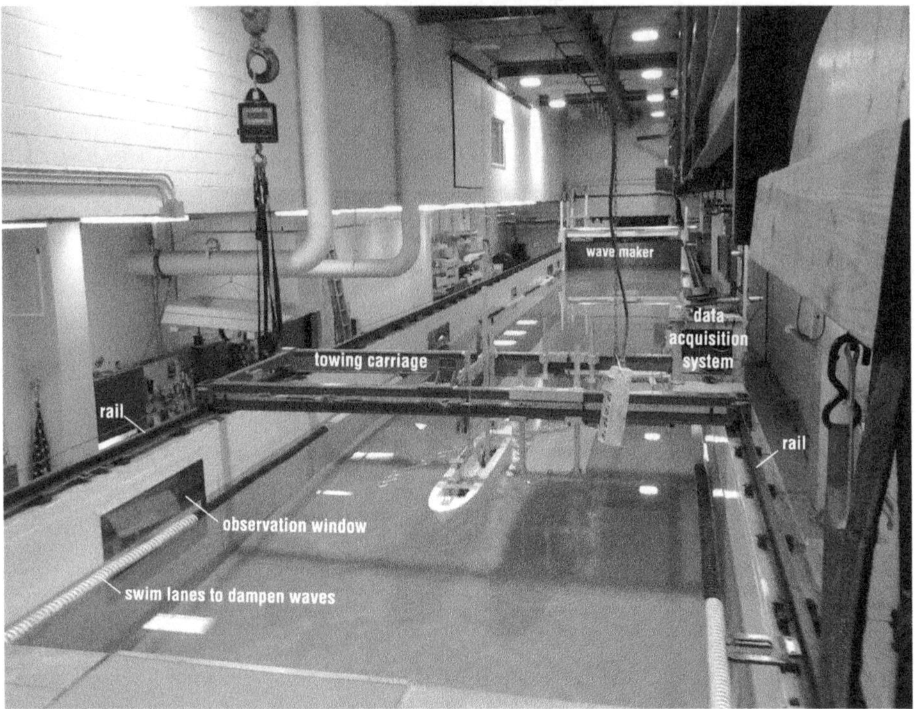

Figure 26.2 Towing tank at the School of Naval Architecture and Marine Engineering of the University of New Orleans

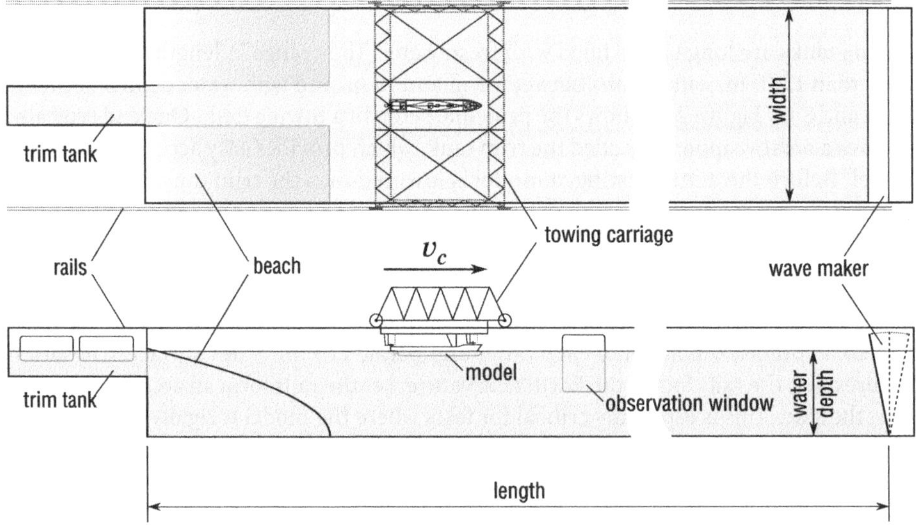

Figure 26.3 Schematic view of a towing tank

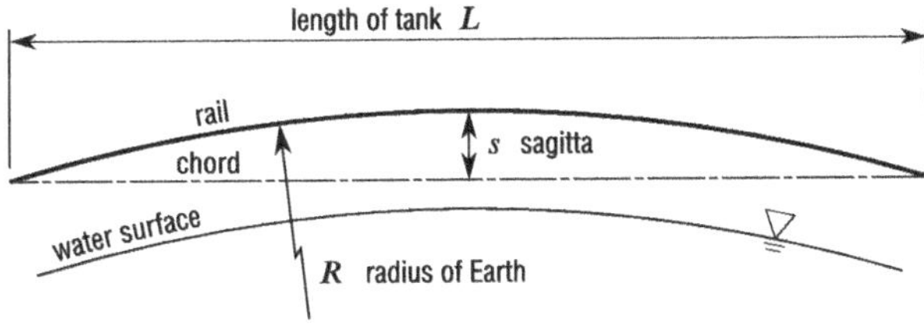

Figure 26.4 Definition of rail sagitta s

The height or sagitta s of a circular arc is a function of its radius R and the length L of the chord.

$$s = R - \sqrt{R^2 - \frac{L^2}{4}} \tag{26.1}$$

According to the World Geodetic System 1984 (WGS 84), the mean radius of the Earth is R =6 371 008.7714 m (NIMA, 2000). For the towing tank at the Hamburg Ship Model Basin (HSVA, Hamburg, Germany) with a length of L =300 m, the rails are s = 1.8 mm higher in the middle than at the ends. The L = 1324 m long tank at the Krylov State Research Center (St. Petersburg, Russia) needs a sagitta of s =34.3 mm (1 3/8 in).

Towing carriage

A towing carriage is a moving laboratory in itself. It carries power supplies and electronics to process data from the sensors. In large basins the carriage is manned and has its own drive motors. In smaller model basins, the towing carriage is often pulled by a cable system and operated from a fixed outside station. Towing carriages must be free of vibrations that could be transferred into the resistance measurement system. This places great demands on the speed control of the carriage motors, roundness of wheels, alignment of rails, and stiffness of the carriage structure.

Wave maker

Most basins feature a wave maker at the end opposite to the trim tank. Wave makers consist of a hydraulically actuated flap or board. A computer controls the board movement to create waves with a wide range of frequencies and wave amplitudes. Typically, wave forces and motions of stationary or moving models can be measured only in head seas. However, model basins with wide tanks might also feature segmented wave makers on the long side of the tank capable of creating waves that propagate at an angle to the model's course. A beach at the trim tank end of the basin dampens the waves generated by models and wave maker. An effective beach will reflect very little wave energy back into the tank.

Advantages and disadvantages

Towing a ship model through the basin has the advantage that it reflects the real ship–waterway system: the water is at rest and the vessel moves. A model creates a wave pattern dynamically similar to the wave system of the full scale vessel, if the tests are conducted at the same Froude number. Moving the model at a constant speed is much easier and cheaper than creating a uniform flow over the entire cross section of a tank by pumping water around. Limited length of a basin, however, restricts the usable observation time, especially for high speed vessels. Another difficulty for towing tanks is the simulation of cavitation. Very few model basins have the necessary capability to lower the ambient air pressure above the water surface. Such a tank must be enclosed in an air tight bunker and carriage operation has to be fully automated. The interested reader might look for the depressurized wave tank of the Dutch model basin MARIN in Wageningen as an example.

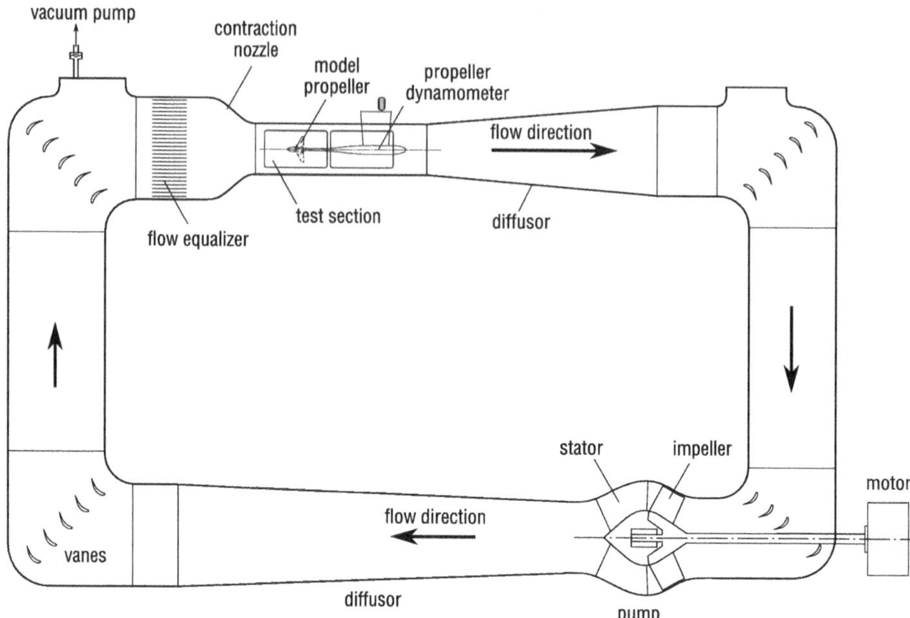

Figure 26.5 Schematic of a cavitation tunnel without free surface

26.1.2 Cavitation tunnel

Circulating water tunnel

In a circulating water tunnel the model is stationary, and a pump moves the water past the model. Most circulating water tunnels can reduce the static pressure in the fluid by reducing the effect of atmospheric pressure. Pressure distributions and not just pressure forces become dynamically similar, which is necessary to simulate propellers and foils at model scale that are subject to cavitation (see Chapter 28). For this reason, circulating water tunnels are more commonly known as cavitation tunnels.

Cavitation tunnel

Figure 26.5 shows a schematic view of a cavitation tunnel. In contrast to wind tunnels, circulating water tunnels are built upright: in small tunnels the test section might be as little as 2 m height above the ciculation pump. Large cavitation tunnels easily span several stories of a building. Although construction is more expensive, the height difference between measuring section and circulation pump increases the static pressure at the latter, thus forestalling cavitation of the pump itself. The fluid accelerated by the circulation pump is slowed down in a diffusor to prevent cavitation at the vanes. Foil shaped vanes help to redirect the flow around corners in the tunnel.

Flow equalizer and test section

Before the nozzle accelerates the fluid into the test section, the flow passes through a grid which acts as a flow equalizer. Together, grid and nozzle aim to make the flow as uniform as possible across the test section. Test sections vary in shape and size considerably. Some are circular but most are rectangular, with widths from a few inches to several meters.

Propeller models are mounted to a dynamometer which rotates the propeller and records thrust, torque, and rate of revolution (see Chapter 41). In tunnels with large

Figure 26.6 A model is prepared for cutting on the five axis mill. Photo courtesy of Hamburgische Schiffbau-Versuchsanstalt GmbH (HSVA), www.hsva.de

cross section, models of the submerged hull of a ship may be mounted in front of the propeller model to simulate the wake field.

Advantages and disadvantages

In a cavitation tunnel, the flow into the test section is never truly uniform. However, deviations from an average flow velocity will be small for a carefully designed tunnel. Circulating water tunnels allow measurements for prolonged periods of time. It is relatively easy to adjust the ambient pressure to the correct scale of the model which enables the detailed investigation of cavitation phenomena. Most tunnels are completely filled with water. Therefore, effects of waves cannot be studied due to the lack of a free surface. Some of the larger tunnels can be operated with a free surface so that the total resistance of a ship model may be measured. Since the cross sections of cavitation tunnels are usually much smaller than the cross section of towing tanks, effects due to blockage and walls have to be carefully studied.

26.2 Ship and Propeller Models

Scale

Ship models for resistance and propulsion tests are scale replicas of the actual vessels' outer hull shape. All linear dimensions are reduced by the same scale factor λ.

$$\lambda = \frac{L_S}{L_M} = \frac{B_S}{B_M} = \frac{T_S}{T_M} \tag{26.2}$$

Consequently, surfaces and volumes scale with λ^2 and λ^3, respectively.

$$S_M = \frac{S_S}{\lambda^2} \qquad V_M = \frac{V_S}{\lambda^3} \tag{26.3}$$

The subscript $_S$ marks properties of the full scale vessel and the subscript $_M$ stands for model properties. In general, models do not feature deck houses, equipment, or interior structure.

Construction Models are made as long as possible given the restrictions of towing tank or circulating water tunnel. Some of the larger model basins test models up to 12 m in length. Most models are made of wood. Wood is relatively cheap and easily machined or shaped with hand tools. The resulting models are dimensionally stable if the model is properly treated and sealed. First, thick boards are roughly sawed according to water line contours and glued together into a block (see Figure 26.6). The actual shaping is done with industrial five axis mills. The massive computer controlled router can be seen hanging from its gantry in the right half of Figure 26.6. As a final step, a model is sanded with 300 or 400 grit sand paper and painted to achieve a hydraulically smooth surface. Cheaper than wooden models are models made of paraffin wax. The wax can be remelted and reused. However, the models are more sensitive to temperature changes and will not keep their shape as long as wooden models.

Accuracy Models are manufactured according to specifications in the ITTC recommended procedures (ITTC, 2017b). The length of a model should be accurate within ± 0.05 % of length between perpendiculars or at least within ± 1 mm of its target value. The latter limit also applies to all other linear dimensions. The locations of knuckle lines and transoms in particular have to be modeled accurately because they may significantly influence the resistance.

Appendages If possible, the ITTC recommends conducting tests with and without appendages to gauge their effects on the resistance. Movable appendages, like rudders and stabilizer fins, should not be included. However, fixed appendages like bilge keels, bossings, and struts are quite often omitted in resistance tests as well. The larger the scale factor λ, the smaller a model will be and the more difficult it becomes to accurately represent the appendages. They might get too flimsy and would easily be bent out of shape during testing. In addition, appendages mostly affect the viscous resistance which is not scaled correctly in model tests due to the lower Reynolds numbers (see Chapter 28).

26.2.1 Turbulence generation

Sand strips or trip wires Lower Reynolds numbers at model scale are also responsible for the requirement to have turbulence generators. On a full scale vessel, the boundary layer will transition from laminar to turbulent flow a couple of centimeters behind the bow. As a consequence, the boundary layer will be turbulent over essentially the entire hull. On a model this transition would happen at a position considerably further aft relative to its length. To ensure that most of the model boundary layer is turbulent, the flow is tripped by a wire (0.5–0.8 mm diameter) or a strip of rough sand at a station 5% of the length behind the forward perpendicular. A bulbous bow receives its own turbulence generator about one third of its length from the tip.

Some model basins prefer a series of studs parallel to the bow contour. Cylindrical studs stick out into the flow and create a turbulent wake behind them. Diameter, height, and separation should follow the guidance given in the ITTC recommended procedures (ITTC, 2017b).

Enforcing a turbulent boundary layer allows us to use the same model–ship correlation line in the computation of the frictional resistance of model and ship (see Chapter 29). It also prevents problems with separation, which is more likely in a laminar boundary layer and may lead to an exaggerated viscous pressure resistance for the model. As a consequence, the performance prediction would fail for the full scale vessel.

26.2.2 Loading condition

Ballast based on volumetric displacement

Typically, a model is tested over a range of speeds at different loading conditions (design draft, fully loaded, ballast, etc.). The model is ballasted according to its volumetric displacement for the loading condition in question. Waterline markings on the model are solely for checking heel and trim angles. The distribution of ballast in the model is not important for resistance tests as long as displacement and attitude are correct. However, dynamically similar moments of inertia and centers of gravity must be maintained for maneuvering and seakeeping tests. Ballast must be stored accordingly on the model.

Ballast calculation

Calculation of the necessary ballast is straightforward:

1. Determine volumetric displacement V for the targeted loading condition, e.g. from the ship's curves of form. Equation (26.3) provides the relationship between model volume V_M and ship volume V_S.

2. Weigh the model as built and record its mass m_M. This mass should include all the equipment that will be placed on the model during testing. The equipment includes half of the mass of the yoke which connects model and towing carriage (see Figure 29.1).

3. Measure the temperature t_W of the tank water at a depth of half the model draft.

4. Measure the tank water density ρ_M with a calibrated hydrometer or, if this is not available, use the value from the water properties tables in ITTC (2011).

5. Determine the total model mass for the selected loading condition

$$\Delta_M = \rho_M V_M \tag{26.4}$$

6. Compute the ballast mass m_{ballast} by subtracting the model mass m_M from the total mass $\rho_M V_M$ of the model for the target loading condition

$$m_{\text{ballast}} = \Delta_M - m_M \tag{26.5}$$

7. Adjust the ballast mass if tests are conducted over a longer time period where water temperature and density may change.

Figure 26.7 The beginnings of a five bladed propeller model. Photo courtesy of Hamburgische Schiffbau-Versuchsanstalt GmbH (HSVA), www.hsva.de

26.2.3 Propeller models

Fabrication of propeller models requires high precision (Figure 26.7). In most cases computer controlled milling machines are used; however, 3D-printing may also be effective. The diameter of model propellers manufactured according to ITTC (2017a) may deviate by at most ± 0.1 mm from its nominal value. Other important propeller parameters like pitch and shape of blade sections also have tight tolerances. The blade surfaces are polished to less than 0.5 μm mean roughness. Turbulence generation is not used on propeller models because it would negatively affect their lift to drag ratio. Relevant Reynolds numbers for the propeller at model scale must be high enough to ensure that the boundary layer flow is mostly turbulent.

26.3 Model Basins

Table 26.1 lists model basins and their websites. This list is definitely incomplete and my apologies if your favorite institution is missing. Emphasis has been placed on model basins whose websites provide information and photos of their facilities. All reported web links were active in September 2018. The ITTC offers an extensive list of facilities on its website at www.ittc.info. Some of the institutions, like the David Taylor Model Basin in the United States, do not provide a lot of information because they mostly perform classified research for their respective naval forces.

Table 26.1 A selection of model basins around the world, sorted alphabetically according to their commonly used abbreviations

Name	Location	Website
China Ship Scientific Research Center (CSSRC)	Wuxi, Jiangsu, P.R. China	www.cssrc.com/
David Taylor Model Basin (DTMB), Naval Surface Warfare Center at Carderock	West Bethesda, MD, USA	www.navsea.navy.mil/Home/WarfareCenters/NSWCCarderock.aspx
Canal de Experiencias Hidrodinámicas de El Pardo (El Pardo Model Basin)	Madrid, Spain	www.cehipar.es/cehiparweb/index.php?lang=english
Marine Technology Research Institute (INSEAN)	Rome, Italy	www.insean.cnr.it/
Hamburg Ship Model Basin (HSVA)	Hamburg, Germany	www.hsva.de
Korea Research Institute of Ships and Ocean Engineering (KRISO)	Daejeon, South Korea	www.kriso.re.kr/eng/
Krylov State Research Center	St. Petersburg, Russia	krylov-center.ru/eng
Maritime Research Institute Netherlands (MARIN)	Wageningen, The Netherlands	www.marin.nl
SINTEF Ocean (formerly Marintek)	Trondheim, Norway	www.sintef.no/en/ocean/laboratories2/
National Maritime Research Institute (NMRI)	Tokyo, Japan	www.nmri.go.jp/index_e.html
National Research Council Canada	St. John's, Newfoundland, Canada	www.nrc-cnrc.gc.ca/eng/solutions/facilities/marine_performance/towing_tank.html
QinetiQ Maritime Test Facilities	Haslar, United Kingdom	www.qinetiq.com/services-products/maritime/platform-readiness/Pages/maritime-test-facilities.aspx
SSPA Sweden AB	Gothenburg, Sweden	http://www.sspa.se
Schiffbau-Versuchsanstalt Potsdam (SVA)	Potsdam, Germany	www.sva-potsdam.de/home.html
Vienna Model Basin (SVA)	Vienna, Austria	www.sva.at

References

ITTC (2011). *Fresh water and seawater properties.* International Towing Tank Conference, Recommended Procedures and Guidelines 7.5-02-01-03. Revision 2.

ITTC (2017a). *Propeller model accuracy.* International Towing Tank Conference, Recommended Procedures and Guidelines 7.5-01-02-02. Revision 1.

ITTC (2017b). *Ship models.* International Towing Tank Conference, Recommended Procedures and Guidelines 7.5-01-01-01. Revision 4.

NIMA (2000). Department of Defense World Geodetic System 1984. Technical Report NIMA TR8350.2, third edition, Amendment 1, National Imagery and Mapping Agency.

Self Study Problems

1. What are the advantages of conducting a resistance test in a towing tank compared with a resistance test in a cavitation tunnel?
2. Why must the rails follow the Earth's curvature?
3. What is necessary to conduct model tests with cavitation conditions similar to the full scale vessel?
4. Which geometric property of a hull determines how much ballast must be added to a ship model before testing?
5. Research the dimensions of the towing tank of the Vienna Model Basin.

27

Dimensional Analysis

Dimensional analysis is an important tool in the search for functional relationships in experimental as well as theoretical models. Typically, this topic is part of courses in fluid mechanics. However, for the sake of completeness this chapter provides a short summary, explaining the purpose and application of dimensional analysis. The Buckingham π-theorem is recaptured and applied to the ship resistance problem. We rediscover the characteristic numbers already found by making the Navier-Stokes equations dimensionless.

Learning Objectives

At the end of this chapter students will be able to:

- apply dimensional analysis
- define dimensionless characteristic numbers

27.1 Purpose of Dimensional Analysis

Experiments represent an obvious means to obtain more insight into the flow processes around a ship hull and the resultant resistance. However, before we conduct any experiments, two important questions need to be answered:

(i) which parameters influence the resistance, and

(ii) how can we obtain the most information from the smallest set of experiments?

Need for dimensional analysis (DA)

Let us assume that the flow around a submerged sphere is governed by four variables:

(a) water density ρ (type of fluid),

(b) speed v (operation),

(c) diameter D (geometry), and

(d) dynamic viscosity μ (another fluid property).

Fundamentals of Ship Hydrodynamics: Fluid Mechanics, Ship Resistance and Propulsion, First Edition.
Lothar Birk.

Companion website: www.wiley.com/go/birk/hydrodynamics

Then the sphere's resistance R will be a yet unknown function $\mathcal{F}$ of ρ, v, D, and μ.

$$R = \mathcal{F}(\rho, v, D, \mu) \tag{27.1}$$

To determine the function $\mathcal{F}$, we may select 10 different values for each variable. The total number of experiments to cover all combinations of variable values would be $N = 10^4 = 10\,000$ experiments. Even if we manage to perform 50 experiments per day, it would take 200 days, or more than half a year, to assemble the results. We rarely have this much time.

27.2 Buckingham π-Theorem

Buckingham π-theorem

Dimensional analysis helps to reduce the number of relevant variables by combining them into dimensionless parameters. The theory is based on the Buckingham π-theorem, named after the American physicist Edgar Buckingham (*1867 – †1940), although it was first proved by the French mathematician Joseph Bertrand (*1867 – †1940) (Bertrand, 1878). Assume we have a physically correct formula that involves n physical variables. The variables may be expressed in terms of k independent fundamental physical quantities, e.g. time, length, and force. Then the Buckingham π-theorem postulates that the physically correct formula is equivalent to an equation involving a set of $m = n - k$ dimensionless parameters, so-called π-terms. The dimensionless parameters are constructed from the original variables.

Considering the example in the previous section, the formula for the resistance of a sphere has $n = 4$ variables. Independent physical quantities are force measured in Newtons [N], length [m], and time [s] or simply $k = 3$. Therefore, a single dimensionless parameter $m = n - k = 4 - 3 = 1$ can be used to describe the relationship $\mathcal{F}$ resulting in a significant reduction in the number of necessary experiments. We go from $\mathcal{F}(\rho, v, D, \mu)$ to $\mathcal{F}(Re)$. Thus, 10 instead of 10 000 experiments would yield the same information about the drag of spheres.

27.3 Dimensional Analysis of Ship Resistance

The resistance R_T of a ship may serve as an illustration of the steps involved in a dimensional analysis (DA).

Collection of relevant variables

1. First, we collect all variables known (or believed) to affect ship resistance. This is the most difficult task of DA, often supported by observation, experience, and trial and error.

   Let us assume R_T depends on density of water ρ, velocity v of the ship, dimension of the ship represented by its length L, dynamic viscosity μ of the water, acceleration of gravity g, and pressure p:

   $$R_T = \mathcal{F}(\rho, v, L, \mu, g, p) \tag{27.2}$$

   There is a total of six variables ($n = 6$) which involve the following fundamental units:

(a) ρ water density $\left[\frac{\text{kg}}{\text{m}^3}\right]$ or $\left[\frac{\text{Ns}^2}{\text{m}^4}\right]$ using $\text{N} = \frac{\text{kg m}}{\text{s}^2}$

(b) v ship speed $\left[\frac{\text{m}}{\text{s}}\right]$

(c) L ship length [m]

(d) μ dynamic viscosity $\left[\frac{\text{Ns}}{\text{m}^2}\right]$ relevant to viscous forces

(e) g acceleration of gravity $\left[\frac{\text{m}}{\text{s}^2}\right]$ relevant to gravity forces

(f) p pressure $\left[\frac{\text{N}}{\text{m}^2}\right]$ relevant to pressure forces

Therefore, the fundamental physical quantities involved are force, length, and time with the independent units Newton [N], meter [m], and seconds [s], i.e. $k = 3$. Mass [kg] could be used instead of force, but the former is more convenient in this case. The Buckingham π-Theorem tells us that the number of dimensionless parameters m is equal to the number of variables minus the number of independent units

$$m = n - k = 6 - 3 = 3$$

2. The unknown resistance function $R_T = \mathcal{F}(\rho, v, L, \mu, g, p)$ will contain products of the variables, possibly raised to yet unknown powers a through f, and a constant factor C.

Product of variables

$$R_T = C\,\rho^a v^b L^c \mu^d g^e p^f \tag{27.3}$$

In physical units the equation requests [N] = [N]. We will derive the values of the powers in subsequent steps.

3. The equation above can only be satisfied if the product on the right-hand side yields the independent unit of a force [N]. We substitute the variables with their corresponding independent units in Equation (27.3).

Comparison of coefficients

$$\left[\text{N}\right]^1 = \left[\frac{\text{Ns}^2}{\text{m}^4}\right]^a \cdot \left[\frac{\text{m}}{\text{s}}\right]^b \cdot \left[\text{m}\right]^c \cdot \left[\frac{\text{Ns}}{\text{m}^2}\right]^d \cdot \left[\frac{\text{m}}{\text{s}^2}\right]^e \cdot \left[\frac{\text{N}}{\text{m}^2}\right]^f \tag{27.4}$$

From Equation (27.4) follow three conditions, one for each physical quantity. The sum of powers on the left- and right-hand side have to be equal for each unit separately, because they are independent of each other.

$$\text{for force } \left[\text{N}\right]: \quad 1 = a + d + f \tag{27.5}$$

$$\text{for length } \left[\text{m}\right]: \quad 0 = -4a + b + c - 2d + e - 2f \tag{27.6}$$

$$\text{for time } \left[\text{s}\right]: \quad 0 = 2a - b + d - 2e \tag{27.7}$$

The equations are derived by collecting corresponding terms from Equation (27.4). On the left-hand side of Equation (27.4) only N^1 appears. The power is equal to 1 for the force unit [N] (27.5) and zero for unit length [m] (27.6) and time [s] (27.7). On the right-hand side we collect powers of all appearances of the independent unit. Powers which are part of a denominator are preceded with minus signs.

Reduction

4. In total, we derive three equations for six unknown powers a, b, c, d, e, and f. Although we cannot solve for all of them, we are able to reduce the number of unknown powers to three by expressing powers a, b, and c in terms of powers d, e, and f. From Equation (27.5) follows

$$a = 1 - d - f \tag{27.8}$$

and from Equation (27.7) we get

$$b = 2a + d - 2e$$

Substituting the result for a into the latter yields

$$\begin{aligned} b &= +2 - 2d - 2f + d - 2e \\ &= 2 - d - 2e - 2f \end{aligned} \tag{27.9}$$

Using Equations (27.8) and (27.9) allows us to derive c from Equation (27.6):

$$\begin{aligned} 0 &= -4 + 4d + 4f + 2 - d - 2e - 2f + c - 2d + e - 2f \\ -c &= -2 + d - e \\ c &= 2 - d + e \end{aligned} \tag{27.10}$$

Dimensionless parameters

5. We replace a, b, and c in (27.3) with the results from Equations (27.8), (27.9), and (27.10). Subsequently, we sort the terms with respect to equal power coefficients.

$$\begin{aligned} R_T &\sim \rho^{1-d-f} \cdot v^{2-d-2e-2f} \cdot L^{2-d+e} \cdot \mu^d \cdot g^e \cdot p^f \\ &\sim \rho v^2 L^2 \cdot \left[\frac{\mu^d}{\rho^d v^d L^d}\right] \cdot \left[\frac{g^e L^e}{v^{2e}}\right] \cdot \left[\frac{p^f}{\rho^f v^{2f}}\right] \\ &\sim \frac{1}{2}\rho v^2 L^2 \cdot \left[\frac{\mu}{\rho v L}\right]^d \cdot \left[\frac{gL}{v^2}\right]^e \cdot \left[\frac{p}{\frac{1}{2}\rho v^2}\right]^f \end{aligned} \tag{27.11}$$

The terms in brackets are indeed dimensionless. Introducing the kinematic viscosity $\nu = \mu/\rho$, the first pair of square brackets (power d) is the reciprocal value of the Reynolds number Re.

$$\text{Reynolds number} \qquad Re = \frac{vL}{\nu} \tag{27.12}$$

The second pair of brackets (power e) contains the reciprocal value of the squared Froude number Fr.

$$\text{Froude number} \qquad Fr = \frac{v}{\sqrt{gL}} \tag{27.13}$$

The third set of brackets yields a form of the Euler number or pressure coefficient.

$$\text{Euler number} \qquad Eu = \frac{p}{\frac{1}{2}\rho v^2} \tag{27.14}$$

Dimensional analysis reveals that the calm water resistance of a ship depends on the type of fluid (density ρ), ship speed υ, ship geometry (represented by length L or wetted surface S), and the three dimensionless parameters: Reynolds number Re, Froude number Fr, and Euler number Eu.

$$R_T = \frac{1}{2}\rho\, \upsilon^2\, S \cdot \mathcal{F}(Re, Fr, Eu) \tag{27.15}$$

We often express this in the form of a dimensionless resistance coefficient C_T

$$C_T = \frac{2R_T}{\rho\, \upsilon^2\, S} = \mathcal{F}(Re, Fr, Eu) \tag{27.16}$$

The actual function $\mathcal{F}$ will have to be determined by experiments.

References

Bertrand, J. (1878). Sur l'homogénéité dans les formules de physique. *Comptes rendus*, 86(15):916–920.

Self Study Problems

1. Assume that thrust T of a propeller depends on the seven physical variables density ρ, kinematic viscosity ν, gravitational acceleration g, diameter D, speed of advance υ_A, pressure p, and rate of revolution n. How many dimensionless characteristic numbers will govern the similarity of model tests measuring thrust?

2. Thrust T of a propeller is assumed to depend on water density ρ, dynamic viscosity of the water μ, acceleration of gravity g, diameter D, speed of advance υ_A, pressure in the fluid p, and rate of revolution n. Perform a dimensional analysis to derive the dimensionless coefficients governing propeller thrust. Look-up and name the dimensionless coefficients.

28

Laws of Similitude

Need for model testing

Even with cutting-edge CFD tools naval architects cannot reliably compute the resistance of a ship for daily design purposes. Although accuracy and reliability of CFD results have steadily improved, model tests will be used to confirm results for the foreseeable future. Since most ships are one-of-a-kind designs, prototypes cannot be used to improve the design which further increases the importance of model tests.

However, model tests are useful only if they are accompanied by methods which transfer physical quantities measured at model scale to the full scale design. This chapter explains the fundamental concepts of similarity and points out the difficulties that arise if only partial dynamic similarity is possible.

Learning Objectives

At the end of this chapter students will be able to:

- understand the concepts of geometric, kinematic, and dynamic similarity
- interpret Froude number, Reynolds number, and Euler number in the context of dynamic similarity
- identify the limitations of similarity in ship model tests
- discuss Froude's hypothesis and the ITTC form factor method

28.1 Similarities

Required similarities

Three basic conditions regulate the conversion of model test results into full scale results:

1. geometric similarity,
2. kinematic similarity, and
3. dynamic similarity.

Fundamentals of Ship Hydrodynamics: Fluid Mechanics, Ship Resistance and Propulsion, First Edition.
Lothar Birk.

Companion website: www.wiley.com/go/birk/hydrodynamics

The individual similarities will be discussed below. Unfortunately, we often cannot maintain all three similarities. Reliable approximations have to be introduced which provide corrections for the condition of partial similarity in the model test.

28.1.1 Geometric similarity

Geometric similarity

The necessity for geometric similarity is the most obvious. If you want to know the resistance of a sphere, you most likely will not get usable results by testing a box. We require that our ship is converted into a model by applying a fixed scale factor λ to all linear dimensions. Any length L_S at the ship will be λ times the corresponding length on the model L_M or

Scale of length, area and volume

$$\lambda = \frac{L_S}{L_M} \qquad \text{scaling of lengths} \tag{28.1}$$

Areas are of dimension (length unit) × (length unit) and volumes are of dimension (length unit)^3. Consequently, areas and volumes will scale according to

$$\lambda^2 = \frac{S_S}{S_M} \qquad \text{scaling of areas} \tag{28.2}$$

$$\lambda^3 = \frac{V_S}{V_M} \qquad \text{scaling of volumes} \tag{28.3}$$

As a result of the constant scale in length, angles are the same for full scale ship and model.

The recommended ITTC procedure on model ships (ITTC, 2011) states that the model should be accurate within ±1 mm in linear dimensions for all surfaces that get in contact with the water. The surfaces should be smoothly sanded with 300 to 400 grit sand paper. Small appendages, like struts or bilge keels, may not be reproduced in many cases, because members may become too thin at model scale. Details of the surface like weld seems or rivets are not reproduced at model scale.

Tank dimensions

In the context of geometric similarity, it is easily forgotten that we should scale the ocean environment as well. Due to space limitations, however, depth and especially width of our model oceans are limited. A careful assessment has to be performed to determine whether restricted water depth and tank width influence the model test results or not. If you are a stickler for detail, water particles should be scaled too, but let us file this under impossible.

28.1.2 Kinematic similarity

Time scale

Kinematic similarity in fluid mechanics is often referred to as the similarity of streamline patterns. Fluid particles should not only take geometric similar paths, but, in addition, should maintain a constant time scale τ at full scale and model scale.

$$\tau = \frac{\Delta T_S}{\Delta T_M} \tag{28.4}$$

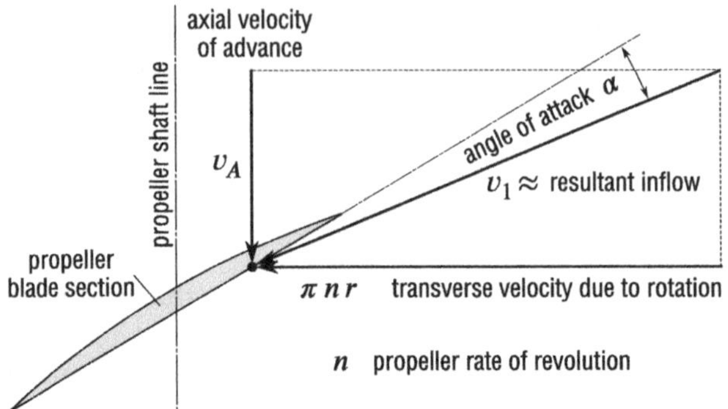

Figure 28.1 Simplified flow pattern at a propeller blade section. Kinematic similarity requires that the angle of attack remains the same for full scale and model propeller blade section

As a consequence, ratios of velocities are constant between full and model scale.

Similarity of velocities A section of a rotating propeller blade may serve as an example (Figure 28.1). In order to have the same flow pattern, the angle of attack α between blade section and inflow must be equal in full scale and model scale. This is achieved if the ratio of axial flow velocity (speed of advance v_A) and tangential velocity (transverse speed due to propeller rotation, $1/2\pi\, n\, D$ for blade tip) is constant. For convenience we omit the factor π.

$$\left(\frac{v_{AM}}{n_M\, D_M}\right) = \left(\frac{v_{AS}}{n_S\, D_S}\right) = J \tag{28.5}$$

The fraction $v_A/(n\, D)$ is called advance coefficient J. Obviously, kinematic similarity cannot be achieved without geometric similarity.

Scale of velocities and acceleration Together the scales for geometry λ and time τ define the scales of velocities and accelerations.

$$\text{velocity scale} \qquad \frac{v_S}{v_M} = \frac{\dfrac{L_S}{\Delta T_S}}{\dfrac{L_M}{\Delta T_M}} = \frac{\dfrac{L_S}{L_M}}{\dfrac{\Delta T_S}{\Delta T_M}} = \frac{\lambda}{\tau} \tag{28.6}$$

$$\text{acceleration scale} \qquad \frac{a_S}{a_M} = \frac{\dfrac{L_S}{\Delta T_S^2}}{\dfrac{L_M}{\Delta T_M^2}} = \frac{\dfrac{L_S}{L_M}}{\dfrac{\Delta T_S^2}{\Delta T_M^2}} = \frac{\lambda}{\tau^2} \tag{28.7}$$

28.1.3 Dynamic similarity

Types of forces Dynamic similarity requires that forces in full and model scale have the same direction and the ratio κ of magnitudes full scale to model scale is constant. According to the Navier-Stokes equations for incompressible flow, four classes of forces act on our ship:

1. inertia force
2. gravity force (volume force)

3. friction force (surface force, shear stress)
4. pressure force (surface force, normal stress)

Below we determine the force ratios and the requirements for dynamic similarity.

Inertia forces

Inertia forces may be expressed as mass times acceleration: $F_i = m\,a$. Mass is usually replaced by volume and density $m = \rho\, V$. As a first part of dynamic similarity, the ratio of inertia forces must be constant. **Inertia forces**

$$\kappa_i = \frac{F_{iS}}{F_{iM}} = \frac{\rho_S\, V_S\, a_S}{\rho_M\, V_M\, a_M} = \frac{\rho_S}{\rho_M}\,\lambda^3\,\frac{\lambda}{\tau^2} = \frac{\rho_S}{\rho_M}\,\frac{\lambda^4}{\tau^2} \tag{28.8}$$

Whatever geometric scale and time scale have been chosen, if only inertia forces would act on a system, they would scale by the ratio of the fluid densities times λ^4/τ^2. Of course, whenever there are inertia forces, at least one other class of force will act too. Otherwise no accelerations could occur according to Newton's second law.

Inertia and gravity forces

The ratio of gravity forces $F_g = m\,g$ would be **Gravity forces**

$$\kappa_g = \frac{F_{gS}}{F_{gM}} = \frac{\rho_S\, V_S\, g_S}{\rho_M\, V_M\, g_M} = \frac{\rho_S}{\rho_M}\,\frac{g_S}{g_M}\,\lambda^3 \tag{28.9}$$

In most practical cases, gravitational acceleration at full and model scale will be the same: $g = g_S = g_M$. However, if you some day conduct experiments on the moon ...

If *both* inertia and gravity forces act, dynamic similarity requires that the force ratios κ_i and κ_g are equal. There can be only one force ratio. If $\kappa_i = \kappa_g$ then **Inertia and gravity forces**

$$\frac{\rho_S}{\rho_M}\,\frac{\lambda^4}{\tau^2} = \frac{\rho_S}{\rho_M}\,\frac{g_S}{g_M}\,\lambda^3$$

or

$$\tau^2 = \frac{g_M}{g_S}\,\lambda. \tag{28.10}$$

If $g = g_S = g_M$, time scale must be equal to the square root of the geometric scale

$$\tau = \sqrt{\lambda} \tag{28.11}$$

Enforcing dynamic similarity of both inertia and gravity forces makes the time scale dependent on the geometric scale.

The scale of velocities (28.6) changes to

$$\frac{v_S}{v_M} = \frac{\lambda}{\tau} = \frac{\lambda}{\sqrt{\lambda}} = \sqrt{\lambda} = \sqrt{\frac{L_S}{L_M}} \tag{28.12}$$

Considering the first and last terms in Equation (28.12), we find the relation of *corresponding velocities*

$$\frac{v_S}{\sqrt{L_S}} = \frac{v_M}{\sqrt{L_M}} \tag{28.13}$$

Froude number

This relation was published by the French marine engineer Ferdinand Reech. However, Froude (1868) was the first to put it to practical use in ship model testing. If we include the ratio of gravitational acceleration in Equation (28.10), we get

$$\frac{v_S}{v_M} = \frac{\lambda}{\tau} = \frac{\lambda}{\sqrt{\frac{g_M}{g_S}\lambda}} = \sqrt{\frac{g_S\,L_S}{g_M\,L_M}} \tag{28.14}$$

or again combining the first and last terms

$$\frac{v_S}{\sqrt{g_S\,L_S}} = \frac{v_M}{\sqrt{g_M\,L_M}} \tag{28.15}$$

This is obviously the definition of the Froude number Fr. Consequently, conducting model tests where the Froude number is equal for full scale ship and model ensures dynamic similarity of inertia and gravity forces (Froude similarity).

In the case of Froude similarity, the velocity at model scale is obtained by dividing full scale velocity by the square root of the geometric scale (assuming equal $g = g_S = g_M$).

$$v_M = \frac{v_S}{\sqrt{\lambda}} \quad \text{and} \quad v_S = v_M\sqrt{\lambda} \tag{28.16}$$

This velocity ratio is very practical. For example, with a typical geometric scale of $\lambda = 25$, a full scale ship speed of 10 m/s (almost 20 kn) translates into a model speed of 2 m/s. *In cases where only inertia and gravity forces are relevant,* full scale forces F_S are obtained from model scale forces F_M by the following simple conversion:

$$F_S = \frac{\rho_S}{\rho_M}\frac{g_S}{g_M}\lambda^3 F_M \quad \text{for purely inertia and gravity forces} \tag{28.17}$$

Inertia and friction forces

Friction forces

In real fluids a more or less noticeable friction force exists. Friction forces can be modeled as shear stress times wetted surface: $F_\nu = \mu \frac{\partial u}{\partial y} S$. Here, μ is the dynamic viscosity of the fluid and $\frac{\partial u}{\partial y}$ is the change in velocity perpendicular to the flow direction. Thus, the friction forces for ship and model form the ratio

$$\kappa_\nu = \frac{F_{\nu S}}{F_{\nu M}} = \frac{\mu_S \frac{\partial u_S}{\partial y_S} S_S}{\mu_M \frac{\partial u_M}{\partial y_M} S_M} = \frac{\mu_S}{\mu_M}\frac{\frac{\lambda}{\tau}}{\lambda}\lambda^2 = \frac{\mu_S}{\mu_M}\frac{\lambda^2}{\tau} \tag{28.18}$$

Inertia and friction forces

Assuming inertia and friction forces are present, their respective force ratios must be equal to maintain dynamic similarity. For $\kappa_i = \kappa_\nu$ we get

$$\frac{\rho_S}{\rho_M}\frac{\lambda^4}{\tau^2} = \frac{\mu_S}{\mu_M}\frac{\lambda^2}{\tau} \tag{28.19}$$

or by introducing the kinematic viscosity $\nu = \mu/\rho$

$$\frac{\lambda^2}{\tau} = \frac{\mu_S}{\rho_S}\frac{\rho_M}{\mu_M} = \frac{\dfrac{\mu_S}{\rho_S}}{\dfrac{\mu_M}{\rho_M}} = \frac{\nu_S}{\nu_M}$$

Thus the necessary time scale is equal to

$$\tau = \frac{\nu_M}{\nu_S}\lambda^2 \tag{28.20}$$

In contrast to the similarity of combined inertia and gravity forces, the time scale is now proportional to the squared geometric scale.

$$\frac{v_S}{v_M} = \frac{\lambda}{\tau} = \frac{\lambda}{\dfrac{\nu_M}{\nu_S}\lambda^2} = \frac{\nu_S}{\nu_M}\frac{1}{\lambda} = \frac{\nu_S}{\nu_M}\frac{L_M}{L_S} \tag{28.21}$$

This yields

$$\frac{v_M\,L_M}{\nu_M} = \frac{v_S\,L_S}{\nu_S} \tag{28.22}$$

The Reynolds number $Re = vL/\nu$ governs the dynamic similarity of inertia and friction forces. Equal Reynolds number results in a very impractical velocity ratio. If model tests are conducted in the same fluid, i.e. $\nu_S = \nu_M$, model speed is equal to full scale speed times geometric scale

$$v_M = \lambda\, v_S \tag{28.23}$$

For a geometric scale of $\lambda = 25$ and a ship speed of 10 m/s, a model speed of $v_M =$ 250 m/s would be necessary to maintain dynamic similarity of inertia and friction forces. This is in the range of the cruising speed of a commercial airliner and very unlikely to be realized in a towing tank!

Inertia, gravity, and friction forces

Our ship resistance problem includes inertia, gravity, *and* friction forces. In order to maintain dynamic similarity, the scales of all three classes of forces have to be equal:

Inertia, gravitational and friction forces

$$\kappa_i = \kappa_g = \kappa_\nu \tag{28.24}$$

The second part yields with $\kappa_g = \kappa_\nu$

$$\begin{aligned}\frac{\rho_S}{\rho_M}\frac{g_S}{g_M}\lambda^3 &= \frac{\mu_S}{\mu_M}\frac{\lambda^2}{\tau}\\ \frac{g_S}{g_M}\lambda &= \frac{\nu_S}{\nu_M}\frac{1}{\tau}\end{aligned} \tag{28.25}$$

We enforce similarity of inertia and gravitational forces by selecting the time scale proportional to the square root of the geometric scale – see Equation (28.10). This also leads to more practical model speeds. Substituting $\tau = \sqrt{\lambda\, g_M / g_S}$ in (28.25) yields the condition

$$\lambda\sqrt{\lambda} = \frac{\nu_S}{\nu_M}\frac{g_S}{g_M} \tag{28.26}$$

With $g_S = g_M$, kinematic viscosity of the model fluid must be

$$\nu_M = \frac{\nu_S}{\lambda^{3/2}} \tag{28.27}$$

in order to maintain similarity of inertia, gravity, and friction forces. For a model scale of $\lambda = 25$, kinematic viscosity ν_M in the model test would have to be as small as $0.008\ \nu_S$, i.e. eight thousandths of the kinematic viscosity of seawater.

A fluid with a significant smaller kinematic viscosity than water is mercury. Not that anyone would seriously consider filling a towing tank with this extremely dangerous substance, but let us play with some numbers. The ratio of kinematic viscosities of water H_2O and mercury Hg is $\dfrac{\nu_S}{\nu_M} = \dfrac{\nu_{H_2O}}{\nu_{Hg}} \approx 8$. The resultant model scale would be as small as $\sqrt{\lambda^3} = 8$ or $\lambda = 4$. Unfortunately, the models would still be way too large for common ship sizes and available test facilities.

Partial dynamic similarity In summary, we will not be able to conduct our ship resistance model test with complete dynamic similarity. We will have only *partial dynamic similarity* for inertia and gravity forces. All friction force components measured in the model test will have to be carefully corrected for scale effects before they can be applied to the full scale ship. For this we make use of the ITTC 1957 model–ship correlation line, which allows as to *compute* the frictional part of ship resistance.

$$C_F = \frac{0.075}{\left[\log_{10}(Re) - 2\right]^2} \tag{2.18}$$

We apply this equation to the model as well as to the ship. Of course, the Reynolds numbers will be quite different.

Inertia and pressure forces

Pressure forces Remaining are the pressure forces which can be modeled as pressure times area $F_p = p\,S$. The ratio of pressure forces is

$$\kappa_p = \frac{p_S\,S_S}{p_M\,S_M} = \frac{p_S}{p_M}\lambda^2 \tag{28.28}$$

Inertia and pressure forces Enforcing dynamic similarity for inertia and pressure forces requires $\kappa_i = \kappa_p$. Then

$$\frac{\rho_S}{\rho_M}\frac{\lambda^4}{\tau^2} = \frac{p_S}{p_M}\lambda^2$$

$$\frac{p_S}{p_M} = \frac{\rho_S}{\rho_M}\left(\frac{\lambda}{\tau}\right)^2 \tag{28.29}$$

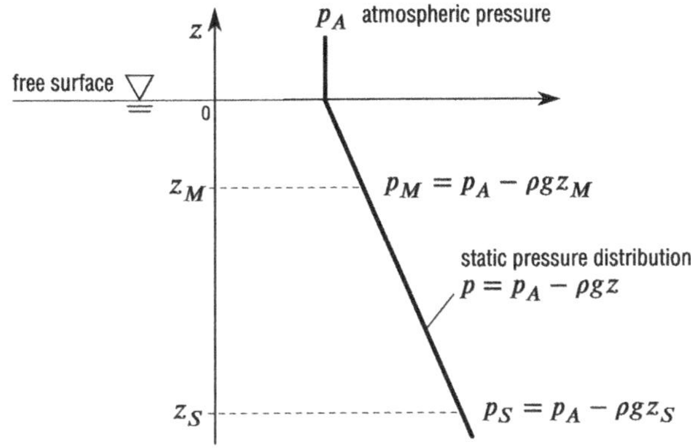

Figure 28.2 Relationship of total static pressures for model and full scale ship

The ratio λ/τ is the the scale of velocities, i.e.

$$\frac{p_S}{p_M} = \frac{\rho_S}{\rho_M}\frac{v_S^2}{v_M^2} \tag{28.30}$$

$$\frac{p_S}{\frac{1}{2}\rho_S\, v_S^2} = \frac{p_M}{\frac{1}{2}\rho_M\, v_M^2} \tag{28.31}$$

An Euler number or a dimensionless pressure coefficient $\overline{C}_P = 2p/(\rho v^2)$ can be used to characterize dynamic similarity for systems subject to inertia and pressure forces.

Inertia, gravity and pressure forces

We already concluded that we cannot simultaneously attain dynamic similarity for inertia, gravity, and friction forces. What about the combination of inertia, gravity, and pressure forces? Again we choose the time scale according to Froude's law of similarity, i.e. $\tau = \sqrt{\lambda\, g_M/g_S}$. The required ratio of pressures becomes

$$\frac{p_S}{p_M} = \frac{\rho_S}{\rho_M}\lambda \tag{28.32}$$

Is this possible? Figure 28.2 compares the total static pressure for a depth z_M related to a model and the corresponding depth z_S of the full scale ship. Certainly $z_S/z_M = \lambda$. In this case the ratio of total static pressures yields

$$\frac{p_S}{p_M} = \frac{p_A - \rho_S\, g\, z_S}{p_A - \rho_M\, g\, z_M} \neq \frac{\rho_S}{\rho_M}\lambda \tag{28.33}$$

Since the atmospheric pressure is equal for full scale ship and model, we do *not* have the required dynamic similarity for the total pressures.

However, pressure forces are the result of pressure differences. The atmospheric pressure acts equally on all parts of the ship surface above and below the waterline. Hence, its resultant force vanishes. Comparing only the hydrostatic pressures for model and full scale reveals that the necessary similarity for pressure differences is satisfied.

$$\frac{\Delta p_S}{\Delta p_M} = \frac{-\rho_S\, g\, z_S}{-\rho_M\, g\, z_M} = \frac{\rho_S}{\rho_M}\lambda \tag{28.34}$$

As a consequence, we are able to maintain simultaneous dynamic similarity for inertia, gravity, and pressure forces.

Cavitation

For some flow problems, similarity of absolute pressure is needed. *Cavitation* is one of these flow phenomena. Cavitation occurs when the pressure in the fluid reaches the vapor pressure level. Water changes from liquid into gaseous phase forming bubbles with low pressure gas in the flow. The bubbles are transported with the flow and collapse when they reach areas of higher pressure. This process can destroy propellers, rudders, or any other material surface nearby.

The inception of cavitation depends on the absolute pressure level. It is usually higher in the model than it is in the full scale ship, because we do not normally scale back the atmospheric pressure (equal for both). Therefore models appear cavitation free whereas the full scale vessel will show cavitation which can have significant effects on performance. To investigate this phenomenon, special towing tanks or circulating water tunnels are used that allow the atmospheric pressure above the water surface to be changed.

28.1.4 Summary

Summary

The discussion of the requirements for dynamic similarity discloses one of the greatest difficulties of marine model testing: *We are unable to conduct our model tests under complete dynamic similarity.* Given the available conditions, we are not able to maintain similarity of friction forces. Elaborate procedures have been developed over the past 140 years to account for this deficiency. Although practical solutions have been found, you should always be aware of their limitations. Inertia, gravity, and pressure forces are modeled correctly – within the limitations regarding geometric and kinematic similarity – if the time scale is selected based on equal Froude numbers for full scale ship and model $Fr_M = Fr_S$.

28.2 Partial Dynamic Similarity

As we have learned in the previous section, full dynamic similarity is *not* achievable in practical ship model testing. We are unable to satisfy the requirements of equal Froude number and equal Reynolds number at the same time. The lack of full dynamic similarity has severe consequences on our model testing procedures. To emphasize the problem we first look at the *hypothetical* case of full dynamic similarity.

28.2.1 Hypothetical case: full dynamic similarity

Hypothetical full dynamic similarity

Assuming the requirements of geometric similarity and kinematic similarity are satisfied, a hypothetical full dynamic similarity is achieved when all force scales are equal:

$$\kappa_i = \kappa_g = \kappa_f = \kappa_p \tag{28.35}$$

Putting Equation (28.35) into words, the ratios of full scale to model scale forces have to be the same for all type of forces important to the problem at hand.

$$\frac{F_{iS}}{F_{iM}} = \frac{F_{gS}}{F_{gM}} = \frac{F_{fS}}{F_{fM}} = \frac{F_{pS}}{F_{pM}} \tag{28.36}$$

Here, scales would have to be equal for inertia, gravity (representing body forces), frictional (representing surface forces due to viscous stress), and pressure force (representing surface forces due to normal pressure).

Hypothetical full scale predictions

Making full scale predictions from model test results would be easy, *if* we could have full dynamic similarity. Since all forces would scale the same way, the actual force scale could be derived from the physical force unit.

$$\mathrm{N} = \mathrm{kg}\,\frac{\mathrm{m}}{\mathrm{s}^2} \quad \longrightarrow \quad \kappa = \frac{\rho_S}{\rho_M}\lambda^3\,\frac{\lambda}{\tau^2} \tag{28.37}$$

Applying – for practical reasons – the time scale that follows from similarity of inertia and gravity forces $\tau = \sqrt{\lambda}$ (follows from $Fr_S = Fr_M$) yields:

$$\kappa = \frac{\rho_S}{\rho_M}\lambda^3 \tag{28.38}$$

Therefore, all model scale forces would scale up to full scale like volumes (λ^3) multiplied by the scale of densities.

$$\kappa_i = \kappa_g = \kappa_f = \kappa_p = \frac{\rho_S}{\rho_M}\lambda^3 \quad \text{hypothetical} \tag{28.39}$$

Hypothetical full dynamic similarity example

Assume that we are testing a model of geometric scale $\lambda = 40$ in fresh water of density $\rho_M = 1000.0\,\mathrm{kg/m^3}$ under the condition that $Fr_M = Fr_S$. In the hypothetical case of full dynamics similarity, a drag force of $F_{DM} = 25.21\,\mathrm{N}$ measured at model scale would be equivalent to a full scale drag force of

$$\begin{aligned}
\frac{F_{DS}}{F_{DM}} &= \frac{\rho_S}{\rho_M}\lambda^3 \\
F_{DS} &= \frac{\rho_S}{\rho_M}\lambda^3 \cdot F_{DM} \\
F_{DS} &= \frac{1026.0}{1000.0}\,40^3 \cdot 25.21\,\mathrm{N} \\
&= 1\,655\,389.44\,\mathrm{N} = 1655.4\,\mathrm{kN}
\end{aligned}$$

in salt water of density $\rho_S = 1026.0\,\mathrm{kg/m^3}$.

Dimensionless coefficients

It is common practice to make experimental results dimensionless. The selection of an appropriate normalizing force is up to the user. In ship hydrodynamics forces are mostly normalized by dynamic pressure $1/2\rho v^2$ multiplied with the wetted surface at rest S. The result is a dimensionless drag coefficient:

$$C_D = \frac{F_{DM}}{\frac{1}{2}\rho_M v_M^2 S_M} \tag{28.40}$$

The coefficient would be the same for model and full scale vessel because with

$$F_{DM} = C_D \frac{1}{2} \rho_M v_M^2 S_M \quad \text{and} \quad F_{DS} = C_D \frac{1}{2} \rho_S v_S^2 S_S \tag{28.41}$$

we get

$$\begin{aligned} \frac{F_{DS}}{F_{DM}} &= \frac{C_D \frac{1}{2} \rho_S v_S^2 S_S}{C_D \frac{1}{2} \rho_M v_M^2 S_M} \\ &= \frac{\rho_S}{\rho_M} \frac{v_S^2}{v_M^2} \frac{S_S}{S_M} \\ &= \frac{\rho_S}{\rho_M} \frac{\lambda^4}{\tau^2} \end{aligned} \tag{28.42}$$

Again, assuming we performed the model tests under the condition of equal Froude number, the time scale is fixed to $\tau = \sqrt{\lambda}$. Substituting the time scale τ into Equation (28.42) results in

$$\frac{F_{DS}}{F_{DM}} = \frac{\rho_S}{\rho_M} \lambda^3 \tag{28.43}$$

which is the force scale factor (28.39) we derived above.

Consequently, dimensionless coefficients would apply to both model and full scale, if the quantities (forces) could satisfy the conditions of full dynamic similarity.

28.2.2 Real world: partial dynamic similarity

As we have pointed out in the previous section, we cannot achieve full dynamic similarity. Conducting experiments under the requirement of equal Froude number achieves dynamic similarity for inertia, gravity, and pressure forces, but not for friction forces.

$$\kappa = \kappa_i = \kappa_g = \kappa_p \tag{28.44}$$

$$\kappa \neq \kappa_f \tag{28.45}$$

This complicates ship model testing tremendously. The dimensionless total calm water resistance coefficient

$$C_{TM} = \frac{R_{TM}}{\frac{1}{2} \rho_M v_M^2 S_M} \tag{28.46}$$

cannot be applied to the full scale vessel without further corrections.

$$C_{TS} \neq C_{TM} \tag{28.47}$$

28.2.3 Froude's hypothesis revisited

William Froude was the first to come up with a practical solution to the scaling problem (Froude, 1872). He postulated that the resistance has to be divided into two parts: **Froude's hypothesis**

$$R_T = R_R + R_F \tag{28.48}$$

or in dimensionless form

$$C_T = C_R + C_F \tag{28.49}$$

1. The first part contains the forces with full dynamic similarity if the tests are carried out for corresponding speeds which is equivalent to equal Froude number with $g_S = g_M$. **Residuary resistance**

$$\frac{v_S}{\sqrt{L_S}} = \frac{v_M}{\sqrt{L_M}} \tag{28.50}$$

   This resistance part became known as the residuary resistance R_R. Its dimensionless coefficient applies to both model and full scale vessel:

$$C_R = \frac{R_{RM}}{\frac{1}{2}\rho_M v_M^2 S_M} = C_{RM} = C_{RS} \tag{28.51}$$

2. The second part is the frictional resistance R_F which is a function of the Reynolds number. Since Reynolds numbers for ship and model are different $Re_S \neq Re_M$, viscous forces are not dynamically similar. The frictional resistance coefficient will differ for model and full scale vessel. Consequently, a method is needed to scale the model test results. **Frictional resistance**

At model scale the total resistance R_{TM} is measured and used to find the residuary resistance coefficient C_R which applies to both the model and the full scale vessel:

$$C_R = C_{TM} - C_{FM} \tag{28.52}$$

The full scale resistance coefficient is derived by summing up the residuary resistance, the frictional resistance (full scale), and some more or less empirical corrections:

$$C_{TS} = C_R + C_{FS} + \text{corrections} \tag{28.53}$$

For this procedure to work, we need a method to derive frictional resistance coefficients C_F for the model and the full scale vessel. They cannot be directly measured with the ship model. Froude determined a curve for C_F experimentally by testing flat plates of different sizes and roughness. Nowadays we employ the ITTC 1957 model–ship correlation line (2.18).

Froude knew that his hypothesis was flawed because the residuary resistance still contains the viscous pressure resistance R_{VP}. Since R_{VP} is influenced by the viscosity of the fluid it depends at least somewhat on the Reynolds number. Therefore, the residuary resistance C_R coefficient does not reflect full dynamic similarity. **Improving dynamic similarity**

The ITTC introduced the form factor k to eliminate most of the viscous effects from the force determined under Froude similarity. The dynamically similar force is the wave resistance.

$$C_W = C_{TM} - (1+k)C_{FM} \tag{28.54}$$

This is better in theory but still hard to accomplish in practice, because accurate determination of the form factor is difficult. We will study the procedures recommended by the ITTC in the following chapter.

References

Froude, W. (1868). Observations and suggestions on the subject of determining by experiment the resistance of ships. Correspondence with the British Admiralty. Published in *The Papers of William Froude*, A.D. Duckworth, The Institution of Naval Architects, London, United Kingdom, pp. 120–128, 1955.

Froude, W. (1872). Experiments on surface-friction experienced by a plane moving through water. Read before the British Association for the Advancement of Science at Brighton. Published in *The Papers of William Froude*, A.D. Duckworth, The Institution of Naval Architects, London, United Kingdom, pp. 138–147, 1955.

ITTC (2011). *Ship models.* International Towing Tank Conference, Recommended Procedures and Guidelines 7.5-01-01-01. Revision 3.

Self Study Problems

1. Provide definitions of Froude and Reynolds number (according to the ITTC).
2. What is the greatest difficulty in marine model testing?
3. Under the assumption of equal Froude number for model and full scale vessel, how do the following quantities scale if the scale of the model is $\lambda = L_S/L_M = 50$?
   (a) velocity v_S/v_M =?
   (b) acceleration a_S/a_M =?
   (c) force F_S/F_M =?
   (d) torque Q_S/Q_M =?
   (e) wave period T_S/T_M =?
   (f) wave frequency ω_S/ω_M =?

29

Resistance Test

The resistance test is the first in a series of experiments which together provide the basis for a performance prediction of the full scale hull–propulsor system. This chapter summarizes the procedures for measuring the resistance of a scale model. It discusses test conditions, arrangements, data acquisition, and processing of the test results.

Learning Objectives

At the end of this chapter students will be able to:

- understand the purpose of resistance tests
- prepare and set up resistance tests
- process and interpret test results
- calculate form factor and residuary resistance coefficient

29.1 Test Procedure

A resistance test serves two purposes: **Purpose**

- determine the wave resistance coefficient C_W as a function of Froude number, and
- ascertain a form factor k.

Resistance tests are routinely performed to confirm the estimates of early ship design. However, strictly speaking, performance predictions could be done without a resistance test. Results of a load variation test may be used to substitute for the resistance test data. This will be discussed later in Section 43.4.

ITTC 1978 performance prediction method

The resistance test is part of the ITTC 1978 performance prediction method. The procedure has been revised several times since 1978 and the reader is advised to consult the latest issues of the ITTC recommended procedures. The description here is based on the documents issued in 2017 (ITTC, 2011, 2017a,c,d).

Fundamentals of Ship Hydrodynamics: Fluid Mechanics, Ship Resistance and Propulsion, First Edition.
Lothar Birk.

Companion website: www.wiley.com/go/birk/hydrodynamics

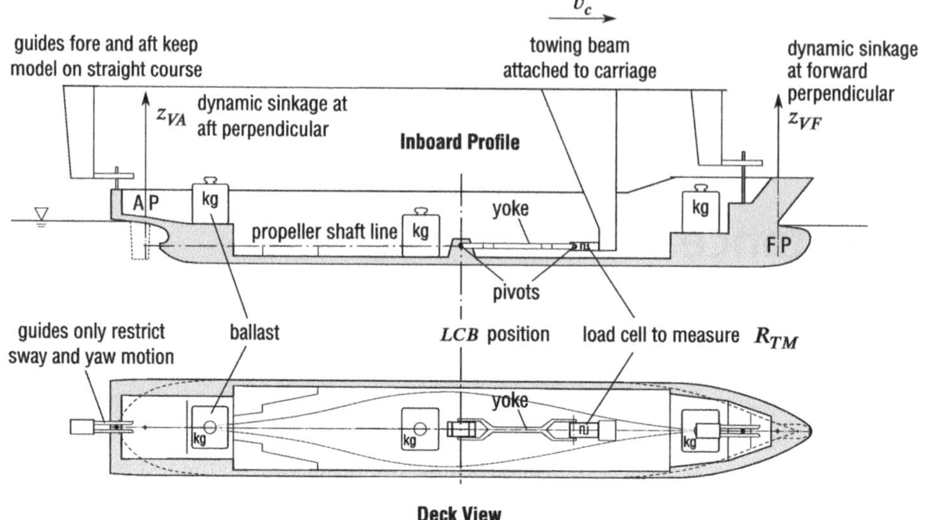

Figure 29.1 Ship model set up for resistance test

Model setup Figure 29.1 shows a typical model setup for a resistance test. After water temperature and density have been measured, the model is ballasted to the desired volumetric displacement (see Section 26.2).

Nowadays, resistance is measured with a load cell attached to the lower end of a stiff column which itself is fastened to the towing carriage. The towing force is transferred from the load cell to the model through a yoke. The yoke pivots in the vertical plane and allows the model to heave and pitch freely. The towing point is located at the intersection of the vertical line through the longitudinal center of buoyancy of the model and the extended line of the propeller shaft. That way, the towing force will not unduly change the attitude of the model. In some cases, shaft inclination might be too steep and alternative approaches are needed, e.g. use a line through the center of the thrust bearing as the height of the towing force attachment point.

The pressure distribution over the hull of a ship underway changes compared to its hydrostatic equilibrium. A large area of lower pressure over the ship bottom causes most ships to sink deeper into the water. They also might trim slightly aft or forward depending on speed and hull shape. Linear transducers or heave potentiometers are used to measure sinkage close to the fore and aft perpendiculars. The two sinkage measurements enable us to compute mean sinkage and running trim angle.

Guides at the forward and aft ends of the model keep the model in line with the center line of the tank. It is important that the guides do not impose any force in longitudinal direction (resistance) or vertical direction (sinkage, trim). They restrict only sway and yaw motion of the vessel.

Data flow Figure 29.2 shows the principal data flow. During the tests environmental data related to the tank and the model are directly measured. See ITTC (2017c) for details.

- Tank water temperature t_W is measured at least at the beginning and the end of a series of tests. Water temperature is taken at half the draft of the model with

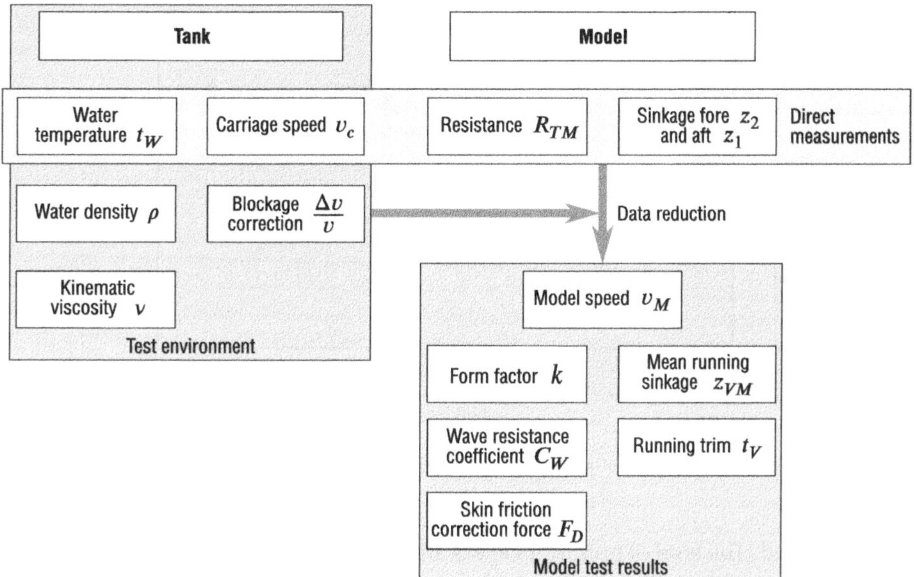

Figure 29.2 Measured and derived data in the resistance test

at least 0.1 °C of accuracy. In a long tank, temperature should be measured at several locations.

The temperature is needed to determine water density ρ and kinematic viscosity ν. See the tables in the ITTC Recommended Procedures (ITTC, 2011). If a hydrometer is available, the density should be measured directly.

- The carriage speed v_C is measured to within 0.1% of the maximum speed or 3 mm/s whichever is larger. However, what we really need for our analysis is the model speed v_M through the water. Unfortunately, v_M is impractical to measure. Since we conduct our test in confined channels, influences of the tank walls and the tank bottom may have to be considered. Details will be discussed in the paragraphs on blockage correction below.

- Resistance R_{TM} is measured as a *horizontal* towing force in line with the propeller shaft and at the longitudinal center of buoyancy location LCB. As mentioned, this arrangement avoids trim effects. The tow force should be measured within 0.2% of the maximum capacity of the dynamometer (load cell) or 0.05 N, whichever is larger (ITTC, 2017c).

- Changes in vessel attitude are recorded as linear changes in vertical position z_1 and z_2 toward the aft and fore ends of the vessel. Mean dynamic sinkage z_{VM} at $L_{PP}/2$ and trim angle t_V are calculated from z_1 and z_2 (see page 350).

Maintaining preturbulence

Tests are conducted for several towing velocities covering the whole range of Froude numbers of interest for the full scale vessel. It is important to keep a relatively constant level of preturbulence in the towing tank water throughout the tests. After long periods of inactivity in the tank, water has settled down and its naturally appearing turbulence

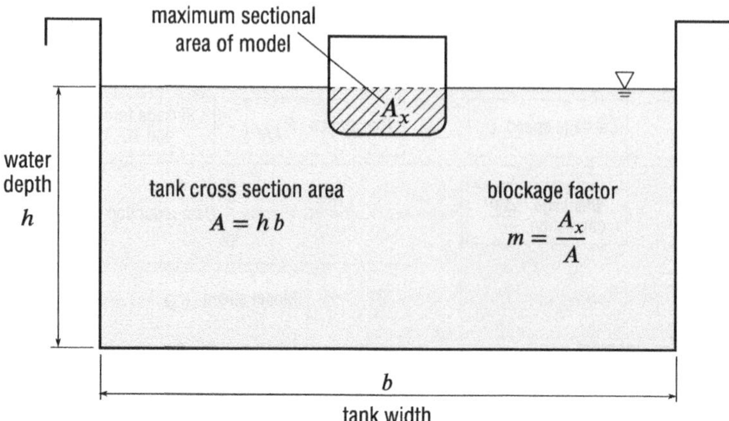

Figure 29.3 Tank cross section area A and blockage factor m

has diminished. The level of preturbulence is affecting the laminar–turbulent transition in the boundary layer. Resistance from repeated tests at the same speed will be in disagreement, if the levels of preturbulence have changed significantly. Therefore, it is custom to perform a higher speed run (without data recording) or drag a special fork through the water to attain preturbulence before the actual test sequence commences. Afterward, carriage speeds should not be simply increased from test to test, but alternate between high and low speeds.

Steady state data In each run the model has to be accelerated from zero to the actual speed. Due to the transient acceleration phase, certain fluctuations of the resistance can be expected even once a steady speed has been reached. The resistance R_{TM} should be taken as the mean value of at least 5 cycles. The time window for averaging should consist of an integer number of full cycles.

29.2 Reduction of Resistance Test Data

Test results As a result of the tests, we obtain a table with total resistance $R_{TM}(v_C)$, sinkage measurements $z_1(v_C)$, $z_2(v_C)$ as a function of carriage speed v_C. In a first step, the actual model speed through the water has to be determined from carriage speed and tank dimensions. Once the model speed v_M is known, form factor and residuary resistance coefficient may be determined.

Blockage correction Obviously, a model moves at a speed relative to the towing tank walls which matches the carriage speed v_C. However, for the performance prediction the speed of a model relative to the surrounding fluid is the determining factor.

As Figure 29.3 illustrates, the cross section of the tank $A = bh$ narrows at the model slightly to $A - A_x$. A_x is the maximum submerged sectional area of the model. Comparable to a flow in a narrowing pipe, the water will speed up around the model. You can also imagine the model pushing the water, which creates an indiscernible rise of the water ahead of the model, and thus causing a flow from bow to stern which is superimposed to the model movement. As a result of blockage, the relative velocity between model and the surrounding body of water is commonly higher than the carriage speed by a small fraction. How much faster mainly depends on the blockage factor $m = A_x/A$ and the depth Froude number $Fr_h = v/\sqrt{gh}$.

Blockage correction and model speed

Most model basins have developed their own blockage correction methods considering their tank dimensions and the size of models they typically investigate. Blockage corrections are expressed as relative increase of model speed over carriage speed $\Delta v/v$. For typical towing tank cross sections and model sizes, the increase in velocity is typically less than 1%. The carriage velocity v_C is augmented by the fractional increase $\Delta v/v$ to obtain the actual model speed v_M relative to the surrounding water.

$$v_M = v_C\left(1+\frac{\Delta v}{v}\right) \tag{29.1}$$

The former model basin Versuchsanstalt für Wasserbau und Schiffbau (VWS) in Berlin sometimes simply used $\Delta v/v = m$ which accounts for conservation of mass effects only (Schuster, 1952).

Schuster's blockage correction

ITTC suggests three different blockage correction methods, two of which are fairly simple to implement (ITTC, 2017c). Schuster (1956) suggests to compute the blockage correction as a function of the blockage factor m, the depth Froude number Fr_h, and the ratio of viscous and total resistance.

$$\frac{\Delta v}{v} = \frac{m}{1-m-Fr_h^2} + \frac{2}{3}\left(1-\frac{R_{VM}}{R_{TM}}\right)Fr_h^{10} \tag{29.2}$$

The depth Froude number is defined as

$$Fr_h = \frac{v_M}{\sqrt{g\,h}} \tag{29.3}$$

h is the water depth in the towing tank at the time of the test. R_{VM} is the total viscous resistance of the model.

This obviously involves a 'chicken or egg' dilemma. At this point, we only know the carriage speed v_C, but we should use the actual model speed v_M to compute the depth Froude number Fr_h (29.3) and to estimate the viscous model resistance R_{VM}. Due to the small differences in speed, it will not matter much, if we initially estimate the depth Froude number based on the carriage speed instead of the model speed.

$$\tilde{Fr}_h = \frac{v_C}{\sqrt{g\,h}} \tag{29.4}$$

Using this approximation, a first estimate for the model velocity is derived utilizing the first and dominant part of Schuster's formula (29.2):

$$\tilde{v}_M = v_C\left(1+\frac{m}{1-m-\tilde{Fr}_h^2}\right) \tag{29.5}$$

Equipped with this approximate model speed, the depth Froude number is updated and the viscous resistance of the model is estimated:

$$\text{Second depth Froude number estimate:} \quad Fr_h = \frac{\tilde{v}_M}{\sqrt{g\,h}} \tag{29.6}$$

$$\text{Reynolds number estimate:} \quad \tilde{Re} = \frac{\tilde{v}_M L_{OS_M}}{\nu_M} \tag{29.7}$$

$$\text{Friction coefficient:} \quad \tilde{C}_{FM} = \frac{0.075}{\left[\log_{10}(\tilde{Re}) - 2\right]^2} \tag{29.8}$$

$$\text{Viscous resistance estimate:} \quad R_{VM} = \frac{1}{2}\rho_M\,\tilde{v}_M^2 S_M \tilde{C}_{FM} \tag{29.9}$$

The latter does not include a form factor as it is unknown at this point. However, since the depth Froude number should satisfy $Fr_h < 0.7$, the factor for the correction is small ($Fr_h^{10} < 0.029$). All quantities are now known, and Equations (29.2) and (29.1) may be employed to compute the speed correction. This process works well for moderate speeds with $Fr < 0.35$.

Tamura's blockage correction Another simple blockage correction method has been proposed by Tamura (1972). It computes the fractional increase in speed from the following formula:

$$\frac{\Delta v}{v} = 0.67\,m \left[\frac{L_{WL}}{B}\right]^{\frac{3}{4}} \frac{1}{\left(1 - Fr_h^2\right)} \tag{29.10}$$

Again, model speed follows from Equation (29.1). Tamura's blockage correction seems to overpredict the speed for the relatively small models of about 3 m length used in the towing tank at the University of New Orleans.

Total resistance coefficient With the actual model speed v_M known, the measured resistance is converted into a dimensionless total resistance coefficient:

$$C_{TM} = \frac{R_{TM}}{\frac{1}{2}\rho\,v_M^2\,S_M} \tag{29.11}$$

If we would have been able to conduct the test under *full dynamic similarity*, C_{TM} would also apply to the full scale vessel. Unfortunately, this is *not* the case.

Dynamic trim and sinkage During the test run, sinkage measurements are taken near stern and bow of the vessel. Figure 29.4 shows the general setup. The lengths L_1 and L_2 mark the distances of the measurement points from midships $L_{PP}/2$.

Trim angle The trim angle t_V follows from the ratio of difference in sinkage fore and aft and the distance between the measurement points.

$$t_V = \tan^{-1}\left(\frac{-(z_2 - z_1)}{L_1 + L_2}\right) \tag{29.12}$$

If we use a right handed $< x, y, z >$-coordinate system with x-axis pointing forward and the z-axis pointing upwards, a positive trim angle $t_V > 0$ will cause the bow to trim down.

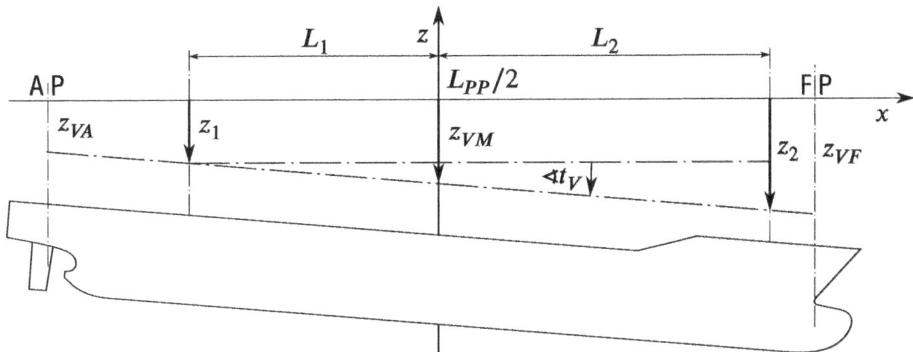

Figure 29.4 Measurements and length definitions for the computation of sinkage and trim (sinkage and trim are exaggerated)

Mean sinkage at midships follows from basic geometrical relationships. **Mean sinkage**

$$z_{VM} = z_1 + \left(\frac{z_2 - z_1}{L_2 + L_1} \right) L_1 \qquad (29.13)$$

Most vessels will have negative sinkage, i.e. they settle deeper into the water when underway. This effect is even more pronounced in shallow water and has to be considered when sailing fast in shallow waterways in order to prevent grounding.

29.3 Form Factor *k*

The form factor k has been introduced to augment the flat plate frictional resistance coefficient C_F to include the Reynolds number dependent viscous pressure resistance and form effects on the boundary layer itself. Once a form factor is known, the wave resistance coefficient[1] C_W is found by solving **Form factor**

$$C_{TM} = (1 + k) C_{FM} + C_W \qquad (29.14)$$

for C_W:

$$C_W = C_{TM} - (1 + k) C_{FM} \qquad (29.15)$$

C_{FM} is the ITTC 1957 model–ship correlation line coefficient for the Reynolds number Re_M of the model test (2.18). Note that the Froude number is computed based on the waterline length L_{WL}, but the Reynolds number is computed based on the length over wetted surface L_{OS}.

$$Fr = \frac{v_M}{\sqrt{g L_{WL_M}}} \qquad Re_M = \frac{v_M L_{OS_M}}{\nu_M} \qquad (29.16)$$

[1]The residuary resistance coefficient C_R was used instead of C_W in editions of the ITTC recommended procedures and guidelines before 2017.

Difficulties Form factors k are not universally applied because the uncertainty in their values is high. Among others, Gross (1975) lists the following difficulties:

- low model speed means low Reynolds number. The boundary layer may be laminar and the danger of separation increases (which will not happen at the full scale vessel). Tripping the boundary layer into turbulent flow may be more difficult. If separation occurs, the increased resistance will cause the form factor to be too large. On the other hand, if the boundary flow becomes mostly laminar without separation form factors may become too small. This problem will be compounded for short models (large scale λ).
- wave breaking may disturb the linearity of the data points used to find k. This is especially true for vessels with high block coefficient.
- side wall effects may influence the form factor
- bulbous bow and transom affect the linearity of the Prohaska data points
- hull–propeller interaction may influence the form factor

If the form factor is set too high, resistance of the full scale vessel will be underpredicted, because the wave resistance estimate will be too low and vice versa. ITTC is discussing formulas for the form factor, but so far no formula has been found that consistently improves the performance prediction (see Gross (1975) and Section 18.1). The most reliable method to determine the form factor was suggested by Prohaska (1966) and is recommended by the ITTC.

Prohaska's method If no boundary layer separation occurs, the difference between total and viscous resistance is the wave resistance. Linear wave resistance theory predicts that for slow speeds $Fr < 0.2$ wave resistance is growing with the fourth power of the Froude number with a constant factor α.

$$C_W \approx \alpha Fr^4 \qquad \text{for } Fr < 0.2 \tag{29.17}$$

This is confirmed by experimental results for Froude numbers $Fr < 0.2$. We substitute assumption (29.17) into Equation (29.14) and divide by the ITTC 1957 model–ship correlation coefficient C_{FM}.

$$\frac{C_{TM}}{C_{FM}} = (1+k)\frac{C_{FM}}{C_{FM}} + \alpha\frac{Fr^4}{C_{FM}} \tag{29.18}$$

Equation (29.18) may be interpreted as the equation of a straight line, where $b = (1+k)$ becomes the y-intercept, α is the slope, and $x = \frac{Fr^4}{C_{FM}}$ is the variable.

$$y = b + \alpha x \tag{29.19}$$

We plot left- and right-hand side of (29.18) as data points $(x_i, y_i) = \left(\frac{Fr^4}{C_{FM}}, \frac{C_{TM}}{C_{FM}}\right)$. According to Equations (29.18) and (29.19), the data points should align in a straight line for model speeds with $Fr < 0.2$ (Figure 29.5). ITTC recommends to use only data points in the range of $Fr \in [0.1, 0.2]$. If the speed is too low ($Fr < 0.1$), errors in the resistance measurements may lead to incorrect form factor predictions.

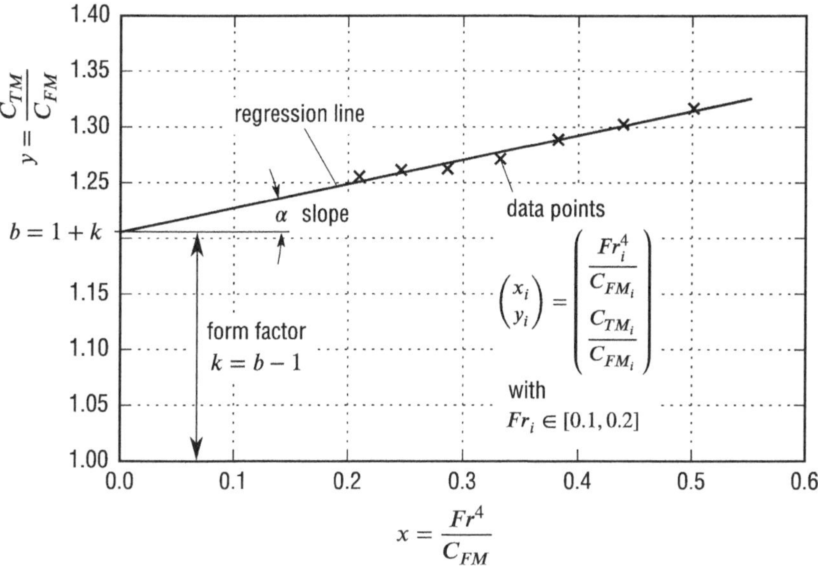

Figure 29.5 Method of Prohaska to determine the form factor k

Holtrop (2001) suggests to carefully select the data used to compute the form factor. Data points $(C_{TM}/C_{FM}, Fr^4/C_{FM})$ which drop sharply for small Fr^4/C_{FM} are an indication of possible laminar flow effects. If the Reynolds number becomes too small, the flow may become laminar again downstream of the turbulence generator (relaminarization). If data points $(C_{TM}/C_{FM}, Fr^4/C_{FM})$ rise sharply for small Fr^4/C_{FM} values, it might be an indication of flow separation. However, rising values may also be caused by a transom that is too deeply submerged for low speeds and, consequently, creates significant pressure drag. The same is true for a partially submerged bulbous bow. Careful observation of flow and wave patterns during the tests is required to choose a suitable data set.

Linear regression

The form factor can be determined graphically from the y-intercept or better by linear regression. Depending on how slow the model could be towed without flow separation, data may have to be extrapolated to $Fr \longrightarrow 0$ quite a bit. First, compute the mean values $\bar{x}$ and $\bar{y}$ of the used x_i and y_i values (M pairs).

$$\bar{x} = \frac{1}{M}\sum_{i=1}^{M} x_i = \frac{1}{M}\sum_{i=1}^{M} \frac{Fr_i^4}{C_{FM_i}} \tag{29.20}$$

$$\bar{y} = \frac{1}{M}\sum_{i=1}^{M} y_i = \frac{1}{M}\sum_{i=1}^{M} \frac{C_{TM_i}}{C_{FM_i}} \tag{29.21}$$

The slope α is given by

$$\alpha = \frac{\sum_{i=1}^{M}(x_i - \bar{x})(y_i - \bar{y})}{\sum_{i=1}^{M}(x_i - \bar{x})^2} \tag{29.22}$$

The y-intercept follows from

$$b = \bar{y} - \alpha\bar{x} \tag{29.23}$$

and the form factor k is

$$k = b - 1 \tag{29.24}$$

In some cases, the assumption $C_W = \alpha Fr^4$ does not work out. If the data points do not show a linear relationship, higher powers than 4, i.e. 5 or 6, may be tried.

Full scale form factor The form factor k found for the model is also used for the full scale vessel. After all, k expresses a form effect, and the shapes of model and full scale vessel ought to be identical. The boundary layers are, however, not similar. It is relatively thicker on the model. There is some evidence that the assumption $k_M = k_S$ is not entirely correct, but so far no consistent estimate for the scale effects on the form factor has been developed. Form factor values typically fall into the range $k \in [0.1, 0.4]$. Slender vessels tend to have smaller values than vessels with high block coefficient. The form factor will vary for the same vessel from loading condition to loading condition.

29.4 Wave Resistance Coefficient C_W

Wave resistance coefficient Finally, with model speed v_M and form factor k known, we can extract the wave resistance coefficient C_W. If not yet done, compute the Reynolds number Re_M for all model velocities v_M

$$Re_M = \frac{v_M L_{OS_M}}{\nu_M} \tag{29.25}$$

which leads to the ITTC 1957 model–ship correlation coefficient

$$C_{FM} = \frac{0.075}{\left[\log_{10}(Re_M) - 2\right]^2} \tag{2.18}$$

Finally, the wave resistance coefficient follows from Equation (29.15).

$$C_W = C_{TM} - (1 + k)C_{FM} \tag{29.15}$$

Residuary resistance coefficient In cases where a form factor cannot be found experimentally and the estimates do not apply, the residuary resistance coefficient C_R replaces the wave resistance coefficient.

$$C_R = C_{TM} - C_{FM} \qquad \text{without form factor} \tag{29.26}$$

Full scale wave resistance coefficient Based on Froude's fundamental assumption that the residuary resistance depends only on the Froude number Fr, the wave resistance coefficient C_W (or C_R) derived for the model is applied to the full scale vessel for the same Froude number and corresponding speed.

$$C_W = C_{WM} = C_{WS} \tag{29.27}$$

A numerical example is presented at the end of the following chapter.

29.5 Skin Friction Correction Force F_D

Friction deduction force

An additional result of the resistance test is the skin friction correction force F_D (ITTC, 2017b). For each run of the propulsion test, a skin friction correction F_D is precomputed and applied to the model as an additional towing force. It corrects the otherwise excessive loading of the model propeller during the propulsion test.

$$F_D = \frac{1}{2} \rho v_M^2 S_M \left[(1+k)(C_{FM} - C_{FS}) - \Delta C_F \right] \quad (29.28)$$

Omission of the skin friction correction ΔC_F results in incorrect values for the hull–propeller interaction parameters (see Chapter 34).

Of course, if the residuary resistance coefficient is determined with Froude's method ($k = 0$), the form factor is also omitted in the calculation of the skin friction deduction.

$$F_D = \frac{1}{2} \rho v_M^2 S_M \left[(C_{FM} - C_{FS}) - \Delta C_F \right] \quad (29.29)$$

An alternative to applying a specific friction deduction force is to do tests with load variations on the propeller. We will get back to this when we discuss propulsion.

References

Gross, A. (1975). Form factor. In *Proceedings of the 14th ITTC*, pages 576–590, Ottawa, Canada. International Towing Tank Conference. Report of the Performance Committee, Appendix 4.

Holtrop, J. (2001). Extrapolation of propulsion tests for ships with appendages and complex propulsors. *Marine Technology*, 38(3):145–157.

ITTC (2011). *Fresh water and seawater properties*. International Towing Tank Conference, Recommended Procedures and Guidelines 7.5-02-01-03. Revision 2.

ITTC (2017a). *1978 ITTC performance prediction method*. International Towing Tank Conference, Recommended Procedures and Guidelines 7.5-02-03-01.4. Revision 4.

ITTC (2017b). *Propulsion/bollard pull test*. International Towing Tank Conference, Recommended Procedures and Guidelines 7.5-02-03-01.1. Revision 5.

ITTC (2017c). *Resistance test*. International Towing Tank Conference, Recommended Procedures and Guidelines 7.5-02-02-01. Revision 4.

ITTC (2017d). *Ship models*. International Towing Tank Conference, Recommended Procedures and Guidelines 7.5-01-01-01. Revision 4.

Prohaska, C. (1966). A simple method for the evaluation of the form factor and the low speed wave resistance. In *Proceedings of the 11th ITTC*, pages 65–66, Tokyo, Japan. International Towing Tank Conference. Resistance Session, Written Contributions.

Schuster, S. (1952). Untersuchungen über Strömungs- und Widerstandsverhältnisse bei der Fahrt von Schiffen in beschränktem Wasser. In *Jahrbuch der Schiffbautechnischen Gesellschaft*, volume 46, pages 244–288. Springer-Verlag, Berlin.

Schuster, S. (1956). Beitrag zur Frage der Kanalkorrektur bei Modellversuchen. *Schiffstechnik/Ship Technology Research*, 3:93–96.

Tamura, K. (1972). Study on the blockage correction. *Journal of the Society of Naval Architects of Japan*, 1972(131):17–28.

Self Study Problems

1. Summarize the data acquired during a resistance test and state the quantities derived from the measured data.
2. Discuss why models are ballasted based on weight rather than draft.
3. Why is the speed of the model relative to the surrounding water higher than the actual carriage speed?
4. Explain why the attitude changes for a moving ship compared to its position at rest.
5. Why is the form factor determined for the model also used for the full scale vessel?
6. What is the major difference between the residuary resistance coefficient and the wave resistance coefficient? What do they have in common?

30

Full Scale Resistance Prediction

This chapter continues the discussion of the resistance test. We show how the resistance values found in the model test are used to predict the resistance of the full scale vessel. The description follows the ITTC recommended procedure 7.5-02-02-01 for the resistance test and procedure 7.5-02-03-01.4 for the 1978 ITTC performance prediction method (ITTC, 2017a,b). Additions and corrections to the resistance coefficients of the model test are explained and discussed. An example closes the chapter.

Learning Objectives

At the end of this chapter students will be able to:

- understand the issues associated with scaling model test results to full scale
- make full scale resistance predictions based on model test results
- understand and apply necessary corrections

30.1 Model Test Results

Wave resistance and form factor

The data of resistance tests result in the form factor k and a set of wave resistance coefficients C_W as a function of Froude number Fr. In the case a form factor is not determined, a set of residuary resistance coefficients C_R replaces the wave resistance coefficients.

Wave resistance coefficient (or residuary resistance coefficient) and form factor are used for the full scale vessel without corrections. This is based on the assertions that model and ship are geometrically similar and that C_W only considers forces that are dynamically similar.

Ship speed

Resistance tests are conducted under Froude similarity, i.e. the full scale ship speed equivalent to the model speed is given by

$$v_S = Fr\sqrt{gL_{WL_S}} = v_M\frac{\sqrt{gL_{WL_M}}}{\sqrt{gL_{WL_S}}} = v_M\sqrt{\lambda} \qquad (30.1)$$

Fundamentals of Ship Hydrodynamics: Fluid Mechanics, Ship Resistance and Propulsion, First Edition.
Lothar Birk.

Companion website: www.wiley.com/go/birk/hydrodynamics

The last term uses the model scale λ and assumes equal gravitational acceleration for model and full scale ship.

Total resistance coefficient of model

The total resistance for the model has been measured and used to derive the total resistance coefficient C_{TM}. During the model test, only partial dynamic similarity is possible. Therefore, the dimensionless total resistance coefficient of the model cannot be applied to the full scale vessel. The total resistance of the model is split into viscous and wave resistance $C_{TM} = (1+k)C_{FM} + C_W$ in order to overcome the problem of partial dynamic similarity. Although the wave resistance coefficient C_W applies to both, model and full scale ship, simply replacing the model frictional resistance coefficient C_{FM} with the equivalent full scale coefficient C_{FS} in the formula is not quite sufficient. Additional corrections are needed, which are explained in the next section.

30.2 Corrections and Additional Resistance Components

Roughness allowance

As stated in our discussion of the frictional resistance, the finished surfaces of models are hydraulically smooth. However, the full scale ship hull is not. Due to the high Reynolds number and the resulting thin viscous sublayer, even smooth weld seams and ever present surface imperfections call for an upward correction of the resistance compared to the smooth model. Therefore, a roughness allowance ΔC_F is applied to account for the relative increase of frictional resistance from model to full scale vessel. ITTC (2017a) recommends the following correction for surface roughness.

$$\Delta C_F = 0.044\left[\left(\frac{k_s}{L_{WL_S}}\right)^{\frac{1}{3}} - 10\left(Re_S\right)^{-\frac{1}{3}}\right] + 0.000125 \tag{30.2}$$

In this formula k_s is the equivalent sand roughness of the hull and *not* the form factor! The standard value for it is $k_s = 150 \cdot 10^{-6}$ m. For advanced hull coatings available today, lower values might be more appropriate.

Full scale Reynolds number

Re_S is the Reynolds number for the full scale vessel.

$$Re_S = \frac{v_S L_{OS_S}}{\nu_S} \tag{30.3}$$

By convention, a kinematic viscosity of $\nu_S = 1.1892\ 10^{-6}\mathrm{m}^2/\mathrm{s}$ for seawater of 15 °C temperature is used for the full scale prediction.

Correlation allowance

The correlation allowance accounts for remaining differences between model and full scale vessel which may be attributed to the partial dynamic similarity of the model test. Model basins usually develop their own formula by comparing their predictions with data gathered during ship trials, but ITTC (2017a) suggests using the following equation in combination with the roughness allowance (30.2) above.

$$C_A = \left[5.68 - 0.6\log_{10}(Re_S)\right] \cdot 10^{-3} \tag{30.4}$$

Surface corrections, bilge keels

The sum of 3D frictional resistance $(1+k)C_{FS}$, roughness allowance ΔC_F, and correlation allowance C_A represent the full scale viscous resistance. It is mainly a function of Reynolds number Re_S. Since the viscous resistance is proportional to the wetted surface area S, an additional correction factor $\frac{S_S+S_{BK_S}}{S_S}$ is applied to account for extra hull surfaces not replicated at model scale. Here, S_{BK_S} is the wetted surface of the bilge keels.

Air resistance

Ship models are usually built without deck house and other superstructure like for example cargo containers. Therefore, the full scale resistance prediction contains an addition for the resistance of a ship moving through air.

$$C_{AAS} = C_{DA} \frac{\rho_A \, A_{VS}}{\rho_S \, S_S} \tag{30.5}$$

For some vessels wind tunnel test data may be available, but in general, $C_{DA} = 0.8$ is the suggested default value for the air drag coefficient. $\rho_A = 1.225 \text{ kg/m}^3$ is the standard density of dry air at a temperature of 15 °C. A_{VS} is the projected area above the waterline perpendicular to the direction of motion. It has to be emphasized that Equation (30.5) captures only the resistance of a ship moving through air at rest (trial conditions). Significantly higher forces due to steady wind and wind gusts can occur in service conditions.

Additional corrections

Additional resistance components may have to be considered as needed. Examples include the additional viscous resistance due to appendages (struts, bossings for twin propulsion systems, foils, etc.). They are collected in an appendage resistance coefficient C_{APP_S}. Estimates of appendage resistances are provided by Holtrop and Mennen (1982). Additional corrections may be needed for spray resistance, steering resistance, or other flow phenomena not captured anywhere else.

30.3 Total Resistance and Effective Power

Frictional resistance

From the ship speed follows the full scale Reynolds number Re_S (30.3) and the corresponding ITTC 1957 model–ship correlation line coefficient C_{FS} for the ship.

$$C_{FS} = \frac{0.075}{\left[\log_{10}(Re_S) - 2\right]^2} \tag{2.18}$$

Full scale total resistance coefficient

The full scale resistance coefficient C_{TS} is build on C_{FS} and C_W including the additional corrections discussed above.

$$C_{TS} = \frac{S_S + S_{BK_S}}{S_S} \left[(1+k)\,C_{FS} + \Delta C_F + C_A\right] + C_W + C_{AAS} + C_{APP_S} \tag{30.6}$$

Total resistance

The actual, total resistance of the ship without a propulsor in calm water is obtained by multiplying the total resistance coefficient C_{TS} with dynamic pressure $q = \rho v_S^2/2$ and wetted surface S_S.

$$R_{TS} = \frac{1}{2} \rho \, v_S^2 \, S_S \, C_{TS} \tag{30.7}$$

Effective power

At steady state, work is equal to force multiplied by distance. The effective power

Table 30.1 Particulars of full scale vessel and model used in the prediction example

Main particular		Unit	Vessel	Model
model number		–	[–]	0001
model scale	λ	–	[–]	25
length in waterline	L_{WL}	[m]	172.440	6.8976
length over wetted surface	L_{OS}	[m]	178.040	7.1216
beam	B	[m]	26.529	1.0612
draft	T	[m]	10.204	0.4081
maximum sectional area	A_X	[m^2]	264.466	0.4231
transverse area above WL	A_{VS}	[m^2]	464.000	–
wetted surface	S	[m^2]	5 540.508	8.8648
bilge keel surface	S_{BK}	[m^2]	85.000	–
displacement (volume)	V	[m^3]	27 285.788	1.7463
displacement (mass)	Δ	[t]	27 995.791	1.7444
distance midship–aft sinkage sensor	L_1	[m]	–	2.400
distance midship–forward sinkage sensor	L_2	[m]	[–]	2.700
standard sand roughness	k_S	[μm]	150.0	–

needed to tow the vessel is defined by the work done per unit time which is equivalent to force times velocity.

$$P_{ES} = R_{TS}\, v_S \tag{30.8}$$

Make sure you use consistent units! Especially, do not use the ship speed expressed in knots with these equations. The whole procedure is best implemented as a spread sheet or program.

30.4 Example Resistance Prediction

Summary of vessel data A resistance test has been conducted for a single screw container ship model at scale $\lambda = 25$. Table 30.1 shows an excerpt of geometric particulars for vessel and model. Each model should have a unique identifier within the institution performing the tests (here the fictitious number 0001). Typically, a resistance test report will include a full set of hydrostatic data and potentially a lines plan. ITTC (2017b) lists the minimum set of data to be included. The model discussed here features a sand strip as a turbulence generator and is towed at the LCB–propeller shaft line intersection point.

Displacement Note that the displacement mass Δ_M of the model is not derived by scaling the full scale mass. Only the volumetric displacement V is scaled. Then, the mass at model scale follows from Equation (26.4).

Test environment Important for the evaluation of any model test is the testing environment. Temperature of the tank water t_W determines density and kinematic viscosity of the water. Water depth h and tank cross section area A influence the blockage correction. Full scale resistance predictions are usually performed for standard seawater at temperature 15 °C. The relevant environmental data for this example are presented in Table 30.2. Values are taken from ITTC's recommended procedures on fresh and seawater properties (ITTC, 2011).

Table 30.2 Testing and full scale environments for resistance prediction

gravitational acceleration at test site	g	9.793167 m/s^2
water depth in towing tank	h	4.450 m
cross section area of towing tank	A	35.600 m^2
blockage factor	m	0.01189
tank water temperature	t_W	16.300 °C
tank water density	ρ_M	998.8968 kg/m^3
tank water kinematic viscosity	ν_M	1.1007 10^{-6} m^2/s
seawater temperature	t_{WS}	15.000 °C
seawater density	ρ_S	1026.021 kg/m^3
seawater kinematic viscosity	ν_S	1.1892 10^{-6} m^2/s
standard air density	ρ_A	1.225 kg/m^3

Table 30.3 Measured total resistance and sinkage of model; blockage correction (Schuster, 1956), dynamic sinkage, and trim

Measured data				Blockage correction, dynamic sinkage, and trim						
v_C [m/s]	R_{TM} [N]	z_1 [cm]	z_2 [cm]	Fr_h [–]	$\Delta v/v$ [%]	v_M [m/s]	Fr [–]	Re_M [–]	z_{VM} [cm]	t_V [deg]
1.296	28.41	−1.080	−0.644	0.1988	1.2530	1.3125	0.1597	8 491 701.0	−0.875	−0.049
1.347	30.61	−1.214	−0.600	0.2065	1.2572	1.3635	0.1659	8 822 152.6	−0.925	−0.069
1.396	32.72	−1.286	−0.628	0.2141	1.2615	1.4136	0.1720	9 146 167.8	−0.976	−0.074
1.446	35.16	−1.343	−0.675	0.2219	1.2660	1.4647	0.1782	9 476 709.5	−1.028	−0.075
1.496	37.92	−1.421	−0.718	0.2295	1.2707	1.5148	0.1843	9 800 817.5	−1.091	−0.079
1.546	40.73	−1.547	−0.780	0.2372	1.2756	1.5659	0.1905	10 131 458.6	−1.195	−0.084
1.596	43.63	−1.669	−0.886	0.2448	1.2807	1.6160	0.1966	10 455 668.9	−1.301	−0.088
1.645	46.71	−1.780	−0.907	0.2524	1.2859	1.6661	0.2027	10 779 933.8	−1.369	−0.098
1.694	49.99	−1.895	−0.960	0.2600	1.2914	1.7163	0.2088	11 104 255.9	−1.455	−0.105
1.745	53.59	−1.965	−1.004	0.2677	1.2972	1.7674	0.2150	11 435 126.3	−1.513	−0.108
1.794	57.62	−2.073	−1.085	0.2753	1.3031	1.8175	0.2211	11 759 572.9	−1.608	−0.111
1.846	62.07	−2.192	−1.141	0.2832	1.3096	1.8697	0.2275	12 097 067.9	−1.698	−0.118
1.895	66.46	−2.378	−1.301	0.2908	1.3160	1.9199	0.2336	12 421 653.8	−1.871	−0.121
1.943	71.59	−2.519	−1.389	0.2983	1.3226	1.9690	0.2396	12 739 819.5	−1.987	−0.127

Recorded data

Carriage speed v_C, total resistance R_{TM}, and dynamic sinkage measurements z_1 (aft) and z_2 (forward) recorded during the tests are summarized in Table 30.3. Note that the total resistance is less than 0.5% of the weight force for the model even at the highest speed. Therefore, carriage speed has to be kept as steady as possible because even small fluctuations in velocity may cause accelerations big enough to generate inertia forces of sufficient magnitude to distort the measurement.

Blockage correction, sinkage, and trim

As a first step, a blockage correction is selected and applied in order to find the actual speed of the model through the water (see Section 29.2). Here, the blockage correction by Schuster (1956) is applied using Equations (29.4) through (29.9), and Equations (29.2) and (29.1). The velocity correction $\Delta v/v$ ranges from 1.25% to 1.32% above the carriage velocity v_C in this case. With the model velocity v_M known, Froude and Reynolds number are computed with Equation (29.16). Mean sinkage z_V and running trim t_V are

Data for regression line

Fr	C_{FM_i}	C_{TM_i}	$\frac{Fr_i^4}{C_{FM_i}}$	$\frac{C_{TM_i}}{C_{FM_i}}$
[−]	[10^{-3}]	[10^{-3}]	[−]	[−]
0.1597	3.0871	3.7257	0.2106	1.2069
0.1659	3.0664	3.7182	0.2471	1.2126
0.1720	3.0471	3.6987	0.2872	1.2138
0.1782	3.0282	3.7021	0.3331	1.2225
0.1843	3.0105	3.7324	0.3833	1.2398
0.1905	2.9932	3.7514	0.4402	1.2533
0.1966	2.9769	3.7736	0.5021	1.2676

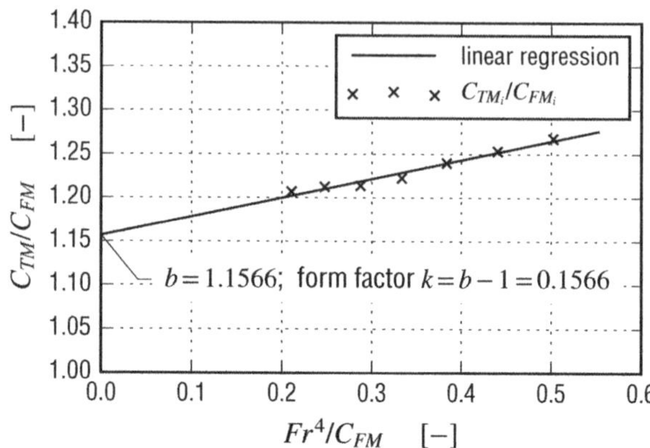

Figure 30.1 Finding the form factor with Prohaska's method; only data points with $0.1 \leq Fr \leq 0.2$ are used

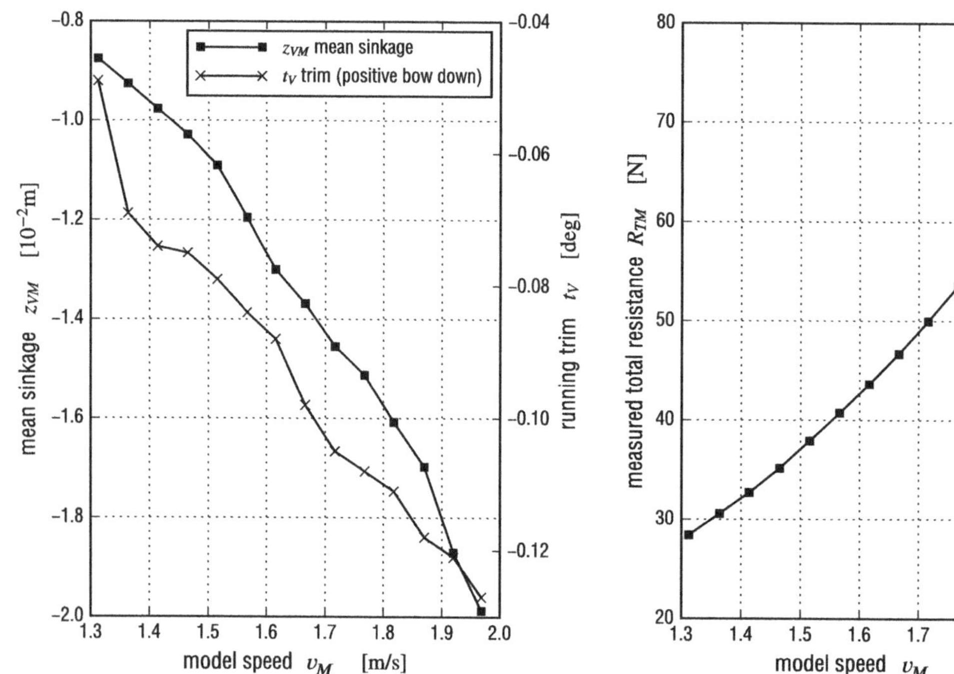

Figure 30.2 Measured mean sinkage and running trim angle of model

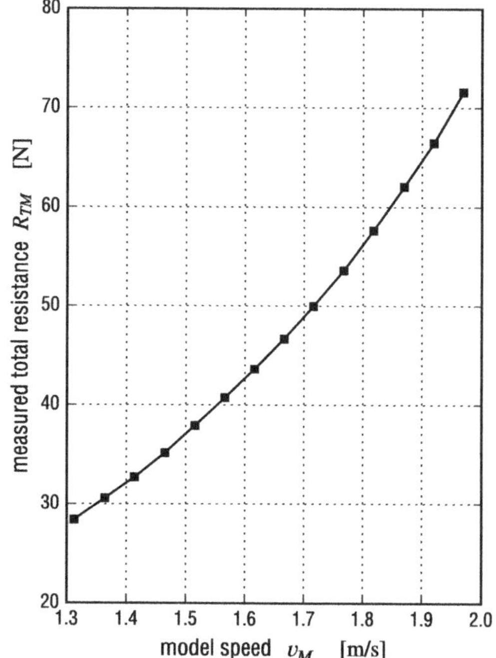

Figure 30.3 Measured total resistance of model as function of model speed

found using Equations (29.13) and (29.12). Negative values for z_V and t_V in Table 30.3 indicate that the model is settling down and trimming aft. Both quantities increase with increasing velocity (see Figure 30.2).

Form factor, Prohaska's method

Based on the seven tests with Froude numbers in the range $0.1 \leq Fr \leq 0.2$, Prohaska's method from Section 29.3 is applied to find the form factor with result $k = 0.1566$. Data points and regression line are shown in Figure 30.1.

Table 30.4 Resistance coefficients for the model ($k = 0.1566$)

v_M [m/s]	Fr [–]	Re_M [–]	R_{TM} [N]	C_{TM} [10^{-3}]	C_{FM} [10^{-3}]	C_W [10^{-3}]
1.3125	0.1597	8 491 701.0	28.41	3.7257	3.0871	0.1552
1.3635	0.1659	8 822 152.6	30.61	3.7182	3.0664	0.1715
1.4136	0.1720	9 146 167.8	32.72	3.6987	3.0471	0.1743
1.4647	0.1782	94 76 709.5	35.16	3.7021	3.0282	0.1996
1.5148	0.1843	9 800 817.5	37.92	3.7324	3.0105	0.2503
1.5659	0.1905	10 131 458.6	40.73	3.7514	2.9932	0.2894
1.6160	0.1966	10 455 668.9	43.63	3.7736	2.9769	0.3304
1.6661	0.2027	10 779 933.8	46.71	3.8004	2.9612	0.3754
1.7163	0.2088	11 104 255.9	49.99	3.8334	2.9461	0.4258
1.7674	0.2150	11 435 126.3	53.59	3.8749	2.9313	0.4844
1.8175	0.2211	11 759 572.9	57.62	3.9394	2.9173	0.5652
1.8697	0.2275	12 097 067.9	62.07	4.0102	2.9032	0.6522
1.9199	0.2336	12 421 653.8	66.46	4.0727	2.8901	0.7299
1.9690	0.2396	12 739 819.5	71.59	4.1706	2.8777	0.8422

Resistance coefficients of model

Model speed v_M through the water, total resistance R_{TM}, and form factor k make up the input for the wave resistance coefficient computation. Following the procedure outlined in Section 29.4, Equation (29.15) yields the wave resistance coefficient by subtracting the viscous model resistance $C_{VM} = (1 + k)C_{FM}$ from the total resistance coefficient C_{TM}. The results are shown in Table 30.4 and Figure 30.4. Note that the wave resistance coefficient C_W of the model is also valid for the full scale vessel, because it is based on a dynamically similar force.

Full scale resistance coefficients

The resistance coefficients for the full scale vessel are derived from Equations (30.1) through (30.6). Again, full scale resistance predictions are done for seawater at temperature 15 °C. The standard seawater properties are included in Table 30.2. First, ship velocities v_S are computed for the Froude numbers of the resistance test. From v_S follow Reynolds numbers and the ITTC 1957 model–ship correlation line coefficient C_{FS} (2.18) for the full scale vessel. Combined with the unchanged form factor k and the roughness allowance ΔC_F, the viscous resistance coefficient C_{VS} of the full scale vessel is built:

$$C_{VS} = (1 + k)C_{FS} + \Delta C_{FS} \tag{30.9}$$

The viscous resistance of the full scale vessel is augmented by the correlation allowance C_A and by the increase in wetted surfaces due to areas not included in the model test. Here, the additional surfaces of the bilge keels lead to the following correction factor:

$$\frac{S_S + S_{BK_S}}{S_S} = \frac{5540.508\ \text{m}^2 + 85.0003\text{m}^2}{5540.508\ \text{m}^2} = 1.0153$$

The total resistance coefficient C_{TS} (30.6) for the full scale vessel is obtained by adding wave resistance coefficient C_W and air resistance coefficient C_{AAS} to the augmented viscous resistance coefficient. The example vessel does not feature additional appendages. Consequently, we set C_{APP_S} to zero. All full scale resistance coefficients are presented

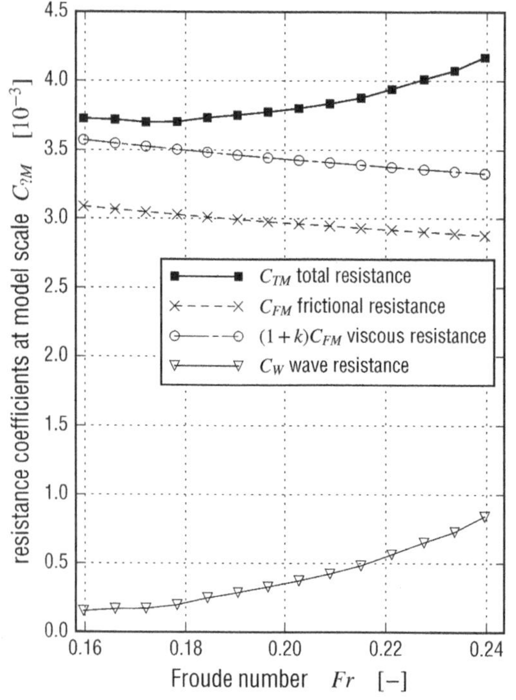

Figure 30.4 Resistance coefficients of model

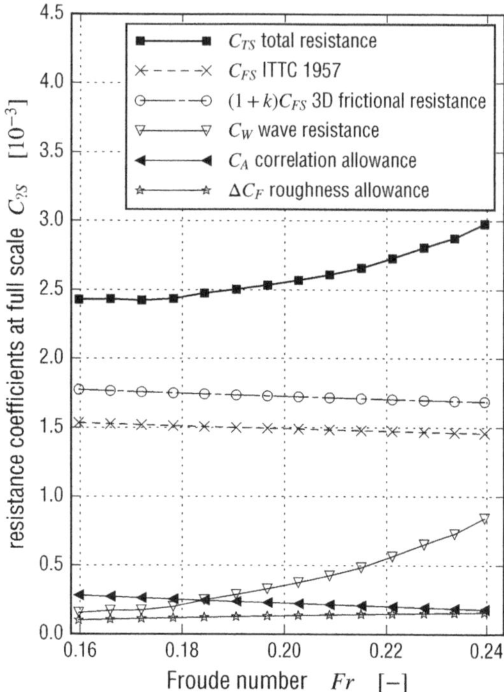

Figure 30.5 Resistance coefficients of full scale ship

Table 30.5 Predicted resistance coefficients for the full scale vessel

Fr [–]	v_S [m/s]	Re_S [–]	C_{FS} [10^{-3}]	$(1+k)C_{FS}$ [10^{-3}]	ΔC_F [10^{-3}]	C_A [10^{-3}]	C_W [10^{-3}]	C_{TS} [10^{-3}]
0.1597	6.562	982 468 812.4	1.5340	1.7742	0.1024	0.2846	0.1552	2.4296
0.1659	6.818	1 020 701 245.3	1.5267	1.7659	0.1080	0.2747	0.1715	2.4330
0.1720	7.068	1 058 189 005.1	1.5199	1.7580	0.1132	0.2653	0.1743	2.4236
0.1782	7.324	1 096 431 860.8	1.5133	1.7503	0.1183	0.2560	0.1996	2.4368
0.1843	7.574	1 133 930 354.9	1.5070	1.7431	0.1231	0.2472	0.2503	2.4761
0.1905	7.829	1 172 184 711.5	1.5009	1.7360	0.1277	0.2386	0.2894	2.5039
0.1966	8.080	1 209 695 047.8	1.4951	1.7293	0.1321	0.2304	0.3304	2.5342
0.2027	8.331	1 247 211 689.6	1.4895	1.7228	0.1363	0.2224	0.3754	2.5688
0.2088	8.581	1 284 734 951.6	1.4841	1.7166	0.1403	0.2147	0.4258	2.6092
0.2150	8.837	1 323 015 844.6	1.4788	1.7104	0.1442	0.2071	0.4844	2.6577
0.2211	9.088	1 360 553 510.5	1.4738	1.7046	0.1479	0.1998	0.5652	2.7290
0.2275	9.348	1 399 600 847.2	1.4687	1.6987	0.1517	0.1924	0.6522	2.8064
0.2336	9.599	1 437 154 639.4	1.4640	1.6933	0.1551	0.1855	0.7299	2.8750
0.2396	9.845	1 473 965 621.8	1.4595	1.6881	0.1584	0.1789	0.8422	2.9787
for all speeds: factor $(S_S + S_{BK_S})/S_S = 1.0153$; air resistance coefficient is $10^3 C_{AAS} = 0.08$								

Table 30.6 Full scale resistance R_{TS} and effective power P_{ES}

Fr [–]	v_S [m/s]	v_S [kn]	C_{TS} [10^{-3}]	R_{TS} [kN]	P_{ES} [kW]
0.1597	6.562	12.76	2.4296	297.38	1951.53
0.1659	6.818	13.25	2.4330	321.43	2191.42
0.1720	7.068	13.74	2.4236	344.14	2432.43
0.1782	7.324	14.24	2.4368	371.48	2720.54
0.1843	7.574	14.72	2.4761	403.74	3057.89
0.1905	7.829	15.22	2.5039	436.27	3415.79
0.1966	8.080	15.71	2.5342	470.27	3799.84
0.2027	8.331	16.19	2.5688	506.71	4221.24
0.2088	8.581	16.68	2.6092	546.11	4686.30
0.2150	8.837	17.18	2.6577	589.92	5213.09
0.2211	9.088	17.67	2.7290	640.60	5821.53
0.2275	9.348	18.17	2.8064	697.12	6517.00
0.2336	9.599	18.66	2.8750	753.00	7228.31
0.2396	9.845	19.14	2.9787	820.64	8079.37

in Table 30.5 and Figure 30.5. Finally, employing Equations (30.7) and (30.8), the dimensional total resistance R_{TS} and effective power P_{ES} can be predicted for the full scale vessel based on the total resistance coefficient C_{TS}. Results are shown in Table 30.6 and Figures 30.6 and 30.7.

References

Holtrop, J. and Mennen, G. (1982). An approximate power prediction method. *International Shipbuilding Progress*, 29(335):166–170.

ITTC (2011). *Fresh water and seawater properties*. International Towing Tank Conference, Recommended Procedures and Guidelines 7.5-02-01-03. Revision 2.

ITTC (2017a). *1978 ITTC performance prediction method*. International Towing Tank Conference, Recommended Procedures and Guidelines 7.5-02-03-01.4. Revision 4.

ITTC (2017b). *Resistance test*. International Towing Tank Conference, Recommended Procedures and Guidelines 7.5-02-02-01. Revision 4.

Schuster, S. (1956). Beitrag zur Frage der Kanalkorrektur bei Modellversuchen. *Schiffstechnik/Ship Technology Research*, 3:93–96.

Self Study Problems

1. Why is a roughness allowance applied to the full scale vessel but not to the model?

2. Explain the purpose of the correlation allowance.

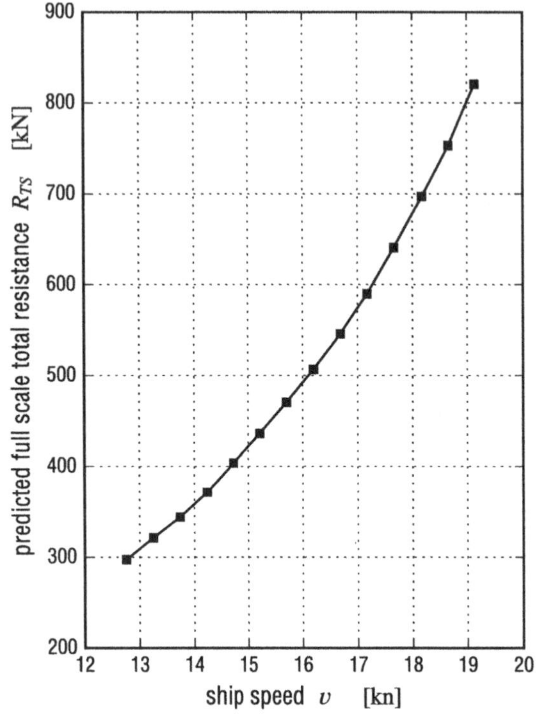

Figure 30.6 Full scale total resistance prediction (calm water)

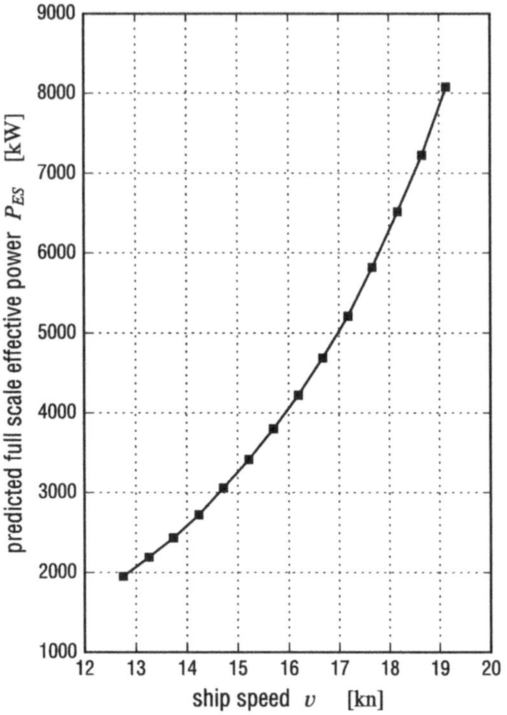

Figure 30.7 Full scale effective power

3. Write a program or create a spread sheet to perform the evaluation of a resistance test using the latest version of the ITTC recommended procedures. Use the ship's geometric data, carriage speed v_C, and measured total resistance R_{TM} of the model as input. Implement Schuster's blockage correction.

4. Add an option for Tamura's blockage correction to your program/spread sheet. How much does the predicted resistance change for the lowest and highest speed of the example in Section 30.4?

31

Resistance Estimates – Guldhammer and Harvald's Method

Data on resistance and powering requirements are required in the ship design process long before a model test can be conducted. Therefore, methods have been developed which allow the naval architect to estimate resistance and powering requirements based on ship speed and a few principal geometric characteristics. Traditionally, data of model tests with systematically varied hull forms and resulting design charts have been used. However, the advent of the personal computer has mostly replaced these with formulas derived from regression analysis of vast sets of model test and trial data. As an example for such a method, this chapter illustrates a resistance estimate which was originally developed by Guldhammer and Harvald (1974) and later amended in collaboration with other researchers (Andersen and Guldhammer, 1986; Kristensen and Lützen, 2012).

Learning Objectives

At the end of this chapter students will be able to:

- understand the basics of resistance estimates
- program and apply Guldhammer and Harvald's method
- study the relationships between main particulars and calm water resistance

31.1 Historical Development

Admiralty coefficient

One of the earliest attempts to compare and predict performance of ships is based on the admiralty coefficient:

$$A_C = \frac{\Delta^{2/3} v_S^3}{P} \tag{31.1}$$

In order to estimate the effective power $P_E = R_T v_S$ for a new ship design, the admiralty coefficient of a *very* similar ship is multiplied by the third root of displacement Δ squared and the design speed v_S cubed, i.e. $P_E = \sqrt[3]{\Delta^2}\, v_S^3\, A_C$.

Fundamentals of Ship Hydrodynamics: Fluid Mechanics, Ship Resistance and Propulsion, First Edition.
Lothar Birk.

Companion website: www.wiley.com/go/birk/hydrodynamics

Unfortunately, admiralty coefficients A_C are dimensional and their values depend on the units used. Harvald (1983) suggests the following formula for preliminary design:

$$A_C = 3.7\left(\sqrt{L} + \frac{75}{v_S}\right) \tag{31.2}$$

Here, length of the ship L is entered in meter and the speed v_S in meter per second. The power is obtained in kilowatt, if the displacement Δ is entered in metric tons, i.e. $1\ \mathrm{t} = 1000\ \mathrm{kg}$. As ship length L the length in waterline L_{WL} is applied for vessels without a bulbous bow, and length over wetted surface L_{OS} is used for vessels with bulbous bow.

Admiralty coefficients capture fundamental relationships between speed, size of the vessel, and the power required to move them. Due to the simplicity of the formula (31.1), power estimates are only good for ships with similar geometry and speed and the results encompass high uncertainty. If at all, they should be used only as a rough first estimate.

Systematic series With Froude's hypothesis and the availability of model testing facilities, naval architects could analyze the resistance of an existing hull form. Obviously, it is desirable to predict the resistance in advance to eliminate impractical hull forms before time consuming model tests are conducted. Early on in the 20th century, model basins developed basic hull designs and then systematically varied important geometric characteristics like the prismatic coefficient C_P, position of the longitudinal center of buoyancy LCB, beam–draft ratio, and more.

Taylor standard series Already in 1906 the Experimental Model Basin, which became later known as the David Taylor Model Basin, developed a parent hull model from the lines of the British cruiser Leviathan. 38 models were derived from this hull and tested to explore the effect of the prismatic coefficient on resistance. Over the following five decades more parent hulls and overall more than 200 models were added to the data base which is referred to as Taylor Standard Series. Like the model basin he founded, the Taylor Standard Series is named after Admiral David W. Taylor (*1864 – †1920), naval architect, engineer, and Chief Constructor of the United States Navy during World War I. With the continuous improvement of model testing procedures, the data have been reanalyzed and published by Gertler (1954) as residuary resistance curves based on the Schoenherr friction line Equation (13.29).

Series 60 In the 1950s, a new series was developed for merchant vessels covering a wide range of block coefficients. It became known as Series 60, and the data have been widely used in ship design (Todd, 1963). Several other systematic series exist, some of which address specific ship types like trawlers and high speed planing boats (Lewis, 1988). All series provide useful insight into the relationships between resistance and key geometric characteristics. However, most of the series are decades old and do not reflect modern hull designs.

Estimates based on regression Some series data are available as regression formulas, but most are only available as charts that do not fit very well into today's computerized engineering work flow. Over the past 40 years several researchers have developed resistance estimates which condensed the results of available model test and ship trial data into relatively simple formula which are easily programmed. Three of these that are applicable to displacement type vessels are discussed in this book. As a first example, the following section presents the method by Guldhammer and Harvald (1974) in its latest form. Methods by Holtrop

Table 31.1 Range of parameters suitable for Guldhammer and Harvald's method

Permissible range of main dimensions			
Parameter	**Symbol**	**Unit**	**Range**
ship speed	v	[m/s]	$Fr < 0.33$
length-beam ratio	L/B	[–]	5.0 … 8.0
block coefficient	C_B	[–]	0.55 … 0.85
length-displacement ratio	$M = L/\sqrt[3]{V}$	[–]	4.0 … 6.0

and Mennen (1982) and Hollenbach (1999) are described in some detail after we have covered ship propulsion and propeller selection.

31.2 Guldhammer and Harvald's Method

Origin and computerization

The method was developed in the late 60s and early 70s when most ships did not yet have a bulbous bow. In later publications, Harvald (1983) and Andersen and Guldhammer (1986) updated the procedure and introduced a *computation length* L which made the method applicable to ships with bulbous bows. Andersen and Guldhammer (1986) provide regression formulas to replace the traditional C_R-charts and extend the method to include powering estimates. The regression formulas are the basis for this section, as they are more applicable to today's computerized work flow than H.E. Guldhammer and S.A. Harvald's original design charts. However, the design charts provide an important tool to perform basic error checks during implementation.

Updates to the method

In addition, more recent updates reported by Kristensen and Lützen (2012) are referenced. The report provides formulas for wetted surfaces of common vessel types and an updated bulbous bow correction for the residuary resistance, which promises to give more accurate results for modern hull forms. Kristensen and Lützen (2012) also include additional regression formulas for the residuary resistance of ships with high prismatic coefficient C_P.

31.2.1 Applicability

Single and twin screw vessels

The method of Guldhammer and Harvald is applicable to single and twin screw displacement type vessels. Based on the regression formulas provided by Andersen and Guldhammer (1986) for the residuary resistance, limits of application for the method are shown in Table 31.1. The original charts allow predictions for velocities up to Froude number $Fr = 0.45$.

31.2.2 Required input

Input

Guldhammer and Harvald's method requires only the hull form parameters listed in Table 31.2. A special computation length L plays a prominent role for modern hull forms.

Table 31.2 Required and optional input parameters for Guldhammer and Harvald's method

Parameter	Symbol	Remarks
	required parameters	
length between perpendiculars	L_{PP}	
length of aft overhang in waterline	L_{aft}	usually $L_{WL} = L_{PP} + L_{\text{aft}}$
extension of S beyond fore perpend.	L_{fore}	
computation length	L	total extension of wetted surface (usually equal to L_{OS})
maximum molded beam in waterline	B	
molded draft	T	
block coefficient	C_B	or volumetric displacement V
prismatic coefficient	C_P	or midship section area A_M
transverse cross section area of bulb	A_{BT}	at forward perpendicular FP
	optional parameters	
propeller diameter	D_P	for propulsion analysis
longitudinal center of buoyancy	LCB	or assume optimum position
wetted surface (hull+rudder)	S	
wetted surface of appendages	S_{APP}	bilge keels, stabilizer fins, bossings, etc.
form factors for fore and aft body	F_F, F_A	$-3 \leq F_A, F_F \leq +3$

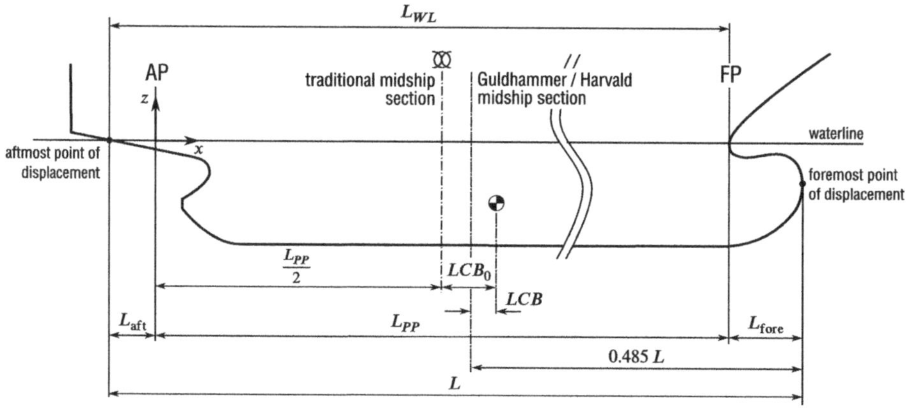

Figure 31.1 Definition of the midship section and the computational length L for Guldhammer and Harvald's resistance estimate (Andersen and Guldhammer, 1986)

Computation length

Before 1970, the longitudinal extension of a submerged hull was usually equal to its length in the waterline L_{WL}. Therefore, the waterline length often served as basis for resistance computations. However, with the introduction of bulbous bows, submerged volume is stretched over a longer length. The increased length has to be considered in the resistance estimate. Andersen and Guldhammer (1986) called it the computation length L (see Figure 31.1). It is used as reference length throughout the method described here.

For most modern ships, the computation length will be equal to the length over wetted surface L_{OS}. It is derived from the length between perpendiculars L_{PP} by adding the length L_{fore} of the additional wetted parts protruding forward of the forward perpendicular FP (mostly for bulbous bows) and the length L_{aft} of wetted surface extending past the aft perpendicular AP.

$$L = L_{PP} + L_{\text{aft}} + L_{\text{fore}} \tag{31.3}$$

L_{aft} is positive, if the submerged hull extends *aft* of the aft perpendicular. L_{fore} is positive, if the submerged hull extends *forward* of the fore perpendicular. In unusual cases, L_{fore} or L_{aft} may be negative.

If the length in waterline or the aft overhang are unknown, Kristensen and Lützen (2012) suggest to use the following relationships:

$$\begin{aligned} L_{WL} &= 1.02\, L_{PP} \quad \text{for tanker and bulk carriers} \\ L_{WL} &= 1.01\, L_{PP} \quad \text{for container ships} \end{aligned} \tag{31.4}$$

Definition of midship location

Guldhammer and Harvald (1974) assumed that the waterline extends about 3% of L_{PP} aft of the aft perpendicular AP. Therefore, midship was located 48.5% of the length in waterline L_{WL} aft of the foremost point of displacement (originally equal to FP). Within the method, this definition of midship location is kept for vessels with bulbous bow. Nowadays, the foremost point of displacement is the tip of the bulb. As a result, the midship location moves forward of the usual position at $L_{PP}/2$ (see Figure 31.1).

Longitudinal center of buoyancy

Consequently, the longitudinal location of the center of buoyancy has to be stated with respect to the midship position as defined by Andersen and Guldhammer (1986).

If the position of the center of buoyancy is given as length LCB_0 with respect to the traditional midship position ($L_{PP}/2$), its location with respect to Guldhammer's midship definition is

$$LCB = LCB_0 - 0.015 L_{PP} + 0.485 L_{\text{aft}} - 0.515 L_{\text{fore}} \tag{31.5}$$

Obviously, all quantities have the dimension of the selected length unit. The equation in Andersen and Guldhammer (1986) corresponding to Equation (31.5) has a misprint: $0.15\, L_{PP}$ instead of $0.015\, L_{PP}$. Note that in contrast to Andersen and Guldhammer (1986), a positive LCB indicates that the center of buoyancy is forward of midships throughout this book.

Wetted surface

In early design stages, the wetted surface S is often not yet known. Andersen and Guldhammer (1986) suggest Mumford's formula (31.6) but it is likely to underpredict the wetted surface of modern ship hulls with bulbous bow.

$$\begin{aligned} S &= 1.025\, L_{PP}(C_B\, B + 1.7\, T) \\ &= 1.025 \left(\frac{V}{T} + 1.7\, L_{PP}\, T \right) \quad \text{Mumford's formula} \end{aligned} \tag{31.6}$$

Kristensen and Lützen (2012) modified the constants in Mumford's formula to improve the wetted surface prediction for modern hull forms. The formulas employ the length in waterline instead of the length between perpendiculars.

$$\begin{aligned} S &= 0.99 \left(\frac{V}{T} + 1.9\, L_{WL}\, T \right) \quad \text{bulk carriers, tankers, vessels with high } C_B \\ S &= 0.995 \left(\frac{V}{T} + 1.9\, L_{WL}\, T \right) \quad \text{container ships (vessels with low to medium } C_B) \end{aligned} \tag{31.7}$$

$$S = 1.53\left(\frac{V}{T} + 0.55\, L_{WL}\, T\right) \quad \text{twin screw ships (Ro-Ro vessels) with open shaft lines}$$
$$S = 1.2\left(\frac{V}{T} + 1.5\, L_{WL}\, T\right) \quad \text{twin skeg ships (Ro-Ro vessels with twin rudders)} \qquad (31.8)$$
$$S = 1.11\left(\frac{V}{T} + 1.7\, L_{WL}\, T\right) \quad \text{double ended ferries}$$

Kristensen and Lützen (2012) also provide two useful equations to estimate the change in wetted surface for changes in draft. Assuming that the wetted surface S_1 at draft T_1 is known, the wetted surface S_2 at draft T_2 is approximately equal to

$$\begin{aligned} S_2 &= S_1 + 2.0(T_2 - T_1)(L_{WL} + B) && \text{tanker, bulk carrier} \\ S_2 &= S_1 + 2.4(T_2 - T_1)(L_{WL} + B) && \text{container ships} \end{aligned} \qquad (31.9)$$

31.2.3 Resistance estimate

Total resistance

Within the resistance estimation method by Guldhammer and Harvald the total resistance coefficient is defined as

$$C_T = C_R + C_F' + C_A + C_{AA} + \left[C_{AS}\right] \qquad (31.10)$$

C_R is the residuary resistance coefficient. The frictional resistance coefficient C_F' is based on the ITTC 1957 model–ship correlation line augmented by the effect of appendage surfaces not included in the wetted surface S. C_A stands for a combined correlation and roughness allowance. As usual, C_{AA} is the air resistance coefficient. The steering resistance coefficient C_{AS} originally included in Equation (31.10) has been dropped by Kristensen and Lützen (2012).

All resistance coefficients have been rendered dimensionless by division with the dynamic pressure $\frac{1}{2}\,\rho\, v^2$ and the wetted surface S of the hull at rest. Therefore, the total resistance is computed as usual:

$$R_T = \frac{1}{2}\rho\, v^2\, S\, C_T \qquad (31.11)$$

Residuary resistance

Residuary resistance

Key to all resistance estimates is the derivation of wave or residuary resistance. The method does not employ a form factor, hence, the residuary resistance coefficient C_R is used. In Guldhammer and Harvald's method, it is a function of Froude number Fr, prismatic coefficient C_P, and the length–displacement ratio (31.12).

$$M = \frac{L}{\sqrt[3]{V}} \qquad (31.12)$$

For consistency, the prismatic coefficient C_P should be computed on the basis of the computation length L.

Residuary resistance charts

Guldhammer and Harvald (1974) provided charts for C_R similar to the ones shown in Figures 31.5 and 31.6. Charts are selected based on the length–displacement ratio M, and values for C_R read from the curve with the correct prismatic coefficient C_P as a function of Froude number Fr.

Residuary resistance formula

In order to accommodate today's computerized work flow, Andersen and Guldhammer (1986) provided regression formulas for C_R which replace the charts. The example charts have been redrawn based on the regression formula (31.13). They represent the original charts fairly well. Although for Froude numbers above 0.3 and higher prismatic coefficients ($C_P > 0.7$), resistance coefficients tend to be larger as found in the original charts, Andersen and Guldhammer (1986) state that the regression formulas for C_R are usable for Froude numbers smaller than $Fr < 0.33$. Consequently, curves for $Fr > 0.33$ are printed in light gray in Figures 31.5 and 31.6. For higher Froude numbers, the original charts should be consulted which provide residuary resistance coefficients up to $Fr = 0.45$ (Guldhammer and Harvald, 1974).

$C_{R_{\mathrm{std}}}$ value for standard form

Andersen and Guldhammer (1986) stated the following regression formula for a standard residuary resistance coefficient $C_{R_{\mathrm{std}}}$:

$$10^3\, C_{R_{\mathrm{std}}} = E + G + H + K \tag{31.13}$$

The four contributions depend on Froude number Fr, length–displacement ratio M, and prismatic coefficient C_P:

$$E = \left(A_0 + 1.5 Fr^{1.8} + A_1 Fr^{N_1}\right)\left[0.98 + \frac{2.5}{(M-2)^4}\right] + (M-5)^4 (Fr-0.1)^4 \tag{31.14}$$

with variables

$$A_0 = 1.35 - 0.23\, M + 0.012\, M^2 \qquad A_1 = 0.0011\, M^{9.1}$$

$$N_1 = 2\, M - 3.7 \qquad M = \frac{L}{\sqrt[3]{V}}$$

$$G = \frac{(7 - 0.09\, M^2)(5\, C_P - 2.5)^2}{\left[600(Fr - 0.315)^2 + 1\right]^{1.5}} \tag{31.15}$$

$$H = \mathrm{e}^{H_1} \quad \text{with } H_1 = 80\left[Fr - (0.04 + 0.59\, C_P) - 0.015(M-5)\right] \tag{31.16}$$

$$K = 180\, Fr^{3.7}\, \mathrm{e}^{(20\, C_P - 16)} \tag{31.17}$$

Standard hull form

The residuary resistance coefficient $C_{R_{\mathrm{std}}}$ is valid only for what was considered a standard hull form in the 1960s:

1. beam–draft ratio of $B/T = 2.5$
2. location of LCB at its optimal position (see later)
3. no bulbous bow
4. no appendages like bossings, struts, etc.

C_R corrections

For ship hulls that deviate from the standard shape, five corrections have been introduced by Guldhammer and Harvald (1974):

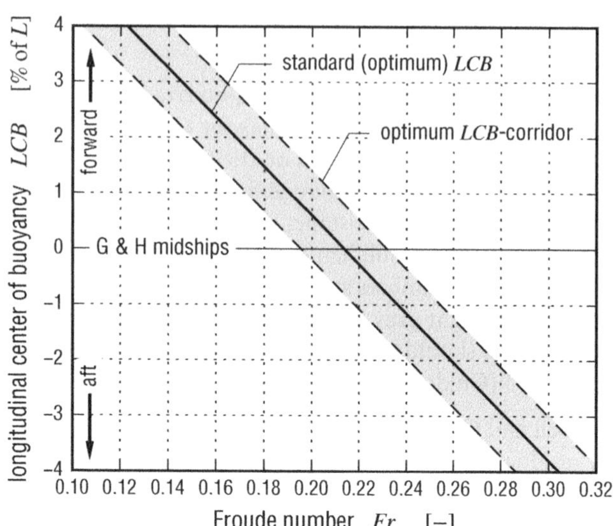

Figure 31.2 Guidance for the optimum location of LCB as a function of Froude number Fr. Here, negative LCB values indicate a location aft of midship

1. $\Delta C_{R(B/T)}$ for beam–draft ratios other than 2.5
2. $\Delta C_{R(LCB)}$ for LCB positions deviating from optimum
3. $\Delta C_{R(\text{form})}$ for section shapes at bow and stern that deviate from normal hull forms
4. $\Delta C_{R(\text{bulb})}$ for bulbous bows
5. $\Delta C_{R(\text{app})}$ for appendages other than the typical single rudder and bilge keels

The adjusted residuary resistance coefficient will be

$$C_R = C_{R_{\text{std}}} + \Delta C_{R(B/T)} + \Delta C_{R(\text{form})} + \Delta C_{R(\text{bulb})} + \Delta C_{R(LCB)} + \Delta C_{R(\text{app})} \quad (31.18)$$

Beam–draft ratio correction For vessels with $B/T \neq 2.5$ a simple speed independent correction is applied.

$$10^3 \Delta C_{R(B/T)} = 0.16\left(\frac{B}{T} - 2.5\right) \quad (31.19)$$

Longitudinal center of buoyancy correction In their original work, Guldhammer and Harvald (1974) provided two charts to estimate a correction for LCB-locations which deviate from the presumed optimum. This correction bears considerable uncertainties, and Kristensen and Lützen (2012) suggest to drop it altogether. For the sake of completeness, we detail the procedure outlined in Andersen and Guldhammer (1986).

Standard LCB Figure 31.2 provides an LCB-location corridor which is considered optimal for a given design speed. A regression formula for the mean line is reported in Andersen and Guldhammer (1986). The formula below has an additional minus sign, reflecting the convention that a positive LCB is *forward* of midships.

$$LCB_{\text{std}}(Fr) = -\left(0.44Fr - 0.094\right) L \quad (31.20)$$

Table 31.3 Bulbous bow corrections to the standard residuary resistance coefficient (Andersen and Guldhammer, 1986)

C_P	Fr	0.15	0.18	0.21	0.24	0.27	0.30	0.33
0.5		+0.2	+0.2	+0.2	0.0	−0.2	−0.4	−0.4
0.6	$\Delta C_{R(\text{table})}$	+0.2	+0.2	+0.2	0.0	−0.2	−0.3	−0.3
0.7		+0.2	+0.2	0.0	−0.2	−0.3	−0.3	−0.3
0.8		+0.1	0.0	−0.2	−0.2	−0.2	−0.2	−0.2

Based on the actual longitudinal center of buoyancy location LCB_{actual} of the ship project, the relative deviation from the standard value for each speed is determined:

$$\Delta LCB(Fr) = \frac{LCB_{\text{actual}} - LCB_{\text{std}}(Fr)}{L} \tag{31.21}$$

The actual LCB_{actual} is of course considered constant. Effects due to sinkage and trim are ignored.

A correction for the residuary resistance coefficient is computed, if, for a given Froude number Fr, the actual LCB is forward of the standard LCB location, i.e. $\Delta LCB > 0$. *LCB-correction*

$$10^3 \Delta C_{R(LCB)} = \begin{cases} 0 & \text{if } Fr < (-C_P^2 + 1.1C_P - 0.0875) \text{ or } \Delta LCB \leq 0 \\ 90\left(\dfrac{Fr}{-C_P^2 + 1.1C_P - 0.0875} - 1\right)\Delta LCB & \text{else} \end{cases} \tag{31.22}$$

Again, this correction has been dropped from the newest version of the Guldhammer and Harvald method (Kristensen and Lützen, 2012).

A standard hull form has neither pronounced V- nor pronounced U-shaped stations in entrance and run of the hull. A raked stem without a bulbous bow and a moderate cruiser stern are considered normal. Modern bulk carriers and tankers with high block coefficient often feature extreme U-shaped stations in the fore body (entrance). Fast container ships often have entrances with pronounced V-shaped stations. Guldhammer and Harvald (1974) assigned fore and aft body shape factors F_F and F_A between −3 and +3, respectively. The value −3 is for extreme V-shaped stations, 0 for normal stations, and +3 for extreme U-shaped stations. **Shape factors**

Once values have been assigned to F_F and F_A, a correction to residuary resistance coefficient is computed from **Form correction**

$$10^3 \Delta C_{R(\text{form})} = -\frac{0.1}{3}\left(F_F - F_A\right) \tag{31.23}$$

Clearly, U-shaped stations in the fore body are deemed beneficial whereas V-shaped stations increase resistance. Vice-versa for the aft body.

The influence of a bulbous bow on residuary resistance is derived by interpolation in Fr- and C_P-directions between the values provided in Table 31.3 (Andersen and Guldhammer, 1986). **Bulbous bow correction**

Original bulbous bow correction

The full correction is applied, if the transverse cross section area of the bulb at the forward perpendicular A_{BT} is larger than 10% of the midship section area A_M. If $A_{BT} < 0.1A_M$, the correction is linearly scaled down to zero.

$$10^3 \Delta C_{R(\text{bulb})} = \begin{cases} 10^3 \Delta C_{R(\text{table})} & \text{from Table 31.3 if } A_{BT} \geq 0.1A_M \\ 10^3 \Delta C_{R(\text{table})} \dfrac{10A_{BT}}{A_M} & \text{if } A_{BT} < 0.1A_M \end{cases} \tag{31.24}$$

New bulbous bow correction

Kristensen and Lützen (2012) analyzed model tests of 27 newer tanker and bulk carrier hull forms and 21 newer hull forms of vessels with lower block coefficients. As indicated by Andersen and Guldhammer (1986), they concluded that the original bulbous bow correction is too pessimistic. Experience in the design of bulbs and optimization of their shape has made them more effective. The effectiveness of a bulbous bow depends on different factors and its exact shape should be determined by shape optimization later in the design process. Design guidance can be found in Kracht (1978).

To reflect these improvements, Kristensen and Lützen (2012) proposed new corrections making a distinction between vessels with high ($C_B \approx 0.8$) and low block coefficients ($C_B \approx 0.6$).

$$10^3 \Delta C'_{R(\text{bulb})} = \begin{cases} \max\left(-0.4, -(0.1 + 1.6Fr)\right) & \text{for tankers and bulk carriers (high } C_B) \\ \dfrac{(250Fr - 90)}{100} 10^3 C_{R(\text{no bulb})} & \text{for container ships (vessels with low } C_B) \end{cases} \tag{31.25}$$

For vessels with high block coefficient (and $M \in [4.4, 5.2]$ and $C_P \in [0.78, 0.87]$), the correction is mostly speed dependent. More slender vessels have larger variations of the length–displacement ratio M and prismatic coefficient C_P. Therefore, the correction is expressed as a fraction of the residuary resistance coefficient without a bulb.

$$10^3 C_{R(\text{no bulb})} = 10^3 \left(C_{R_{\text{std}}} + \Delta C_{R(B/T)} + \Delta C_{R(LCB)} + \Delta C_{R(\text{form})}\right) \tag{31.26}$$

Although not specifically mentioned, it seems safe to assume that the new correction values (31.25) have to be scaled for bulbous bows with A_{BT} less than 10% of A_M.

$$10^3 \Delta C_{R(\text{bulb})} = \begin{cases} 10^3 \Delta C'_{R(\text{bulb})} & \text{from Equation (31.25) if } A_{BT} \geq 0.1A_M \\ 10^3 \Delta C'_{R(\text{bulb})} \dfrac{10A_{BT}}{A_M} & \text{if } A_{BT} < 0.1A_M \end{cases} \tag{31.27}$$

Once all adjustments have been evaluated, the final residuary resistance coefficient can be computed from Equation (31.18).

Appendage correction for C_R

Guldhammer and Harvald (1974) also include a correction for appendages. Effects of single rudder and bilge keels are included in the standard C_R value. However, if the vessel features bossings, exposed shafts, and shaft brackets they may affect wave and viscous pressure resistance which warrants additional corrections to the residuary resistance coefficient.

- bossings: add 3%–5% of $C_{R_{\text{std}}}$ for full ship forms
- shafts and brackets: add 5%–8% of $C_{R_{\text{std}}}$ for fine hull forms

Andersen and Guldhammer (1986) caution that the corrections should be small compared to the overall C_R value.

Frictional resistance

Frictional resistance coefficient

The frictional resistance coefficient is computed with the ITTC 1957 model–ship correlation line.

$$C_F = \frac{0.075}{\left[\log_{10} Re - 2\right]^2} \tag{2.18}$$

Again, a form factor is not used in Guldhammer and Harvald's method.

In case the wetted surface S does not include the surface of appendages, the frictional coefficient should be augmented by the ratio of total wetted surface $S + S_{APP}$ and wetted surface of the hull S.

$$C_F' = \frac{S + S_{APP}}{S} C_F \tag{31.28}$$

Additional resistance components

Incremental resistance coefficient

The incremental resistance coefficient represents a combined correlation and roughness allowance C_A. Guldhammer and Harvald (1974, (page 5)) stated only a few discrete values depending on vessel length. In their update to the procedure, Andersen and Guldhammer (1986) provided the formula (31.29).

$$10^3\, C_A = 0.5 \log_{10}(V) - 0.1\left(\log_{10}(V)\right)^2 \tag{31.29}$$

It considers that the roughness effect tends to decrease with increasing ship length and size. The input is unfortunately dimensional, and the displacement V must be entered in cubic meter [m^3]. This equation should potentially be replaced with the roughness allowance ΔC_F from the ITTC 1978 performance prediction procedure (see Equation (30.2)).

Air and steering resistance

Air resistance coefficient C_{AA} and steering resistance coefficient C_{AS} are assumed to be constant in Guldhammer and Harvald (1974) and Andersen and Guldhammer (1986). Suggested values are

$$10^3\, C_{AA} = 0.07 \tag{31.30}$$

$$10^3\, C_{AS} = 0.04 \tag{31.31}$$

The air resistance represents the resistance caused by moving the superstructure of the vessel through air at rest. Resistance in high winds must be considered separately for the selection of a service margin and maneuvering assessment. Steering resistance may occur in vessels which need constant rudder action to counter the wheel effect of a single propeller.

Table 31.4 Air resistance coefficients for different types of vessels (Kristensen and Lützen, 2012)

Vessel type	$10^3 C_{AA}$
small tanker	0.07
Handysize tanker	0.07
Handymax tanker	0.07
Panamamax tanker	0.05
Aframax tanker	0.05
Suezmax tanker	0.05
very large crude carrier (VLCC)	0.04
container ship	see Equation (31.32)

Table 31.5 Principal dimensions for resistance estimate example

1000 TEU container ship		
length between perpendiculars	L_{PP}	145.0 m
extension of bulb forward of FP	L_{fore}	3.3 m
extension of waterline aft of AP	L_{aft}	2.7 m
beam	B	24.0 m
draft	T	8.2 m
displacement	V	18 872.0 m^3
estimated wetted surface	S	4400.0 m^2
bulbous bow cross section area at FP	A_{BT}	14.0 m^2
wetted area of appendages	S_{APP}	52.0 m^2
prismatic coefficient (based on L_{PP})	C_P	0.6783
location of (with respect to $L_{PP}/2$)	LCB_0	+0.4 % of L_{PP}
design speed	v_S	17.5 kn
standard shaped fore and aft body		

Kristensen and Lützen (2012) analyzed data for different types of vessels and recommend the constants in Table 31.4. For container ships they developed a simple formula based on the container carrying capacity of the vessel.

$$10^3\, C_{AA} = \max\left((0.28\,\text{TEU}^{-0.126}), 0.09\right) \tag{31.32}$$

TEU stands for Twenty-foot Equivalent Unit, i.e. a 20 ft long container. The steering resistance is no longer used in their update to the Guldhammer and Harvald method.

Andersen and Guldhammer (1986) complemented the resistance estimate with equations for the hull–propeller interaction parameters. We will discuss these in Chapter 34.

31.3 Extended Resistance Estimate Example

Example input data

In this section, the Guldhammer and Harvald method in its latest form is illustrated with an example. Table 31.5 summarizes data selected for the preliminary design of a 1000 TEU container ship. The task at hand is to determine total resistance R_T and effective power P_E as a function of ship speed v.

31.3.1 Completion of input parameters

As a first step we have to complete our set of input data. To perform the resistance estimate we have to know

- computation length L,
- length–displacement ratio $M = L/V^{1/3}$,
- midship section area A_M,
- prismatic coefficient based on computation length, and
- longitudinal center of buoyancy position in the reference frame defined by Guldhammer and Harvald

Computation length

The computation length is equal to the total length of the wetted surface $L = L_{OS}$. With Equation (31.3) and the values from Table 31.5 we get

$$L = L_{PP} + L_{\text{aft}} + L_{\text{fore}} = 145.0\,\text{m} + 2.7\,\text{m} + 3.3\,\text{m} = 151.0\,\text{m}$$

Length–displacement ratio

The length–displacement ratio M for this vessel is

$$M = \frac{L}{V^{1/3}} = \frac{151.0\,\text{m}}{\left(18\,872.0\,\text{m}^3\right)^{1/3}} = 5.6716$$

Midship section area

From displacement and given prismatic coefficient C_P (based on L_{PP}) follows the midship section area.

$$A_M = \frac{V}{L_{PP}\,C_P} = \frac{18\,872.0\,\text{m}^3}{145.0\,\text{m} \cdot 0.6783} = 191.88\,\text{m}^2$$

Prismatic coefficient

For the assessment of the residuary resistance coefficient and its corrections, the prismatic coefficient has to be stated on the basis of the computational length L.

$$C_P = \frac{V}{L\,A_M} = 0.6513 \qquad \text{based on computational length } L$$

***LCB* position**

Redefinition of the longitudinal center of buoyancy position is required only if the LCB-correction (31.22) is employed. Although it is not included in the new version of the method, an example is provided to illustrate the confusing procedure.

As explained by Andersen and Guldhammer (1986), midships is considered 48.5% of L aft of the foremost point of displacement (usually the tip of bulbous bow). Based on Equation (31.5) the LCB position with respect to the redefined midship section is:

$$\begin{aligned} LCB &= LCB_0 - 0.015L_{PP} + 0.485L_{\text{aft}} - 0.515L_{\text{fore}} \\ &= 0.004 \cdot 145\,\text{m} - 0.015 \cdot 145\,\text{m} + 0.485 \cdot 2.7\,\text{m} - 0.515 \cdot 3.3\,\text{m} \\ &= -1.985\,\text{m} \\ &= -1.315\,\%\text{ of } L \text{ aft of G+H midship} \end{aligned}$$

Table 31.6 Selected ship speeds and resulting Froude and Reynolds number

v	v	$Fr = \frac{v}{\sqrt{g\,L}}$	$Re = \frac{v\,L}{\nu}$
[kn]	[m/s]	[–]	[–]
15.00	7.717	0.2005	979832380
15.50	7.974	0.2072	1012493460
16.00	8.231	0.2139	1045154539
16.50	8.488	0.2205	1077815618
17.00	8.746	0.2272	1110476698
17.50	9.003	0.2339	1143137777
18.00	9.260	0.2406	1175798856
18.50	9.517	0.2473	1208459936
19.00	9.774	0.2540	1241121015

31.3.2 Range of speeds

The service speed of our vessel is $v_S = 17.5$ kn. In our example we will perform the estimates for speeds in the range of $v = 15$ kn to 19 kn. In a full application, estimates would be done for more speeds and different loading conditions if applicable.

Table 31.6 summarizes ship speed in knots and meter per second and the corresponding Froude and Reynolds numbers. We made use of the gravitational acceleration $g = 9.807\ \mathrm{m/s^2}$ in SI-units. Density and kinematic viscosity for salt water at 15 °C are taken from the ITTC Recommended Procedure 7.5-02-01-03 (ITTC, 2011). Density is $\rho = 1026.021\ \mathrm{kg/m^3}$ and kinematic viscosity is $\nu = 1.1892 \cdot 10^{-6}\ \mathrm{m^2/s}$. Note that, in contrast to ITTC recommendation, the method derives Froude number Fr and Reynolds number Re based on the computation length L rather than length in waterline and length over wetted surface.

31.3.3 Residuary resistance coefficient

The range of speeds is well within the limit of $Fr < 0.33$. There is no chart of our length-displacement ratio of $M = 5.6716$. If charts were to be used, a linear interpolation is performed between the standard $C_{R_{\mathrm{std}}}$ values read from the charts with the next lower and the next higher length–displacement ratio.

$$10^3\,C_{R_{\mathrm{std}}}\left(\tfrac{L}{\nabla^{1/3}}{=}5.6716\right) = 10^3\,C_{R_{\mathrm{std}}}\left(\tfrac{L}{\nabla^{1/3}}{=}5.5\right) + \frac{(5.6716-5.5)}{(6.0-5.5)}\cdot\left[10^3\,C_{R_{\mathrm{std}}}\left(\tfrac{L}{\nabla^{1/3}}{=}6.0\right) - 10^3\,C_{R_{\mathrm{std}}}\left(\tfrac{L}{\nabla^{1/3}}{=}5.5\right)\right] \quad (31.33)$$

$C_{R_{\mathrm{std}}}$ values The $C_{R_{\mathrm{std}}}$ values presented in Table 31.7 have been derived with the regression formula (31.13). The contribution H (31.16) is negligibly small in this case (of order 10^{-8}).

The last column in Table 31.7 lists the residuary resistance coefficients for the presumed standard hull form. Now, the somewhat tedious process of adjusting the $C_{R_{\mathrm{std}}}$ values for the actual hull shape begins.

Table 31.7 Residuary resistance value computation for the standard hull form

v [kn]	Fr [–]	E [–]	G [–]	H [–]	K [–]	$C_{R_{std}}$ [10^{-3}]
15.00	0.2005	0.548147	0.089096	0.000000	0.024105	0.6613
15.50	0.2072	0.563610	0.104452	0.000000	0.027215	0.6953
16.00	0.2139	0.581698	0.123384	0.000000	0.030607	0.7357
16.50	0.2205	0.602898	0.146934	0.000000	0.034298	0.7841
17.00	0.2272	0.627774	0.176503	0.000000	0.038303	0.8426
17.50	0.2339	0.656977	0.213986	0.000000	0.042640	0.9136
18.00	0.2406	0.691249	0.261950	0.000000	0.047324	1.0005
18.50	0.2473	0.731443	0.323880	0.000000	0.052374	1.1077
19.00	0.2540	0.778524	0.404461	0.000000	0.057805	1.2408

1. **B/T–correction:** The $C_{R_{std}}$ values are only valid for a beam–draft ratio of $B/T = 2.5$. Our ship project has a shallower hull with *B/T–correction*

$$\frac{B}{T} = \frac{24.0\,\text{m}}{8.2\,\text{m}} = 2.9268. \tag{31.34}$$

With Equation (31.19), the speed independent correction becomes

$$10^3 \Delta C_{R(B/T)} = 0.16\left(\frac{B}{T} - 2.5\right) = 0.16\,(2.9268 - 2.5) = 0.0683 \tag{31.35}$$

2. **LCB–correction:** If an LCB-correction is computed at all, the LCB-location has to be expressed in the method's reference frame first. We completed this above and found that LCB is -1.315% of the computation length L aft of midships (48.5% of L aft or foremost point of displacement). *LCB–correction*

   Computation of the correction requires monitoring both factors in Equation (31.22)

$$\Delta LCB = \frac{LCB_{\text{actual}} - LCB_{\text{std}}(Fr)}{L} \overset{!}{>} 0$$

$$C_1 = \left(\frac{Fr}{-C_P^2 + 1.1C_P - 0.0875} - 1\right) \overset{!}{>} 0$$

   The correction is zero if either or both values are negative for a given Froude number. ΔLCB is positive only if the actual LCB-position is aft of the standard position. The optimum LCB-position moves aft with increasing Froude number. Therefore, corrections are likely to occur only for higher vessel speeds. In our example, a positive correction is applied for the two highest velocities 18.5 kn and 19.0 kn.

   Table 31.8 presents the values for the LCB-correction of C_R for the example at hand.

3. **Hull form correction:** As stated in Table 31.5, we assume standard section shapes ($F_A = F_F = 0$). As a consequence, there is no contribution from this adjustment. *Hull form correction*

$$10^3 C_{R(\text{form})} = 0$$

Table 31.8 Computation of the LCB-correction for the residuary resistance coefficient

v [kn]	Fr [–]	LCB_{std} [% of L]	LCB_{actual} [% of L]	ΔLCB [% of L]	C_1 [–]	$\Delta C_{R(LCB)}$ $[10^{-3}]$
15.00	0.2005	0.5768	−1.3146	−1.8914	−0.0205	0.0000
15.50	0.2072	0.2827	−1.3146	−1.5973	0.0121	0.0000
16.00	0.2139	−0.0114	−1.3146	−1.3032	0.0448	0.0000
16.50	0.2206	−0.3055	−1.3146	−1.0091	0.0774	0.0000
17.00	0.2273	−0.5996	−1.3146	−0.7149	0.1101	0.0000
17.50	0.2339	−0.8937	−1.3146	−0.4208	0.1427	0.0000
18.00	0.2406	−1.1878	−1.3146	−0.1267	0.1754	0.0000
18.50	0.2473	−1.4819	−1.3146	0.1674	0.2080	0.0313
19.00	0.2540	−1.7760	−1.3146	0.4615	0.2407	0.1000

Table 31.9 Comparison of old and updated bulbous bow correction to the residuary resistance coefficient

		old method interpolated, Table 31.3		updated method Equations (31.25), (31.27)	
v [kn]	Fr [–]	unscaled $[10^{-3}]$	$\Delta C_{R(\text{bulb})}$ $[10^{-3}]$	unscaled $[10^{-3}]$	$\Delta C_{R(\text{bulb})}$ $[10^{-3}]$
15.00	0.2005	0.1297	0.0947	−0.2909	−0.2122
15.50	0.2072	0.1069	0.0780	−0.2917	−0.2128
16.00	0.2139	0.0713	0.0521	−0.2937	−0.2143
16.50	0.2206	0.0268	0.0195	−0.2971	−0.2168
17.00	0.2273	−0.0178	−0.0130	−0.3023	−0.2205
17.50	0.2339	−0.0623	−0.0455	−0.3094	−0.2258
18.00	0.2406	−0.1058	−0.0772	−0.3190	−0.2327
18.50	0.2473	−0.1389	−0.1014	−0.3401	−0.2482
19.00	0.2540	−0.1721	−0.1255	−0.3734	−0.2724

Bulbous bow correction

4. **Bulbous bow correction:** The necessary corrections $\Delta 10^3 C_{R(\text{bulb})}$ can be found by interpolating between the values given in Table 31.3 (original method) or by employing Equation (31.25) (updated method, Kristensen and Lützen, 2012). Table 31.9 compares the results for both methods. The values of the updated correction are employed for this example.

   The bulbous bow has a cross section area of less than 10% of the midship cross section area. Therefore, the values derived from Equation (31.25) are scaled with factor $10\,A_{BT}/A_M = 0.7296$.

Appendage correction

5. **Appendage correction:** Effects of standard appendages like rudder and bilge keels are already included in the C_R values above. We have a single screw vessel which usually does not feature bossings, brackets, or exposed shafts. Thus, a correction considering appendages is not necessary here.

$$10^3 \Delta C_{R(\text{app})} = 0$$

Table 31.10 Estimate of residual resistance coefficient

v [kn]	Fr [–]	$C_{R_{std}}$ [10^{-3}]	$\Delta C_{R(B/T)}$ [10^{-3}]	$\Delta C_{R(LCB)}$ [10^{-3}]	$\Delta C_{R(\text{form})}$ [10^{-3}]	$\Delta C_{R(\text{bulb})}$ [10^{-3}]	C_R [10^{-3}]
15.00	0.2005	0.6613	0.0683	0.0000	0.0000	−0.2122	0.5174
15.50	0.2072	0.6953	0.0683	0.0000	0.0000	−0.2128	0.5508
16.00	0.2139	0.7357	0.0683	0.0000	0.0000	−0.2143	0.5897
16.50	0.2206	0.7841	0.0683	0.0000	0.0000	−0.2168	0.6356
17.00	0.2273	0.8426	0.0683	0.0000	0.0000	−0.2205	0.6903
17.50	0.2339	0.9136	0.0683	0.0000	0.0000	−0.2258	0.7561
18.00	0.2406	1.0005	0.0683	0.0000	0.0000	−0.2327	0.8361
18.50	0.2473	1.1077	0.0683	0.0313	0.0000	−0.2482	0.9592
19.00	0.2540	1.2408	0.0683	0.1000	0.0000	−0.2724	1.1366

Table 31.11 Frictional resistance estimate

v [kn]	Fr [–]	v [m/s]	Re [–]	C_F [10^{-3}]	C'_F [10^{-3}]
15.00	0.2005	7.717	979832380	1.5345	1.5526
15.50	0.2072	7.974	1012493460	1.5283	1.5463
16.00	0.2139	8.231	1045154539	1.5223	1.5402
16.50	0.2205	8.488	1077815618	1.5165	1.5344
17.00	0.2272	8.746	1110476698	1.5109	1.5288
17.50	0.2339	9.003	1143137777	1.5055	1.5233
18.00	0.2406	9.260	1175798856	1.5003	1.5180
18.50	0.2473	9.517	1208459936	1.4953	1.5129
19.00	0.2540	9.774	1241121015	1.4904	1.5080

Corrected residuary resistance

Pulling together the standard value and its corrections, we obtain the residuary resistance coefficient C_R for our vessel as listed in Table 31.10.

31.3.4 Frictional resistance coefficient

Based on the ITTC 1957 model–ship correlation line (2.18) and the scaling factor

$$\frac{S + S_{APP}}{S} = \frac{4400\,\text{m}^2 + 52\,\text{m}^2}{4400\,\text{m}^2} = 1.01182$$

we obtain the friction coefficients stated in Table 31.11.

31.3.5 Additional resistance coefficients

Additional resistance contributions

Minor contributions to the total resistance result from the correlation allowance, Equation (31.29), and the air resistance C_{AA}. The displacement has to be entered in cubic meter.

$$\begin{aligned} 10^3\,C_A &= 0.5\,\log_{10}(V) - 0.1\big(\log_{10}(V)\big)^2 \\ &= 0.5\,\log_{10}(18\,872) - 0.1\big(\log_{10}(18\,872.0)\big)^2 = 0.3096 \end{aligned}$$

Table 31.12 Resistance coefficients computed with Guldhammer and Harvald's method using input from Table 31.5

v [kn]	v [m/s]	Fr [–]	C'_F [10^{-3}]	C_R [10^{-3}]	C_A [10^{-3}]	C_{AA} [10^{-3}]	C_{AS} [10^{-3}]	C_T [10^{-3}]
15.00	7.717	0.2005	1.5526	0.5174	0.3096	0.1173	0.0000	2.4969
15.50	7.974	0.2072	1.5463	0.5508	0.3096	0.1173	0.0000	2.5240
16.00	8.231	0.2139	1.5402	0.5897	0.3096	0.1173	0.0000	2.5569
16.50	8.488	0.2206	1.5344	0.6356	0.3096	0.1173	0.0000	2.5970
17.00	8.746	0.2273	1.5288	0.6903	0.3096	0.1173	0.0000	2.6460
17.50	9.003	0.2339	1.5233	0.7561	0.3096	0.1173	0.0000	2.7064
18.00	9.260	0.2406	1.5180	0.8361	0.3096	0.1173	0.0000	2.7811
18.50	9.517	0.2473	1.5129	0.9592	0.3096	0.1173	0.0000	2.8990
19.00	9.774	0.2540	1.5080	1.1366	0.3096	0.1173	0.0000	3.0715

Kristensen and Lützen (2012) suggest using Equation (31.32) for the air drag coefficient C_{AA} of container ships.

$$\begin{aligned} 10^3\,C_{AA} &= \max\left((0.28\mathrm{TEU}^{-0.126}), 0.09\right) = \max\left((0.28 \cdot 1000^{-0.126}), 0.09\right) \\ &= \max\left(0.1173, 0.09\right) = 0.1173 \end{aligned}$$

The steering resistance is zero $C_{AS} = 0$ in this example.

31.3.6 Total resistance coefficient

Resistance estimate

Finally, we assemble all resistance parts into the total resistance coefficient C_T according to Equation (31.10). Results are collected in Table 31.12 and shown in Figure 31.3.

31.3.7 Total resistance and effective power

Required total resistance R_T and the effective power P_E at trial conditions (new hull, calm water) follow from

$$R_T = \frac{1}{2}\rho\, v^2\, S\, C_T \tag{31.36}$$

$$P_E = R_T\, v \tag{31.37}$$

Results are presented in Table 31.13 and Figure 31.4.

References

Andersen, P. and Guldhammer, H. (1986). A computer-oriented power prediction procedure. In *Proc. of Int. Conf. on Computer Aided Design, Manufacture, and Operation in the Marine and Offshore Industries (CADMO '86)*, Washington, DC, USA.

Table 31.13 Total resistance and effective power computed with Guldhammer and Harvald's method using input from Table 31.5

v [kn]	Fr [–]	v [m/s]	C_T [10^{-3}]	R_T [kN]	P_E [kW]
15.00	0.2005	7.717	2.4969	335.62	2589.85
15.50	0.2072	7.974	2.5240	362.25	2888.54
16.00	0.2139	8.231	2.5569	391.03	3218.58
16.50	0.2206	8.488	2.5970	422.36	3585.17
17.00	0.2273	8.746	2.6460	456.82	3995.13
17.50	0.2339	9.003	2.7064	495.13	4457.51
18.00	0.2406	9.260	2.7811	538.28	4984.49
18.50	0.2473	9.517	2.8990	592.72	5641.07
19.00	0.2540	9.774	3.0715	662.39	6474.54

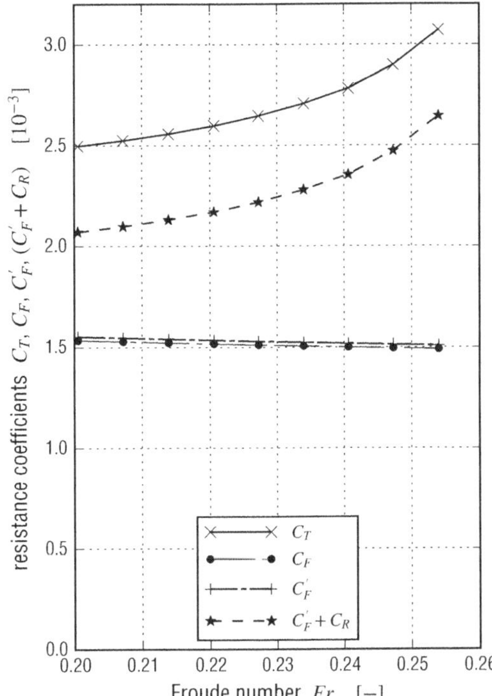

Figure 31.3 Resistance coefficients for the Guldhammer and Harvald method example

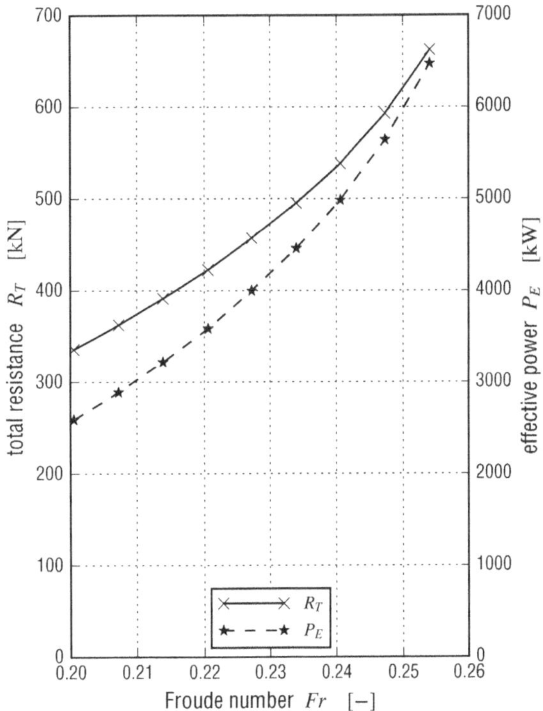

Figure 31.4 Total resistance and effective power for the Guldhammer and Harvald method example

Gertler, M. (1954). A reanalysis of the original test data for the Taylor Standard Series. Report 806, David Taylor Model Basin, Washington, DC.

Guldhammer, H. and Harvald, S. (1974). *Ship resistance – effect of form and principal dimensions (revised).* Akademisk Forlag, Copenhagen.

Harvald, S. (1983). *Resistance and propulsion of ships.* John Wiley & Sons, New York, NY.

Hollenbach, K. (1999). Estimating resistance and propulsion for single-screw and twin-screw ships in the preliminary design. In *Proc. of 10th Int. Conference on Computer Applications in Shipbuilding (ICCAS '99).*

Holtrop, J. and Mennen, G. (1982). An approximate power prediction method. *International Shipbuilding Progress,* 29(335):166–170.

ITTC (2011). *Fresh water and seawater properties.* International Towing Tank Conference, Recommended Procedures and Guidelines 7.5-02-01-03. Revision 2.

Kracht, A. (1978). Design of bulbous bows. *SNAME Transactions,* 86:197–217.

Kristensen, H. and Lützen, M. (2012). Prediction of resistance and propulsion power of ships. Technical Report Project No. 2010-56, Work Package 2, Report No. 04, University of Southern Denmark and Technical University of Denmark.

Lewis, E., editor (1988). *Principles of naval architecture,* volume II – Resistance, propulsion and vibration. The Society of Naval Architects and Marine Engineers, Jersey City, NJ, second edition.

Todd, F. (1963). SERIES 60 – Methodical experiments with models of single-screw merchant ships. Technical Report Techn. Report 1712, David Taylor Model Basin.

Self Study Problems

1. Implement Guldhammer and Harvald's method for resistance estimates and compute the total resistance for the PANMAX container ship specified below. Check the standard C_R-value against Figures 31.5 and 31.6. The total resistance at design speed should be $R_T = 1114.96$ kN and $R_T = 1532.00$ kN at 24 kn if a correction for LCB is included.

Principal dimensions		
length between perpendiculars	L_{PP}	201.1 m
length of bulb forward of FP	L_{fore}	5.6 m
length of waterline aft of AP	L_{aft}	4.92 m
beam	B	32.2 m
draft	T	11.3 m
displacement	V	43171.63 m^3
container capacity	TEU	2800
block coefficient (based on L_{PP})	C_B	0.5900
prismatic coefficient (based on L_{PP})	C_P	0.6244
longitudinal center of buoyancy	LCB	0.4 % of L_{PP} aft of $L_{PP}/2$
wetted surface	S	7316.81 m^2
wetted surface of bilge keels	S_{APP}	120.00 m^2
bulbous bow cross section area at FP	A_{BT}	31.0 m^2
ship design speed	v	22.0 kn
ship speeds to consider	v	from 15 kn to 24 kn

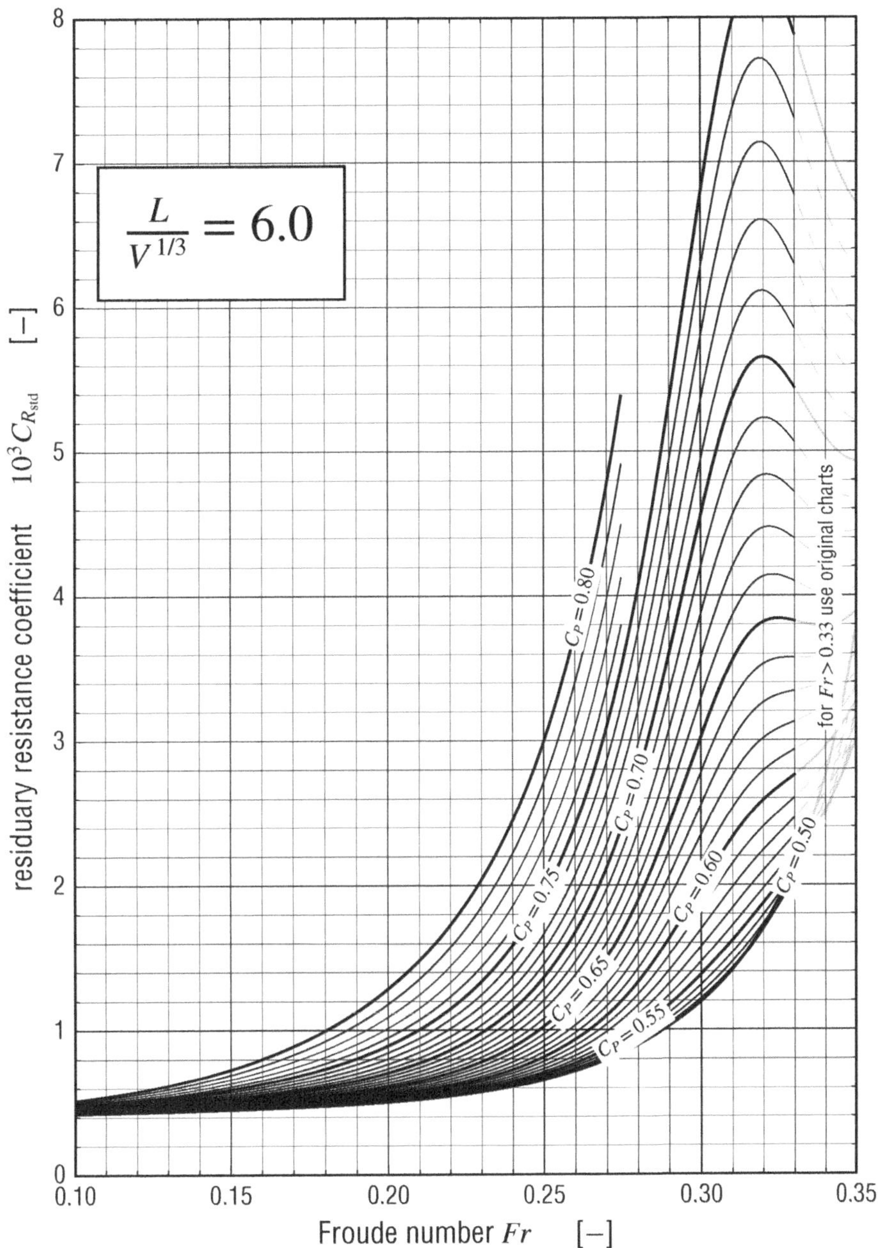

Figure 31.5 Charts for standard residuary resistance coefficients $C_{R_{std}}$ after Guldhammer and Harvald (1974) for vessels with length-speed ratio $L/V^{1/3} = 6.0$. The values have been computed and redrawn based on the regression formula provided by Andersen and Guldhammer (1986)

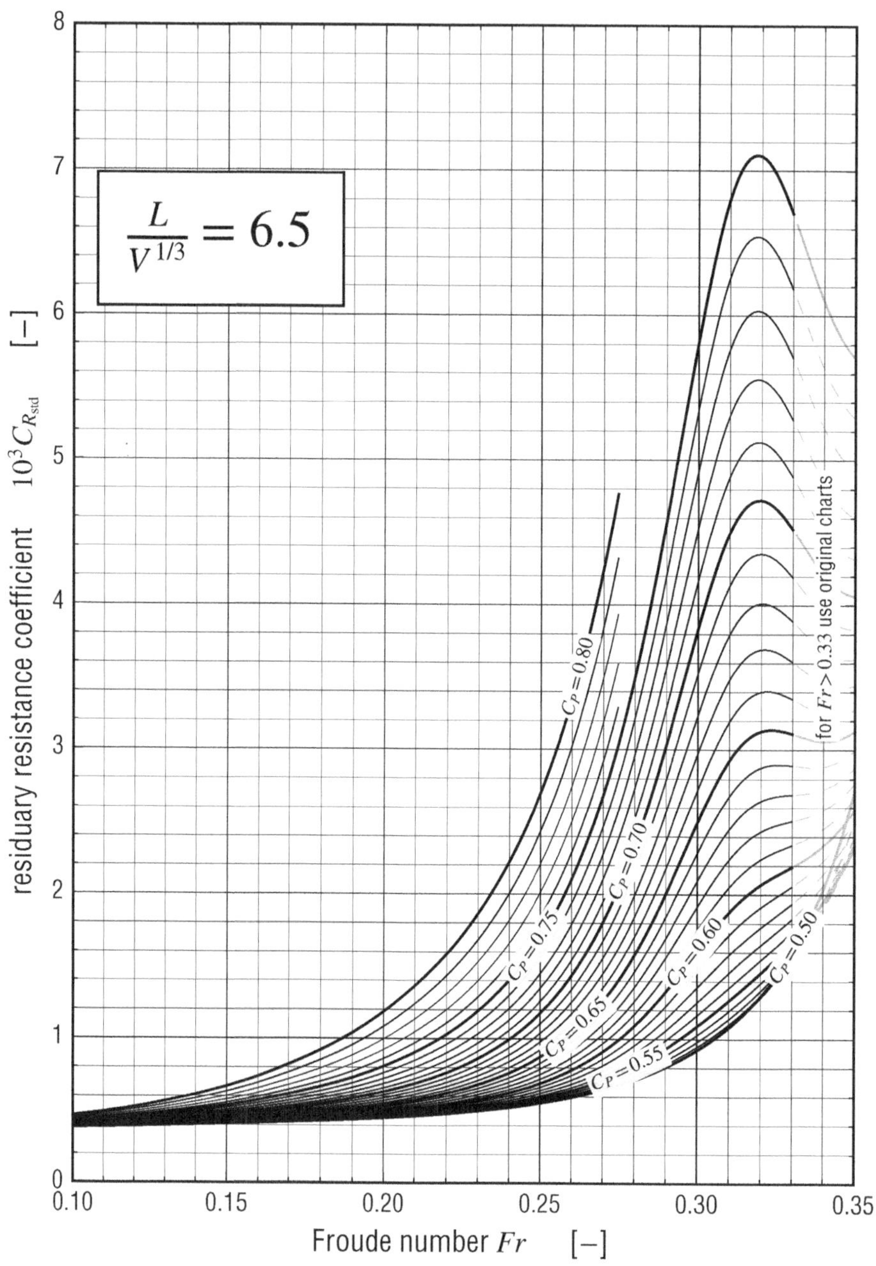

Figure 31.6 Charts for standard residuary resistance coefficients $C_{R_{std}}$ after Guldhammer and Harvald (1974) for vessels with length-speed ratio $L/V^{1/3} = 6.5$. The values have been computed and redrawn based on the regression formula provided by Andersen and Guldhammer (1986)

32

Introduction to Ship Propulsion

The discussion of ship hydrodynamics in this book started out with separating the ship–propulsor system into ship hull and the propulsor itself. We discussed the flow around the ship hull and the resultant resistance. Obviously, a propulsion force is needed that acts in the opposite direction of the resistance in order to maintain a constant speed.

This section provides a general overview on how these systems work together, and generates a road map for developing the necessary information on propulsor and hull–propulsor interaction. Basic definitions of power and power ratios are provided.

Learning Objectives

At the end of this chapter students will be able to:

- define power and efficiencies related to ship propulsion
- perform computations with efficiencies
- discuss advantages and disadvantages of propulsors

32.1 Propulsion Task

Effective power

In our discussion so far, the ship hull miraculously moves through the water at constant speed v without any propulsion system (Figure 32.1(a)). We defined the force acting on the hull as the total resistance R_T. An imaginary tow force of equal strength is introduced to balance the external forces acting on the ship. The effective power P_E is spent towing the hull.

$$P_E = R_T\, v \qquad \text{effective power (hull)} \tag{32.1}$$

Thrust and thrust deduction

The imaginary tow force is ultimately replaced with the thrust generated by a propulsion system. In Figure 32.1(b) a marine propeller was added to the ship hull. Based on Newton's first law one would expect that the propeller thrust is equal to the resistance but acts in the opposite direction so that the forces are in equilibrium, i.e. $T = R_T$. However, adding a propulsor changes the flow regime around the hull. In general, the

Fundamentals of Ship Hydrodynamics: Fluid Mechanics, Ship Resistance and Propulsion, First Edition.
Lothar Birk.

Companion website: www.wiley.com/go/birk/hydrodynamics

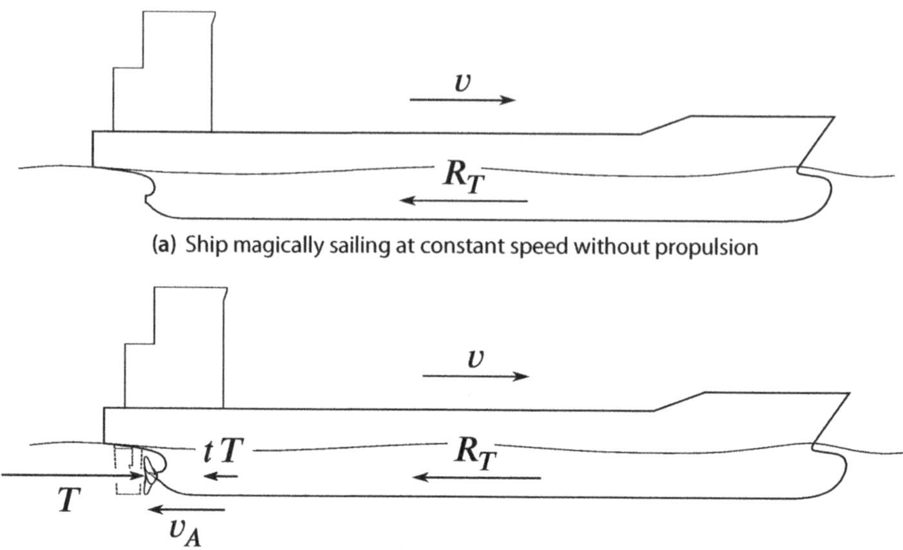

(a) Ship magically sailing at constant speed without propulsion

(b) Ship sailing at constant speed with propulsion

Figure 32.1 Forces acting on ship without and with propulsion system

drag force acting on the hull–propulsor system is larger than the total resistance R_T of the hull alone. A fraction t of the propulsor thrust T is spent compensating for the additional resistance (Figure 32.1(b)).

$$T = R_T + tT \tag{32.2}$$

The number t is called the thrust deduction fraction and defined as the ratio of increase in resistance and thrust.

$$t = \frac{T - R_T}{T} \tag{32.3}$$

The thrust deduction fraction and other parameters quantifying the interaction between hull and propulsor are discussed in detail in Chapter 34.

Wake fraction For any outside observer a propulsor is moving with speed v of its ship. Otherwise it would fall off! However, due to changes in the flow field caused by the moving body, a propeller at the stern of a vessel moves with an advance velocity v_A relative to the surrounding fluid that is smaller than the ship's speed. The difference in ship speed and speed of advance is represented by the dimensionless wake fraction w.

$$w = \frac{v - v_A}{v} \tag{32.4}$$

Again, details will be illustrated in Chapter 34.

Thrust power We introduce the thrust power P_T as the product of thrust and speed of advance. It represents the power transferred by a propulsor into the fluid.

$$P_T = T\, v_A \qquad \text{thrust power (hull–propeller system)} \tag{32.5}$$

A major objective of the work of naval architects is to design a hull–propulsor system that accomplishes this transfer with the highest possible efficiency. The design process will start with the selection of a suitable propulsion system.

32.2 Propulsion Systems

In this section we will discuss briefly the properties, advantages, and disadvantages of the most common mechanical propulsion systems. For a more in depth review see Carlton (2007).

Most vessels today are driven by one of these systems:

- marine propeller
- water jet propulsion
- Voith Schneider (cycloidal) propeller

32.2.1 Marine propeller

The most common propulsor is the marine propeller. It consists of a hub and 2–7 radially oriented blades. Marine propellers come in many different variations:

- fixed pitch propellers (Figure 32.2),
- controllable pitch propellers,
- ducted propellers, and
- azimuthing propellers.

Advantages

Propellers provide fairly high efficiency and reliable operation. Changes in draft or motions of the vessel have only minor impact on propeller performance as long as it is fully submerged. This is one of the reasons that propellers superseded paddle wheels so quickly for seagoing vessels. Combined with the appropriate reduction gears, propellers can be driven by a wide variety of engines. Typically located at the stern of the vessel, they are protected and not easily damaged.

Disadvantages

Their complex shape poses manufacturing challenges and makes a propeller an expensive piece of ship machinery. Propeller and shaft make up about 7% of the cost for the whole engine system. The principal characteristics of a propeller have a major impact on its performance. Therefore, careful tuning of its parameters is required for a given propulsion task.

Fixed pitch propellers

Fixed pitch propellers (FPPs) are the most common propulsor because of their high reliability and comparatively low price. Fixed pitch refers to the fact that the blades and hub are rigidly connected. In most cases, hub and blades are cast as one continuous piece.

Controllable pitch propellers

The blades of controllable pitch propellers (CPPs) are not rigidly fixed to the hub but can rotate about their radial axis. Ship speed may be changed by increasing or decreasing the angle of the blades rather than changing the engine rate of revolution. This is beneficial if the propeller shaft is also driving a generator which works best at a constant rate of revolution. It also allows for faster changes in speed and thrust reversal resulting in higher maneuverability of the vessel. However, best efficiency is only

Figure 32.2 A five-bladed fixed pitch propeller with a Schneekluth nozzle to improve propeller inflow

achieved at the design point of the propeller. At all other pitch angles efficiency drops and cavitation becomes more likely. Complexity of the control mechanism increases cost compared with a fixed pitch propeller. Controllable pitch propellers also require larger hub diameters, reducing the efficiency. If thrust reversal is desired, expanded area ratios are limited to values below 0.75.

Ducted propellers For applications where high thrust is needed at low speeds of advance, efficiency can be improved by placing the propeller in a short duct. In some applications the duct is fixed but usually the direction (azimuth) of ducted propellers can be changed to generate thrust in varying directions.Most active dynamic positioning systems feature azimuthing ducted propellers.

Azimuthing propellers Not all azimuthing propellers feature a duct. In recent years podded propulsion systems have become popular for diesel–electric propulsion. On vessels with high demand for hotel power, like cruise ships, electric motors are used to drive the propeller. The electric motor can be placed in an azimuthing pod providing a high level of maneuverability.

32.2.2 Water jet propulsion

Water jet propulsion might provide better efficiency as a marine propeller in high speed applications where the propeller is likely to experience significant cavitation. The raised internal pressure in the pump delays the onset of cavitation to higher speeds. Surface piercing propellers might be an alternative. In essence, a water jet is a pump typically designed to create a high mass flow (Figure 32.3). Water is diverted from the bottom of the vessel through the inlet to the impeller. The impeller accelerates the fluid and

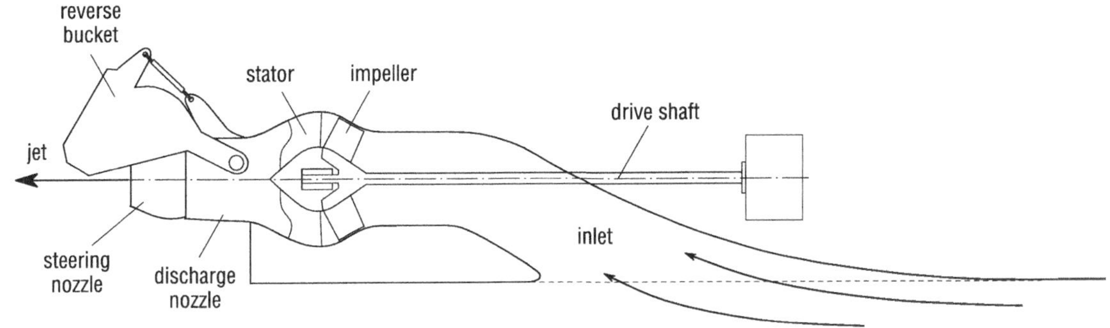

Figure 32.3 Schematic of a water jet

pushes it out at high speed above the transom. Steering nozzle and a reverse thrust bucket can be used to direct and reverse the thrust.

Water jets eliminate appendages that stick out beneath the hull which typically come with marine propeller installations for high speeds (e.g. exposed shafts, bossings, struts, and rudders). Delicate machinery is protected inside the hull and can be serviced without docking the vessel. The inlet and the grating preventing larger objects from being sucked into the drive require careful design to minimize their impact on resistance.

32.2.3 Voith Schneider propeller (VSP)

Cycloidal propeller

Whereas a water jet could still be considered as a type of propeller stuck into a pipe, the Voith Schneider propeller has a completely different shape. It consists of a rotating disk with a set of four to five high aspect ratio blades oriented perpendicular to the disk. The blades themselves rotate about their vertical axes. The angle of attack is controlled by a mechanism inside the ship. Voith Schneider propellers were invented in the 1920s, originally conceived as hydrokinetic turbines. They are also known as cycloidal propellers. Voith Schneider propellers combine propulsion with steering. This makes them applicable to vessels which require high maneuverability like harbor tugs and ferries.

Voith Schneider propellers can generate thrust in any direction, which gives the vessels their high level of maneuverability. Typically, they are installed in pairs, which provides a high level of redundancy. While propelling the ship, VSPs may also be employed to increase roll damping in heavy seas and thus eliminate the need for other means of roll stabilization. They are also suitable for dynamic positioning. Another advantage is comparatively low noise levels.

VSPs are mechanically complex and servicing the blades can be done only while the ship is in dry dock. The rotating disk and blades all need watertight seals. Due to the multitude of moving parts installable, delivered power is limited to about 4000 kW.

32.3 Efficiencies in Ship Propulsion

Definition of efficiency

Efficiency is commonly defined as the ratio of output (usable) power to input (total) power:

$$\text{efficiency} = \frac{\text{output power}}{\text{input power}} = \frac{\text{usable power}}{\text{total power spent}} \qquad (32.6)$$

Conservation of energy demands that efficiency η in a closed system is smaller than $\eta \leq 1$. When you switch on a light only a small portion of the energy flowing to the light bulb (total energy) is converted into light (usable energy). Most of the energy is converted into heat. We consider this a loss, although the energy is not gone. It has been converted into anenergy form we cannot use for the intended purpose of lighting a space. A system that has more usable power than total power spent would violate the first and second laws of thermodynamics. Ergo, no such system exists!

Hull efficiency

As we will see later, some of the efficiencies defined here take values larger than one. This apparent contradiction is a result of comparing power output from two different systems rather than a closed system. For instance, forming the ratio of power needed to tow the vessel (without propulsion system) and the power the propulsor transfers into the water yields the hull efficiency η_H. Hull efficiency is the ratio of effective power and thrust power (32.5):

$$\eta_H = \frac{P_E}{P_T} = \frac{R_T\, v}{T\, v_A} \qquad (32.7)$$

Hull efficiency for single screw vessels takes values typically in the range from 1.05 to 1.1. Of course our hull–propeller system is not a perpetual motion machine. Remember that effective power is defined for the hull alone whereas thrust power refers to the combined hull–propeller system. Therefore, hull efficiency is not defined in a closed system and one may argue that the application of the term efficiency might not be the best choice.

Delivered power

Rotation of the propeller requires application of torque Q to the shaft. At the propeller we measure delivered power P_D with

$$P_D = \omega Q = 2\pi\, n\, Q \qquad \text{delivered power} \qquad (32.8)$$

The angular velocity ω of the propeller is two times π times the rate of revolution n. Beware that if SI-units are used, rate of revolution has to be entered in 1/s, although engine revolutions are commonly stated in rpm (revolutions per minute).

Behind efficiency

The efficiency of the propeller working behind the ship is the ratio of thrust power (usable power) and delivered power (total power spent).

$$\eta_B = \frac{P_T}{P_D} = \frac{T\, v_A}{2\pi\, n\, Q} \qquad \text{behind efficiency} \qquad (32.9)$$

Open water efficiency

In Chapter 41 we discuss propeller characteristics under open water conditions, i.e. the propeller is working in undisturbed parallel flow instead of the wake of the ship. For this condition we define the propeller open water efficiency.

$$\eta_O = \frac{P_{TO}}{P_{DO}} = \frac{T_O\, v_A}{2\pi\, n\, Q_O} \qquad \text{open water efficiency} \qquad (32.10)$$

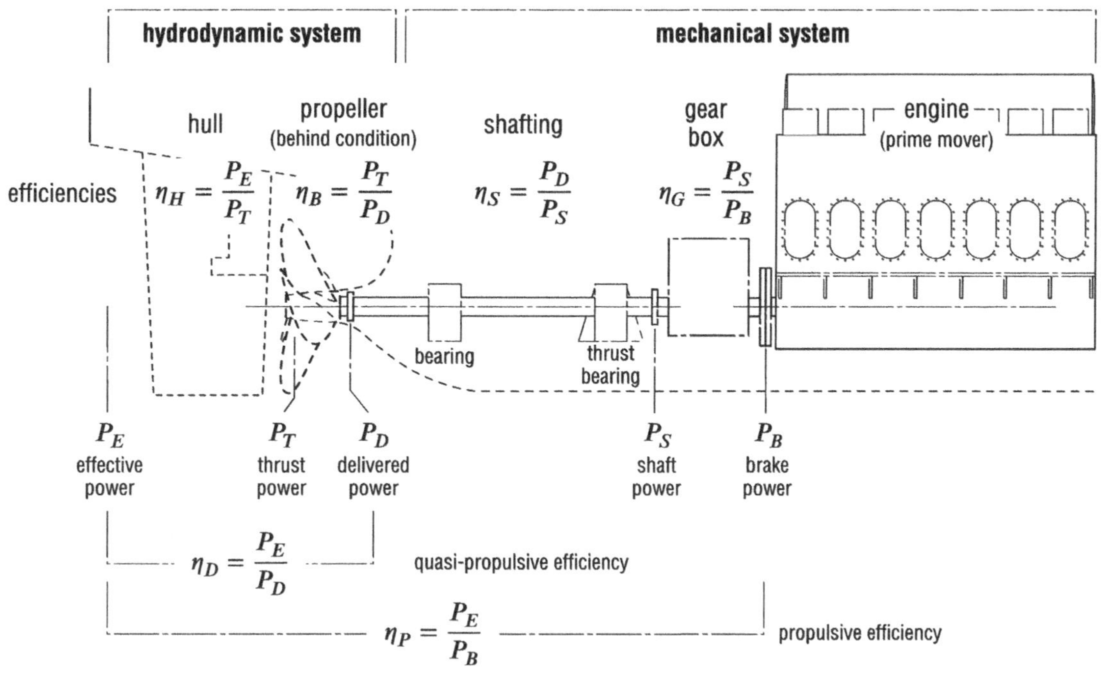

Figure 32.4 The propulsion system with transmission powers and efficiencies

The difference from the behind efficiency is that thrust T_O and torque Q_O refer to the values measured in parallel flow.

Relative rotative efficiency

The ratio of behind and open water efficiency is known as relative rotative efficiency.

$$\eta_R = \frac{\eta_B}{\eta_O} \qquad \text{relative rotative efficiency} \tag{32.11}$$

This is a ratio rather than an efficiency. As will be discussed in Chapter 34, relative rotative efficiency is typically a bit larger than one for single screw vessels and a bit smaller than one for twin screw vessels.

Quasi-propulsive efficiency

The complete hydrodynamic system of propeller and hull uses only the effective power to overcome the resistance. The ratio of effective and delivered power defines the quasi-propulsive efficiency η_D (Figure 32.4).

$$\eta_D = \frac{P_E}{P_D} \qquad \text{quasi-propulsive efficiency} \tag{32.12}$$

$$= \frac{P_E}{P_T}\frac{P_T}{P_D} = \eta_H\,\eta_B \tag{32.13}$$

The quasi-propulsive efficiency represents the efficiency of the hydrodynamic system and is a naval architect's main responsibility.

Maximum continuous rating

To better understand the whole propulsion system, we follow the power to its ship based source (Figure 32.4). Each combustion engine, like for example a Diesel engine or a gas turbine, is rated by the maximum power output it may deliver over prolonged periods of operation without being damaged beyond normal wear and tear. The engine manufacturer calls this the maximum continuous rating (MCR).

Brake power A motor or turbine converts the energy contained in its fuel into rotation of the engine shaft. An engine without any load will turn faster and faster until its controls kick in and prevent a runaway and failure of the engine. Any load (power take off) attached to the coupling flange brakes (slows down) the engine. If you take your bicycle from a flat stretch into a steep climb you will slow down or maintain speed by pedaling harder. The power necessary to maintain a constant rate of revolution of the engine is known as the brake power P_B. Brake power increases with the rate of revolution n. In their typical range of operation brake power of large internal combustion engines is approximately proportional to the rate of revolution cubed ($P_B \sim n^3$).

Shaft power and gearing efficiency Very large diesel engines rotate slowly enough (between 70 and 150 rpm) to be connected directly to the propeller shaft without any gearing. Without a gearbox, shaft power is equal to brake power.

$$P_S = P_B \quad \text{direct drive without gearbox} \tag{32.14}$$

Medium and small engines rotate faster than the rate of revolution for best propeller performance. Therefore, a reduction gear is necessary to transform the engine rate of revolution down to the optimum propeller rate of revolution. The power take off at the gearbox becomes the new shaft power P_S. The ratio of shaft power and brake power defines the gearing efficiency.

$$\eta_G = \frac{P_S}{P_B} \quad \text{with gearbox} \tag{32.15}$$

Losses in gears are usually smaller than 5%.

Shafting efficiency Frictional losses in the shaft bearings reduce the power that the shaft delivers to the propeller. The ratio of shaft and delivered power provides the shafting efficiency.

$$\eta_S = \frac{P_D}{P_S} \tag{32.16}$$

In most vessels, engines are installed toward the stern, with short shafts supported by very few bearings. As a consequence, losses are small and shafting efficiency is only slightly smaller than one ($\eta_S \approx 0.98$).

Total propulsive efficiency Power is transmitted from the engine coupling flange to the propeller through a series of subsystems: engine $\longrightarrow$ gearbox $\longrightarrow$ shafting $\longrightarrow$ propeller $\longrightarrow$ hull. The propulsive efficiency η_P combines the efficiencies of all subsystems (Figure 32.4).

$$\eta_P = \frac{P_E}{P_B} = \frac{P_E}{P_T}\frac{P_T}{P_D}\frac{P_D}{P_S}\frac{P_S}{P_B} = \eta_H\,\eta_B\,\eta_S\,\eta_G \tag{32.17}$$

Design of the propulsion system aims to make the propulsive efficiency as high as possible. Gearing and shafting efficiency are topics for studies in marine engineering. In the following chapters, we will concentrate on the efficiency η_D of the hydrodynamic system consisting of propeller and hull. We will study which properties affect the quasi-propulsive efficiency η_D and how it may be determined through model testing and how it may be estimated for early design purposes.

References

Carlton, J. (2007). *Marine propellers and propulsion.* Butterworth-Heinemann, Oxford, United Kingdom.

Self Study Problems

1. What is thrust power?
2. Where do you measure brake power?
3. Explain in your own words the difference between propulsive and quasi-propulsive efficiency.
4. What are the advantages and disadvantages of a Voith Schneider propeller compared with a fixed pitch propeller?
5. Explain in your own words the difference between open water and behind efficiency.
6. The quasi-propulsive efficiency is estimated to be $\eta_D = 69\%$ at a speed of 18 kn. If the resistance of the vessel at this speed is $R_T = 350$ kN, what is its effective and delivered power? In addition, which quantity or quantities would you have to know to also estimate the thrust power?
7. For a single screw vessel design project an engine has been selected and a preliminary propulsion analysis resulted in the engine and propulsion data below. Estimate the quasi-propulsive efficiency η_D by means of the data. Assume typical values for data *not* provided, making especially sure that you do not overestimate quasi-propulsive efficiency. Also state torque, thrust power, effective power, and resistance of the vessel.

Ship and engine data	
ship design speed	$v_S = 24.0$ kn
engine brake power	$P_B = 25\,000.0$ kW
rate of revolution	$n = 90$ rpm
hull efficiency	$\eta_H = 1.100$
propeller open water efficiency	$\eta_O = 0.62$
shaft efficiency	$\eta_S = 0.98$

33

Momentum Theory of the Propeller

Propellers are by far the most common mechanically driven propulsors. Before we take a closer look at their geometry and hydrodynamic performance, we apply the fundamental principles of conservation of mass and momentum to an idealized model of the propulsion system.

When a swimmer leaves the starting block, he generates forward momentum by pressing the block backwards. The block does not move and provides a reaction force equal to the force exerted by the swimmer's feet. A propulsor presses against the water to generate the required forward momentum for the ship. However, unlike the starting block for the swimmer, water will not stay in place. Therefore, some of the energy employed to create the forward momentum is lost by accelerating water in the opposite direction. This physical principle can be shown in the momentum theory for a propeller (Glauert, 1947). This basic mathematical model of a propulsor ignores actual geometry and flow details, but nonetheless yields valuable insight into the workings of a propeller and what is necessary to generate thrust with high efficiency.

Learning Objectives

At the end of this chapter students will be able to:

- apply conservation of mass and momentum principles
- understand the basic workings of a propulsor
- relate thrust loading and propulsor efficiency

33.1 Thrust, Axial Momentum, and Mass Flow

Actuator disk

In the model a propulsor is represented as a thin disk of area A_0 (see Figure 33.1). The disk is oriented perpendicular to the flow and located at position ② between inlet ① and outlet ③. The coordinate system is fixed to the disk. Its positive x-axis points in the direction of the ship's forward movement. Fluid flows through the disk without interruption. However, the disk 'magically' creates a pressure difference Δp between its inflow and outflow side.

$$\Delta p = p_2' - p_2 \tag{33.1}$$

Fundamentals of Ship Hydrodynamics: Fluid Mechanics, Ship Resistance and Propulsion, First Edition.
Lothar Birk.

Companion website: www.wiley.com/go/birk/hydrodynamics

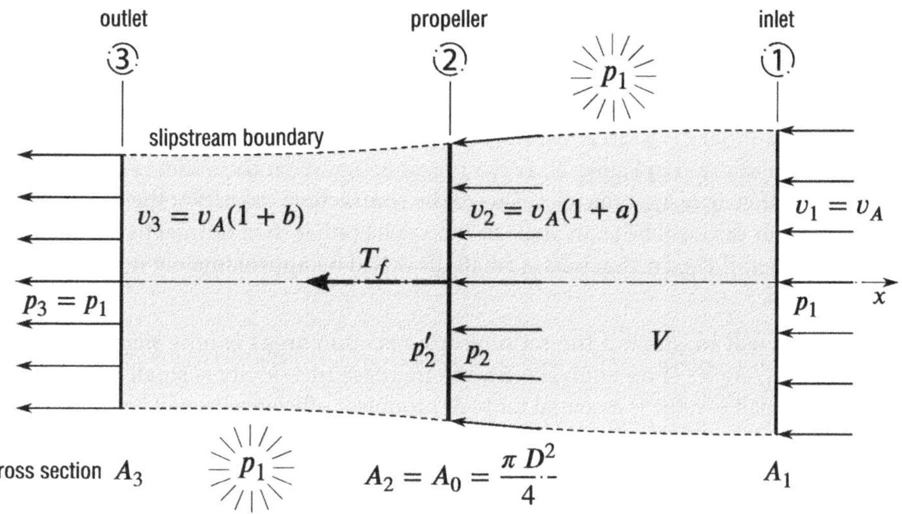

Figure 33.1 Fixed control volume around an idealized propeller (actuator disk)

For the time being, we ignore how this is actually accomplished. The pressure p_2' on the outflow side will be higher than the pressure p_2 on the inflow side. As a result, a thrust T acts on the disk, which is also known as an actuator disk.

Propulsor thrust

If we assume that the jump in pressure is constant across the disk area, the thrust is given as the product of area times pressure difference.

$$T = A_0 \, \Delta p \tag{33.2}$$

Next, conservation of momentum and conservation of mass principles are applied to the control volume, which will reveal the relationship between flow properties and thrust.

Input and assumptions

The propeller disk area A_0 and the mean inflow velocity at ① $v_1 = v_A$ are given as input to the problem. A known pressure p_1 acts all around the control volume. To proceed, we make the following additional assumptions:

(i) The fluid is inviscid (ideal), incompressible, and steady, i.e. independent of time.

(ii) The race boundary is a stream surface, i.e. no fluid enters or escapes the control volume V through the race boundary.

(iii) The flow velocity is perpendicular to the actuator disk and is treated as constant across A_0.

(iv) The flow is purely axial in x-direction throughout the control volume.

Lightly loaded propeller

The assumptions (ii) and (iii) above generate something of a conflict. As mentioned in the introduction, a propulsor – represented by the actuator disk – pushes the fluid in negative x-direction. Consequently, one must expect that the flow velocity will be higher at the outlet ③ than at the inlet ①. In Figure 33.1 this is indicated by expressing the velocity increase by a yet unknown fraction b.

$$v_3 = v_A(1+b) \tag{33.3}$$

If the race boundary is a stream surface, conservation of mass requires that the race contracts as indicated in Figure 33.1. The cross section A_3 at the outlet is smaller than the cross section area A_1 at the inlet. Due to the contraction, velocity at the outer ranges of the actuator disk will be at an angle to the x-axis rather than normal to the actuator disk. However, if the contraction is small, flow will be approximately normal to the actuator disk.

Contraction will be small if the ratio of cross section areas is only slightly smaller than one ($A_3/A_1 \approx 1$) or, equivalently, the increase in velocity is small ($b \ll 1$). In addition, a small b-value is essential for high propulsor efficiency, as will be shown later. However, smaller b-values also mean less thrust if all other parameters remain the same. A propeller that produces low thrust compared with its disk size is lightly loaded. For now, we will assume that our propulsor satisfies this.

Application of conservation of momentum

From the conservation of momentum Equation (9.1) equation follows that the rate of change of momentum is equal to the forces acting on the control volume V.

$$\frac{\partial \underline{M}}{\partial t} = \sum_{\text{external}} \underline{F} \tag{33.4}$$

The only external force acting on the fluid in V is the reaction T_f to the propulsor thrust T.

$$\sum_{\text{external}} \underline{F} = T_f \underline{i} = -T \underline{i} \tag{33.5}$$

The vector $\underline{i} = (1,0,0)^T$ is the unit length vector in positive x-direction.

The rate of change of momentum in V is given by

$$\frac{\partial \underline{M}}{\partial t} = \iiint_V \frac{\partial(\rho \underline{v})}{\partial t}\,\mathrm{d}V + \iint_S \rho \underline{v}\left(\underline{v}^T \underline{n}\right)\mathrm{d}S \tag{33.6}$$

The local time derivative of the density is zero since the flow is steady. As a consequence, the volume integral vanishes. Momentum changes in the control volume only by momentum transport across its boundaries. Momentum transport only happens on inlet and outlet disks, respectively, and we therefore may restrict the integration to the cross sections A_1 and A_3.

$$\frac{\partial \underline{M}}{\partial t} = \iint_{A_1} \rho \underline{v}_1\left(\underline{v}_1^T \underline{n}_1\right)\mathrm{d}S + \iint_{A_3} \rho \underline{v}_3\left(\underline{v}_3^T \underline{n}_3\right)\mathrm{d}S \tag{33.7}$$

Using the normal and velocity vectors pictured in Figure 33.1 and considering that the velocities are assumed to be constant over the cross sections, we have:

$$\begin{aligned}\frac{\partial \underline{M}}{\partial t} &= -\rho v_A \underline{v}_1 \iint_{A_1} \mathrm{d}S + \rho v_A(1+b)\,\underline{v}_3 \iint_{A_3} \mathrm{d}S \\ &= -\rho v_A A_1\,\underline{v}_1 + \rho v_A(1+b) A_3\,\underline{v}_3 \end{aligned} \tag{33.8}$$

The quantity $Q_1 = \rho v_A A_1$ represents the mass flow in [kg/s] through the inlet and $Q_3 = \rho v_A(1+b)A_3$ the mass flow through the outlet.

Application of conservation of mass

$$\frac{\partial \underline{\underline{M}}}{\partial t} = -Q_1 \underline{v}_1 + Q_3 \underline{v}_3 \tag{33.9}$$

The continuity equation requires that

$$\iiint_V \frac{\partial \rho}{\partial t}\,\mathrm{d}V = -\iint_S \rho\,\underline{v}^T \underline{n}\,\mathrm{d}S \tag{5.18}$$

Again, the volume integral vanishes for steady flow. Since we postulated no flow through the race boundary, the surface integral reduces to the fact that inflow and outflow must be equal. No fluid can be lost in between.

$$Q_1 = Q_3 \quad \longrightarrow \quad \rho v_A A_1 = \rho v_A (1+b)\,A_3 \tag{33.10}$$

The mass flows of inlet or outlet also apply to the actuator disk:

$$Q_2 = \rho\, v_A(1+a)A_0 = Q_1 = Q_3 \tag{33.11}$$

In this expression only the factor a is unknown.

Next, the result for the mass flow Equation (33.11) is substituted into the rate of change of momentum Equation (33.9).

Combine conservation of mass and momentum

$$\frac{\partial \underline{\underline{M}}}{\partial t} = -Q_2 \begin{pmatrix} -v_A \\ 0 \\ 0 \end{pmatrix} + Q_2 \begin{pmatrix} -v_A(1+b) \\ 0 \\ 0 \end{pmatrix} = -b\,Q_2 \begin{pmatrix} v_A \\ 0 \\ 0 \end{pmatrix} \tag{33.12}$$

Together with the result for the external force (33.5), we get:

$$\frac{\partial \underline{\underline{M}}}{\partial t} = -T\underline{i}$$

$$-b\,Q_2 \begin{pmatrix} v_A \\ 0 \\ 0 \end{pmatrix} = -T \begin{pmatrix} 1 \\ 0 \\ 0 \end{pmatrix} \quad \longrightarrow \quad T = b\,Q_2\, v_A \tag{33.13}$$

From this expression it becomes clear that a propulsor produces forward thrust ($T > 0$) only if the factor b is positive. In other words, thrust is generated only if fluid is accelerated aft.

Apply Bernoulli equation

The Bernoulli equation provides a relationship between the velocities v_2 and v_3. For that we apply Bernoulli's equation between stations ① and ② right in front of the actuator disk and from right behind the disk ②' to the outlet ③.

$$① \longrightarrow ② : \quad \frac{1}{2}\rho\, v_A^2 + p_1 = \frac{1}{2}\rho\left(v_A(1+a)\right)^2 + p_2 \tag{33.14a}$$

$$②' \longrightarrow ③ : \quad \frac{1}{2}\rho\left(v_A(1+a)\right)^2 + p_2' = \frac{1}{2}\rho\left(v_A(1+b)\right)^2 + p_1 \tag{33.14b}$$

Note that the velocity through the actuator disk must be continuous, i.e. $v_2 = v_2'$ in order to satisfy the conservation of mass principle.

Adding Equations (33.14a) and (33.14b) eliminates $1/2\,\rho\,(1+a)^2 v_A^2 + p_1$ on both sides:

$$\frac{1}{2}\rho\, v_A^2 + p_2' = \frac{1}{2}\rho\left(v_A(1+b)\right)^2 + p_2 \tag{33.15}$$

By expansion and rearrangement of the terms we find:

$$\Delta p = p_2' - p_2 = \rho\, b\, v_A^2\left(1 + \frac{b}{2}\right) \tag{33.16}$$

From Equation (33.2) follows with Equations (33.13) and (33.11) a relationship for the jump in pressure and the increase of velocity in the slipstream.

$$\Delta p = \frac{T}{A_0} = \frac{b\, Q_2\, v_A}{A_0} = \frac{b\left[\rho\, v_A(1+a)\, A_0\right] v_A}{A_0} = b\,\rho\,(1+a)\, v_A^2 \tag{33.17}$$

Finally, comparing the right-hand sides of Equations (33.17) and (33.16) reveals

$$(1+a) = \left(1 + \frac{b}{2}\right) \quad \text{or} \quad a = \frac{b}{2} \tag{33.18}$$

Axially induced velocity Some authors prefer to use an axially induced velocity u_A instead of the factors a and b. The velocities at propeller and outlet are then given as

$$\begin{aligned} v_2 &= \left(1 + \frac{b}{2}\right) v_A = v_A + u_A \\ v_3 &= (1+b)\, v_A = v_A + 2u_A \end{aligned} \tag{33.19}$$

Therefore $2u_A = bv_A$. In some references you may also find that the axial induced velocity has been defined as the whole axially induced velocity $u_A = bv_A$.

Pressure distribution The propulsor accelerates fluid aft. Half of the acceleration takes already place in front of the actuator disk (Figure 33.2). As the flow velocity increases in front of the actuator disk, pressure is decreasing to the pressure p_2:

$$p_2 = p_1 - \frac{1}{2}\rho\, v_A^2\left(b + \frac{b^2}{4}\right) \tag{33.20}$$

Surprisingly, the acceleration continues behind the disk as more potential energy (high pressure) is converted into kinetic energy until the pressure returns to the pressure level p_1 of the undisturbed flow. The propulsor raises the pressure by Δp. This eliminates the pressure drop in front of the disk and adds

$$p_2' - p_1 = \frac{1}{2}\rho\, v_A^2\left(b + \frac{3b^2}{4}\right) \tag{33.21}$$

to the pressure level. Note that the pressure difference $p_1 - p_2$ in front of the actuator disk is smaller than the pressure difference $p_2' - p_1$ behind the disk. Thus, the pressure curve shown in Figure 33.2 does not feature point symmetry with respect to the p_1 level at position ②.

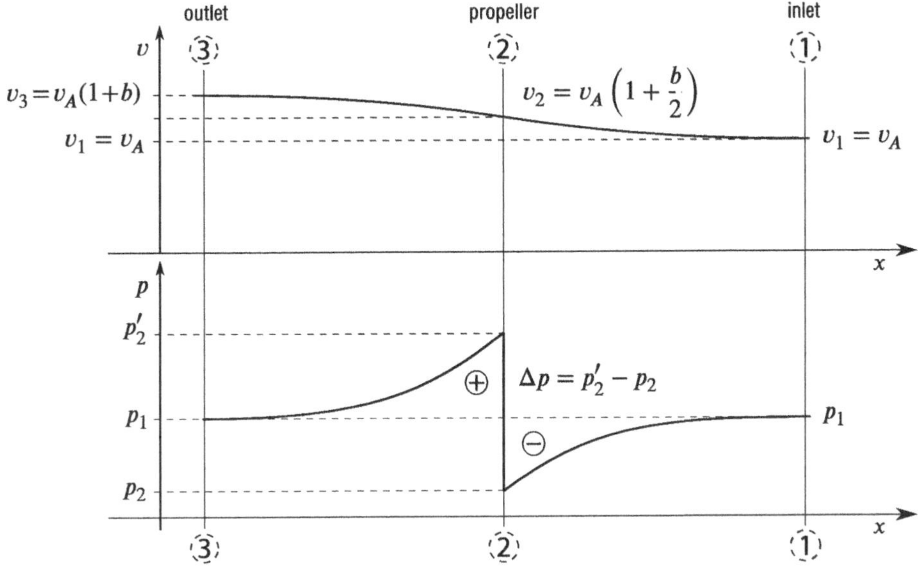

Figure 33.2 Velocity and pressure distribution according to momentum theory

Mass flow Q_2 and thrust T may now be written as functions of the advance velocity v_A and the factor b which expresses the fraction by which the velocity is increased in the slipstream.

Mass flow and thrust

$$\text{mass flow} \qquad Q_2 = \rho\, v_A \left(1 + \frac{b}{2}\right) A_0 \qquad (33.22)$$

$$\text{thrust} \qquad T = \frac{1}{2}\rho\, v_A^2 \left(2b + b^2\right) A_0 \qquad (33.23)$$

Equation (33.23) shows that there are two ways to increase thrust T for a given constant speed of advance v_A.

(i) increase the fluid flow velocity at the outlet, i.e. increase the factor b, or

(ii) increase the propeller disk area A_0 by selecting a larger propeller diameter.

The question arises: which of the two methods yields a higher efficiency?

33.2 Ideal Efficiency and Thrust Loading Coefficient

Answering the question above requires a closer look at the energy balances in the momentum theory model. As usual, efficiency is defined as the ratio of useful power to total power spent. Since we neglect viscous losses, it is called ideal efficiency η_{id}. It is also known as jet efficiency η_{TJ} in the ITTC's nomenclature.

Useful power

The useful work per unit time obtained from the actuator is the thrust power P_T (see Section 32.3).

$$P_T = T\, v_A \qquad (32.5)$$

For an outside observer the fluid is at rest outside of the propeller slipstream and the disk is moving with the speed of advance v_A relative to the observer. Since our coordinate system is fixed to the moving disk, water is streaming toward it.

Total power spent

The actuator disk adds kinetic energy to the flow. For rigid bodies, kinetic energy is equal to one half times mass times velocity squared. In case of continua, an integration over the control volume is required. However, since the flow is incompressible, mass flow is the same in every cross section. Within a time interval Δt, the mass $m = Q_2\,\Delta t$ flows through any given cross section. Since the velocity is assumed to be constant across each cross section, we find that a kinetic energy of

$$E_{\text{kin}_1} = \frac{1}{2} Q_2\, \Delta t\, v_1^2 \tag{33.24}$$

is streaming into the control volume. At the outlet, kinetic energy E_{kin_3} is leaving V:

$$E_{\text{kin}_3} = \frac{1}{2} Q_2\, \Delta t\, v_3^2 \tag{33.25}$$

Consequently, the total increase in kinetic energy between inlet and outlet per time period Δt is

$$\begin{aligned}\frac{E_{\text{total}}}{\Delta t} = \frac{\Delta E_{\text{kin}}}{\Delta t} &= \frac{E_{\text{kin}_3} - E_{\text{kin}_1}}{\Delta t}\\ &= \frac{1}{2} Q_2 \left[v_3^2 - v_1^2 \right]\end{aligned}$$

Substituting the definitions for velocities v_1 and v_3

$$\frac{E_{\text{total}}}{\Delta t} = \frac{1}{2} Q_2 \left[v_A^2 (1+b)^2 - v_A^2 \right]$$

and replacing the mass flow Q_2 with the right-hand side of Equation (33.22) yields:

$$\frac{E_{\text{total}}}{\Delta t} x = \frac{1}{2} \rho\, v_A \left(1 + \frac{b}{2}\right) A_0\, v_A^2 (2b + b^2)$$

Finally, we utilize Equation (33.23) and collect some factors into the thrust T.

$$\frac{E_{\text{total}}}{\Delta t} = T\, v_A \left(1 + \frac{b}{2}\right) \tag{33.26}$$

Ideal efficiency

The ideal efficiency of an actuator disk is equal to the ratio of thrust power and the total energy transferred into the flow.

$$\begin{aligned}
\eta_{TJ} &= \frac{\text{useful energy per unit time}}{\text{total energy spent per unit time}} \\
&= \frac{P_T}{\dfrac{E_{\text{total}}}{\Delta t}} \\
&= \frac{T\, v_A}{T\, v_A \left(1 + \dfrac{b}{2}\right)} \\
&= \frac{1}{1 + \dfrac{b}{2}} = \frac{2}{2+b}
\end{aligned} \tag{33.27}$$

Equation (33.27) provides an answer to the initial question. Clearly, the denominator of the fraction in Equation (33.27) is larger than the numerator for all $b > 0$. Therefore, ideal efficiency reaches its maximum value for $b = 0$. Unfortunately, this would be a useless propulsor because thrust will vanish too (check Equation (33.23)). For a given thrust, the best efficiency is obtained if as much fluid as possible (high Q_2) is accelerated aft by as little as possible (low b). Thus increasing the propeller disk area A_0 results in better efficiency than increasing b.

From this follows a general propeller design rule: for a given thrust, a larger propeller yields better efficiency. In practice, however, propeller diameters are limited due to ever present viscous losses, which we ignored in our simple propulsor model. If the propeller diameter is increased beyond its optimum value, expected gains in efficiency are eaten up by increased frictional losses.

Axial and radial losses

The propeller generates thrust by transferring energy into the fluid which is subsequently lost to the system. It exits the slipstream. Besides adding kinetic energy in axial direction, propellers also induce a rotational motion in the fluid, which causes an additional drop in efficiency.

Thrust loading coefficient

Another important number with respect to propeller efficiency is the thrust loading coefficient.

$$C_{Th} = \frac{T}{\frac{1}{2}\,\rho\, v_A^2 A_0} \tag{33.28}$$

It expresses how much thrust is generated per unit pressure and propeller disk area.

Most ship propellers are lightly to moderately loaded with thrust loading coefficients in the order of magnitude one. In heavy weather or bollard pull conditions, where high thrust is generated at low speeds of advance v_A, propellers are highly loaded with high C_{Th} values.

Effect of thrust loading on efficiency

Next, we investigate the effect of thrust loading on the ideal efficiency. As a first step, we establish a relationship between thrust loading and the increase in velocity b. Substituting the result for the thrust (33.23) into the definition of the thrust loading coefficient (33.28) yields:

$$C_{Th} = \frac{T}{\frac{1}{2}\,\rho\, v_A^2 A_0} = \frac{\frac{1}{2}\,\rho\, v_A^2 \left(2b + b^2\right) A_0}{\frac{1}{2}\,\rho\, v_A^2 A_0} = 2b + b^2 \tag{33.29}$$

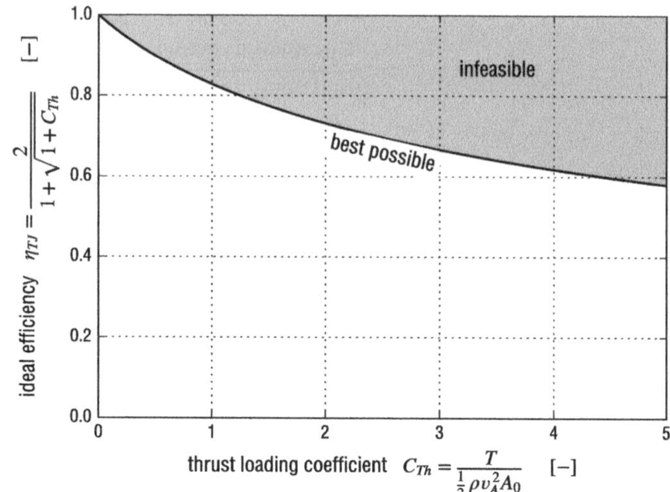

Figure 33.3 Ideal efficiency (jet efficiency) of propulsor momentum theory as a function of thrust loading coefficient

Equation (33.29) may be rewritten as a quadratic equation to express the factor b as a function of the thrust loading coefficient.

$$0 = b^2 + 2b - C_{Th} \tag{33.30}$$

This equation has the solutions

$$b_{1,2} = -1 \pm \sqrt{1 + C_{Th}} \tag{33.31}$$

The negative solution does not provide positive thrust. Therefore we only keep case 1.

$$b = -1 + \sqrt{1 + C_{Th}} \tag{33.32}$$

Substituting this relationship into the equation for the ideal efficiency (33.27) results in

$$\begin{aligned} \eta_{TJ} &= \frac{2}{2+b} \\ &= \frac{2}{1 + \sqrt{1 + C_{Th}}} \end{aligned} \tag{33.33}$$

Figure 33.3 pictures the relationship between thrust loading coefficient and ideal efficiency. The curve may be considered as the theoretical limit of attainable efficiency for a propulsor. Values above the line are not feasible. Considering the neglected radial and frictional losses, the actual attainable efficiency will fall short of the ideal efficiency. For that reason any propulsion system should be sceptically scrutinized if it promises efficiencies close or even above the ideal efficiency.

References

Glauert, H. (1947). *The elements of airofoil and airscrew theory.* Cambridge University Press, Cambridge, United Kingdom, second edition.

Self Study Problems

1. Compute the ratio of the pressure drop $p'_2 - p_1$ (33.21) behind the actuator disk to the pressure drop $p_1 - p_2$ (33.20) in front of the actuator disk as a function of the factor b which represents the increase in fluid velocity far behind the propulsor.
2. Compute the slipstream contraction ratio A_3/A_1 as a function of propeller disk area A_0 and the ideal efficiency η_{TJ}.

34

Hull–Propeller Interaction

Based on our discussions of the fluid flow around a ship hull, it should be obvious that the hull affects the inflow to the propeller. A propeller, in turn, affects how fluid flows around the hull. It almost always increases the drag of the hull–propulsor system beyond the bare hull resistance. These mutual effects are commonly referred to as hull–propeller interaction.

Effects of a hull on the propeller are represented by wake fraction w and relative rotative efficiency η_R. The former expresses changes in inflow velocity and the latter the change of propeller efficiency in comparison to the open water condition. The effect of a propeller on the hull is captured in the thrust deduction fraction t.

Separating the mutual effects is an imperfect model of the real word. It helps to explain and, in part, quantify the effects, but keep in mind that we are dealing with a single, integrated system.

Learning Objectives

At the end of this chapter students will be able to:

- explain hull–propeller interaction phenomena
- define interaction parameters and their effects
- apply hull–propeller interaction parameters

34.1 Wake Fraction

Effect of hull on propeller

The presence of the ship hull causes non-uniform inflow to a propeller behind the ship. It will significantly differ from the open water condition with its undisturbed parallel inflow. The propeller is now working in the wake of the hull. As a consequence, the speed of advance v_A of a propeller will differ from the ship speed v. At first glance this seems impossible. For an outside observer, both hull and propeller move with the same speed and the propeller, hopefully, stays attached to the ship. So how can its speed differ from that of the ship? The speed of advance v_A is the velocity of a propeller relative to the surrounding water. At the stern of a vessel, the fluid flow slows down relative to the

Fundamentals of Ship Hydrodynamics: Fluid Mechanics, Ship Resistance and Propulsion, First Edition.
Lothar Birk.

Companion website: www.wiley.com/go/birk/hydrodynamics

hull, in part to rebuild pressure and in part due to viscous losses in the boundary layer which is thickest at the stern.

Wake fraction

The difference in speed of advance and ship speed is expressed in the wake fraction w.

$$\text{wake fraction:} \quad w = \frac{\text{wake speed}}{\text{ship speed}} = \frac{v - v_A}{v} \tag{34.1}$$

This definition is attributed to Admiral David W. Taylor.[1] The difference between ship and propeller speed is also referred to as propeller slip. Later we will learn how to estimate or experimentally determine the wake fraction.

Speed of advance

Once the wake fraction w is known, the speed of advance v_A can be computed for the propeller.

$$v_A = (1 - w)\, v \tag{34.2}$$

For most ships the wake fraction is positive, which means speed of advance of the propeller is smaller than ship speed. For example, if $w = 0.28$ and the ship speed is 20 kn, the speed of advance becomes

$$v_A = (1 - w)\, v = (1 - 0.28)\, v = 0.72\, v$$

For the speed of advance a representation in meter per second is more appropriate, since it is used in several equations later on.

$$v = 20\ \text{kn} = 20\ \frac{\text{sm}}{\text{h}} = 20\ \frac{1852\ \text{m}}{3600\ \text{s}} = 10.289\ \frac{\text{m}}{\text{s}} \quad \text{ship speed}$$

Thus the speed of advance in the example is

$$v_A = 0.72 \cdot 10.289\ \frac{\text{m}}{\text{s}} = 7.408\ \frac{\text{m}}{\text{s}} \quad \text{propeller speed of advance}$$

Local wake fraction and wake field

The wake fraction for a ship is an average of inflow speed over the propeller disk.

$$w = \frac{1}{\left(A_0 - \pi r_h^2\right)} \iint_{A_0\setminus\text{hub}} w(y, z)\, \mathrm{d}A = \frac{1}{\left(A_0 - \pi r_h^2\right)} \int_0^{2\pi} \int_{r_h}^{R} w(r, \varphi)\, r\, \mathrm{d}r\, \mathrm{d}\varphi \tag{34.3}$$

Figure 34.1 shows isolines of local wake fraction as it would be typically found for a slender single screw vessel with V-type frames in the run. The local wake fraction is based on the local axial inflow $v_A(y, z)$ into the propeller plane:

$$w(y, z) = \frac{v - v_A(y, z)}{v} \tag{34.4}$$

[1] Some old publications use a definition by Froude which divides the wake speed $(v - v_A)$ with the speed of advance instead of ship speed v.

$$w_{\text{Froude}} = \frac{v - v_A}{v_A} \quad \text{(this is no longer used)}$$

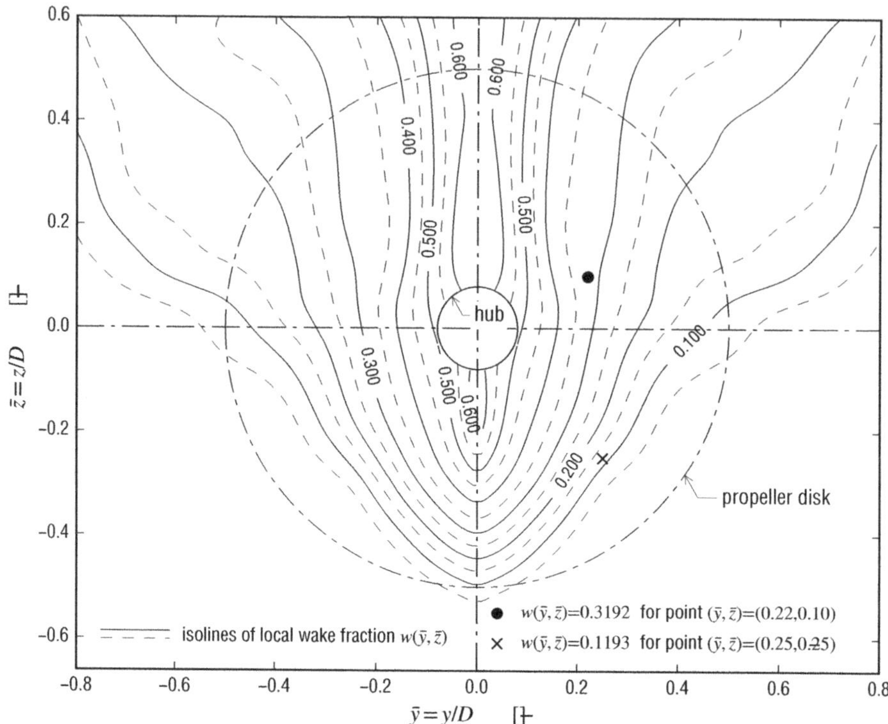

Figure 34.1 Example of a nominal wake field of a single propeller ship with moderate block coefficient

For example, at point $(y, z) = (0.22, 0.1)$, marked with a dot in Figure 34.1, the local wake fraction is $w(0.22, 0.1) = 0.3192$. Therefore, the difference between local axial inflow velocity and a ship speed of $v = 9$ m/s at point $(0.22, 0.1)$ is

$$v - v_A(0.22, 0.1) \;=\; v\,w(0.22, 0.1) \;=\; 9\text{ m/s} \cdot 0.3192 \;=\; 2.873\text{ m/s}$$

Consequently, the local axial inflow speed is

$$v_A(0.22, 0.1) \;=\; v\big[1 - w(0.22, 0.1)\big] \;=\; 9\text{ m/s} - 2.873\text{ m/s} \;=\; 6.127\text{ m/s}$$

Wake hook Figure 34.2 shows a wake fraction distribution typical for a full bodied vessel with U-type sections in the run. The wake field in each half of the propeller disk is formed like a hook. This feature is caused by a pronounced vortex emanating from the turn of the bilge. Vessels with high block coefficient tend to have a higher wake fraction, but parameters like beam–draft ratio B/T, mid ship section coefficient C_M, shape of the run, and surface roughness influence the wake distribution as well.

Nominal and effective wake Data for plots like Figure 34.1 and 34.2 stems from flow speed measurements or flow speed calculations in the propeller plane. Note that a propeller is not present and its influence on the flow is not accounted for. The wake field determined in the absence of a propeller is called nominal wake. Measurements of the wake field are difficult in the plane of a working propeller and also difficult to interpret because the wake field is

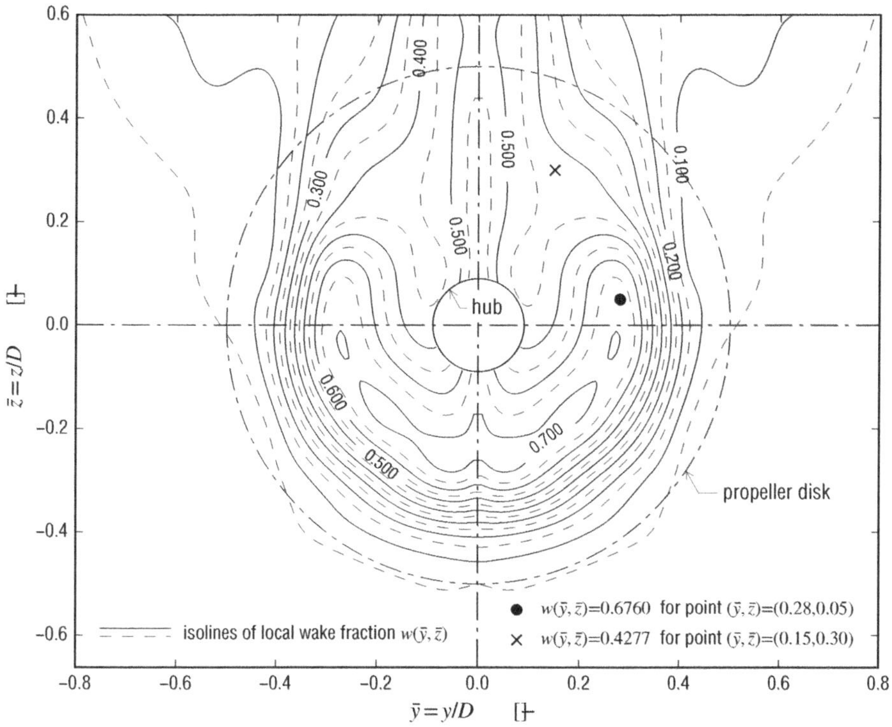

Figure 34.2 Example of a nominal wake field of a single propeller ship with high block coefficient

locally time dependent. A wake field that includes the influence of a working propeller is called effective wake.

Phenomena influencing wake

For design and analysis purposes, it is useful to study contributions to wake or wake fraction individually. Once the dominant influences are identified, procedures and measures may be conceived for wake prediction and control. Remind yourself that wake is an influence of the hull on the flow pattern in the vicinity of the propeller and that the wake fraction states how much the flow has slowed down compared to the ship speed. In our recapture of basic flow phenomena, we came across three flow patterns that cause local changes of flow speed:

- boundary layers on walls
- potential flow around bodies
- particle movement in waves

Friction between water and ship hull causes fluid to slow down within the boundary layer. Even in an inviscid flow around a body, molecules slow down toward the aft stagnation point, converting kinetic into potential energy. Underneath a wave crest, particles move in the direction of wave propagation, whereas underneath a wave trough they move in the opposite direction.

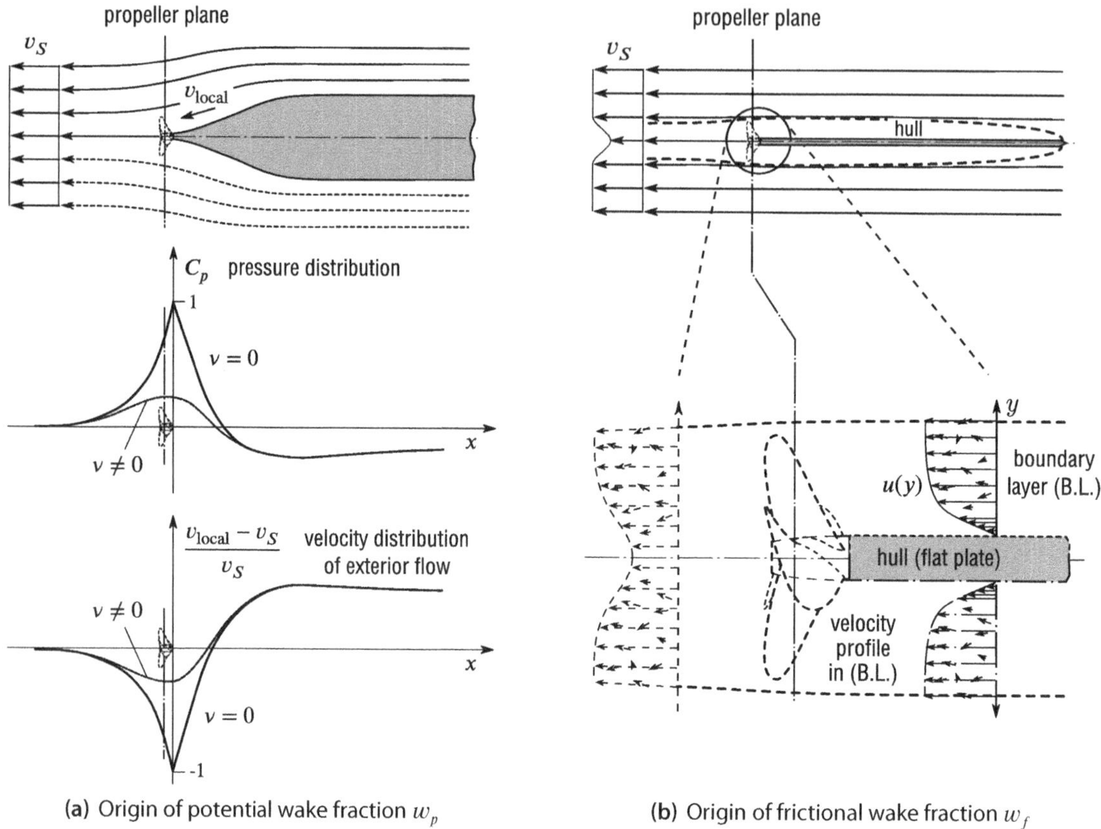

(a) Origin of potential wake fraction w_p

(b) Origin of frictional wake fraction w_f

Figure 34.3 Major contributions to the nominal wake fraction

Theoretical wake model The phenomena give rise to a simple three part model of the wake fraction (Harvald, 1983).

$$w = w_p + w_f + w_w \tag{34.5}$$

The total wake fraction is split into a potential wake fraction w_p, a frictional wake fraction w_f, and a wave wake fraction w_w. Note that the model (34.5) is not based on a conclusive theory. For starters, none of the three parts can be truly measured individually and the additive model expressed by Equation (34.5) suggests that the parts have no influence on each other, which is definitely not true. Nevertheless, the model will prove helpful in explaining what influences the wake and how to deal with scale effects in model tests.

Potential wake fraction w_p In our discussion of potential flow around a cylinder, we learned that stagnation points exist at the front and back of a body. Thus, a propeller at the stern of a vessel operates close to the virtual stagnation point in a region with small inflow velocity (Figure 34.3(a)). This is also applicable to viscous flows. For most vessels, the exterior flow will not come to a complete stop but will slow down nonetheless. This conversion of kinetic energy into potential energy causes the potential wake which is captured as the wake fraction component w_p.

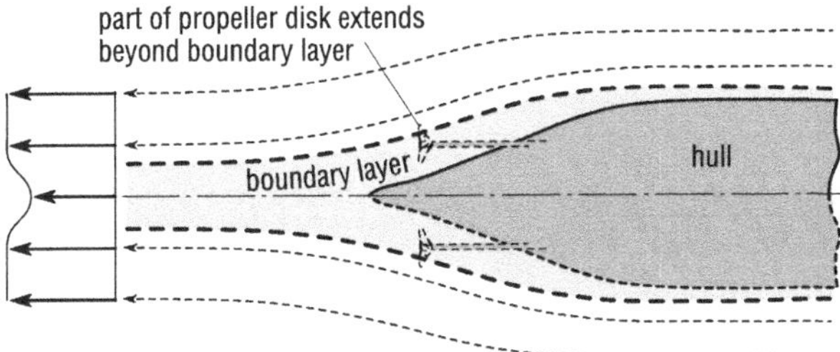

Figure 34.4 Frictional wake for twin screw vessels

Frictional wake fraction w_f

The frictional portion of the wake fraction is attributed to the momentum loss caused by viscous forces within a boundary layer. The propeller of single screw vessels is completely immersed in the boundary layer of the hull and experiences its decelerated flow over the entire disk. Note that the propeller is drawn with dashed lines in Figure 34.3(b) to indicate that we are studying the nominal wake, i.e. the influence of the hull on a propeller without flow effects caused by the working propeller.

Wave wake fraction w_w

The wave wake fraction w_w is negligibly small for most vessels. The idea is that, depending on the propeller location relative to the ship wave pattern, the water particle velocities within the stern wave add to or subtract from the wake. Under the crest of a stern, wave particles move in the direction of the ship, which reduces the local speed of advance. A slower inflow causes a higher wake fraction. Under a wave trough, particles move against the motion of the ship and slightly increase the velocity of advance (lower wake fraction).

Wake of twin screw vessels

In twin screw arrangements, propellers are only partially immersed in the hull's boundary layer (Figure 34.4). As a consequence, wake fractions of twin screw vessels tend to be smaller than for single screw vessels.

Model and full scale wake fraction

Potential and frictional part are the major contributions to the wake fraction. Chapter 43 discusses how the wake fraction is determined through model tests. Since the frictional part of the wake fraction is of considerable size, and the Reynolds number is a lot smaller in the model test than for the full scale vessel, wake fractions determined from model test data have to be corrected before they can be applied to the ship.

Typical wake fraction values for single screw vessels are in the range from 0.2 to 0.35. Ships with high block coefficients or short runs tend to have higher values toward the upper end of the range, slender vessels have lower wake fractions.

Wake fraction estimate

The resistance estimate by Guldhammer and Harvald (see Chapter 31) has been extended by Andersen and Guldhammer (1986) to provide estimates for the hull–propeller interaction parameters. For the wake fraction w of single screw vessels, Andersen and Guldhammer (1986) suggest the following method:

(i) First, compute the parameters a, b, and c:

$$\begin{aligned} a &= 0.1\frac{B}{L} + 0.149 \\ b &= 0.05\frac{B}{L} + 0.449 \\ c &= 585 - 5027\frac{B}{L} + 11700\left(\frac{B}{L}\right)^2 \end{aligned} \tag{34.6}$$

Again, L stands for the computation length introduced in Section 31.2.2.

(ii) The wake fraction is split into three parts: w_1 captures the influence of the hull form, w_2 represents a correction for the stern shape, and w_3 marks the influence of the propeller.

$$\begin{aligned} w_1 &= a + \frac{b}{c(0.98 - C_B)^3 + 1} \\ w_2 &= \frac{0.025 F_A}{100(C_B - 0.7)^2 + 1} \\ w_3 &= -0.18 + \frac{0.00756}{\left(\frac{D}{L} + 0.002\right)} \qquad \text{however, } w_3 \leq 0.1 \end{aligned} \tag{34.7}$$

The parameter F_A is the shape factor for the aft body. D stands for the propeller diameter. If the computed value for w_3 exceeds 0.1 it is set to 0.1.

(iii) The estimate for the model scale wake fraction is the sum of w_1, w_2, and w_3. 70% of the model scale value is assigned to the full scale wake fraction for trial conditions.

$$w_M = w_1 + w_2 + w_3 \qquad \text{model} \tag{34.8}$$

$$w_S = 0.7 w_M \qquad \text{full scale, trial condition} \tag{34.9}$$

For in service conditions, Andersen and Guldhammer (1986) suggest using the model scale value for the full scale ship. This has to be coordinated with the selected service margin.

Based on comparisons with model test data for contemporary bulk carrier and tanker hull designs, Kristensen and Lützen (2012) suggest to correct the value obtained from Equation (34.8) based on the length–displacement ratio M:

$$w_{S_{\text{corrected}}} = 0.7 w_M - 0.45 + 0.08 M \qquad \text{bulk carriers and tankers} \tag{34.10}$$

Other methods for wake fraction estimates are presented in Chapters 50 and 51.

34.2 Thrust Deduction Fraction

Thrust deduction

A working propeller affects the flow around a ship hull, especially in the stern region. The propeller causes an additional resistance which we expressed as a fraction of thrust (see Equation (32.2)). Since part of the propeller thrust is used to compensate for the increase in resistance, we called the effect thrust deduction.

Let us look into the mechanisms which cause the increase in resistance. As we discussed in Chapter 33 on momentum theory, propellers accelerate the fluid aft even before it reaches the propeller. Changes in flow velocity over the hull surface will affect the boundary layer development and will change pressure and shear stress distribution. As a consequence, the drag force on the hull will differ from the case without a propeller. In addition to changes in hull drag, we have to consider the propeller's own drag. Experience shows that the resistance of a hull with propeller is larger than the resistance

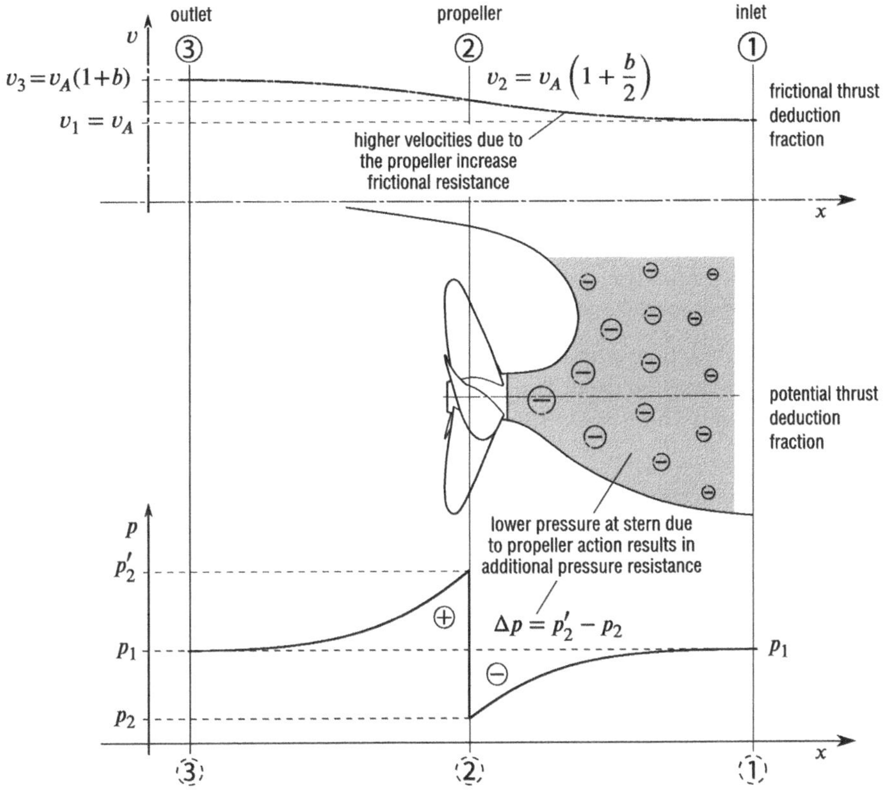

Figure 34.5 Effect of propeller on pressure and velocity distribution at the stern

of a hull alone. Therefore, thrust T needs to be larger than the calm water total resistance R_T.

Thrust deduction fraction

The effect is quantified by the thrust deduction $T - R_T$ and a corresponding thrust deduction fraction t:

$$t = \frac{\text{thrust deduction}}{\text{thrust}} = \frac{T - R_T}{T} \tag{34.11}$$

Like the wake fraction, the thrust deduction fraction is subdivided into three parts for a discussion of influencing factors.

$$t = t_p + t_f + t_w \tag{34.12}$$

Potential thrust deduction

The potential part of the thrust deduction fraction t_p is by far the largest contribution to t. From momentum theory we learned that the propeller is accelerating water through the propeller disk. The increasing velocity causes a drop in pressure compared with the flow around the bare hull without propeller (see Figure 34.5). Lower pressure at the stern results in additional resistance which has to be compensated by thrust larger than the total resistance R_T of the hull alone.

Frictional thrust deduction

The frictional part t_f may be attributed to the increased flow velocities in the stern region (Figure 34.5). Higher velocities will result in higher wall shear stress values. Since efficient propellers keep the increase in velocity small, the frictional part of the thrust deduction fraction is negligibly small for propellers with moderate thrust loading.

Wave thrust deduction

The wave portion t_w of the thrust deduction fraction captures the small effect that the propeller has on the wave resistance. The working propeller changes the pressure distribution in its vicinity which in turn will affect the contribution of the stern wave to the wave pattern. The effect on the wave resistance may be positive or negative. In general this contribution is also negligible.

Scale effects are negligible

Reynolds number effects are very small in the thrust deduction fraction. Therefore, the value determined in model tests is applied to the full scale ship without corrections.

Thrust deduction fraction values are typically lower than the wake fraction. As a rough estimate $t = 0.7w$ may be used. Hollenbach (1996) uses a fixed value of $t = 0.19$ for single screw vessels and $t = 0.15$ for a standard twin screw arrangement with two rudders.

Thrust deduction estimate

For the thrust deduction fraction t of single screw vessels Andersen and Guldhammer (1986) suggest the following procedure:

(i) Compute three parameters d, e, and f:

$$\begin{aligned} d &= 0.625\frac{B}{L} + 0.08 \\ e &= 0.165 - 0.25\frac{B}{L} \\ f &= 825 - 8060\frac{B}{L} + 20300\left(\frac{B}{L}\right)^2 \end{aligned} \tag{34.13}$$

L represents the computation length introduced in Section 31.2.2.

(ii) The thrust deduction is split into three parts: t_1 captures the influence of the hull form, t_2 represents a correction for the stern shape, and t_3 marks the influence of the propeller.

$$\begin{aligned} t_1 &= d + \frac{e}{f(0.98 - C_B)^3 + 1} \\ t_2 &= -0.01F_A \\ t_3 &= 2\left(\frac{D}{L} - 0.04\right) \end{aligned} \tag{34.14}$$

The parameter F_A is the shape factor for the aftbody. D is the propeller diameter.

(iii) The estimate for the thrust deduction fraction is the sum of t_1, t_2, and t_3 (Andersen and Guldhammer, 1986):

$$t = t_1 + t_2 + t_3 \tag{34.15}$$

Similar to the correction for the wake fraction, Kristensen and Lützen (2012) suggest correcting the thrust deduction fraction obtained from Equation (34.15) for bulk carriers and tankers.

$$t_{\text{corrected}} = t - 0.26 + 0.04M \qquad \text{bulk carriers and tankers} \tag{34.16}$$

Since the influence of viscosity on the thrust deduction fraction is small, we assume the same value for model and ship.

34.3 Relative Rotative Efficiency

Open water condition

During open water tests, thrust T_O and torque Q_O are determined as a function of the advance coefficient J. The subscript $_O$ indicates the open water condition with uniform parallel inflow into the propeller. The ratio of thrust power and delivered power determines the open water efficiency η_O.

$$\eta_O = \frac{P_T}{P_D} = \frac{T_O v_A}{2\pi\, n\, Q_O} \tag{34.17}$$

It expresses how effectively the torque Q_O delivered via the shaft to the propeller is converted into thrust T_O. Increasing thrust requires increased torque.

Behind condition

As mentioned, a propeller in the behind condition is affected by the hull wake. Provided we use the same advance coefficient J, the ratio of average speed of advance v_A and rate of revolution n will be the same for open water and behind condition. However, generated thrust and required torque will probably differ in open water and behind condition. This also affects the efficiency of the propeller. We defined the behind efficiency η_B as

$$\eta_B = \frac{P_T}{P_D} = \frac{T\, v_A}{2\pi\, n\, Q} \tag{34.18}$$

This is seemingly the same as the open water efficiency. However, the flow conditions are vastly different: uniform (open water) versus nonuniform flow (behind condition).

Thrust identity

This scenario has many influencing factors. To simplify matters we introduce the assumption of thrust identity: we postulate that the propeller produces the same thrust $T = T_O$ in open water and behind condition for the same advance coefficient J. Only the torque differs $Q \neq Q_O$. We will discuss this further in Chapter 43.

Relative rotative efficiency

Under the condition of thrust identity, the ratio of behind and open water efficiency yields:

$$\eta_R = \frac{\eta_B}{\eta_O} = \left.\frac{\dfrac{T\, v_A}{2\pi\, n\, Q}}{\dfrac{T_O\, v_A}{2\pi\, n\, Q_O}}\right|_{T=T_O} = \frac{Q_O}{Q} \quad \text{assuming thrust identity} \tag{34.19}$$

The ratio η_R is known as relative rotative efficiency. It expresses the change in propeller efficiency from open water (uniform flow) to behind condition (nonuniform flow). Surprisingly, relative rotative efficiency tends to be slightly larger than one for single screw vessels ($\eta_R \in [1.0, 1.05]$), i.e. the propeller is more efficient in the nonuniform wake than in the uniform inflow. For twin screw vessels, relative rotative efficiency is usually smaller than one ($\eta_R \in [0.96, 1.0]$). The difference in η_R values between single and twin screw vessels is attributed to the fact that propellers of twin screw vessels are less affected by the hull wake. A portion of the propeller disks usually falls outside of the boundary layer of the hull.

Torque identity

The definition of the relative rotative efficiency provided with Equation (34.19) is not unique. We could also define relative rotative efficiency under the assumption of torque identity: at the same advance coefficient J the same torque in behind and open water condition $Q = Q_O$ results in different thrust $T \neq T_O$.

$$\eta_R = \frac{\eta_B}{\eta_O} = \left.\frac{\frac{T\,v_A}{2\pi\,n\,Q}}{\frac{T_O\,v_A}{2\pi\,n\,Q_O}}\right|_{Q=Q_O} = \frac{T}{T_O} \quad \text{assuming torque identity} \tag{34.20}$$

Thrust identity is preferred

Torque identity has the advantage that performance predictions are more easily verified in ship trials. Torque can be measured on board a ship without great effort, thrust less so. In general, however, thrust identity seems to produce more accurate performance predictions. Therefore, the ITTC recommends the use of thrust identity.

Estimates

In ship design we often know the open water efficiency of a propeller from series data or propeller design calculations like lifting line and lifting surface computations. Prediction of the vessel performance requires the efficiency for the behind condition. Consequently, estimates for the relative rotative efficiency are necessary, which enables us to convert the measured or computed open water efficiency into the behind efficiency:

$$\eta_B = \eta_O\,\eta_R \tag{34.21}$$

As a first conservative estimate $\eta_R = 1$ may be used for single screw vessels and $\eta_R = 0.96$ for twin screw vessels. Hollenbach (1998) suggests the following values:

Relative rotative efficiency	η_R
single screw vessel at design draft	1.009
single screw vessel in ballast condition	1.000
twin screw vessel at design draft	0.981

See also Chapters 50 and 51 on the performance prediction methods by Holtrop (1984) and Hollenbach (1996). Chapter 43 explains how the hull–propeller interaction parameters are determined during model tests.

References

Andersen, P. and Guldhammer, H. (1986). A computer-oriented power prediction procedure. In *Proc. of Int. Conf. on Computer Aided Design, Manufacture, and Operation in the Marine and Offshore Industries (CADMO '86)*, Washington, DC, USA.

Harvald, S. (1983). *Resistance and propulsion of ships*. John Wiley & Sons, New York, NY.

Hollenbach, K. (1996). Verfahren zur Abschätzung von Widerstand und Propulsion von Ein- und Zweischraubenschiffen im Vorentwurf. In *Jahrbuch der Schiffbautechnischen Gesellschaft*, volume 90, pages 119–133. Schiffbautechnische Gesellschaft (STG).

Hollenbach, K. (1998). Estimating resistance and propulsion for single-screw and twin-screw ships. *Schiffstechnik/Ship Technology Research*, 45(2):72–76.

Holtrop, J. (1984). A statistical re-analysis of resistance and propulsion data. *International Shipbuilding Progress*, 31(363):272–276.

Kristensen, H. and Lützen, M. (2012). Prediction of resistance and propulsion power of ships. Technical Report Project No. 2010-56, Work Package 2, Report No. 04, University of Southern Denmark and Technical University of Denmark.

Self Study Problems

1. Name and explain the three hull–propeller interaction parameters.
2. Discuss the major contributions of the wake fraction.
3. Discuss the effect the propeller has on the hull.
4. What is the relative rotative efficiency?
5. Explain the nature of the potential part of the thrust deduction fraction.
6. Given a ship speed of $v_S = 18.2$ kn and a wake fraction of $w = 0.242$, compute the speed of advance of the propeller.
7. A ship is sailing at a speed of $v_S = 18.2$ kn. Resistance and propulsion estimates yield the following characteristics for the service speed:

Ship and engine data	
total resistance	$R_T = 316.00$ kN
wake fraction	$w = 0.242$
thrust deduction fraction	$t = 0.175$
open water efficiency	$\eta_O = 0.62$
rate of revolution	$n = 124.00$ rpm
propeller diameter	$D = 5.24$ m

Compute the effective power, calm water thrust, speed of advance, advance coefficient, hull efficiency, and perform an estimate of quasi-propulsive efficiency, delivered power, and propeller torque. Assume conservative values for data not provided.

35

Propeller Geometry

Before we discuss the hydrodynamic characteristics of a propeller, it is necessary to define and explain its principal geometric characteristics: number of blades, diameter, pitch–diameter ratio, and expanded area ratio, as well as rake, skew-back, and other properties. The general impact of main propeller characteristics on performance is discussed.

Learning Objectives

At the end of this chapter students will be able to:

- identify and describe propeller parts
- define principal geometric characteristics of propellers
- understand impact of geometric characteristics on performance

35.1 Propeller Parts

Fixed versus controllable pitch

A propeller consists of a hub and two or more blades. Figure 35.1 shows two views of a fixed pitch propeller with four blades. The term 'pitch' will be explained shortly. The adjective 'fixed' refers to the fact that the blades are cast together with the hub as a single piece. The position of the blades is immutable. Controllable pitch propellers have blades which can be rotated around a radial axis. About one third of installed propellers on commercial vessels with more than 2000 horse power are controllable pitch propellers (Carlton, 2007). They are commonly used on ships which require high maneuverability like ferries, tugs, or supply vessels. Fixed pitch propellers are cheaper and due to the smaller hub often achieve a higher efficiency at their design point.

Hub

The hub provides a connection to the propeller shaft. Propellers on pleasure boats might have a key. On larger vessels the hub has a conical bore which is pressed on the conical end of the propeller shaft and secured with bolts. The hub diameter is typically about one fifth of the total propeller diameter. Controllable pitch propellers have hub diameters of about 25% of the propeller diameter.

Fundamentals of Ship Hydrodynamics: Fluid Mechanics, Ship Resistance and Propulsion, First Edition.
Lothar Birk.

Companion website: www.wiley.com/go/birk/hydrodynamics

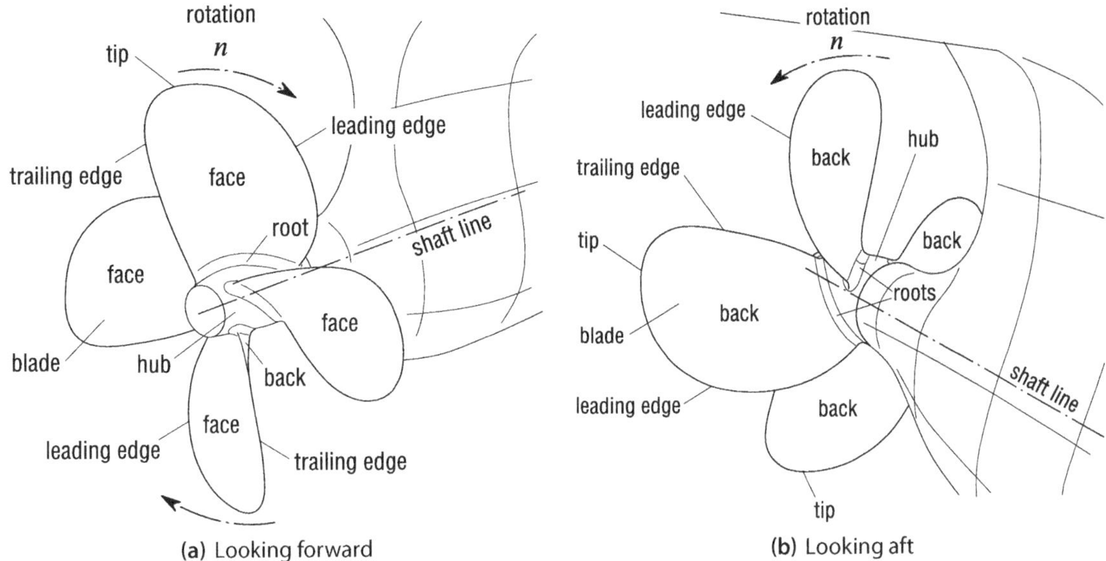

Figure 35.1 Parts of a propeller. Shown is a right-handed, fixed pitch propeller with four blades

Blades

Like the wings of an aircraft, propeller blades generate a 'lift' force. A portion of the lift force points in direction of the propeller shaft and contributes to the propeller thrust. The connection of a propeller blade to the hub is called the blade root (Figure 35.1). Structurally, blades are cantilevered beams. Stress concentrations at the blade root are avoided by casting fillets. The outer, free end of the blade is the blade tip. Looking forward, we see the blade faces (Figure 35.1(a)). This is the high pressure side of the blade, equivalent to the lower side of an aircraft wing. Looking aft, we see the blade backs (Figure 35.1(b)), which represent the low pressure (suction) side of the blade, equivalent to the upper side of an aircraft wing. Note that blade faces are oriented aft relative to the ship and blade backs are oriented forward.

Leading and trailing edge

The edge of a blade pointing in the direction of rotation is called the leading edge. By convention, single screw vessels operate with right turning (right-handed) propellers, i.e. looking forward along the shaft they turn clockwise when they produce forward thrust. Therefore, the leading edge is found on the right side of the blade (Figure 35.1(a)). The downstream edge of the blade is the trailing edge.

Left- and right-handed

Single propellers are always right-handed, i.e. they rotate clockwise when locking forward. Due to the clockwise rotation of the propeller, single screw vessels tend to steer to port if left uncontrolled. This is known as the wheel effect. Imagine the propeller being a rolling wheel which shifts the stern to starboard; the vessel will steer to port. Twin screw vessels use one right-handed (starboard) and one left-handed propeller (port) to compensate for the wheel effect.

Dynamic balance

Blades are always arranged in a regular pattern because the propeller needs to be dynamically balanced. Any imbalance due to damage in grounding, collision, or high seas will cause subsequent damage to shaft, bearings, and possibly the engine.

35.2 Principal Propeller Characteristics

Principal characteristics Ship power prediction in early design requires at a minimum the selection of four geometric and one basic operational characteristic of the propeller:

- number of blades Z,
- diameter D,
- pitch–diameter ratio P/D,
- expanded area ratio A_E/A_0, and
- rate of revolution n

The overall objective of matching propeller and ship is to maximize the propulsive efficiency of the vessel.

The four geometric parameters are explained below. The rate of revolution n should be self-explanatory. We will come back to it when we discuss propeller selection. As a first step in the design process, principal parameters are selected on the basis of published propeller series data and later confirmed and refined by lifting line and lifting surface calculations. We will discuss the selection of propeller characteristics based on series data in later chapters. For a discussions of lifting line and lifting surface methods the reader may consult Kerwin and Hadler (2010) and Breslin and Andersen (1994).

Number of blades Z The number of blades ranges from two to seven. Cost of a propeller increases with the number of blades. Therefore, lower blade numbers are preferable. Auxiliary drives for sailing yachts might feature only two blades. Power boats and small vessels typically use threebladed propellers. Most propellers for larger merchant vessels have four or five blades. More than five blades are used only when it is necessary to avoid cavitation or to reduce vibrations and noise signatures. An important factor in the selection of the number of blades Z is the avoidance of resonances with the main engine. Propellers develop cyclical forces with frequencies that are multiples of $Z\,n$. These frequencies should not match up with natural frequencies of the main engine. For example, a four bladed propeller should not be used in conjunction with an engine which has an even number of cylinders.

Diameter D The diameter D of a propeller is defined by the outermost circle which circumscribes the blade tips (Figure 35.2). The enclosed area is the propeller disk area A_0.

$$A_0 = \frac{\pi\,D^2}{4} \tag{35.1}$$

The distance between shaft center and tip of a blade is the blade radius R. It is equal to half the diameter and is equivalent to the span of an aircraft wing. In many propeller drawings and calculation methods a dimensionless radial coordinate x is used with

$$x = \frac{r}{R} \qquad \text{with } x \in [0, 1] \tag{35.2}$$

For conical or streamlined hubs, the hub radius r_h is usually measured where the blade reference line intersects the hub. This point also defines the longitudinal location of the propeller plane (see also Figure 35.12).

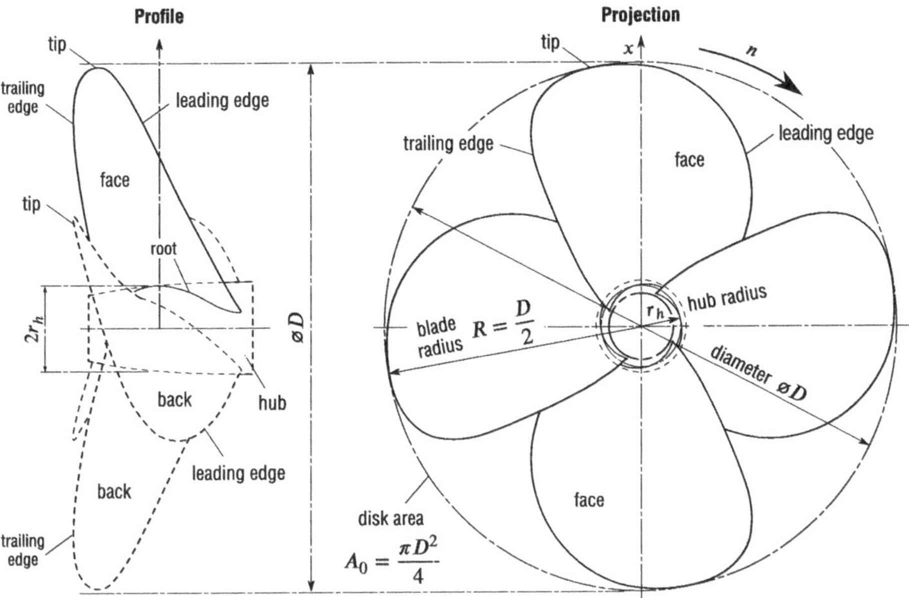

Figure 35.2 Definition of propeller diameter D, blade radius R, hub radius r_h, and disk area A_0

The diameter has a major impact on propeller performance. As we learned in the axial momentum theory of propellers, a larger diameter is usually more efficient. However, viscous losses and the wake field of the hull usually define an optimum value for the diameter which has to be found during the design process.

Pitch P

The complicated twisted shape of propeller blades is due to the pitch P. Every point on a propeller travels on a helical curve controlled by the rate of revolution n and the forward movement of the propeller. The geometric pitch P is the distance which a point on the propeller travels in direction of the shaft during a single rotation. This warrants further explanation.

Blade sections

First, we take a look at the foil sections that make up a propeller blade. On an aircraft wing or a rudder we commonly find the foil section by cutting straight through the wing/rudder perpendicular to its span. Finding the foil section of a propeller blade is a bit more complicated. Take a look at Figure 35.3. It shows a four bladed propeller whose top blade is intersected with a cylinder (shaded surface) of radius $x = r/R = 0.7$. The axis of the cylinder coincides with the propeller shaft. The intersection of blade and cylinder reveals the blade foil section (dark gray) at the selected radius.

If we repeat the intersection process for cylinders with different radii and unroll the resulting foils onto a flat surface, we see the foils as they are shown in Figure 35.4 for the radii $x = 0.2, 0.7$, and 0.975. Notice that the pitch angle ϕ is highest close to the hub ($x = 0.2$) and decreases going outward toward the blade tip. This change in pitch angle gives the propeller blade its characteristic twisted shape.

Constant pitch

Now remember that every point on the propeller travels on a helical path. If we view the helical paths from the top, they look like sine curves – assuming we start with the blade in a vertical position (Figure 35.5). Since the arc length of the inner radii is much

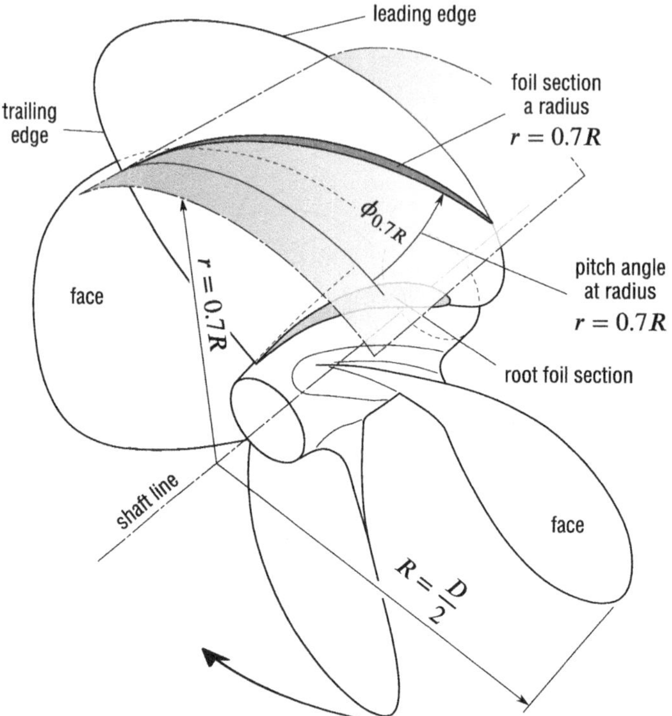

Figure 35.3 Hydrofoil section within a propeller blade

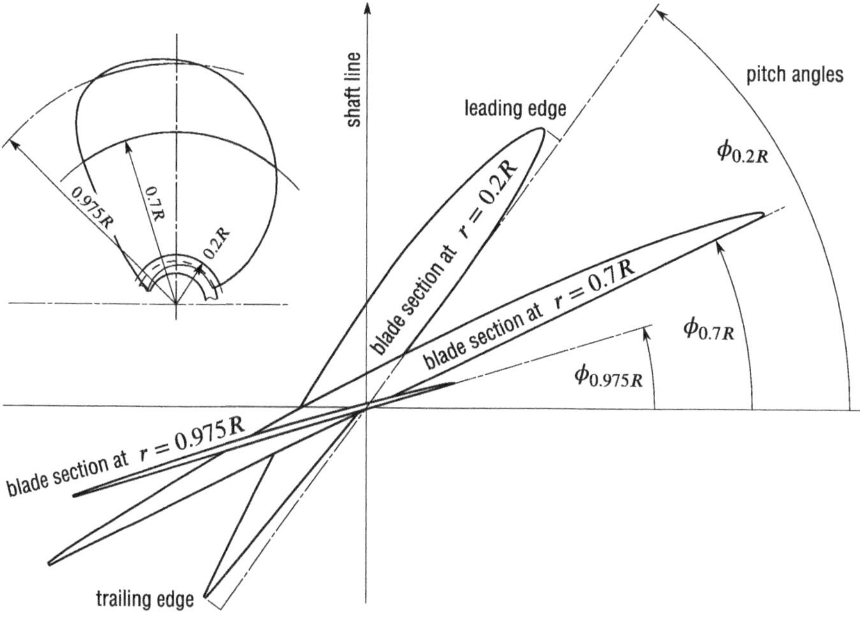

Figure 35.4 Pitch angle variation for a propeller with constant pitch P

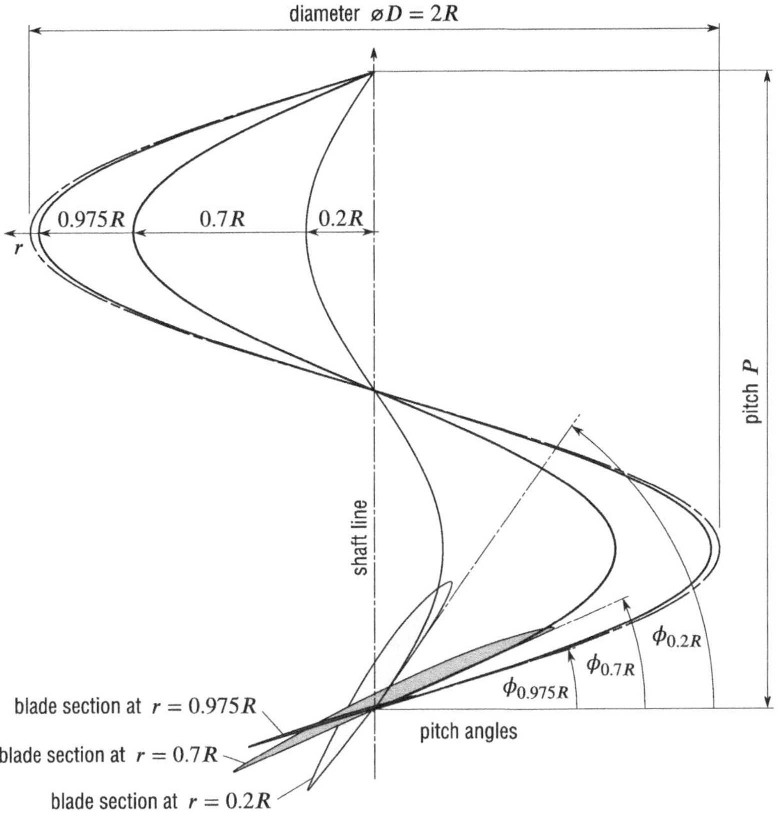

Figure 35.5 Helical paths for propeller with constant pitch P

shorter than the helix arc length toward the tip, inner radii must have a greater pitch angle in order to travel the same pitch distance P as the outer radii in one revolution. If the pitch is independent of the radii, i.e.

$$P(r) = P = \text{const.} \tag{35.3}$$

we have a propeller with constant pitch.

Pitch–diameter ratio P/D

Pitch P is a length and measured in meters. For design purposes the dimensionless pitch–diameter ratio P/D is used to specify the propeller pitch. Pitch–diameter ratios vary from 0.5 to 1.4 and sometimes higher for racing boats. High vessel speed is usually combined with a high pitch–diameter ratio which provides better efficiency compared with a higher rate of revolution. However, propellers with high P/D values need engines capable of generating high torque at low rate of revolutions.

Design impact of P/D

Besides the diameter, pitch–diameter ratio is also an important parameter in propeller selection and design. P/D is fine-tuned to provide the highest possible efficiency for a given vessel speed and wake field. Errors in finding the optimal pitch–diameter ratio will lead to poor performance of the vessel.

Pitch angle

The connection between pitch angle ϕ and pitch P is found by unrolling the helix on a flat surface (Figure 35.6). Pitch P and circumference $2\pi r$ form a right-angled triangle.

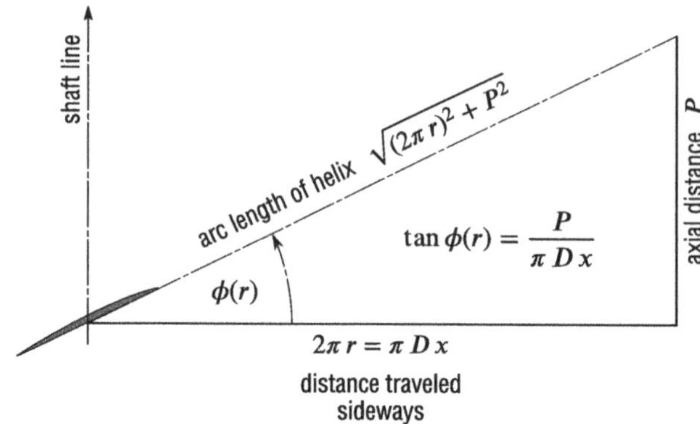

Figure 35.6 Relationship between pitch P and pitch angle ϕ

The hypotenuse is equal to the arc length of the helix.

$$\tan\phi(r) = \frac{P}{2\pi r} \quad \text{or} \quad \phi(r) = \tan^{-1}\left(\frac{P}{2\pi r}\right) \tag{35.4}$$

Pitch angle example Some examples might help to get comfortable with these geometric relationships. Consider a propeller of diameter $D = 3.2$ m and a constant pitch distribution with pitch–diameter ratio $P/D = 0.8$. Therefore, pitch is equal to

$$P = D\frac{P}{D} = 3.2 \text{ m} \cdot 0.8 = 2.56 \text{ m}$$

Using Equation (35.4), the following pitch angle distribution is found:

x [–]	r [m]	$\tan\phi$ [–]	ϕ [rad]	ϕ [deg]
0.20	0.32	1.2732	0.9050	51.854
0.30	0.48	0.8488	0.7038	40.326
0.40	0.64	0.6366	0.5669	32.482
0.50	0.80	0.5093	0.4711	26.990
0.60	0.96	0.4244	0.4014	22.997
0.70	1.12	0.3638	0.3489	19.991
0.80	1.28	0.3183	0.3082	17.657
0.90	1.44	0.2829	0.2757	15.798
1.00	1.60	0.2546	0.2493	14.287

Note that the pitch angle distribution is nonlinear and that the pitch angle at the tip may vanish only if the pitch is zero (unlikely case).

Variable pitch Variable pitch means that the pitch is a function of the radius r. The helix of different radii no longer end in the same point (Figure 35.7). This is admittedly confusing because the propeller is certainly not disintegrating. A steel bolt and corresponding thread in a steel plate with variable pitch would make no sense. The bolt would not be able to turn. Again, we have to remind ourselves that the propeller moves through a fluid which gives way. Therefore, blade sections of propellers with variable pitch may travel different distances relative to the water but an outside observer still sees the propeller moving with ship speed.

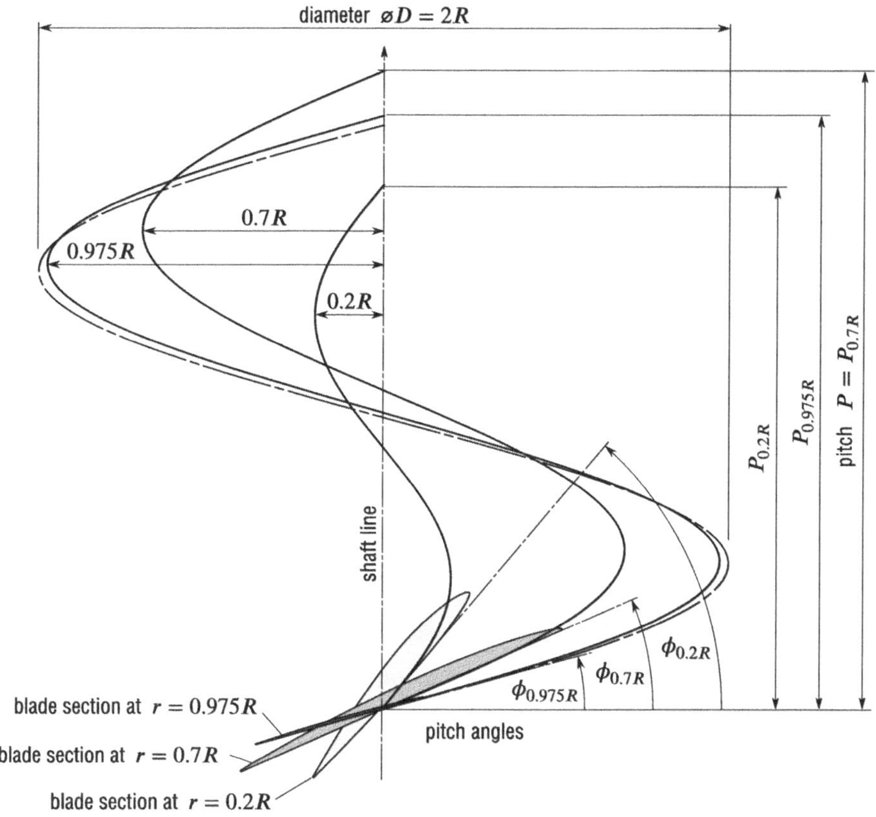

Figure 35.7 Helical paths for propeller with variable pitch P

It is common practice to reduce the pitch for inner and outer radii to adjust angles of attack to the wake field and to lower the risk of cavitation. The reduction of pitch reduces the angle of attack which in turn lowers the amount of lift produced by the foil sections with reduced pitch angle. Lower lift force requires a smaller drop in pressure which means less cavitation.

Variable pitch $P(r)$ should not be confused with controllable pitch where the whole blade can be rotated. Both a fixed pitch and a controllable pitch propeller may feature variable pitch distributions. The pitch $P_{0.7}$ at radius $x = 0.7$ is often stated as a representative pitch value for variable pitch propellers.

The distance actually traveled through the water is smaller than the geometric pitch P. The difference is known as propeller slip.

Propeller slip

Last but not least, we explain the expanded area ratio A_E/A_0, which is arguably the most abstract of the propeller parameters.

Expanded area ratio A_E/A_0

$$\frac{A_E}{A_0} = \frac{Z\,R \int\limits_{x_h}^{1} c(x)\,\mathrm{d}x}{\frac{\pi\,D^2}{4}} \tag{35.5}$$

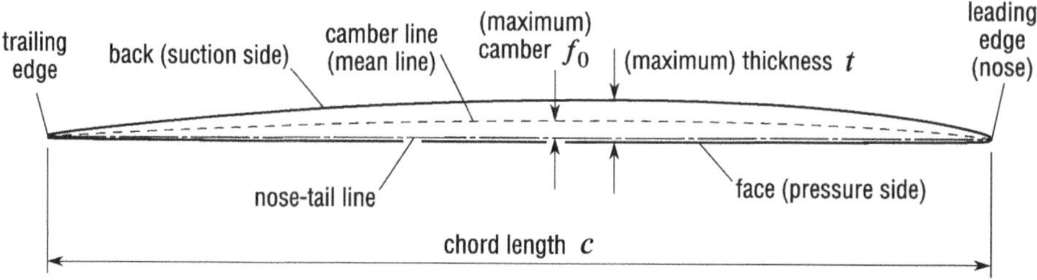

Figure 35.8 Basic geometric properties of a lifting hydrofoil

Foil sections We take another look at the foil sections derived by intersecting the blade with a cylinder (Figure 35.3). Figure 35.8 shows a foil section laid out flat and rotated to zero pitch angle. The distance from leading to trailing edge of a foil section is called chord length c. Most foils used in propellers have a camber f_0, i.e. their center line is slightly bent toward the back side of the foil. Camber is usually just a few percent (low single digits) of the chord length. The bigger the camber, the more lift a foil produces at zero angle of attack. Strength requirements determine the foil section thickness t. Classification rules define minimum requirements. Thickness is highest at the root of the blade and declines toward the tip where bending stresses in the blade vanish.

Expanded blade If all foil sections are extracted, rolled flat, and rotated to zero pitch angle, the expanded blade is obtained as shown in Figure 35.9. Foil sections would normally appear as thin lines in this view but the sections have been rotated upright to make them visible. The expanded blade reveals the true chord length distribution $c(x)$ as a function of the dimensionless blade radius $x = r/R$. The blade reference line marks the position of the mid chord. For manufacturing purposes points on the blade surfaces are specified with respect to the generator line. The generator line bisects the chord length on the smallest radius used to define the blade shape. The corresponding chord length at $x = 0.15$ is bisected in Figure 35.9. The expanded view also shows a curve marking the location of maximum foil section thickness. Both reference line and maximum thickness location are fair curves.

The expanded area of a single blade is obtained by integration of the chord length distribution from hub radius to blade tip.

$$\frac{A_E}{Z} = R \int_{x_h}^{1} c(x)\, \mathrm{d}x \qquad \text{expanded area of propeller blade} \qquad (35.6)$$

The result is equal to the gray shaded area in Figure 35.10. Multiplying Equation (35.6) by the number of blades Z results in the expanded area of the propeller:

$$A_E = Z\,R \int_{x_h}^{1} c(x)\, \mathrm{d}x \qquad \text{expanded area of propeller} \qquad (35.7)$$

Division by the propeller disk area A_0 yields the dimensionless expanded area ratio A_E/A_0.

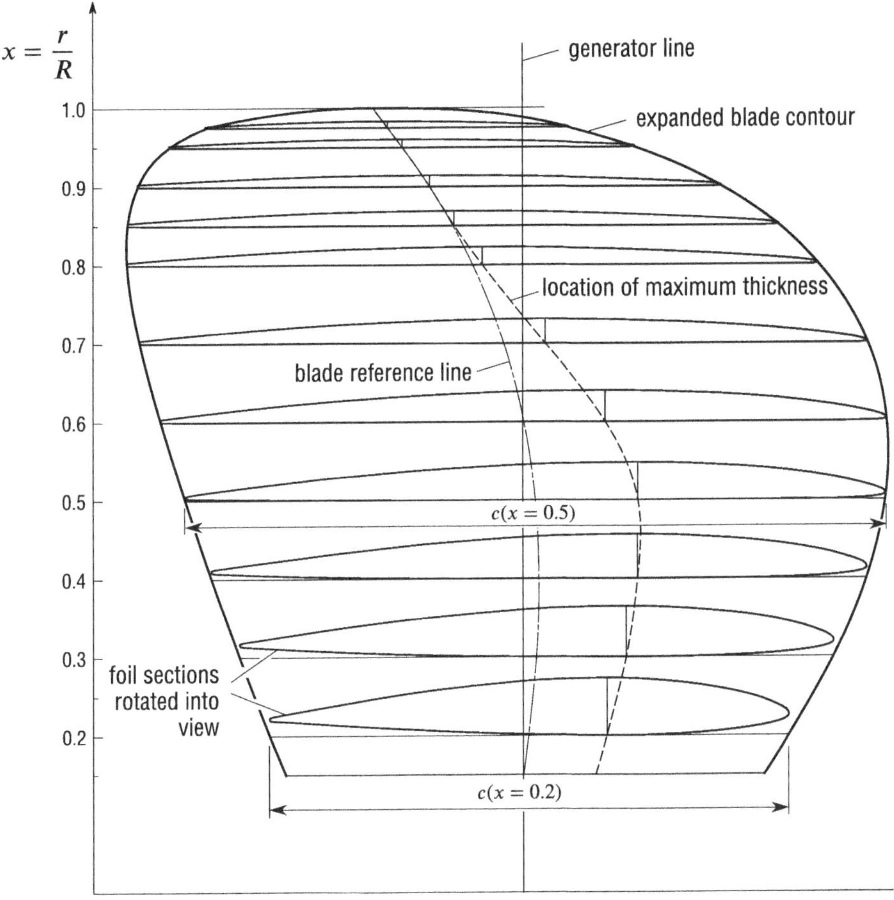

Figure 35.9 The expanded blade for a Wageningen B-Series propeller with four blades and $A_E/A_0 = 0.85$. Drawing is based on data from Oosterveld and van Oossanen (1975)

Design impact of A_E/A_0

The expanded area ratio plays a key role in cavitation prevention. A higher expanded area ratio means that thrust generation is spread over a larger blade area. This in turn requires smaller pressure differences on the blade compared with a propeller with lower expanded blade area ratio. Smaller pressure differences lowers the danger of cavitation (see Chapter 45 on cavitation). On the other hand, larger blade area increases frictional losses and diminishes propeller efficiency. Therefore, as a general rule, expanded area ratio A_E/A_0 is chosen as low as possible for high efficiency but as high as necessary to prevent cavitation. Excessive cavitation would lead to even higher efficiency losses and might ultimately damage the propeller.

Typically, expanded area ratios for fixed pitch propellers vary between 0.5 and 1.1. The lower the thrust loading coefficient, the lower is the necessary expanded area ratio. Fast merchant vessels like container ships feature expanded area ratios between 0.8 and 1.0. Expanded area ratios for tankers and bulk carriers range between 0.5 and 0.7. Controllable pitch propellers are available up to about $A_E/A_0 = 0.75$. For higher expanded area ratios the capability to reverse thrust by turning the blades to negative pitch angles is lost because the blades will make contact during the rotation. Figure 35.11 illustrates two typical expanded area ratios for propellers with four blades.

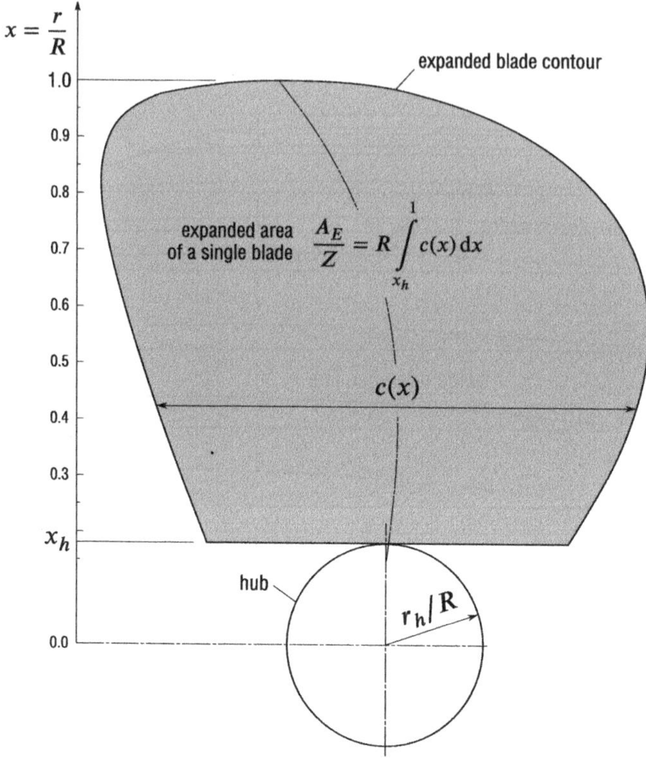

Figure 35.10 Definition of the expanded area A_E/Z of a propeller blade

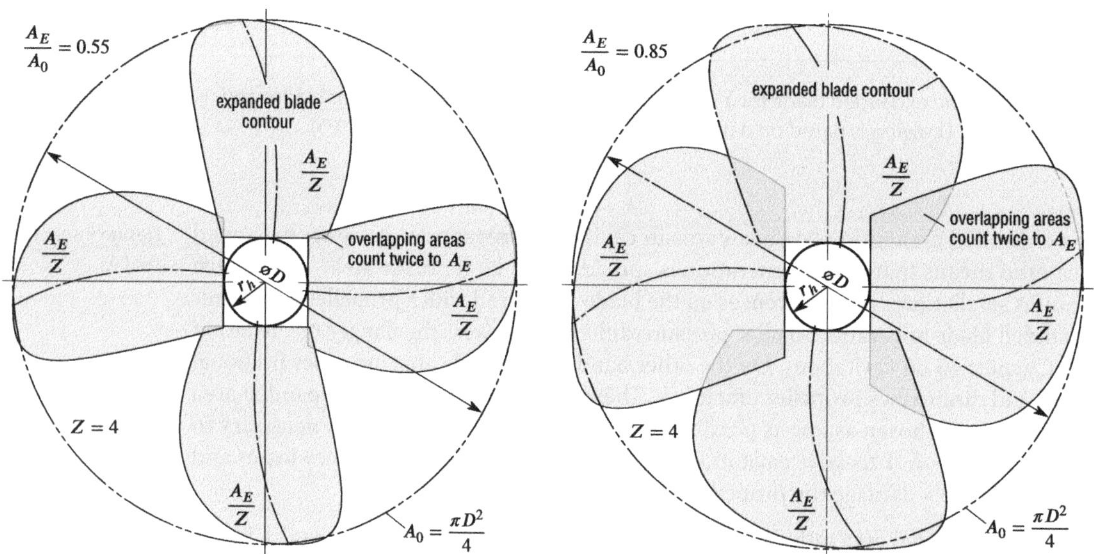

Figure 35.11 Two examples of expanded area ratios. $A_E/A_0 = 0.55$ on the left and $A_E/A_0 = 0.85$ on the right

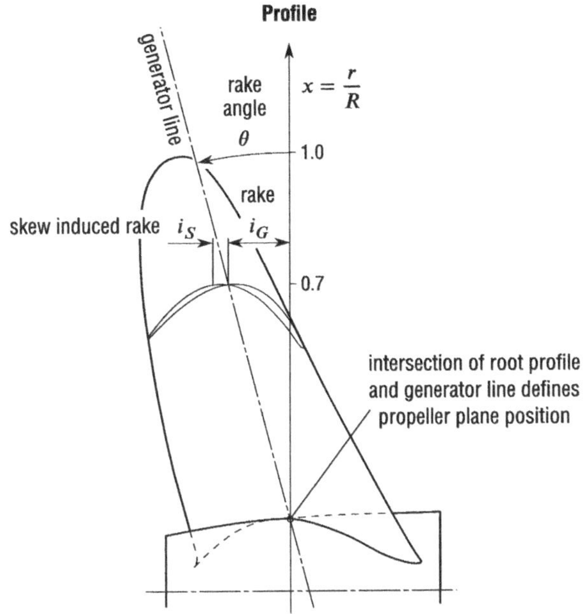

Figure 35.12 Side elevation of a propeller blade and definition of propeller rake $i_G(x)$ and rake angle θ

35.3 Other Geometric Propeller Characteristics

Rake and rake angle

The generator line is drawn upwards on the expanded view and used to define the blade in its 12 o'clock position. For most propellers the generator line is inclined aft by some degrees when viewed in the side elevation (Figure 35.12). The rake angle θ of the generator line is measured in degrees from a normal to the propeller shaft. Older propeller designs, like the Wageningen B-Series, have rake angles θ of up to 15 degrees. The rake was used to increase clearance between propeller and hull. Its influence on performance is small.

More contemporary designs tend to have smaller rake angles. In some cases the angle of the generator line to the vertical axis is reduced toward the tip. The generator line is no longer straight. Due to the rake angle, blade sections are shifted backward parallel to the propeller shaft direction. The distance is known as the rake i_G and measured in meters. The subscript $_G$ stands for the generator line influence. Rake is considered positive if blade sections are shifted aft. This is the usual case.

Skew-back

Additional rake of blade sections is created by propeller skew or, more appropriately, skew-back. Figure 35.13 illustrates the definitions. A blade is skewed if the projection of the blade reference line into the propeller plane deviates from the generator line. Positive skew-back shifts the blade section aft along its pitch helix, i.e. against the direction of rotation for forward thrust. The skew-back is equal to the arc length of the backward shift measured along the helix between points ① and ② in Figure 35.13. The backward shift along the helix causes an additional skew induced rake i_S which is equal to the component of the skew-back that is parallel to the shaft line. The skew-back is a function of the radius r.

Skew angle

Skew-back is commonly measured in degrees as the angle between the intersections of the generator line ① and intersection of the blade reference line ② with the cylinder at

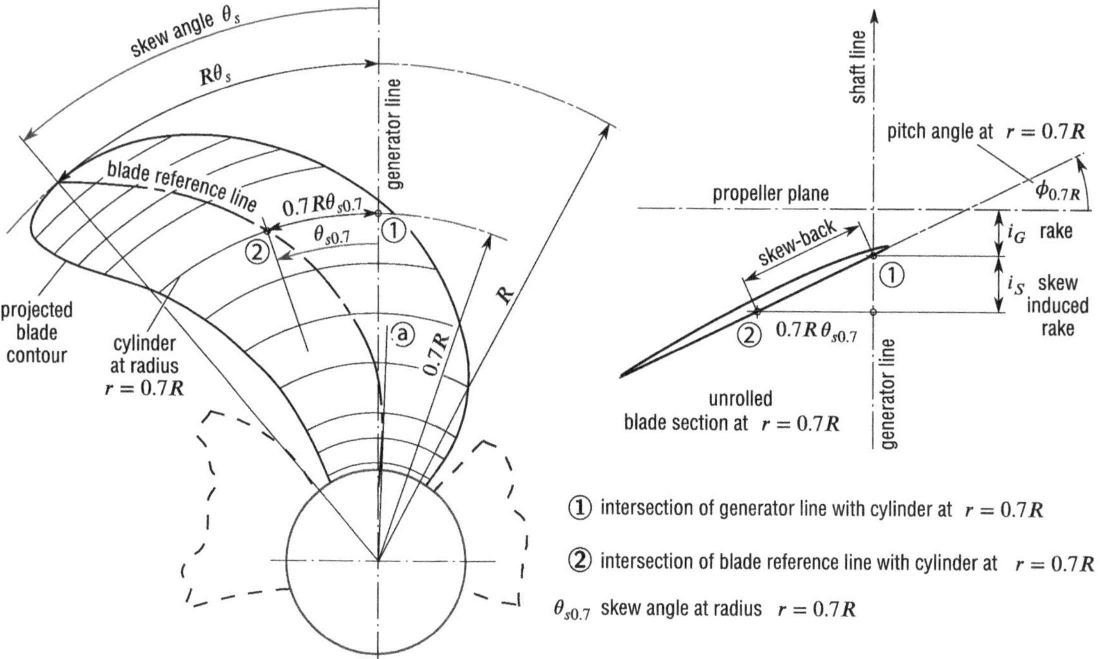

Figure 35.13 Definition of propeller skew-back and skew angle θ_s

radius r. The skew angle is sometimes also referred to as warp. Use of the alternative definition ⓐ is discouraged. This alternative replaces point ① with the intersection of the cylinder and a radial which is tangent to the blade reference line.

The skew angle θ_s for the blade tip is used to quantify the skew of a propeller. However, each blade section has its own skew angle as indicated for the radius $x = 0.7$ in Figure 35.13. Skew angles may be negative for radii close to the hub, before the reference line bends backwards for positive propeller skew angles toward the blade tip. In the projected view (looking forward along the propeller shaft) only the arc $r\,\theta_s(r)$ is visible. The actual skew-back is equal to the length $\sqrt{(r\,\theta_s(r))^2 + i_S^2(r)}$.

High skew propeller Propellers with a skew angle of more than 25 degrees at the tip are considered high skew propellers. The blade pictured in Figure 35.13 falls into this category.

Impact of skew-back Blades with high skew-back are considerably more expensive. Assessment of propeller strength of highly skewed blades requires detailed analysis. Skew-back is applied to minimize a propeller's noise signature. Passage of the propeller blades underneath the stern causes pressure fluctuations with a pronounced peak at the hull plating closest to the tip. Those pressure fluctuations excite vibrations in the hull structure which may be strong enough to be felt and definitely can be heard by sensitive underwater microphones. Skew-back of blades flattens the pressure peak and helps to reduce vibrations. Skew-back may also be used to influence cavitation patterns on propeller blades. The impact of skew-back on the efficiency is fairly small for forward motion. However, high skew propellers may be quite inefficient for operations with reversed direction of rotation.

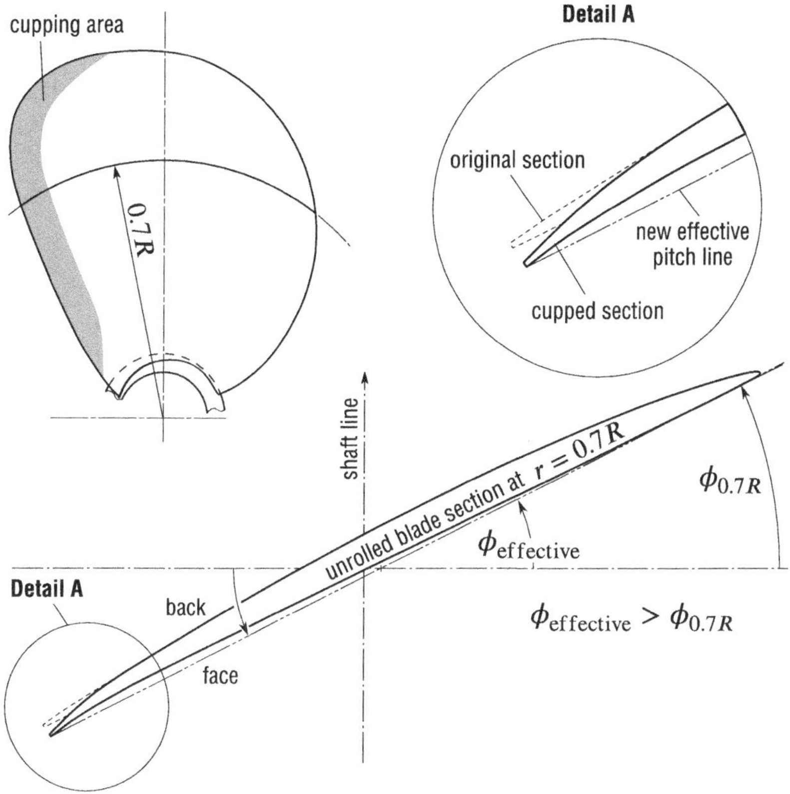

Figure 35.14 Cupping of a propeller blade

A feature seemingly popular with powerboat enthusiasts is to cup the propeller blades. The blade sections are bent toward the pressure side (face) close to the trailing edge (Figure 35.14). Similar to flaps at the trailing edge of a wing, cupping increases the camber of a foil, which results in a higher lift coefficient for the same angle of attack. Cupping may also be interpreted as a correction of the effective pitch of the propeller. Increases in effective pitch angle result in a propeller that produces more thrust, but also requires more torque. In many cases, efficiency of the propulsion system does not increase. On the contrary, if the effective pitch is increased too much the engine torque might be insufficient to bring the propeller up to its design rate of revolution, with detrimental effects on performance. This is especially true in service with additional weather related resistance components. A properly designed propeller will not need cupping. **Cupping**

References

Breslin, J. and Andersen, P. (1994). *Hydrodynamics of ship propellers.* Cambridge University Press, Cambridge, United Kingdom.

Carlton, J. (2007). *Marine propellers and propulsion.* Butterworth-Heinemann, Oxford, United Kingdom.

Kerwin, J. and Hadler, J. (2010). *Propulsion*. The Principles of Naval Architecture Series. Society of Naval Architects and Marine Engineers (SNAME), Jersey City, NJ.

Oosterveld, M. and van Oossanen, P. (1975). Further computer-analyzed data of the Wageningen B-screw series. *International Shipbuilding Progress*, 22:251–261.

Self Study Problems

1. State the four most important geometric propeller characteristics.
2. Consider the example about pitch angles on page 426. Assume the data describes a controllable pitch propeller. Compute the radial pitch distribution $P(x)$ if the blade is rotated by 12 degrees.
3. Consider a propeller with the following radial pitch distribution:

x	r	$P(x)$
[–]	[m]	[m]
0.20	0.40	2.4000
0.30	0.60	2.7200
0.40	0.80	2.9500
0.50	1.00	3.0000
0.60	1.20	3.0000
0.70	1.40	3.0000
0.80	1.60	3.0000
0.90	1.80	2.9700
1.00	2.00	2.8000

   Plot the pitch distribution as a function of x. Make x the vertical axis of the plot. State pitch–diameter ratio and diameter of the propeller. Compute the pitch angle distribution $\phi(x)$.

4. A standard ceiling fan uses inclined but flat boards as wings. Is this a variable pitch or constant pitch propeller?
5. Why is the pitch of propeller blades sometimes reduced toward hub and tip of the blade?
6. Which criterion is used to select the expanded area ratio of the propeller?

36

Lifting Foils

A basic element of a propeller blade is the foil section. Figure 35.3 visualizes how foil sections are embedded in a blade. Foil sections are streamlined shapes designed to generate a lift force perpendicular to the onflow, hence the name lifting foil. Lifting foils are also used in rudders, stabilizer fins, and sails. In this chapter we discuss geometry and basic flow patterns across lifting foils. We also introduce thin foil theory which is one of the most widely used tools in foil design.

Learning Objectives

At the end of this chapter students will be able to:

- explain geometry and flow patterns of lifting foils
- define lift and drag coefficients
- understand the assumptions of thin foil theory

36.1 Foil Geometry and Flow Patterns

Foil section geometry

The extent of a foil from trailing edge to leading edge is called chord length c (Figure 36.1). The straight tail–nose line connects trailing and leading edge and serves as the x-axis of the foil coordinate system. Here we place the origin in the middle of the chord length c. Chord length c together with thickness distribution $t(x)$ and camber distribution $f_m(x)$ defines the geometry of a foil section. For the leading edge, a radius r_{LE} (nose radius) may be specified to better define the shape of the hydrodynamically important nose. The trailing edge of foils is comparatively sharp in most cases. However, thickness at the trailing edge has to be finite for strength and manufacturing reasons. Superscripts $^+$ and $^-$ are used to identify points or properties of upper and lower foil contour, respectively. Points $\underline{f}^+$ on the upper foil side are obtained by moving a distance $t(x)$ upward in a direction normal to the camber line. The lower foil side is obtained by moving $t(x)$ downward normal to the camber line. The following equations define upper $^+$ and lower $^-$ foil contour.

$$x^{\pm} = x \mp t(x)\sin\big(\theta(x)\big) \tag{36.1}$$

$$f^{\pm} = f_m(x) \pm t(x)\cos\big(\theta(x)\big) \tag{36.2}$$

Fundamentals of Ship Hydrodynamics: Fluid Mechanics, Ship Resistance and Propulsion, First Edition.
Lothar Birk.

Companion website: www.wiley.com/go/birk/hydrodynamics

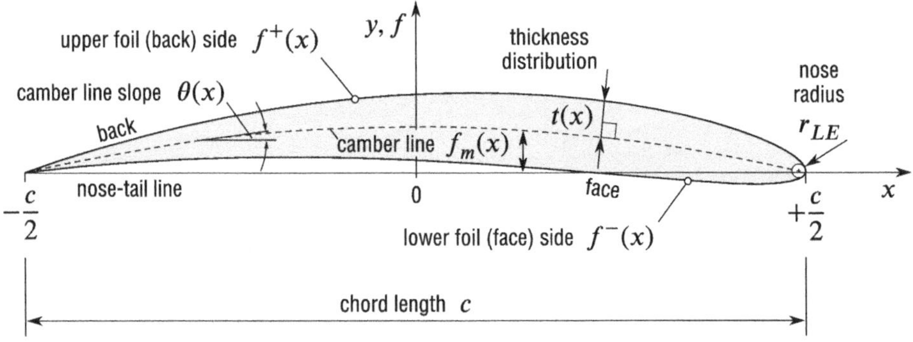

Figure 36.1 Definition of foil geometry

The minus sign in the right-hand side of Equation (36.1) is listed on top because it applies to the upper side of the foil. $\theta(x)$ measures the angle of the camber distribution $f_m(x)$ with respect to the tail–nose line.

$$\theta(x) = \tan^{-1}\left(\frac{\mathrm{d}f_m}{\mathrm{d}x}\right) \tag{36.3}$$

Typically, the ratio f_0/c of maximum camber f_0 and chord length c is less than 2%. Consequently, the angle θ is small, which justifies the simplifications $\sin(\theta) \approx 0$ and $\cos(\theta) \approx 1$. This is equivalent to defining the foil contours by moving up or down from the line perpendicular to the tail–nose line rather than normal to the camber line.

$$x^{\pm} = x \tag{36.4}$$

$$f^{\pm} = f_m(x) \pm t(x) \tag{36.5}$$

The documentation of the foil shape should specify which version of the coordinate definition is employed.

With the simplified foil contour description (36.4) and (36.5), thickness and camber distribution are equivalent to half the difference and half the sum of upper and lower contour y-coordinates respectively.

$$t(x) = \frac{1}{2}\left(f^+ - f^-\right) \qquad f_m(x) = \frac{1}{2}\left(f^+ + f^-\right) \tag{36.6}$$

Owing to the latter equation, the camber line is also referred to as the mean line.

Flow patterns Figure 36.2 shows a typical flow pattern and the associated pressure distribution. The foil is rotated anticlockwise by the angle of attack α with respect to the onflow $\underline{v}_0$. On the upper (back) side of the foil, streamlines bunch together. The continuity equation tells us that a narrowing of streamlines indicates an acceleration of the flow. Thus, much of the upper side of the foil sees flow speeds $|\underline{v}|$ higher than the speed of the onflow $|\underline{v}_0|$. According to Bernoulli's equation, increased flow speed causes a reduction in pressure. This is clearly visible in the negative pressure coefficient C_p^+ values for the back of the foil. As stated before, a negative pressure coefficient does not mean the pressure is negative; rather, it implies that local pressure p is smaller than the reference pressure p_0. Therefore, the alternative name for the back of the foil is suction side. Especially

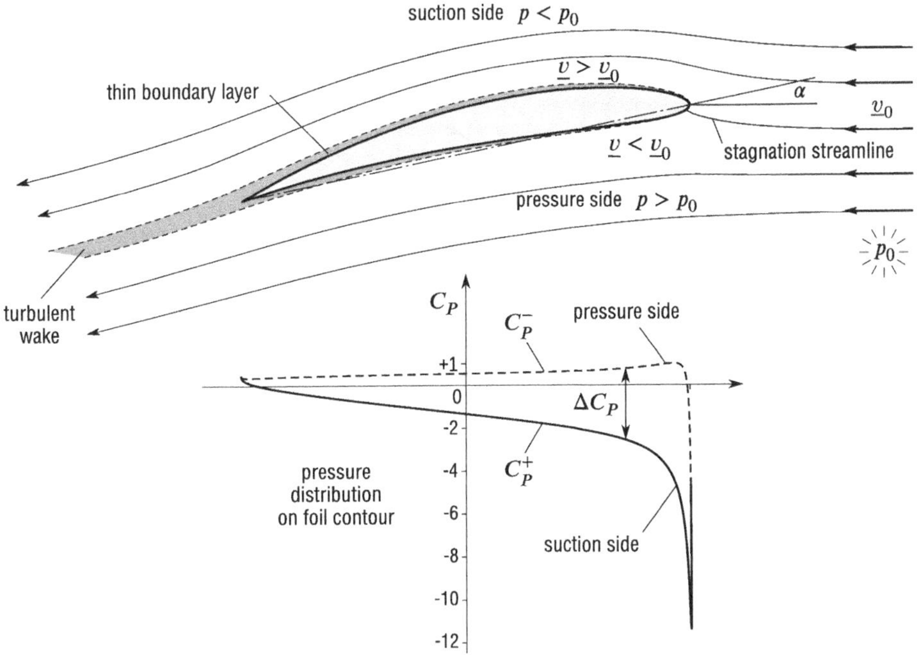

Figure 36.2 Flow pattern and pressure distribution for a 2D foil section at angle of attack α

toward the leading edge, we encounter low pressure and high flow speeds when the foil is operating at large angles of attack.

The opposite happens on the underside of the foil, the pressure side. Flow is decelerating to a stop at the stagnation point and is slower than the onflow across almost all of the face side of the foil. The pressure coefficient C_p^- is positive. The pressure difference between upper and lower side of a foil determines its lift force.

$$\Delta C_p = C_p^+ - C_p^- \tag{36.7}$$

Friction between fluid and foil develops into a thin boundary layer. The resulting wall shear stress creates a drag force.

Stall

Note that in Figure 36.2, the depicted angle of attack α is rather high with about 12 degrees. Typical operational angles of foils are much smaller. Increases in angle of attack will cause the foil to stall. During a stall event, the flow separates already at the leading edge, as illustrated in Figure 36.3. Instead of a flow with high speed and low pressure, the foils produce a large wake filled with large eddies. The lift force all but vanishes and drag becomes the dominant force.

Stall causes a sudden loss of lift and flight control in airplanes. If the pilots fail to counteract immediately, the plane will crash. The stalling of marine propeller blades has less severe consequences but will also result in loss of performance and increased vibrations.

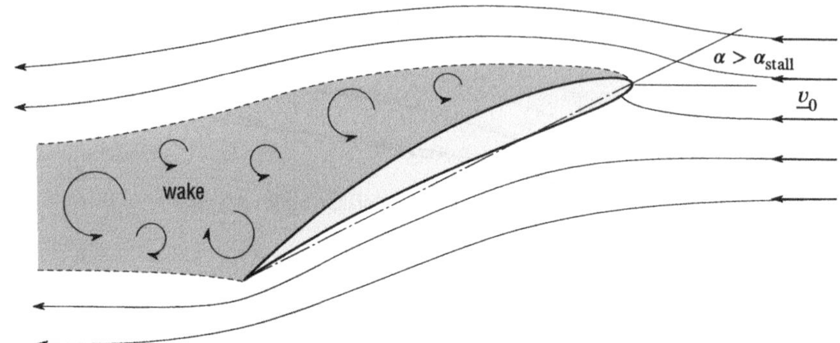

Figure 36.3 Foil in stalled flow condition

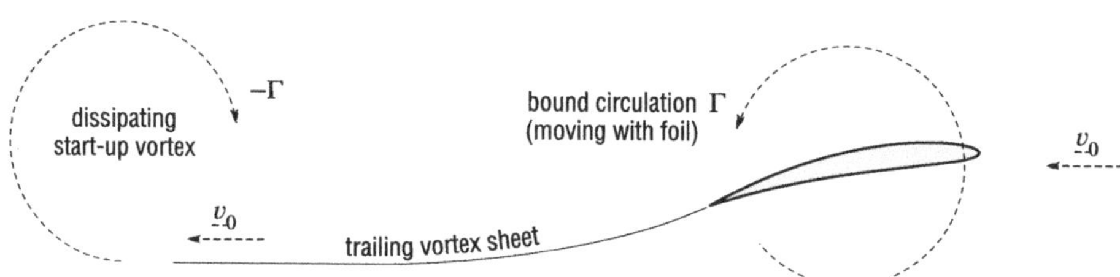

Figure 36.4 Complete vortex system of the foil section

36.2 Lift and Drag

Kutta-Joukowsky's lift theorem

The German mathematician Martin Wilhelm Kutta (*1867–†1944) and the Russian scientist Nikolay Yegorovich Joukowsky (*1847–†1921, name is anglicanized from Russian) derived independently from each other the Kutta-Joukowsky's lift theorem.

$$L = \rho \, |\underline{v}_0| \, \Gamma \tag{36.8}$$

Note that the section lift L has the dimension of force per unit length [N/m]. Kutta is most famous as one of the inventors of the Runge-Kutta method for solving ordinary differential equations and for the Kutta condition which is necessary to describe lifting flows with potential flow theory. Joukowsky is one of the founding fathers of aerodynamics.

Lift may be accurately computed with potential flow theory assuming an inviscid, incompressible, and irrotational fluid. A nonzero circulation $\Gamma \neq 0$ requires the curl of the velocity vector to vanish as well, i.e. $\underline{\nabla} \times \underline{v} \neq \underline{0}$. This seems to contradict the application of potential flow. However, as explained in Chapter 15, the condition of vanishing curl only applies to simply connected regions. If we include the foil, the flow region becomes multiply connected.

The effect of circulation on a lifting flow can be modeled with vortices. According to Helmholtz's theorems, vortex strength must be constant in inviscid fluids. So, where did the circulation come from if it was not there already (foil at rest)? When the foil is accelerating and circulation is established around the foil, a start-up vortex is left behind (Figure 36.4). The start-up vortex is of equal strength but rotates in the opposite

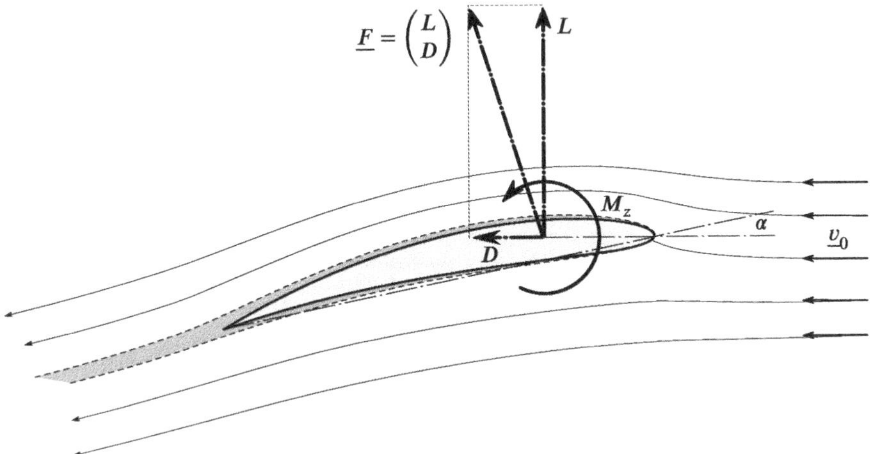

Figure 36.5 Forces acting on the foil section at angle of attack α

direction compared with the circulation that is traveling with the foil. Therefore, the sum of vorticity in the entire flow field is still zero.

Although this is a crude model of the physics, start-up vortices are real. The start-up vortex left behind by a big commercial jet at takeoff may easily flip over a small single propeller aircraft that happens to fly through it. It is one of the reasons why aircraft of different sizes need longer time periods between landing and takeoff than aircraft of equal size. The start-up vortex is dissolved by friction over time, and the vortices are blown away by the wind.

Lift and drag force

Figure 36.5 shows lift force L and drag force D acting on a foil section. Lift acts perpendicular to the onflow. However, an additional drag force D exists in viscous (real) fluids which acts in the direction of the onflow $\underline{v}_0$. Viscous pressure drag for thin foils is small as long as the angle of attack is small. It increases with increasing angle of attack α. Frictional drag is the major drag component and also depends on the surface roughness of the foil. Increased surface roughness due to fouling or damage is especially detrimental to performance if it affects the leading edge and the first quarter of the suction side. See Section 13.5 for details.

Pitch moment

Besides lift and drag forces there is also a rotational moment which aspires to flip the foil over. Aircraft have a tail unit to counterbalance the moment. The moment is of interest for the design of controllable pitch propellers.

Lift and drag coefficients

Hydrodynamic properties of foil sections are often provided as lift and drag coefficients C_l and C_d respectively.

$$C_l = \frac{L}{\frac{1}{2}\rho v_0^2 c} \tag{36.9}$$

$$C_d = \frac{D}{\frac{1}{2}\rho v_0^2 c} \tag{36.10}$$

These two-dimensional foil section coefficients use the chord length c as reference length. Note that this is different from the three-dimensional resistance coefficients

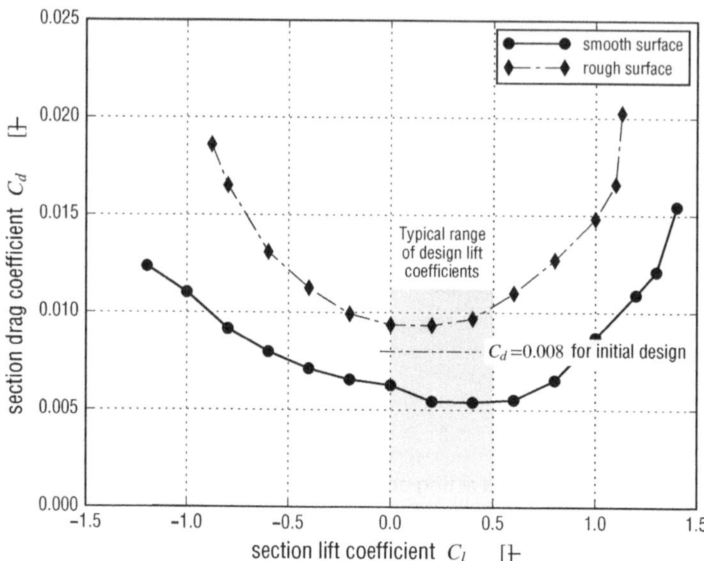

Figure 36.6 Typical lift–drag curves for a thin, cambered foil

used in ship hydrodynamics which use the wetted surface of the hull as reference area rather than a projection onto a plane.

Lift–drag ratio The quality of a foil section may be expressed as the lift–drag ratio L/D or C_l/C_d. The more lift is generated per unit drag, the more effective is the foil. Form drag of the foil section increases with the angle of attack α and the lift–drag ratio decreases. This may be best visualized in a plot of drag coefficient C_d over lift coefficient C_l as illustrated in Figure 36.6. The graph shows typical section drag coefficient curves for a smooth and a rough foil section. The foil has a small camber and its most efficient range of operation is for section lift coefficients between 0.0 and 0.5. Lift–drag curves for many standard foil section shapes can be found in Abbott and von Doenhoff (1959).

In the following we will discuss a method suitable for analyzing the potential flow around two-dimensional foil sections. It is surprisingly accurate in predicting the lift force of foil sections. Drag has to be determined from model tests or by semi-empirical formulas which combine theory and experimental results. A section drag coefficient of $C_d = 0.008$ is often used for initial propeller design.

36.3 Thin Foil Theory

Conformal mapping Conformal mapping is the most accurate method for the analysis of two-dimensional potential flow around foil sections. Conformal mapping transforms the known solution of the flow around a cylinder into the flow around a foil. The geometries of cylinder and foil are represented in complex planes $z = x + \mathrm{i}y$ and $\zeta = \xi + \mathrm{i}\eta$ respectively. At the heart of the procedure is a complex transformation function which maps points of the z-plane (cylinder) into points of the ζ-plane (foil). Finding the appropriate transformation

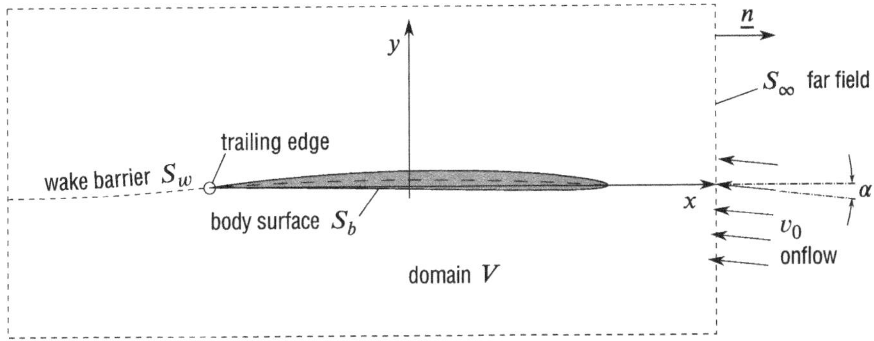

Figure 36.7 Setup of boundary value problem for a thin foil operating at angle of attack α

function for a given geometry is the most difficult task. Readers interested in lifting flow are encouraged to study conformal mapping. Detailed descriptions are found in Abbott and von Doenhoff (1959) and Katz and Plotkin (2001).

Thin foil theory assumptions

Although conformal mapping is a powerful tool for the analysis of two-dimensional forms, it is less well suited for design purposes. In conformal mapping, hydrodynamic properties are evaluated for a given shape. During design of lifting foils we would like to specify desired dynamic characteristics first, e.g. a section lift coefficient, and then find the geometry which produces this characteristic. A suitable tool for this purpose is thin foil theory. We assume that the foil geometry is given in the form of a mean line $f_m(x)$ and a thickness distribution $t(x)$. Offsets of upper and lower foil contour $f^{\pm}(x)$ may be derived from Equations (36.4) and (36.5). A foil is considered thin if its slope with respect to the x-axis (tail–nose line) is small.

$$\frac{\partial f^{\pm}}{\partial x} \ll 1 \qquad \text{for } x \in [-c/2, c/2] \tag{36.11}$$

In addition, thin foil theory assumes that the angle of attack α is small.

$$\alpha \ll 1 \qquad \text{angle of attack in radians} \tag{36.12}$$

36.3.1 Thin foil boundary value problem

Analogous to the analysis of the flow around a cylinder, we seek the velocity potential Φ as the solution of a boundary value problem. We use a body fixed $< x, y >$ coordinate system, with the x-axis falling onto the tail–nose line and pointing from trailing edge to leading edge (Figure 36.7). The velocity potential Φ is initially comprised of two parts:

- a uniform parallel flow in the far field which is oriented at the angle of attack α with respect to x-axis. It has the velocity potential

$$\phi_\infty(x, y) = -v_0\, x \cos(\alpha) + v_0\, y \sin(\alpha) \tag{36.13}$$

  The velocity v_0 is the speed of the foil relative to the fluid in the far field.

- a disturbance caused by the foil which fades away in the far field. Its potential $\phi(x, y)$ has yet to be determined.

We exploit again the superposition principle grounded in the linearity of the Laplace equation. The total flow potential is simply the sum of its parts.

$$\Phi(x, y) = \phi_\infty(x, y) + \phi(x, y) \tag{36.14}$$

The boundary of the fluid domain V consists of the foil contour S_b, the far field S_∞, and the wake barrier S_w (Figure 36.7). The latter is needed to determine the circulation Γ. As before, the normal vector $\underline{n}$ is pointing out of the fluid domain. That means $\underline{n}$ is pointing into the foil contour on S_b.

Boundary value problem The boundary value problem consists of the Laplace equation as the field equation representing conservation of mass of an ideal, irrotational fluid, and the boundary condition for each boundary.

$$\begin{aligned} \Delta\Phi &= 0 && \text{in } V \\ \frac{\partial \Phi}{\partial n} &= 0 && \text{body boundary condition for } S_b \\ \frac{\partial \phi}{\partial n} &= 0 && \text{far field condition for } S_\infty \end{aligned} \tag{36.15}$$

plus a Kutta condition for the wake S_w. We will specify a Kutta condition in more detail when we solve the lifting flow problem in Chapter 38. The body boundary condition requires that no fluid flows through the foil contour S_b. This means that the velocity component normal to the contour has to vanish and the contour itself becomes a streamline. The far field condition requires that the disturbance potential ϕ vanishes far away from the foil.

36.3.2 Thin foil body boundary condition

Physical interpretation The body boundary condition holds the key to a significant simplification of the problem. Therefore, it is discussed in greater detail. The body boundary condition in the boundary value problem (36.15) requests that the normal derivative of the potential vanishes. The normal derivative is the component of the gradient of the potential that points in the direction of the normal vector. It is computed via the dot product, and the body boundary condition requires that this component is of zero length, i.e. the two vectors, normal and gradient of the potential, are perpendicular to each other.

$$\frac{\partial \Phi}{\partial n} = \underline{n}^T \underline{\nabla} \Phi = 0 \tag{36.16}$$

Since the gradient of the velocity potential is equal to the velocity vector, the latter must be tangent to the foil contour. No flow passes through the foil surface.

Normal vector First, we need to relate the normal vector to the geometry of the foil contour. As for the discussion of the free surface condition in linear wave theory, we make use of implicit functions. As before, we use superscripts $^+$ and $^-$ to identify properties of upper and lower foil contour, respectively.

$$\text{upper foil surface} \quad F^+(x, y) = y - f^+(x) = 0 \tag{36.17}$$

$$\text{lower foil surface} \quad F^-(x, y) = y - f^-(x) = 0 \tag{36.18}$$

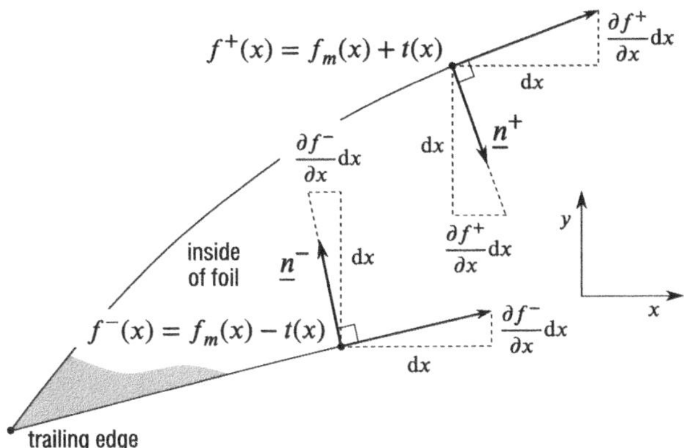

Figure 36.8 Definition of normal vectors for upper and lower foil surface

The implicit definition represents an isoline of $F^{\pm}$, here $F^{\pm} = \text{const.} = 0$. The gradient of a function is pointing in the direction of steepest ascent which is equivalent to being normal to the isolines of the function. Then, the normal vector of an implicit function must be parallel to its gradient.

$$\underline{n}^{\pm} = \mp \frac{\underline{\nabla}\, F^{\pm}}{|\underline{\nabla}\, F^{\pm}|} \tag{36.19}$$

Since the contours $f^{\pm}$ are only functions of x, the gradients of F^+ and F^- are equal to

$$\underline{\nabla}\, F^{\pm} = \begin{pmatrix} \dfrac{\partial F^{\pm}}{\partial x} \\ \dfrac{\partial F^{\pm}}{\partial y} \end{pmatrix} = \begin{pmatrix} -\dfrac{\partial f^{\pm}}{\partial x} \\ 1 \end{pmatrix} \tag{36.20}$$

As a consequence, the normal vector $\underline{n}^+$ on the suction side becomes

$$\underline{n}^+ = \frac{1}{\sqrt{1 + \left(\dfrac{\partial f^+}{\partial x}\right)^2}} \begin{pmatrix} \dfrac{\partial f^+}{\partial x} \\ -1 \end{pmatrix} \tag{36.21}$$

and the normal vector on the pressure side is

$$\underline{n}^- = \frac{1}{\sqrt{1 + \left(\dfrac{\partial f^-}{\partial x}\right)^2}} \begin{pmatrix} -\dfrac{\partial f^-}{\partial x} \\ 1 \end{pmatrix} \tag{36.22}$$

Note the different signs to make the normal vector point into the foil (out of the fluid domain V) on both suction and pressure side (see Figure 36.8).

Suction side

Substitution of the normal vector (36.21) into the body boundary condition (36.16) yields:

$$(\underline{n}^+)^T \underline{\nabla}\,\Phi = \left(\frac{\partial f^+}{\partial x}, -1\right)\begin{pmatrix}\dfrac{\partial \Phi}{\partial x}\\ \dfrac{\partial \Phi}{\partial y}\end{pmatrix} = 0 \tag{36.23}$$

$$\frac{\partial f^+}{\partial x}\frac{\partial \Phi}{\partial x} - \frac{\partial \Phi}{\partial y} = 0 \qquad \text{on } f^+ \tag{36.24}$$

In addition, we split the total velocity potential into parallel flow and disturbance potential

$$\frac{\partial f^+}{\partial x}\left(-v_0\cos(\alpha) + \frac{\partial \phi^+}{\partial x}\right) - v_o\sin(\alpha) - \frac{\partial \phi^+}{\partial y} = 0 \tag{36.25}$$

For a flat plate the change of velocity $\partial\phi/\partial x$ along the plate would vanish. For a thin foil it will not be zero but significantly smaller than the onflow velocity v_0. Together with the basic assumptions (36.11) and (36.12) of thin foil theory we introduce the following simplifications:

$$\frac{\partial f^+}{\partial x} \ll 1, \qquad \cos(\alpha) \approx 1, \qquad \sin(\alpha) \approx \alpha, \text{ and} \qquad \frac{\partial \phi^+}{\partial x} \ll v_0 \tag{36.26}$$

As a consequence, the first parenthesis in Equation (36.25) reduces to $-v_0$.

$$-\frac{\partial f^+}{\partial x} v_0 - \alpha\, v_o - \frac{\partial \phi^+}{\partial y} = 0 \tag{36.27}$$

We rearrange the terms to have the unknown y-derivative of the disturbance potential on the left-hand side.

$$\frac{\partial \phi^+}{\partial y} = -\frac{\partial f^+}{\partial x} v_0 - \alpha\, v_o \tag{36.28}$$

This is sometimes called the linearized body boundary condition for thin foil theory, although technically the body boundary condition (36.16) is already linear. Nonetheless, the body boundary condition in its form (36.28) for the upper foil surface represents a significant simplification. All terms on the right-hand side are known.

Pressure side Repeating the conversion of Equations (36.23) through (36.28) for the pressure side f^- results in the same simplified body boundary condition except that f^+ is replaced by f^-.

$$\frac{\partial \phi^-}{\partial y} = -\frac{\partial f^-}{\partial x} v_0 - \alpha\, v_o \tag{36.29}$$

or in short the body boundary condition for a thin foil now reads as

$$\frac{\partial \phi^\pm}{\partial y} = -\frac{\partial f^\pm}{\partial x} v_0 - \alpha\, v_o \qquad \text{for } y = \pm 0,\ x \in [-c/2, c/2] \tag{36.30}$$

We satisfy this condition on the upper and lower sides of the x-axis with $y = \pm 0$ because the right-hand side is a univariate function of x.

Introducing the definition (36.5) of foil geometry into the thin foil body boundary condition (36.30) allows us to split the right-hand side into three distinct parts. With $f^{\pm} = f_m + t$ we get:

Thin foil body boundary condition

$$\begin{aligned} \frac{\partial \phi^{\pm}}{\partial y} &= -\left(\frac{\partial f_m}{\partial x} \pm \frac{\partial t}{\partial x}\right) v_0 - \alpha\, v_o \\ &= \mp \frac{\partial t}{\partial x} v_0 - \frac{\partial f_m}{\partial x} v_0 - \alpha\, v_0 \qquad \text{for } y = \pm 0 \end{aligned} \tag{36.31}$$

This condition has to be satisfied on the tail–nose line between trailing and leading edge, i.e. $x \in [-c/2, c/2]$.

36.3.3 Decomposition of disturbance potential

Each term on the right-hand side of Equation (36.31) depends on a different foil characteristic.

Decomposition of disturbance potential

(i) The first term is a function of the slope of the thickness distribution $\partial t(x)/\partial x$.

(ii) The second term is defined by the slope of the camber distribution $\partial f_m(x)/\partial x$.

(iii) The last term depends on the angle of attack α.

The superposition principle for solutions of the Laplace equation enables us to subdivide the disturbance potential into three parts.

$$\phi = \phi_1 + \phi_2 + \phi_3 \tag{36.32}$$

The original boundary value problem is split into three separate boundary value problems. Table 36.1 summarizes their key points. In practice, the two lifting flow problems are solved together. The potential ϕ_1 depends only on the thickness distribution $t(x)$ and describes the flow around a symmetric foil. This is called the displacement flow problem or thickness problem. For zero angle of attack the flow will be symmetric as well and we do not expect a resultant force acting normal to the flow. The potentials ϕ_2 and ϕ_3 capture the flow effects caused by camber and the angle of attack, respectively. In both cases, the flow will be asymmetric to the x-axis and a lift force will be generated. Together, ϕ_2 and ϕ_3 describe the lifting flow. In the following two chapters we will solve displacement and lifting flow problems under the condition that the foil is thin. We will start with the displacement flow. For the lifting flow we also have to address the question of the circulation strength Γ.

Displacement flow

Lifting flow

References

Abbott, I. and von Doenhoff, A. (1959). *Theory of wing sections – including a summary of airfoil data.* Dover Publications, Inc., New York, NY, Dover edition.

Katz, J. and Plotkin, A. (2001). *Low-speed aerodynamics.* Cambridge Aerospace Series. Cambridge University Press, New York, NY, second edition.

Table 36.1 The three subtasks of thin foil theory

displacement flow	**lifting flow**	
Symmetric foil at $\alpha = 0$	**Cambered plate at $\alpha = 0$**	**Flat plate at angle of attack α**
$\Delta\phi_1 = 0$	$\Delta\phi_2 = 0$	$\Delta\phi_3 = 0$
$\frac{\partial \phi_1^\pm}{\partial y} = \mp \frac{\partial t}{\partial x} v_0 \quad \text{on } y = \pm 0$	$\frac{\partial \phi_2^\pm}{\partial y} = -\frac{\partial f_m}{\partial x} v_0 \quad \text{on } y = \pm 0$	$\frac{\partial \phi_3^\pm}{\partial y} = -\alpha\, v_0 \quad \text{on } y = \pm 0$
$\underline{\nabla}\,\phi_1 \longrightarrow \underline{0}$ for $x \rightarrow \pm\infty$	$\underline{\nabla}\,\phi_2 \longrightarrow \underline{0}$ for $x \rightarrow \pm\infty$	$\underline{\nabla}\,\phi_3 \longrightarrow \underline{0}$ for $x \rightarrow \pm\infty$
	Kutta condition for Γ	Kutta condition for Γ
Neither camber nor angle of attack is present in this problem. It describes the flow around a symmetric foil with thickness distribution $t(x)$.	The BVP for ϕ_2 involves neither thickness distribution nor the angle of attack. Only the mean line $f_m(x)$ is required. The problem defines the flow over a cambered plate at zero angle of attack.	The geometry of the foil ($f_m(x)$ and $t(x)$) does not appear in this problem. The potential ϕ_3 represents the flow over a flat plate at angle of attack α.

Self Study Problems

1. Explain stall in your own words. When does it occur and what are its effects?
2. Summarize the underlying assumptions of thin foil theory.
3. The body boundary condition for the pressure side of a foil requires that the velocity normal vanishes.

$$-\frac{\partial f^-}{\partial x}\left(-v_0 \cos(\alpha) + \frac{\partial \phi^-}{\partial x}\right) + v_o \sin(\alpha) + \frac{\partial \phi^-}{\partial y} = 0$$

   Apply the assumptions of thin foil theory to derive the body boundary condition for thin foil theory (36.31).

4. Research the geometry of the NACA 4-digit series and the NACA 16-xxx series of foils. Learn how to extract important foil characteristics from the specification, i.e. which specifications are embedded in the designation NACA 2412. Use the provide equations to plot the shape of NACA 2412.

37

Thin Foil Theory – Displacement Flow

In this chapter we continue to discuss the potential flow around a thin foil. The upper and lower sides of the foil are represented by a mean line f_m and a symmetric thickness distribution.

$$f^{\pm}(x) = f_m(x) \pm t(x) \qquad \text{for } x \in [-c/2, c/2] \tag{37.1}$$

By introducing the requirements that

- the angle of attack α is small,
- the foil represented by $f^{\pm}(x)$ is thin with $\dfrac{\partial f^{\pm}}{\partial x} \ll 1$, and
- any velocities induced by the disturbance potential ϕ are small compared to the parallel stream velocity $\left|\dfrac{\partial \phi}{\partial x}\right| \ll v_0$

we have been able to split the flow problem into three separate boundary value problems. Here, we solve the first of these boundary value problems which describes the flow around a symmetric foil with finite but small thickness–chord length ratio t/c, no camber, and at zero angle of attack.

Learning Objectives

At the end of this chapter students will be able to:

- state and solve boundary value problems
- understand and apply singularity distributions
- discuss and compute the pressure distributions around symmetric foils

37.1 Boundary Value Problem

The flow around the symmetric foil at zero angle of attack may be represented by a velocity potential Φ which consists of two parts: A flow potential ϕ_∞, which describes

Fundamentals of Ship Hydrodynamics: Fluid Mechanics, Ship Resistance and Propulsion, First Edition.
Lothar Birk.

Companion website: www.wiley.com/go/birk/hydrodynamics

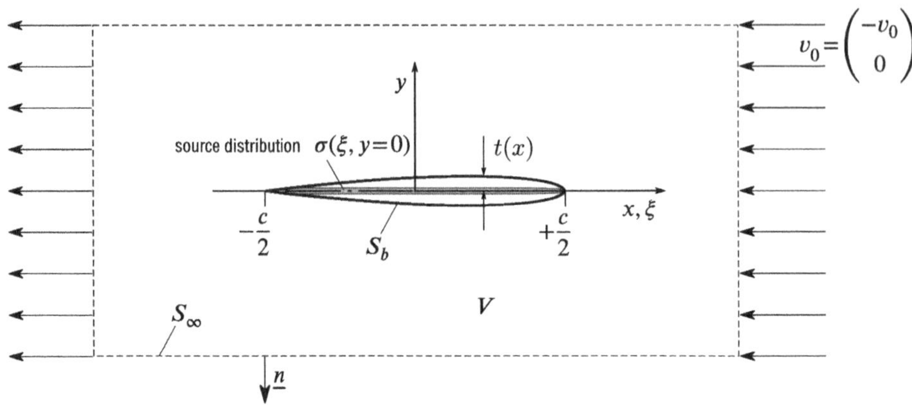

Figure 37.1 The boundary value problem of a symmetric thin foil with finite thickness and zero angle of attack

the flow in the far field, and a velocity potential ϕ_1, which describes the disturbance caused by the foil. The potentials are functions of the planar coordinates x and y.

$$\Phi(x, y) = \phi_\infty(x, y) + \phi_1(x, y) \tag{37.2}$$

Figure 37.1 shows the control volume and its boundaries. The boundary value problem for the disturbance potential consists of the Laplace equation and appropriate boundary conditions.

$$\begin{aligned} \Delta\phi_1(x, y) &= 0 && \text{for } (x, y) \in V \\ \frac{\partial \phi_1}{\partial y} &= \mp v_0 \frac{\mathrm{d}t}{\mathrm{d}x} && \text{for } \quad x \in \left[-\frac{c}{2}, \frac{c}{2}\right] \text{ and } y = 0^\pm \\ \phi_1(x, y) &\longrightarrow \text{const.} && \text{for } \sqrt{x^2 + y^2} \to \infty \end{aligned} \tag{37.3}$$

Note that we simplified the body boundary condition according to thin foil theory assumptions.

This problem is similar to the flow around a cylinder or a Rankine oval. The flow around the latter was described by the potentials of a source, a sink, and parallel. This lends credit to the idea that more complicated shapes may be represented by a multitude of sources and sinks. As long as the sum of their strengths vanishes we will obtain a closed contour.

Line distribution of sources

Going one step further, we use a line of distributed sources instead of a set of discrete sources and sinks. This is the same concept as a distributed load in mechanics of materials. The source strength σ is now given as a 'volume' flow per unit length. Note that the 'volume' flow in the plane is measured in length units squared per second. We place the sources on the chord line $y = 0$ between $x = -c/2$ and $x = +c/2$. The total disturbance potential is obtained by integration along the length of the source distribution.

$$\phi_1(x, y) = \int_{-c/2}^{+c/2} \frac{\sigma(\xi)}{2\pi} \ln\left(\sqrt{(x-\xi)^2 + y^2}\right) \mathrm{d}\xi \tag{37.4}$$

A pair of coordinates (x, y) marks the point where we want to know the potential or velocity. The integration variable ξ marks the position of sources along the tail–nose line ($\eta = 0$). We have to adjust the yet unknown source strength $\sigma(\xi)$ so that the closed contour looks like our thin foil and represents a streamline of the flow. Therefore we will employ the linearized body boundary condition for the displacement flow from Table 36.1 to find the source strength.

Transverse velocity

The derivative of the disturbance potential (37.4) with respect to y is part of the body boundary condition. In the potential, both the limits of the integral and the integration variable are independent of y, which allows us to interchange integration and differentiation.

$$\frac{\partial \phi_1(x, y)}{\partial y} = \int_{-c/2}^{+c/2} \frac{\sigma(\xi)}{2\pi} \frac{y}{\left((x-\xi)^2 + y^2\right)} \, \mathrm{d}\xi \tag{37.5}$$

Evaluating the integral

The linearized body boundary condition is satisfied over the tail–nose line on the x-axis with $y = 0$ instead of the actual foil contour. The integrand in (37.5) vanishes everywhere for $y = 0$ except for the position $x = \xi$ for which we obtain an undefined expression of the type '0/0.' Without loss of accuracy, we may limit the extent of the integral to some finite interval of length 2δ around x. δ is a small positive number. The regions outside of this area will make no contribution to the integral because the integrand vanishes ($y = 0$).

$$\frac{\partial \phi_1(x, y)}{\partial y} = \int_{x-\delta}^{x+\delta} \frac{\sigma(\xi)}{2\pi} \frac{y}{\left((x-\xi)^2 + y^2\right)} \, \mathrm{d}\xi \tag{37.6}$$

We make δ so small that it is safe to assume that the source strength takes the value at x, i.e. $\sigma(\xi) = \sigma(x)$ for $\xi \in [x-\delta, x+\delta]$. The source strength becomes independent of the integration variable and can be brought in front of the integral sign.

$$\frac{\partial \phi_1(x, y)}{\partial y} = \frac{\sigma(x)}{2\pi} \int_{x-\delta}^{x+\delta} \frac{y}{\left((x-\xi)^2 + y^2\right)} \, \mathrm{d}\xi \tag{37.7}$$

We will proceed with caution and search for the value of integral (37.7) by letting $y = 0 \pm \varepsilon$ and compute the limits for $\varepsilon \to +0$ and $\varepsilon \to -0$ if they exist.

$$\frac{\partial \phi_1(x, y = \pm 0)}{\partial y} = \lim_{\varepsilon \to \pm 0} \frac{\sigma(x)}{2\pi} \int_{x-\delta}^{x+\delta} \frac{\varepsilon}{\left((x-\xi)^2 + \varepsilon^2\right)} \, \mathrm{d}\xi \tag{37.8}$$

The following variable substitution simplifies the integration:

$$\lambda = (x - \xi) \qquad \frac{\mathrm{d}\lambda}{\mathrm{d}\xi} = -1 \quad \text{i.e.} \quad \mathrm{d}\xi = -\mathrm{d}\lambda$$

The limits of the integral transform into

$$\lambda_l = x - (x - \delta) = +\delta \qquad \lambda_u = x - (x + \delta) = -\delta$$

Figure 37.2 The inverse tangent function

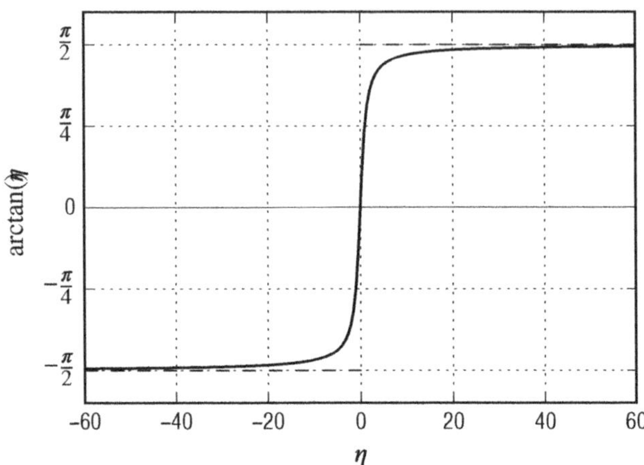

Substituting λ for ξ changes the integral changes into the following form:

$$\frac{\partial \phi_1(x, y = \pm 0)}{\partial y} = \lim_{\varepsilon \to \pm 0} \frac{\sigma(x)}{2\pi} \int_{+\delta}^{-\delta} \frac{\varepsilon}{\left(\lambda^2 + \varepsilon^2\right)} (-\mathrm{d}\lambda) \tag{37.9}$$

We switch upper and lower limits and consequently remove the minus sign.

$$\frac{\partial \phi_1(x, y = \pm 0)}{\partial y} = \frac{\sigma(x)}{2\pi} \lim_{\varepsilon \to \pm 0} \int_{-\delta}^{+\delta} \frac{\varepsilon}{\left(\lambda^2 + \varepsilon^2\right)} \mathrm{d}\lambda \tag{37.10}$$

The antiderivative of the integral of type (37.10) is the inverse tangent function arctan() (Figure 37.2).

$$\begin{aligned} \frac{\partial \phi_1(x, y = \pm 0)}{\partial y} &= \frac{\sigma(x)}{2\pi} \lim_{\varepsilon \to \pm 0} \left[\varepsilon \frac{1}{\varepsilon} \arctan\left(\frac{\eta}{\varepsilon}\right) \right]_{-\delta}^{+\delta} \\ &= \frac{\sigma(x)}{2\pi} \lim_{\varepsilon \to \pm 0} \left(\arctan\left(\frac{\delta}{\varepsilon}\right) - \arctan\left(\frac{-\delta}{\varepsilon}\right) \right) \end{aligned}$$

If ε approaches zero from the positive side, the first arctan() function's argument will grow toward $+\infty$. Then the function value becomes $+\pi/2$ (see Figure 37.2). The argument of the second arctan() grows toward $-\infty$ and the tangent value approaches $-\pi/2$.

$$\begin{aligned} \frac{\partial \phi_1(x, y = +0)}{\partial y} &= \frac{\sigma(x)}{2\pi} \lim_{\varepsilon \to +0} \left(\arctan\left(\frac{\delta}{\varepsilon}\right) - \arctan\left(\frac{-\delta}{\varepsilon}\right) \right) \\ &= \frac{\sigma(x)}{2\pi} \left(\frac{\pi}{2} - \frac{-\pi}{2} \right) \\ &= + \frac{\sigma(x)}{2} \end{aligned} \tag{37.11}$$

For the case where ε approaches zero from the negative side, the signs reverse for the arguments of the arctan() functions.

$$\begin{aligned}\frac{\partial\phi_1(x,y=-0)}{\partial y} &= \frac{\sigma(x)}{2\pi}\lim_{\varepsilon\to-0}\left(\arctan\left(\frac{\delta}{\varepsilon}\right)-\arctan\left(\frac{-\delta}{\varepsilon}\right)\right)\\ &= \frac{\sigma(x)}{2\pi}\left(\frac{-\pi}{2}-\frac{\pi}{2}\right)\\ &= -\frac{\sigma(x)}{2}\end{aligned} \tag{37.12}$$

In summary:

$$v_1^+ = \frac{\partial\phi_1(x,+0)}{\partial y} = +\frac{\sigma(x)}{2} \quad \text{and} \quad v_1^- = \frac{\partial\phi_1(x,-0)}{\partial y} = -\frac{\sigma(x)}{2} \tag{37.13}$$

Normal velocity induced by line sources

This result needs some physical interpretation:

- On the upper side of the foil with $y = +0$ the source distribution generates an upward oriented transverse disturbance velocity of $v_1^+ = +\sigma(x)/2$.
- On the lower side of the foil where $y = -0$ the transverse velocity $v_1^- = -\sigma(x)/2$ is oriented downward with the same magnitude.

Thus the velocity discontinuously changes when crossing from one side of the foil to the other. The total change in velocity is equal to the source strength $\sigma(x)$. Although this was derived here for a line distribution of sources it is a general feature of source distributions.

Exploiting the body boundary condition

Finally, we introduce the result (37.13) into the linearized body boundary condition (37.3).

$$\frac{\partial\phi_1}{\partial y} = \mp v_0\frac{\mathrm{d}t}{\mathrm{d}x} \tag{37.14}$$

Therefore, the source strength is proportional to the slope of the thickness distribution:

$$\sigma(x) = -2v_0\frac{\mathrm{d}t(x)}{\mathrm{d}x} \tag{37.15}$$

Both the upper and lower sides yield the same result which is expected since we have only one source distribution.

Source strength distribution

Figure 37.3 shows a symmetric foil section and the distribution of source strength over the chord of the foil. The source strength will be negative toward the trailing edge; the sources are actually sinks. At the position of maximum thickness with $\mathrm{d}t/\mathrm{d}x = 0$ the source strength is zero and then grows positive toward the leading edge. If the nose is round and meets the chord line ($y = 0$) at a right angle the source strength distribution will have a singularity. Clearly, $|\mathrm{d}t/\mathrm{d}x| \to \infty$ is in violation of our thin foil assumption. A correction is needed to represent the flow correctly in the vicinity of the leading edge (see e.g. Lighthill, 1951).

With the source strength distribution known, we can restate the disturbance potential.

$$\phi_1(x,y) = -\frac{v_0}{\pi}\int_{-c/2}^{+c/2}\frac{\mathrm{d}t(\xi)}{\mathrm{d}\xi}\ln\left(\sqrt{(x-\xi)^2+y^2}\right)\mathrm{d}\xi \tag{37.16}$$

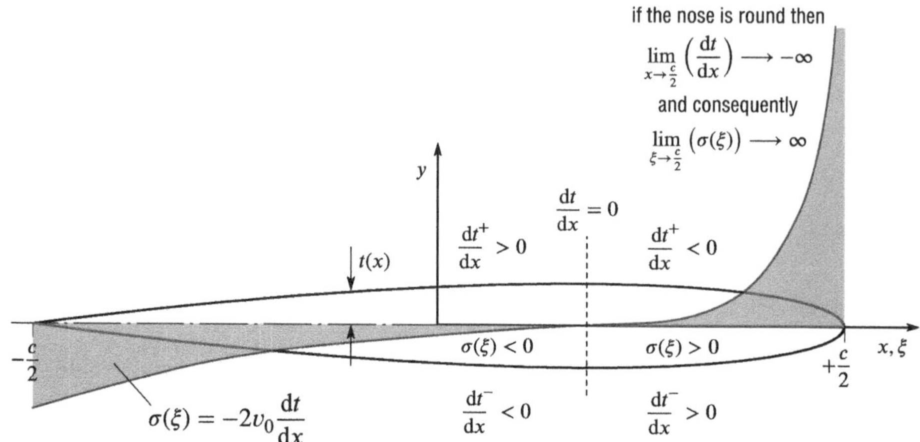

Figure 37.3 The source strength distribution $\sigma(\xi)$ as a function of the slope $\mathrm{d}t/\mathrm{d}x$ of the foil surface

Since the effect of sources declines with growing distance, the disturbance potential will satisfy the far field condition of the boundary value problem (37.3).

Total potential for thickness problem

Together with the parallel flow we have a velocity potential for the flow around our thin, symmetric foil at zero angle of attack.

$$\Phi(x,y) = -v_0 x - \frac{v_0}{\pi}\int_{-c/2}^{+c/2} \frac{\mathrm{d}t(\xi)}{\mathrm{d}\xi}\ln\left(\sqrt{(x-\xi)^2+y^2}\right)\mathrm{d}\xi \tag{37.17}$$

37.2 Pressure Distribution

Bernoulli equation for thin foils

With the potential (37.17) the velocity distribution in the domain around the foil is known. The Bernoulli equation for steady flow allows us to compute the pressure distribution. We ignore the gravity forces assuming that foil and fluid are in equilibrium when at rest.

$$\frac{1}{2}\rho|\underline{\nabla}\,\Phi|^2 + p = \frac{1}{2}\rho\, v_0^2 + p_0 \tag{37.18}$$

The magnitude of the velocity squared $|\underline{\nabla}\,\Phi|^2$ is given by

$$\begin{aligned}|\underline{\nabla}\,\Phi|^2 &= \left(\frac{\partial\Phi}{\partial x}\right)^2 + \left(\frac{\partial\Phi}{\partial y}\right)^2\\ &= v_0^2 - 2v_0\frac{\partial\phi_1}{\partial x} + \left(\frac{\partial\phi_1}{\partial x}\right)^2 + \left(\frac{\partial\phi_1}{\partial y}\right)^2\end{aligned}$$

The latter two terms must be very small if the foil is thin because they are a function of $(\mathrm{d}t/\mathrm{d}x)^2$. These small, quadratic terms will be ignored from hereon forward.

$$|\underline{\nabla}\,\Phi|^2 = v_0^2 - 2v_0\frac{\partial\phi_1}{\partial x} \tag{37.19}$$

Substitution of this result into (37.18) yields

$$p - p_0 = \rho v_0 \frac{\partial \phi_1}{\partial x} \tag{37.20}$$

The pressure coefficient $C_p(x)$ will be

$$C_p(x) = \frac{p - p_0}{\frac{1}{2}\rho v_0^2} = \frac{2}{v_0}\frac{\partial \phi_1}{\partial x} \tag{37.21}$$

The x-derivative of the disturbance potential (37.16) is

$$\frac{\partial \phi_1}{\partial x} = -\frac{v_0}{\pi}\int_{-c/2}^{+c/2} \frac{\mathrm{d}t(\xi)}{\mathrm{d}\xi}\frac{(x-\xi)}{\left((x-\xi)^2 + y^2\right)}\,\mathrm{d}\xi \tag{37.22}$$

The pressure coefficient on the foil surfaces, approximated by $y = \pm 0$, has to be evaluated carefully because the integrand has a singularity for $x = \xi$.

$$C_p(x) = -\frac{2}{\pi}\,⨍_{-c/2}^{+c/2} \frac{\mathrm{d}t(\xi)}{\mathrm{d}\xi}\frac{1}{(x-\xi)}\,\mathrm{d}\xi \tag{37.23}$$

This is called a Cauchy principal value integral, and indicated by the dash through the middle of the integral sign.

Cauchy principal value integral

In this case, it can be shown that the integral exists. We abbreviate $\mathrm{d}t/\mathrm{d}\xi = t'$ and split the integral into three parts.

$$⨍_{-c/2}^{+c/2} t'(\xi)\frac{1}{(x-\xi)}\,\mathrm{d}\xi$$

$$= \int_{-c/2}^{x-\varepsilon} t'(\xi)\frac{1}{(x-\xi)}\,\mathrm{d}\xi + ⨍_{x-\varepsilon}^{x+\varepsilon} t'(\xi)\frac{1}{(x-\xi)}\,\mathrm{d}\xi + \int_{x+\varepsilon}^{c/2} t'(\xi)\frac{1}{(x-\xi)}\,\mathrm{d}\xi \tag{37.24}$$

The first and last integral on the right-hand side do not contain singularities. The case $x = \xi$ is not possible. Therefore, the two integrals will exist if $t' = \mathrm{d}t/\mathrm{d}\xi$ is bounded.

The middle integral in Equation (37.24) needs further investigation. We assume that ε is so small that we can extract the derivative of the thickness distribution $t'(\xi)$ from the integral. It will take the value $t'(x)$ at position x along the tail–nose line. The remaining integral is a Cauchy principal value integral, and it has a value, if the limit on the right-hand side exists.

$$⨍_{x-\varepsilon}^{x+\varepsilon} \frac{1}{(x-\xi)}\,\mathrm{d}\xi = \lim_{\delta\to+0}\left(\int_{x-\varepsilon}^{x-\delta} \frac{1}{(x-\xi)}\,\mathrm{d}\xi + \int_{x+\delta}^{x+\varepsilon} \frac{1}{(x-\xi)}\,\mathrm{d}\xi\right)$$

$$= \lim_{\delta\to+0}\left(\int_{x-\varepsilon}^{x-\delta} \frac{1}{(x-\xi)}\,\mathrm{d}\xi - \int_{x+\delta}^{x+\varepsilon} \frac{1}{(\xi-x)}\,\mathrm{d}\xi\right) \tag{37.25}$$

During this conversion the operands switched position in the denominator of the second integral. The resulting minus sign has been extracted from the integral. This avoids negative arguments in the integrand, since $\xi > x$ for the second integral. This is helpful in the subsequent integration because the natural logarithm is undefined for negative arguments.

The remaining integrals can now be solved.

$$\int_{x-\varepsilon}^{x+\varepsilon} \frac{1}{(x-\xi)}\,\mathrm{d}\xi = \lim_{\delta\to+0}\left(\Big[-\ln(x-\xi)\Big]_{x-\varepsilon}^{x-\delta} - \Big[\ln(\xi-x)\Big]_{x+\delta}^{x+\varepsilon}\right)$$

The additional minus sign in the first antiderivative results from the inner derivative. Finally, we substitute the limits for the integration variable ξ:

$$\begin{aligned}\int_{x-\varepsilon}^{x+\varepsilon} \frac{1}{(x-\xi)}\,\mathrm{d}\xi &= \lim_{\delta\to+0}\left(\Big[-\ln(\delta)+\ln(\varepsilon)\Big] - \Big[\ln(\varepsilon)-\ln(\delta)\Big]\right) \\ &= \lim_{\delta\to+0}\left(-\ln(\delta)+\ln(\delta)+\ln(\varepsilon)-\ln(\varepsilon)\right) \\ &= \lim_{\delta\to+0}(0) \\ &= 0\end{aligned} \tag{37.26}$$

The limit for $\delta \to 0$ obviously exists. The value of the integral is zero, i.e. the middle integral in Equation (37.24) makes no contribution to the pressure coefficient. Thus we have proven that the Cauchy principal value integral for the pressure coefficient (37.23) exists, if the derivative of the thickness distribution is well behaved.

37.3 Elliptical Thickness Distribution

Example: elliptical foil section

As an example, we investigate the elliptical thickness distribution.

$$t(\xi) = \frac{t_{\max}}{2}\sqrt{1-\left(\frac{2\xi}{c}\right)^2} \qquad \text{for } \xi \in \left[-\frac{c}{2},\frac{c}{2}\right] \tag{37.27}$$

Elliptical thickness distributions are not often used, although they are less prone to cavitation than other profile shapes.

Variable substitution

Before substituting the elliptical thickness distribution (37.27) into the integral for the pressure coefficient (37.23), we initiate a variable substitution. The integration variable ξ will be replaced by a polar coordinate φ such that

$$\xi = \frac{c}{2}\cos(\varphi) \qquad \frac{\mathrm{d}\xi}{\mathrm{d}\varphi} = -\frac{c}{2}\sin(\varphi) \qquad \mathrm{d}\xi = -\frac{c}{2}\sin(\varphi)\,\mathrm{d}\varphi \tag{37.28}$$

The integration limits become:

$$\xi_l = -\frac{c}{2} \longrightarrow \varphi_l = \pi \qquad \xi_u = \frac{c}{2} \longrightarrow \varphi_u = 0 \tag{37.29}$$

The point $(x, 0)$, where we evaluate the pressure coefficient along the foil, is transformed in the same way, but it has its own angle θ, independent of the integration variable.

$$x = \frac{c}{2}\cos(\theta) \tag{37.30}$$

with $\theta \in [0, \pi]$.

Over the φ-axis, the elliptical thickness distribution (37.27) is represented by a sine curve.

$$t(\varphi) = \frac{t_{\max}}{2}\sqrt{1-\cos^2(\varphi)} = \frac{t_{\max}}{2}\sin(\varphi) \tag{37.31}$$

Derivative of thickness distribution

We also need the derivative of the thickness distribution with respect to the integration variable ξ.

$$\frac{dt}{d\xi} = \frac{dt}{d\varphi}\frac{d\varphi}{d\xi} = \frac{t_{\max}}{2}\cos(\varphi)\frac{1}{\left(\frac{d\xi}{d\varphi}\right)}$$

and with the derivative $d\xi/d\varphi$ from (37.28)

$$\begin{aligned}\frac{dt}{d\xi} &= \frac{t_{\max}}{2}\frac{\cos(\varphi)}{\left(-\frac{c}{2}\sin(\varphi)\right)}\\ &= -\frac{t_{\max}}{c}\frac{\cos(\varphi)}{\sin(\varphi)}\end{aligned} \tag{37.32}$$

Pressure coefficient

We introduce the variable substitution (37.28), the transformed field point coordinate (37.30), and the derivative of the thickness distribution (37.32) into Equation (37.23) for the pressure coefficient.

$$C_p(\theta) = -\frac{2}{\pi}\int_{\pi}^{0} -\frac{t_{\max}}{c}\frac{\cos(\varphi)}{\sin(\varphi)}\frac{1}{\left[\frac{c}{2}\cos(\theta)-\frac{c}{2}\cos(\varphi)\right]}\left(-\frac{c}{2}\sin(\varphi)\right)d\varphi \tag{37.33}$$

We carefully consolidate the three minus signs into one and extract all constants from the integral.

$$C_p(\theta) = -\frac{2\,t_{\max}}{\pi\,c}\int_{\pi}^{0}\frac{\cos(\varphi)}{\sin(\varphi)}\frac{\sin(\varphi)}{\left[\cos(\theta)-\cos(\varphi)\right]}d\varphi \tag{37.34}$$

Next, we eliminate the $\sin(\varphi)$ in numerator and denominator and extract a factor -1 from the denominator which we convert into a reversal of the integration limits.

$$C_p(\theta) = -\frac{2\,t_{\max}}{\pi\,c}\int_{0}^{\pi}\frac{\cos(\varphi)}{\left[\cos(\varphi)-\cos(\theta)\right]}d\varphi \tag{37.35}$$

The Cauchy principal value integral in equation (37.35) is not quite so easy to solve.

Glauert integrals Luckily for us, Glauert (1947, pp. 92–93)[1] solved this as part of a whole group of integrals, which are known in hydrodynamics as Glauert integrals.

$$I_n(\theta) = \fint_0^\pi \frac{\cos(n\varphi)}{\left[\cos(\varphi) - \cos(\theta)\right]} \, d\varphi = \pi \frac{\sin(n\theta)}{\sin(\theta)} \qquad (37.36)$$

In our case $n = 1$, and the value of the integral $I_1 = \pi$. For the pressure coefficient of an elliptical thickness distribution we obtain a constant negative value:

$$C_p(\theta) = -\frac{2\,t_{\mathrm{max}}}{c} \qquad (37.37)$$

This is somewhat surprising because we expect to see the value $C_p = 1$ at the aft and forward stagnation points. But then, close to the stagnation points our foil shape violates the primary condition of thin foil theory. At the rounded ends of the elliptical foil section, the derivative $\frac{\partial f}{\partial x} = \frac{\partial t}{\partial x}$ of the shape is anything but small. In fact, it does not exist and its values approach $\pm\infty$.

Examples Overall, the pressure distribution matches the exact potential theory result obtained by conformal mapping quite well. Figure 37.4 shows the results for an elliptical foil with 10% thickness to chord length ratio. The minimum C_p value is slightly under predicted by thin foil theory, but it is close to its conformal mapping value. As expected, thin foil theory fails to predict the correct pressure toward trailing and leading edges of the foil.

One could argue that a foil with $t_{\mathrm{max}}/c = 0.10$ is not really a thin foil. Figure 37.5 shows the result for an elliptical foil section with a thickness to chord length ratio of just 5%. The minimum C_p value is fairly accurate and the deviations are much smaller toward leading and trailing edges.

Figures 37.4 and 37.5 also show the velocity magnitude as derived by conformal mapping. If one plots the magnitude of the velocity derived from thin foil theory, huge discrepancies are found toward the ends of the foil if the transverse velocity component $v_1 = \partial\phi_1/\partial y$ is included. As listed in Equation (37.14), v_1 is a function of the slope of the thickness distribution. At the ends of the foil, the slope is $\pm\infty$ which renders the velocity prediction as incorrect. Hence, it is common practice to compute the magnitude of the velocity as in Equation (37.19) without the parts u_1^2 and v_1^2.

$$|\underline{\nabla}\,\Phi| = v_0\sqrt{1 - \frac{2}{v_0}\frac{\partial\phi_1}{\partial x}} \qquad (37.38)$$

References

Glauert, H. (1947). *The elements of airofoil and airscrew theory*. Cambridge University Press, Cambridge, United Kingdom, second edition.

Lighthill, M. (1951). A new approach to thin foil theory. *The Aeronautical Quarterly*, 3(3):193–210.

[1] Hermann Glauert (*1892 – †1934), British aerodynamicist

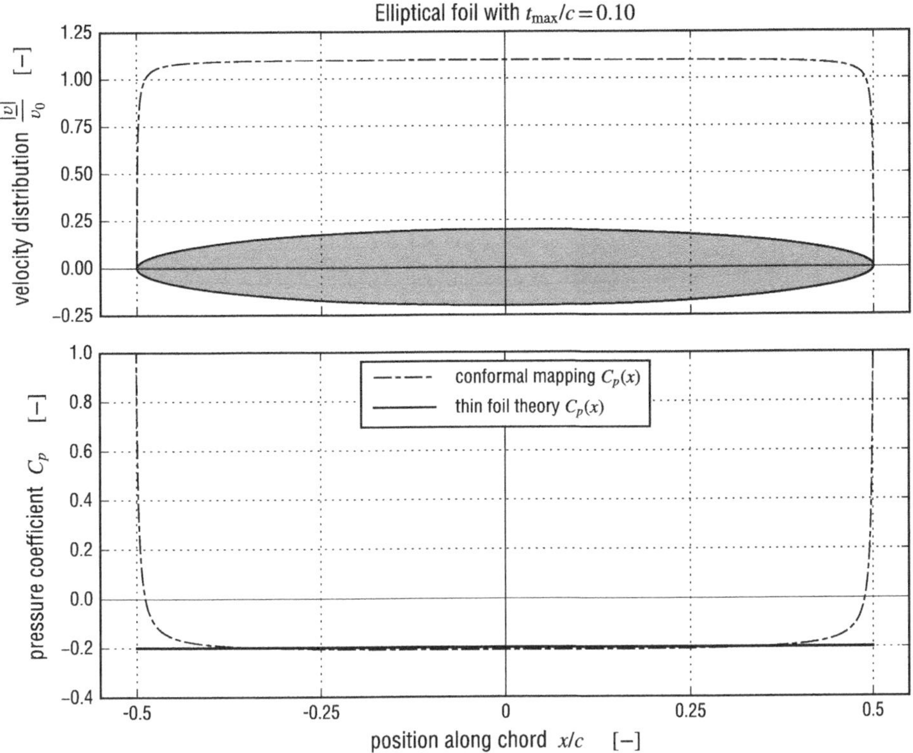

Figure 37.4 Comparison of thin foil theory and conformal mapping (exact) pressure coefficient for an elliptical foil with thickness to chord length ratio of $t_{max}/c = 0.10$

Self Study Problems

1. Summarize the equations that define the linearized boundary value problem for the displacement flow. State the physical meaning of each equation.
2. Given is a line distribution of sources along the tail–nose line with strength $\sigma(x)$. By how much changes the transverse velocity v as you move from the underside $(y = 0^-)$ to the upper side of the source distribution $(y = 0^+)$?
3. Explain in your own words why the source strength at the leading edge is unbounded in Figure 37.3.
4. Compute and plot the pressure coefficient over the chord length of an elliptical foil with 7% thickness–chord length ratio. Use the thin foil approximation.
5. Compute the value of the following integral for $\theta = \pi/3$:

$$I_3(\theta) = \fint_0^\pi \frac{\cos(3\varphi)}{\left[\cos(\varphi) - \cos(\theta)\right]} \, \mathrm{d}\varphi \qquad (37.39)$$

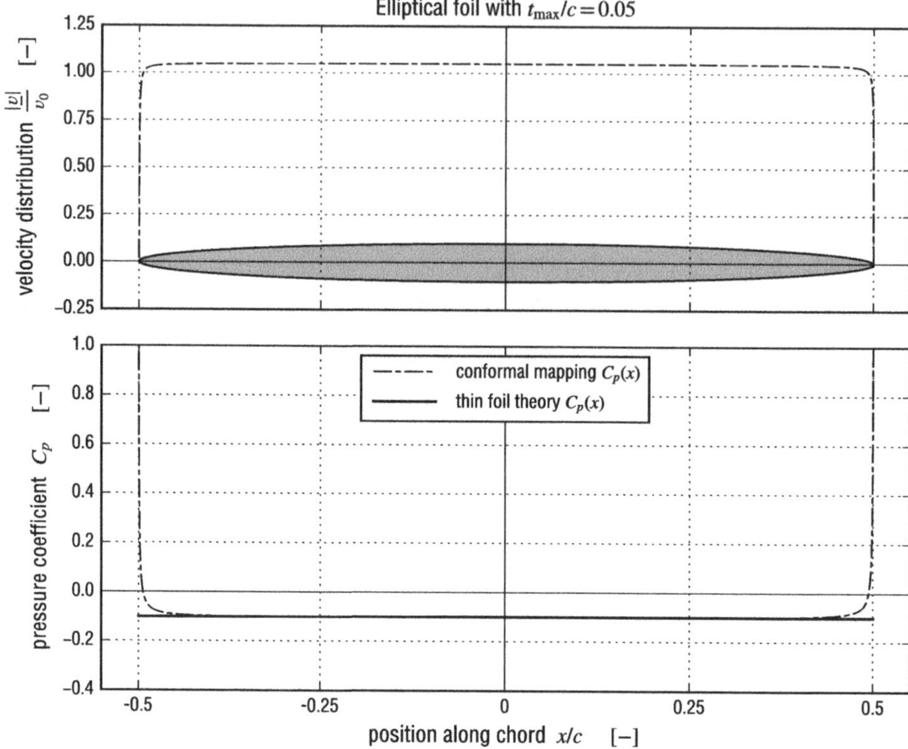

Figure 37.5 Comparison of thin foil theory and conformal mapping (exact) pressure coefficient for an elliptical foil with thickness to chord length ratio of $t_{max}/c = 0.05$

6. Derive the velocity distribution induced on a thin foil ($x \in [-c/2, c/2]$) by a parabolic thickness distribution

$$t(x) = \frac{t_{max}}{2}\left[1 - \left(\frac{2x}{c}\right)^2\right] \qquad \text{for} \quad -\frac{c}{2} \leq x \leq \frac{c}{2}$$

The parabolic thickness distribution is used to approximate the flow around ogival foil sections. Compare the velocity and pressure distribution with the elliptical thickness distribution.

38

Thin Foil Theory – Lifting Flow

The remaining task in solving the flow around thin foils is to find expressions for the disturbance potentials ϕ_2 and ϕ_3 which describe the lifting flow. We will use the results to study basic properties of lifting foils. First, we define a combined boundary value problem for lifting flow which will be solved for the combined lifting flow potential ϕ_{23}.

$$\phi_{23} = \phi_2 + \phi_3 \tag{38.1}$$

The disturbance potential ϕ_{23} represents the effects of camber and angle of attack on the flow. The solution is based on a series expansion developed by Glauert (1947).

Learning Objectives

At the end of this chapter students will be able to:

- understand and apply line distributions of vortices to simulate lifting flows
- comprehend the physical meaning of the Kutta condition
- compute lift coefficients and ideal angle of attack for thin foils

38.1 Lifting Foil Problem

Boundary value problem

The lifting problem considers an infinitely thin plate with camber operating at an angle of attack α (Figure 38.1). We will solve it under the assumption that the fluid is incompressible, inviscid, and irrotational in simply connected regions. Accordingly, we may apply potential flow theory. The total velocity potential consists of parallel flow and the disturbance potential ϕ_{23} since the thickness is assumed to be zero $t(x) = 0$ for the lifting problem.

$$\Phi(x, y) = \phi_\infty + \phi_{23} = -v_0 \cos(\alpha) + v_0 \sin(\alpha) + \phi_{23} \tag{38.2}$$

The generation of lift requires a nonzero circulation $\Gamma \neq 0$. Therefore, the flow must have a component which rotates around the foil (see also Figure 14.6). Obviously, all velocity potentials must satisfy the Laplace equation.

$$\Delta\Phi = 0 \quad \text{and} \quad \Delta\phi_{23} = 0 \tag{38.3}$$

Fundamentals of Ship Hydrodynamics: Fluid Mechanics, Ship Resistance and Propulsion, First Edition.
Lothar Birk.

Companion website: www.wiley.com/go/birk/hydrodynamics

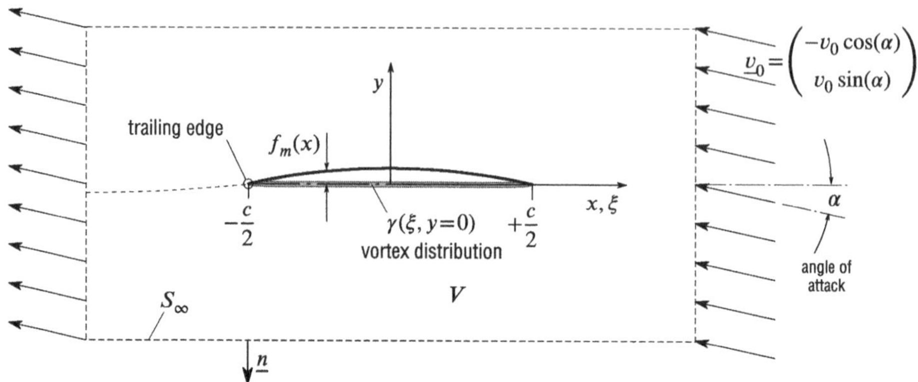

Figure 38.1 Boundary value problem for an infinitely thin cambered plate at angle of attack α

The body boundary condition for the foil contour consists of the parts of Equation (36.31) which involve camber f_m and angle of attack α.

$$\frac{\partial \phi_{23}^{\pm}}{\partial y} = -\frac{\partial f_m}{\partial x} v_0 - \alpha v_0 \qquad \text{for } y = \pm 0 \tag{38.4}$$

As consequence of our thin foil assumptions, the body boundary condition is satisfied along the x-axis between trailing and leading edges $x \in [-c/2, c/2]$. The disturbance of the parallel flow caused by the foil must vanish far away from the body. This can be expressed as

$$\phi_{23} = \text{const.} \quad \text{for } r \longrightarrow \infty \qquad \text{or} \qquad \underline{\nabla}\,\phi_{23} = \underline{0} \quad \text{for } r \longrightarrow \infty. \tag{38.5}$$

Kutta condition In addition to the body boundary and far field conditions, we also have to implement the Kutta condition. The Kutta condition enforces the physical observation that the flow is leaving the trailing edge smoothly. If, hypothetically, the fluid would flow around the sharp trailing edge, it would create infinitely high flow velocities. The Kutta condition may be enforced in different ways.

- One way to enforce the Kutta condition is to request equal pressure on upper and lower sides of the foil right at the trailing edge i.e. $p^+ = p^-$ at $x = -c/2$. Differences in pressure would force the fluid to flow around the trailing edge from the high pressure side to the side with lower pressure.
- Alternatively, we can impose the Kutta condition as a kinematic requirement, i.e. the velocity vector at the trailing edge must be equal for upper and lower sides $\underline{v}^+ = \underline{v}^-$ at $x = -c/2$.

Lifting flow disturbance potential We know from our first discussion of the lift force (page 438) that a circulation is needed to create faster flow speeds on the upper side and slower flow speeds on the lower side of the foil. In essence, a flow pattern, like the one created by a vortex, must be included in the flow. A distribution of sources has been used to represent the disturbance potential associated with the thickness distribution. Likewise, the disturbance potential for the lifting flow may be based on a distribution of vertices. With the potential of a 2D-vortex from Chapter 16, we define the disturbance potential as

$$\phi_{23}(x,y) = \int_{-c/2}^{+c/2} \frac{\gamma(\xi)}{2\pi} \arctan\left(\frac{y}{x-\xi}\right) \mathrm{d}\xi \tag{38.6}$$

Note that the coordinate ξ marks the position of a vortex. Its effect is probed at the field point (x, y). Unknown in the potential is the vortex strength $\gamma(\xi)$. By design, the disturbance potential ϕ_{23} will satisfy the Laplace equation because the integral is a combination of many vortex potentials which all satisfy the Laplace equation. The potential (38.6) also satisfies the far field condition because the effect of a vortex vanishes far away from its origin.

Velocity field of a line vortex distribution

Before we solve the boundary value problem, we take a look at the velocity field defined by the disturbance potential ϕ_{23} (38.6). The velocity vector is equal to the gradient of the potential.

$$\underline{v}_{23} = \underline{\nabla}\,\phi_{23} = \begin{pmatrix} \dfrac{\partial \phi_{23}}{\partial x} \\ \dfrac{\partial \phi_{23}}{\partial y} \end{pmatrix} = \begin{pmatrix} u_{23} \\ v_{23} \end{pmatrix} \tag{38.7}$$

Horizontal velocity component u_{23}

For the horizontal velocity component we have to compute the x-derivative of Equation (38.6).

$$\frac{\partial \phi_{23}}{\partial x} = \frac{\partial}{\partial x} \int_{-c/2}^{+c/2} \frac{\gamma(\xi)}{2\pi} \arctan\left(\frac{y}{x-\xi}\right) \mathrm{d}\xi \tag{38.8}$$

The integration limits are constant and the integration variable is ξ, i.e. the position of the vortex element rather than the location x of the field point. Hence, differentiation and integration may be interchanged. In the integrand, only the inverse tangent function arctan() is a function of x.

$$\frac{\partial \phi_{23}}{\partial x} = \int_{-c/2}^{+c/2} \frac{\gamma(\xi)}{2\pi} \frac{\partial}{\partial x} \left[\arctan\left(\frac{y}{x-\xi}\right)\right] \mathrm{d}\xi \tag{38.9}$$

The first order derivative of the inverse tangent function is equal to

$$\frac{\mathrm{d}}{\mathrm{d}z} \arctan(z) = \frac{1}{1+z^2}$$

Applying this and the chain rule for $z = y/(x-\xi)$ yields

$$u_{23}^{\pm}(x,y) = \frac{\partial \phi_{23}}{\partial x} = \int_{-c/2}^{+c/2} \frac{\gamma(\xi)}{2\pi} \left(\frac{-y}{\left[(x-\xi)^2 + y^2\right]}\right) \mathrm{d}\xi \tag{38.10}$$

Except for the minus sign, and the vortex strength $\gamma(\xi)$ standing in for the source strength, this integral is identical to the y-derivative of the source distribution (37.5) from the preceding chapter.

Between trailing edge ($x = -c/2$) and leading edge ($x = c/2$), the integrand in Equation (38.10) vanishes for $y = \pm 0$, except for the point $\xi = x$. To evaluate the resulting undefined expression, we exercise the same limiting process used in Equations (37.5) through (37.13). As a result, the horizontal velocity on the x-axis $y = \pm 0$ is equal to half the vortex strength.

$$u_{23}^{\pm}(x, y = \pm 0) \; = \; \frac{\partial \phi_{23}(x, y{=}\pm 0)}{\partial x} \; = \; \mp \frac{\gamma(x)}{2} \qquad \text{for } x \in [-c/2, c/2] \tag{38.11}$$

On the upper side of the foil $y = +0$, the additional horizontal velocity u_{23}^{+} is pointing in negative x-axis direction and, thus, augments the horizontal onflow $-v_0 \cos \alpha$. On the pressure side of the foil, u_{23}^{-} points in positive x-axis direction and reduces the overall horizontal flow speed.

Across a line source strength distribution, the *vertically* induced velocity component jumps by the source strength σ. If we pass from the lower side to the upper side of a line vortex distribution, the *horizontal* velocity changes by the negative of the vortex strength γ.

Kutta condition revisited

Equation (38.11) bears some significance for the Kutta condition. Since the vortex distribution induces horizontal velocities which point in opposite directions on the upper and lower foil surfaces, a Kutta condition must enforce that the vortex strength vanishes at the trailing edge.

$$\gamma(x = -c/2) \; = \; 0 \qquad \text{Kutta condition} \tag{38.12}$$

Otherwise, we cannot be sure that the flow velocity at the trailing edge is equal on the upper and lower sides as the Kutta condition requests.

Transverse velocity component v_{23}

The transverse velocity component follows from the y-derivative of Equation (38.6).

$$\frac{\partial \phi_{23}}{\partial y} \; = \; \frac{\partial}{\partial y} \int_{-c/2}^{+c/2} \frac{\gamma(\xi)}{2\pi} \arctan\left(\frac{y}{x - \xi}\right) \mathrm{d}\xi \tag{38.13}$$

Again, interchanging differentiation and integration yields together with the basic arctan derivative and the chain rule the transverse disturbance velocity

$$v_{23} \; = \; \int_{-c/2}^{+c/2} \frac{\gamma(\xi)}{2\pi} \left(\frac{x - \xi}{(x - \xi)^2 + y^2}\right) \mathrm{d}\xi \tag{38.14}$$

This is the same type of integral which has been found for the x derivative of the source distribution (37.22).

The body boundary condition is evaluated on the x-axis. From Equation (38.14) follows a Cauchy principal value integral for $y = \pm 0$.

$$v_{23}(x, y = \pm 0) \; = \; \fint_{-c/2}^{+c/2} \frac{\gamma(\xi)}{2\pi} \frac{1}{(x - \xi)} \mathrm{d}\xi \tag{38.15}$$

For $x = \xi$ the integrand does not exist, but on pages 453 and 454 it has been shown that the integral has a finite value, as long as the vortex strength is bounded.

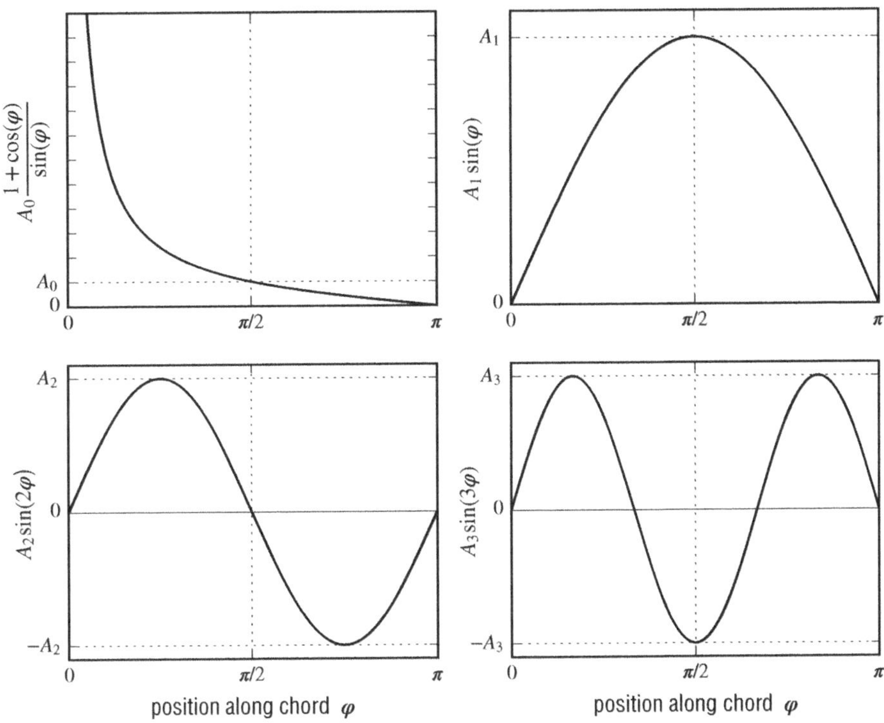

Figure 38.2 The first four elements of Glauert's trigonometric series for the vortex strength γ

38.2 Glauert's Classical Solution

Exploiting the body boundary condition

The next step is to compute the unknown vortex strength. Since our disturbance potential satisfies all but the body boundary condition, we can only employ the latter to find the vortex strength as a function of foil geometry. Combining Equations (38.4) and (38.15) links the vortex strength with camber distribution and angle of attack.

$$\int_{-c/2}^{+c/2} \frac{\gamma(\xi)}{2\pi} \frac{1}{(x-\xi)} \, \mathrm{d}\xi = -\frac{\partial f_m}{\partial x} v_0 - \alpha v_0 \tag{38.16}$$

At this point, we may replace the partial derivative of the camber distribution with an ordinary derivative because only x remains as variable.

Glauert's series for vortex strength

Glauert (1947) proposed a trigonometric series to approximate the unknown vortex strength.

$$\gamma(\varphi) = 2 v_0 \left[A_0 \frac{1+\cos(\varphi)}{\sin(\varphi)} + \sum_{n=1}^{\infty} A_n \sin(n\varphi) \right] \tag{38.17}$$

Figure 38.2 shows the shape of the first four elements of the series expansion (38.17). Note that the series implicitly satisfies the Kutta condition $\gamma(x = -c/2) = 0$. All functions in the series vanish at the trailing edge $\varphi = \pi$.

$$\gamma(\varphi = \pi) \;=\; 0 \qquad \text{Kutta condition} \tag{38.18}$$

Glauert's series (38.17) uses the same variable transformation as was employed in Section 37.3 for the elliptical thickness distribution. The integration variable ξ is replaced by a polar coordinate φ such that

$$\xi \;=\; \frac{c}{2}\cos(\varphi) \qquad \frac{\mathrm{d}\xi}{\mathrm{d}\varphi} \;=\; -\frac{c}{2}\sin(\varphi) \qquad \mathrm{d}\xi \;=\; -\frac{c}{2}\sin(\varphi)\,\mathrm{d}\varphi \tag{37.28}$$

Consequently, the integration limits become:

$$\xi_l = -\frac{c}{2} \longrightarrow \varphi_l \;=\; \pi \qquad\qquad \xi_u = \frac{c}{2} \longrightarrow \varphi_u \;=\; 0 \tag{38.19}$$

The leading edge is $\varphi = 0$ and the trailing edge is $\varphi = \pi$. The field point $(x, 0)$ is rewritten as

$$x \;=\; \frac{c}{2}\cos(\theta) \tag{38.20}$$

The position angle θ also takes values from the interval $[0, \pi]$.

Body boundary condition for lifting flow

Substituting (37.28), (38.19), and (38.20) into Equation (38.16) converts the body boundary condition into the following form:

$$-\fint_0^{\pi} \frac{\gamma(\varphi)}{2\pi}\,\frac{\sin(\varphi)}{\big(\cos(\varphi) - \cos(\theta)\big)}\,\mathrm{d}\varphi \;=\; -\frac{\mathrm{d}f_m}{\mathrm{d}x}\,v_0 \;-\; \alpha\, v_0 \tag{38.21}$$

We replace the vortex strength γ with its series representation (38.17), and divide the equation by the onflow velocity v_0.

$$-\frac{1}{\pi}\fint_0^{\pi}\left[A_0\,\frac{1+\cos(\varphi)}{\sin(\varphi)} + \sum_{n=1}^{\infty} A_n\,\sin(n\varphi)\right]\frac{\sin(\varphi)}{\big(\cos(\varphi)-\cos(\theta)\big)}\,\mathrm{d}\varphi \;=\; -\frac{\mathrm{d}f_m}{\mathrm{d}x} \;-\; \alpha \tag{38.22}$$

The goal is to define the unknown series coefficients A_n as functions of the slope of the camber line and the angle of attack. To this end, the integral on the left-hand side of Equation (38.22) is broken up into individual integrals, which each contains only one unknown coefficient A_n.

$$\begin{aligned} I \;&=\; -\frac{1}{\pi}\fint_0^{\pi}\left[A_0\,\frac{1+\cos(\varphi)}{\sin(\varphi)} + \sum_{n=1}^{\infty} A_n\,\sin(n\varphi)\right]\frac{\sin(\varphi)}{\big(\cos(\varphi)-\cos(\theta)\big)}\,\mathrm{d}\varphi \\ &=\; \sum_{n=0}^{\infty} I_n \;=\; I_0 + I_1 + I_2 + \ldots \end{aligned} \tag{38.23}$$

Integral I_0

The first integral involves only the coefficient A_0.

$$
\begin{aligned}
I_0 &= -\frac{1}{\pi} \int_0^{\pi} A_0 \frac{\left(1+\cos(\varphi)\right)}{\sin(\varphi)} \frac{\sin(\varphi)}{\left(\cos(\varphi)-\cos(\theta)\right)} \mathrm{d}\varphi \\
&= -\frac{A_0}{\pi} \int_0^{\pi} \frac{1+\cos(\varphi)}{\left(\cos(\varphi)-\cos(\theta)\right)} \mathrm{d}\varphi \\
&= -\frac{A_0}{\pi} \left[\int_0^{\pi} \frac{1}{\left(\cos(\varphi)-\cos(\theta)\right)} \mathrm{d}\varphi + \int_0^{\pi} \frac{\cos(\varphi)}{\left(\cos(\varphi)-\cos(\theta)\right)} \mathrm{d}\varphi \right]
\end{aligned}
$$

The two resulting integrals are both Glauert integrals (37.36) (see Section 37.3) and result in 0 and π respectively.

$$
I_0 = -\frac{A_0}{\pi}\left[0+\pi\right] = -A_0 \tag{38.24}
$$

That is a surprisingly simple result. Of course, it is connected to the specific choice of function which accompanies the coefficient A_0 in the series expansion.

Integrals I_n, $n > 0$

The remaining integrals can also be reduced to Glauert integrals.

$$
\begin{aligned}
I_n &= -\frac{1}{\pi} \int_0^{\pi} \left[A_n \sin(n\varphi)\right] \frac{\sin(\varphi)}{\left(\cos(\varphi)-\cos(\theta)\right)} \mathrm{d}\varphi \\
&= -\frac{A_n}{\pi} \int_0^{\pi} \frac{\sin(n\varphi)\sin(\varphi)}{\left(\cos(\varphi)-\cos(\theta)\right)} \mathrm{d}\varphi
\end{aligned}
$$

Further progress needs the help of a trigonometric identity:

$$
\sin(n\varphi)\sin(\varphi) = \frac{1}{2}\left[\cos\left((n-1)\varphi\right) - \cos\left((n+1)\varphi\right)\right] \tag{38.25}
$$

This one is hardly in any books but can be derived by using Euler's formulas:

$$
\sin(x) = \frac{\mathrm{e}^{\mathrm{i}x}-\mathrm{e}^{-\mathrm{i}x}}{2i} \quad \text{and} \quad \cos(x) = \frac{\mathrm{e}^{\mathrm{i}x}+\mathrm{e}^{-\mathrm{i}x}}{2} \tag{38.26}
$$

Employing the formula for the sine function with arguments $n\varphi$ and φ yields:

$$
\begin{aligned}
\sin(n\varphi)\sin(\varphi) &= \frac{1}{2\mathrm{i}}\left(\mathrm{e}^{\mathrm{i}nx}-\mathrm{e}^{-\mathrm{i}nx}\right)\frac{1}{2\mathrm{i}}\left(\mathrm{e}^{\mathrm{i}x}-\mathrm{e}^{-\mathrm{i}x}\right) \\
&= -\frac{1}{4}\left[\mathrm{e}^{\mathrm{i}(n+1)x}-\mathrm{e}^{\mathrm{i}(n-1)x}-\mathrm{e}^{-\mathrm{i}(n-1)x}+\mathrm{e}^{-\mathrm{i}(n+1)x}\right] \\
&= -\frac{1}{2}\left[\frac{\mathrm{e}^{\mathrm{i}(n+1)x}+\mathrm{e}^{-\mathrm{i}(n+1)x}}{2} - \frac{\mathrm{e}^{\mathrm{i}(n-1)x}+\mathrm{e}^{-\mathrm{i}(n-1)x}}{2}\right]
\end{aligned}
$$

The formula for the cosine allows us to transform the expression back into the real domain which proves the identity (38.25).

$$
\begin{aligned}
\sin(n\varphi)\sin(\varphi) &= -\frac{1}{2}\left[\cos\left((n+1)\varphi\right) - \cos\left((n-1)\varphi\right)\right] \\
&= \frac{1}{2}\left[\cos\left((n-1)\varphi\right) - \cos\left((n+1)\varphi\right)\right]
\end{aligned}
$$

With the help of Equation (38.25) the integral I_n is split into two Glauert integrals.

$$\begin{aligned} I_n &= -\frac{A_n}{2\pi}\left[\int_0^\pi \frac{\cos\left((n-1)\varphi\right)}{\left(\cos(\varphi)-\cos(\theta)\right)}\,\mathrm{d}\varphi - \int_0^\pi \frac{\cos\left((n+1)\varphi\right)}{\left(\cos(\varphi)-\cos(\theta)\right)}\,\mathrm{d}\varphi\right] \\ &= -\frac{A_n}{2}\left[\frac{\sin\left((n-1)\theta\right)}{\sin(\theta)} - \frac{\sin\left((n+1)\theta\right)}{\sin(\theta)}\right] \end{aligned}$$

These, in turn, can be reduced to

$$I_n = A_n\left[\frac{\cos(n\theta)\sin(\theta)}{\sin(\theta)}\right] = A_n\cos(n\theta) \tag{38.27}$$

This conversion employs the trigonometric identity:

$$\sin\left((n-1)\theta\right) - \sin\left((n+1)\theta\right) = -2\cos(n\theta)\sin(\theta) \tag{38.28}$$

Series coefficients Collecting the results from Equations (38.24) and (38.27), the integral I (38.23) is reduced to

$$I = I_0 + \sum_{n=1}^{\infty} I_n = -A_0 + \sum_{n=1}^{\infty} A_n\cos(n\theta) \tag{38.29}$$

Going back to the body boundary condition in its form (38.22), the integral on the left-hand side is replaced with the right-hand side of Equation (38.29).

$$-A_0 + \sum_{n=1}^{\infty} A_n\cos(n\theta) = -\frac{\mathrm{d}f_m}{\mathrm{d}x} - \alpha \tag{38.30}$$

This is a purely geometric relationship, which may be exploited to compute the coefficients A_n with $n = 0, 1, 2, \ldots$.

Coefficient A_0 In order to gain explicit equations for the coefficients, we first multiply Equation (38.30) with $\cos(k\theta)$ on both sides.

$$-A_0\cos(k\theta) + \sum_{n=1}^{\infty} A_n\cos(n\theta)\cos(k\theta) = -\frac{\mathrm{d}f_m}{\mathrm{d}x}\cos(k\theta) - \alpha\cos(k\theta) \tag{38.31}$$

k is an arbitrary integer. Next, we formally integrate both sides with respect to the position of the field point θ from leading edge $\theta = 0$ to trailing edge $\theta = \pi$.

$$\begin{aligned} &-A_0\int_0^\pi \cos(k\theta)\,\mathrm{d}\theta + \sum_{n=1}^{\infty} A_n\int_0^\pi \cos(n\theta)\cos(k\theta)\,\mathrm{d}\theta \\ &\qquad = -\int_0^\pi \frac{\mathrm{d}f_m}{\mathrm{d}x}\cos(k\theta)\,\mathrm{d}\theta - \alpha\int_0^\pi \cos(k\theta)\,\mathrm{d}\theta \end{aligned} \tag{38.32}$$

Equation (38.32) contains two integrals which may be solved directly:

$$\int_0^\pi \cos(k\theta)\,\mathrm{d}\theta = \begin{cases} \pi & \text{for } k = 0 \\ 0 & \text{for } k > 0 \end{cases} \tag{38.33}$$

$$\int_0^\pi \cos(n\theta)\cos(k\theta)\,\mathrm{d}\theta = \begin{cases} \frac{\pi}{2} & \text{for } k = n \\ 0 & \text{for } k \neq n \end{cases} \tag{38.34}$$

If one chooses $k = 0$, our thin foil body boundary condition (38.32) for the lifting problem becomes

Coefficient A_0

$$-A_0\,\pi + \sum_{n=1}^{\infty} A_n\,0 = -\int_0^\pi \frac{\mathrm{d}f_m}{\mathrm{d}x}\,\mathrm{d}\theta - \alpha\,\pi \tag{38.35}$$

The integrals in the sum are all zero because $k = 0$ and $n \geq 1$ and, as a consequence, $k \neq n$. This effectively eliminates all coefficients but A_0, and we obtain

$$A_0 = \alpha + \frac{1}{\pi}\int_0^\pi \frac{\mathrm{d}f_m}{\mathrm{d}x}\,\mathrm{d}\theta \tag{38.36}$$

For any integer $k > 0$, Equation (38.32) reduces to

Coefficients A_n

$$-A_0\,0 + A_k\,\frac{\pi}{2} = -\int_0^\pi \frac{\mathrm{d}f_m}{\mathrm{d}x}\cos(k\theta)\,\mathrm{d}\theta - \alpha\,0 \tag{38.37}$$

or

$$A_n = -\frac{2}{\pi}\int_0^\pi \frac{\mathrm{d}f_m}{\mathrm{d}x}\cos(n\theta)\,\mathrm{d}\theta \qquad \text{for } n = 1, 2, 3, \ldots \tag{38.38}$$

Thus, for a known foil geometry and angle of attack we can compute as many coefficients A_n as we need to accurately define the disturbance potential ϕ_{23} for the lifting flow. Equations (38.36) and (38.38) may be utilized in numerical approximations. If the foil mean line is discretized, slope $\mathrm{d}f_m/\mathrm{d}x$ may be approximated by finite differences between offsets. Subsequently, the integrals can be solved numerically with the trapezoidal rule or possibly more accurate methods.

References

Glauert, H. (1947). *The elements of airofoil and airscrew theory.* Cambridge University Press, Cambridge, United Kingdom, second edition.

Self Study Problems

1. Starting with Equation (38.10), explicitly derive the horizontal velocity

$$u_{23}(x, y = \pm 0) = \mp \frac{\gamma(x)}{2}$$

induced by the line distribution of vortices on the upper and lower side of the x-axis between trailing and leading edge.

2. Given is a line distribution of vortices along the tail–nose line with strength $\gamma(x)$. By how much changes the horizontal velocity u as you move from the underside ($y = 0^-$) to the upper side of the vortex distribution ($y = 0^+$)?

3. Prove the trigonometric identity (38.28) using Euler's formula (38.26).

4. You are part of a survivor team and as the team's naval architect you are tasked with building a hydrofoil supported boat. You have three straight steel plates that you can weld into the crude foil shape shown below. The function f_m describes the mean line and shape of the foil.

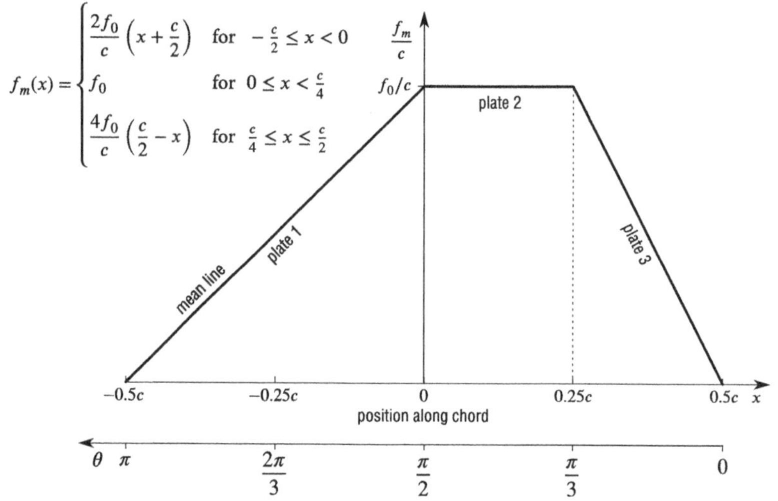

Using thin foil theory results, estimate the section lift coefficient C_l as a function of the maximum camber ratio f_0/c and the angle of attack α. *Hint: integrals can be split up into integrals over parts of the integration range.*

39

Thin Foil Theory – Lifting Flow Properties

With the coefficients A_n $(n = 0, 1, 2, \dots)$ known, the vortex strength may be computed with Equation (38.17) at all points along the chord of a foil. That in turn defines the disturbance potential ϕ_{23} for the lifting flow. We are most interested in the lift force acting on the foil. This chapter exploits the solutions of the thin foil boundary value problems to define and quantify important lifting foil properties.

Learning Objectives

At the end of this chapter students will be able to:

- understand pressure distributions and shock free entry
- compute lift coefficients
- compute the ideal angle of attack

39.1 Lift Force and Lift Coefficient

Linearized Bernoulli equation

In our discussion of the flow over a symmetric thin foil (see Section 37.2), we derived the pressure difference between local pressure p and the reference p_0 from a linearized Bernoulli equation.

$$p - p_0 = \rho\, v_0 \frac{\partial \phi_1}{\partial x} \qquad \text{displacement flow} \qquad (37.20)$$

Following the same arguments that led to Equation (37.20), it may be concluded that, in the lifting flow problem, the pressure difference is a function of the horizontal disturbance velocity $\partial\phi_{23}/\partial x$.

$$p - p_0 = \rho\, v_0 \frac{\partial \phi_{23}}{\partial x} \qquad \text{lifting flow} \qquad (39.1)$$

Pressure over the foil contour

Within the limits of thin foil theory, the horizontal velocity u_{23} induced on the foil is

Fundamentals of Ship Hydrodynamics: Fluid Mechanics, Ship Resistance and Propulsion, First Edition.
Lothar Birk.

Companion website: www.wiley.com/go/birk/hydrodynamics

equal to half the vortex strength.

$$u_{23}^{\pm}(x, y = \pm 0) = \frac{\partial \phi_{23}}{\partial x} = \mp \frac{\gamma(x)}{2} \qquad \text{for } x \in [-c/2, c/2] \tag{38.11}$$

Based on this result, the pressure on upper and lower foil surface is derived from Equation (39.1)

$$p^+ - p_0 = -\rho\, v_0 \frac{\gamma}{2} \qquad \text{for } y = +0 \tag{39.2}$$

$$p^- - p_0 = +\rho\, v_0 \frac{\gamma}{2} \qquad \text{for } y = -0 \tag{39.3}$$

The difference in pressure between the lower and upper sides becomes

$$\Delta p = p^- - p^+ = +\rho\, v_0\, \gamma \qquad \text{for } y = 0 \tag{39.4}$$

Pressure coefficients A division of the equations above by the dynamic pressure yields the pressure coefficients for upper and lower foil side as well as the pressure difference.

$$C_p^+ = \frac{p^+ - p_0}{\frac{1}{2}\rho\, v_0^2} = -\frac{\gamma}{v_0} \qquad \text{for } y = +0 \tag{39.5}$$

$$C_p^- = \frac{p^- - p_0}{\frac{1}{2}\rho\, v_0^2} = +\frac{\gamma}{v_0} \qquad \text{for } y = -0 \tag{39.6}$$

$$\Delta C_p = C_p^- - C_p^+ = +2\frac{\gamma}{v_0} \qquad \text{for } y = 0 \tag{39.7}$$

Lift force from pressure Integrating the pressure difference along the foil yields the lift force.

$$\underline{F}_p = \int_{-c/2}^{c/2} \Delta p\, \underline{n}\, \mathrm{d}\xi \tag{39.8}$$

The normal vector $\underline{n}$ is derived from the slope of the camber line (see Figure 36.8).

$$\underline{n} = \frac{1}{\sqrt{1 + \left(\frac{\mathrm{d}f_m}{\mathrm{d}x}\right)^2}} \begin{pmatrix} -\frac{\mathrm{d}f_m}{\mathrm{d}x} \\ 1 \end{pmatrix} \tag{39.9}$$

Force on flat plate Consider the case of a flat plate. Its camber is zero and the normal vector simplifies to $\underline{n} = (0, 1)^T$. The x-component of the pressure force vanishes, and the y-component is perpendicular to the plate.

$$F_y = \int_{-c/2}^{c/2} \Delta p\, \mathrm{d}\xi \tag{39.10}$$

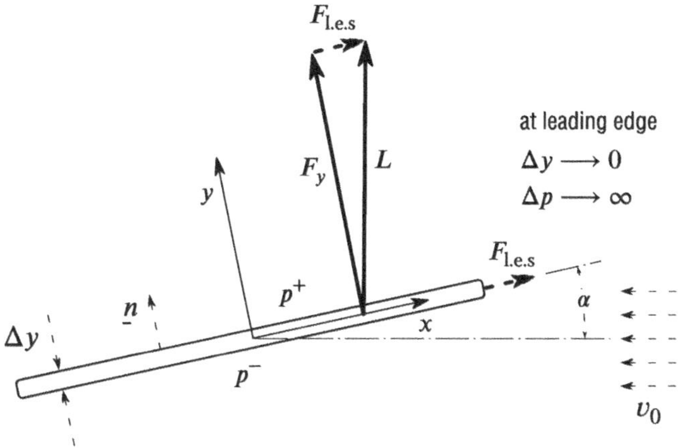

Figure 39.1 The effect of leading edge suction for a thin plate at angle of attack α

Introducing Equation (39.4) yields

$$F_y = \rho v_0 \int_{-c/2}^{c/2} \gamma(\xi)\, d\xi \tag{39.11}$$

If the plate operates at angle of attack α, we find a lift force $d\tilde{L}$ perpendicular to the onflow and a drag force $d\tilde{D}$ parallel to the onflow v_0.

$$\tilde{D} = F_y \sin(\alpha) \tag{39.12}$$

$$\tilde{L} = F_y \cos(\alpha) \tag{39.13}$$

Leading edge suction

This appears to be in contradiction to Kutta-Joukowsky's lift theorem. For inviscid flow, we expect the drag force to vanish, and the lift force is supposed to be perpendicular to the onflow. However, we derived a pressure force (39.11) which acts perpendicular to the plate. Something is obviously amiss. Either Kutta-Joukowsky's lift theorem does not apply, or our pressure integration is missing a component of the force. At first glance, you could argue that there cannot be a pressure force along the infinitely thin plate edge because there is no area with normal vector along the edge for the pressure to act on. This argument, however, is flawed. Study the top left graph in Figure 38.2! The vortex strength γ becomes infinite at the leading edge ($\varphi = 0$), if the series coefficient A_0 is nonzero. An infinite vortex strength results in an infinite pressure difference $\Delta p \longrightarrow \infty$ at the leading edge (Figure 39.1). An infinitely small area multiplied by an infinitely large pressure may create a finite force $F_{\text{l.e.s}}$, which acts along the plate and completely compensates for the apparent drag force $\tilde{D}$. This effect is known as leading edge suction. The actual leading edge suction force $F_{\text{l.e.s}}$ may be computed with conformal mapping. This is beyond the scope of this book, but the interested reader may find this in standard aerodynamics texts like Katz and Plotkin (2001).

Lift force

Considering the corrective leading edge suction force, the lift force (perpendicular to onflow) is equal to the transverse pressure force F_y.

$$L = \rho v_0 \int_{-c/2}^{c/2} \gamma(\xi)\, d\xi \tag{39.14}$$

This in turn must correspond to Kutta-Joukowsky's lift theorem (15.14).

$$L = \rho\, v_0\, \Gamma \tag{39.15}$$

Circulation A comparison of Equations (39.14) and (39.15) leads to the connection between circulation Γ and vortex strength γ:

$$\Gamma = \int_{-c/2}^{c/2} \gamma(\xi)\,\mathrm{d}\xi \tag{39.16}$$

In Chapter 38 we expressed the vortex distribution as a Glauert series and linked the series coefficients to the foil geometry. We replace the integration variable ξ in Equation (39.16) with φ (37.28), adjust the integration limits according to Equation (38.19), and substitute the series expansion (38.17) for the vortex strength $\gamma(\xi)$.

$$\begin{aligned}\Gamma &= \int_0^\pi \gamma(\varphi)\frac{c}{2}\sin(\varphi)\,\mathrm{d}\varphi \\ &= \int_0^\pi \left[A_0\,\frac{1+\cos(\varphi)}{\sin(\varphi)} + \sum_{n=1}^{\infty} A_n\,\sin(n\varphi)\right]\sin(\varphi)\mathrm{d}\varphi\end{aligned}$$

Next, the integral is split into parts.

$$\Gamma = c\,v_0\,A_0 \int_0^\pi \left[1+\cos(\varphi)\right]\mathrm{d}\varphi + c\,v_0 \sum_{n=1}^{\infty} A_n \int_0^\pi \sin(n\varphi)\sin(\varphi)\mathrm{d}\varphi$$

The remaining integrals are of standard form and yield the following results:

$$\int_0^\pi \left[1+\cos(\varphi)\right]\mathrm{d}\varphi = \pi \tag{39.17}$$

$$\int_0^\pi \sin(n\varphi)\sin(\varphi)\mathrm{d}\varphi = \begin{cases} \dfrac{\pi}{2} & \text{for } n = 1 \\ 0 & \text{for } n \neq 1 \end{cases} \tag{39.18}$$

Because all integrals vanish for $n > 1$, only the first two terms of the series expansion remain.

$$\Gamma = c\,v_0\,A_0\,\pi + c\,v_0\,A_1\,\frac{\pi}{2} = c\,v_0\,\pi\left[A_0 + \frac{A_1}{2}\right] \tag{39.19}$$

Lift coefficient As a consequence, Kutta-Joukowsky's lift theorem (39.15) yields a lift force which depends on the coefficients A_0 and A_1 only.

$$L = \pi\,\rho\,c\,v_0^2\left[A_0 + \frac{A_1}{2}\right] \tag{39.20}$$

The dimensionless section lift coefficient C_l follows suit:

$$C_l = \frac{L}{\frac{1}{2}\rho v_0^2 c} = 2\pi \left[A_0 + \frac{A_1}{2} \right] \tag{39.21}$$

We recapture the series coefficients A_0 and A_1 from Equations (38.36) and (38.38).

$$A_0 = \alpha + \frac{1}{\pi} \int_0^{\pi} \frac{\mathrm{d}f_m}{\mathrm{d}x}\, \mathrm{d}\theta \qquad A_1 = -\frac{2}{\pi} \int_0^{\pi} \frac{\mathrm{d}f_m}{\mathrm{d}x} \cos(\theta)\, \mathrm{d}\theta \tag{39.22}$$

Inserting them into the equation for the lift coefficient, splits C_l into two distinct parts, as we would expect considering our body boundary condition (38.4).

$$C_l = 2\pi\,\alpha + 2\pi \left(\frac{1}{\pi} \int_0^{\pi} \frac{\mathrm{d}f_m}{\mathrm{d}x} \left[1 - \cos(\theta) \right] \mathrm{d}\theta \right) \tag{39.23}$$

(second term: contribution of camber line; first term: contribution of flat plate at angle α)

- The first term represents the lift coefficient of a flat plate at angle of attack α.

$$C_l = 2\pi\,\alpha \qquad \text{flat plate} \tag{39.24}$$

  The slope of the camber is zero for a flat plate and, as a consequence, the second term vanishes. The exact lift coefficient derived from conformal mapping is $C_{l_{c.m.}} = 2\pi \sin(\alpha)$. Equation (39.24) is equivalent to the exact result for small angles of attack for which $\sin(\alpha) \approx \alpha$.

- The second term in Equation (39.23) captures the effect of camber. Obviously, a cambered foil may produce a lift force even if it operates at zero angle of attack.

Note that the overall slope of the lift coefficient curve is 2π for thin foils. Beware that the angle of attack in these formulas has to be entered in radians rather than in degrees!

Zero lift angle α_0

The lift coefficient vanishes for the so called zero lift angle α_0, i.e. $C_l(\alpha_0) = 0$. Setting $C_l = 0$ in Equation (39.23) yields:

$$C_l = 0 = 2\pi\,\alpha_0 + 2\pi \left(\frac{1}{\pi} \int_0^{\pi} \frac{\mathrm{d}f_m}{\mathrm{d}x} \left[1 - \cos(\theta) \right] \mathrm{d}\theta \right)$$

or

$$\alpha_0 = -\frac{1}{\pi} \int_0^{\pi} \frac{\mathrm{d}f_m}{\mathrm{d}x} \left[1 - \cos(\theta) \right] \mathrm{d}\theta \qquad \text{zero lift angle} \tag{39.25}$$

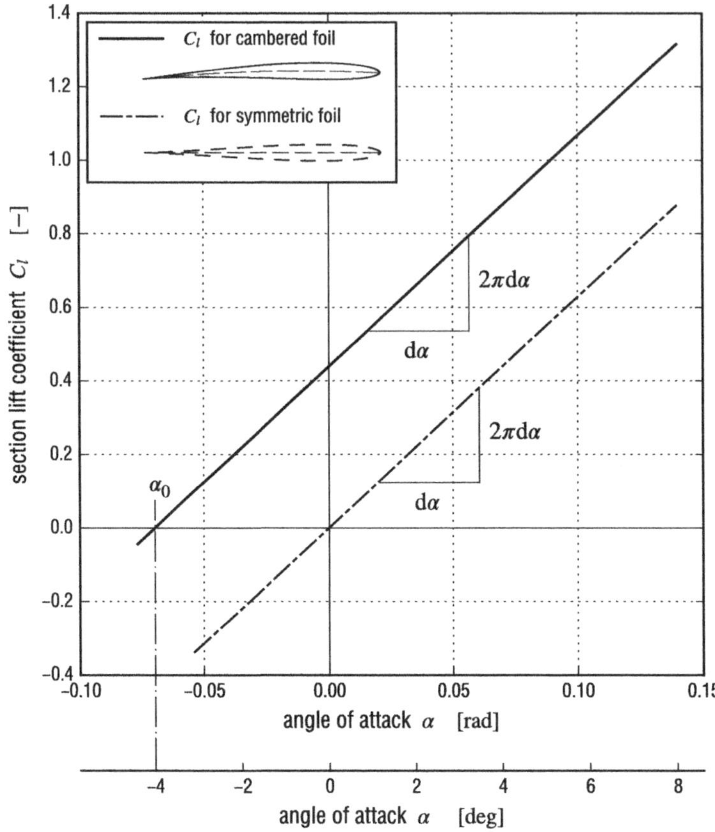

Figure 39.2 Section lift coefficient C_l of thin, symmetric foil sections and a thin, cambered foil section with zero lift angle $\alpha_0 = -4$ degree

Zero lift angles are negative for foil sections with positive camber.

Figure 39.2 pictures the lift coefficient as a function of the angle of attack. The results of thin foil theory hold up very well in practice as long as the foils are actually thin ($t_{\max}/c < 0.1$) and the angle of attack is small ($|\alpha| < 5$ deg). Thicker foils, however, have a slightly steeper slope for small angles of attack.

39.2 Moment and Center of Effort

Besides the lift force and lift coefficient, it is useful to know the center of effort for the lift force. It is important for balancing control surfaces like rudders or stabilizers and also affects the spindle moment of the blades in a controllable pitch propeller.

Definition of Moment M_z

In general, a moment is a vector and computed by integration of the cross product of lever $\underline{r}$ and differential force vector $d\underline{F}$.

$$\underline{M} = \int_{-c/2}^{c/2} \underline{r} \times d\underline{F} \tag{39.26}$$

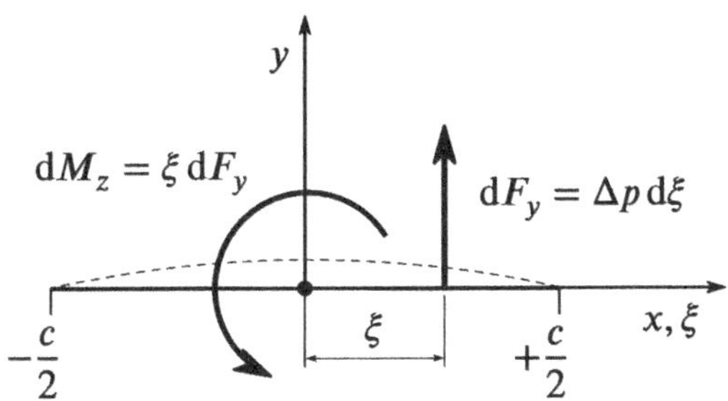

Figure 39.3 Definition of the moment M_z created by the pressure force acting on a thin foil

Our flow problem is two-dimensional and only a moment M_z with axis normal to the x-y-plane exists (Figure 39.3).

Resultant pressure force distribution

The differential pressure force $\mathrm{d}F_p^{\pm} = p^{\pm}\,\mathrm{d}\xi$ acts on upper and lower foil contours. However, in accordance with thin foil theory assumptions, we perform the integration along the tail–nose line from trailing edge to leading edge rather than integrating over the actual contour. The force acting at a position ξ along the chord is a result of the pressure difference between lower and upper side:

$$\mathrm{d}F_y = \mathrm{d}F_p^- - \mathrm{d}F_p^+ = \Delta p\,\mathrm{d}\xi \tag{39.27}$$

Its lever with respect to the z-axis is equal to ξ. Therefore, the moment M_z becomes

$$M_z = \int_{-c/2}^{c/2} \xi\,\mathrm{d}F_y = \int_{-c/2}^{c/2} \xi\,\Delta p\,\mathrm{d}\xi \tag{39.28}$$

Subsequently, the pressure difference is expressed with Equation (39.4) in terms of the vortex strength γ. Again, the now familiar variable substitution $\xi \longrightarrow \varphi$ (37.28) is employed.

$$\begin{aligned} M_z &= \rho\,v_0 \int_{-c/2}^{c/2} \xi\,\gamma(\xi)\,\mathrm{d}\xi \\ &= \rho\,v_0 \int_{\pi}^{0} \frac{c}{2}\cos(\varphi)\,\gamma(\varphi)\left[-\frac{c}{2}\sin(\varphi)\right]\mathrm{d}\varphi \end{aligned}$$

Reversal of the integration limits eliminates the minus sign in the integrand.

$$M_z = \rho\,v_0 \frac{c^2}{4} \int_0^{\pi} \gamma(\varphi)\,\sin(\varphi)\,\cos(\varphi)\,\mathrm{d}\varphi \tag{39.29}$$

Glauert's series

Finally, Glauert's series expansion (38.17) is used to represent the vortex strength $\gamma(\varphi)$.

$$M_z = \rho\,v_0^2\,\frac{c^2}{2} \int_0^{\pi} \left[A_0 \frac{1+\cos(\varphi)}{\sin(\varphi)} + \sum_{n=1}^{\infty} A_n \sin(n\varphi)\right] \sin(\varphi)\,\cos(\varphi)\,\mathrm{d}\varphi \tag{39.30}$$

Separating the integrals yields:

$$M_z = \rho v_0^2 \frac{c^2}{2} \left[A_0 \int_0^{\pi} \left(\cos(\varphi) + \cos^2(\varphi) \right) d\varphi \right.$$

$$\left. + \sum_{n=1}^{\infty} A_n \int_0^{\pi} \sin(n\varphi) \sin(\varphi) \cos(\varphi) \, d\varphi \right] \quad (39.31)$$

Solving the first integral

Once more, we are left with a set of integrals involving products of trigonometric functions. The first integral is of a standard type.

$$\int_0^{\pi} \left(\cos(\varphi) + \cos^2(\varphi) \right) d\varphi = \left[-\sin(\varphi) \right]_0^{\pi} + \left[\frac{\varphi}{2} + \frac{1}{4} \sin(2\varphi) \right]_0^{\pi}$$

$$= 0 - 0 + \frac{\pi}{2} - 0 = \frac{\pi}{2} \quad (39.32)$$

Solving the second integral

For the second integral in Equation (39.31), we exploit yet another identity for trigonometric functions.

$$\sin(\varphi) \cos(\varphi) = \frac{1}{2} \sin(2\varphi) \quad (39.33)$$

Therefore

$$\int_0^{\pi} \sin(n\varphi) \sin(\varphi) \cos(\varphi) \, d\varphi = \frac{1}{2} \int_0^{\pi} \sin(n\varphi) \sin(2\varphi) \, d\varphi \quad (39.34)$$

The integrand of the remaining integral on the right-hand side may be replaced by a generalization of the trigonometric identity in Equation (38.25).

$$\sin(n\varphi) \sin(k\varphi) = \frac{1}{2} \left[\cos\left((n-k)\varphi\right) - \cos\left((n+k)\varphi\right) \right] \quad (39.35)$$

Keeping in mind that $k = 2$, the integral identity (39.35) converts our integral (39.34) into a difference of integrals:

$$\frac{1}{2} \int_0^{\pi} \sin(n\varphi) \sin(k\varphi) \, d\varphi = \frac{1}{4} \int_0^{\pi} \left[\cos\left((n-k)\varphi\right) - \cos\left((n+k)\varphi\right) \right] d\varphi$$

$$= \frac{1}{4} \int_0^{\pi} \cos\left((n-k)\varphi\right) d\varphi - \frac{1}{4} \int_0^{\pi} \cos\left((n+k)\varphi\right) d\varphi \quad (39.36)$$

Two cases may occur:

(i) For $k \neq n$ we set $k - n = m$. The definite integral of a cosine function vanishes with the limits 0 to π.

$$\int_0^{\pi} \cos(m\varphi) \, d\varphi = \left[\frac{1}{m} \sin(m\varphi) \right]_0^{\pi} = 0 \quad (39.37)$$

The same will be true for the second integral with $k + n = m$.

(ii) If $k = n$, the argument of the first integrand becomes the constant one.

$$\cos\left((n-k)\varphi\right) = \cos(0) = 1 \qquad \text{for } n = k$$

Therefore, the integral takes the value

$$\int_0^\pi \cos\left((n-k)\varphi\right) \mathrm{d}\varphi = \int_0^\pi 1 \,\mathrm{d}\varphi = \Big[\varphi\Big]_0^\pi = \pi \tag{39.38}$$

For $k = n$ the second integral turns into

$$\int_0^\pi \cos\left((n+k)\varphi\right) \mathrm{d}\varphi = \int_0^\pi \cos(2n\varphi)\,\mathrm{d}\varphi = 0 \tag{39.39}$$

which vanishes like the integral in Equation (39.37).

In summary, the first integral in Equation (39.31) for the moment M_z is equal to $\pi/2$ (39.32). The integrals in the sum over n of Equation (39.31) vanish except for the case $n = 2$. Based on the results (39.38) and (39.37), we get

$$\int_0^\pi \sin(n\varphi)\,\sin(\varphi)\,\cos(\varphi)\,\mathrm{d}\varphi = \begin{cases} \dfrac{\pi}{4} & \text{for } n = 2 \\ 0 & \text{for } n \neq 2 \end{cases} \tag{39.40}$$

Result for moment M_z

Like in the case of the lift force, only two of the Glauert series coefficients are required to compute the moment M_z. The coefficient A_0 is known from the lift force. In addition, we need the coefficient A_2.

$$M_z = \rho\, v_0^2\, \frac{c^2}{2} \left[\frac{\pi}{2} A_0 + \frac{\pi}{4} A_2\right] = \pi\, \rho\, v_0^2\, \frac{c^2}{2} \left[A_0 + \frac{A_2}{2}\right] \tag{39.41}$$

Center of effort x_l

The pressure distribution responsible for the moment M_z creates the lift L as a resultant force. Based on the moment M_z, we may compute the center of effort x_l for the lift force. Using Equation (39.20) for the lift force and Equation (39.41) for the moment, we obtain the coordinate

$$x_l = \frac{M_z}{L} = \frac{\pi\, \rho\, v_0^2\, \dfrac{c^2}{2} \left[A_0 + \dfrac{A_2}{2}\right]}{\pi\, \rho\, c\, v_0^2 \left[A_0 + \dfrac{A_1}{2}\right]} = \frac{c}{4}\, \frac{\left[A_0 + \dfrac{A_2}{2}\right]}{\left[A_0 + \dfrac{A_1}{2}\right]} \tag{39.42}$$

Center of effort for flat plate

For the special case of a flat plate, all coefficients $A_n = 0$ vanish for $n > 0$ because the derivative of the nonexisting camber is zero $\mathrm{d}f_m/\mathrm{d}x = 0$. Therefore, the center of effort is

$$x_l = \frac{c}{4} \qquad \text{for a flat plate} \tag{39.43}$$

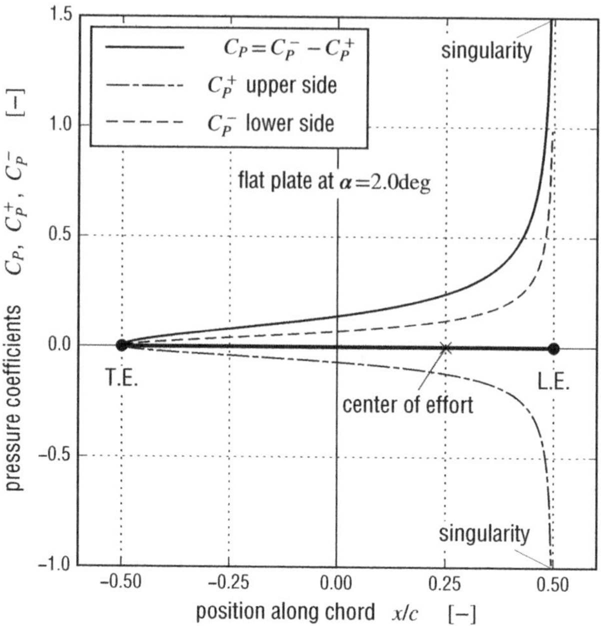

Figure 39.4 Pressure distribution for a flat plate at 2 degrees angle of attack

The value is positive, which means the center of effort is upstream, halfway between origin and leading edge. Result (39.43) also applies to thin symmetric foils which have a small, finite thickness, but no camber. With increasing thickness, the center of effort tends to move aft toward the origin at mid-chord.

Thus, if you plan to use a small rudder engine, your shaft should be attached to the rudder at a quarter chord length from its leading edge. In practice, you might place the shaft a bit forward, closer to the leading edge. This will reduce possible vibrations, and the rudder will tend to return to midships if left alone.

39.3 Ideal Angle of Attack

Pressure distribution on flat plate

We continue to study the pressure distribution over a flat plate in more detail. Of the expansion series coefficients $A_0, A_1, A_2, \ldots,$ from Equations (38.36) and (38.38) only A_0 has a finite value, because the integrals over the slope of the camber line vanish.

$$\begin{aligned} A_0 &= \alpha + \frac{1}{\pi}\int_0^{\pi} \frac{\mathrm{d}f_m}{\mathrm{d}x}\,\mathrm{d}\theta = \alpha \\ A_n &= -\frac{2}{\pi}\int_0^{\pi} \frac{\mathrm{d}f_m}{\mathrm{d}x}\cos(n\theta)\,\mathrm{d}\theta = 0 \qquad \text{for } n > 0 \end{aligned} \tag{39.44}$$

The pressure coefficients for upper and lower side of the foil follow from Equations (39.5) and (39.6).

$$
\begin{aligned}
C_p^{\pm} &= \mp \frac{\gamma}{v_0} = \mp \frac{1}{v_0} 2 v_0 \left[A_0 \frac{1+\cos(\varphi)}{\sin(\varphi)} + \sum_{n=1}^{\infty} A_n \sin(n\varphi) \right] \\
&= \mp 2\,\alpha \frac{1+\cos(\varphi)}{\sin(\varphi)}
\end{aligned} \tag{39.45}
$$

Figure 39.4 shows the pressure coefficients of upper and lower foil contour for an angle of attack of 2 degrees along with the resulting dimensionless pressure difference ΔC_p.

$$
\Delta C_p = C_p^- - C_p^+ = 4\,\alpha \frac{1+\cos(\varphi)}{\sin(\varphi)} \tag{39.46}
$$

With the help of the identity $\sin(\varphi) = \sqrt{1-\cos^2(\varphi)}$ and the inverse of the variable substitution (37.28) $\cos(\varphi) = 2x/c$, we express ΔC_p as a function of x.

$$
\begin{aligned}
\Delta C_p &= 4\,\alpha \frac{1+\cos(\varphi)}{\sin(\varphi)} = 4\,\alpha \frac{1+\cos(\varphi)}{\sqrt{1-\cos^2(\varphi)}} \\
&= 4\,\alpha \sqrt{\frac{1+\cos(\varphi)}{1-\cos(\varphi)}} = 4\,\alpha \sqrt{\frac{\frac{c}{2}+x}{\frac{c}{2}-x}}
\end{aligned} \tag{39.47}
$$

Clearly, the pressure distribution has a singularity at the leading edge. **Pressure shock**

$$
\Delta C_p \longrightarrow \infty \qquad \text{for } x \longrightarrow +c/2 \text{ and } \alpha \neq 0 \tag{39.48}
$$

Of course, in a real fluid, the pressure would not rise that far and there is no infinitely thin plate which could support such a high pressure difference. However, even foils with a rounded nose will experience a peak in the pressure distribution. The sudden rise and fall in pressure over the chord wise position has a similar signature as the pressure shock wave over time after an explosion. Therefore, the peak of the pressure distribution on a foil operating at a nonzero angle of attack $\alpha \neq 0$ is referred to as pressure shock. We will see later that this pressure shock increases the danger of cavitation on a foil and should be avoided whenever possible.

Mathematically, the singularity stems from the term associated with coefficient A_0 in the expansion series for the vortex strength γ (38.17). The position angle becomes zero $\varphi \longrightarrow 0$ at the leading edge with $x \longrightarrow +c/2$. As a consequence, the term associated with A_0 grows toward infinity. **Ideal angle of attack**

$$
\lim_{\varphi \to 0} \left(A_0 \frac{1+\cos(\varphi)}{\sin(\varphi)} \right) \longrightarrow \infty \tag{39.49}
$$

The only option to avoid the pressure shock is to design and operate the foil such that the coefficient A_0 vanishes. Recapturing Equation (38.36) for the coefficient

$$
A_0 = \alpha + \frac{1}{\pi} \int_0^{\pi} \frac{\mathrm{d}f_m}{\mathrm{d}x}\,\mathrm{d}\varphi \tag{38.36}
$$

and enforcing $A_0 = 0$, leads to a specific angle of attack $\alpha = \alpha_{id}$:

$$0 = \alpha_{id} + \frac{1}{\pi}\int_0^\pi \frac{\mathrm{d}f_m}{\mathrm{d}x}\,\mathrm{d}\varphi$$

$$\alpha_{id} = -\frac{1}{\pi}\int_0^\pi \frac{\mathrm{d}f_m}{\mathrm{d}x}\,\mathrm{d}\varphi \qquad (39.50)$$

The ideal angle of attack α_{id} is important because it provides shock free entry for the leading edge. If the foil is operating at an angle of attack $\alpha = \alpha_{id}$, pressure is bounded everywhere and sharp pressure peaks near the leading edge are avoided. It should be clear that a flat plate is shock free only if $\alpha = \alpha_{id} = 0$ because there is no camber. Ideally, a foil will operate at its ideal angle of attack to delay cavitation inception.

39.4 Parabolic Mean Line

Parabolic camber

As an example for the use of Glauert's series expansion (38.17), we investigate the parabolic mean line.

$$f_m(x) = f_0\left[1 - \left(\frac{2x}{c}\right)^2\right] \qquad (39.51)$$

The camber line is symmetric to the middle of the chord length and has a maximum camber value of f_0 for $x = 0$. The parabolic mean line is often used as a stand-in for the mean line of a circular segment type foil, where the pressure side is flat and the suction side is a circular arc. This section type was standard for propellers in the first half of the 20th century and is still used in off-the-shelf boat propellers.

Coefficients A_0, A_1, and A_2 for parabolic mean line

We begin by computing the first three coefficients of the series expansion (38.17).

$$A_0 = \alpha + \frac{1}{\pi}\int_0^\pi \frac{\mathrm{d}f_m}{\mathrm{d}x}\,\mathrm{d}\varphi \qquad (39.52)$$

$$A_1 = -\frac{2}{\pi}\int_0^\pi \frac{\mathrm{d}f_m}{\mathrm{d}x}\cos(\varphi)\,\mathrm{d}\varphi \qquad (39.53)$$

$$A_2 = -\frac{2}{\pi}\int_0^\pi \frac{\mathrm{d}f_m}{\mathrm{d}x}\cos(2\varphi)\,\mathrm{d}\varphi \qquad (39.54)$$

In all three cases, we need the derivative of the parabolic mean line.

$$\frac{\mathrm{d}f_m}{\mathrm{d}x} = \frac{\mathrm{d}}{\mathrm{d}x}\left(f_0\left[1 - \left(\frac{2x}{c}\right)^2\right]\right) = -\frac{8\,f_0}{c^2}\,x \qquad (39.55)$$

Substituting the slope into the formula for A_0 and employing the variable substitution $x = c\cos(\varphi)/2$ (37.28) yields:

$$\begin{aligned} A_0 &= \alpha + \frac{1}{\pi}\int_0^{\pi}\left(-\frac{8\,f_0}{c^2}\right)\frac{c}{2}\cos(\varphi)\,\mathrm{d}\varphi \;=\; \alpha - \frac{4\,f_0}{\pi\,c}\Big[\sin(\varphi)\Big]_0^{\pi} \\ &= \alpha - \frac{4\,f_0}{\pi\,c}\big(0-0\big) \;=\; \alpha \end{aligned} \tag{39.56}$$

The integral vanishes, which is actually not surprising. The parabolic mean line is symmetric to the middle of the chord length. Its derivative must be antimetric, and the integral over antimetric functions is zero, when it is computed over a symmetric interval. This result obviously holds true for all symmetric mean lines.

Ideal angle of attack

As a consequence, the ideal angle of attack is zero for the parabolic mean line or any other symmetric mean line.

$$A_0 = 0 \quad\longrightarrow\quad \alpha_{id} = 0 \qquad \text{for symmetric mean lines} \tag{39.57}$$

The coefficient A_1 becomes

$$\begin{aligned} A_1 &= -\frac{2}{\pi}\int_0^{\pi}\left(-\frac{8\,f_0}{c^2}\right)\frac{c}{2}\cos(\varphi)\,\cos(\varphi)\,\mathrm{d}\varphi \\ &= \frac{8\,f_0}{\pi\,c}\int_0^{\pi}\cos^2(\varphi)\,\mathrm{d}\varphi \;=\; \frac{8\,f_0}{\pi\,c}\left[\frac{1}{2}\varphi + \frac{1}{4}\sin(2\varphi)\right]_0^{\pi} \\ &= 4\,\frac{f_0}{c} \end{aligned} \tag{39.58}$$

And we get for the coefficient A_2:

$$\begin{aligned} A_2 &= -\frac{2}{\pi}\int_0^{\pi}\left(-\frac{8\,f_0}{c^2}\right)\frac{c}{2}\cos(\varphi)\,\cos(2\varphi)\,\mathrm{d}\varphi \\ &= \frac{8\,f_0}{\pi\,c}\int_0^{\pi}\cos(\varphi)\,\cos(2\varphi)\,\mathrm{d}\varphi \end{aligned}$$

The remaining integral is found in tables of standard integrals (e.g. Gradshteyn and Ryshik, 2000).

$$\int_0^{\pi}\cos(\varphi)\,\cos(2\varphi)\,\mathrm{d}\varphi = \left[\frac{\sin(-\varphi)}{-2} + \frac{\sin(3\varphi)}{6}\right]_0^{\pi} = 0 \tag{39.59}$$

In fact, the integrals vanish for all coefficients A_n with $n \geq 2$.

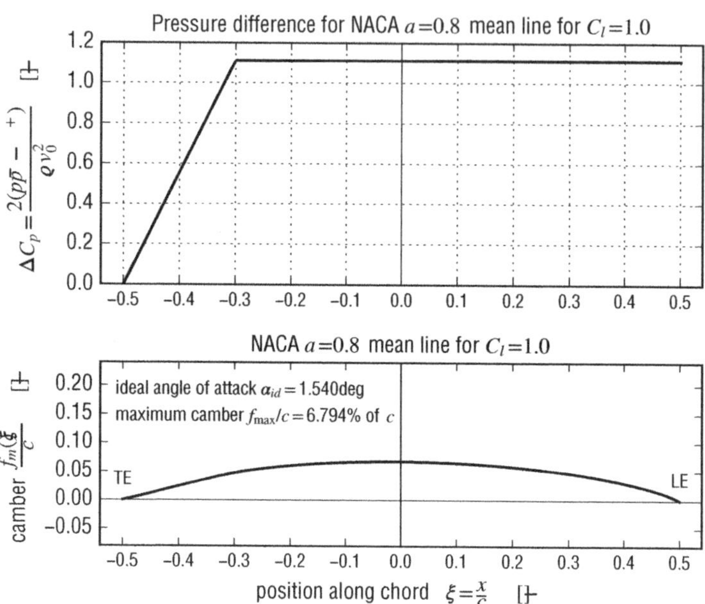

Figure 39.5 A prescribed pressure distribution resulting in the NACA $a = 0.8$ mean line

Section lift coefficient C_l

According to Equation (39.21), the lift coefficient depends on coefficients A_0 and A_1 only:

$$C_l = 2\pi \left[A_0 + \frac{A_1}{2}\right] = 2\pi \left[\alpha + 2\frac{f_0}{c}\right] \tag{39.60}$$

As stated before, the slope of the lift coefficient curve $C_l(\alpha)$ is 2π for thin foils.

Zero lift angle

For a foil with positive camber, the curve is shifted to the left. The zero lift angle α_0 is derived from the condition $C_l = 0$.

$$C_l = 0 \quad \longrightarrow \quad 0 = 2\pi \left[\alpha_0 + 2\frac{f_0}{c}\right] \quad \longrightarrow \quad \alpha_0 = -2\frac{f_0}{c} \tag{39.61}$$

The zero lift angle (39.61) for the parabolic mean line is computed in radians.

NACA $a = 0.8$ mean line

In the practical design of foil sections, engineers start with a desired pressure distribution rather than an equation for the mean line. An example is the NACA $a = 0.8$ mean line. NACA stands for National Advisory Committee for Aeronautics and was the predecessor to today's National Aeronautics and Space Administration (NASA). The a-mean lines prescribe a constant pressure difference from the leading edge of a foil to the fraction a of the chord length c toward the trailing edge. In the case of $a = 0.8$, the pressure difference between lower and upper foil contour will be constant over 80% of the chord length starting at the leading edge. From the end of the constant pressure distribution, the pressure difference linearly declines to zero at the trailing edge. This pressure distribution and the resulting mean line are presented in Figure 39.5 for $a = 0.8$, which is the most widely used value. Pressure distribution and camber line will be scaled with the actual design lift coefficient. Usually, lift coefficients for propeller foil sections are in the range from $C_l = 0.1$ to 0.5. The actual equations are quite cumbersome and go beyond the scope of this book. Details can be found in the classical text by Abbott and von Doenhoff (1959).

References

Abbott, I. and von Doenhoff, A. (1959). *Theory of wing sections – including a summary of airfoil data*. Dover Publications, Inc., New York, NY, Dover edition.

Gradshteyn, I. and Ryshik, I. (2000). *Tables of integrals, series and products*. Academic Press, San Diego, CA, sixth edition. p. 1132.

Katz, J. and Plotkin, A. (2001). *Low-speed aerodynamics*. Cambridge Aerospace Series. Cambridge University Press, New York, NY, second edition.

Self Study Problems

1. Derive Equation (39.1) for the pressure difference in a lifting flow from Bernoulli's equation.

$$\frac{1}{2}\rho|\underline{\nabla}\,\Phi|^2 + p = \frac{1}{2}\rho\, v_0^2 + p_0 \tag{37.18}$$

   Ignore gravity effects and assume the foil has no thickness. Apply appropriate thin foil theory assumptions.

2. Plot the curve for the section lift coefficient C_l as a function of the angle of attack α for a thin foil section with parabolic mean line and a camber–chord length ratio of $f_0/c = 1.5\%$.

3. Compute and plot the dimensionless pressure distribution over the chord length for upper and lower side of a flat plate at an angle of attack $\alpha = 3$ degree.

4. Revisit Problem 4 at the end of Chapter 38. What will be the ideal angle of attack α_{id} of the foil if the maximum camber ratio is $f_0/c = 0.03$? State the result in degrees.

40

Lifting Wings

So far, our discussion of lifting flow has been restricted to bodies that stretch to infinity in one of the three spatial coordinates. This allowed us to treat the flow around cylinders or foil sections as two-dimensional flow problems. This chapter discusses the effects of a finite span on the flow patterns. We still assume that the fluid is inviscid, incompressible, and irrotational.

Learning Objectives

At the end of this chapter students will be able to:

- discuss flow patterns of wings with finite span
- understand the origin of downwash
- know the Biot–Savart Law
- explain induced drag

40.1 Effects of Limited Wingspan

From two- to three-dimensional flow

The length of a wing from tip to tip is called span. In the middle of a finite wing the flow pattern may still be planar, and fluid flows from leading edge to trailing edge. However, the closer you get to the wing tips, the more you will see three-dimensional flow patterns.

Changes in flow pattern

Study the wing of finite span s in Figure 40.1. Like the two-dimensional flow over a foil section, the flow across a wing is governed by the onflow v_0 and the effect of circulation. In order to generate lift, the pressure on the upper side of a wing must be lower, and the pressure on the underside of a wing must be higher compared to the free stream pressure far away from the wing. In general, fluid flows from high pressure areas to low pressure areas attempting to equalize the pressures. At the wing tips, fluid will move from the underside with high pressure to the upper side of the wing with low pressure. Therefore, an additional transverse flow component is initiated at the wing tips for wings with finite span. This transverse flow component will bend velocity

Fundamentals of Ship Hydrodynamics: Fluid Mechanics, Ship Resistance and Propulsion, First Edition.
Lothar Birk.

Companion website: www.wiley.com/go/birk/hydrodynamics

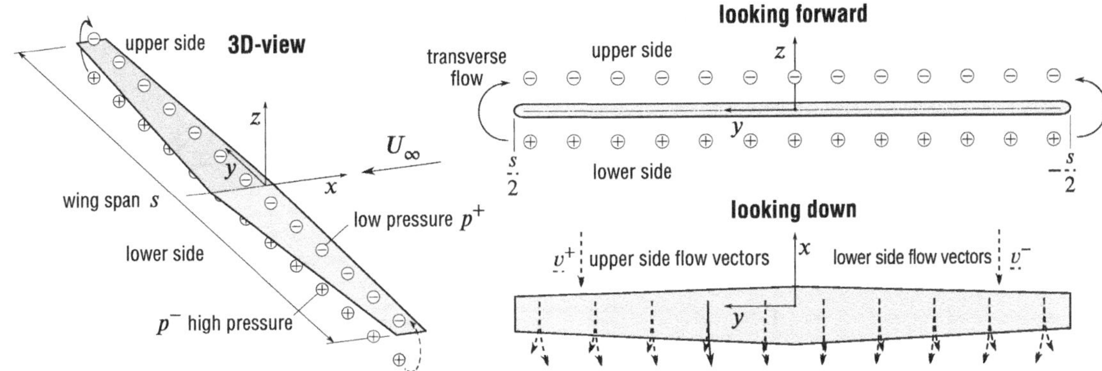

Figure 40.1 Pressure distribution on a wing of finite span s

vectors toward the wing tips on the underside and toward the wing center on the upper side, as indicated by the flow arrows in Figure 40.1.

Lifting line

According to Kutta-Joukowsky's lift theorem (39.15), the lifting effect of a foil section may be represented by a single line vortex. The vortex is also known as lifting line, or bound vortex, because it travels with the wing. For two-dimensional flow, we assume that the vortex is perpendicular to the flow plane and stretches to infinity on both sides. The vortex in three-dimensional cases may not extend past the wing tips. However, it cannot simply end at the wing tips either. Under the condition that all forces acting on the fluid have a potential, the German physician and physicist Hermann von Helmholtz (*1821 – †1894) concluded in 1858 that the motion of an inviscid fluid is governed by three important theorems which are quoted here from Uwe Parpart's translation (Helmholtz, 1978).

Helmholtz's theorems

(i) No water particle that was not originally in rotation is made to rotate.

(ii) The water particles that at any given time belong to the same vortex line, however they may be translated, will continue to belong to the same vortex line.

(iii) The product of cross section and the velocity of rotation of an infinitely thin vortex filament is constant along the entire length of the filament and retains the same value during all displacements of the filament. Hence, the vortex filaments must run back into themselves in the interior of the fluid or else must end at the bounding surface of the fluid.

These are collectively known as Helmholtz's theorems of fluid mechanics.

Horseshoe vortex

From the third theorem follows that the bound vortex cannot simply end at the wing tips. It finds its continuation in the tip vortices, which stretch downstream from the tips of the wing to the start-up vortex far behind the lifting wing (Figure 40.2). The bound vortex, tip vortices, and start-up vortex form a closed vortex ring in the fluid with constant circulation, as requested by the third theorem.

In modeling a wing, the influence of the start-up vortex is often neglected because it is assumed to be very far away and its effect has subsided at the location of the wing. This is a realistic assumption because the viscosity of a real fluid will dissipate the kinetic

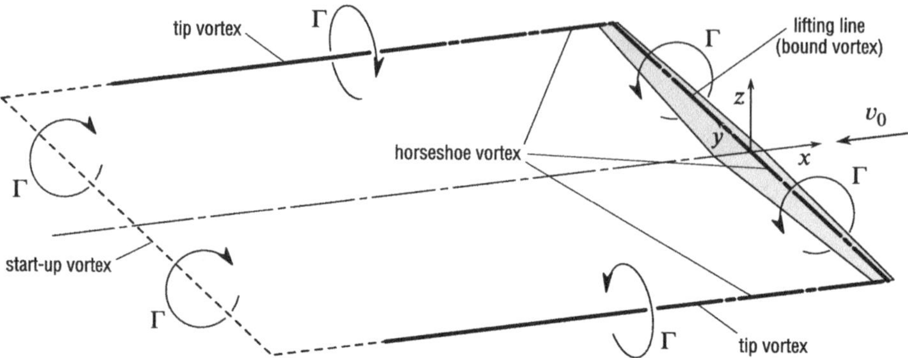

Figure 40.2 Simplest model of the vortex system of a wing

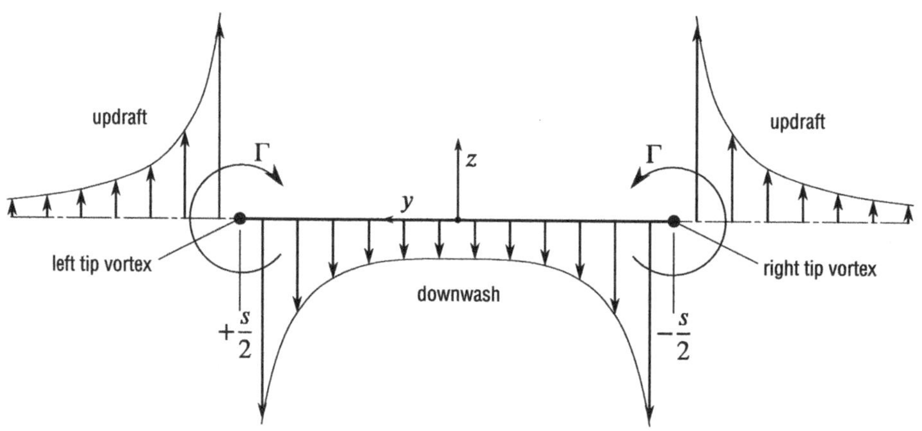

Figure 40.3 Cross section through the velocity field of the wing tip vortices revealing downwash and updraft

energy of a start-up vortex causing it to vanish over time. Because of its shape, the remaining effective vortex system is known as a horseshoe vortex and consists of bound vortex and tip vortices.

Downwash and updraft

The simple line vortex model of a wing helps to explain why migrating birds fly in V-shaped formations. Bicyclists ride in the slipstream of the riders in front of them to safe energy. However, birds seem to avoid flying directly behind another bird. Figure 40.3 shows a cross section of the velocity field generated by the tip vortices. Left and right tip vortex rotate in opposite directions. Both contribute a downward oriented velocity component in the space between them. This is known as downwash. A bird avoids flying in the downwash of birds in front because it would have to work harder maintaining altitude. It rather flies in the updraft, which the tip vortices create in the zone outside of the wing tips. This transverse shift leads to the V-shaped flight formations.

Trailing vortex sheet

A single horseshoe vortex is a rather crude model of a lifting wing. Like a fluid will not flow around a sharp corner without shedding eddies, a vortex filament will not suddenly change direction at the wing tips, like it is depicted in Figure 40.2. Also, the bound circulation will not be constant over the wing span and actually will vanish at the wing

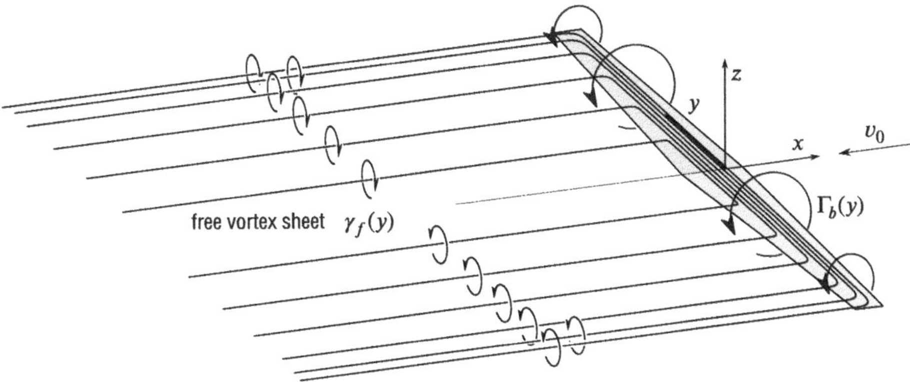

Figure 40.4 A model of a wing with varying bound circulation $\Gamma_b(\eta)$ and the resulting free vortex sheet

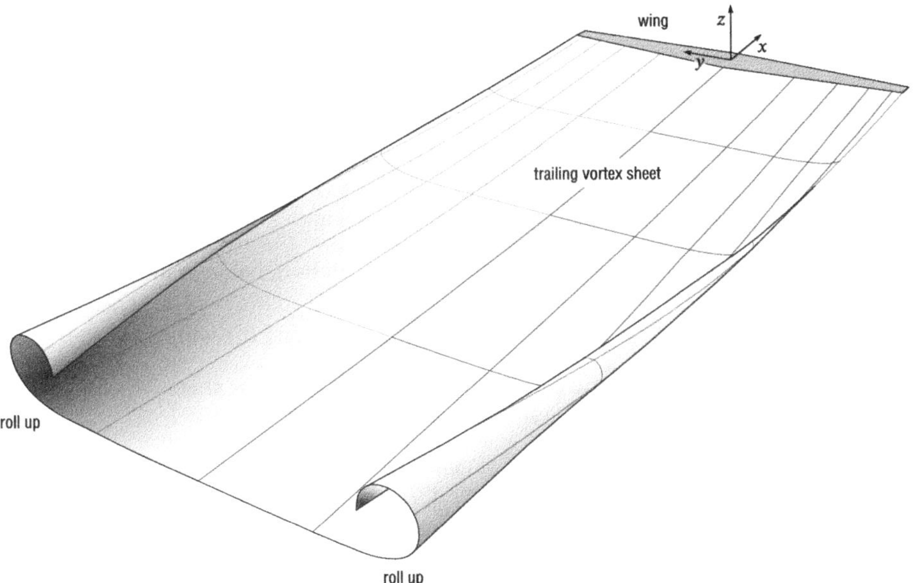

Figure 40.5 Actual shape and roll up of trailing vortex sheet

tips. In order to comply with Helmholtz' theorem that the vorticity may not change along a single vortex filament, we picture the bound vortex as the superposition of an infinite number of vortex filaments which stretch over varying portions of the wing span (Figure 40.4). Their continuations in the wake of the wing form a continuous sheet of vortices. It is known as the free vortex sheet or trailing vortex sheet. The vorticity $\gamma_f(y)$ in the free vortex sheet is a function of the rate of change of bound circulation (see Section 40.2).

Roll up of free vortex sheet

Figure 40.4 shows the free vortex sheet as a planar, rectangular area. This is a useful and often employed approximation. In reality, however, the edges of a trailing vortex sheet roll up behind the wing deforming the vortex sheet (Figure 40.5). Trailing vortex

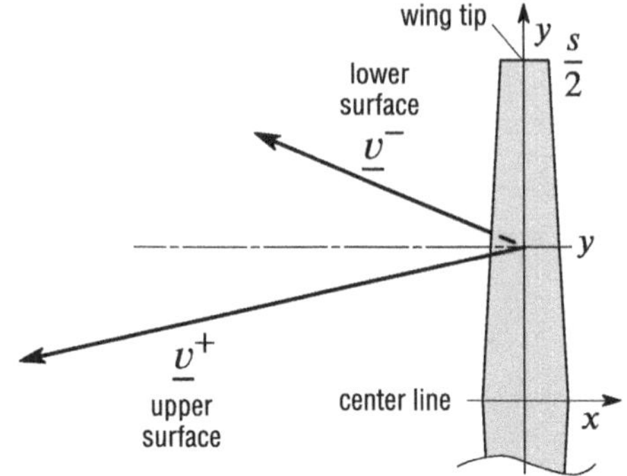

Figure 40.6 Velocity vectors on upper and lower wing surfaces

sheets are invisible and seem a fairly abstract concept. However, they do exist and can be quite strong. In early 2017, a small private jet flipped over several times and almost crashed because it accidentally flew through the trailing vortex sheet of a giant Airbus A380 before it had dissolved (Jamieson, 2017).

Summary

In summary, we may conclude:

- Fluid flows from the high pressure area on the lower side of a wing to the low pressure area on the upper side of a wing. This causes a transverse velocity component, which is oriented toward the wing tips on the lower side and oriented toward the center on the upper side of a lifting wing.
- The overall lifting effect may, like that of a foil section, be described by a single line vortex which we call bound vortex. In contrast to the two-dimensional foil section, the bound circulation strength is varying along the wing span.
- The variation of the bound circulation creates a free vortex sheet which extends behind the wing to the origin of the circulation (start-up vortex).
- The trailing vortex sheet induces a downwash, a velocity component which points roughly into the opposite direction of the lift force.

In the following section, the relationship between free and bound vorticity is explored in more mathematical terms.

40.2 Free and Bound Vorticity

Mean velocity

Figure 40.6 compares the velocities on upper and lower side at an off-center position y along a wing of finite span s. We assume the wing is thin so that we may employ results of thin foil theory. As explained above, the velocity $\underline{v}^-$ on the underside is rotated toward the wing tips and the velocity $\underline{v}^+$ on the upper side is pointing toward the center line of the wing. We define the mean velocity $\underline{v}_m$ (Figure 40.7):

$$\underline{v}_m = \frac{1}{2}\left(\underline{v}^+ + \underline{v}^-\right) \tag{40.1}$$

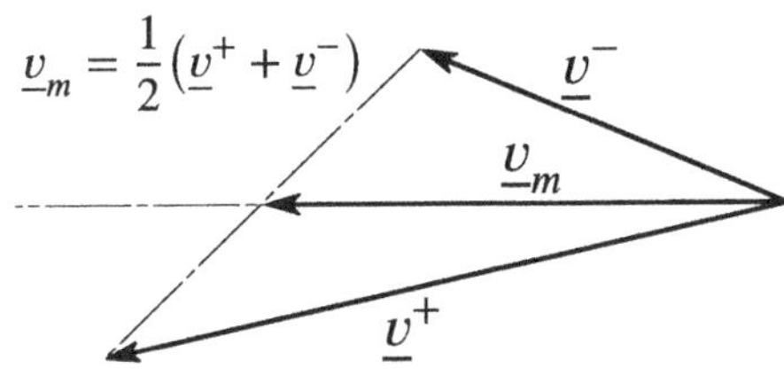

Figure 40.7 Velocity vectors on upper and lower wing surface and mean velocity $\underline{v}_m$

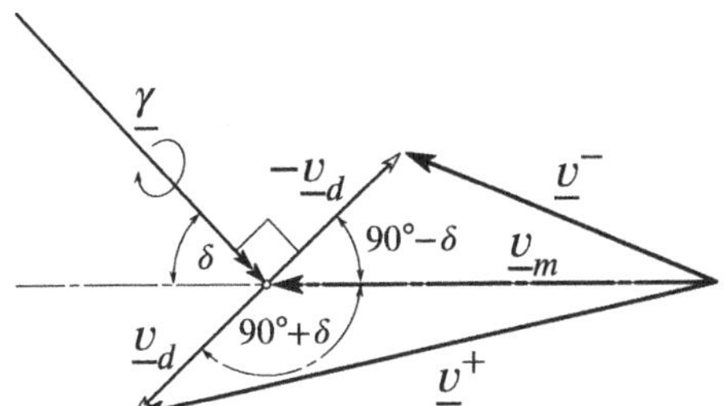

Figure 40.8 Vorticity vector $\underline{\gamma}$ and the resulting difference velocity $\underline{v}_d$

Difference velocity

Upper and lower velocity are now connected via the mean velocity and a difference velocity $\underline{v}_d$ (Figure 40.8). With

$$2\underline{v}_d = \underline{v}^+ - \underline{v}^- \tag{40.2}$$

we may write for upper and lower velocity:

$$\underline{v}^+ = \underline{v}_m + \underline{v}_d \qquad \underline{v}^- = \underline{v}_m - \underline{v}_d \tag{40.3}$$

In two-dimensional foil theory, a distribution of line vortices is applied to model the difference in upper and lower side speed (see Section 38). Two times the difference velocity u_{23} turned out to be equal to the vortex strength $\gamma = 2|u_{23}|$ (see Equation (38.11)). The axes of a two-dimensional vortex distribution is perpendicular to the flow plane, and upper and lower velocity are parallel to each other.

Vortex filament

In the three-dimensional case, we may also employ a vortex to model the difference velocity $\underline{v}_d$. Analog to the two-dimensional case, the strength of the vortex must be equal to two times the magnitude of the difference velocity:

$$\gamma = |\underline{\gamma}| = 2|\underline{v}_d| \tag{40.4}$$

Upper and lower velocity are no longer parallel though. Straight vortex filaments generate velocities perpendicular to their axis (see Section 16.3). Therefore, the axis of the vortex filament $\underline{\gamma}$ must be perpendicular to the difference velocity $\underline{v}_d$. This is shown in Figure 40.8.

Velocities

Let the magnitudes of the velocity vectors be:

$$v^+ = |\underline{v}^+| \qquad v^- = |\underline{v}^-| \tag{40.5}$$

$$v_m = |\underline{v}_m| \qquad v_d = |\underline{v}_d| \tag{40.6}$$

Using the angles depicted in Figure 40.8, relationships between the velocities follow from the law of cosines.

$$(v^+)^2 = v_m^2 + v_d^2 - 2v_m v_d \cos\left(\frac{\pi}{2} + \delta\right) \tag{40.7}$$

$$(v^-)^2 = v_m^2 + v_d^2 - 2v_m v_d \cos\left(\frac{\pi}{2} - \delta\right) \tag{40.8}$$

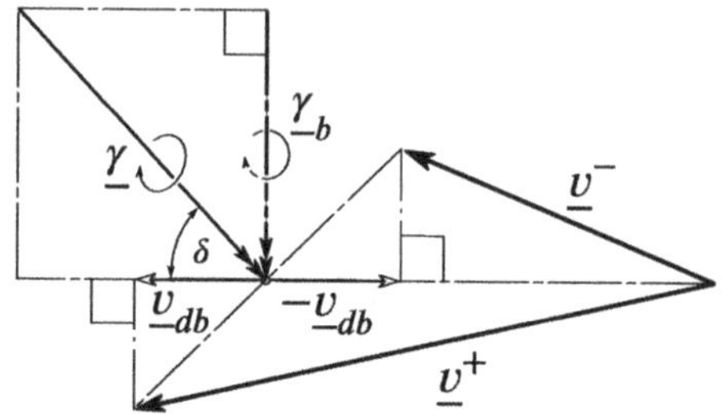

Figure 40.9 Bound vorticity $\underline{\gamma}_b$ and its effect

The angle enclosed by mean velocity $\underline{v}_m$ and vortex filament $\underline{\gamma}$ is marked by δ.

Pressure difference

The Bernoulli equation (37.18) provides a connection between velocity and pressure. The pressure difference between lower and upper wing side is

$$\Delta p = p^- - p^+ = \frac{1}{2}\rho\left[(v^+)^2 - (v^-)^2\right] \tag{40.9}$$

Substituting Equations (40.7) and (40.8) for the squared velocities yields:

$$\Delta p = -\rho v_m v_d \left[\cos\left(\frac{\pi}{2} + \delta\right) - \cos\left(\frac{\pi}{2} - \delta\right)\right] \tag{40.10}$$

Both cosine terms phase shifted by ±90 degrees are identical to a negative sine curve:

$$-\sin(\delta) = \cos\left(\frac{\pi}{2} + \delta\right) = -\cos\left(\frac{\pi}{2} + \delta\right) \tag{40.11}$$

Therefore, the equation for the pressure difference at the wing reduces to

$$\Delta p = \rho v_m 2 v_d \sin(\delta)$$

Equation (40.4) links difference velocity v_d and vortex strength γ. Replacing $2v_d$ with γ expresses the pressure difference as a function of density, mean velocity, and the angle δ.

$$\Delta p = \rho v_m \gamma \sin(\delta) \tag{40.12}$$

We split the vortex filament vector $\underline{\gamma}$ into two components, so that

$$\underline{\gamma} = \underline{\gamma}_b + \underline{\gamma}_f \tag{40.13}$$

Bound vorticity

(i) The first component $\underline{\gamma}_b$ is perpendicular to the mean velocity $\underline{v}_m$. Thus, it is parallel to the wing axis y (Figure 40.9). We call this the bound vorticity, or bound circulation, because it is tied to the wing. The geometry implies that

$$\gamma_b = \gamma \sin(\delta) \tag{40.14}$$

A comparison with Equation (40.12) reveals that the pressure difference depends only on the strength of the bound vorticity.

$$\Delta p = \rho v_m \gamma_b \tag{40.15}$$

This formula is equivalent to Equation (39.4), which we obtained for the two-dimensional foil section where $\gamma = \gamma_b$ ($\delta = 90°$).

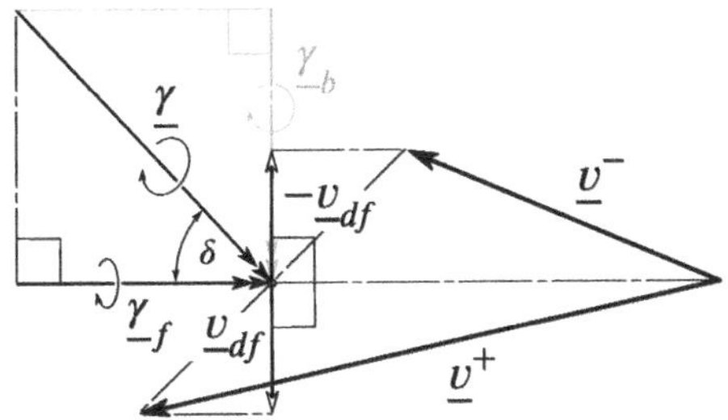

Figure 40.10 Free vorticity $\underline{\gamma}_f$ and its effect

(ii) The second component $\underline{\gamma}_f$ of the vorticity vector is parallel to the mean velocity. From geometry in Figure 40.10, we gather that

Free vorticity

$$\gamma_f = \gamma \cos(\delta) \tag{40.16}$$

This component is called free vorticity because the vortex filament exists free of the wing in the trailing vortex sheet.

Effect of bound vorticity

Bound vorticity creates a difference velocity $\pm\underline{v}_{db}$ parallel to the mean velocity. Thus, the bound vorticity is responsible for the pressure difference between lower and upper wing side and, hence, the lift force.

Effect of free vorticity

The free vorticity induces a difference velocity $\underline{v}_{df}$ perpendicular to the mean velocity. Therefore, free vorticity is responsible for the three-dimensional flow effects that occur on wings with finite span. The lift force may be found with knowledge of the bound vorticity alone. However, we must know the free vorticity to quantify three-dimensional flow effects. Consequently, a connection has to be established between the strengths of bound and free vorticity.

Bound circulation

In two-dimensional foil theory, we computed the circulation created by a distribution of line vortices by integrating the magnitude of the vorticity γ along the chord length. Formally, this is equivalent to the last surface integral in Equation (14.41). The vortex strength $\gamma(x)$ is equal to $\underline{\gamma}^T\underline{n}$ along the chord with ($x \in [-{}^c/_2, {}^c/_2]$) and is zero everywhere else, which reduces the surface integral to the line integral in Equation (39.16). For the three-dimensional wing, we select a path B from trailing edge to leading edge that is always parallel to the mean velocity. By definition, the bound vorticity is perpendicular to this path. Therefore, the bound circulation of a wing at the spanwise location y is equal to

$$\Gamma_b(y) = \int_B \gamma_b \, d\ell \tag{40.17}$$

Link between bound and free vorticity

Now, imagine a closed path C which consists of two almost closed circles connected by an infinitesimally thin bridge. The circles are roughly perpendicular to each other, as depicted in Figure 40.11. This oddly shaped path encloses a simply connected region S. Consequently, the circulation over the area S vanishes for an inviscid and irrotational flow according to Equation (15.15).

The path C provides us with a tool to establish the desired connection between bound and free vorticity. We shift the path so that the area in the first circle lines up with path B across the wing (Figure 40.11). One edge of the thin bridge is just above, the

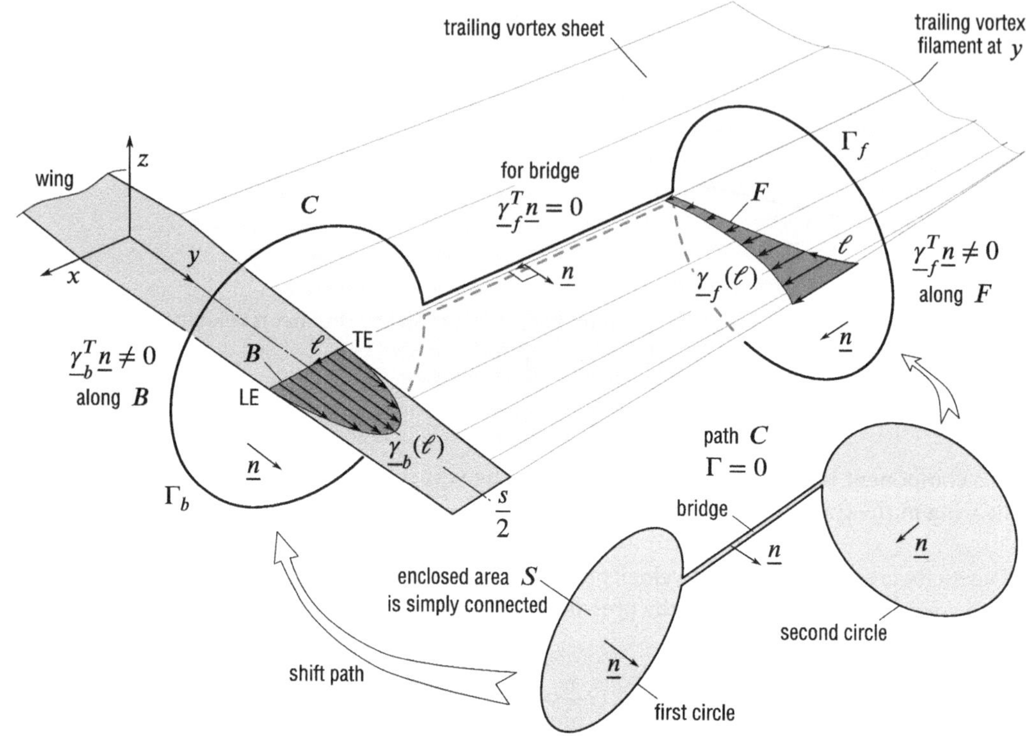

Figure 40.11 Establishing a connection between bound and free vorticity by integration over the boundary of a simply connected region

other just below the trailing vortex sheet and follows the vortex filament that leaves the trailing edge of the wing at the end of path B. The area in the second circle is oriented perpendicular to the trailing vortex filaments along path F. We assume that path F is close enough to the wing, so that we may ignore the roll up of the trailing vortex sheet. The shifted path C is still closed and still encloses a simply connected region S. Therefore, the total circulation is zero.

Total circulation

$$\Gamma = \oint_C \underline{v}^T \mathrm{d}\underline{s} = \iint_S \underline{\gamma}^T \underline{n} \, \mathrm{d}S = 0 \tag{40.18}$$

Free circulation

However, as is evident from Figure 40.11, both circular areas enclose lines with nonzero vorticity. The circulation enclosed by the first circle is equal to the bound circulation of Equation (40.17). An integral over the second circle encloses the free vorticity from the edge of the trailing vortex sheet to the vortex filament at spanwise position y. Integration along path F yields the free circulation:

$$\Gamma_f(y) = \int_F \gamma_f \, \mathrm{d}\ell = \int_{s/2}^{y} \gamma_f \, \mathrm{d}\ell \tag{40.19}$$

The area of the thin bridge does not contribute any circulation since it is aligned parallel to the free vortex filament $\underline{\gamma}_f$ and consequently $\underline{\gamma}_f^T \underline{n} = 0$.

In summary, the total circulation in the enclosed area S consists of three parts.

$$\Gamma = \Gamma_b(y) + 0 + \Gamma_f(y) = 0 \tag{40.20}$$

└ contribution from 2nd circle, Equation (40.19)
└ contribution from bridge
└ contribution from 1st circle, Equation (40.17)

The contributions must add up to zero according to Equation (40.18). Consequently, the free circulation Γ_f, which is contained in the trailing vortex sheet from its outer edge to the vortex filament that left the wing at position y, is equal to the negative bound circulation $\Gamma_b(y)$.

$$\Gamma_f(y) = -\Gamma_b(y) \tag{40.21}$$

Free vorticity

Next, we take the derivative of this equation with respect to the position y.

$$\frac{\mathrm{d}}{\mathrm{d}y}\Gamma_f(y) = -\frac{\mathrm{d}}{\mathrm{d}y}\Gamma_b(y) \tag{40.22}$$

With the integral from Equation (40.19) we have:

$$\frac{\mathrm{d}}{\mathrm{d}y}\int_{s/2}^{y} \gamma_f \,\mathrm{d}\ell = -\frac{\mathrm{d}}{\mathrm{d}y}\Gamma_b(y) \tag{40.23}$$

For the left-hand side, we make use of a rule for the derivative of an integral: if a function g and its antiderivative G are given, then the derivative of the integral of g with respect to the upper bound is equal to g evaluated at the upper bound.

$$\frac{\mathrm{d}}{\mathrm{d}y}\left[\int_{a}^{y} g(t)\,\mathrm{d}t\right] = \frac{\mathrm{d}}{\mathrm{d}y}\big[G(y) - G(a)\big] = \frac{\mathrm{d}G(y)}{\mathrm{d}y} - \frac{\mathrm{d}G(a)}{\mathrm{d}y} = g(y) \tag{40.24}$$

Note that the value $G(a)$ is a constant in this context. Therefore, its derivative with respect to y vanishes. This rule is now applied to the integral in Equation (40.23):

$$\gamma_f(y) = -\frac{\mathrm{d}}{\mathrm{d}y}\Gamma_b(y) \tag{40.25}$$

Thus, the free vorticity γ_f in the trailing vortex sheet depends on the spanwise change of the bound circulation.

This corresponds to our model of the vortex system of a wing from Figure 40.4. In order to comply with Helmholtz' theorems, the vortex system sheds any change of bound circulation as a vortex filament downstream. As a consequence, the filaments do not end in the fluid domain and maintain constant strength.

40.3 Biot–Savart Law

Now that we have a means to compute the strength of the trailing vortex filaments, we may analyze their effects on the flow across the wing. In Section 16.3, the potential

Figure 40.12 The velocity induced by an element of a vortex filament

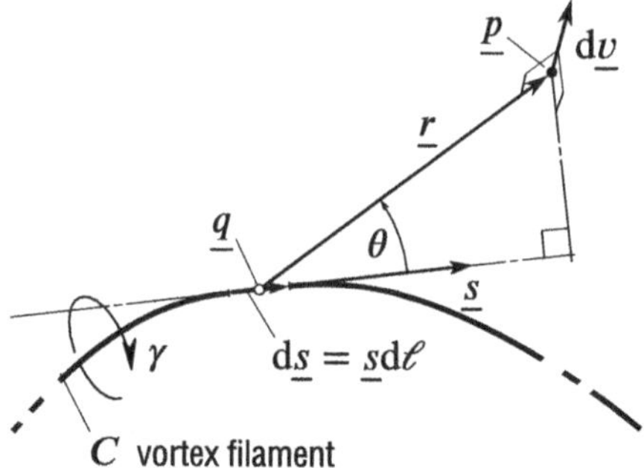

of a two-dimensional vortex has been introduced. Its planar flow field is shown in Figure 16.4. The velocity field is expected to be more complex in the three-dimensional case.

Biot–Savart Law The task at hand is to find the velocity $\underline{v}$ at a point $\underline{p}$ induced by a vortex filament of strength γ_f. Point $\underline{p}$ is also known as field point or collocation point. Figure 40.12 sets up the domain. Note that, unlike the filament shown in the figure, a vortex filament forms a closed loop or stretches to the limits of the flow domain. In an unbounded fluid, it stretches to infinity in both directions. An infinitesimal piece $d\underline{s}$ of the vortex filament of strength γ is oriented in the direction of a unit length tangent vector $\underline{s}$. It is located at point $\underline{q}$, which is also known as a source point.

We want to know the velocity $d\underline{v}$ at a field point $\underline{p}$ induced by the infinitesimal piece $d\underline{s}$ of the vortex filament. A solution to this problem was found by the French scientists Jean-Baptiste Biot (*1774–†1862) and Félix Savart (*1791–†1841), who investigated the magnetic field created by a current through a conductor (Biot and Savart, 1820). This problem may also be described by potential theory. Accordingly, many cross references exist between magneto-electrodynamics and fluid dynamics. The Biot–Savart Law in fluid mechanics computes the velocity induced by a vortex filament:

$$d\underline{v} = -\frac{\gamma}{2\pi}\frac{(\underline{r}\times d\underline{s})}{|\underline{r}|^3} \qquad \text{Biot–Savart Law} \tag{40.26}$$

Induced velocity The result of a cross product is perpendicular to the plane of the vectors, here $\underline{r}$ and $d\underline{s}$. The vector $\underline{r}$ marks the relative position of field point $\underline{p}$ with respect to the source point $\underline{q}$. The influence of a vortex filament declines fast with the inverse of the distance between $\underline{p}$ and $\underline{q}$ squared.[1] Integration along the path C of a vortex filament produces the velocity induced at the field point $\underline{p}$ by the whole vortex filament.

$$\underline{v}(\underline{p}) = -\frac{\gamma}{2\pi}\int_C \frac{(\underline{r}\times d\underline{s})}{|\underline{r}|^3} \tag{40.27}$$

The integral has to be solved anew for each field point.

Straight vortex filament Unfortunately, the integral has no explicit solution unless the geometry of the path C is known and of reasonably simple shape. One such case is a straight vortex filament like

[1] Don't get fooled by the $|\underline{r}|^3$ in the denominator. $\underline{r}$ also appears in the numerator.

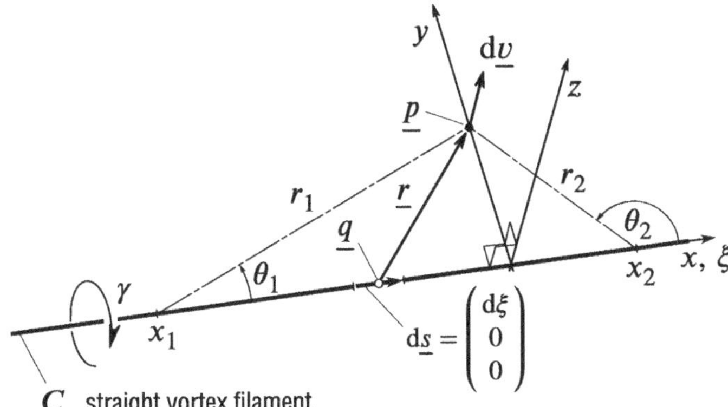

Figure 40.13 The velocity induced by a straight vortex filament

it is shown in Figure 40.13. For convenience, the filament is aligned with the x-axis of our coordinate system, and the field point $\underline{p}$ lies on the y-axis. Therefore, the vectors $\underline{p}$ and $\underline{q}$ and their difference $\underline{r} = \underline{p} - \underline{q}$ are given by:

$$\underline{p} = \begin{pmatrix} 0 \\ y \\ 0 \end{pmatrix} \qquad \underline{q} = \begin{pmatrix} \xi \\ 0 \\ 0 \end{pmatrix} \qquad \underline{r} = \begin{pmatrix} -\xi \\ y \\ 0 \end{pmatrix} \tag{40.28}$$

The distance between field and source point is

$$r = |\underline{r}| = \sqrt{\xi^2 + y^2} \tag{40.29}$$

We also need the cross product between $\underline{r}$ and the line element $d\underline{s}$ for the integrand in Equation (40.27). $d\underline{s}$ is stated in Figure 40.13.

$$\underline{r} \times d\underline{s} = \begin{vmatrix} \underline{i} & \underline{j} & \underline{k} \\ -\xi & y & 0 \\ d\xi & 0 & 0 \end{vmatrix} = \begin{pmatrix} y \cdot 0 - 0 \\ 0 \cdot d\xi - (-\xi) \cdot 0 \\ -\xi \cdot 0 - y \cdot d\xi \end{pmatrix} = \begin{pmatrix} 0 \\ 0 \\ -y\, d\xi \end{pmatrix} \tag{40.30}$$

Both vectors $\underline{r}$ and $d\underline{s}$ lie in the x-y-plane. Consequently, the result of their cross product points into the z-direction.

Induced velocity

We substitute the cross product and the distance into the integral of Equation (40.27).

$$\underline{v}_{12} = -\frac{\gamma}{2\pi} \int_{x_1}^{x_2} \frac{\begin{pmatrix} 0 \\ 0 \\ -y \end{pmatrix}}{\left(\sqrt{\xi^2 + y^2}\right)} d\xi \tag{40.31}$$

Our vortex filament stretches to infinity in both directions of the x-axis. For the moment, we consider only the velocity induced by the piece located between x_1 and x_2.

Effect of partial filament

Only the vertical component w_{12} of the resulting velocity will have a nonzero value:

$$w_{12} = +\frac{\gamma}{2\pi} \int_{x_1}^{x_2} \frac{y}{\left(\xi^2 + y^2\right)^{3/2}} d\xi \tag{40.32}$$

This integral is of standard type, and we get:

$$w_{12} = \frac{\gamma}{4\pi} \left| \frac{\xi}{\left(\xi^2 + y^2\right)^{1/2}} \right|_{x_1}^{x_2}$$

$$= \frac{\gamma}{4\pi} \left[\frac{x_2}{\sqrt{x_2^2 + y^2}} - \frac{x_1}{\sqrt{x_1^2 + y^2}} \right] \tag{40.33}$$

The denominators in the result are equal to the distances r_2 and r_1 from the positions x_2 and x_1 to the field point, respectively.

$$w_{12} = \frac{\gamma}{4\pi} \left[\frac{x_2}{r_2} - \frac{x_1}{r_1} \right] \tag{40.34}$$

From the geometry depicted in Figure 40.13, it follows that the fractions x_1/r_1 and x_2/r_2 are related to the cosines of the angles θ_1 and θ_2. Note that for a negative $x_1 < 0$, the fraction x_1/r_1 becomes negative and that for a positive $x_2 > 0$, the angle θ_2 becomes larger than 90 degrees. Therefore, $x_1/r_1 = -\cos(\theta_1)$ and $x_2/r_2 = -\cos(\theta_2)$ in order to preserve the signs.

$$w_{12} = \frac{\gamma}{4\pi} \left[-\cos(\theta_2) + \cos(\theta_1) \right] \tag{40.35}$$

Effect of complete filament According to Helmholtz's theorems the vortex filament cannot simply start at x_1 and end at x_2. Beginning and end must be at physical limits of the domain, like a wall, or the vortex filament stretches to infinity. We investigate the latter. For $x_1 \to -\infty$, the angle θ_1 shrinks to zero: $\lim_{x_1 \to \infty}(\theta_1) = 0$. For $x_2 \to \infty$, the angle θ_2 approaches 180 degrees: $\lim_{x_2 \to \infty}(\theta_2) = \pi$. Therefore, the limits are:

$$\lim_{x_1 \to \infty} \cos(\theta_1) = \cos(0) = +1$$

$$\lim_{x_2 \to \infty} \cos(\theta_2) = \cos(\pi) = -1$$

The resulting induced vertical velocity w becomes

$$w = \lim_{\substack{x_1 \to -\infty \\ x_2 \to \infty}} (w_{12}) = \frac{\gamma}{4\pi} \left[-(-1) + 1 \right] = \frac{\gamma}{2\pi} \tag{40.36}$$

This looks familiar, and it is indeed the magnitude of the velocity created by a two-dimensional vortex at a distance $r = y$ from the location of the vortex (see Section 16.3).

Effect of trailing vortex filament The trailing vortex filaments of our wing do not stretch forward of the wing. They merge into the bound vortex at the wing. If we place the wing at position $x = 0$, the trailing vortex filaments stretch from minus infinity to zero: $-\infty < x \neq 0$. For $x_2 \to 0$ the angle θ_2 becomes 90 degrees and, as a result, $\cos(\theta_2) = 0$. The induced vertical velocity becomes exactly half of the velocity induced by the vortex filament which stretches to infinity in both directions.

$$w = \lim_{\substack{x_1 \to -\infty \\ x_2 \to \infty}} (w_{12}) = \frac{\gamma}{4\pi} \left[-(-1) + 0 \right] = \frac{\gamma}{4\pi} \tag{40.37}$$

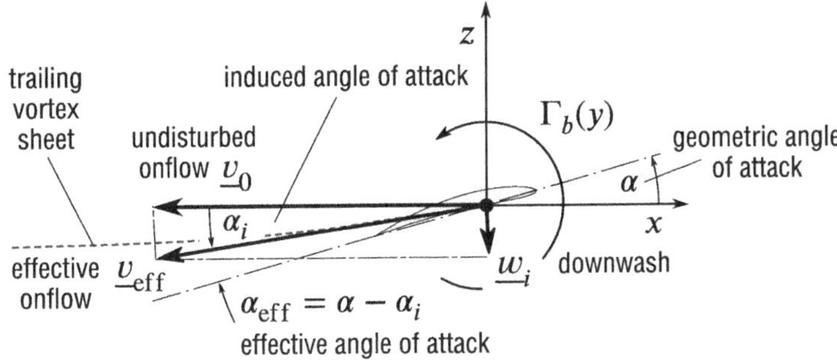

Figure 40.14 The effect of downwash on the angle of attack

40.4 Lifting Line Theory

Lifting line theory

Lifting line theory is the basic mathematical tool used to compute the flow field for wings of finite span s. The wing is modeled as a single vortex, whose bound circulation $\Gamma_b(y)$ varies over the span of the wing. The circulation must become zero at the wing tips to comply with Helmholtz's theorems. The resulting changes in bound circulation create a trailing vortex sheet.

Downwash

Each filament of the trailing vortex sheet induces a velocity at the position of the wing. The precise direction and strength of the velocity depends on the shape of the trailing vortex sheet and the rate of change of bound circulation, which governs the strength γ_f of the trailing vortex filaments. In general, the effect of a trailing vortex sheet looks similar to the velocity distribution in Figure 40.3. There is a downward oriented velocity $\underline{w}_i$ across the whole span of the wing, which we call downwash.

Induced angle of attack

Downwash has a noticeable effect on the flow across a wing. Figure 40.14 shows a cross section through a wing and the velocity vectors at a spanwise position y. The downwash velocity $\underline{w}_i$, induced by the trailing vortex sheet, is superimposed with the onflow $\underline{v}_0$ and creates the effective onflow vector $\underline{v}_{\text{eff}}$. As a result, the geometric angle of attack α of the foil section is reduced to the effective angle of attack α_{eff}. The change in angle of attack is known as induced angle of attack α_i, which should not be confused with the ideal angle of attack α_{id} from the previous chapter.

Typically, the magnitude of downwash $w_i = |\underline{w}_i|$ is considerably smaller than the onflow velocity $v_0 = |\underline{v}_0|$ and roughly perpendicular to the onflow, as long as the sectional lift coefficients remain small. As a consequence, the magnitude of effective velocity v_{eff} is about the same as the undisturbed onflow speed.

$$v_{\text{eff}} \approx v_0 \tag{40.38}$$

The main effect of downwash is a change in the direction of flow speeds. For cases with $w_i \ll v_0$, we can state that the induced angle of attack is approximately

$$\alpha_i \approx \tan(\alpha_i) = \frac{w_i}{v_0} \tag{40.39}$$

Lift force

The change in angle of attack caused by the downwash has substantial consequences. From Kutta-Joukowsky's lift theorem we know that circulation and onflow create a force perpendicular to the onflow. Figure 40.15 shows the force vectors. The total force

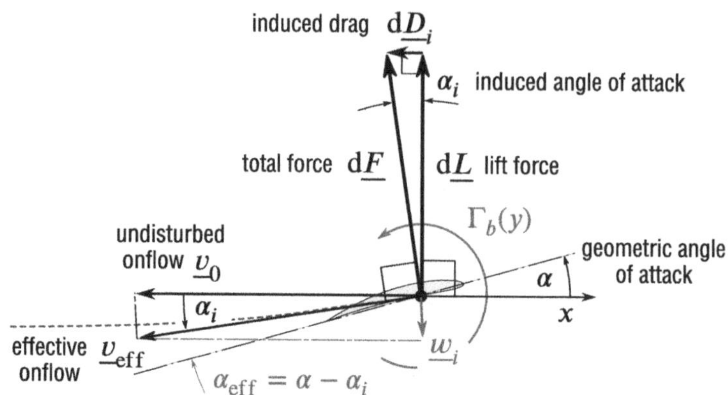

Figure 40.15 Induced drag caused by the downwash

dF on a foil section of the wing is now acting normal to the effective velocity.

$$dF = \rho v_{eff} \Gamma_b dy \approx \rho v_0 \Gamma_b dy \tag{40.40}$$

The usable lift force dL is the force component perpendicular to the original, undisturbed onflow v_0.

$$dL = df \cos(\alpha_i) = \rho v_0 \Gamma_b \cos(\alpha_i) dy \approx \rho v_0 \Gamma_b dy \tag{40.41}$$

Again, since the induced angle is small, the cosine of α_i is approximately one. Therefore, total and lift force are essentially equal in magnitude but point in different directions.

Induced drag Due to the induced angle of attack α_i, a second force component dD_i exists. It points in the direction of the undisturbed onflow. Since it acts against the motion of the wing, it is experienced as an additional drag. This force component is called induced drag.

$$dD_i = dL \tan(\alpha_i) \approx \rho v_0 \Gamma_b \alpha_i dy \tag{40.42}$$

Note that induced drag occurs even in an inviscid fluid. In a real fluid with nonzero viscosity, an additional viscous drag must be considered.

Small induced angles of attack may be replaced by Equation (40.39). Substituting it into Equation (40.42) yields for the induced drag:

$$dD_i = dL \tan(\alpha_i) \approx \rho v_0 \Gamma_b \frac{w_i}{v_0} dy = \rho \Gamma_b w_i dy \tag{40.43}$$

The last term emphasizes that the higher the downwash at a wing section, the higher becomes the induced drag.

Summary This short introduction to wing theory may suffice for the purpose of our basic discussion of ship propulsion. Obviously, further studies are needed to apply lifting line theory to compute downwash, lift, and induced drag for foils and propeller blades. Lifting line theory is covered in all texts on aerodynamics, e.g. Moran (1984). A comprehensive description for lifting line theory in the context of propellers is found in Breslin and Andersen (1994). A working implementation and an introduction to the theory is provided by the OpenProp project (Epps, 2010). Its home page can be found at https://sage-newt.cloudvent.net/ (last visited in December 2018).

References

Biot, J. and Savart, N. (1820). Note sur le magnétisme de la pile de Volta. *Annales de chimie et de physique*, 15:222–223.

Breslin, J. and Andersen, P. (1994). *Hydrodynamics of ship propellers.* Cambridge University Press, Cambridge, United Kingdom.

Epps, B. (2010). *An impulse framework for hydrodynamic force analysis: fish propulsion, water entry of spheres, and marine propellers.* PhD thesis, Massachusetts Institute of Technology (MIT), Cambridge, MA.

Helmholtz, H. (1978). On the integrals of the hydrodynamic equations that correspond to vortex motions. Translation by Uwe Parpart. *Int. Journal of Fusion Energy*, 1(3–4):41–68. Original published in German in Journal für die reine und angewandte Mathematik, 56:25–55, 1858.

Jamieson, A. (2017). Private jet flipped over in wake turbulence from Airbus A380. NBC News at http://www.cnbc.com/2017/03/22/private-jet-flipped-over-in-wake-turbulence-from-airbus-a380.html. Last visited June 4, 2018.

Moran, J. (1984). *An introduction to theoretical and computational aerodynamics.* John Wiley & Sons, New York, NY.

Self Study Problems

1. Explain why migrating birds fly in a V-shaped formation.

2. Given is the following elliptical distribution of bound circulation.

$$\Gamma_b(y) = \Gamma_0\sqrt{1-\left(\frac{2y}{s}\right)^2}$$

   Γ_0 is the maximum bound circulation, s is the wingspan, and y is the coordinate along the wingspan. Γ_0 and s are constants and known. Compute the strength of the free vorticity γ_f in the trailing vortex sheet.

3. State the Biot–Savart Law.

4. Use the solution of Problem 2 to compute the distribution of downwash w_i.

$$w_i(y) = \frac{1}{4\pi}\int\limits_{-s/2}^{s/2}\frac{\gamma_f(\eta)}{(y-\eta)}\,\mathrm{d}\eta$$

   This formula is valid under the assumptions that the trailing vortex sheet is straight, oriented in the direction of the undisturbed onflow, and does not show noticeable effects due to roll up. Hint: the integral can be converted into a Glauert integral with the substitution $\eta = s\cos(\varphi)/2$. Details can be found in Section 37.3.

5. Explain to a friend the origin of induced drag.

41

Open Water Test

Main objective of propeller design is the creation of a hull–propeller propulsion system with high efficiency. The system must produce as much thrust as possible for as little power spent as possible. Although great strides has been made in the computational assessment of propeller characteristics, model tests with propellers still play an important part of design verification. Like the ship hull alone is evaluated in the resistance test, we first evaluate the propeller without the influence of the hull. These open water tests are discussed in this chapter.

Learning Objectives

At the end of this chapter students will be able to:

- understand hydrodynamic propeller characteristics
- conduct open water tests and interpret their results
- determine open water efficiency

41.1 Test Conditions

Open water test In Chapter 1 we subdivided the hull–propeller system into parts to simplify the analysis of their hydrodynamic characteristics. The ship hull alone and its resistance when moving with constant speed through deep, calm water are investigated with a resistance test. The thrust created by a propeller and the required torque to generate that thrust are analyzed in an open water test. It is called open water test because the propeller operates in a uniform flow. This may be achieved in two ways:

- a rotating propeller is moved at constant speed through a towing tank where the water itself is initially at rest
- the test is conducted in a cavitation tunnel where the water streams past the propeller rotating in place

More details are provided in Section 41.3.

Fundamentals of Ship Hydrodynamics: Fluid Mechanics, Ship Resistance and Propulsion, First Edition.
Lothar Birk.

Companion website: www.wiley.com/go/birk/hydrodynamics

Before measurement equipment is set up, laws of similarity have to be considered which govern the model test. Maintaining geometric, kinematic, and dynamic similarity are critical if the results are used for performance prediction of the full scale propeller.

Geometric, kinematic, and dynamic similarity

The thrust T of a propeller will depend on water density ρ and kinematic viscosity ν. As usual we assume that they are constant throughout the fluid and over the duration of a test. The propeller diameter D may serve a characteristic length. Flow conditions are represented by the speed of advance v_A and the rate of revolution n of the propeller. Gravitational acceleration g and pressure p indicate the magnitude of external forces. In summary, the propeller thrust T is assumed to be a yet unknown function $\mathcal{F}$ of the seven physical variables $(\rho, \nu, g, D, v_A, p, n)$:

Dimensional analysis

$$T = \mathcal{F}(\rho, \nu, g, D, v_A, p, n) \tag{41.1}$$

Function $\mathcal{F}$ uses three fundamental physical quantities: force, length, and time with units Newton [N], meter [m], and seconds [s]. Consequently, according to dimensional analysis and the Buckingham π-theorem the flow problem is characterized by $7 - 3 = 4$ dimensionless numbers.

$$T = \frac{1}{2}\rho\, v_A^2\, D^2 \cdot C \cdot f\left(\left[\frac{g\,D}{v_A^2}\right], \left[\frac{p}{\frac{1}{2}\rho\, v_A^2}\right], \left[\frac{\nu}{v_A\, D}\right], \left[\frac{n\,D}{v_A}\right]\right) \tag{41.2}$$

The unknown function f is essentially equal to the thrust loading coefficient C_{Th}. With the constant factor $C = \pi/4$ we may write:

$$C_{Th} = \frac{T}{\frac{1}{2}\rho\, v_A^2\, \frac{\pi\, D^2}{4}} = f\left(\left[\frac{g\,D}{v_A^2}\right], \left[\frac{p}{\frac{1}{2}\rho\, v_A^2}\right], \left[\frac{\nu}{v_A\, D}\right], \left[\frac{n\,D}{v_A}\right]\right) \tag{41.3}$$

The unknown function f may be found by conducting experiments by systematically varying the four dimensionless numbers. The numbers represent different laws of similarity (see Chapter 28).

The first dimensionless parameter in Equation (41.3) is equal to the inverse of a squared propeller Froude number

Propeller Froude number

$$\left[\frac{g\,D}{v_A^2}\right] = \frac{1}{Fr^2}$$

with

$$Fr = \frac{v_A}{\sqrt{g\,D}} \tag{41.4}$$

Ship speed and length in waterline for the Froude number of the ship are replaced by speed of advance v_A and propeller diameter D, respectively.

The Froude number represents the similarity of inertia and gravity forces. During resistance tests it ensured that the wave pattern and wave resistance are similar to the full scale vessel. Most propellers operate fully submerged, in which case, wave making and gravity forces are of lesser concern.

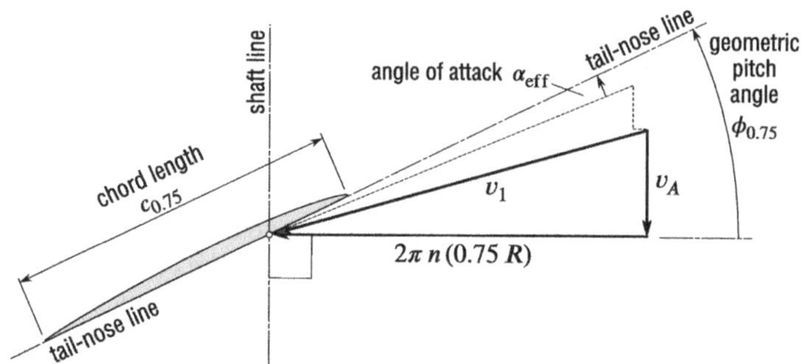

Figure 41.1 Simplified velocities triangle for an unrolled propeller blade section at radius $x = 0.75$ (induced velocities have been ignored)

Propeller pressure coefficient The second dimensionless parameter in Equation (41.3) is an Euler number and ensures similarity of inertia and pressure forces.

$$Eu = \left[\frac{2p}{\rho\, v_A^2} \right] \tag{41.5}$$

We already learned that in order to keep Eu the same for model and full scale, air pressure above the tank water surface has to be reduced for model tests. This is essential for cavitation tests (see Chapter 45). Fortunately, the thrust generated by the propeller stems from pressure differences on the propeller blades. Therefore, static pressure components may be neglected and thrust at model and full scale are similar if the pressure coefficient is the same, i.e. $C_{PM} = C_{PS} = 2\Delta p/(\rho v_A^2 A_0)$. This will be the case, if the angles of attack are the same for model and full scale propeller blade sections.

Propeller Reynolds number A form of Reynolds number Re is given with the third dimensionless parameter in Equation (41.3).

$$\left[\frac{\nu}{v_A\, D} \right] = \frac{1}{Re} \quad \text{with} \quad Re = \frac{v_A\, D}{\nu} \tag{41.6}$$

Again, maintaining both, equal Froude number *and* equal Reynolds number for model and full scale vessel, is practically impossible. Since gravity forces are of lesser importance for a deeply submerged propeller, the Reynolds number is kept as high as is feasible with the available testing equipment in order to ensure turbulent boundary layer flow over the propeller blades. However, full similarity of viscous forces ($Re_M = Re_S$) is generally not attainable for model scale propellers.

Since Reynolds numbers are typically lower during open water tests for models than for the full scale propellers, model test results have to be corrected for the full scale propeller.

Propeller advance coefficient Finally, the fourth and last dimensionless parameter in Equation (41.3) is the inverse of the advance coefficient J. J is also referred to as advance ratio.

$$\left[\frac{n\, D}{v_A} \right] = \frac{1}{J} \quad \text{with} \quad J = \frac{v_A}{n\, D} \tag{41.7}$$

v_A is again the speed of advance, n the rate of revolution, and D the diameter of the propeller.

Figure 41.1 shows a simplified velocity triangle for a propeller blade section. The transverse velocity due to rotation of the propeller grows linearly from zero at the shaft center to its maximum value of $2\pi nR$ at the blade tip. At radius r the transverse velocity is $2\pi nr$. The denominator (nD) in Equation 41.7 is proportional to the transverse speed of the blade tip. The factor π has been dropped for simplicity.

The ratio of speed of advance v_A and the transverse velocity $(n\ D)$ in Equation (41.7) is a measure for the hydrodynamic pitch of the blade section. If the velocity triangles are similar at model and full scale, blade sections will operate at the same angle of attack. Therefore, keeping advance coefficient equal for model and full scale propeller ensures kinematic similarity.

$$J_M = J_S \quad \longrightarrow \quad \frac{v_{AM}}{n_M\ D_M} = \frac{v_{AS}}{n_S\ D_S} \tag{41.8}$$

Dynamic similarity cannot be obtained without kinematic similarity for a scale model propeller. Consequently, maintaining an equal advance coefficient for model scale and full scale propeller is of utmost importance to ensure results of the model test can be used to make a full scale performance prediction.

Hypothetical full dynamic similarity

If all four parameters Fr, Re, C_p, and J are equal for two geometrically similar propellers, the propellers have the same thrust coefficient K_T. However, as it is the case with the resistance test, full dynamic similarity is usually not attainable in propeller open water tests for practical model scales and test conditions. As a consequence, scaling of the model test results to the full scale propeller requires additional corrections. The corresponding procedure recommended by the ITTC is discussed in Chapter 42.

41.2 Propeller Models

Propeller model accuracy

Except for small boat propellers, testing of full scale propellers would be unrealistically expensive. However, due to their complicated shape and tight manufacturing tolerances, model propellers are typically more expensive than ship models. Accurate leading edge shapes and thickness distributions are especially difficult to attain. Modern CAD/CAM and CNC manufacturing tools like 3D printing help to improve the accuracy of propeller models. The ITTC distinguishes two classes of model propellers governed by separate recommended procedures:

- propellers for open water and propulsion tests (ITTC, 2011)
- propellers for cavitation tests (ITTC, 2005).

Model propellers for open water tests

Typical model propellers have diameters in the range of 150 mm $\leq D \leq$ 300 mm. For open water tests, models should adhere to the following tolerances of their design values (ITTC, 2011):

diameter	D	± 0.10 mm
thickness	t	± 0.10 mm
blade width	c	± 0.20 mm
mean pitch at each radius	P/D	± 0.50 %

Surfaces of blades and hub should be polished.

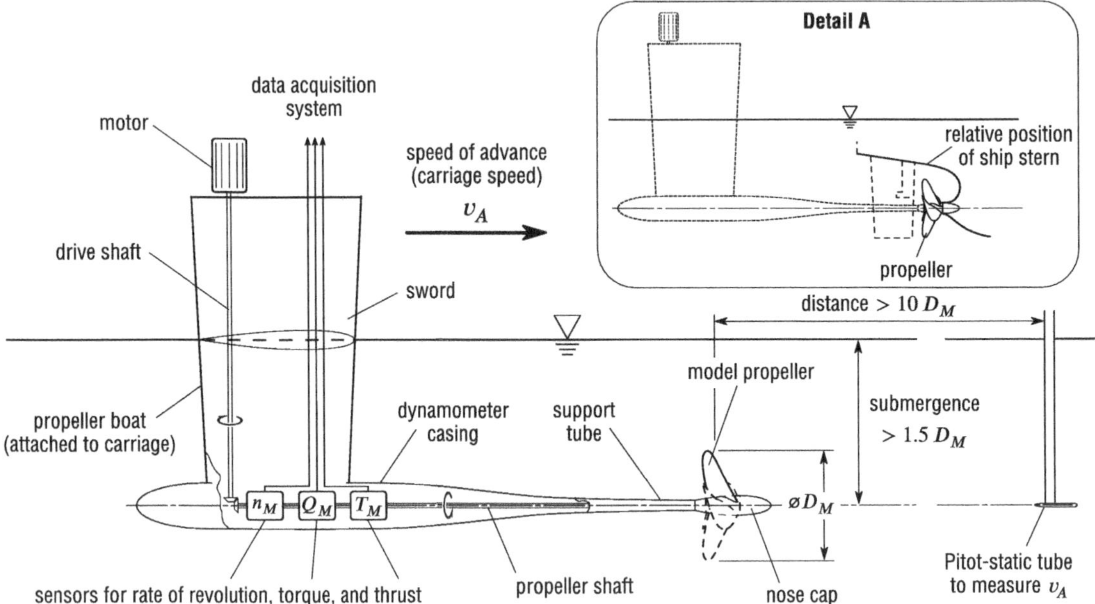

Figure 41.2 Open water test of model propeller with propeller boat in towing tank

Model propellers for cavitation tests Because cavitation phenomena are sensitive to small changes in test parameters, model propellers for cavitation tests must adhere to more detailed requirements which are documented in ITTC (2005). Special attention has to be placed on leading edge geometry and surface quality.

41.3 Test Procedure

The purpose of the open water test is to determine thrust under different operating conditions and the associated torque necessary. Open water refers to the fact that propellers are tested in undisturbed onflow conditions. Tests should be conducted in accordance with ITTC (2014). Two different types of testing facilities may be employed:

- towing tanks
- cavitation or circulating water tunnels

In towing tanks undisturbed parallel inflow is generated by moving the rotating propeller through the water at speed of advance v_A. In cavitation tunnels water streams with velocity v_A past the propeller which is rotating in place.

Towing tank For open water tests in towing tanks, the propeller is mounted on a propeller dynamometer, colloquially called propeller boat (Figure 41.2). The propeller boat consists of a vertical sword with a foil shaped cross section and the streamlined casing, which contains the propeller shafting and the actual propeller dynamometer. The propeller dynamometer has sensors for thrust T (force in line with shaft), torque Q, and rate of revolution n. Advanced propeller dynamometers may feature additional sensors for transverse bearing forces in the support tube which provide data on propeller side forces (transverse to axis) and the center of thrust.

Note that the propeller shaft in the open water test is driving the propeller from behind. The ship hull would be in front of the propeller (see Detail A in Figure 41.2). This allows the propeller dynamometer to trail the propeller and avoids disturbing the onflow. The more voluminous parts of the dynamometer should be at least two diameters away from the propeller to avoid influencing the flow through the propeller disk. During testing the propeller shaft is submerged by at least 1.5 times the propeller diameter. This avoids wave generation and other free surface effects like ventilation.

The speed of advance v_A is measured in front of the propeller at the same depth as the propeller shaft (see Figure 41.2). In general, it should correspond to the carriage speed.

Before tests with a propeller are conducted, effects of the nose cap are measured by running the dynamometer with a dummy hub without any blades. Measured forces and torque are later subtracted from the values found with the actual propeller model.

Selection of Reynolds number

As mentioned above, full dynamic similarity is usually not attained. In the towing tank open water test, the rate of revolution is set to a fixed value n so that the local propeller Reynolds number Re_{c0} is as high as possible.

$$Re_{c0} = \frac{v_1\, c_{0.75}}{\nu} \tag{41.9}$$

with the approximation of the local blade section velocity v_1

$$v_1 = \sqrt{v_A^2 + (0.75\pi\, n\, D)^2} \tag{41.10}$$

and the chord length $c_{0.75}$ at radius $r = 0.75R$ of the propeller (Figure 41.1). The radius $x = 0.75$ is in the blade region where most of the thrust is generated. Usually the rate of revolution is limited by the range of the thrust sensor in the propeller dynamometer. At $J = 0$ a propeller will generate the largest thrust. The advance coefficient is varied by varying the speed of advance. For each speed of advance setting, values for thrust, torque, and rate of revolution are measured (see Table 42.1 for an example).

Comparison with propulsion test

If the same propeller is also used in a propulsion test, a second set of open water tests should be conducted for the Reynolds numbers occurring in the propulsion test. In practice, this means the propeller is run at the same Froude numbers and the same rate of revolutions as the self-propelled model. At a minimum, Re_{c0} must be larger than 200 000 to ensure turbulent flow.

Cavitation tunnel

For open water tests in a circulating water tunnel or cavitation tunnel, the propeller dynamometer is fixed to the test section of the tunnel (Figure 41.3). The water is circulated by a pump. In contrast to the open water test in a towing tank, we keep the flow speed constant because it is more difficult to control than the propeller rate of revolution. The advance coefficient is varied by modifying the rate of revolution of the model propeller. As a consequence, the bollard pull condition ($J = 0$) is unattainable. Depending on the diameter of the model propeller and the size of the cross section of the tank, wall corrections have to be applied to the test data (Glauert, 1947).

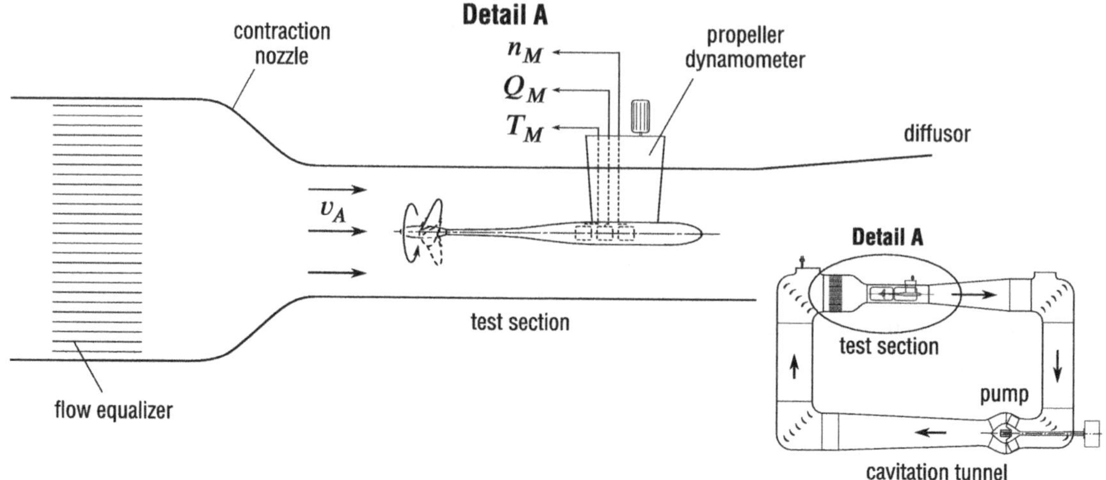

Figure 41.3 Open water test of model propeller in a circulating water tunnel / cavitation tunnel

41.4 Data Reduction

Thrust and torque coefficients Thrust T, torque Q, and rate of revolution n are measured at different speeds of advance and converted into three dimensionless coefficients:

$$\text{advance coefficient} \qquad J = \frac{v_A}{n\,D} \tag{41.11}$$

$$\text{thrust coefficient} \qquad K_T = \frac{T}{\rho\,n^2\,D^4} \tag{41.12}$$

$$\text{torque coefficient} \qquad K_Q = \frac{Q}{\rho\,n^2\,D^5} \tag{41.13}$$

Open water efficiency The efficiency of the propeller in open water condition η_O is given by the ratio of usable power and total power spent, i.e. the ratio of thrust power and delivered power.

$$\eta_O = \frac{P_T}{P_D} \tag{41.14}$$

Introducing the definitions for thrust power and delivered power and replacing thrust and torque using the dimensionless coefficients K_T (41.12), K_Q (41.13), and J (41.11) yields:

$$\begin{aligned} \eta_O &= \frac{v_A\,T}{2\pi\,n\,Q} = \frac{v_A\,\rho\,n^2\,D^4\,K_T}{2\pi\,n\,\rho\,n^2\,D^5\,K_Q} \\ &= \frac{v_A\,K_T}{2\pi\,n\,D\,K_Q} = \frac{1}{2\pi}\,\frac{v_A}{n\,D}\,\frac{K_T}{K_Q} \\ &= \frac{J}{2\pi}\,\frac{K_T}{K_Q} \end{aligned} \tag{41.15}$$

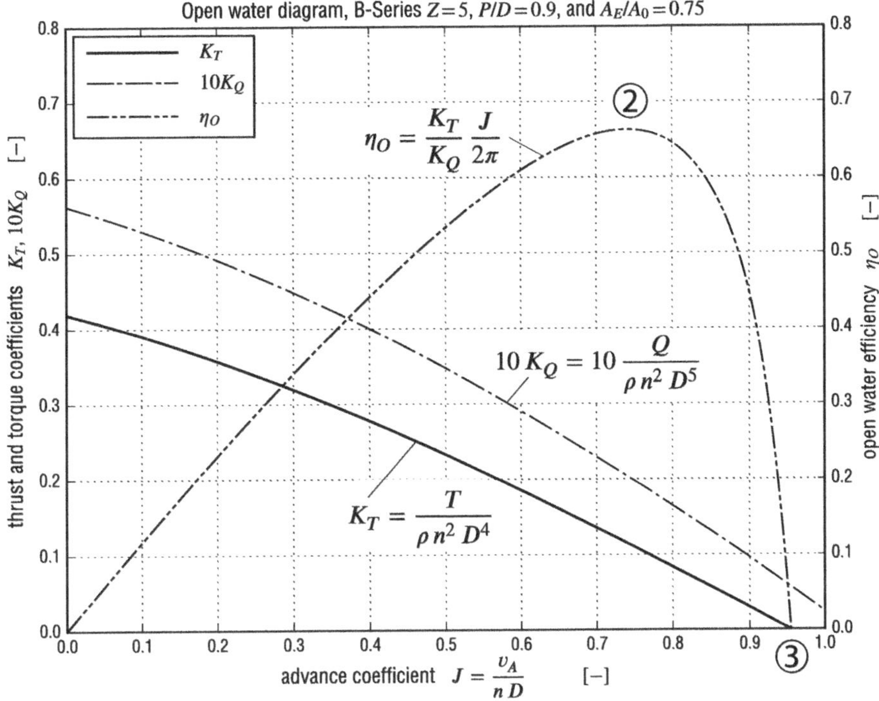

Figure 41.4 Typical propeller open water diagram

Open water diagram

The results of the open water test are presented in an open water diagram (Figure 41.4). Thrust coefficient K_T, torque coefficient K_Q, and open water efficiency η_O (41.15) are plotted as a function of the advance coefficient J. The torque coefficient is usually one order of magnitude smaller than the thrust coefficient. Therefore it is common practice to plot ten times the torque coefficient. Keep this in mind when you read torque coefficients off open water diagrams and divide the value by ten before computing dimensional torque. In Figure 41.4 open water efficiency has the same scale as thrust and torque coefficient but in general they may differ.

Three points in Figure 41.4 are marked by circled numbers because they represent notable operational conditions for the propeller.

① Bollard pull: the propeller produces maximum thrust for advance coefficient $J = 0$, i.e. initially there is no speed of advance although the propeller is rotating ($n > 0$). This reflects the working condition at the beginning of a tow or simply when a boat is getting underway. The angles of attack for the blade sections are very high and close to the geometric pitch angles. High angles of attack will create high lift forces but also require high torque values. The efficiency is zero because thrust power vanishes for $v_A = 0$. Since the propeller is accelerating fluid in its onflow it creates its own speed of advance even if the ship is not moving, i.e. $J > 0$ as soon as the propeller starts producing thrust.

② Maximum efficiency: the efficiency of the propeller grows steadily with increasing advance coefficient until it reaches a maximum. The propeller shown attains its maximum open water efficiency max(η_O) at $J = 0.75$. Obviously, we would like the propeller to work close to its maximum efficiency in normal operating conditions. For practical reasons, the operational advance coefficient is typically slightly to the left of the maximum efficiency.

③ Zero thrust: at this point the propeller stops producing thrust. Although no thrust is generated the engine is still providing torque to rotate the propeller. The torque at zero thrust is a reflection of the propeller's rotational drag. You may compare this condition to a bicyclist going downhill. At some point the bike will roll faster than he or she can pedal. At point ③ the ship is moving so fast that the propeller is no longer able to accelerate water aft. This condition typically occurs during stop maneuvers. The engine is powered down but the ship is still going forward. If the advance coefficient is increased beyond point ③, the propeller is generating negative thrust and helps to slow down the vessel.

The open water diagram shows the thrust generating capability of a propeller for all possible working conditions (advance coefficients J). It plays an important role in matching a propeller to a given ship hull.

References

Glauert, H. (1947). *The elements of airofoil and airscrew theory.* Cambridge University Press, Cambridge, United Kingdom, second edition.

ITTC (2005). *Model manufacture, propeller models, propeller model accuracy.* International Towing Tank Conference, Recommended Procedures and Guidelines 7.5-01-02-02. Revision 0.

ITTC (2011). *Ship models.* International Towing Tank Conference, Recommended Procedures and Guidelines 7.5-01-01-01. Revision 3.

ITTC (2014). *Open water test.* International Towing Tank Conference, Recommended Procedures and Guidelines 7.5-02-03-02.1. Revision 3.

Self Study Problems

1. Why is the advance coefficient J used as the equivalent speed of model and full scale propellers instead of Froude or Reynolds number?

2. Discuss the advantages and disadvantages of conducting open water tests in a towing tank versus conduction them in a circulating water tunnel.

3. Explain in your own words the characteristics of the points marked ①, ②, and ③ in Figure 41.4.

42

Full Scale Propeller Performance

The dimensionless thrust and torque coefficients K_{TM} and K_{QM}, determined in the open water test, cannot be directly applied to the full scale vessel. Like the resistance test, open water tests are conducted under partial dynamic similarity. Characteristic Reynolds numbers for model propellers are smaller than those for the full scale propeller. Therefore viscous effects are not replicated to scale in open water tests. In this chapter we discuss the forces acting on propeller blade sections and the differences between model scale and full scale caused by the difference in Reynolds numbers. The chapter closes with a description of the recommended ITTC procedure for full scale predictions from open water tests.

Learning Objectives

At the end of this chapter students will be able to:

- identify forces acting on propeller blade sections
- understand scale effects in open water tests
- predict full scale propeller performance from open water test results

42.1 Comparison of Model and Full Scale Propeller Forces

Figure 42.1 shows the forces acting on a foil section of the model propeller blade. Typically, sections at radius $r = 0.7R$ or $r = 0.75R$ are used as representatives for the blade because sectional forces are largest in this range of the blade radius.

The total force $\mathrm{d}F_M$ on the blade section may be decomposed into two components:

Lift and drag force on model blade section

(i) a drag force $\mathrm{d}D_M$ in the direction of the resultant onflow velocity v_r, and

(ii) a lift force $\mathrm{d}L_M$ perpendicular to v_r.

The center of effort of blade section forces is located approximately one quarter of the chord length c_M from the leading edge. The lift force follows from Kutta-Joukowsky's theorem (39.15), and the drag force is computed from section properties.

Fundamentals of Ship Hydrodynamics: Fluid Mechanics, Ship Resistance and Propulsion, First Edition.
Lothar Birk.

Companion website: www.wiley.com/go/birk/hydrodynamics

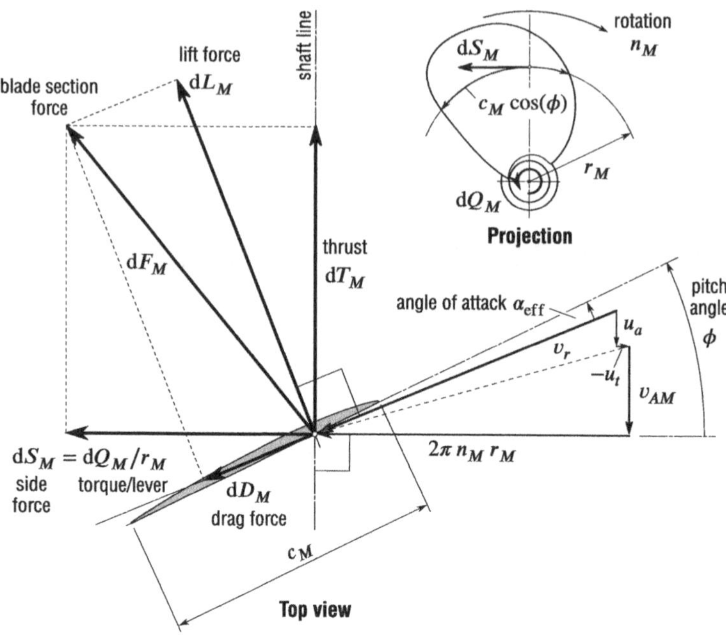

Figure 42.1 Fluid forces acting on a propeller blade section at model scale

Thrust and torque of model blade section

We are, however, more interested in the thrust generated by the blade section. For this the total blade section force dF_M is split into a component dT_M parallel to the shaft and a side force dS_M acting perpendicular to the shaft. The force dT_M is the section's contribution to the propeller thrust. The component dS_M perpendicular to the shaft line is responsible for the torque a propeller exerts on the shaft.

$$dS_M = \frac{dQ_M}{r_M}$$

Forces of the full scale blade section

The same decompositions may be applied to the full scale section force dF_S. The decisive difference between the flows over model scale and full scale blade section is the much higher Reynolds number for the full scale flow. Consequently, the drag coefficient of the full scale blade section is smaller than for the model scale section.

$$C_{DS} < C_{DM} \qquad \text{for propeller blade sections} \tag{42.1}$$

Figure 42.2 shows vectors of both model and full scale section forces. The full scale forces have been scaled down so that the lengths of the lift force vectors dL_M and dL_S have the same length in the sketch. Since $C_{DS} < C_{DM}$, the drag force at full scale is comparatively smaller than at model scale, i.e. the full scale blade section has a better lift–drag ratio.

$$\frac{dL_S}{dD_S} > \frac{dL_M}{dD_M} \tag{42.2}$$

Due to the better lift–drag ratio, the full scale resultant force dF_S points more in the direction of the propeller shaft than the resultant model force dF_M.

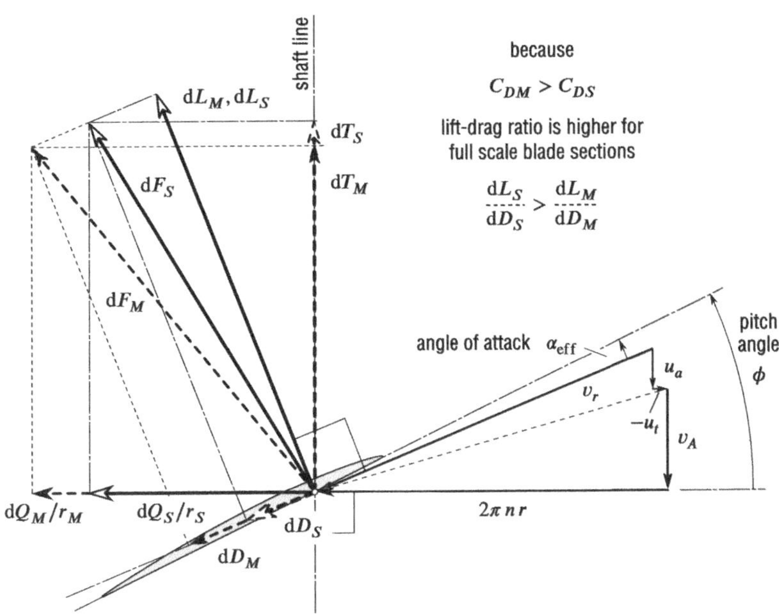

Figure 42.2 Comparison of model scale and full scale forces acting on a propeller blade section

Thrust and torque of the full scale blade section

As a consequence of the forward rotation, the projection of dF_S onto the shaft line becomes longer and the projection onto the transverse direction is shorter (Figure 42.2). Subdividing the full scale blade section force into thrust dT_S and torque dQ_S shows that the full scale propeller for the equivalent onflow conditions produces relatively more thrust with less torque than the model propeller.

Summary of scale effects

In summary, we can expect that a full scale propeller is slightly more efficient than the model propeller.

$$\begin{aligned} K_{TM} &< K_{TS} \\ K_{QM} &> K_{QS} \\ \eta_{OM} &< \eta_{OS} \end{aligned} \tag{42.3}$$

42.2 ITTC Full Scale Correction Procedure

As part of its 1978 Performance Prediction Method, ITTC provides a procedure to scale the open water model test results (ITTC, 2014). The equations yield corrections for thrust and torque coefficient of the model test. The semi-empirical method is applied to the test data in Table 42.1 to illustrate the procedure.

Reduction of model test data

As outlined in the previous section on open water tests, test results are made dimensionless using Equations (41.11), (41.12), and (41.13).

$$J = \frac{v_{AM}}{n_M\, D_M} \qquad K_{TM} = \frac{T_M}{\rho\, n_M^2\, D_M^4} \qquad K_{QM} = \frac{Q_M}{\rho\, n_M^2\, D_M^5} \tag{41.11–41.13}$$

Table 42.1 Example results of an open water test conducted in a towing tank

Model propeller data		Test data			
number of blades	$Z = 4$	v_A	T_M	Q_M	n_M
model scale	$\lambda_M = 25$	[m/s]	[N]	[Nm]	[1/s]
diameter	$D_M = 0.2200$ m	0.000	236.74	6.347	15.02
		0.330	213.86	5.803	14.96
Section data at radius $x = 0.75$		0.661	191.69	5.304	15.01
chord length	$c_{0.75_M} = 0.0733$ m	0.990	166.45	4.742	15.02
max. thickness	$t_{0.75_M} = 0.0031$ m	1.321	142.54	4.218	15.04
pitch–diameter ratio	$P/D_{0.75} = 0.8260$	1.651	117.40	3.671	15.04
		1.982	91.57	3.071	15.01
Test conditions (fresh water)		2.312	65.75	2.406	14.98
temperature	$t_w = 19.5$ °C	2.641	37.07	1.639	14.98
density	$\rho = 998.309$ kg/m^3	2.972	5.87	0.794	15.00
kin. viscosity	$\nu = 1.0157 \cdot 10^{-6}$ m^2/s	3.137	−12.21	0.311	15.02

Table 42.2 Open water characteristics of model propeller (see Table 42.1)

J [–]	K_{TM} [–]	$10\,K_{QM}$ [–]	η_{OM} [–]
0.0000	0.44872	0.54683	0.0000
0.1003	0.40861	0.50398	0.1294
0.2002	0.36382	0.45758	0.2533
0.2996	0.31549	0.40855	0.3682
0.3498	0.29409	0.38727	0.4228
0.4500	0.24404	0.33775	0.5175
0.5514	0.19600	0.28916	0.5948
0.6490	0.15113	0.23816	0.6555
0.7486	0.09977	0.17787	0.6683
0.8483	0.04340	0.10784	0.5434
0.9493	−0.02314	0.02679	−1.3050

The dimensionless open water characteristics for the propeller from Table 42.1 are listed in Table 42.2.

Model section drag coefficient

The corrections for thrust and torque coefficients are based on the difference in drag coefficients for model and full scale blade sections. Data for the blade section at radius $x = 0.75$ is used to represent the propeller. First, a Reynolds number is computed for the model blade section using the chord length at radius $x = 0.75$ as reference length.

$$Re_{c0} = \frac{v_1\, c_{0.75_M}}{\nu_M} \tag{42.4}$$

v_1 is the approximated onflow velocity, which excludes the axial and tangential induced velocities u_a and u_t:

$$v_1 = \sqrt{v_A^2 + (0.75\pi\, n\, D)^2} \tag{42.5}$$

Table 42.3 Intermediate results for scaling open water characteristics of model propeller (see Table 42.1)

J [–]	v_1 [m/s]	Re_{c0} [–]	C_{DM} [10^{-3}]	C_{DS} [10^{-3}]	ΔC_D [10^{-3}]	ΔK_T [–]	$10\,\Delta K_Q$ [–]
0.0000	7.786	561878	12.0692	7.4928	4.5764	−0.001511	0.01525
0.1003	7.762	560140	12.0779	7.4928	4.5851	−0.001514	0.01528
0.2002	7.809	563527	12.0610	7.4928	4.5682	−0.001509	0.01522
0.2996	7.848	566402	12.0468	7.4928	4.5539	−0.001504	0.01517
0.3498	7.871	568038	12.0387	7.4928	4.5331	−0.001497	0.01510
0.4500	7.921	571653	12.0210	7.4928	4.5114	−0.001490	0.01503
0.5514	7.970	575139	12.0040	7.4928	4.4905	−0.001483	0.01496
0.6490	8.081	583192	11.9654	7.4928	4.4655	−0.001475	0.01488
0.7486	8.180	590341	11.9318	7.4928	4.4316	−0.001464	0.01477
0.8483	8.286	597984	11.8964	7.4928	4.3910	−0.001450	0.01463
0.9493	8.394	605771	11.8610	7.4928	4.3681	−0.001443	0.01455

The drag coefficient for the blade section at model scale is a function of this Reynolds number.

$$C_{DM} = 2\left(1+2\frac{t}{c}\right)\left[\frac{0.044}{\left(Re_{c0}\right)^{1/6}} - \frac{5}{\left(Re_{c0}\right)^{2/3}}\right] \tag{42.6}$$

The Reynolds number Re_{c0} should be larger than 200 000.

Full scale section drag coefficient

The formula for the drag coefficient of the full scale blade section assumes that the Reynolds number (full scale) is so high that even a newly manufactured blade surface operates in the fully rough flow regime where the drag coefficient becomes independent of the Reynolds number and solely depends on thickness ratio (t/c) and roughness ratio ($c_{0.75_S}/k_p$) (see Section 13.5).

$$C_{DS} = 2\left(1+2\frac{t}{c}\right)\left[1.89 + 1.62\log_{10}\left(\frac{c_{0.75_S}}{k_p}\right)\right]^{-2.5} \tag{42.7}$$

The roughness k_p of a new propeller is assumed to be 0.00003 m ($k_p = 30 \cdot 10^{-6}$ m). Note that the full scale chord length $c_{0.75_S}$ must be used in Equation (42.7).

Difference in drag coefficients

The difference between model (42.6) and full scale (42.7) section drag coefficients is always positive for typical values of the blade roughness k_p.

$$\Delta C_D = C_{DM} - C_{DS} \tag{42.8}$$

A negative ΔC_D may occur if the model scale chord length $c_{0.75_M}$ is erroneously used in Equation (42.7) for the full scale section drag coefficient.

Corrections for thrust and torque

With the difference in drag coefficient the following corrections are determined for thrust and torque coefficients.

$$\Delta K_T = -\Delta C_D\, 0.3\,\frac{P}{D}\frac{c\,Z}{D} \tag{42.9}$$

$$\Delta K_Q = \Delta C_D\, 0.25\,\frac{c\,Z}{D} \tag{42.10}$$

Table 42.4 Predicted open water characteristics of full scale propeller (see Table 42.1)

J [–]	K_{TS} [–]	$10\,K_{QS}$ [–]	η_{OS} [–]
0.0000	0.45023	0.53158	0.0000
0.1003	0.41013	0.48870	0.1339
0.2002	0.36533	0.44236	0.2631
0.2996	0.31700	0.39338	0.3843
0.3498	0.27095	0.34733	0.4957
0.4500	0.22342	0.30040	0.5906
0.5514	0.17528	0.24997	0.6698
0.6490	0.12676	0.19352	0.7314
0.7486	0.07210	0.12720	0.7230
0.8483	0.01261	0.05396	0.3349
0.9493	−0.02170	0.01224	−2.6786

Note that the correction ΔK_T is negative for the thrust coefficient. Table 42.3 summarizes the intermediate data, drag coefficients, and thrust and torque coefficient corrections for the example data from Table 42.1.

Full scale thrust and torque coefficients

Finally, the full scale thrust and torque coefficients are obtained by subtracting the corrections from the coefficients obtained in the open water test.

$$K_{TS} = K_{TM} - \Delta K_T \qquad (42.11)$$

$$K_{QS} = K_{QM} - \Delta K_Q \qquad (42.12)$$

Make sure to apply the torque correction to the torque coefficient itself instead of $10\,K_{QM}$. Because ΔK_T is negative, the full scale thrust coefficient is larger than its model scale counter part. As expected, the full scale torque coefficient is smaller. The full scale open water efficiency η_{OS} (42.13) grows by a few percent.

$$\eta_{OS} = \frac{J}{2\pi}\frac{K_{TS}}{K_{QS}} \qquad (42.13)$$

Table 42.4 shows the predicted full scale open water characteristics for the propeller and open water test data from Table 42.1. Figure 42.3 compares model and full scale open water characteristics. Note that the thrust and torque coefficient curves for the full scale propeller lie within the pair of curves for the model. Differences are small and barely visible for the thrust coefficient. The effect on the torque coefficient is amplified because we plot ten times its value. An increase in efficiency is clearly visible. In this example maximum efficiency is 6% higher at full scale which is significant.

References

ITTC (2014). *1978 ITTC performance prediction method*. International Towing Tank Conference, Recommended Procedures and Guidelines 7.5-02-03-01.4. Revision 3.

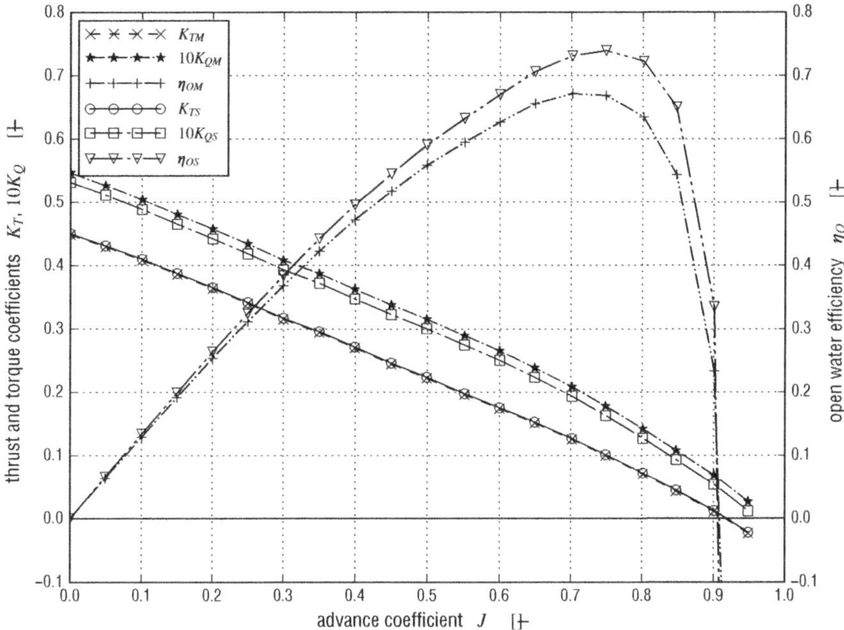

Figure 42.3 Comparison of measured open water characteristics and predicted full scale propeller performance

Self Study Problems

1. Explain with a sketch of the forces acting on a blade section why the efficiency of a full scale propeller is slightly higher than that of its model.

2. Given is the following open water test data set. Predict and plot the full scale open water characteristics.

Model propeller data

number of blades	$Z = 5$
model scale	$\lambda_M = 24$
diameter	$D_M = 0.2500$ m

Section data at radius $x = 0.75$

chord length	$c_{0.75_M} = 0.08281$ m
max. thickness	$t_{0.75_M} = 0.00311$ m
pitch–diameter ratio	$P/D_{0.75} = 0.900$

Test conditions (fresh water)

temperature	$t_w = 18.5$ °C
density	$\rho = 998.505$ kg/m^3
kin. viscosity	$\nu = 1.0410 \cdot 10^{-6}$ m^2/s

Test data

v_A [m/s]	T_M [N]	Q_M [Nm]	n_M [1/s]
0.000	320.50	10.741	14.01
0.387	296.09	10.047	14.00
0.774	267.26	9.226	13.99
1.161	234.50	8.286	14.02
1.548	198.32	7.234	14.00
1.935	159.20	6.076	13.99
2.322	117.67	4.820	14.00
2.709	74.21	3.473	14.02
3.095	29.32	2.042	14.01
3.482	−16.48	0.534	13.99

43

Propulsion Test

After discussion of resistance, propeller open water characteristics, and hull–propeller interaction in general, it is time to explain how the hull–propeller interaction parameters are determined for a given ship. Deriving values for thrust deduction fraction, wake fraction, and relative rotative efficiency is the purpose of the propulsion test. In a propulsion test, propeller thrust, torque, and rate of revolution are measured with a self-propelled model. This chapter explains the setup of models, methods to conduct the actual propulsion test, data reduction, and evaluation of hull–propeller interaction parameters for model and full scale vessel. The chapter closes with an alternative form of the propulsion test called the load variation test.

Learning Objectives

At the end of this chapter students will be able to:

- understand the purpose of propulsion tests
- prepare and set up propulsion tests
- process and interpret test results
- determine hull–propeller interaction parameters for the model

43.1 Testing Procedure

The propulsion test setup and procedure should follow the guidelines of the International Towing Tank Conference (ITTC, 2011).

Model setup First, the model used in the resistance test is equipped with a complete drive system of propeller, shaft, and motor. A propeller dynamometer capable of measuring propeller thrust, torque, and rate of revolution is integrated into the shaft line. The ship model will feature the same turbulence generator as during the resistance test. Again, self-propelled models will be free in surge, heave, roll, and pitch, but sway and yaw motion are prevented by guides at bow and stern. Typically, sinkage fore and aft is also recorded in the propulsion test. Figure 43.1 shows the typical setup of a model.

Fundamentals of Ship Hydrodynamics: Fluid Mechanics, Ship Resistance and Propulsion, First Edition.
Lothar Birk.

Companion website: www.wiley.com/go/birk/hydrodynamics

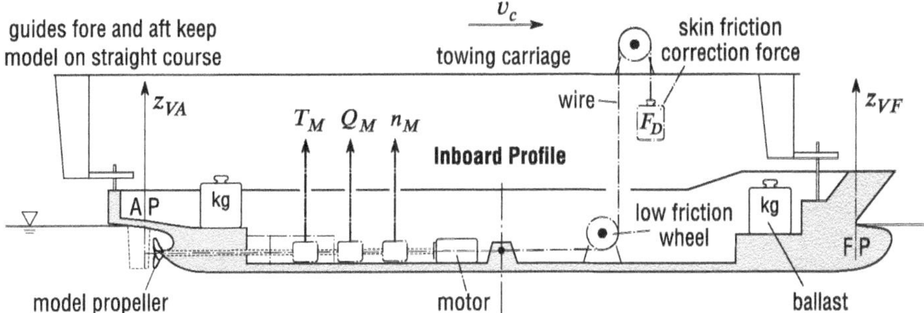

Figure 43.1 Setup of model for propulsion test with single skin friction correction force (continental method)

Even when conducted with great care, differences occur in dynamic trim between resistance and propulsion test. Although displacement and velocity are the same, the propeller will change flow field and pressure distribution especially at the stern. This introduces considerable uncertainties in thrust deduction fraction and wake fraction which are determined from a combination of resistance and propulsion test data.

Stock propeller

The propeller is often not an exact replica of the actual ship propeller. Model propellers are quite expensive due to the complicated shape and required tight manufacturing tolerances. Therefore it is common practice to use a 'stock propeller', i.e. a propeller with the correct general characteristics but not necessarily the exact blade shape and section profiles. Most importantly, it must have the correct diameter D according to the geometric scale λ of the hull. In fact, the availability of a stock propeller often dictates the scale λ of the model. Furthermore, the stock propeller's pitch–diameter ratio should match the pitch–diameter ratio of the actual propeller design. Typically number of blades Z and expanded area ratio A_E/A_0 are of lesser concern but should be matched whenever possible. If the actual propeller has special features, like for example a highly varying pitch distribution to reduce tip and hub cavitation or high skew blades, for which a suitable stock propeller is not readily available, manufacturing a suitable model should be strongly considered.

Effect of propeller loading

We conduct the propulsion test to find the hull–propeller interaction parameters wake fraction w, thrust deduction fraction t, and relative rotative efficiency η_R. Besides other parameters these quantities depend on the propeller loading (Harvald, 1983).

$$C_{Th} = \frac{T}{\frac{1}{2}\rho\, v_A\, \frac{\pi D^2}{4}} \tag{43.1}$$

A smaller propeller diameter D for the same ship will increase the thrust loading coefficient C_{Th} (43.1). Experiments have shown that a larger thrust loading will result in a higher wake fraction w. This is somewhat confusing because we introduced the wake fraction as the influence of a hull onto the propeller. You have to remind yourself that the hull–propeller interaction parameters are just a model to explain and to quantify the effects. However, they are not fundamental laws of physics and have to be interpreted with care.

Partial dynamic similarity

Ideally, we would conduct model tests under full dynamic similarity. This is, unfortunately, not possible as was explained in Chapter 28. Like resistance tests, propulsion

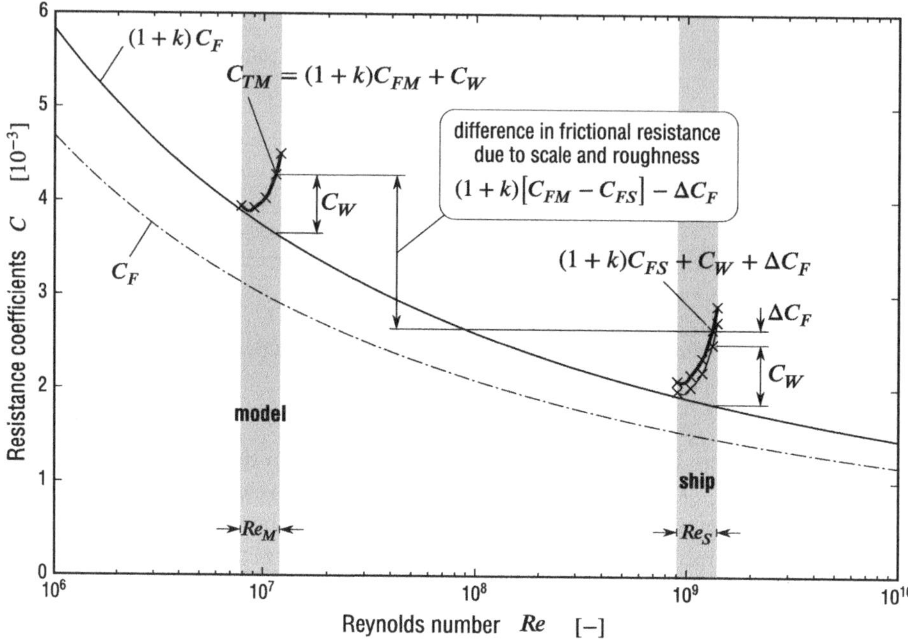

Figure 43.2 The relative difference of resistance for model and full scale vessel

tests are performed satisfying Froude's law. For practical reasons the same Froude numbers are run as in the resistance test. Consequently, the Reynolds numbers in the model test are lower than at full scale. Due to the lower Reynolds numbers at model scale, viscous effects in the resistance are disproportionately larger compared to the full scale vessel. Thus, the model propeller would have a higher thrust loading than the full scale propeller because it has to overcome a comparatively higher resistance.

Thrust loading difference Figure 43.2 illustrates the difference in thrust loading between model and full scale vessel. It shows resistance coefficients of model and full scale vessel as a function of the Reynolds number Re. Model and ship are connected by the ITTC 1957 model–ship correlation line coefficient C_F and the form factor k. They share the same wave resistance coefficient C_W. Compared to the full scale propeller, a model propeller has to overcome an additional frictional resistance of $(1+k)(C_{FM} - C_{FS}) - \Delta C_F$. Note that the roughness allowance ΔC_F increases the loading only for the full scale propeller. As a result, the roughness allowance lowers the difference in frictional resistance between model and full scale vessel.

Skin friction correction force The difference in loading condition is compensated in the propulsion test by applying an additional towing force F_D. It corrects the otherwise excessive loading of the model propeller during the propulsion test. The computation of this skin friction correction force F_D is documented in Section 29.5.

$$F_D = \frac{1}{2}\rho v_M^2 S_M \left[(1+k)\left(C_{FM} - C_{FS}\right) - \Delta C_F\right] \tag{29.28}$$

F_D has to be precomputed for each model speed v_M. This is known as the constant loading method. In Europe this is also called the continental method. Alternatively, a series of tests may be conducted with the same model speed but varying external tow force F_D. This load variation method is described in Section 43.4.

Correction for appendage drag

For models with extensive appendages like exposed shafts and struts, it is recommended to augment the skin friction correction by an appendage drag correction (ITTC, 2011).

$$F_D = \frac{1}{2}\rho v_M^2 S_M \left[(1+k)\left(C_{FM} - C_{FS}\right) - \Delta C_F\right] + (1-\beta)\left[R_{TMA} - R_{TM}\right] \quad (43.2)$$

R_{TMA} is the total resistance of the model with appendages and R_{TM} the total resistance without appendages. The correction factor β for the appendage drag takes values between 0.6 and 1.0 with a typical value of 0.75 for conventional twin screw vessels with bossings, shafts, and brackets.

Model test

During testing the model is propelled by its propeller (or propellers). For each test speed the rate of revolution of the propeller is adjusted so that the model maintains the same speed as the towing carriage. Only when any initial difference in speed has vanished is the recording of thrust T_M, torque Q_M, and rate of revolution n_M started. In the past, the skin friction correction force was applied via a pulley and calibrated weights. Nowadays, the towing staff of the resistance test is used to measure the remaining tow force and the propeller rate of revolution is adjusted until the remaining tow force (resistance) is equal to F_D. This has the additional advantage that the model speed is accurately defined by the carriage speed.

43.2 Data Reduction

At the end of a test series, model thrust T_M, torque Q_M, and rate of revolution n_M have been measured for given sets of model speed v_M and skin friction correction values F_D. The model speed has been derived from the carriage speed by applying appropriate blockage corrections (see Section 29.2).

Thrust and torque coefficients

Thrust and torque are rendered dimensionless like the results of an open water test.

$$K_{TM} = \frac{T_M}{\rho_M\, n_M^2\, D_M^4} \qquad K_{QM} = \frac{Q_M}{\rho_M\, n_M^2\, D_M^5} \quad (43.3)$$

The difference here is that thrust and torque coefficient in Equation (43.3) have been measured in the behind condition.

Behind condition

The open water characteristics of the propeller employed in the propulsion test have been plotted in an open water diagram which shows K_{T_O} and $10\,K_{Q_O}$ as a function of the advance coefficient $J = v_A/(nD)$ (see Chapter 41). Although we measured thrust, torque, and rate of revolution in the propulsion test, we are missing a crucial piece of information. During an open water test the speed of advance is equal to the carriage speed. Obviously, we cannot make the same determination for the propeller working in the wake of a hull (behind condition). However, combining results of open water and propulsion test will allow us to determine the advance ratio and thus speed of advance of the propeller in the behind condition.

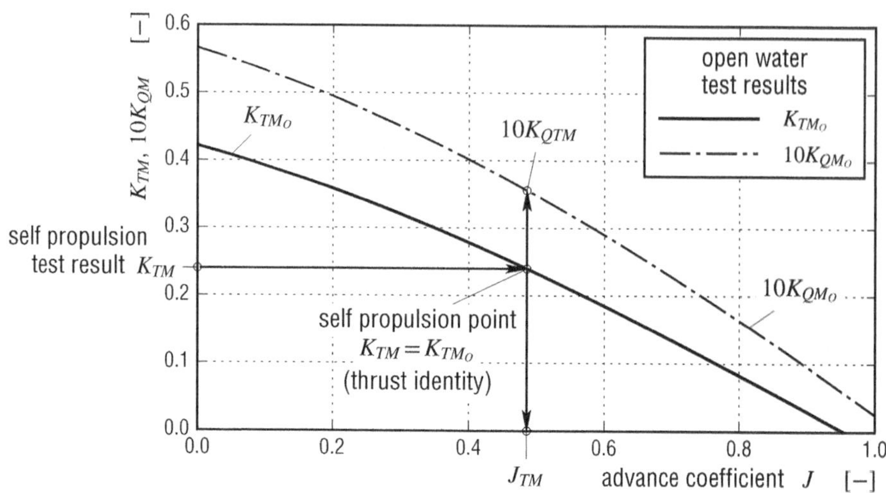

Figure 43.3 Self propulsion point of model propeller under the assumption of thrust identity

43.3 Hull–Propeller Interaction Parameters

Self propulsion point From the results of the propulsion test and from the open water test of the stock propeller we have to determine the operating condition of the propeller behind the ship model, the self propulsion point. Further progress requires us to make an assumption:

(i) We could assume that the model propeller is producing the same thrust in behind and open water condition for the same advance coefficient: $K_{TM} = K_{TMO}$. This is called thrust identity.

(ii) Alternatively, we could assume that the model propeller requires the same torque in behind and open water condition for the same advance coefficient J: $K_{QM} = K_{QMO}$. This is called torque identity.

Probably both, thrust and torque, differ between open water and behind condition but we do not know exactly by how much. In practice, the assumption of thrust identity is preferred as it tends to provide more accurate performance predictions for the full scale vessel. For one, thrust is less affected by different model scales and its measurement is less susceptible to errors caused by friction in bearings and on propeller blades. Working under the assumption of torque identity has the advantage that torque is more easily determined during ship trials to confirm the powering prediction.

Thrust identity Working under the assumption of thrust identity, we take the thrust coefficient K_{TM} (43.3) from the propulsion test and search for the corresponding thrust coefficient K_{TMO} determined under open water condition. Figure 43.3 explains the process. We employ the open water diagram of the model propeller. It represents the capabilities of a propeller in uniform flow, i.e. open water condition. The intersection of open water thrust coefficient curve K_{TMO} and a horizontal line drawn through the thrust coefficient K_{TM} of the propulsion test (43.3) marks the self propulsion point. A vertical

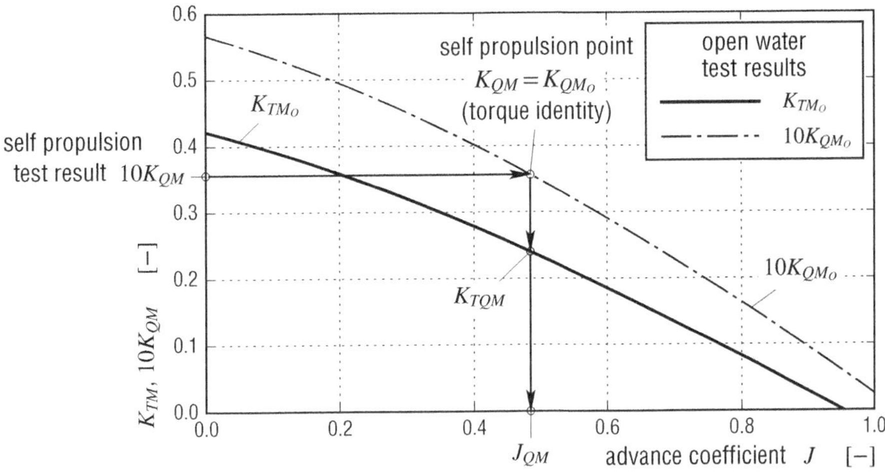

Figure 43.4 Self propulsion point of model propeller under the assumption of torque identity

line drawn through the self propulsion point defines advance coefficient J_{TM} and open water torque coefficient 10 K_{QTM}. The additional subscript $_T$ refers to the assumption of thrust identity.

Note that the self propulsion point changes with model speed!

Torque identity

For the sake of completeness, Figure 43.4 shows the equivalent process under the assumption of torque identity. The self propulsion point is defined by the intersection of the open water torque coefficient curve 10 $K_{QM\,O}$ and the horizontal line drawn through ten times the torque coefficient 10 K_{QM} in the behind condition from Equation (43.3). Advance coefficient J_{QM} and thrust coefficient K_{TQM} for the open water condition are read off the open water diagram. Obviously, the additional subscript $_Q$ refers to the assumption of torque identity.

43.3.1 Model wake fraction

We continue our data analysis assuming thrust identity. The self propulsion point is defined by the combination of advance coefficient J_{TM}, thrust coefficient K_{TM}, and torque coefficient K_{QTM}.

Speed of advance

Wake fraction for the model is defined as the difference between model speed v_M and the speed of advance v_{AM}. The latter may now be derived from the advance coefficient of the self propulsion point. From $J_{TM} = v_{AM}/(n_M D_M)$ follows

$$v_{AM} = J_{TM}\, n_M\, D_M \tag{43.4}$$

The diameter D_M of the propeller is of course known and n_M is the rate of revolution recorded in the propulsion test for the thrust coefficient x value measured for the model velocity v_M.

Wake fraction The definition of the wake fraction (34.1) provides the path to compute it for the self-propelled model:

$$w_{TM} = 1 - \frac{J_{TM}\, n_M\, D_M}{v_M} \tag{43.5}$$

The subscript $_T$ is again used to emphasize the assumption of thrust identity. The wake fraction w_{QM} based on torque identity is obtained by exchanging J_{TM} with J_{QM}.

The wake fraction is subject to significant viscous effects (see Chapter 34). Since the Reynolds number is too small at model scale, the model wake fraction has to be corrected before it is applied to the full scale vessel. The full scale wake fraction is expected to be smaller than the model wake fraction because viscous effects are proportionally larger at model scale. The necessary correction of the model wake fraction for the full scale vessel is explained in Section 43.3.4.

43.3.2 Thrust deduction fraction

Definition In Chapter 34 thrust deduction fraction t was defined as the portion of thrust that must be spent to overcome the difference in resistance of the vessel with and without a propeller.

$$t = \frac{T - R_T}{T} \tag{34.11}$$

Adaption of test results We have to adapt this formula to the specific conditions of the propulsion test.

- During the propulsion test, we reduced the actual model resistance by the skin friction correction F_D (29.28) in order to correct the propeller loading.
- The resistance from the resistance test R_{TM} may have to be corrected for changes in water temperature between the dates of the resistance and propulsion tests. The correction procedure is documented in the ITTC 1978 Performance Prediction Method (ITTC, 2014).

$$R_C = \left[\frac{(1+k)C_{FMC} + C_R}{(1+k)C_{FM} + C_R}\right] R_{TM} \tag{43.6}$$

C_{FM} is the ITTC 1957 model–ship correlation coefficient from the resistance test. C_{FMC} is the equivalent friction coefficient for the water temperature of the propulsion test. A change in water temperature results in a different kinematic viscosity ν_M. Thus, the Reynolds number changes despite equal model speeds.

Prediction With the corrections above, the thrust deduction fraction is computed as

$$t = \frac{T_M - (R_{TM} - F_D)}{T_M} \tag{43.7}$$

You may have noticed that the symbol for the thrust deduction fraction t does not feature a subscript $_M$ for the model. In contrast to the wake fraction, viscous effects on the thrust deduction fraction are small. In lieu of a better prediction, the thrust deduction fraction (43.7) of the model is also used for the full scale ship.

43.3.3 Relative rotative efficiency

Relative rotative efficiency η_R is defined as the ratio of propeller efficiency in the behind condition η_B to the open water efficiency η_O (uniform flow) at the same advance coefficient J.

Change in propeller efficiency

$$\eta_R = \frac{\eta_B}{\eta_O} \tag{34.19}$$

Propeller efficiency is equal to the ratio of thrust power to delivered power. For the behind and open water condition we have

$$\eta_B = \frac{P_T}{P_D} = \frac{T\, v_A}{2\,\pi\, n\, Q} \qquad \eta_O = \frac{P_{To}}{P_{Do}} = \frac{T_O\, v_A}{2\,\pi\, n\, Q_O} \tag{43.8}$$

Noting that speed of advance v_A and rate of revolution n are the same in both cases (equal J), relative rotative efficiency is equal to

$$\eta_R = \frac{\dfrac{T\, v_A}{2\,\pi\, n\, Q}}{\dfrac{T_O\, v_A}{2\,\pi\, n\, Q_O}} = \frac{T\, Q_O}{T_O\, Q} = \frac{K_T\, K_{Q_O}}{K_{T_O}\, K_Q} \tag{43.9}$$

We apply this to the results of the propulsion test. Consistent with the selection of the self propulsion point, we use either thrust identity (assuming $T = T_O$) or torque identity (assuming $Q = Q_O$). Under the assumption of thrust identity, we determined the open water torque K_{QTM}. In the propulsion test we measured the torque K_{QM} for the behind condition. We substitute both into Equation (43.9), and with thrust identity $T = T_O$ we obtain the relative rotative efficiency.

Assuming thrust identity

$$\eta_R = \frac{K_{QTM}}{K_{QM}} \tag{43.10}$$

If the propulsion test would be analyzed using torque identity the relative rotative efficiency would be $\eta_R = K_{TM}/K_{TQM}$.

Assuming torque identity

43.3.4 Full scale hull–propeller interaction parameters

Results of a propulsion test are wake fraction w_{TM}, thrust deduction fraction t, and relative rotative efficiency η_R of the model. The latter two show no significant scale effects and the values found at model scale are also applied to the full scale vessel.

Thrust deduction fraction, relative rotative efficiency

$$t = t_M = t_S \qquad \text{and} \qquad \eta_R = \eta_{RM} = \eta_{RS} \tag{43.11}$$

The wake fraction, however, is heavily influenced by viscous effects. As stated before, viscous effects are more pronounced at model scale than at full scale due to the lower Reynolds number. Therefore, the model wake fraction w_{TM} (w_{QM}) found under thrust (torque) identity is typically larger than the full scale wake fraction w_{TS}.

Full scale wake fraction

The ITTC recommended procedures supply the following formula to predict the full scale wake fraction (ITTC, 2014).

$$w_{TS} = (t + w_r) + \left[w_{TM} - (t + w_r)\right] \frac{(1+k)C_{FS} + \Delta C_F}{(1+k)C_{FM}} \qquad (43.12)$$

$$\text{if } w_{TS} > w_{TM} \text{ use } w_{TS} = w_{TM}$$

This equation needs some explaining.

Components of wake fraction

Chapter 34 (page 34.1) described the wake fraction as the sum of three parts: potential wake w_p, frictional wake w_f, and wave wake w_w.

$$w = w_p + w_f + w_w \qquad (34.5)$$

The wave wake fraction is small compared to the other two parts. Since Froude number for model and ship are equal, we expect the wave systems of model and ship to be geometrically similar. Omitting the wave wake fraction, we are left with two parts:

$$w = w_p + w_f \qquad (43.13)$$

This is also the form of Equation (43.12). The first part is the potential wake fraction of the ship and the second part is its frictional wake fraction.

$$w_{pS} = t + w_r \qquad (43.14)$$

$$w_{fS} = \left[w_{TM} - (t + w_r)\right] \frac{(1+k)C_{FS} + \Delta C_F}{(1+k)C_{FM}} \qquad (43.15)$$

Potential wake fraction

Now, what has the thrust deduction fraction t to do with the potential wake fraction? This is best explained with a thought experiment taken from Harvald (1983). Imagine a flow around the ship without any energy losses. We may model this condition as potential flow. In this hypothetical case, the energy $P_E = R_T\, v$ spent moving the vessel through the fluid must be equal to the work $P_T = T\, v_A$ done by the propulsor.

$$R_T\, v = T\, v_A$$

Resistance may be replaced by $R_T = (1 - t_p)T$. The thrust deduction fraction is reduced to its potential part $t = t_p$ because there is no friction in potential flow, and the effect of the ship wave pattern on the thrust deduction fraction is negligible. Ship speed multiplied by $(1 - w_p)$ may replace the speed of advance v_A. The wake fraction is also equal to its potential part $w = w_p$ because there is no frictional wake in potential flow, and we ignored the small wave wake. We substitute both into the energy balance above.

$$(1 - t_p)T\, v = T\,(1 - w_p)v$$

Dividing by $T\, v$ yields

$$(1 - t_p) = (1 - w_p)$$

For this equation to hold true, it is necessary that wake fraction and potential thrust deduction fraction are of the same size.

$$t_p = w_p$$

As outlined in Chapter 34, the potential part is the largest contribution to the overall thrust deduction fraction even in viscous flows. Thus, $t_p \approx t$ and it is reasonable to substitute the thrust deduction fraction for the potential wake fraction w_p.

$$w_p = t \tag{43.16}$$

The additional term w_r in Equation (43.14) represents the wake effect of the rudder.

$$w_{pS} = w_{pM} = t + w_r \tag{43.14}$$

If it is not determined separately, a value of $w_r = 0.04$ may be assumed (ITTC, 2014).

Frictional wake

In Equation (43.15) for the frictional wake fraction of the ship, the first term in brackets is equal to the difference of model wake fraction w_{TM} and $(t + w_r)$. The latter, we just determined, is the potential wake fraction which applies to model and ship alike (no scale effects). The difference $w_{TM} - (t + w_r)$ is therefore the frictional wake fraction of the model.

$$w_{fM} = w_{TM} - w_{pM} = w_{TM} - (t + w_r) \tag{43.17}$$

The frictional wake of the model cannot be applied to the full scale vessel without correction. A suitable correction factor is represented by the final fraction in Equation (43.15). It represents the ratio of frictional resistances of ship and model including the form effect. Since C_{FS} is considerably smaller than C_{FM} for most scale models, the ratio should be smaller than one.

$$\frac{(1+k)C_{FS} + \Delta C_F}{(1+k)C_{FM}} < 1 \tag{43.18}$$

Note: $w_{TS} \leq w_{TM}$

The full scale wake fraction must be smaller than the model wake fraction. In some rare cases, Equation (43.12) produces a w_{TS} larger than w_{TM}. This makes no sense physically since the frictional effects should always be smaller on the full scale vessel. Therefore, if the result of Equation (43.12) is larger than w_{TM}, the model wake fraction should be applied to the full scale vessel instead.

43.4 Load Variation Test

Problems with continental method

In the continental method described above, a single pre-calculated skin friction correction force is employed to ensure a propeller loading which is dynamically similar to the loading of the full scale propeller (continental method). Values of the hull propeller interaction parameters are quite sensitive to propeller loading. Harvald and Hee (1988) tested a single hull with different model propellers. Significant scatter was found in the resulting thrust deduction fraction. Thus, uncertainties in the estimate of the skin friction correction F_D will produce uncertainties in the prediction of the hull–propeller interaction parameters.

British method

Alternatively, propulsion tests may be repeated at a specific speed each time varying the pulling forces F_P which replaces the skin friction correction force. This varies the propeller loading and is also known as a load variation test or for short the British method.

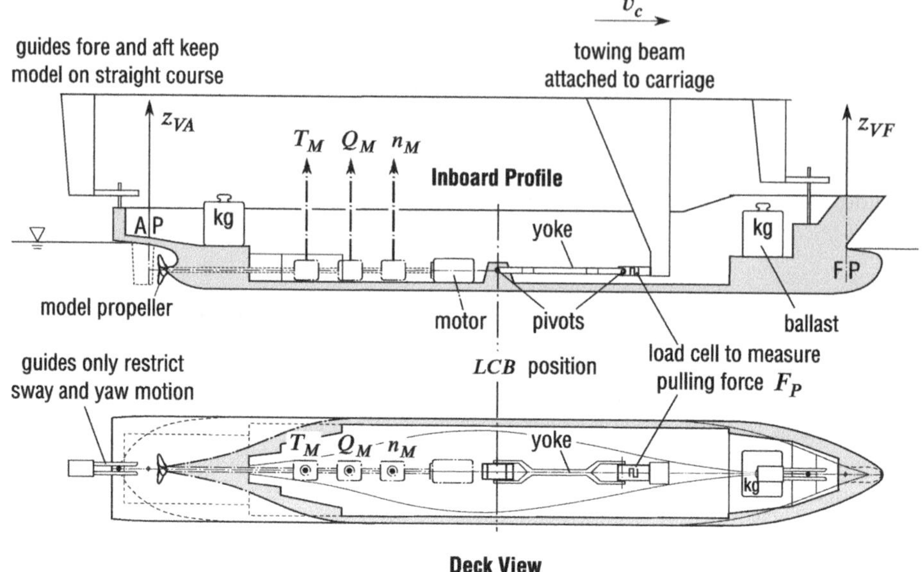

Figure 43.5 Setup of model for propulsion test with load variation (British method)

Setup of load variation test

Figure 43.5 shows the required measurement setup. Essentially, the model equipped for propulsion is hooked up to the resistance dynamometer. For a selected velocity multiple runs are made with the self-propelled model each time increasing the propeller rate of revolution n_M. In the first run, the propeller might be standing still ($n_M = 0$ rpm). The resistance dynamometer will register a pulling (tow) force F_P which will be larger than the resistance R_{TM} of the hull alone. The propeller with its blades perpendicular to the flow generates an additional drag. The propeller dynamometer will measure a negative thrust because the drag force on the propeller is pulling at the shaft.

In a second run, the propeller RPM is set to a small positive value. Let us say it turns just fast enough to produce the necessary thrust to compensate for the additional resistance caused by the propeller. The measured thrust is zero ($T_M = 0$ N). The propeller is turning but not yet pushing the model. The resistance dynamometer measures a towing force which may be interpreted as the total resistance of the complete hull–propeller system.

$$R_{PM0} = F_P(T_M = 0\ \text{N}) \tag{43.19}$$

This point is marked by ① in Figure 43.6. The propeller is essentially idling along. Since no thrust is produced, the propeller loading is zero.

Kracht (1991) reports that the ratio of total resistance with running propeller R_{PM0} and total resistance R_{TM} of hull alone is slightly larger than one for single screw vessels and slightly smaller than one for slender twin screw vessels. Holtrop (1990) states that for single screw vessels $R_{PM0}/R_{TM} \in [1.01, 1.04]$ in 95% of the investigated cases.

In third and subsequent runs, the rate of revolution is increased until the remaining pulling force F_P becomes negative. The model starts pushing the towing carriage. What is astonishing at first is that the relationship between the towing force F_P and the thrust T_M is fairly linear. This is true over a wide range of speeds and ship hull forms. Slight

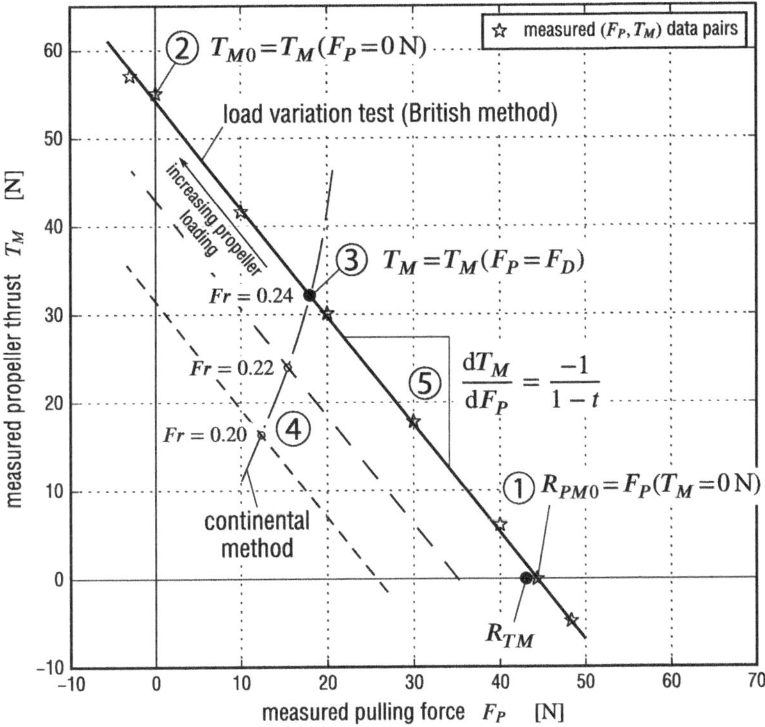

Figure 43.6 Typical results of a load variation test (British method)

nonlinearities seem only to occur for very full vessels with small propellers and in the range of negative F_P forces (Kracht, 1991; Holtrop, 2001).

Model self propulsion point

When the pulling force vanishes ($F_p = 0$ N), the model is completely self-propelled. It produces the thrust $T_{M0} = T_M(F_P = 0\text{ N})$ necessary to overcome the resistance of the model hull–propeller system at the given speed. ② marks this point in Figure 43.6. The thrust T_{M0} must be equal to the resistance of the hull without effects due to the propeller R_{PM0} plus the increase in resistance caused by the propeller. The latter we defined as the thrust deduction $t\,T_{M0}$ (see Section 34.2).

$$T_{M0} = R_{PM0} + t\,T_{M0} \qquad \text{or} \qquad T_{M0} = \frac{R_{PM0}}{1-t} \tag{43.20}$$

The corresponding self propulsion point from the continental method with fixed skin friction correction force F_D is marked with ③.

The slope of the F_p–T_M line shows no or only insignificant variation across the range of ship speeds (Holtrop, 2001). Lines for two smaller Froude numbers have been added into Figure 43.6 at ④.

Balance of longitudinal forces

How can we determine the thrust deduction fraction t based on the result of the load variation test? Considering the forces acting in longitudinal direction on the self-propelled model, we have

- thrust T_M and pulling force F_P acting in the direction of motion, and

- resistance of vessel with running propeller R_{PM0} and the thrust deduction $t\,T_M$ acting against the forward movement of the model.

The thrust deduction tT_M captures the effects on flow pattern and pressure distribution at the stern.

For a model running at constant velocity these forces must be in equilibrium.

$$\sum F_x = 0 = T_M + F_P - R_{PM0} - t\,T_M \tag{43.21}$$

We solve Equation (43.21) for the thrust T_M of the propeller.

$$T_M = \frac{R_{PM0} - F_P}{1 - t} \tag{43.22}$$

This is the equation that describes the line in Figure 43.6. Substituting $F_P = 0$ N into Equation (43.22) yields the self propulsion point:

$$T_M(F_P = 0\,\text{N}) = T_{M0} = \frac{R_{PM0}}{1 - t} \quad \text{(see Equation (43.20))} \tag{43.23}$$

For vanishing thrust $T_M = 0$ N Equation (43.22) resolves into $F_P = R_{PM0}$ as expected.

Thrust deduction fraction Taking the derivative of the thrust T_M (Equation (43.22)) with respect to the pulling force F_P results in the expression shown at ⑤ in Figure 43.6.

$$\frac{\mathrm{d}T_M}{\mathrm{d}F_P} = \frac{-1}{1 - t} \tag{43.24}$$

The thrust deduction fraction defines the slope of the line between the self propulsion point ② and the point of vanishing thrust ①. The fact that the slope does not change for different Froude numbers indicates that the thrust deduction fraction t is independent of speed which is confirmed by practical experience (Holtrop, 2001). In fact, one pair of (R_{PM0}, T_{M0}) values from a load variation test for a single speed is sufficient to compute the thrust deduction fraction. Rearrangement of Equation (43.20) results in:

$$t = 1 - \frac{R_{PM0}}{T_{M0}} \tag{43.25}$$

Unlike the continental method, load variation tests do not require results from a resistance test. Both, R_{PM0} and T_{M0} are results from the same series of tests. Uncertainties from differences in dynamic trim are eliminated. Thrust deduction fractions derived via Equation (43.25) show less spread and are considered more physically sound.

The wake fraction is typically increasing with decreasing propeller loading (higher pulling force F_P). Based on the results of the load variation tests, effects of propeller loading on the wake fraction may be investigated. Again, the concept of thrust identity will be employed to derive the advance coefficient from the propeller open water data. The process is described in Section 43.3 above.

The following chapter summarizes the ITTC 1978 performance prediction method which combines results of resistance test, open water test, and propulsion test to predict the necessary delivered power for trial conditions.

References

Harvald, S. (1983). *Resistance and propulsion of ships.* John Wiley & Sons, New York, NY.

Harvald, S. and Hee, J. (1988). The components of the propulsive efficiency of ships in relation to the design procedure. In *13th Ship Technology and Research Symposium (STAR)*, Pittsburgh, PA.

Holtrop, J. (1990). Are model resistance tests indispensable? In *19th International Towing Tank Conference (ITTC)*, volume 2, pages 450–454, Madrid, Spain.

Holtrop, J. (2001). Extrapolation of propulsion tests for ships with appendages and complex propulsors. *Marine Technology*, 38(3):145–157.

ITTC (2011). *Propulsion/bollard pull test.* International Towing Tank Conference, Recommended Procedures and Guidelines 7.5-02-03-01.1. Revision 4.

ITTC (2014). *1978 ITTC performance prediction method.* International Towing Tank Conference, Recommended Procedures and Guidelines 7.5-02-03-01.4. Revision 3.

Kracht, A. (1991). Load variation tests improve the reliability of ship power predictions based on model test results. *Schiffstechnik / Ship Technology Research*, 38(4):181–190.

Self Study Problems

1. Explain why a skin friction correction force is needed in the propulsion test.
2. Discuss the difference between thrust identity and torque identity.
3. Why can the thrust deduction fraction of the model be applied to the full scale vessel but not the model wake fraction.
4. In your own words, state the difference between the continental and the British method.
5. Why is the thrust deduction fraction derived from load variation tests considered more reliable than the thrust deduction derived from resistance and propulsion test using the continental method?

44

ITTC 1978 Performance Prediction Method

The ITTC 1978 performance prediction method is a procedure to predict the necessary delivered power of a vessel for trial conditions. It is part of the extensive set of model testing procedures recommended by the International Towing Tank Conference (ITTC, www.ittc.info). The power prediction is based on the results of resistance test, open water test, and propulsion test. The individual tests have been discussed in detail in Chapters 29, 30, 41, 42, and 43. Even though only very few naval architects conduct model tests themselves, in order to interpret and use the results in ship design it is important to understand the process and its inherent limitations.

Learning Objectives

At the end of this chapter students will be able to:

- discuss the prediction of delivered power based on model tests
- determine delivered power for trial conditions based on model tests

44.1 Summary of Model Tests

The results of three different model tests contribute to the prediction of the full scale vessel performance in the ITTC 1978 performance prediction method (ITTC, 2017a):

(i) Resistance test: a hull without a propeller is towed through the water. Total resistance R_{TM} is measured as a function of model speed (Froude number). From the model test results follow form factor k and total resistance R_{TS} of the full scale vessel.

(ii) Open water test: the performance characteristics of a propeller are observed while it operates in undisturbed parallel flow. Thrust T_M, torque Q_M, and rate of revolution n_M are measured as a function of advance ratio J. Full scale propeller characteristics are derived from the model test results.

Fundamentals of Ship Hydrodynamics: Fluid Mechanics, Ship Resistance and Propulsion, First Edition.
Lothar Birk.

Companion website: www.wiley.com/go/birk/hydrodynamics

(iii) Propulsion test: a complete hydrodynamics system of hull and propeller is moving under its own power through the water. Thrust T_M, torque Q_M, and rate of revolution n_M are measured as a function of model speed.

44.2 Full Scale Power Prediction

With the full scale thrust deduction fraction and wake fraction as the final puzzle pieces we can predict the delivered power for the ship. During self propulsion tests, thrust and rate of revolution are measured for the model. This enabled us to determine the self propulsion point of the model propeller: first the advance ratio under the assumption of thrust (or torque) identity and then the speed of advance from advance ratio and rate of revolution. Obviously, there are no equivalent full scale measurements available before the ship is built.

Required thrust

From total resistance and thrust deduction fraction we compute the thrust that the ship requires for each tested speed:

$$T_S = \frac{R_{TS}}{(1-t)} \tag{44.1}$$

The wake fraction allows us to estimate the speed of advance at which the propeller operates.

$$v_{AS} = (1 - w_{TS})v_S \tag{44.2}$$

Both required thrust and speed of advance will vary with ship speed v_S. Next, the thrust requirement of the ship has to be matched with the thrust capability of the full scale propeller in order to find the ship's self propulsion point.

Propeller capabilities

The capabilities of the propeller are recorded in its open water diagram. Specifically, the thrust coefficient curve K_{TSO} represents the thrust capability of the propeller under open water conditions (Figure 44.1).

Shortened thrust loading coefficient

Finding the rate of revolution n_S when propeller diameter D_S, speed of advance v_{AS}, and thrust T_S are known is task 4 of the propeller design problems (Chapter 48). In task 4 the self propulsion point is found by intersecting the parabola $C_S\, J^2$ with the thrust coefficient curve K_{T_O} of the propeller. The shortened thrust loading coefficient C_S is derived from the definition of the thrust coefficient K_T. The unknown rate of revolution n is eliminated by dividing Equation (41.12) with J^2:

$$C_S = \frac{K_{TS}}{J^2} = \frac{T_S}{\rho_S\, D_S^2\, v_{AS}^2} \tag{44.3}$$

All quantities on the right-hand side are known. Applying Equations (44.1) and (44.2) yields

$$C_S = \frac{R_{TS}}{\rho_S\, D_S^2 (1-t)\left(1 - w_{TS}\right)^2 v_S^2} \tag{44.4}$$

Finally, we replace the resistance with the dimensionless total resistance coefficient C_{TS}.

$$C_S = \frac{S_S}{2\, D_S^2}\frac{C_{TS}}{(1-t)\left(1 - w_{TS}\right)^2} \tag{44.5}$$

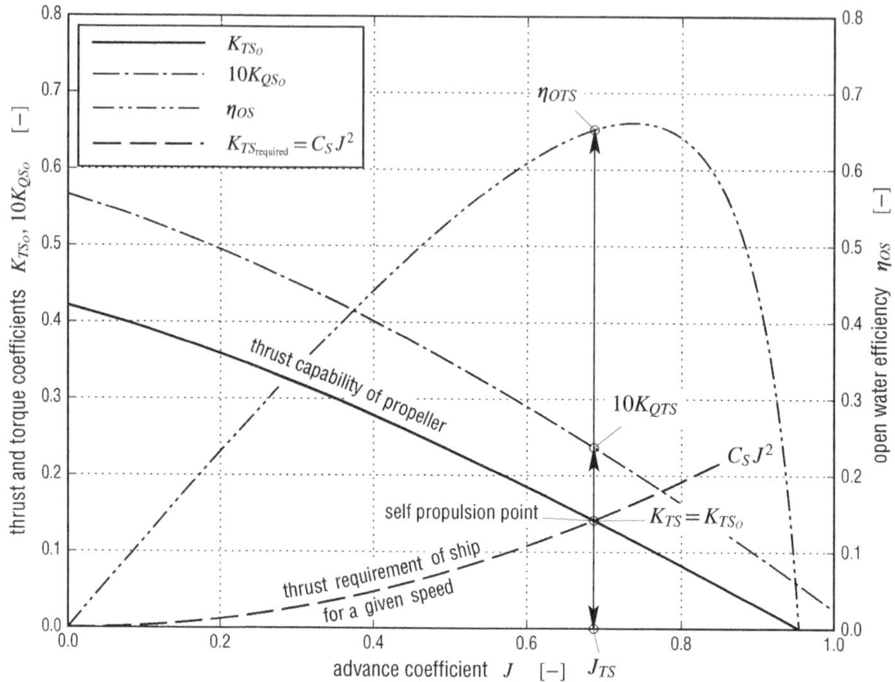

Figure 44.1 Matching propeller thrust K_{TSO} with the thrust requirement of the ship assuming thrust identity

The shortened thrust loading coefficient will vary with ship speed because the resistance changes.

Self propulsion point of ship A match of thrust requirements and thrust capabilities is found by plotting the parabola $C_S\, J^2$ into the propeller open water diagram (see Figure 44.1). To that end, we multiply C_S with increasing values of J squared and mark the points in the open water diagram. The parabola $K_{TS} = C_S\, J^2$ represents the thrust requirement for a given vessel speed. The intersection of the parabola with the propeller K_{TSO}-curve marks the self propulsion point. Note that we again employed the concept of thrust identity assuming that the propeller produces the same thrust for the same rate of revolution and speed of advance in both, open water and behind condition.

Advance ratio under thrust identity The advance coefficient J_{TS} is found via a vertical line drawn into Figure 44.1. Where the vertical line intersects the open water torque curve $10K_{QS_O}$ and the open water efficiency curve η_{O_S} we read off the open water torque $10K_{QTS}$ and the propeller efficiency $\eta_{O_{TS}}$ for the given speed v_S. Section 44.4 explains how to find the intersection numerically if the open water diagram is given as a set of discrete data points.

J_{TS}, K_{QTS}, and $\eta_{O_{TS}}$ have the additional subscript $_T$ because they have been obtained under the assumption of thrust identity. Computing C_S and matching thrust requirements of the ship with the propeller thrust have to be repeated for each speed.

Rate of revolution From the advance coefficient J_{TS}, we retrieve the necessary rate of revolution n_S.

$$n_S = \frac{v_{AS}}{J_{TS}\, D_S} \tag{44.6}$$

With the relative rotative efficiency the required torque K_{QS} for the propeller in the behind condition and the behind efficiency η_B are found.

Torque coefficient and behind efficiency

$$K_{QS} = \frac{K_{QTS}}{\eta_R} \tag{44.7}$$

$$\eta_B = \eta_{O_{TS}} \eta_R \tag{44.8}$$

Finally, required torque Q_S and delivered power P_{DS} may be determined.

Delivered power and quasi-propulsive efficiency

$$\begin{aligned} Q_S &= \rho\, n_S^2 D_S^5\, K_{QS} && \text{torque} \\ P_{DS} &= 2\pi\, n_S\, Q_S = 2\pi\, \rho\, n_S^3\, D_S^5\, K_{QS} && \text{delivered power} \end{aligned} \tag{44.9}$$

The quasi propulsive efficiency η_{DS} of the ship becomes:

$$\begin{aligned} \eta_{DS} &= \frac{P_{ES}}{P_{DS}} \\ &= \eta_{HS}\, \eta_{O_{TS}}\, \eta_R && \text{quasi propulsive efficiency} \end{aligned} \tag{44.10}$$

The predicted values for rate of revolution n_S (44.6) and delivered power P_{DS} (44.9 are usually adjusted for the ship trial. Three different methods are described in the ITTC recommended procedures (ITTC, 2017a):

Trial corrections

- correction factors
- alternative self propulsion point based on estimates of changes in friction ΔC_F and changes in wake fraction Δw_c
- using torque identity

The first and simplest method is correction factors. Power and rate of revolution at trial are

$$P_{DT} = P_{DS}\, C_P \tag{44.11}$$

$$n_T = n_S\, C_N \tag{44.12}$$

New editions of the ITTC 1978 performance prediction method (ITTC, 2017a) do not provide values for the correction factors C_P and C_N. It is left to the model basin to select appropriate values based on its own experience. In older editions values are stated as $C_P = 1.01$ and $C_N = 1.02$.

The brake power P_B of the engine is slightly higher than the delivered power due to losses in shaft bearings and gear boxes if present. Brake power is given with the mechanical efficiency η_S of the shafting as

Brake power

$$P_B = \frac{P_D}{\eta_S} \tag{44.13}$$

Note that delivered power and brake power have been computed for trial conditions, i.e. deep water, calm seas, and no wind or current. However, propeller and engine selection have to take into account weather and sea state conditions for the intended service area of the vessel. A discussion of powering margins is provided in ITTC (2017b).

44.3 Summary

This completes our discussion of power prediction from model tests. Development of this method started with William Froude and is still ongoing today. Every three years researchers meet at the International Towing Tank Conference to summarize and discuss the newest developments and to propose and update the recommended procedures. The current procedure is the essence of decades of experience in power prediction and produces in most cases accurate and reliable design data. However, it definitely has its shortcomings which are briefly discussed below.

A fundamental flaw of the power prediction based on resistance, open water, and propulsion test is the decomposition of the hull-propeller system into individual parts. Resistance test and open water test investigate hull and propeller in hypothetical conditions. The hull is not able to move without propeller and the propeller cannot be driven without a hull housing the prime mover. A possible solution would be to solely rely on load variation tests (see Section 43.4) which at least would be a complete hydrodynamic system. This still leaves the problem of partial dynamic similarity. Unless a fluid is found whose kinematic viscosity is considerably lower than that of fluids known today, Reynolds numbers will be significantly lower at model scale than at full scale. Thus, data derived from load variation tests with models must still be scaled up to the full scale vessel. Possible solutions have been proposed in the past but so far the international community has not been able to replace the current performance prediction procedure with a method more soundly grounded in physics. Some examples of alternatives can be found in Grigson (1990), Holtrop (2001), and most notably in Schmiechen (1991, 2009). Professor M. Schmiechen provides updates and in depth discussions about his rational theory of propulsion on his webpage at http://www.m-schmiechen.homepage.t-online.de.

Less fundamental, but still important, are problems within the current method.

- As mentioned before, the ITTC 1957 model–ship correlation line is not a true flat plate friction coefficient but rather a compromise to facilitate the scaling of model test results to full scale vessels. This is a field of ongoing research. A proposal by Grigson (1993, 2000) seems most promising to eventually replace the ITTC 1957 line with a planar friction coefficient which is based on the best available physical model.
- Resistance and propulsion tests are evaluated assuming the model has reached a steady state. That is however rarely the case, and small fluctuations of velocity do occur. Measured values for resistance, thrust, torque, and rate of revolution are averaged over time to eliminate the fluctuations. Admittedly, the changes in velocity are very small, but they are accompanied by small accelerations. The latter should be less than 1% of the gravitational acceleration g in carefully conducted tests. However, considering how small the resistance values are compared to the large mass of ships (plus additional added mass), inertia forces (mass times acceleration) may not be negligible in the equilibrium of forces. This is addressed by Schmiechen (1991, 2009) but not accounted for in the present performance prediction method.

This being primarily a textbook for future naval architects and marine engineers, the short discussion above should not be construed as a dismissal of current model testing

Table 44.1 Propeller open water characteristics as a set of discrete data points

i	J_i [–]	K_{T_i} [–]	$10K_{Q_i}$ [–]	η_{O_i} [–]
0	0.0000	0.45626	0.65882	0.0000
1	0.0242	0.45021	0.65116	0.0267
2	0.0485	0.44385	0.64318	0.0533
3	0.0727	0.43720	0.63487	0.0797
4	0.0970	0.43026	0.62625	0.1060
5	0.1212	0.42304	0.61730	0.1322
⋮	⋮	⋮	⋮	⋮
$N-2$	1.0424	0.00299	0.06021	0.0823
$N-1$	1.0545	−0.00361	0.05032	−0.1204

practices. Rather, shortcomings of the performance prediction procedure should be understood as an enticement to study these problems and eventually contribute to their solution.

44.4 Solving the Intersection Problem

Various numerical methods may be used to solve for the self propulsion point where the parabola of the ship thrust requirements $C_S J^2$ and the open water thrust coefficient of the propeller intersect. For now, we assume that the curves of the open water diagram of the full scale propeller are given as discrete points (Table 44.1).

The data points may be derived from open water tests, regression formulas of a suitable propeller series, or the results of a lifting line or panel code. The numbers in Table 44.1 have been computed with the Wageningen B-Series regression formulas omitting the additional Reynolds number correction (Oosterveld and van Oossanen, 1975).

To avoid unnecessary clutter, we omit the subscripts for open water and full scale ship. It is important to check if the open water torque coefficient itself was stored or ten times the torque coefficient.

Linear interpolation

The simplest computation of the self propulsion point approximates the open water curves as polylines. Data points are connected by straight line segments. The intersection point is then found by linear interpolation. This will be accurate enough if your data points are close together. A quadratic interpolation over three data points may be more appropriate for a limited number of data points. Here we assume that a sufficient number of data points is available.

We perform a sequential search through the data starting with $i = 0$ and we stop, once the product of design constant and squared advance coefficient $C_S J_i^2$ becomes larger than K_{T_i}. The intersection point is located between J_{i-1} and J_i, i.e. $J_{i-1} \leq J_{TS} \leq J_i$.

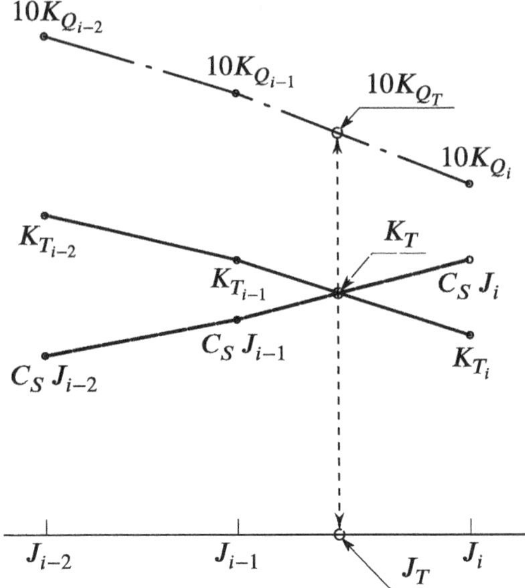

Figure 44.2 Setup for solving the intersection problem with discrete open water data

We describe the K_T curve between $K_{T_{i-1}}$ and K_{T_i} as a linear function:

$$K_T = K_{T_{i-1}} + \frac{K_{T_i} - K_{T_{i-1}}}{J_i - J_{i-1}}\left(J - J_{i-1}\right) \qquad \text{for } J \in [J_{i-1}, J_i] \tag{44.14}$$

A corresponding linear function is created for the thrust requirement of the ship:

$$C_S\,J^2 = C_S\,J_{i-1}^2 + \frac{C_S\,J_i^2 - C_S\,J_{i-1}^2}{J_i - J_{i-1}}\left(J - J_{i-1}\right) \qquad \text{for } J \in [J_{i-1}, J_i] \tag{44.15}$$

In order to find J_{TS}, we begin by subtracting Equation (44.15) from Equation (44.14):

$$\begin{aligned} K_T - C_S\,J^2 = {} & K_{T_{i-1}} - C_S\,J_{i-1}^2 + \frac{K_{T_i} - K_{T_{i-1}}}{J_i - J_{i-1}}\left(J - J_{i-1}\right) \\ & - \frac{C_S\,J_i^2 - C_S\,J_{i-1}^2}{J_i - J_{i-1}}\left(J - J_{i-1}\right) \qquad \text{for } J \in [J_{i-1}, J_i] \end{aligned} \tag{44.16}$$

We simplify the resulting equation and make use of the fact that for $J = J_{TS}$ we get the intersection point $K_{TS} = C_S\,J_{TS}^2$, i.e. the left-hand side of Equation (44.16) vanishes.

$$\begin{aligned} 0 = {} & K_{T_{i-1}} - C_S\,J_{i-1}^2 \\ & + \left[\frac{K_{T_i} - K_{T_{i-1}}}{J_i - J_{i-1}} - \frac{C_S\,J_i^2 - C_S\,J_{i-1}^2}{J_i - J_{i-1}}\right]\left(J_{TS} - J_{i-1}\right) \end{aligned} \tag{44.17}$$

Multiplying with $(J_i - J_{i-1})$ eliminates the fractions.

$$\begin{aligned} 0 = & \left[K_{T_{i-1}} - C_S\, J_{i-1}^2\right](J_i - J_{i-1}) \\ & + \left[K_{T_i} - K_{T_{i-1}} - C_S\, J_i^2 + C_S\, J_{i-1}^2\right](J_{TS} - J_{i-1}) \\ & \text{for } J_{TS} \in [J_{i-1}, J_i] \end{aligned} \tag{44.18}$$

Finally, we solve the equation for the desired advance ratio J_{TS} at which the propeller delivers the required thrust.

$$J_{TS} = J_{i-1} - \frac{\left[K_{T_{i-1}} - C_S\, J_{i-1}^2\right](J_i - J_{i-1})}{\left[K_{T_i} - K_{T_{i-1}} - C_S\, J_i^2 + C_S\, J_{i-1}^2\right]} \tag{44.19}$$

Substituting the value of J_{TS} into Equation (44.14) yields the thrust coefficient for the full scale propeller.

$$K_{TS} = K_{T_{i-1}} + \frac{K_{T_i} - K_{T_{i-1}}}{J_i - J_{i-1}}(J_{TS} - J_{i-1}) \tag{44.20}$$

Substituting J_{TS} into the corresponding intervals of the polynomials for torque coefficient and open water efficiency allows us to compute torque and efficiency for the full scale propeller in open water condition.

$$K_{QTS} = K_{Q_{i-1}} + \frac{K_{Q_i} - K_{Q_{i-1}}}{J_i - J_{i-1}}(J_{TS} - J_{i-1}) \tag{44.21}$$

$$\eta_{O_{TS}} = \eta_{O_{i-1}} + \frac{\eta_{O_i} - \eta_{O_{i-1}}}{J_i - J_{i-1}}(J_{TS} - J_{i-1}) \tag{44.22}$$

Again, if only a few data points define the open water characteristics, a higher order interpolation method is warranted.

44.5 Example

Let us study the process using an example. As input we need the data listed in Table 44.2 which may be derived from model test results or from a resistance and propulsion estimate like we discussed.

With the given input data we first compute the required thrust:

$$T_{\text{required}} = \frac{R_T}{(1-t)} = \frac{1450250\ \text{N}}{1 - 0.1924} = 1795.71\ \text{kN}$$

The speed of advance for $v = 11.472$ m/s follows from equation (44.2).

$$v_A = (1 - w)\,v = (1 - 0.2309)\,11.472\ \text{m/s} = 8.823\ \text{m/s}$$

Table 44.2 Example input data for the calculation of the self propulsion point for a single ship speed

Resistance and propulsion data		
ship speed	v	11.472 m/s
total resistance	R_T	1450.250 kN
thrust deduction fraction	t	0.1924
wake fraction	w	0.2309
relative rotative efficiency	η_R	0.9939
propeller diameter	D	7.200 m
density of salt water	ρ	1026.021 kg/m^3

Propeller open water data			
i	J_i	K_{T_i}	$10K_{Q_i}$
		…	
28	0.6182	0.22329	0.36665
29	0.6303	0.21744	0.35901
30	0.6424	0.21155	0.35131
31	0.6545	0.20563	0.34354
32	0.6667	0.19968	0.33570
33	0.6788	0.19369	0.32779
34	0.6909	0.18767	0.31981
35	0.7030	0.18162	0.31177
36	0.7152	0.17554	0.30366
		…	

Table 44.3 Data for required thrust parabola at $v = 11.472$ m/s and $C_S = 0.43372$ and propeller open water thrust coefficient

Required and provided thrust coefficient				
i	J_i	J_i^2	$C_S\,J_i^2$	K_{T_i}
		…		
28	0.6182	0.38217	0.16577	0.22329
29	0.6303	0.39728	0.17232	0.21744
30	0.6424	0.41268	0.17900	0.21155
31	0.6545	0.42837	0.18579	0.20563
32	0.6667	0.44449	0.19280	0.19968
33	0.6788	0.46077	0.19986	0.19369
34	0.6909	0.47734	0.20705	0.18767
35	0.7030	0.49421	0.21437	0.18162
		…		

With these results we compute the shortened thrust loading coefficient C_S (44.3).

$$C_S = \frac{T_{\text{required}}}{\rho D^2 v_A^2} = \frac{1795710\ \text{N}}{1026.021\ \text{kg/m}^3\ 7.2^2\text{m}^2\ 8.823^2(\text{m/s})^2} = 0.43372$$

Next, we compute values of the required thrust parabola $C_S\,J^2$ for the ship using the J_i values from the open water data in Table 44.2. Table 44.3 summarizes the parabola data and repeats the open water thrust coefficient K_{T_i}. Comparing $C_S\,J_i^2$ and K_{T_i} we find that for $i = 33$ the required thrust coefficient exceeds the provided thrust coefficient for the first time, i.e. $C_S\,J_{33}^2 > K_{T_{33}}$. For $i - 1 = 32$ we still have $C_S\,J_{32}^2 < K_{T_{32}}$. Consequently, the self propulsion point J_T is located between $J_{32} = 0.6667$ and $J_{33} = 0.6788$. With the help of the interpolation formula (44.19) and data for $i = 33$, we find

$$J_T = 0.6731$$

Equipped with the advance ratio J_T, Equations (44.20), (44.21), and (44.22) deliver thrust coefficient, torque coefficient, and open water efficiency for the self propulsion point.

$$\begin{aligned} J_T &= 0.6731 \\ K_T &= 0.1965 \\ K_{QT} &= 0.03315 \\ \eta_{OT} &= 0.6350 \end{aligned}$$

With the data of the propeller at the self propulsion point it is easy to complete the power prediction for the given speed. We use Equations (44.6) through (44.10) and compute

rate of revolution	$n = \dfrac{v_A}{J_T D} = 1.8205\ \mathrm{s}^{-1} = 109.231\ \mathrm{rpm}$
torque in behind condition	$K_Q = \dfrac{K_{QT}}{\eta_R} = 0.033353$
propeller behind efficiency	$\eta_B = \eta_{OT}\eta_R = 0.6311$
propeller torque	$Q = \rho\, n^2\, D^5\, K_Q = 2194.54\ \mathrm{kNm}$
delivered power	$P_D = 2\pi\, n\, Q = 25102.56\ \mathrm{kW}$
quasi-propulsive efficiency	$\eta_D = \eta_H\, \eta_{OT}\, \eta_R = 0.6627$

The process described in this example will be repeated for each considered ship speed.

Based on the predicted delivered power, the required engine brake power may be derived depending on shafting and gearing efficiency as well as suitable engine and sea margins.

References

Grigson, C. (1990). Screws working in behind and prediction of the performance of full ships. *Journal of Ship Research*, 34(4):262–282.

Grigson, C. (1993). An accurate smooth friction line for use in performance prediction. In *Transactions of The Royal Institution of Naval Architects (RINA)*, volume 135, pages 149–162. The Royal Institution of Naval Architects.

Grigson, C. (2000). A planar friction algorithm and its use in analysing hull resistance. In *Transactions of The Royal Institution of Naval Architects (RINA)*, volume 142, pages 76–115. The Royal Institution of Naval Architects.

Holtrop, J. (2001). Extrapolation of propulsion tests for ships with appendages and complex propulsors. *Marine Technology*, 38(3):145–157.

ITTC (2017a). *1978 ITTC performance prediction method*. International Towing Tank Conference, Recommended Procedures and Guidelines 7.5-02-03-01.4. Revision 4.

ITTC (2017b). *Predicting powering margins*. International Towing Tank Conference, Recommended Procedures and Guidelines 7.5-02-03-01.5. Revision 2.

Oosterveld, M. and van Oossanen, P. (1975). Further computer-analyzed data of the Wageningen B-screw series. *International Shipbuilding Progress*, 22:251–261.

Schmiechen, M. (1991). The method of quasisteady propulsion and its trial on board the Meteor. Technical Report Report No. 1184/91, Versuchsanstalt für Wasserbau und Schiffbau, Berlin, Germany.

Schmiechen, M. (2009). 50 Years rational theory of propulsion – Recent results and perspectives. In *International Symposium on Marine Propulsion smp'09*, Trondheim, Norway.

Self Study Problems

1. Compute the self propulsion point data for the data in Table 44.2 applying the following ship speed and total resistance combinations:

$$\begin{aligned} v &= 20\text{kn} & R_T &= 1072.76\,\text{kn} \\ v &= 24\text{kn} & R_T &= 1818.65\,\text{kn} \end{aligned}$$

2. What are the advantages of a load variation test compared with a traditional propulsion test (continental method)?

45

Cavitation

The substance known as water (H_2O) exists in three phases: as a liquid (water), solid (ice), or gas (vapor). The sudden transition from its liquid to its gaseous phase is called cavitation. Cavitation may occur in all flows which have high local flow velocities and associated low pressures. In marine systems, propellers and lifting foils especially are subject to the cavitation phenomena. Cavitation may also occur in pumps and piping systems. This chapter discusses the cavitation phenomenon, where it is likely to occur, and which negative effects it can have. Design measures to prevent cavitation are discussed in the following chapter.

Learning Objectives

At the end of this chapter students will be able to:

- explain and identify cavitation
- quantify the likelihood of cavitation with cavitation numbers
- know about detrimental effects of cavitation

45.1 Cavitation Phenomenon

Phases of Water

Water (H_2O) exists in three different phases:

(i) solid as ice,

(ii) liquid as water, and

(iii) gaseous as vapor.

Which phase is present depends on the combination of temperature and pressure. The relationship is visualized in a phase diagram like the one shown in Figure 45.1. Solid ice exists for temperatures below freezing (0 °C) and (not shown here) at higher temperatures under very high pressure. For temperatures above 0 °C ice becomes liquid. If the combination of pressure and temperature falls below the vapor pressure, ice or water turn into vapor. Vapor pressure is also called saturation pressure.

Fundamentals of Ship Hydrodynamics: Fluid Mechanics, Ship Resistance and Propulsion, First Edition.
Lothar Birk.

Companion website: www.wiley.com/go/birk/hydrodynamics

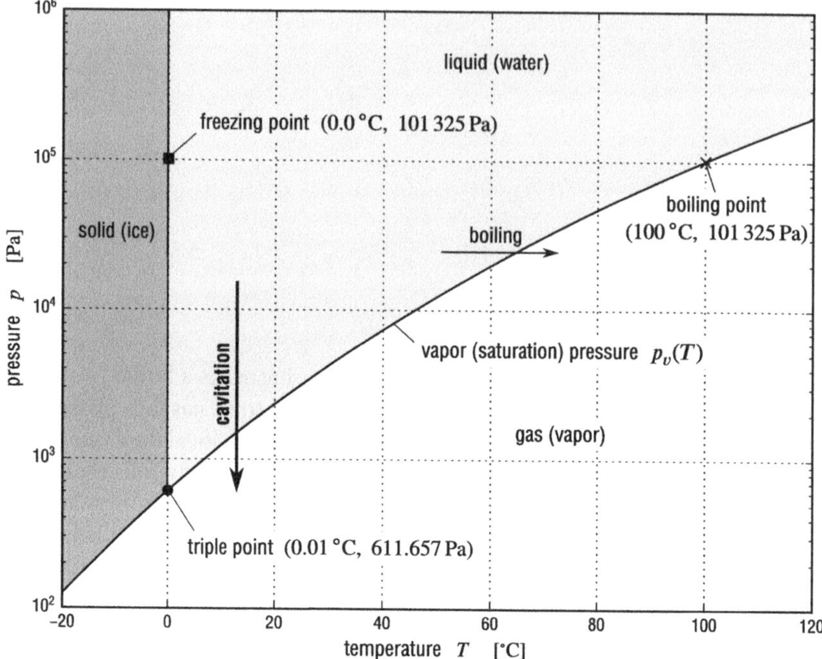

Figure 45.1 Simplified phase diagram of fresh water

Boiling or cavitating? We are interested in the transition from the liquid to the gaseous phase. According to the phase diagram Figure 45.1, transition from liquid to gaseous phase may happen in two different ways:

- If temperature is increased at constant pressure, water will change from liquid to vapor. On the phase diagram we move horizontally from left to right. At the standard atmospheric pressure of 101 325 Pa, transition from liquid to vapor happens at a temperature of 100 °C. This is known as the boiling point. You observe bubbles popping up from the bottom of a saucepan with boiling water. The bubbles burst at the free surface and release the vapor into the air. As mentioned, temperature is considered constant for flow patterns around ships and propellers. Thus this first method of transition from liquid to gas does not apply.
- For constant temperature, the transition into the gaseous phase may be triggered by a drop in pressure. We move vertically in the phase diagram. At the level of the vapor pressure p_v, water changes into its gaseous phase. This is called cavitation. Vapor pressure p_v increases with temperature, i.e. cavitation is more likely in warm than in cold water.

Cavitation When the pressure in a flow approaches vapor pressure, the fluid vaporizes in an instant. Cavitation is the formation of vapor filled cavities in the liquid. Typically, dissolved air or tiny sand grains serve as nuclei for the bubble. The cavities move along with the flow and collapse when they reach areas with higher ambient pressure. This phenomenon occurs in any hydrodynamic or hydraulic system where low pressure is generated by moving fluids or bodies.

Cavitation is a highly unsteady process and depends on numerous fluid properties and flow features, which makes accurate prediction of cavitation and its impact on performance difficult. This difficulty applies to numerical as well as experimental methods.

45.2 Cavitation Inception

Cavitation number

One objective in the design of hydrodynamic systems is the avoidance or at least the delay of cavitation. Cavitation numbers σ are used as a measure to quantify the likelihood of cavitation to occur.

$$\sigma = \frac{p_0 - p_v}{\frac{1}{2}\rho\, v^2} \tag{45.1}$$

The numerator $p_0 - p_v$ measures the difference between the static pressure in the flow p_0 and the vapor pressure p_v. The denominator is the dynamic pressure q.

$$q = \frac{1}{2}\rho v^2 \tag{45.2}$$

Cavitation numbers express the resistance of a flow against cavitation. The larger the cavitation number the less likely is cavitation. High static pressure p_0 lowers the danger of cavitation (higher cavitation number σ). This can be achieved by deeply submerging a foil or propeller. High flow velocities increase dynamic pressure which increases the likelihood of cavitation (smaller cavitation number σ).

Pressure drop

Consider the two-dimensional flow of an inviscid fluid around a foil section shown in Figure 45.2. Two points, A and B, are marked along the same streamline. At point A the fluid has a velocity v_0 and a total static pressure p_0. According to Bernoulli's equation (14.25) for steady flow, the sum of total static pressure and dynamic pressure is constant.

$$p_0 + \frac{1}{2}\rho v_0^2 = p_1 + \frac{1}{2}\rho v_1^2 = \text{const.} \tag{45.3}$$

At point B on the back side of the foil, velocity v_1 will be larger than v_0, and, as a consequence, the pressure p_1 smaller than p_0.

The drop in pressure Δp between points A and B is given by

$$\Delta p = p_1 - p_0 = \frac{1}{2}\rho\left(v_0^2 - v_1^2\right) \tag{45.4}$$

In this case, pressure decreases from A and B, and the drop in pressure Δp is negative. Rewriting the left part of the Equation (45.4) above yields:

$$p_1 = p_o + \Delta p \tag{45.5}$$

Increasing the angle of attack α will lower the pressure p_1 further and $|\Delta p(\alpha_2)| > |\Delta p(\alpha_1)|$ when $\alpha_2 > \alpha_1$. It may happen that Δp becomes equal to $-p_0$, and the pressure at point B reaches zero.

$$p_1 = p_0 + \Delta p = p_0 - p_0 = 0 \qquad \text{for } \Delta p = -p_0 \tag{45.6}$$

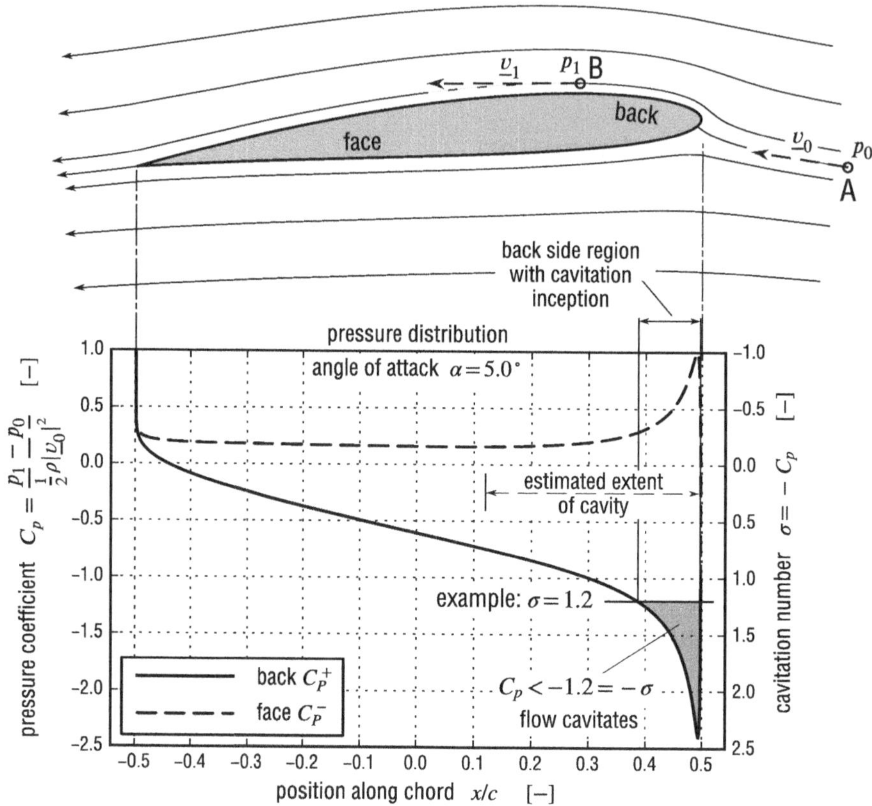

Figure 45.2 Flow around a cavitating foil section and the associated pressure distribution for back (upper) and face (lower) side

This is a theoretical limit. The pressure at point B cannot fall below zero as water does not support tension. The flow will break down with the formation of cavities. In practice this will already happen when the vapor pressure or saturation pressure $p_v > 0$ is reached:

$$p_1 = p_v = p_0 + \Delta p \qquad \text{for } \Delta p = -p_0 + p_v \tag{45.7}$$

or

$$\Delta p = p_v - p_0 = -(p_0 - p_v) \qquad \text{cavitation inception} \tag{45.8}$$

The vapor pressure p_v at which the fluid changes its phase is temperature dependent. Values have been provided earlier in Chapter 3.

Division by the dynamic pressure q (45.2) renders Equation (45.8) dimensionless:

$$\frac{\Delta p}{q} = \frac{-(p_0 - p_v)}{q} \qquad \text{cavitation inception} \tag{45.9}$$

The absolute value of the right-hand side is equal to the cavitation number σ. p_0 is the total static pressure, and $q = \frac{1}{2}\rho v_0^2$ represents the inflow.

Cavitation inception

The cavitation number σ characterizes the flow's resistance to cavitation. If the negative change of pressure becomes larger than the cavitation number, cavitation will occur.

$$-\frac{\Delta p}{q} > \sigma \qquad \text{cavitation inception} \tag{45.10}$$

This cavitation inception is indicated in Figure 45.2 for an example value of $\sigma = 1.2$. Cavitation bubbles will appear and grow in the region with

$$-C_p > \sigma = 1.2$$

The cavities will stretch downstream of this region but will begin to collapse due to the increased pressure outside of the cavity.

Free stream cavitation number

The ability of a flow to withstand cavitation depends on the combination of static pressure p_0 and flow velocity. High static pressure p_0 will lower the danger of cavitation inception. High velocity will increase the risk of cavitation. Both quantities may vary substantially within a given flow regime. A suitable combination of static pressure and velocity must be selected to define the cavitation number which is characteristic for the flow. Common choices are the free stream cavitation number and the propeller cavitation number. The free stream cavitation number σ_0 is formed with the uniform flow speed v_0 of the far field and the total static pressure at the depth h of the considered object.

$$\sigma_0 = \frac{p_A + \rho g h - p_v}{\frac{1}{2}\rho v_0^2} \tag{45.11}$$

Note that the atmospheric pressure p_A in Equation (45.11) is usually the larger part of the total static pressure.

Propeller cavitation number

Velocity at propeller blades are functions of speed of advance v_A and rate of revolution n. The latter creates a transverse flow speed which linearly increases with growing distance from the shaft center. Highest flow speeds are encountered at the blade tips. Since most of the thrust is created over a region near the radius $r/R = 0.7$, the 70% radius is often used to calculate a representative flow speed v_1.

$$v_1 = \sqrt{v_A^2 + (\pi\, n\, 0.7D)^2} \tag{45.12}$$

Submergence also varies considerably over the diameter of the propeller since its axis is more or less horizontal. Submergence will be smallest for a blade in 12 o'clock position, and cavitation always appears first at the blade in upright position. In most cases submergence of the propeller shaft h_0 forms the basis to compute the static pressure but other definitions have been used in the relevant literature. The propeller cavitation number σ_b is defined as

$$\sigma_b = \frac{p_A + \rho g h_0 - p_v}{\frac{1}{2}\rho\left(v_A^2 + (\pi\, n\, 0.7D)^2\right)} \tag{45.13}$$

We will come back to this in the next chapter to select propeller characteristics which minimize or prevent cavitation.

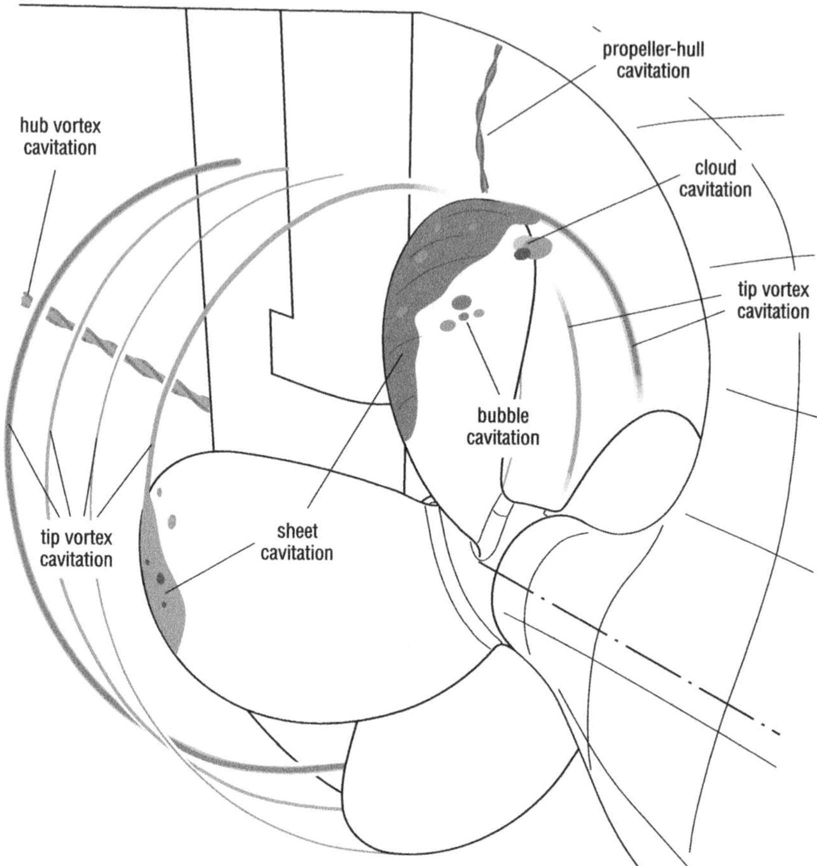

Figure 45.3 Locations and common types of cavitation

45.3 Locations and Types of Cavitation

Cavitation is identified by visual appearance and location. Figure 45.3 depicts common types of cavitation for a four bladed propeller.

Tip vortex and hub vortex cavitation

The highest flow velocities occur near the tip of the propeller blades and especially in the tip vortex. At first, only a portion of the tip vortex of the blade in upright position will cavitate because pressure has dropped to vapor pressure level. The cavitation will occur on each blade passing through the twelve o'clock position and vanish again until the next blade passes through. If flow speeds increase further, tip vortex cavitation will start at all blade tips and persist for sometimes several rotations until decreasing velocities cause the cavitation bubbles to collapse further downstream (see Figure 45.4). Vortices are also shed from the root of each blade. The root vortices combine into a single hub vortex. If the pressure in the hub vortex drops to vapor pressure, a continuous string of cavitation bubbles will appear behind the hub. Hub vortex cavitation is not visible in the photo shown in Figure 45.4. The propeller is mounted in open water condition with the propeller dynamometer downstream of the propeller which disrupts the hub vortex.

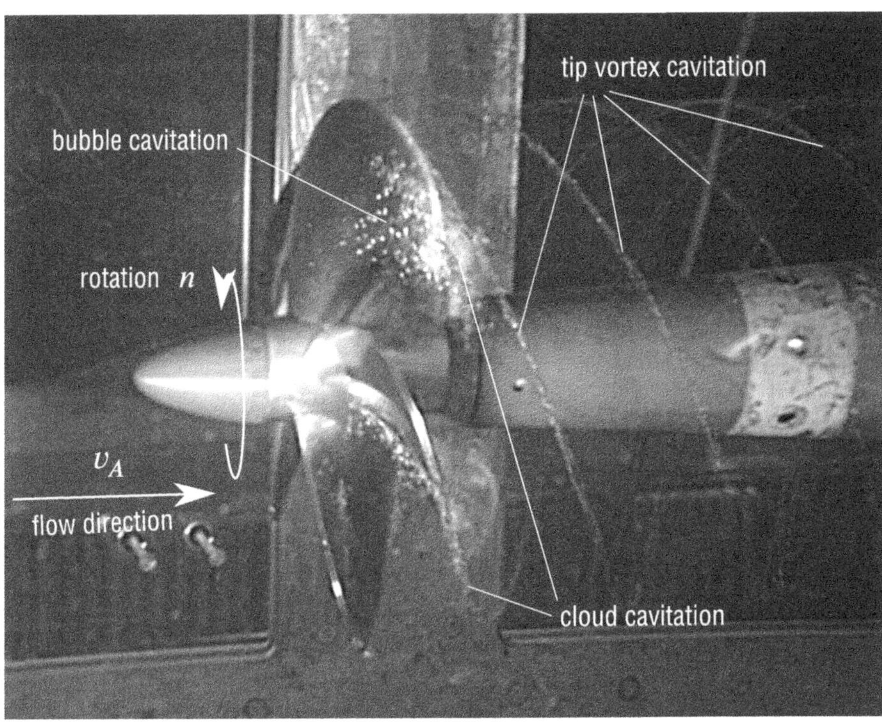

Figure 45.4 Open water test of five-bladed model propeller in a cavitation tunnel; tip vortex and bubble cavitation

Bubble and cloud cavitation

As the cavitation pattern develops, pressure will also drop on the back (suction) side of the blades and cavitation bubbles appear. At first, individual bubbles can be identified as they pop up and collapse pretty soon downstream. Figure 45.4 shows bubble cavitation on the back of the blade. Near the tip of the blades, groups of bubbles form clouds that are washed downstream. These intermittent congregations of bubbles are called cloud cavitation.

Sheet cavitation

Figure 45.3 shows sheet cavitation on the blades in twelve and three o'clock position. Sheet cavitation develops when the pressure drops below vapor pressure over a whole stretch of the leading edge of a blade. It starts near the tip and grows as flow speeds increase and pressure decreases toward the hub and across the blade chord. Due to the smaller static pressure, sheet cavitation extends over a larger area on the blade in 12 o'clock position compared to other blades. Shown is an approximately 10% coverage of the blade surface. This is arguably the worst condition still acceptable for continuous operations. If cavitation is unavoidable, naval architects aim to restrict sheet cavitation patterns to less than 5% as shown on the blade in three o'clock position.

Propeller–hull cavitation

Vessels with relatively large propellers, which leave only a minimum of clearance between blade tip and hull, sometimes experience propeller–hull cavitation. A vortex forms briefly between blade tip and adjacent hull surfaces. If the pressure is low enough, the vortex cavitates like the tip vortices. This highly unsteady process leads to increased noise and vibrations.

Figure 45.5 Open water test of five-bladed model propeller in a cavitation tunnel; propeller blades completely covered in sheet cavitation

Face cavitation Cavitation is mostly seen on the suction side of the blade. However, face cavitation may occur, if the angle of attack of the blade section becomes negative. This often happens when the propeller rate of revolution is winding down but the ship is still going forward close to its original speed. The same effect may occur on controllable pitch propellers when they are operated in off-design conditions.

Fully cavitating propellers Cavitation is a problem for all propellers with high loading (high thrust and small diameter) or high velocities (high ship speed and /or high rate of revolution). As a rule of thumb: if your vessel goes faster than 35 knots you will not be able to avoid cavitation. Figure 45.5 shows a propeller which is fully cavitating. All blades are completely engulfed in sheet cavitation. Serious degradation of performance is to be expected. Propellers designed for this operating condition have blade sections with a sharp leading edge and a blunt trailing edge. This arrangement creates a stable sheet cavitity over the whole chord which collapses downstream of the foil and thus avoids blade erosion.

45.4 Detrimental Effects of Cavitation

Detrimental effects Cavitation has several negative effects. Lifting foils, pumps, and propellers will experience:

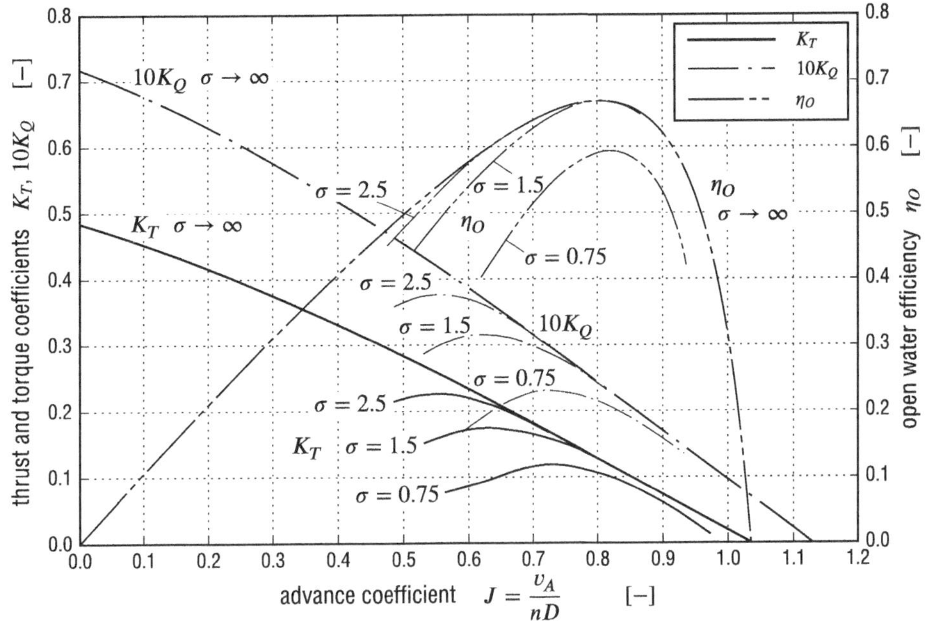

Figure 45.6 Loss of thrust and efficiency due to cavitation

- loss of lift force, thrust, and efficiency
- vibrations and noise
- destruction of blades due to erosion

Loss of thrust and efficiency

In the open water diagram of the propeller effects of cavitation become first visible at lower advance ratios where the propeller loading is high. A distinct loss of thrust is observed. Although torque is decreasing as well, thrust drops faster and results in a loss of propeller efficiency (Figure 45.6). Initially this does not affect the peak efficiency of the propeller. However, with decreasing cavitation number, effects of cavitation spread over larger portions of the propeller blade and affect the working conditions at higher advance ratios J. Eventually loss of thrust in the whole operating range will cause a significant drop in peak propeller efficiency.

Noise

Cavitation can actually be heard before it is visible. Imagine the sound of pebbles hitting a hard surface. As cavitation intensifies, a persistent crackling noise can be heard. Noise itself is of concern for naval vessels. Skilled sonar operators may use the cavitation noise to identify classes of vessels. Today, naval architects should also consider the environmental impact of noise especially its effects on marine mammals.

Vibrations

The highly unsteady nature of cavitation with constant production and collapses of cavities creates high pressure shock waves which are capable of exciting propeller blades, rudder, and nearby hull plating to measurable vibrations. This adds to the signature of a vessel, creates discomfort for crew and passengers, and may lead to fatigue damage in the long term.

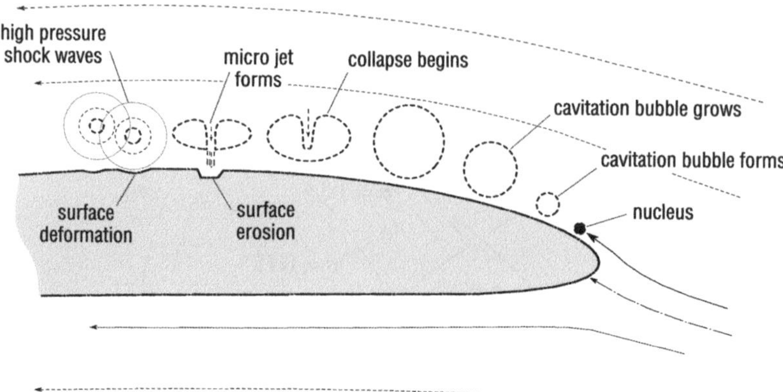

Figure 45.7 Life of cavitation bubble

Erosion The mechanisms leading to erosion of fairly strong materials are still subject of ongoing research. Damage occurs when cavitation bubbles collapse, not when they are created. Two mechanisms have been identified:

- micro-jets and
- high pressure shock waves

Figure 45.7 pictures the short life of a cavitation bubble. It starts with a nucleus moving into an area of pressure equal to the vapor pressure. Air dissolved in the fluid or small solid particles may serve as nuclei. Water surrounding the nucleus evaporates and a cavitation bubble is formed. The bubble is filled with vapor at saturation pressure and grows as long as the pressure stays low. Once the bubble moves into an area with increasing pressure it collapses. Slow motion pictures revealed that the bubbles fold inward which creates a high pressure micro water jet. The water jet is strong enough to break material out of propeller blades and foils. Very hard and polished materials are less susceptible to damage but persistent cavitation will eventually destroy all materials typically used in shipbuilding. The collapsing bubbles also emit high pressure shock waves with short term pressure peaks of several thousand atmospheres. The pressure waves produce the crackling noise and may cause elastic and plastic deformations on propeller blades and hull plating.

Cavitation is still a field of ongoing research. Numerous conference proceedings and a couple of textbooks are available for further study. The book by Franc and Michel (2004) may serve as a starting point for a more scientific discussion of cavitation and the dynamics of cavitation bubbles. In the next chapter, we discuss simple design measures to prevent cavitation under normal operating conditions.

References

Franc, J.-P. and Michel, J.-M. (2004). *Fundamentals of cavitation.* Fluid Mechanics and Its Applications (Book 76). Kluwer Academic Publishers, Dordrecht, The Netherlands. ISBN 978-1402022326.

Self Study Problems

1. Explain in your own words the difference between boiling and cavitation.
2. Discuss the effect of submergence and flow speed on the ability of a flow to withstand cavitation.
3. Which areas of a hydrofoil are likely to cavitate?
4. Summarize the negative effects of cavitation.
5. A propeller is operating under the following conditions.

Propeller data	
propeller diameter	$D = 3.210$ m
rate of revolution	$n = 160.00$ rpm
speed of advance	$v_A = 9.120$ m/s
shaft submergence	$h_0 = 2.900$ m
standard air pressure	$p_A = 1013.250$ hPa
gravitational acceleration	$g = 9.807$ m/s^2
salt water temperature	$t_W = 15.000$ °C

Compute the propeller cavitation number σ_b according to Equation (45.13).

46

Cavitation Prevention

Cavitation has a number of negative effects ranging from noise over performance loss to destruction of the foil or propeller. It is the naval architect's task to prevent or at least limit the negative impacts of cavitation. This chapter introduces the basic design measures necessary to reduce the risk of cavitation. Specifically, the determination of a minimum required area ratio for the propeller is explained using Keller's formula and Burrill's chart. It also discusses the alternatives of fully cavitating and surface piercing propellers in cases where cavitation cannot be avoided.

Learning Objectives

At the end of this chapter students will be able to:

- understand which design measures reduce the risk of cavitation
- compute minimum required expanded area ratios for propellers

46.1 Design Measures

High static pressure

Cavitation occurs when the pressure in a fluid drops below vapor pressure. In ship hydrodynamics this predominantly happens on lifting foils like a propeller. High static pressure and low flow speeds reduce the risk of cavitation as the cavitation number σ introduced in the previous chapter makes clear.

$$\sigma_0 = \frac{p_0 - p_v}{\frac{1}{2}\rho v_0^2} \tag{45.11}$$

High static pressure p_0 may only be achieved by deep submergence of the foil or propeller. This finds its limits in draft restrictions and geometry of the stern in general.

Low speed

Lowering the flow speed v_0 on a foil is possible by lowering the ship speed, reducing the propeller rate of revolution, or selecting foil sections which create a minimum in additional flow speed. The latter requires thin foils in combination with the lowest camber possible to generate the desired lift. We aim to operate foils at their ideal angle of attack, which is often difficult because in early design phases the actual local flow

Fundamentals of Ship Hydrodynamics: Fluid Mechanics, Ship Resistance and Propulsion, First Edition.
Lothar Birk.

Companion website: www.wiley.com/go/birk/hydrodynamics

direction may only be estimated. Paint tests and CFD simulations may help to finalize alignment of struts, bilge keels, and control surfaces.

Local flow speed for propellers is a function of speed of advance v_A, rate of revolution n, and distance r from the shaft center. Speed of advance follows from ship speed and wake fraction. These are typically set when the propeller selection process starts. Consequently, one might lower the rate of revolution n in order to reduce the risk of cavitation. As a rule of thumb, lowering the rate of revolution leads to propellers with higher efficiency. However, smaller rate of revolution needs a bigger propeller and/or increased pitch to generate the same thrust which may counter the effects of lowering n on the risk of cavitation.

Required expanded area ratio

For a given diameter and rate of revolution, a reduction in the risk of cavitation is possible by increasing the propeller blade area. The increased blade area allows the use of foil sections with lower lift coefficients which leads to smaller pressure drops on the suction side of the blade. The necessary blade area to avoid excessive cavitation is specified in the form of a minimum required expanded area ratio $(A_E/A_0)_{\text{req}}$. Care is required not to raise the expanded area ratio too much as this will increase the viscous losses of the propeller and lower its efficiency. In the following sections, two basic methods are introduced to compute the required expanded area ratio of the propeller in early design stages.

46.2 Keller's Formula

A very basic formula for the required expanded area ratio is attributed to auf'm Keller (1966). J. auf'm Keller worked for the model basin in Wageningen, The Netherlands. The original publication is written in Dutch but the formula has been reproduced in many reference books like Carlton (2007).

$$\left(\frac{A_E}{A_0}\right)_{\text{req}} = \frac{(1.3 + 0.3Z)\,T}{(p_0 - p_v)\,D^2} + K \tag{46.1}$$

$$\text{with constant} \quad \begin{aligned} K &= 0.2 && \text{for single screw vessels} \\ K &= 0.1 && \text{for slow twin screw vessels} \\ K &= 0.0 && \text{for fast twin screw vessels} \end{aligned}$$

As the input to Keller's formula serve the number of blades Z, thrust T in service condition for design speed, an estimate for the propeller diameter D, the static pressure head at the propeller shaft p_0, and the vapor pressure p_v. Notable is the absence of any reference to flow velocity. Therefore, we can apply Keller's formula without knowing the rate of revolution or the speed of advance. However, this also limits the accuracy of the minimum expanded area ratio estimate. Keller's formula is suitable to set a starting point, but subsequent design iterations should be checked against Burrill's cavitation chart or more sophisticated assessments.

Initial diameter estimate

Books on ship design feature regression formulas for typical propeller diameters based on ship type and principal hull form parameters (Watson, 1998; Papanikolaou, 2014).

Table 46.1 Regression equations for the limiting lines in the Burrill chart (Figure 46.1)

limiting criteria	regression equation
based on model tests (Burrill and Emerson, 1963)	
30% back cavitation	$\tau_c = 0.910\,\sigma_b^{0.335} - 0.346$
20% back cavitation	$\tau_c = 1.147\,\sigma_b^{0.195} - 0.672$
naval or fast vessels: 10% back cavitation	$\tau_c = 1.267\,\sigma_b^{0.126} - 0.912$
merchant vessels: 5% back cavitation	$\tau_c = 0.715\,\sigma_b^{0.184} - 0.437$
2.5% back cavitation	$\tau_c = 0.611\,\sigma_b^{0.189} - 0.372$
based on experience (Burrill, 1943)	
suggested lower limit for tugs, trawlers, etc.	$\tau_c = 0.527\,\sigma_b^{0.155} - 0.324$
suggested upper limit for merchant vessels	$\tau_c = 0.304\,\sigma_b^{0.497} - 0.034$
suggested upper limit for naval vessels	$\tau_c = 0.464\,\sigma_b^{0.479} - 0.089$

The following formulas are from Kristensen and Lützen (2012):

$$
\begin{aligned}
D &= 0.623\,T_{\max} - 0.16\,\mathrm{m} && \text{for container ships}\\
D &= 0.395\,T_{\max} + 1.3\,\mathrm{m} && \text{for bulk carriers and tankers}\\
D &= 0.713\,T_{\max} - 0.08\,\mathrm{m} && \text{for twin screw Ro-Ro ships}
\end{aligned}
\tag{46.2}
$$

$T_{\max}$ is the maximum draft in meters. If it is not known, simply use the design draft.

46.3 Burrill's Cavitation Chart

Burrill Chart Lennard Constantine Burrill (*1905 – †1965), a British professor of naval architecture at Durham University, UK, conducted cavitation experiments and recorded full scale cavitation observations over many years. The data were converted into a design chart which is known as the Burrill cavitation chart (Burrill, 1943; Burrill and Emerson, 1963).

Figure 46.1 shows a chart similar to the original. However, the cavitation limit curves have been redrawn based on regression formulas. Table 46.1 summarizes the equations. An effort has been made to avoid overpredicting the values derived from the original chart. Regression formulas better fit the modern computerized work flow. Nevertheless, design charts provide an easy way to check computational results.

Local cavitation number σ_b The horizontal axis of the chart is the propeller cavitation number σ_b.

$$\sigma_b = \frac{p_0 - p_v}{q_{0.7}} \tag{46.3}$$

It is a measure for the risk of cavitation in the flow. Low σ_b values indicate high cavitation risks. The propeller cavitation number depends on static pressure head at the propeller

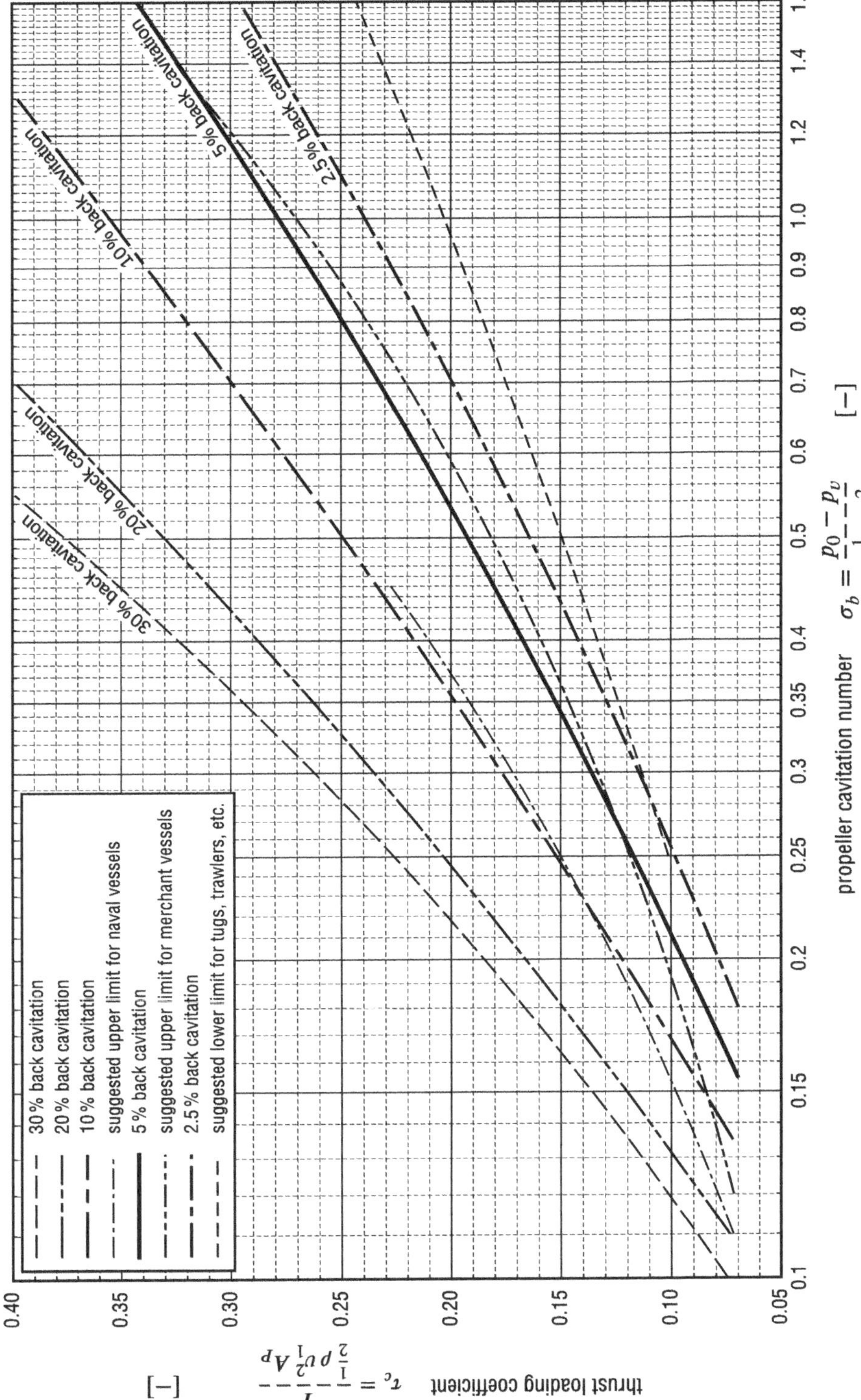

Figure 46.1 Limits for the propeller loading coefficient as a function of cavitation number and acceptable level of cavitation. After Burrill and Emerson (1963), however, the curves represent the regression equations from Table 46.1

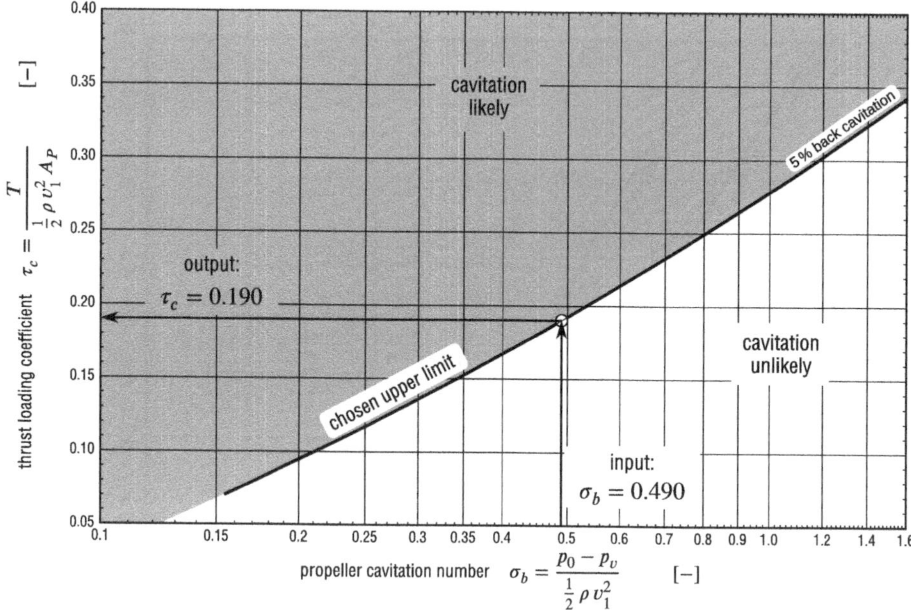

Figure 46.2 Usage of the Burrill chart

shaft p_0, vapor pressure p_v, and dynamic pressure $q_{0.7}$ at blade radius $r = 0.7R$.

$$q_{0.7} = \frac{1}{2}\rho v_1^2 \tag{46.4}$$

The reference velocity v_1 is a combination of axial speed of advance and the transverse velocity at radius $r = 0.7\,R$ which results from the rotation of the propeller.

$$v_1 = \sqrt{v_A^2 + (\pi\, n\, 0.7 D)^2} \tag{45.12}$$

n is the propeller rate of revolution.

Thrust loading coefficient τ_c

The vertical axis of the graph measures a thrust loading coefficient τ_c which is based on the projected area of a propeller A_P.

$$\tau_c = \frac{T_{\text{prop}}}{\frac{1}{2}\rho v_1^2 A_P} \tag{46.5}$$

T_{prop} is the propeller thrust at speed of advance v_A. The thrust loading coefficient represents the working conditions of the propeller.

The Burrill chart shows limiting curves as functions $\tau_c = \tau_c(\sigma_b)$. The percentages indicate how much area of a propeller blade back side is covered in cavitation. A propeller can be expected to be cavitation free or to have cavitation up to the stated extent if the propeller operates at a $(\sigma_b,\ \tau_c)$ combination which falls on or below the limiting curve.

Figure 46.2 explains the use of Burrill's cavitation chart based on the frequently used 5%-cavitation limit. First, the local cavitation number σ_b is computed.

$$\sigma_b = \frac{p_A + \rho g h_0 - p_v}{\frac{1}{2}\rho\left(v_A^2 + (\pi\, n\, 0.7D)^2\right)} \tag{45.13}$$

An example value of $\sigma_b = 0.49$ is marked in Figure 46.2.

For the value of σ_b, the thrust loading coefficient τ_c is read off the limiting curve or computed from its respective regression formula. The 5% back cavitation limit is defined as

$$\tau_c = 0.715\,\sigma_b^{0.184} - 0.437 \tag{46.6}$$

For $\sigma_b = 0.49$ we obtain a value of $\tau_c = 0.19$.

The thrust loading coefficient is connected to the minimum required projected area A_P of the propeller via its definition (46.5). **Projected area**

$$A_P = \frac{T_{\text{prop}}}{\frac{1}{2}\rho\, v_1^2\, \tau_c} \tag{46.7}$$

Using a formula attributed to Admiral David W. Taylor, projected area A_P may be converted into developed area A_D.

$$A_D = \frac{A_P}{\left(1.067 - 0.229\,\frac{P}{D}\right)} \tag{46.8}$$

For propellers with zero or moderate rake, expanded blade area is equal to the developed area.

$$A_E \approx A_D \tag{46.9}$$

Substituting these two results into Equation (46.7) and dividing both sides by the propeller disk area $A_0 = \pi D^2/4$ yields an equation for the required expanded area ratio.

$$\left(\frac{A_E}{A_0}\right)_{\text{req}} = \frac{T_{\text{prop}}}{\frac{1}{2}\rho\, v_1^2\, \tau_c\left(1.067 - 0.229\,\frac{P}{D}\right)\frac{\pi D^2}{4}} \tag{46.10}$$

The propeller will cavitate more than the limit only if the actual expanded area ratio is smaller than $(A_E/A_0)_{\text{req}}$. We will use Burrill's cavitation criteria in the context of propeller design examples (see Chapter 49).

46.4 Other Design Measures

As mentioned above, cavitation can rarely be avoided for vessels sailing faster than 35 knots. If excessive cavitation is expected, the following three alternatives to a conventional propeller should be considered.

- To avoid damage to the propeller and minimize effects on efficiency, special wedge shaped foil sections should be used. They have a sharp leading edge which creates a stable sheet cavitation bubble. The bubbles collapse way downstream of the propeller and thus do not cause erosion of the blades.

- If the machinery arrangements allow a surface piercing propeller may be considered. In this case a part of the propeller disk stays above water. This causes ventilation of the propeller blades. A sheet cavity will develop over the blade but in contrast to sheet cavitation the bubbles are filled with air at atmospheric pressure rather than vapor pressure. As a result damaging micro jets and pressure waves are avoided.

- In some cases water jet propulsion may be a viable solution to extend the speed range of the vessel without experiencing excessive cavitation.

References

auf'm Keller, J. (1966). Enige aspecten bij het ontwerpen van scheepsschroeven. *Schip en Werf*, 33(24):658–663. In Dutch.

Burrill, L. (1943). Developments in propeller design and manufacture for merchant ships. *Transactions of the Institute of Marine Engineers*, 55:148–169.

Burrill, L. and Emerson, A. (1963). Propeller cavitation: further tests on 16in propeller models in the King's College cavitation tunnel. *Transactions of the North East Coast Institution of Engineers and Shipbuilders (NECIES)*, 79:295–320.

Carlton, J. (2007). *Marine propellers and propulsion*. Butterworth-Heinemann, Oxford, United Kingdom.

Kristensen, H. and Lützen, M. (2012). Prediction of resistance and propulsion power of ships. Technical Report Project No. 2010-56, Work Package 2, Report No. 04, University of Southern Denmark and Technical University of Denmark.

Papanikolaou, A. (2014). *Ship design – Methodologies of preliminary design*. Springer, Dordrecht, The Netherlands.

Watson, D. (1998). *Practical ship design*. Elsevier Ocean Engineering Book Series. Elsevier Science, Oxford, United Kingdom.

Self Study Problems

1. Summarize in your own words design measures to decrease the danger of propeller cavitation.

2. A propeller has to be checked for cavitation. It is designed to generate a thrust of $T = 1250.0\,\text{kN}$. Should the expanded area ratio stated in the table below be increased to prevent exceedance of the cavitation limit given by the 5% back cavitation criteria (46.6) or should it be decreased to maximize efficiency? Show all steps to reason your answer.

Propeller data	
number of blades	$Z = 5$
diameter	$D = 6.000$ m
pitch–diameter ratio	$P/D = 0.900$
expanded are ratio	$A_E/A_0 = 0.75$
propeller thrust	$T_{prop} = 1250.0$ kN
speed of advance	$v_A = 9.00$ m/s
rate of revolution	$n = 116.0$ rpm
air pressure	$p_A = 101325$ Pa
vapor pressure	$p_v = 1671$ Pa
submergence of propeller shaft	$h_0 = 6.5$ m
water density	$\rho = 1026.021$ kg/m^3
gravitational acceleration	$g = 9.807$ m/s^2

3. A propeller is operating under the following conditions:

Propeller data	
propeller thrust (service)	$T = 316.000$ kN
propeller diameter	$D = 3.210$ m
pitch–diameter ratio	$P/D = 0.910$
rate of revolution	$n = 160.000$ rpm
speed of advance	$v_A = 9.100$ m/s
shaft submergence	$h_0 = 2.900$ m
standard air pressure	$p_A = 1013.250$ hPa
gravitational acceleration	$g = 9.807$ m/s^2
salt water temperature	$t_W = 15.000$ °C

Compute the propeller cavitation number σ_b according to Equation (45.13) and derive the required expanded area ratio to avoid more than 5% back cavitation.

4. How does the required expanded area ratio in the previous problem change if you employ the 2.5% or the 10% back cavitation limit?

47

Propeller Series Data

The propeller is an essential piece in most ship propulsion systems. Its performance characteristics have to be estimated early in the design because powering requirements of the ship and the resulting engine selection have repercussions on the ship's general layout. Naval architects aim to select a propeller which combines hull and propeller into a highly efficient propulsion system. Propeller series data is a very useful asset in this process. A propeller series is a set of propellers in which key propeller characteristics have been systematically varied. Many different propeller series have been developed over the years, and data has been published to varying degrees. In this chapter we concentrate on the Wageningen B-Series (Oosterveld and van Oossanen, 1975; Kuiper, 1992). The series covers a wide range of principal propeller characteristics. Its data is available in the form of charts and regression polynomials. References for other propeller series are presented at the end of the chapter.

Learning Objectives

At the end of this chapter students will be able to:

- understand the definition of systematic propeller series
- compute propeller characteristics from regression polynomials
- know about important propeller series

47.1 Wageningen B-Series

Origin The Wageningen B-Series was developed in the middle of the 20th century and is the most extensive propeller series (Troost, 1938, 1940, 1951). Its name refers to the town Wageningen, which is the home of the Netherlands Ship Model Basin (NSMB). NSMB is now called MARIN (Maritime Research Institute Netherlands). In older books the series might be referred to as Troost series. The series encompasses a total of 120 propellers with systematic variations of number of blades Z, expanded area ratio A_E/A_0, and pitch–diameter ratio P/D. The Wageningen B-Series propeller had been among the first to use aerofoil shaped section profiles. Circular segment (ogival) sections have only been used from radius $x = 0.6$ to the blade tip. Figure 35.9 shows

Fundamentals of Ship Hydrodynamics: Fluid Mechanics, Ship Resistance and Propulsion, First Edition.
Lothar Birk.

Companion website: www.wiley.com/go/birk/hydrodynamics

Table 47.1 Basic characteristics of the propellers in the Wageningen B-Series. For each combination of Z and A_E/A_0 propellers with $P/D = 0.5, 0.6, 0.8, 1.0, 1.2$, and 1.4 have been tested

Z	expanded area ratios A_E/A_0													
2	0.30													
3		0.35			0.50			0.65			0.80			
4			0.40			0.55			0.70			0.85	1.00	
5				0.45			0.60			0.75				1.05
6					0.50			0.65			0.80			
7						0.55			0.70			0.85		

the blade sections within the expanded view. The propellers feature a 15 degree rake which is rather high by modern standards.

Data range

There are twenty different combinations of Z and A_E/A_0, and for each of the combinations six pitch–diameter ratios have been tested. Table 47.1 summarizes the 20 basic combinations. Over the years propellers have been added to the series and several adjustments have been made. For instance, the first set of propellers with $Z = 4$ blades featured a reduction of the pitch by 20% from radius $x = 0.475$ to the hub. This feature was removed for propellers with less or more than four blades. The original blade contour was widened later to improve cavitation characteristics. This shape was called the BB-Series and equations presented here refer to this form (Kuiper, 1992). The hydrodynamic characteristics of the Wageningen B-Series tests have been reported in graphical form as open water diagrams and as regression formulas. The set of data was extended with open water tests under cavitation conditions.

47.2 Wageningen B-Series Polynomials

Regression analysis

Oosterveld and van Oossanen (1975) present a detailed analysis of the available open water test results for the Wageningen B-Series. Coefficients are provided to retrieve blade and foil section shapes and to compute the thrust and torque coefficients as a function of the basic propeller characteristics diameter D, number of blades Z, expanded area ratio A_E/A_0, pitch–diameter ratio P/D, and the advance ratio J.

For our studies in ship propulsion, we need the open water thrust and torque curves. Oosterveld and van Oossanen (1975) provide regression formulas for thrust and torque coefficients:

$$K_T = \sum_{i=0}^{38} a_i \, J^{b_i} \left(\frac{P}{D}\right)^{c_i} \left(\frac{A_E}{A_0}\right)^{d_i} Z^{e_i} \tag{47.1}$$

$$K_Q = \sum_{i=0}^{46} a_i \, J^{b_i} \left(\frac{P}{D}\right)^{c_i} \left(\frac{A_E}{A_0}\right)^{d_i} Z^{e_i} \tag{47.2}$$

The necessary coefficients a_i and exponents b_i, c_i, d_i, and e_i for the thrust coefficient K_T are listed in Table 47.2. Coefficients and exponents for Equation (47.2) can be found in Table 47.3.

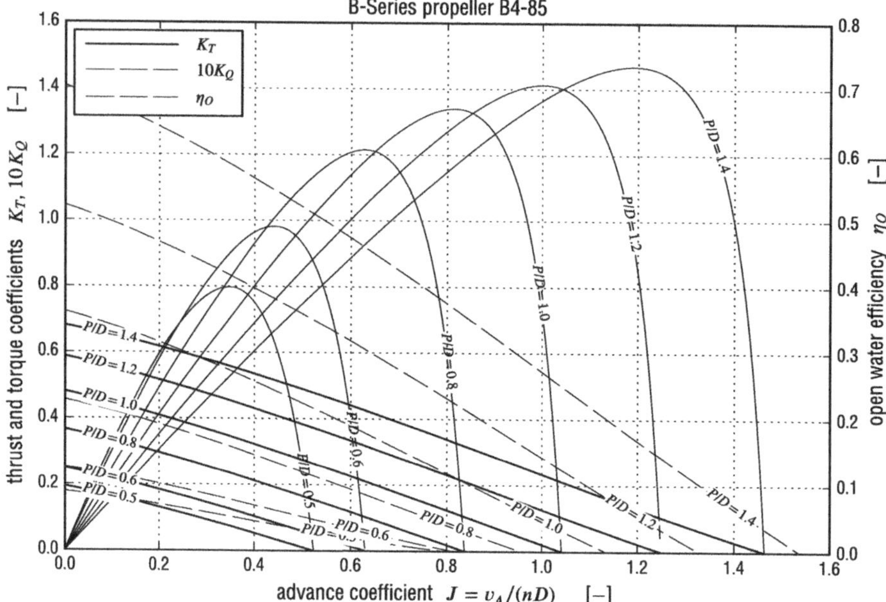

Figure 47.1 Open water chart for a Wageningen B-Series propeller with $Z = 4$ and $A_E/A_0 = 0.85$ derived from K_T and K_Q polynomials Equations (47.1) and (47.2)

The open water efficiency η_O follows from Equation (41.15): $\eta_O = (K_T/K_Q)J/(2\pi)$. Figure 47.1 shows an open water chart based on the polynomials. Note that the open water efficiency has its own scale on the right-hand side of the plot. The key 'B4-85' used in the title identifies the propeller:

- B obviously stands for Wageningen B-Series.
- The number 4 indicates the number of blades $Z = 4$.
- The value after the hyphen (or sometimes a dot) represents the expanded area ratio multiplied by 100, i.e. in Figure 47.1 it is $A_E/A_0 = 0.85$.

For example, B6.105 or B6-105 stand for a Wageningen B-Series propeller with $Z = 6$ blades and an expanded area ratio of $A_E/A_0 = 105/100 = 1.05$.

Studying thrust, torque, and open water efficiency curves in Figure 47.1, it is obvious that thrust and torque increase with increasing pitch–diameter ratio P/D. At the same advance coefficient J, a propeller with higher P/D produces more thrust but also requires more torque, i.e. engine power. The picture is not quite as simple for the open water efficiency η_O. The maximum achievable open water efficiency also grows with the pitch–diameter ratio, at least for the range of P/D-values shown. However, for a given advance coefficient J, a higher pitch–diameter ratio might not necessarily provide a better open water efficiency. Compare the open water efficiency curves in Figure 47.1 at $J = 0.6$! The η_O-curve for $P/D = 0.8$ has the largest open water efficiency whereas the efficiencies for $P/D = 1.0$, 1.2, and 1.4 are smaller. Thus a propeller with high

Table 47.2 Factors and exponents for thrust coefficient polynomials of Wageningen B-Series propellers (Oosterveld and van Oossanen, 1975)

Open Water Thrust Coefficient K_T											
i	a_i	b_i	c_i	d_i	e_i	i	a_i	b_i	c_i	d_i	e_i
0	+0.00880496	0	0	0	0	20	+0.010465	1	6	2	0
1	−0.204554	1	0	0	0	21	−0.00648272	2	6	2	0
2	+0.166351	0	1	0	0	22	−0.00841728	0	3	0	1
3	+0.158114	0	2	0	0	23	+0.0168424	1	3	0	1
4	−0.147581	2	0	1	0	24	−0.00102296	3	3	0	1
5	−0.481497	1	1	1	0	25	−0.0317791	0	3	1	1
6	+0.415437	0	2	1	0	26	+0.018604	1	0	2	1
7	+0.0144043	0	0	0	1	27	−0.00410798	0	2	2	1
8	−0.0530054	2	0	0	1	28	−0.000606848	0	0	0	2
9	+0.0143481	0	1	0	1	29	−0.0049819	1	0	0	2
10	+0.0606826	1	1	0	1	30	+0.0025983	2	0	0	2
11	−0.0125894	0	0	1	1	31	−0.000560528	3	0	0	2
12	+0.0109689	1	0	1	1	32	−0.00163652	1	2	0	2
13	−0.133698	0	3	0	0	33	−0.000328787	1	6	0	2
14	+0.00638407	0	6	0	0	34	+0.000116502	2	6	0	2
15	−0.00132718	2	6	0	0	35	+0.000690904	0	0	1	2
16	+0.168496	3	0	1	0	36	+0.00421749	0	3	1	2
17	−0.0507214	0	0	2	0	37	+0.0000565229	3	6	1	2
18	+0.0854559	2	0	2	0	38	−0.00146564	0	3	2	2
19	−0.0504475	3	0	2	0						

pitch–diameter ratio will be efficient only if it operates at a high advance coefficient. We will learn how to find the best combination of advance coefficient and pitch–diameter ratio in the following chapter.

Viscous scale effects

All model propellers in the series have a diameter of $D = 240$ mm and some scale effects due to the model scale Reynolds numbers are expected. The polynomials (47.1) and (47.2) have been normalized to a Reynolds number $Re_{0.75}$ of $2 \cdot 10^6$. The Reynolds number $Re_{0.75}$ is based on the chord length at radius $x = 0.75$. As resultant flow speed the combined magnitude of axial and rotational speed is used and induced velocities are ignored.

$$Re_{0.75} = \frac{c_{0.75}\sqrt{v_A^2 + (0.75\,\pi\,n\,D)^2}}{\nu} \tag{47.3}$$

The chord length $c_{0.75}$ at radius $x = 0.75$ may be estimated with Equation (47.4).

$$c(x) = C_r(x)\,\frac{D}{Z}\left(\frac{A_E}{A_0}\right) \tag{47.4}$$

Values for the coefficient C_r are listed in Table 47.4.

Full scale thrust and torque

The results from thrust and torque coefficient polynomials (47.1) and (47.2) should be corrected for the full scale Reynolds number. The correction may be computed via

Table 47.3 Factors and exponents for torque coefficient polynomials of Wageningen B-Series propellers (Oosterveld and van Oossanen, 1975)

Open Water Torque Coefficient K_Q													
i	a_i	b_i	c_i	d_i	e_i	i	a_i	b_i	c_i	d_i	e_i
0	+0.00379368	0	0	0	0	24	−0.0397722	0	3	2	0
1	+0.00886523	2	0	0	0	25	−0.00350024	0	6	2	0
2	−0.032241	1	1	0	0	26	−0.0106854	3	0	0	1
3	+0.00344778	0	2	0	0	27	+0.00110903	3	3	0	1
4	−0.0408811	0	1	1	0	28	−0.000313912	0	6	0	1
5	−0.108009	1	1	1	0	29	+0.0035985	3	0	1	1
6	−0.0885381	2	1	1	0	30	−0.00142121	0	6	1	1
7	+0.188561	0	2	1	0	31	−0.00383637	1	0	2	1
8	−0.00370871	1	0	0	1	32	+0.0126803	0	2	2	1
9	+0.00513696	0	1	0	1	33	−0.00318278	2	3	2	1
10	+0.0209449	1	1	0	1	34	+0.00334268	0	6	2	1
11	+0.00474319	2	1	0	1	35	−0.00183491	1	1	0	2
12	−0.00723408	2	0	1	1	36	+0.000112451	3	2	0	2
13	+0.00438388	1	1	1	1	37	−0.0000297228	3	6	0	2
14	−0.0269403	0	2	1	1	38	+0.000269551	1	0	1	2
15	+0.0558082	3	0	1	0	39	+0.00083265	2	0	1	2
16	+0.0161886	0	3	1	0	40	+0.00155334	0	2	1	2
17	+0.00318086	1	3	1	0	41	+0.000302683	0	6	1	2
18	+0.015896	0	0	2	0	42	−0.0001843	0	0	2	2
19	+0.0471729	1	0	2	0	43	−0.000425399	0	3	2	2
20	+0.0196283	3	0	2	0	44	+0.0000869243	3	3	2	2
21	−0.0502782	0	1	2	0	45	−0.0004659	0	6	2	2
22	−0.030055	3	1	2	0	46	+0.0000554194	1	6	2	2
23	+0.0417122	2	2	2	0						

Table 47.4 Coefficients for the estimate of maximum thickness and chord length of Wageningen B-Series propellers (Kuiper, 1992). For convenience values have been added at radii $x = 0.15, 0.25$, and 0.75 by interpolation

$x=r/R$	A_r	B_r	C_r
0.15	0.0588	0.00425	1.473
0.20	0.0526	0.0040	1.600
0.25	0.0495	0.00375	1.719
0.30	0.0464	0.0035	1.832
0.40	0.0402	0.0030	2.023
0.50	0.0340	0.0025	2.163
0.60	0.0278	0.0020	2.243
0.70	0.0216	0.0015	2.247
0.75	0.0185	0.00125	2.208
0.80	0.0154	0.0010	2.132
0.85	0.0123	0.00075	2.005
0.90	0.0092	0.0005	1.798
0.95	0.0061	0.00025	1.434
0.975	0.00455	0.000125	1.122
1.0	0.003	0.000	0.000

the ITTC procedure described in Chapter 42 (ITTC, 2014) or the following regression polynomials (Oosterveld and van Oossanen, 1975).

$$\Delta K_T = \sum_{i=0}^{8} a_i\, J^{b_i} \left(\frac{P}{D}\right)^{c_i} \left(\frac{A_E}{A_0}\right)^{d_i} Z^{e_i}\, \log_{10}\left(Re_{0.75} - 0.301\right)^{f_i} \tag{47.5}$$

$$\Delta K_Q = \sum_{i=0}^{12} a_i\, J^{b_i} \left(\frac{P}{D}\right)^{c_i} \left(\frac{A_E}{A_0}\right)^{d_i} Z^{e_i}\, \log_{10}\left(Re_{0.75} - 0.301\right)^{f_i} \tag{47.6}$$

Coefficients for the Reynolds number corrections to thrust and torque coefficients are provided in Tables 47.5(a) and 47.5(b) respectively.

Thickness to chord length ratio

The ITTC procedure also requires the thickness to chord length ratio t_{max}/c. The maximum thickness of the standard form is given by Equation (47.7).

$$t_{max}(x) = D\big(A_r(x) - B_r(x)Z\big) \tag{47.7}$$

Note that structural strength may require thicker blade sections in some cases. Values for the coefficients A_r and B_r are also found in Table 47.4. The original data from Kuiper (1992) do not contain values for the radii 0.15 (already within hub), 0.25, and 0.75. These values have been found by interpolation through the given data points.

In contrast to the equivalent full scale correction in the ITTC method (see Equations (42.11) and (42.12)), ΔK_T (47.5) and ΔK_Q (47.6) are added to the values determined with Equations (47.1) and (47.2).

$$K_{TS} = K_T + \Delta K_T \qquad K_{QS} = K_Q + \Delta K_Q \tag{47.8}$$

The correction ΔK_Q should be negative and ΔK_T should be positive for Reynolds numbers $Re_{0.75} > 2 \cdot 10^6$.

47.3 Other Propeller Series

This section briefly summarizes other propeller series with references for further study. An extensive summary of different propeller series is given in Carlton (2007).

Gawn series

The Gawn series is a set of 37 three-bladed propellers with pitch–diameter ratios between 0.4 and 2.0 and expanded area ratios from 0.2 to 1.1 (Gawn, 1953). Its extended blade shapes are elliptical and only ogival sections have been used. Although the geometry is outdated, this series is important because its model propellers are exceptionally large with a diameter of 20 in (508 mm). The series data therefore has very little to no scale effects. In addition, the Gawn series also covers a wide pitch–diameter ratio which makes it applicable to high-speed and naval vessels. Blount and Hubble (1981) provide regression polynomials analog to the ones for the Wageningen B-Series. The polynomials are also published in Carlton (2007).

King's College Admiralty (KCA) series

The open water tests for the Gawn series were conducted in a towing tank and performance under cavitation conditions was originally not investigated. However, Gawn and Burrill (1957) tested a set of 30 slightly smaller propellers with 16 in (406.4 mm)

Table 47.5 Factors and exponents for Reynolds number effects on thrust coefficient and torque coefficient of Wageningen B-Series propellers (Oosterveld and van Oossanen, 1975)

(a) Coefficients and exponents for thrust correction ΔK_T

i	a_i	b_i	c_i	d_i	e_i	f_i
0	+0.000353485	0	0	0	0	0
1	−0.00333758	2	0	1	0	0
2	−0.00478125	1	1	1	0	0
3	+0.000257792	2	0	1	0	2
4	+0.0000643192	2	6	0	0	1
5	−0.0000110636	2	6	0	0	2
6	−0.0000276305	2	0	1	1	2
7	+0.0000954	1	1	1	1	1
8	+0.0000032049	1	3	1	2	1

(b) Coefficients and exponents for torque correction ΔK_Q

i	a_i	b_i	c_i	d_i	e_i	f_i
0	−0.000591412	0	0	0	0	0
1	+0.00696898	0	1	0	0	0
2	−0.0000666654	0	6	0	1	0
3	+0.0160818	0	0	2	0	0
4	−0.000938091	0	1	0	0	1
5	−0.00059593	0	2	0	0	1
6	+0.0000782099	0	2	0	0	2
7	+0.0000052199	2	0	1	1	1
8	−0.00000088538	1	1	1	1	2
9	+0.0000230171	0	6	0	1	1
10	−0.00000184341	0	6	0	1	2
11	−0.00400252	0	0	2	0	1
12	+0.000220915	0	0	2	0	2

diameter in the cavitation tunnel of the King's College, Newcastle, UK[1]. The series was conceived for fast naval vessels, hence the name King's College Admiralty or short KCA series. It is also referred to as Gawn-Burrill series. Nowadays it is mostly used as design basis for high-speed small craft.

Newton–Rader series

Additional information for propeller performance under cavitation conditions may be derived from the Newton–Rader series (Newton and Rader, 1961). This series consists of 12 three-bladed propellers. Expanded area ratios of $A_E/A_0 = 0.48, 0.71$, and 0.95 have been tested each with four pitch–diameter ratios between $P/D = 1.04$ and 2.08. This series is of interest for the design of small high-speed craft where significant cavitation is suspected.

DTMB skewed propeller series

The effect of skew on the propeller was investigated by Boswell (1971). Open water and cavitation data is reported for four propellers with skews of 0, 36, 72, and 108 degree at the blade tip. No other propeller characteristics were modified besides the skew. Therefore, this mini-series is not suitable for propeller selection. Nevertheless, the propellers provide important test cases for computer codes that predict effects of skew on propeller performance.

Ducted propellers, Ka-Series

Oosterveld (1970) conducted extensive open water and cavitation tests with ducted propellers, the so called Ka-Series. Regression polynomials are provided for thrust and torque of the propeller as well as the influence of the nozzle (duct). Four different nozzle types have been investigated.

Controllable pitch propellers

Series data are limited for controllable pitch propellers. Data for three-bladed propellers can be found in Gutsche and Schroeder (1963) and Chu et al. (1979). Hansen (1967) reports data for five different controllable pitch propellers with four blades.

[1] In 1963 King's College was renamed into University of Newcastle upon Tyne

MARIN recently added the Wageningen C- and D-Series to its portfolio which provide data for controllable pitch propellers with and without duct. Dang et al. (2013) provide some results in the form of charts but regression polynomials have not yet been released to the public.

In the following two Chapters we will use propeller series data to select a suitable propeller for a ship. Chapter 48 explains the principal propeller selection tasks and process. Chapter 49 provides two extended examples.

References

Blount, D. and Hubble, E. (1981). Sizing segmental section commercially available propellers for small craft. In *Transactions of SNAME*, volume 89. The Society of Naval Architects and Marine Engineers.

Boswell, R. (1971). Design, cavitation performance, and open-water performance of series of research skewed propellers. Report 3339, Naval Ship Research Development Center, Washington, D.C.

Carlton, J. (2007). *Marine propellers and propulsion*. Butterworth-Heinemann, Oxford, United Kingdom.

Chu, C., Chan, Z., She, Y., and Yuan, V. (1979). The 3-bladed JD-CPP series. In *4th Lips Propeller Symposium*, pages 53–82, Drunen, The Netherlands.

Dang, J., van den Boom, H., and Ligtelijn, J. (2013). The Wageningen C- and D-series propellers. In *Proc. of the FAST 2013 Conference*, Amsterdam, The Netherlands. Paper ID 61.

Gawn, R. (1953). Effect of pitch and blade width on propeller performance. In *Transactions of The Royal Institution of Naval Architects (RINA)*, volume 95. The Royal Institution of Naval Architects.

Gawn, R. and Burrill, L. (1957). Effect of cavitation on the performance of a series of 16 in. model propellers. In *Transactions of The Royal Institution of Naval Architects (RINA)*, volume 99. The Royal Institution of Naval Architects.

Gutsche, F. and Schroeder, G. (1963). Freifahrversuche an Propellern mit festen und verstellbaren Flügeln 'voraus' und 'rückwärts'. *Schiffbauforschung*, 2(4):146–154.

Hansen, E. (1967). Thrust and blade spindle torque measurements of five controllable pitch propeller designs for MSO 421. Report 2325, Naval ship Research and Development Center, Washington, D.C.

ITTC (2014). *Open water test*. International Towing Tank Conference, Recommended Procedures and Guidelines 7.5-02-03-02.1. Revision 3.

Kuiper, G. (1992). *The Wageningen propeller series*. Publication 92-001. Maritime Research Institute Netherlands (MARIN), Wageningen, The Netherlands.

Newton, R. and Rader, H. (1961). Performance data of propellers for high speed craft. In *Transactions of The Royal Institution of Naval Architects (RINA)*, volume 103. The Royal Institution of Naval Architects.

Oosterveld, M. (1970). Wake adapted ducted propellers. NSBM Publication 345, Netherlands Ship Model Basin, Wageningen, The Netherlands. NSBM is now MARIN.

Oosterveld, M. and van Oossanen, P. (1975). Further computer-analyzed data of the Wageningen B-screw series. *International Shipbuilding Progress*, 22:251–261.

Troost, L. (1938). Open water test series with modern propeller forms. *Transactions of the North East Coast Institution of Engineers and Shipbuilders (NECIES)*, 54.

Troost, L. (1940). Open water test series with modern propeller forms II. three bladed propellers. *Transactions of the North East Coast Institution of Engineers and Shipbuilders (NECIES)*, 56.

Troost, L. (1951). Open water test series with modern propeller forms III. Two bladed and five bladed propellers– extension of the three and five bladed b-series. *Transactions of the North East Coast Institution of Engineers and Shipbuilders (NECIES)*, 67.

Self Study Problems

1. The Wageningen B-Series was developed more than 50 years ago. Explain why it is still relevant.
2. Why are there polynomials to correct thrust and torque for full scale Reynolds numbers for the Wageningen B-Series but not for the Gawn series?
3. What is the Newton-Rader propeller series known for?
4. Implement the Wageningen B-Series polynomials as a program (use Python, Matlab, or similar) so that you can compute and visualize propeller open water characteristics. The following table provides some test data:

Prop 1	$Z = 4,\ P/D = 0.7000,\ A_E/A_0 = 0.5500$					
J:	0.000000	0.200000	0.400000	0.600000	0.800000	1.000000
K_T:	0.293034	0.235922	0.163956	0.080357	−0.011655	−0.108858
$10K_Q$:	0.314941	0.265489	0.202513	0.125021	0.032021	−0.077478
Prop 2	$Z = 5,\ P/D = 0.7000,\ A_E/A_0 = 0.5500$					
J:	0.000000	0.200000	0.400000	0.600000	0.800000	1.000000
K_T:	0.304817	0.247985	0.173682	0.084871	−0.015485	−0.124422
$10K_Q$:	0.325948	0.279672	0.218642	0.138134	0.033421	−0.100223
Prop 3	$Z = 4,\ P/D = 0.9000,\ A_E/A_0 = 0.5500$					
J:	0.000000	0.200000	0.400000	0.600000	0.800000	1.000000
K_T:	0.382545	0.327866	0.258221	0.176763	0.086647	−0.008971
$10K_Q$:	0.502623	0.442389	0.366125	0.273048	0.162376	0.033326
Prop 4	$Z = 5,\ P/D = 0.9000,\ A_E/A_0 = 0.7000$					
J:	0.000000	0.200000	0.400000	0.600000	0.800000	1.000000
K_T:	0.413463	0.353818	0.276834	0.186159	0.085442	−0.021668
$10K_Q$:	0.550275	0.483373	0.396737	0.290990	0.166755	0.024655
Prop 5	$Z = 6,\ P/D = 1.0000,\ A_E/A_0 = 0.9500$					
J:	0.000000	0.200000	0.400000	0.600000	0.800000	1.000000
K_T:	0.500113	0.434368	0.349707	0.250456	0.140944	0.025498
$10K_Q$:	0.754051	0.666074	0.552772	0.420127	0.274123	0.120741

48

Propeller Design Process

Once estimates of resistance and hull–propeller interaction parameters are completed, a propeller must be selected before propulsive power for a ship can be determined. As a first step in the propeller design process, basic characteristics are selected on the basis of propeller series data. Afterwards, details of the propeller design are determined with additional computational tools like propeller lifting line and lifting surface programs. In this and the following chapter we study the first step in the propeller design process and define principal propeller characteristics based on Wageningen B-series data. Here, the objectives are discussed which govern the selection process and the basic tools are described. Complete examples are provided in the following chapter.

Learning Objectives

At the end of this chapter students will be able to:

- understand and prepare propeller design tasks
- compute required design constants
- understand the process of hull–propeller matching
- find optimum combinations of propeller characteristics for given thrust or powering requirements

48.1 Design Tasks and Input Preparation

Propeller design problems may be subdivided into two groups:

Basic design tasks

(i) optimum diameter design problems, and

(ii) optimum rate of revolution design problems.

In the first group (i) the rate of revolution is known, and we search for the optimum diameter which maximizes efficiency of the propeller. In the second group (ii) the diameter is known, and we look for the best rate of revolution. In each group we have

Fundamentals of Ship Hydrodynamics: Fluid Mechanics, Ship Resistance and Propulsion, First Edition.
Lothar Birk.

Companion website: www.wiley.com/go/birk/hydrodynamics

Table 48.1 The four basic propeller design tasks

	Task	given		wanted		design constant
optimum diameter	1	delivered power	P_D	diameter	D	$\left[\frac{K_Q}{J^5}\right] = \frac{P_D\, n^2\, \eta_R}{2\pi\, \rho\, v_A^5}$
		speed of advance	v_A	pitch–diameter ratio	P/D	
		rate of revolution	n	open water efficiency	η_O	
	2	thrust	T	diameter	D	$\left[\frac{K_T}{J^4}\right] = \frac{T\, n^2}{\rho\, v_A^4}$
		speed of advance	v_A	pitch–diameter ratio	P/D	
		rate of revolution	n	open water efficiency	η_O	
optimum rate of revolution	3	delivered power	P_D	rate of revolution	n	$\left[\frac{K_Q}{J^3}\right] = \frac{P_D\, \eta_R}{2\pi\, \rho\, D^2\, v_A^3}$
		speed of advance	v_A	pitch–diameter ratio	P/D	
		diameter	D	open water efficiency	η_O	
	4	thrust	T	rate of revolution	n	$\left[\frac{K_T}{J^2}\right] = \frac{T}{\rho\, D^2\, v_A^2}$
		speed of advance	v_A	pitch–diameter ratio	P/D	
		diameter	D	open water efficiency	η_O	

two subcategories, depending on whether the delivered power P_D or the thrust T is provided as input. Table 48.1 summarizes the four basic propeller design tasks. Note that the number of blades Z and the expanded area ratio A_E/A_0 are not listed. The number of blades Z is chosen to detune natural frequencies of propeller, shaft, and engine. The expanded area ratio A_E/A_0 is determined by a suitable cavitation criterion (see Chapter 46). Besides Z and A_E/A_0, the input for a propeller design task consists of a combination of speed of advance v_A, thrust T or delivered power P_D, and rate of revolution n or diameter D. The missing parameters pitch–diameter ratio P/D and diameter D or rate of revolution n are chosen to provide the highest possible open water efficiency η_O. How will be discussed below.

Service margin Delivered power P_D and thrust T are commonly determined for trial conditions (calm and deep water, no wind, no current). For propeller design purposes P_D and T have to be augmented with a service margin s_M.

$$T_{\text{service}} = (1 + s_M)T \qquad P_{D_{\text{service}}} = (1 + s_M)\, P_D \tag{48.1}$$

The service or sea margin accounts for increases in ship resistance during operation. Over time fouling will roughen the hull surface and increase its frictional resistance. It will also increase the wake fraction. Wind, current, and waves will increase resistance as well. If the propeller is erroneously optimized for trial conditions, it will have too much pitch. As a consequence, the engine might not reach its design rate of revolution during normal operations.

Service margins are to a large degree a matter of choice. The margin is influenced by

- owner preferences
- target area of operations
- size and trade of the vessel

Resistance of a vessel will be higher than the calm water resistance even during normal operation in fair weather. Wind, waves, and currents may cause the resistance to increase by 15%–35%. In head seas and heavy weather it might be even more; however, the ship will sail more slowly anyway to avoid damage to its structure. Typical weather patterns influence the size of the margin. Winter navigation on the rough North Atlantic between Europe and North America will require a higher service margin than service in the calmer Mediterranean. On one hand, a smaller margin might suffice for large vessels because they are less affected by wind and waves. On the other hand, frictional resistance is the largest resistance component for large and slow vessels. Therefore an increase in surface roughness due to fouling will have a comparatively larger negative effect than on a fast container ship for which wave (making) resistance is a sizable part of resistance. Marine fouling can cause a 25%–50% increase in frictional resistance depending on hull coating and docking intervals. Vessels which sail on a tight schedule, like container ships and ferries, may need larger margins to be able to recover lost time.

In summary, it is important that thrust T and delivered power P_D reflect service conditions rather than trial conditions. Current practice is to adjust trial estimates with a service margin. Service margins are a crude tool considering the effort that goes into estimating, measuring, and computing ship resistance and hull-propeller interaction parameters. Many ships arguably sail across the oceans with underutilized engine power. The ITTC added a manual to its set of recommended procedures with the goal of making service margin estimates more rational ITTC (2008). See also the method proposed by Stasiak (2004).

Engine margin

Engine manufacturers provide for their engines the maximum continuous rating (MCR) and the associated rate of revolution n. The maximum continuous rating is the brake power an engine can deliver during prolonged operation. MCR and n of an engine may be adjusted within narrow limits. The brake power used for propeller selection is typically 10% to 15% lower than MCR. This engine margin allows engine operation with greater fuel economy during normal operation and provides reserve power for maneuvering and bad weather. It comes on top of the service margin.

Design constants

The available input data is combined into a dimensionless design constant specific for each task. It will allow us to find the self propulsion points. This is essentially the same process we used during the full scale power prediction (see Section 44.2).

48.2 Optimum Diameter Selection

Optimum diameter selection tasks

Let us first discuss the optimum diameter selection tasks. In group (i) the rate of revolution n has to be determined first. There are two likely scenarios:

- The owner already operates ships of similar type and size. He may specify that the new ship must have the same engine as an existing vessel. This will simplify crew training and management. It may also lower maintenance cost.
- A preliminary powering estimate has been completed by assuming desired propeller data. An engine is pre-selected based on the estimated delivered power.

48.2.1 Propeller design task 1

Task 1, given: P_D, v_A, n **wanted:** $D, P/D, \eta_O$

Besides the rate of revolution n, we have to know the delivered power P_D and the speed of advance v_A (see Table 48.1). Delivered power may be derived from brake power by means of the mechanical or shafting efficiency: $P_D = P_B \eta_S$. Delivered power and rate of revolution also specify the available propeller torque in behind condition.

$$Q = \frac{P_D}{2\pi n} \tag{48.2}$$

Using the relative rotative efficiency (34.19) yields the torque Q_O for open water conditions:

$$Q_O = Q\,\eta_R \tag{48.3}$$

The next step is to find propellers which absorb this torque at the provided combination of speed of advance and rate of revolution.

Design constant for task 1

The torque absorbed by the propeller is embodied in the torque coefficient K_Q of the open water diagram:

$$K_Q = \frac{Q_O}{\rho\, n^2\, D^5} \tag{48.4}$$

In this equation we miss the propeller diameter D. Dividing Equation (48.4) by the fifth power of the advance ratio $J = v_A/(nD)$ eliminates the diameter from the right-hand side:

$$\frac{K_Q}{J^5} = \frac{\frac{Q_O}{\rho\, n^2\, D^5}}{J^5} = \frac{\frac{Q_O}{\rho\, n^2\, D^5}}{\left(\frac{v_A}{n\,D}\right)^5} = \frac{Q_O\, n^5\, D^5}{\rho\, n^2\, D^5\, v_A^5} = \frac{Q_O\, n^3}{\rho\, v_A^5} \tag{48.5}$$

The last expression in Equation (48.5) is known. Equations (48.3) and (48.2) replace the open water torque with delivered power and relative rotative efficiency η_R. The result is the dimensionless design constant for task 1:

$$\left[\frac{K_Q}{J^5}\right] = \frac{P_D\, n^2 \eta_R}{2\pi\, \rho\, v_A^5} \tag{48.6}$$

Setting the relative rotative efficiency to one is an option if a good estimate for it is unavailable.

Example data

Solution of propeller design task 1 is illustrated based on the data set given in Table 48.2. Input data has to be carefully converted to a consistent set of SI-units:

- delivered power $P_D = 23\,400.00$ kW$= 23\,400\,000$ W,
- rate of revolution $n =$ 108.0 rpm$=$ 1.8 s^{-1}, and
- ship speed $v_S = 21.3$ kn$=$ 0.958 m/s.

The speed of advance is not directly specified but we can compute it from ship speed and wake fraction.

$$v_A = (1 - w)v_S = (1 - 0.24)\, 10.958 \text{ m/s} = 8.295 \text{ m/s}$$

Table 48.2 Input data to illustrate task 1: optimum propeller diameter selection based on delivered power, speed of advance, and rate of revolution

Example ship data – Task 1	
number of blades	$Z = 4$
expanded area ratio	$A_E/A_0 = 0.85$
delivered power (service)	$P_D = 23\,400.00$ kW
rate of revolution	$n = 108.00$ rpm
design ship speed	$v_S = 21.30$ kn
wake fraction	$w = 0.243$
relative rotative efficiency	$\eta_R = 1.03$
salt water density (at 15 °C)	$\rho = 1026.021$ kg/m^3

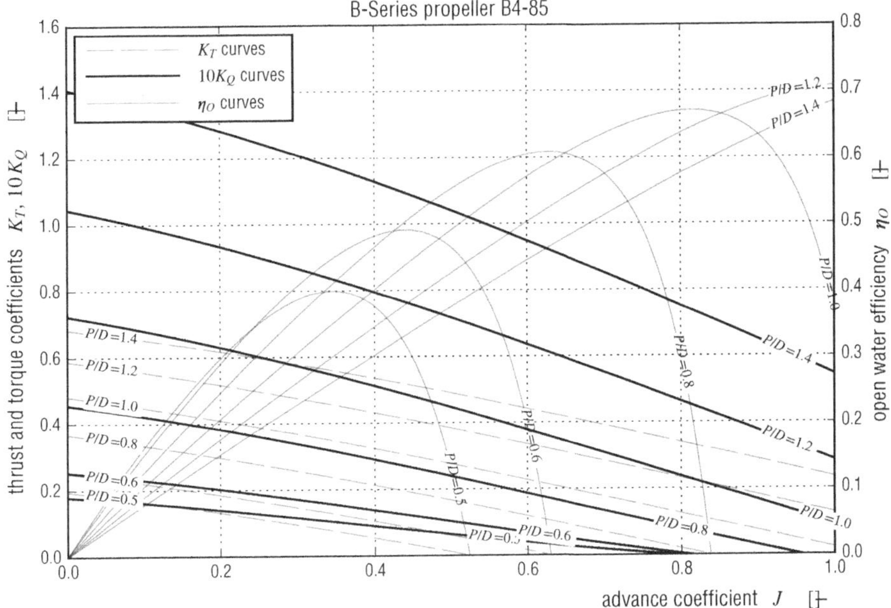

Figure 48.1 Design task 1 – Input: open water diagram for Wageningen B-series propellers with $Z = 4$ and $A_E/A_0 = 0.85$ derived from K_T and K_Q polynomials (47.1) and (47.2). Torque coefficient curves $10K_Q$ are emphasized

Based on the input data above, Equation (48.6) yields the design constant value for task 1. **Step 1: design constant**

$$\left[\frac{K_Q}{J^5}\right] = \frac{P_D\, n^2 \eta_R}{2\pi\, \rho\, v_A^5} = \frac{23400000 \text{ W} \cdot \left(1.8 \text{ s}^{-1}\right)^2 \cdot 1.03}{2\pi\; 1026.021 \text{ kg m}^{-3} \cdot \left(8.295 \text{ m/s}\right)^5} = 0.30846 \qquad (48.7)$$

Check the units to make sure the design constant is dimensionless. The design constant for task 1 measures how much torque is available to turn the propeller.

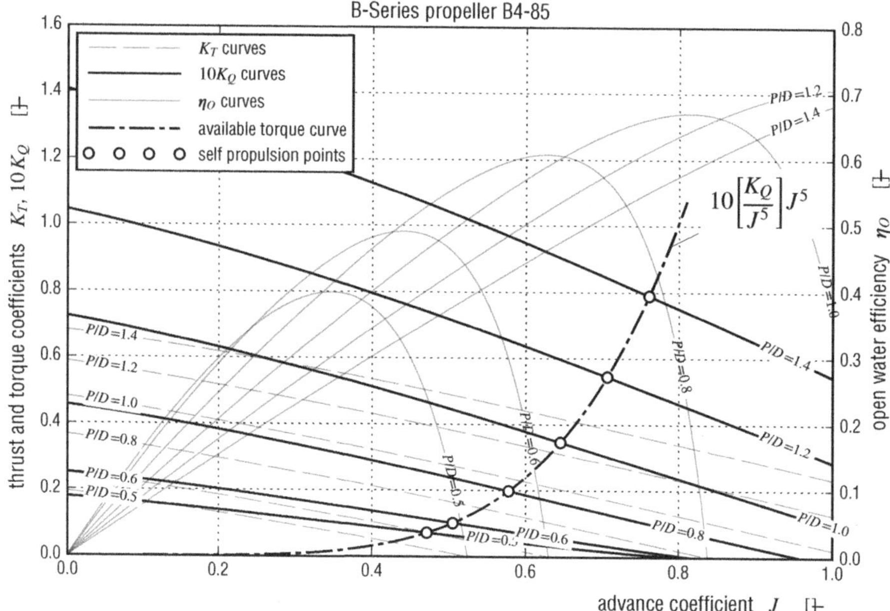

Figure 48.2 Design task 1 – Step 2: locate self propulsion points o at which the propellers absorb the delivered power specified with the design constant $[K_Q/J^5]$ from Equation (48.7)

Step 2: self propulsion points

With this design constant a self propulsion point may be estimated for a given propeller. Of course, we do not yet have a specific one, but we have the open water diagram for the B-Series propellers B4-85 (Figure 48.1). The curves represent six propellers with varying pitch–diameter ratios but all have four blades and an expanded area ratio of $A_E/A_0 = 0.85$. The torque coefficient curves have been emphasized in Figure 48.1 because they represent the torque needed to turn the propellers as a function of the advance ratio J. This is equivalent to saying that the propellers absorb the delivered power $2\pi\, n\, Q_O$.

The dimensionless torque available to the propeller is obtained by multiplying the design constant value (48.7) with J^5.

$$10\,K_{Q_{\text{available}}} = 10\left[\frac{K_Q}{J^5}\right]J^5 = 3.0846\,J^5 \qquad \text{available torque curve} \qquad (48.8)$$

We compute $10K_{Q_{\text{available}}}$ for a set of advance ratios and plot the result as curve into the open water diagram (Figure 48.2). Do not forget that in most cases the open water diagram shows ten times the torque coefficient! The intersections of the available torque curve and the six open water torque curves represent the self propulsion points. They are marked with circles o in Figure 48.2. At the self propulsion points the torque from the propeller shaft is in equilibrium with the torque a propeller requires to turn.

Any of the six propellers is a possible solution to our design problem. For example, for a pitch–diameter ratio $P/D = 0.8$ the intersection point is at the advance ratio

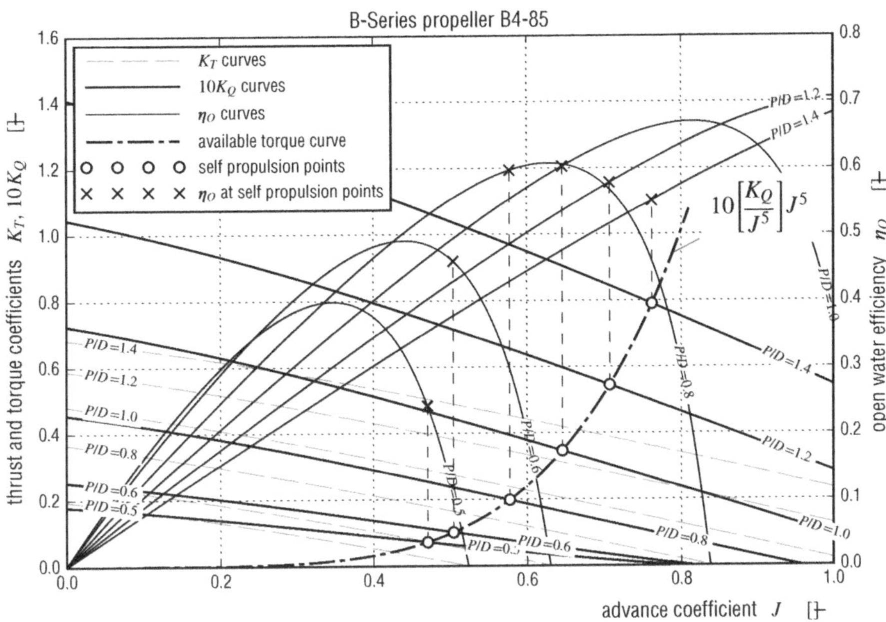

Figure 48.3 Design task 1 – Step 3: find open water efficiencies × for self propulsion points o

$J_{0.8} = 0.5775$. From the advance ratio $J_{0.8}$ we may compute the missing propeller diameter $D_{0.8}$.

$$J_{0.8} = \frac{v_A}{n\, D_{0.8}} \longrightarrow D_{0.8} = \frac{v_A}{n\, J_{0.8}} = \frac{8.295\ \mathrm{m/s}}{1.8\ \mathrm{s}^{-1} \cdot 0.5775} = 7.979\ \mathrm{m}$$

The open water efficiency of the propeller with $P_D = 0.8$ at $J_{0.8} = 0.5775$ is 0.5982. That is not necessarily bad; however, our objective is to find the pitch–diameter ratio which will maximize the efficiency. Therefore, a check of the achievable open water efficiency is needed.

Step 3: open water efficiencies

In Figure 48.3 the open water efficiencies are marked with crosses × for each self propulsion point. Of these, the open water efficiency is highest for $P/D = 1.0$.

Step 4: auxiliary curve

The best possible open water efficiency seems to be found for a pitch–diameter ratio in between $P/D = 0.8$ and 1.0. We draw an interpolation curve through the six open water efficiency points × (see Figure 48.4). This curve is sometimes called the auxiliary curve. It represents the achievable open water efficiency for this set of B4-85 propellers and the given design constant (48.7).

Step 5: optimum propeller

The maximum of the auxiliary curve marks the highest open water efficiency which can be achieved for the chosen data set. Figure 48.5 shows the optimum at $\eta_O = 0.6051$ and the associated self propulsion point □ and optimum advance ratio $J = 0.6202$. The optimum advance ratio yields an optimum diameter of

$$D_O = \frac{v_A}{n\, J} = \frac{8.295\ \mathrm{m/s}}{1.8\ \mathrm{s}^{-1} \cdot 0.6202} = 7.430\ \mathrm{m}$$

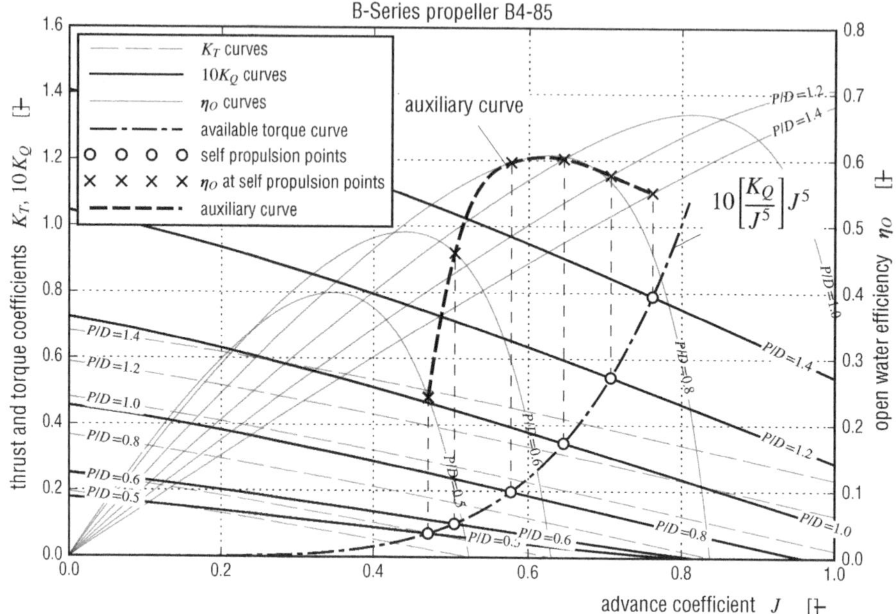

Figure 48.4 Design task 1 – Step 4: draw auxiliary curve through open water efficiency values

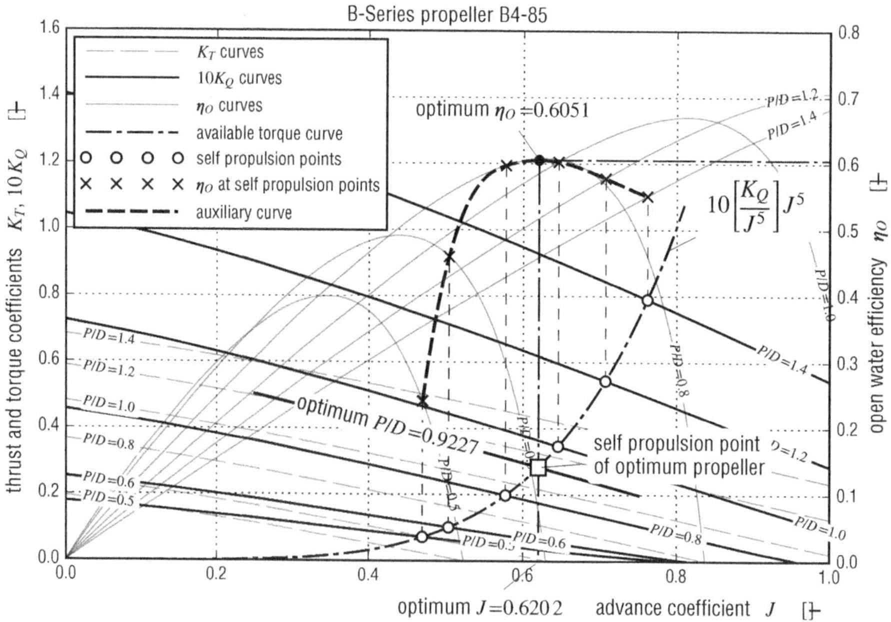

Figure 48.5 Design task 1 – Result: optimum propeller is defined by maximum of auxiliary curve

This diameter is optimal for open water condition only. The optimum propeller diameter is 2%–5% smaller for the behind condition. We will come back to this mystery in the following chapter.

Table 48.3 Results for the propeller selection task 1 example

Optimum open water propeller data – Task 1	
number of blades	$Z = 4$
expanded area ratio	$A_E/A_0 = 0.85$
delivered power (service)	$P_D = 23\,400.00$ kW
rate of revolution	$n = 108.00$ rpm
advance ratio	$J = 0.6202$
pitch–diameter ratio	$P/D = 0.9227$
open water efficiency	$\eta_O = 0.6051$
optimum diameter	$D_O = 7.43$ m

The pitch–diameter ratio of the optimum propeller has to be found by interpolation between the torque coefficient curves. A $10K_Q$ curve is sketched through the self propulsion point □ of the optimum propeller. Figure 48.5 reveals that the optimum pitch–diameter ratio is $P/D = 0.9227$. Table 48.3 summarizes the optimum propeller characteristics based on the input from Table 48.2.

Next steps

In propeller design task 1 we match the propeller torque to an available torque. There is no guarantee that the resulting optimum propeller produces sufficient thrust to achieve the design speed. For that reason, the thrust generated by the optimum propeller has to be checked against the thrust required by the ship in design tasks which use the delivered power as input. If the produced thrust is insufficient, the rate of revolution must be lowered or the available power must be increased.

In addition, the initial value for the expanded area ratio has to be checked against a suitable cavitation limit. A higher expanded area ratio must be chosen if the limit is exceeded. For the sake of brevity these steps are skipped here but we will perform them in the examples of the following chapter.

48.2.2 Propeller design task 2

Task 2, given: T, v_A, n **wanted:** $D, P/D, \eta_O$

The second propeller design task uses the same input as task 1 except that the delivered power is swapped for the required service thrust T. Since the final propeller produces exactly the thrust a ship needs the aforementioned check on ship speed is obsolete. However, one must select an engine which must be capable of generating the required torque at the selected rate of revolution.

Design constant for task 2

Since the thrust is given, the design constant is based on the definition of the thrust coefficient K_T. The equation is divided by the fourth power of the advance ratio to eliminate the unknown diameter D.

$$\left[\frac{K_T}{J^4}\right] = \frac{\dfrac{T}{\rho\, n^2\, D^4}}{J^4} = \frac{\dfrac{T}{\rho\, n^2\, D^4}}{\dfrac{v_A^4}{n^4\, D^4}} = \frac{T\, n^2}{\rho\, v_A^4} \tag{48.9}$$

We use the input data from task 1 again but exchange the delivered power P_D in

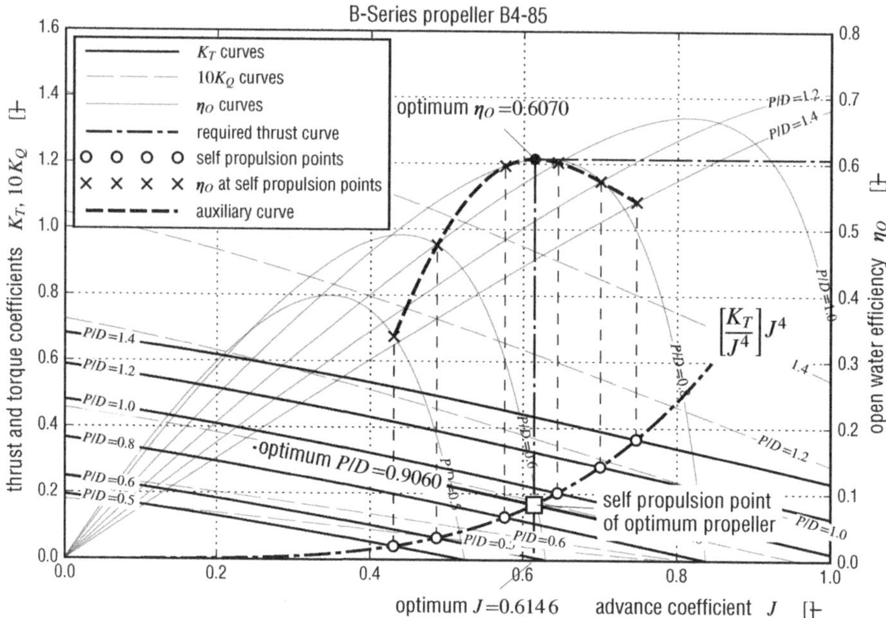

Figure 48.6 Design task 2 – Result: optimum propeller is defined by maximum of auxiliary curve

Table 48.2 for the estimated thrust for the service condition.

$$T = 1762.35\ \text{kN} \qquad \text{new input for task 2} \tag{48.10}$$

The resultant design constant yields the curve of required thrust for the example data.

$$K_{T_\text{required}} = \left[\frac{K_T}{J^4}\right] J^4 = 1.1755\, J^4 \qquad \text{required thrust curve}$$

Self propulsion points for task 2

The self propulsion points are now defined by the intersections of the available thrust curve with the K_T curves of the propellers in the open water diagram. Rather than showing all intermediate steps of the process Figure 48.6 presents the final result. Note that in contrast to Figure 48.5 from task 1 the solid lines are now the thrust coefficient curves K_T. The reader should complete the process as an exercise.

Although a consistent data set has been used for both tasks 1 and 2, it should not surprise that results slightly differ between the tasks. Locating the maximum of the auxiliary curve precisely is hard because the curve may be quite flat around the self propulsion point. In addition, we are working with thrust and torque curves derived from regression polynomials. They themselves have been produced by extensive fairing of the available open water data (Oosterveld and van Oossanen, 1975). In our example, efficiency is minimally higher for the propeller from task 2. The diameter is 1% larger and the pitch–diameter ratio is 2% smaller (see Table 48.4).

Table 48.4 Results for the propeller selection task 2 example

Optimum open water propeller data – Task 2	
number of blades	$Z = 4$
expanded area ratio	$A_E/A_0 = 0.85$
thrust (service)	$T = 1762.35$ kN
rate of revolution	$n = 108.00$ rpm
advance ratio	$J = 0.6146$
pitch–diameter ratio	$P/D = 0.9060$
open water efficiency	$\eta_O = 0.6070$
optimum diameter	$D_O = 7.498$ m
required delivered power	$P_D = 23\,394.60$ kW

48.3 Optimum Rate of Revolution Selection

Propeller design tasks 3 and 4 mirror design tasks 1 and 2, respectively. However, now the diameter D instead of the rate of revolution is known. Consequently, the objective is to find a combination of rate of revolution n and pitch–diameter ratio P/D which provides the best efficiency η_O.

An initial estimate of the diameter D may be derived in two ways:

- In early design stages the diameter may be estimated from regression formulas. See Section 46.2, page 553 and Equation (46.2) for some examples.
- Once preliminary hull lines have been drawn, the shape of the stern defines the largest possible diameter which may be used as a starting point. Observe the clearances! A minimum vertical clearance between propeller tip and hull surface of 25% should be considered. Smaller vessels may have inclined shafts where the propeller disk reaches below the baseline. However, larger commercial vessels maintain a clearance of about 3% of the diameter to the baseline. It protects the propeller and simplifies docking procedures.

48.3.1 Propeller design task 3

In propeller design task 3 we start again with the delivered power P_D as representation of the available torque. The diameter $D = 7.430$ m replaces the rate of revolution on the list of required input. Otherwise the data is taken from Table 48.2.

Task 3, given: P_D, v_A, D **wanted:** n, P/D, η_O

Based on the torque coefficient

Design constant for task 3

$$K_Q = \frac{Q_O}{\rho\, n^2\, D^5} = \frac{P_D\, \eta_R}{2\pi\, \rho\, n^3\, D^5} \tag{48.11}$$

we derive the design constant for task 3. The unknown rate of revolution n is eliminated

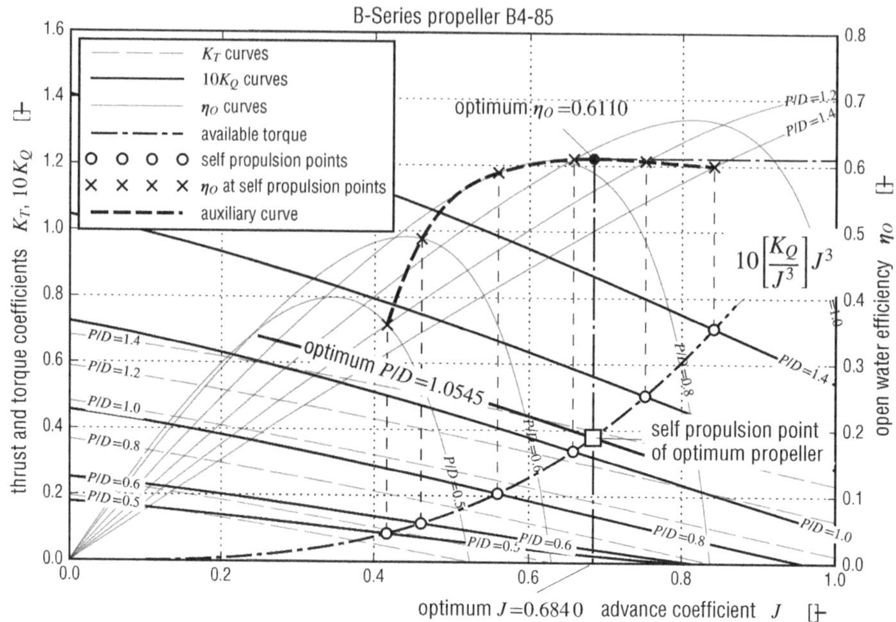

Figure 48.7 Design task 3 – Result: optimum propeller is defined by maximum of auxiliary curve

by dividing Equation (48.11) above by J^3.

$$\left[\frac{K_Q}{J^3}\right] = \frac{\dfrac{P_D}{2\pi\,\rho\,n^3\,D^5}}{\dfrac{v_A^3}{n^3\,D^3}} = \frac{P_D\,\eta_R}{2\pi\,\rho\,D^2\,v_A^3} \tag{48.12}$$

The curve of available torque is represented by a cubic polynomial.

$$10\,K_{Q_{\text{available}}} = 10\left[\frac{K_Q}{J^3}\right] J^3 = 1.1866\,J^3 \tag{48.13}$$

Self propulsion points for task 3

Figure 48.7 shows the polynomial (48.13) and the self propulsion points at the intersections with the open water torque $10K_Q$ curves. The maximum achievable open water efficiency marks the optimum advance ratio $J = 0.6851$. This is about 10% higher than in task 1 although the delivered power is the same as in task 1 and we used the resulting diameter $D = 7.430$ m of task 1 as input. The pitch–diameter ratio is also higher. As a consequence of the higher advance ratio, the optimum rate of revolution n_O is smaller than the input to task 1.

$$n_O = \frac{v_A}{D\,J} = \frac{8.295\ \text{m/s}}{7.430\ \text{m}\cdot 0.6851} = 1.6297\ \text{s}^{-1} = 97.782\ \text{rpm} \tag{48.14}$$

The resultant open water efficiency is 1% higher than the results from tasks 1 and 2.

It should be noted that the maximum of the auxiliary curve depends on the interpolation algorithm used to create the auxiliary curve. Obviously, there is an infinite number of curves of different shape that could be fitted through the six given efficiency points marked with crosses.

48.3.2 Propeller design task 4

Finally, in propeller design task 4 we know thrust T, speed of advance v_A, and diameter D, and we are looking for the rate of revolution n_O and pitch–diameter ratio which provides the highest open water efficiency. We again employ the thrust $T = 1762.35$ kN from task 2 as an example. The input diameter is again $D = 7.430$ m.

Task 4, given: T, v_A, D **wanted:** $n, P/D, \eta_O$

The design constant for task 4 was already introduced when we discussed the prediction of full scale power prediction (see Section 44.2).

Design constant for task 4

$$\left[\frac{K_T}{J^2}\right] = \frac{\frac{T}{\rho n^2 D^4}}{J^2} = \frac{\frac{T}{\rho n^2 D^4}}{\frac{v_A^2}{n^2 D^2}} = \frac{T}{\rho D^2 v_A^2} = C_S \tag{48.15}$$

From the input data we derive the parabola of required thrust:

$$K_{T_{\text{required}}} = \left[\frac{K_T}{J^2}\right] J^2 = 0.4522\, J^2 \qquad \text{required thrust curve}$$

Figure 48.8 shows the resulting optimum propeller. Its characteristics are close to the result of task 3.

optimum open water efficiency	η_O	$= 0.6114$
advance ratio	J	$= 0.6808$
optimum rate of revolution	n_O	$= 1.6369\ \text{s}^{-1} = 98.214$ rpm
required delivered power	P_D	$= 23210.89$ kW

As for the other tasks, this result has to be corrected for the behind condition and double checked against the cavitation limit. Details of this process will be explained in the following chapter.

48.4 Design Charts

Selection of the optimum propeller characteristics based on open water charts is quite cumbersome. Nevertheless, understanding and being able to follow the process described above is a valuable skill. It enables basic checks of computational results and can be applied to propellers for which design charts are not available.

Early on, naval architects converted available open water data into more practical propeller design charts. The idea is to redraw the open water data as a function of the design constant. Easy to work with are the B_P- and B_U-charts. B_P and B_U refer to

B_P- and B_U-charts

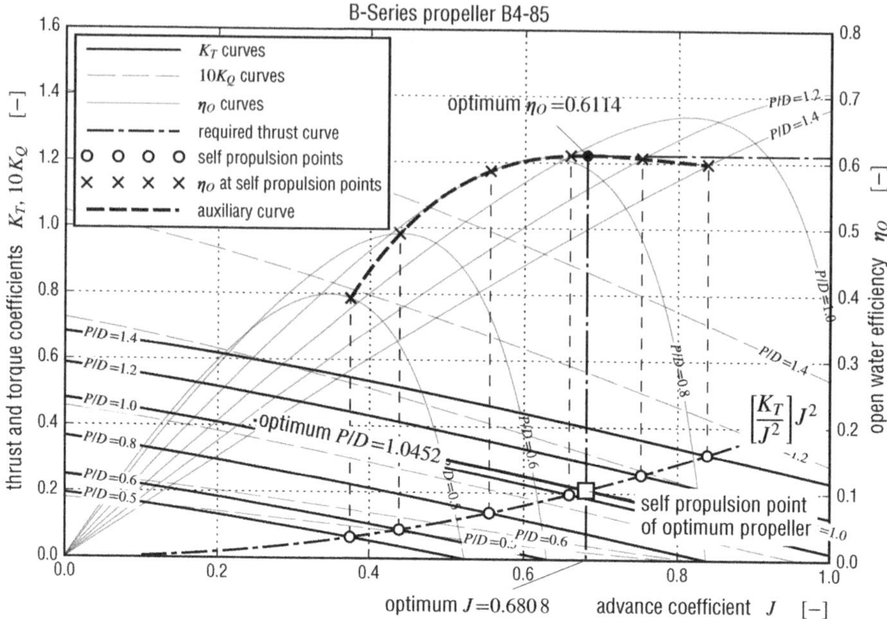

Figure 48.8 Design task 4 – Result: optimum propeller is defined by maximum of auxiliary curve

Admiral D.W. Taylor's propeller coefficients. The original coefficients are dimensional and their values depend on the system of units employed. They should no longer be used. Like open water diagrams, design charts show data for a set of fixed number of blades and expanded area ratio. Figure 48.9 shows a B_{P_1}-chart for task 1 with the data for Wageningen B-series propellers B4-85. The chart has been simplified by omitting intermediate curves.

Propeller design charts are plotted into an axes system of task specific design constant and pitch–diameter ratio. Each task therefore has a specific type of chart. Beware that the horizontal axis uses the fourth root of the design constant! For task 1 we use:

$$\sqrt[4]{\left[\frac{K_Q}{J^5}\right]} = \left(\frac{P_D\, n^2 \eta_R}{2\pi\, \rho\, v_A^5}\right)^{0.25} \tag{48.16}$$

As shown in Figure 48.9, this stretches the horizontal axis and makes it easier to read.

B_P- and B_U-propeller design charts consist of three distinct sets of curves (Figure 48.9):

(i) curves of constant open water efficiency η_O = const.,

(ii) curves of constant inverse advance ratio $1/J$ = const., and

(iii) optimum propeller curves.

B_{P_1}-charts solve design task 1 and B_{P_2}-charts solve design task 3. If the thrust is provided, B_U-charts are used: B_{U_1}-charts for design task 2 and B_{U_2}-charts for design task 4.

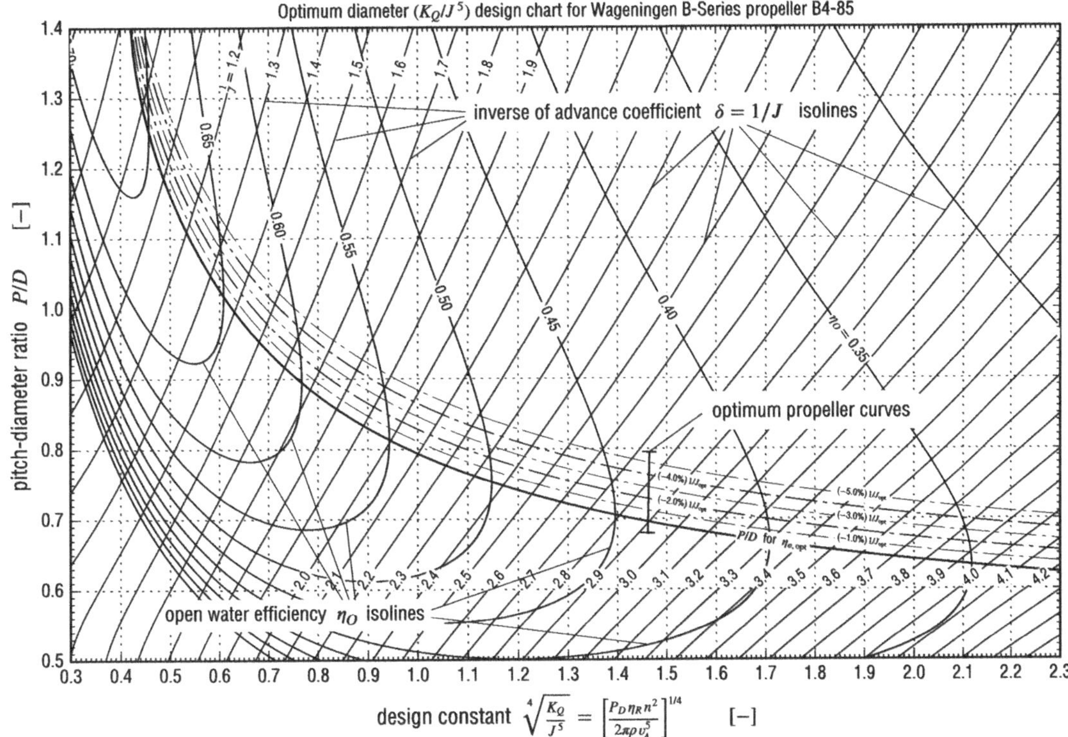

Figure 48.9 Simplified task 1 propeller design B_{P_1}-chart for Wageningen B-Series propeller with $Z = 4$ and $A_E/A_o = 0.85$

Open water efficiency isolines

The highly curved open water efficiency isolines may be interpreted like the lines of constant altitude on a map. Open water efficiency changes faster where the isolines are closer together and it changes slower where the isolines are farther apart. The open water efficiency forms a 'ridge' that curves from the top left corner to the lower right corner of the chart. At the end of a selection process we want to end up somewhere close to the crest of the ridge.

Inverse advance ratio isolines

The lines of constant inverse advance ratio $1/J$ appear like a hatching pattern roughly tracing from the lower left toward the upper right on the chart. In contrast to the open water efficiency, values of $1/J = \text{const.}$ increase from the upper left to the lower right corner of the chart.

Optimum open water propeller curve

With open water diagrams we found the maximum open water efficiency as the highest point on the auxiliary curve. In Figure 48.9 the lower, solid optimum propeller curves reflect the pitch–diameter ratios with the maximum open water efficiency $\eta_{O,\text{opt}}$. The 'P/D for $\eta_{O,\text{opt}}$'-curve marks the points where the open water efficiency isolines have vertical tangents. Any vertical line in the chart corresponds to a specific design constant value. Deviation from the P/D for $\eta_{O,\text{opt}}$ curve upwards or downwards along $\sqrt[4]{K_Q/J^5} = \text{const.}$ causes a decline in open water efficiency. The decline is, however, much smaller for increases in P/D-ratio than for lower P/D-ratios.

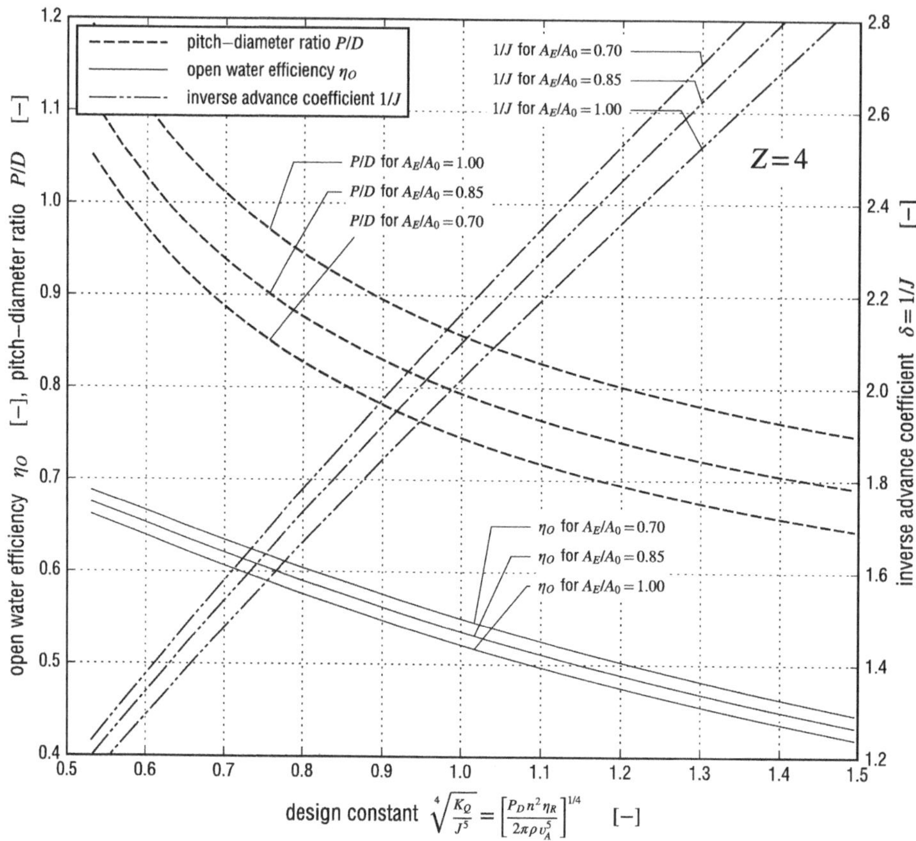

Figure 48.10 Optimum diameter chart for design task 1

Optimum propeller curves

Many published propeller design charts only feature the curve for the optimum efficiency $\eta_{O,\text{opt}}$. However, as mentioned before, this line identifies the best propeller under open water conditions. It does not necessarily mean that this is the best propeller for the behind condition with a nonuniform wake field. Design experience shows that the optimum propeller in the behind condition operates at a reduced $1/J$-ratio. In fact, diameter or rate of revolution of the optimum propeller are 1% to 5% smaller for a nonuniform wake field compared with the optimum propeller for the open water condition. The additional optimum propeller curves (dash-dotted lines) represent the proper pitch–diameter ratio for reductions in $1/J$ from 1% to 5%.

Other propeller design charts

Even simpler than the B_P- and B_U-charts are plots of the data along the optimum open water efficiency curve. Figure 48.10 shows curves for three expanded area ratios. B_P-charts like Figure 48.9 contain data for just one expanded area ratio. However, in many cases A_E/A_0 is adjusted during the selection process according to the chosen cavitation criterion. The middle curve of each triplet represents the same data as the B_p-chart in Figure 48.9. Two reports by Bernitsas and Ray (1982a,b) contain a complete set of design charts like Figure 48.10.

Harvald (1983) explains the use of logarithmic propeller design charts. Their advantage is that all four tasks may be solved with a single design chart. They are, however, fairly complex and data take-off is more difficult compared with the charts discussed above.

48.5 Computational Tools

The propeller design charts above have been derived from the thrust and torque coefficient polynomials of the Wageningen B-Series propellers. It is fairly straight forward to include the polynomials in an optimization tool which solves the selection process computationally rather than with charts. For design task 1, one could minimize the function

$$f\left(P/D,[K_Q/J^5]\right) = 1 - \eta_O(J_{QS}, P/D, A_E/A_0, Z) \tag{48.17}$$

The pitch–diameter ratio serves as a free variable in the optimization process and the design constant $[K_Q/J^5]$ is a parameter which defines the self propulsion point. The open water efficiency is computed from the K_T- and K_Q-polynomials (47.1) and (47.2)

$$\eta_O = \frac{J_{QS}}{2\pi} \frac{K_T(J_{QS}, P/D, A_E/A_0, Z)}{K_Q(J_{QS}, P/D, A_E/A_0, Z)} \tag{48.18}$$

with the advance ratio J_{QS} as the solution of the implicit and nonlinear equilibrium condition for the self propulsion point

$$0 = \left[\frac{K_Q}{J^5}\right] J_{QS}^5 - K_Q(J_{QS}, P/D, A_E/A_0, Z) \tag{48.19}$$

The process is similar for the other design tasks.

The result of the propeller selection based on design charts provides the necessary data to complete the power prediction for a ship design. The result also forms the basis for further computations. At a minimum, a propeller lifting line code should be used for the final design. It helps determining the following details of the propeller design:

- Defines the circulation distribution Γ which in turn determines section lift coefficients and section camber.
- Adjust pitch distribution to account for axial and tangential wake distribution (if available) and unload the extreme ends of the propeller blades to reduce tip and hub vortices.
- Optimize chord length distribution to reduce cavitation issues.

Lifting line codes represent the simplest numerical approximation of hydrodynamic propeller characteristics. Unfortunately, this topic extends beyond the scope of this book. Interested readers should take a look at the OpenProp lifting line code (http://engineering.dartmouth.edu/epps/openprop/, last visited 12/25/2018) and the associated documentation (Epps, 2010a,b). Further analysis with lifting surface and panel codes may be required if cavitation inception and sound signature are of interest. For details see Breslin and Andersen (1994) and Kerwin and Hadler (2010).

References

Bernitsas, M. and Ray, D. (1982a). Optimal diameter B-Series propellers. Technical Report 245, The University of Michigan, Ann Arbor, USA.

Bernitsas, M. and Ray, D. (1982b). Optimal revolution B-Series propellers. Technical Report 244, The University of Michigan, Ann Arbor, USA.

Breslin, J. and Andersen, P. (1994). *Hydrodynamics of ship propellers.* Cambridge University Press, Cambridge, United Kingdom.

Epps, B. (2010a). *An impulse framework for hydrodynamic force analysis: fish propulsion, water entry of spheres, and marine propellers.* PhD thesis, Massachusetts Institute of Technology (MIT), Cambridge, MA.

Epps, B. (2010b). OpenProp v2.4 theory document. Technical report, Department of Mechanical Engineering, Massachusetts Institute of Technology (MIT), Cambridge, MA.

Harvald, S. (1983). *Resistance and propulsion of ships.* John Wiley & Sons, New York, NY.

ITTC (2008). *Predicting powering margins.* International Towing Tank Conference, Recommended Procedures and Guidelines 7.5-02-03-01.5. Revision 1.

Kerwin, J. and Hadler, J. (2010). *Propulsion.* The Principles of Naval Architecture Series. Society of Naval Architects and Marine Engineers (SNAME), Jersey City, NJ.

Oosterveld, M. and van Oossanen, P. (1975). Further computer-analyzed data of the Wageningen B-screw series. *International Shipbuilding Progress,* 22:251–261.

Stasiak, J. (2004). Service margin – solution of the problem or a problem waiting for solution? *Polish Maritime Research,* 11(Special issue):29–33.

Self Study Problems

1. State input and output for all propeller design tasks.
2. Derive the design constant for all propeller design tasks.
3. The engine chosen for a ship generates a delivered power of $P_D = 22\,000$ kW at a propeller rate of revolution $n = 99$ min^{-1}. Use the open water diagram for B4-70 to estimate the optimum propeller diameter D, pitch–diameter ratio P/D, and open water efficiency η_O when the number of blades is $Z = 4$ and the expanded area ratio is $A_E/A_0 = 0.7$. You can assume the propeller produces sufficient thrust to sail at a speed of 22.1 kn with a wake fraction of $w = 0.271$ and relative rotative efficiency η_R. Density of water is $\rho = 1026.021$ kg/m^3 and gravitational acceleration is $g = 9.807$ m/s^2. Do not forget to convert everything to consistent units and that the open water diagram shows $10K_Q$.

49

Hull–Propeller Matching Examples

In this chapter we determine basic propeller characteristics using propeller design charts and a cavitation criterion. Two examples are presented:

- solution of the optimum rate of revolution problem for a container ship (fourth propeller design task)
- solution of the optimum diameter problem for a bulk carrier (first propeller design task)

The examples make use of propeller design charts which have been redrawn on the basis of the Wageningen B-series polynomials (47.1) and (47.2) by Oosterveld and van Oossanen (1975). The 5% back cavitation limit is used as the cavitation criterion.

Learning Objectives

At the end of this chapter students will be able to:

- set up and solve propeller design tasks
- apply criteria for cavitation prevention
- find the optimum propeller for a ship
- predict propulsive power requirements

49.1 Optimum Rate of Revolution Problem

A performance estimate for a container ship yields the basic data listed in Table 49.1. We aim for a slow running diesel engine as the main propulsor. An early design formula for container ships by Kristensen and Lützen (2012) is used to estimate the maximum propeller diameter.

$$D = 0.623\,T_{\text{max}} - 0.16\text{ m} \tag{49.1}$$

With the data from Table 49.1 this yields a propeller diameter of

$$D = 0.623 \cdot 12.5\text{m} - 0.16\text{ m} = 7.6275\text{ m} \tag{49.2}$$

Fundamentals of Ship Hydrodynamics: Fluid Mechanics, Ship Resistance and Propulsion, First Edition.
Lothar Birk.

Companion website: www.wiley.com/go/birk/hydrodynamics

Table 49.1 Optimum rate of revolution problem – example input data for a container ship

Data for design task 4	
length in waterline	$L_{WL} = 211.08$ m
beam	$B = 32.20$ m
draft	$T = 11.00$ m
maximum draft	$T_{\max} = 12.50$ m
wetted surface	$S = 8560.83$ m^2
design speed	$v_S = 20.50$ kn
calm water resistance (design speed)	$R_T = 1149.06$ kN
requested service margin	$\Delta s = 15\%$
nonuniform wake adjustment	$\Delta\delta = 5\%$
relative rotative efficiency	$\eta_R = 1.009$
wake fraction	$w = 0.2865$
thrust deduction fraction	$t = 0.190$
number of blades	$Z = 5$
propeller diameter	$D = 7.6275$ m
shaft submergence	$h_0 = 6.60$ m
air pressure	$p_A = 101325.00$ Pa
vapor pressure (salt water, 15 °C)	$p_v = 1671.00$ Pa
density (salt water, 15 °C)	$\rho_S = 1026.021$ kg/m^3
gravitational acceleration	$g = 9.807$ m/s^2

This diameter should be checked against a preliminary lines plan to ensure that propeller clearances and shaft submergence are sufficient.

49.1.1 Design constant

Identifying the design task

Our first step is to define the design task by analyzing the given data. We are given ship speed, resistance, hull–propeller interaction parameters, and a maximum propeller diameter. Resistance at design speed, service margin, and thrust deduction fraction allow us to calculate the required thrust. From ship speed and wake fraction follows the speed of advance of the propeller. Summarizing, we know:

- required thrust $T_{\text{req}} = \dfrac{R_{T_{\text{service}}}}{(1-t)}$
- speed of advance $v_A = (1-w)v_S$, and
- propeller diameter D.

This indicates we have to solve propeller design task 4. We must select a sufficiently high expanded blade area ratio A_E/A_0 to avoid cavitation and find optimum values for rate of revolution n, pitch–diameter ratio P/D, and the open water efficiency η_O.

For any design task it is important to note that the propeller should be optimized for a realistic service condition and *not* for the ideal calm water conditions at the ship's trial. The ship contract usually specifies a service margin. It depends on the desired service area. Small margins (~ 5%) are used for sheltered waterways, large margins for oceans with severe weather like, for instance, the North Atlantic in winter (suggested margin ~ 35%; Harvald, 1983).

Service condition

With the service margin given in Table 49.1, the resistance for service condition is

Resistance in service condition

$$R_{T_{\text{service}}} = \left(1 + \Delta s\right) R_T = \left(1 + 0.15\right) 1149.06\,\text{kN} \qquad (49.3)$$
$$= 1321.419\,\text{kN}$$

Based on this in service resistance we determine the required thrust:

$$T_{\text{req}} = \frac{R_{T_{\text{service}}}}{(1-t)} = \frac{1321.419\,\text{kN}}{\left(1 - 0.190\right)} \qquad (49.4)$$
$$= 1631.3815\,\text{kN}$$

For the speed of advance we obtain

$$v_A = (1-w)v_S = \left(1 - 0.2865\right) 20.5\,\text{kn} \cdot \frac{1852\,\frac{\text{m}}{\text{M}}}{3600\,\frac{\text{s}}{\text{h}}}$$
$$= 0.7135 \cdot 10.5461\,\frac{\text{m}}{\text{s}} \qquad (49.5)$$
$$= 7.5247\,\frac{\text{m}}{\text{s}}$$

The design constant for task 4 is the shortened thrust loading coefficient which we employed before in the ITTC 1978 performance prediction method.

Design constant

$$C_S = \frac{K_T}{J^2} = \frac{T_{\text{req}}}{\rho\, v_A^2\, D^2} \qquad (49.6)$$
$$= \frac{1631381.5\,\text{N}}{1026.021\,\frac{\text{kg}}{\text{m}^3} \cdot 7.6275^2\text{m}^2 \left(7.5247\,\frac{\text{m}}{\text{s}}\right)^2}$$
$$= 0.482683$$

Do not forget to convert everything into a consistent set of units. In this case, the thrust has to be converted from kilo Newton [kN] into Newton and the speed of advance has to be expressed in m/s.

49.1.2 Initial expanded area ratio

In order to select an appropriate design chart we have to estimate an initial value for the required expanded area ratio. Keller's formula can be used for this.

Keller's formula

$$\left(\frac{A_E}{A_0}\right)_{\text{req}} = \frac{\left(1.3 + 0.3\,Z\right) T_{\text{req}}}{\left(p_0 - p_v\right) D^2} + K \qquad \text{Keller's formula} \qquad (46.1)$$

The parameter K in Keller's formula (46.1) is not the form factor but a constant selected based on ship type. The value $K = 0.2$ is used for single screw merchant vessels. The total static pressure p_0 is taken at the propeller shaft level. It consists of atmospheric pressure and the hydrostatic pressure.

$$\begin{aligned} p_0 &= p_A + \rho\, g\, h_0 \\ &= 101325\,\text{Pa} + 1026.021 \cdot 9.807 \cdot 6.6\,\text{Pa} \\ &= 167735.44\,\text{Pa} \end{aligned} \tag{49.7}$$

Note that the atmospheric pressure represents the larger share of the total static pressure and cannot be ignored.

Substituting all values into Keller's formula (46.1) yields

$$\begin{aligned} \left(\frac{A_E}{A_0}\right)_{\text{req}} &= \frac{\left(1.3 + 0.3 \cdot 5\right) 1631381.5\,\text{N}}{\left(167735.44\,\text{Pa} - 1671\,\text{Pa}\right) 7.6275^2\,\text{m}^2} + 0.2 \\ &= 0.6729 \end{aligned} \tag{49.8}$$

We will start with a design chart for the expanded blade area ratio $A_E/A_0 = 0.60$ (see Figure 49.1).

49.1.3 First iteration

Design chart selection The B_{u2}-chart (Figure 49.1) used here has been calculated and redrawn on the basis of the Wageningen B-Series polynomials by Oosterveld and van Oossanen (1975). Original charts can be obtained from MARIN (http://www.marin.nl).

Design chart application We draw a vertical line for the value of the fourth root of our design constant C_S Equation (49.6).

$$\sqrt[4]{C_S} = \left[\frac{K_T}{J^2}\right]^{\frac{1}{4}} = 0.833519 \tag{49.9}$$

The intersection of the vertical line and the curve of the optimum efficiency $\eta_{O_{\text{opt}}}$ marks the best achievable self propulsion point for open water conditions. We read off the following values:

open water efficiency	$\eta_O = 0.6244$
pitch–diameter ratio	$(P/D)_O = 1.0314$
reciprocal advance coefficient	$\delta_O = \frac{1}{J_O} = 1.4809$

A value of $\delta_O = 1/J_O = 1.4809$ is equivalent to an advance coefficient of

$$J_O = \frac{1}{\delta_O} = 0.6753 \tag{49.10}$$

This, in turn, provides the optimum rate of revolution in open water conditions (uniform wake):

$$n_O = \frac{v_A}{D\, J_O} = \frac{7.5247\,\frac{\text{m}}{\text{s}}}{7.6275\,\text{m} \cdot 0.6753} = 1.4609\,\text{s}^{-1} \tag{49.11}$$

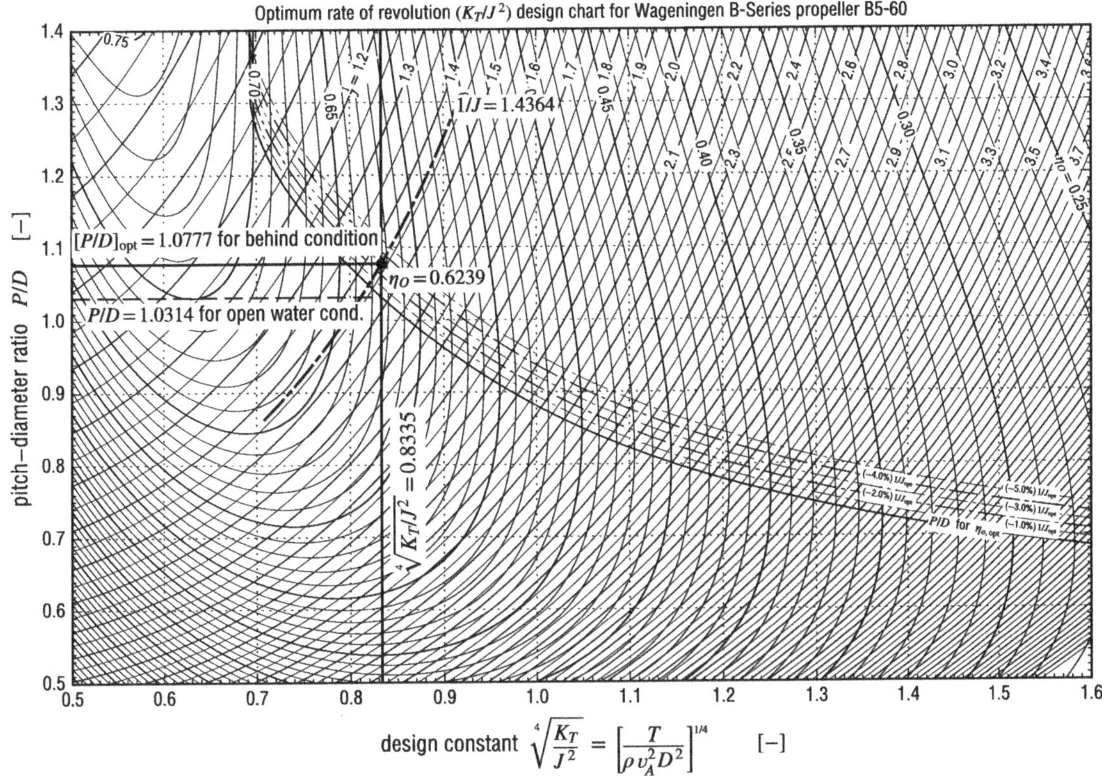

Figure 49.1 Propeller design B_{u2}-chart for Wageningen B-series propeller B5-60. Calculated and plotted based on the K_T, K_Q polynomials by Oosterveld and van Oossanen (1975)

The thrust coefficient for the open water propeller will be

$$K_T = C_S J_O^2 = 0.482683 \cdot 0.6753^2 = 0.2201 \tag{49.12}$$

Of course, it produces the required thrust used as input. However, we are not quite done.

Adjustment for behind condition

Experience shows that the optimum propeller for the behind condition with a nonuniform wake works at a slightly higher advance coefficient. The value of $\delta = 1/J$ is commonly *reduced* by 1% to 5% (Lewis, 1988). Vessels with fine lines or small propeller loading have small reductions, vessels with high block coefficient or high propeller loading have larger reductions in $1/J$. For design tasks where the rate of revolution n is given, a reduction in $1/J$ leads to a lower diameter D. In our case the diameter D is given. Thus, the rate of revolution n is reduced instead.

Advance coefficient for nonuniform wake

The container ship under consideration is sailing reasonably fast and will have normal lines. A medium reduction of $\Delta\delta = 3\%$ is chosen for this example. The new $1/J$ value for the behind condition is

$$\delta = (1 - \Delta\delta)\frac{1}{J_O} = 0.97 \cdot 1.4809 = 1.4364 \tag{49.13}$$

This is equivalent to a reduction in the rate of revolution for the behind condition to

$$n = (1 - \Delta\delta)n_O = 1.4171\,\mathrm{s}^{-1} \tag{49.14}$$

Reducing the rate of revolution for constant speed of advance and constant diameter is equivalent to an increase in advance coefficient J.

$$J = \frac{v_A}{n\,D} = \frac{7.5247\,\frac{\mathrm{m}}{\mathrm{s}}}{1.4171\,\mathrm{s}^{-1}\cdot 7.6275\,\mathrm{m}} = 0.6962 \tag{49.15}$$

Adjustment of pitch–diameter ratio

However, the thrust of a propeller is decreasing with increasing advance coefficient J. In fact, calculating the thrust coefficient K_T for a B-series propeller with $Z = 5$, $P/D = 1.0314$, $A_E/A_0 = 0.60$, and $J = 0.6962$ yields only $K_T = 0.2103$, which is lower than the required value of 0.2201. Together with the reduced rate of revolution, this would cause a significant loss in thrust ($T = 1466.75$ kN). As a consequence, the ship would not reach the desired speed.

To compensate for the loss in thrust caused by the reduction of rate of revolution, the propeller pitch has to be increased. The optimum pitch–diameter ratio for the best self propulsion point in behind condition can be taken off the B_{u2}-chart at the intersection of the $\sqrt[4]{C_S}$ line and the $(-3\%)\ 1/J_{\mathrm{opt}}$ curve plotted above the optimum open water efficiency curve. For this example we get (see Figure 49.1):

$$\frac{P}{D} = 1.0777 \qquad \text{adapted for nonuniform wake} \tag{49.16}$$

Checking with the open water data for the new propeller, we get a thrust coefficient of $K_T = 0.2339$ with a propeller thrust of $T_{\mathrm{prop}} = 1631.382$ kN, which is equal to the required thrust.

Adjustment of open water efficiency

Plotting the current propeller data into the B_{u2}-chart, we moved upward along the vertical line $\sqrt[4]{C_S} = 0.833519$ and away from the optimum efficiency. However, the slope of the efficiency curves is small relative to the line of constant $\sqrt{K_T/J^2}$. As a consequence, the reduction in open water efficiency will be small compared with the case without $1/J$ reduction:

$$\eta_O = 0.6239 \quad \text{for } \frac{P}{D} = 1.0777 \text{ and } J = 0.6962 \quad (\delta = 1.4364)$$

In addition, this reduces the danger of losing efficiency very quickly if the speed of advance is further reduced due to fouling of the hull and the operating point falls below the optimum efficiency line.

First iteration result

In summary, our first iteration results in an optimum propeller with the following characteristics:

Result of 1. iteration	
rate of revolution	$n = 1.4171\ \mathrm{s}^{-1}$
open water efficiency	$\eta_O = 0.6239$
diameter	$D = 7.6275$ m
pitch–diameter ratio	$P/D = 1.0777$
advance coefficient	$J = 0.6962$
expanded area ratio	$A_E/A_0 = 0.60$

49.1.4 Cavitation check for first iteration

Burrill cavitation chart

We now have to check whether the initial guess of the expanded blade area ratio $A_E/A_0 = 0.60$ is sufficient to keep the propeller free from excessive cavitation. For this task we employ the empirical cavitation charts by Burrill and Emerson (1963). Figure 46.1 shows a replication of this chart based on regression curves (see Chapter 46). The curves represent maximum thrust loading coefficients τ_c for different amounts of cavitation on the blade. We will use the curve labeled 5% *back cavitation* which is similar to the suggested upper limit for merchant ship propellers (see Table 46.1). For programming and calculation purposes, it can be expressed as the following regression curve.

$$\tau_c = 0.715\,\sigma_b^{0.184} - 0.437 \qquad \text{5\% back cavitation limit curve} \tag{49.17}$$

Static pressure and flow velocity

The propeller cavitation number σ_b is computed using the total static pressure at the depth of the propeller shaft and a simplified flow velocity v_1 for the blade section at radius $r = 0.7R$. The total static pressure is equal to the atmospheric pressure p_A plus the hydrostatic pressure at the propeller shaft.

$$p_0 = p_A + \rho\, g\, h_0 \tag{49.18}$$

With a standard atmospheric pressure of $p_A = 101325$ Pa and a shaft submergence of $h_0 = 6.6$ m, we get

$$\begin{aligned} p_0 &= 101325\ \text{Pa} + 1026.021\ \text{kg/m}^3\ 9.807\ \text{m/s}^2\ 6.6\ \text{m} \\ &= 167735.44\ \text{Pa} \end{aligned}$$

The simplified flow velocity v_1 is

$$\begin{aligned} v_1 &= \sqrt{v_A^2 + (0.7\,\pi\, n\, D)^2} \qquad (49.19) \\ &= \sqrt{7.5247^2\ (\text{m/s})^2 + \left(0.7\,\pi\ 1.4171\ \text{s}^{-1}\ 7.6275\ \text{m}\right)^2} \\ &= 24.9322\ \text{m/s} \end{aligned}$$

Propeller cavitation number

With this input we can compute the propeller cavitation number

$$\sigma_b = \frac{p_0 - p_v}{\frac{1}{2}\rho\, v_1^2} \tag{49.20}$$

The vapor pressure for seawater at 15 °C is $p_v = 1671$ Pa (ITTC, 2011). Therefore, we get

$$\begin{aligned} \sigma_b &= \frac{167735.44\ \text{Pa} - 1671\ \text{Pa}}{\frac{1}{2}\,1026.021\ \text{kg/m}^3\ 24.9322^2\ (\text{m/s})^2} \\ &= 0.5207 \qquad (49.21) \end{aligned}$$

Required projected area

Based on the cavitation number $\sigma_b = 0.5207$, a thrust loading coefficient of $\tau_c = 0.1971$

is computed with Equation (49.17). The thrust loading coefficient is connected to the minimum required projected area A_P of the propeller by

$$A_{P_{\text{req}}} = \frac{T_{\text{prop}}}{\frac{1}{2}\rho v_1^2 \tau_c} \tag{46.7}$$

Projected area may be converted into developed area A_D using a formula by Admiral David W. Taylor.

$$A_D = \frac{A_P}{\left(1.067 - 0.229 \frac{P}{D}\right)} \tag{49.22}$$

or

$$A_P = \left(1.067 - 0.229 \frac{P}{D}\right) A_D \tag{49.23}$$

For propellers with zero or moderate rake, expanded blade area is equal to the developed area.

$$A_E \approx A_D \tag{49.24}$$

Required expanded area ratio Substituting these two results into Equation (46.7) and dividing both sides by the propeller disk area $A_0 = \pi D^2/4$ yields

$$\left(\frac{A_E}{A_0}\right)_{\text{req}} = \frac{T_{\text{prop}}}{\frac{1}{2}\rho v_1^2 \tau_c \left(1.067 - 0.229 \frac{P}{D}\right) \frac{\pi D^2}{4}} \tag{46.10}$$

With the data for the propeller derived in the first iteration, the Burrill criterion asks for a required minimum expanded area ratio of

$$\left(\frac{A_E}{A_0}\right)_{\text{req}} = 0.6926 \tag{49.25}$$

to prevent excessive cavitation. This is more than the value from our first design chart ($A_E/A_0 = 0.6$). We will have to perform a second iteration using a higher expanded area ratio.

49.1.5 Second iteration

Design chart application In the second iteration, we make use of the B_{u2}-chart for an expanded blade area ratio of $A_E/A_0 = 0.75$. The design constant C_S remains unchanged because our thrust requirements for the ship have not changed. Performing the same steps as above:

- read off values for η_O, $(P/D)_O$, and $1/J_O$
- calculate diameter
- reduce diameter and derive new pitch

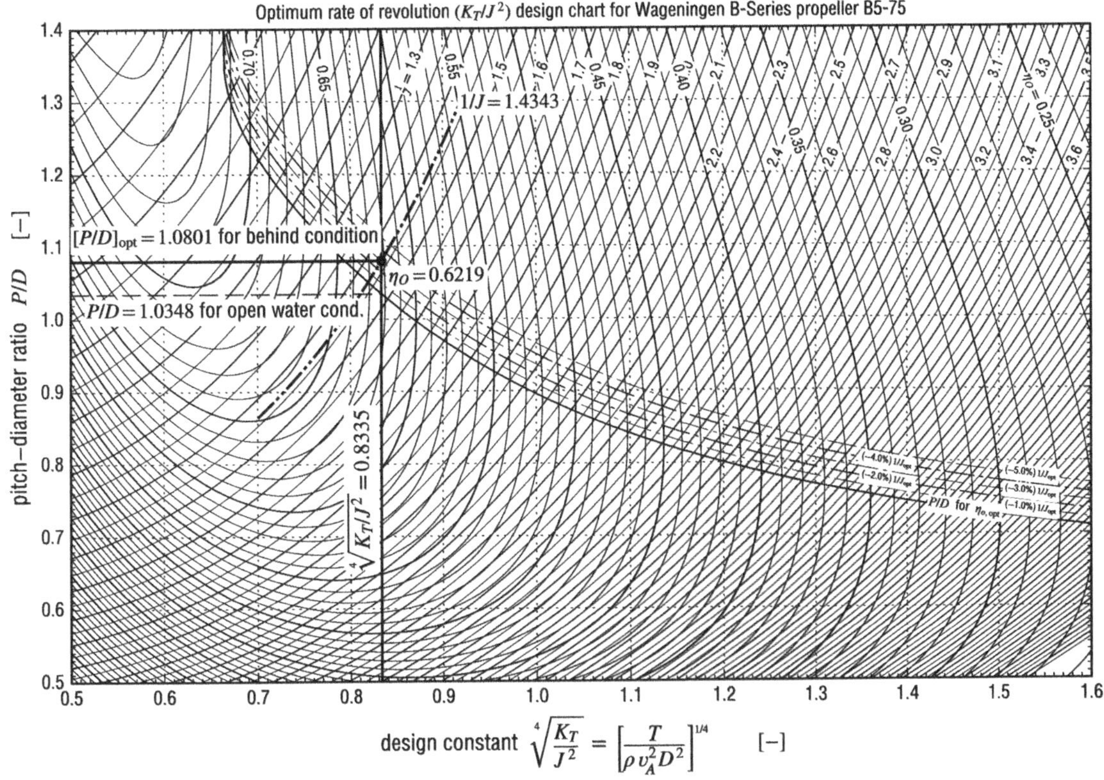

Figure 49.2 Propeller design B_{u2}-chart for Wageningen B-series propeller B5-75. Calculated and plotted based on the K_T, K_Q polynomials by Oosterveld and van Oossanen (1975)

yields the following set of propeller parameters:

$$\begin{aligned} &\text{open water efficiency} & \eta_O &= 0.6224 \\ &\text{pitch–diameter ratio} & (P/D)_O &= 1.0348 \\ &\text{reciprocal advance coefficient} & \delta_O = \frac{1}{J_O} &= 1.4787 \end{aligned}$$

Reduction of $1/J$ by 3% yields the following optimum propeller data for the behind condition:

Second iteration result

Result of 2. iteration	
rate of revolution	$n = 1.4150\ \text{s}^{-1}$
open water efficiency	$\eta_O = 0.6219$
diameter	$D = 7.6275\ \text{m}$
pitch–diameter ratio	$P/D = 1.0801$
advance coefficient	$J = 0.6972$
expanded area ratio	$A_E/A_0 = 0.75$

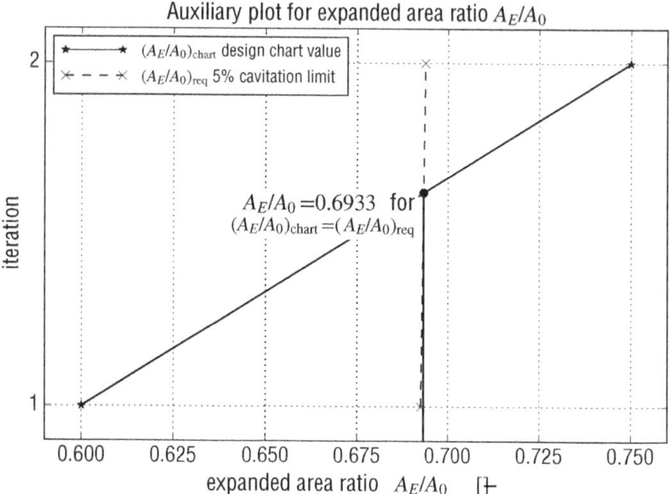

Figure 49.3 Auxiliary plot to determine the expanded area ratio of the final optimum propeller

This propeller produces the required thrust of $T_{prop} = 1631.3815$ kN. A renewed cavitation check with Burrill's chart, i.e. Equation (49.17), yields a required expanded area ratio of

$$\left(\frac{A_E}{A_0}\right)_{req} = 0.6938 \tag{49.26}$$

This is less than the chart value of $A_E/A_0 = 0.75$ we used. A lower expanded area ratio typically results in higher efficiency because the reduced blade surfaces have smaller frictional losses.

49.1.6 Final selection by interpolation

We find the final propeller characteristics by interpolating between the results of the first and second iterations.

Expanded area ratio

First, the expanded area ratio is found by comparing the expanded area ratios we used (from the charts) and the required expanded area ratios according to Equation (46.10). We draw an auxiliary diagram which plots the actually used expanded area ratios and the required expanded area ratios for the two iterations (Figure 49.3). Actual and required expanded blade area ratios are equal where the two lines intersect.

$$\left(\frac{A_E}{A_0}\right)_{chart} = \left(\frac{A_E}{A_0}\right)_{req} \tag{49.27}$$

In our example, the intersection point yields

$$\left(\frac{A_E}{A_0}\right) = 0.6933 \tag{49.28}$$

Pitch–diameter and advance coefficient

For the final optimum propeller characteristics, we plot straight lines connecting the results of the first and second iterations for pitch–diameter ratio P/D, advance coefficient J, open water efficiency η_O, and rate of revolution n as a function of the expanded

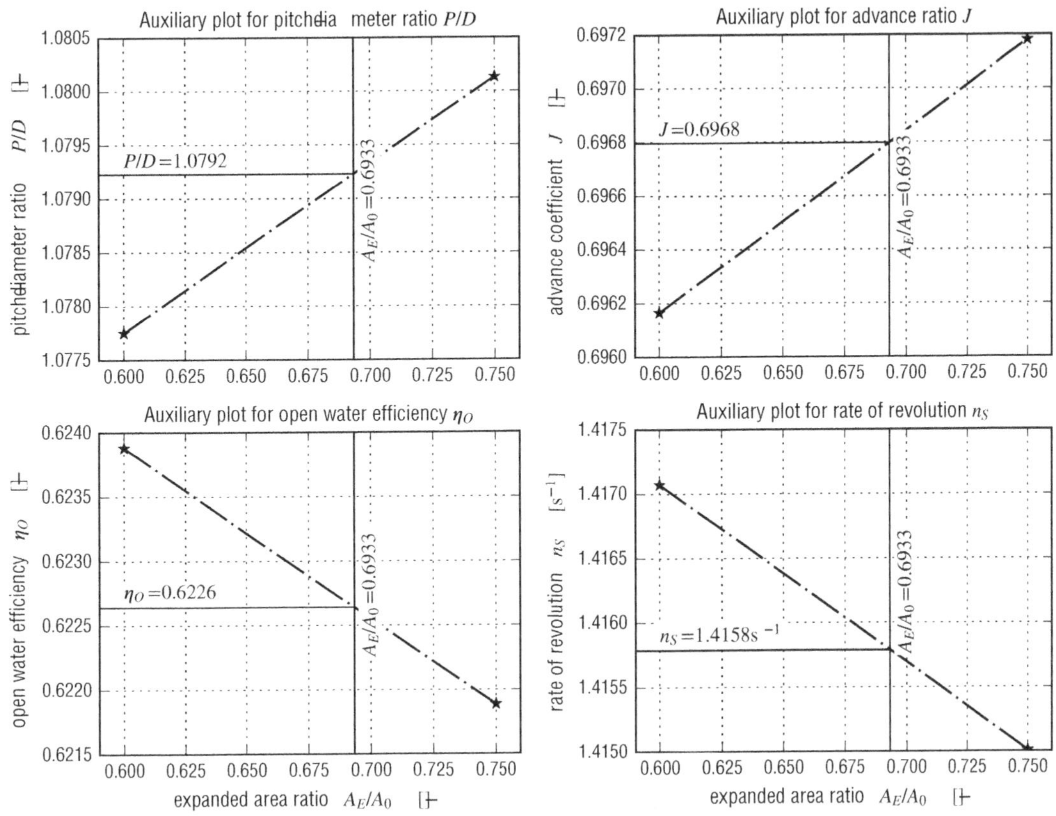

Figure 49.4 Auxiliary plots to determine final optimum propeller characteristics. The ★ mark results of first and second iterations

blade area ratio (Figure 49.4). The graph for the rate of revolution is not really necessary as n could be recomputed from the final advance coefficient J via Equation (49.11).

Final selection

Final values are read off the plots or found by computing the intersection with the vertical line for $A_E/A_0 = 0.6933$.

Result of final selection	
rate of revolution	$n = 1.4158\ \text{s}^{-1}$
open water efficiency	$\eta_O = 0.6226$
diameter	$D = 7.6275\ \text{m}$
pitch–diameter ratio	$P/D = 1.0792$
advance coefficient	$J = 0.6968$
expanded area ratio	$A_E/A_0 = 0.6933$

This propeller provides the required thrust. Checking with the Burrill chart for a third time yields a required expanded area ratio of $A_E/A_0 = 0.6937$ which confirms our final selection.

Table 49.2 Optimum diameter problem – example input data for a bulk carrier

Data for design task 1	
length in waterline	$L_{WL} = 172.700$ m
beam	$B = 28.000$ m
draft	$T = 10.200$ m
maximum draft	$T_{\max} = 10.650$ m
wetted surface	$S = 7079.440$ m^2
design speed	$v_S = 14.500$ kn
requested service margin	$\Delta s = 20\%$
nonuniform wake adjustment	$\Delta\delta = 5\%$
brake power of engine	$P_B = 7650.000$ kW
target rate of revolution	$n = 116.000$ rpm
shafting efficiency	$\eta_S = 0.980$
relative rotative efficiency	$\eta_R = 1.000$
wake fraction	$w = 0.3414$
thrust deduction fraction	$t = 0.2541$
number of blades	$Z = 4$
shaft submergence	$h_0 = 7.000$ m
air pressure	$p_A = 101325.000$ Pa
vapor pressure	$p_v = 1671.000$ Pa
density of salt water at 15 °C	$\rho_S = 1026.021$ kg/m^3
gravitational acceleration	$g = 9.807$ m/s^2

Summary Based on this preliminary propeller selection, the propulsive power estimate may be completed. Typically, wake fraction and thrust deduction fraction depend on some of the propeller characteristics. If the final estimates deviate from the data used for the propeller selection (see Table 49.1), the propeller selection process must be repeated until a converged and consistent solution is obtained. In later design stages the propeller selection should be refined by employing a higher fidelity computational method, for instance a lifting line calculation which would allow to adapt the pitch distribution to the estimated, measured, or computed wake field of the hull. At this point Reynolds number effects may be considered as well.

49.2 Optimum Diameter Problem

Using the Guldhammer and Harvald method, a performance estimate has been completed for a handy size bulk carrier. Table 49.2 lists the basic ship data and Table 49.3 summarizes the resistance estimate in the range of the design speed.

Owner specifies engine The future owner of the bulk carrier already has similar ships in his or her fleet and wants the new design to use the same main engine. Therefore, brake power and rate of revolution are used as input.

49.2.1 Design constant

Our first step is to define the design task by analyzing the given data. We are given speed, brake power, rate of revolution, and hull-propeller interaction parameters. Based on brake power and shaft efficiency we can readily compute the available delivered power. The resistance, service margin, and thrust deduction fraction allow us to calculate the required thrust. From ship speed and wake fraction follows the speed of advance of the propeller. In summary, we know:

Identify Design Task

- delivered power $P_D = P_B \eta_S$
- speed of advance $v_A = (1 - w)v_S$, and
- rate of revolution n.

This indicates we have to solve propeller design task 1. We must select a sufficiently high expanded blade area ratio A_E/A_0 to avoid cavitation and find optimum values for the propeller diameter D, the pitch–diameter ratio P/D, and the open water efficiency η_O.

For any design task it is important to note that the propeller should be optimized for a realistic service condition and *not* for the ideal calm water conditions at the ship's trial. The service margin specified for this design is $\Delta s = 20\%$.

With the data given in Table 49.2, we obtain the available delivered power

$$P_D = P_B \eta_S = 7650\,\text{kW} \cdot 0.98 = 7497.00\,\text{kW} \tag{49.29}$$

The speed of advance is:

$$\begin{aligned} v_A &= (1 - w)v_S \qquad (49.30) \\ &= (1 - 0.3413)\,14.5\,\text{kn} \cdot \frac{1852\,\frac{\text{m}}{\text{M}}}{3600\,\frac{\text{s}}{\text{h}}} \\ &= 0.6587 \cdot 7.4594\,\frac{\text{m}}{\text{s}} \\ &= 4.9135\,\frac{\text{m}}{\text{s}} \end{aligned}$$

The design constant for task 1 is

Design constant

$$\begin{aligned} \left[\frac{K_Q}{J^5}\right] &= \frac{P_D\, n^2\, \eta_R}{2\pi\, \rho\, v_A^5} \qquad (49.31) \\ &= \frac{7497000\,\text{W} \cdot 1.9333^2\,\text{s}^{-2} \cdot 1.0}{2\pi\, 1026.021\,\frac{\text{kg}}{\text{m}^3}\left(4.9135\,\frac{\text{m}}{\text{s}}\right)^5} \\ &= 1.517729 \end{aligned}$$

Do not forget to convert everything into a consistent set of units. In this case, the delivered power has to be converted from kilo Watt (kW) into Watt and the rate of revolution has to be expressed in $1/\text{s}$.

Table 49.3 Resistance estimate for bulk carrier from Table 49.2

Resistance estimate	
v_S [kn]	R_T (trial) [kN]
13.5	427.54
14.0	468.92
14.5	514.62
15.0	565.36
15.5	623.51

49.2.2 Initial expanded area ratio

Keller's formula

In order to select an appropriate design chart, we have to estimate an initial value for the minimum expanded blade area ratio. Keller's formula can be used for this. The parameter K in Keller's formula Equation (46.1) is not the form factor but a constant selected based on ship type. The value $K = 0.2$ is used for single propeller merchant vessels. The total static pressure p_0 is taken at the propeller shaft level. It consists of atmospheric pressure and the hydrostatic pressure.

$$\begin{aligned} p_0 &= p_A + \rho\, g\, h_0 \\ &= 101325\,\text{Pa} + 1026.021 \cdot 9.807 \cdot 7.0\,\text{Pa} \\ &= 171760.32\,\text{Pa} \end{aligned} \tag{49.32}$$

The required thrust T at design speed is estimated using the total resistance for trial conditions R_T, the service margin Δs, and the thrust deduction fraction t.

$$\begin{aligned} T_{\text{req}} &= \frac{(1+\Delta s)R_T}{1-t} \\ &= \frac{(1+0.2)\,514620\,\text{N}}{1-0.2541} \\ &= 827.92\,\text{kN} \end{aligned} \tag{49.33}$$

The total resistance for the design speed $v_S = 14.5$ kn has been taken from Table 49.3.

In cases where a resistance estimate for the design speed is not yet available, one may first assume a quasi-propulsive efficiency η_D (for instance $\eta_D = 0.6$) and then compute effective power from delivered power.

$$P_E = P_D\,\eta_D \tag{49.34}$$

From the effective power follows the total resistance.

$$R_{T_{\text{service}}} = \frac{P_E}{v_S} \tag{49.35}$$

The resistance is considered the resistance for service condition since we started using all available delivered power. At the end of the selection process, we will check whether the propeller provides sufficient thrust.

Initial propeller diameter

An early design formula for bulk carriers by Kristensen and Lützen (2012) is used to estimate an initial propeller diameter.

$$D = 0.395\, T_{max} + 1.3 \text{ m} \tag{49.36}$$

With the data from Table 49.2 this yields a propeller diameter of

$$D = 0.395 \cdot 10.65\text{m} + 1.3 \text{ m} = 5.5068 \text{ m} \tag{49.37}$$

Again, a check of this estimate against a preliminary lines plan is warranted to ensure sufficient propeller clearances.

Substituting all values into Keller's formula (46.1) yields

$$\begin{aligned}\left(\frac{A_E}{A_0}\right)_{req} &= \frac{(1.3 + 0.3 \cdot 4)\, 827918 \text{ N}}{(171735.32 \text{ Pa} - 1671 \text{ Pa})\, 5.5068^2 \text{ m}^2} + 0.2 \\ &= 0.6013\end{aligned} \tag{49.38}$$

We will start with a design chart for the expanded blade area ratio $A_E/A_0 = 0.55$ (see Figure 49.5).

49.2.3 First iteration

Selecting the design chart

The B_{p1}-chart (Figure 49.5) used here has been calculated and redrawn on the basis of the Wageningen B-Series polynomials by Oosterveld and van Oossanen (1975). Original charts can be obtained from MARIN (http://www.marin.nl).

Reading values off design chart

We draw a vertical line for the value of the fourth root out of our design constant C_n Equation (49.31).

$$\sqrt[4]{C} = \left[\frac{K_Q}{J^5}\right]^{\frac{1}{4}} = 1.109938 \tag{49.39}$$

For the intersection of the vertical line and the curve of the optimum efficiency $\eta_{O_{opt}}$ we read off the following values:

open water efficiency	$\eta_O = 0.5330$
pitch–diameter ratio	$(P/D)_O = 0.6876$
reciprocal advance coefficient	$\delta = \frac{1}{J_O} = 2.4053$

The reciprocal advance coefficient is equivalent to an advance coefficient of

$$J_O = 0.4158 \tag{49.40}$$

This, in turn, gives us the optimum diameter for open water conditions (uniform wake):

$$D_O = \frac{v_A}{D\, J_O} = \frac{4.9135 \frac{\text{m}}{\text{s}}}{1.9333 \text{ s}^{-1} \cdot 0.4158} = 6.1130 \text{ m} \tag{49.41}$$

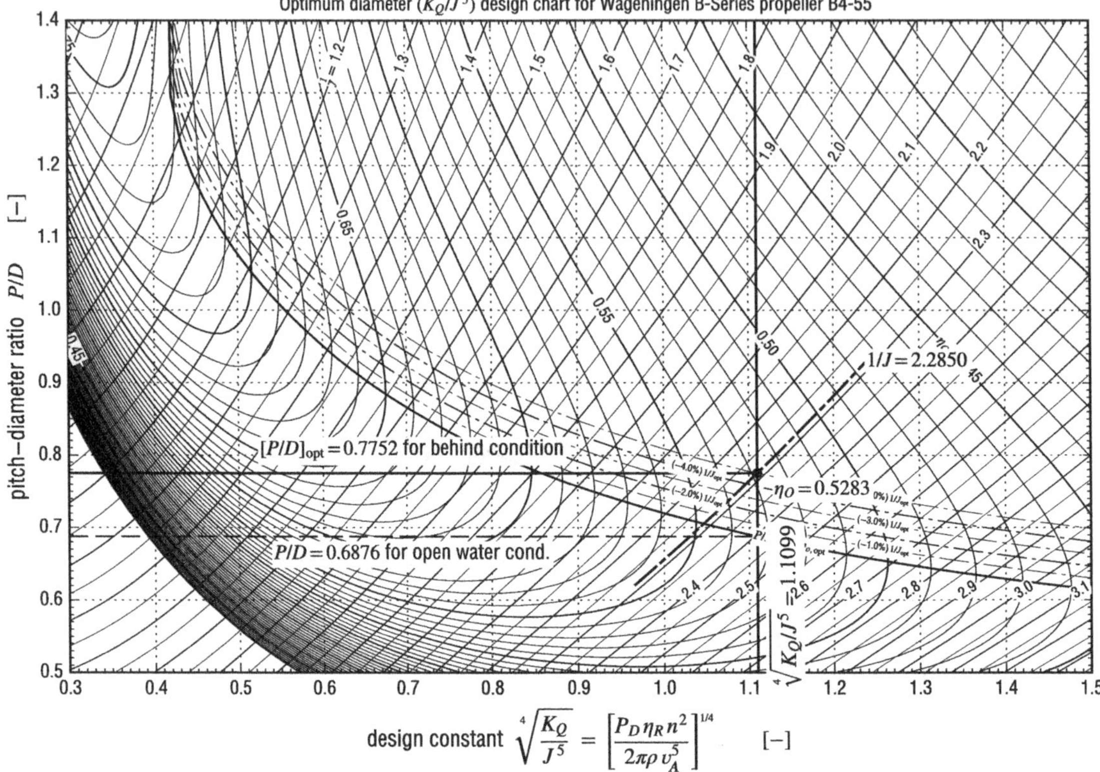

Figure 49.5 Propeller design B_{p1}-chart for Wageningen B-series propeller B4-55. Calculated and plotted based on the K_T, K_Q polynomials by Oosterveld and van Oossanen (1975)

Selection of a nonuniform wake adjustment

As stated before, experience shows that the optimum propeller for the behind condition with a nonuniform wake works at a slightly higher advance coefficient. The value of $\delta = 1/J$ is commonly reduced by 1% – 5% (Lewis, 1988). Vessels with fine lines or small propeller loading have small reductions, vessels with high block coefficient or high propeller loading have larger reductions in $1/J$. For design tasks where the rate of revolution n is given, a reduction in $1/J$ leads to a lower diameter D. In our case, the rate of revolution n is given and the diameter D is reduced. The bulk carrier is a fairly full vessel which usually leads to larger variations in the wake. Therefore, a reduction of 5% is chosen for this example.

Nonuniform wake adjustment

The new $1/J$ value for the behind condition is

$$\delta_B = (1 - \Delta\delta)\frac{1}{J_O} = 0.95 \cdot 2.4053 = 2.2850 \qquad (49.42)$$

This results in an equivalent reduction of the propeller diameter for the behind condition of

$$D = \left(1 - \Delta\delta\right) D_O = 5.8073\,\text{m} \qquad (49.43)$$

Reducing the diameter for constant speed of advance and rate of revolution is equal to an increase in advance coefficient.

$$J = \frac{v_A}{n\,D} = \frac{4.9135\,\frac{\text{m}}{\text{s}}}{1.9333\,1s \cdot 5.8073\,\text{m}} = 0.4376 \tag{49.44}$$

With increasing advance coefficient J, the torque a propeller absorbs is decreasing. As a consequence, the engine would speed up until the torque absorbed by the propeller and the torque delivered by the engine are in equilibrium again. To compensate for the loss in torque caused by the reduction of the diameter, pitch has to be increased. The optimum pitch–diameter ratio for the behind condition can be taken off the B_{p1}-chart at the intersection of the $\sqrt[4]{C}$ line and the (−5%) $1/J_{\text{opt}}$ curve plotted above the optimum open water efficiency curve. Here

$$\frac{P}{D} = 0.7752 \qquad \text{adapted for behind condition} \tag{49.45}$$

Rule of thumb

If you have a propeller design chart without these curves, the following rule of thumb may be applied: When changing diameter or pitch, keep the sum of pitch and diameter constant to achieve the same thrust:

$$P + D = P_O + D_O \tag{49.46}$$

We solve this for the new pitch for the behind condition.

$$P = \left[1 + \left(\frac{P}{D}\right)_O\right] D_O - D \tag{49.47}$$

$$= [1 + 0.6876] \cdot 6.1130\,\text{m} - 5.8073\,\text{m}$$

$$= 4.5090\,\text{m} \tag{49.48}$$

This, in turn, results in a new (increased) pitch–diameter ratio for the propeller of $P/D = 0.7764$ in behind condition. It is close to the value we read off the chart with the (−5%) $1/J_{\text{opt}}$ curve.

By adjusting $1/J$ for nonuniform wake and subsequently increasing the pitch–diameter ratio, we moved away from the optimum efficiency. The new open water efficiency will be slightly smaller:

$$\eta_O = 0.5283 \quad \text{for } \frac{P}{D} = 0.7752 \text{ and } J = 0.4376 \quad (\delta_B = 2.2850)$$

First iteration result

In summary, the first iteration yields the following optimum propeller characteristics:

Result of 1. iteration	
rate of revolution	$n = 1.9333\ \text{s}^{-1}$
open water efficiency	$\eta_O = 0.5283$
diameter	$D = 5.8073\ \text{m}$
pitch–diameter ratio	$P/D = 0.7752$
advance coefficient	$J = 0.4376$
expanded area ratio	$A_E/A_0 = 0.55$

49.2.4 Cavitation check for first iteration

Cavitation limit

The optimum diameter obtained during the first iteration is quite a bit larger than our initial estimate for Keller's formula. Therefore, the initially selected expanded blade area ratio $A_E/A_0 = 0.55$ might be too pessimistic. Again, we employ the empirical cavitation charts by Burrill and Emerson (1963) and use the curve labeled 5% *back cavitation* as limiting criteria (see Table 46.1). For programming and calculation purposes, it can be expressed as a regression curve (see Equation (49.17)).

Static pressure and flow velocity

The local cavitation number σ_b is computed using the total static pressure at the depth of the propeller shaft and a simplified flow velocity v_1 for the blade section at radius $r = 0.7R$. The total static pressure is equal to the atmospheric pressure p_A plus the hydrostatic pressure at the propeller shaft.

$$p_0 = p_A + \rho\, g\, h_0 \tag{49.49}$$

With a standard atmospheric pressure of $p_A = 101325$ Pa and a shaft submergence of $h_0 = 7.0$ m, we get:

$$\begin{aligned} p_0 &= 101325\,\text{Pa} + 1026.021\,\text{kg/m}^3\, 9.807\,\text{m/s}^2\, 7.0\,\text{m} \\ &= 171\,760.32\,\text{Pa} \end{aligned}$$

The simplified flow velocity v_1 is

$$\begin{aligned} v_1 &= \sqrt{v_A^2 + (0.7\,\pi\, n\, D)^2} \qquad (49.50) \\ &= \sqrt{4.9135^2\,(\text{m/s})^2 + \left(0.7\,\pi\, 1.9333\,\text{s}^{-1}\, 5.8073\,\text{m}\right)^2} \\ &= 25.1748\,\text{m/s} \end{aligned}$$

Local cavitation number

With this input we compute the local cavitation number:

$$\sigma_b = \frac{p_0 - p_v}{\frac{1}{2}\rho\, v_1^2} \tag{49.51}$$

With a vapor pressure of $p_v = 1671$ Pa (ITTC, 2011) we get

$$\begin{aligned} \sigma_b &= \frac{171760.32\,\text{Pa} - 1671\,\text{Pa}}{\frac{1}{2}\, 1026.021\,\text{kg/m}^3\, 25.1748^2\,(\text{m/s})^2} \\ &= 0.5231 \end{aligned}$$

Required projected area

Based on the cavitation number $\sigma_b = 0.5231$, we get a thrust loading coefficient of $\tau_c = 0.1976$ from Equation (49.17). The thrust loading coefficient is connected to the minimum required projected area A_P of the propeller by

$$A_{P_{req}} = \frac{T_{prop}}{\frac{1}{2}\rho\, v_1^2\, \tau_c} \tag{49.52}$$

In order to proceed, we need an accurate estimate of the thrust generated by the propeller defined by the first iteration. Given J, P/D, A_E/A_0, and Z, the Wageningen B-series polynomials from Chapter 47.2 can be used to compute the thrust coefficient K_T and thus the propeller thrust. Using the polynomial (47.1), we obtain

$$K_T = 0.1848$$
$$T_{\text{prop}} = 806.10 \text{ kN}$$

If polynomials are not available, direct computational methods like blade element theory or lifting line theory might be employed.

Required expanded area ratio

As before, we use Equation (46.10) to compute the required expanded area ratio

$$\left(\frac{A_E}{A_0}\right)_{\text{req}} = \frac{T_{\text{prop}}}{\frac{1}{2}\,\rho\, v_1^2\, \tau_c \left(1.067 - 0.229\,\frac{P}{D}\right)\frac{\pi D^2}{4}} \tag{49.53}$$

With the data for the propeller derived in the first iteration, the 5% back cavitation criterion asks for a required minimum expanded area ratio of

$$\left(\frac{A_E}{A_0}\right)_{\text{req}} = 0.5325 \tag{49.54}$$

to prevent excessive cavitation. This is less than the value from our first design chart ($A_E/A_0 = 0.55$). A lower expanded area ratio typically results in higher efficiency because the reduced blade surfaces have smaller frictional losses. We will have to perform a second iteration using a lower expanded area ratio.

49.2.5 Second iteration

Design chart application

In the second iteration, we make use of the B_{p1}-chart for an expanded blade area ratio of $A_E/A_0 = 0.40$. The design constant C remains unchanged because the available delivered power for the ship has not changed. Performing the same steps as above, i.e.

- reading off values for η_O, $(P/D)_O$, and $1/J_O$,
- calculating diameter, and
- reducing diameter and deriving new pitch

yields the following set of propeller parameters:

open water efficiency	$\eta_O = 0.5363$
pitch–diameter ratio	$(P/D)_O = 0.6912$
reciprocal advance coefficient	$\delta = \frac{1}{J_O} = 2.3950$

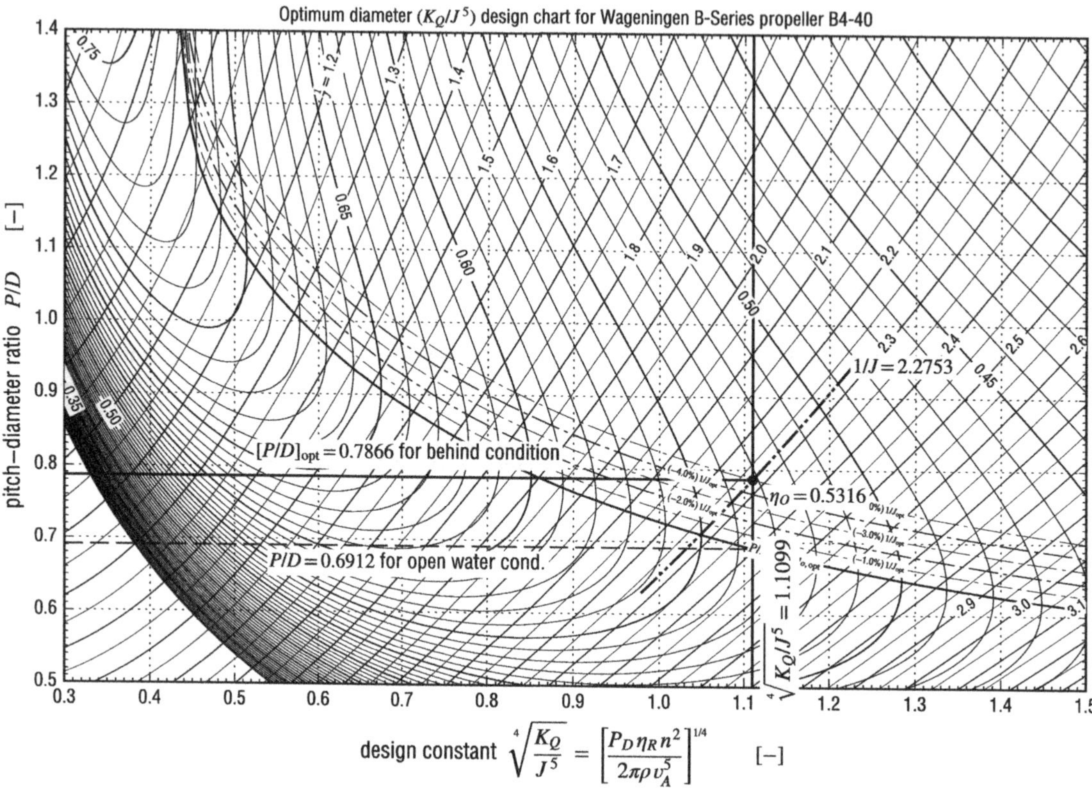

Figure 49.6 Propeller design B_{p1}-chart for Wageningen B-series propeller B4-40. Calculated and plotted based on the K_T, K_Q polynomials by Oosterveld and van Oossanen (1975)

Adjustment for nonuniform wake

Reduction of $1/J$ by $\Delta\delta = 5\%$ yields:

Result of 2. iteration	
rate of revolution	$n = 1.9333\ \text{s}^{-1}$
open water efficiency	$\eta_O = 0.5316$
diameter	$D = 5.7825$ m
pitch–diameter ratio	$P/D = 0.7866$
advance coefficient	$J = 0.4395$
expanded area ratio	$A_E/A_0 = 0.40$

Cavitation check

This propeller produces a thrust of $T_{\text{prop}} = 811.14$ kN. The cavitation check with Burrill's chart, i.e. Equation (49.17), yields a required expanded area ratio of

$$\left(\frac{A_E}{A_0}\right)_{\text{req}} = 0.5438 \tag{49.55}$$

This is more than the $A_E/A_0 = 0.40$ from the chart we used.

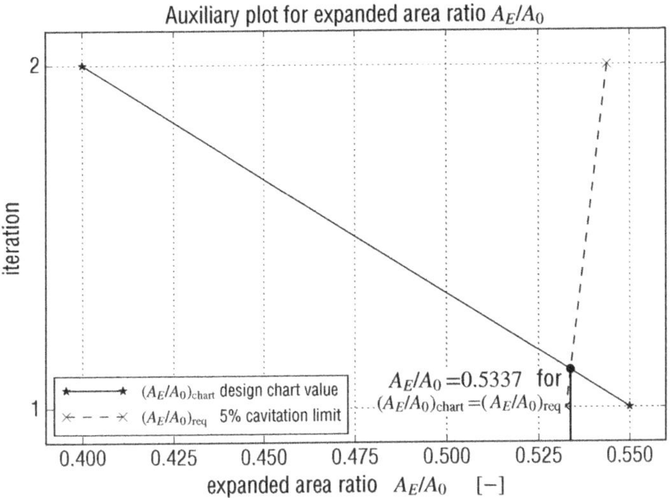

Figure 49.7 Auxiliary plot to determine the expanded blade area ratio of the final optimum propeller

49.2.6 Final selection by interpolation

We find the final propeller characteristics by interpolating between the results of the first and the second iterations.

Expanded area ratio

First, the expanded area ratio is found by comparing the expanded area ratios we used (from the charts) and the required expanded area ratios according to Equation (49.53). The auxiliary plot for the expanded blade area ratio (Figure 49.7) compares the actually used expanded blade area ratios and the required expanded blade area ratios for the two iterations. Where the two lines intersect, we find:

$$\left(\frac{A_E}{A_0}\right) = 0.5337 \tag{49.56}$$

Pitch–diameter ratio, advance coefficient

For the final optimum propeller characteristics, we plot straight lines connecting the results of the first and the second iterations for pitch–diameter ratio P/D, advance coefficient J, open water efficiency η_O, and diameter D as a function of the expanded area ratio (Figure 49.8). The fourth graph for the diameter may be omitted and the diameter recomputed from the final advance coefficient J via Equation (49.41).

Final selection

Final values are read off the plots or found by computing the intersection with the vertical line for $A_E/A_0 = 0.5337$.

Result of final selection	
rate of revolution	$n = 1.9333\ \text{s}^{-1}$
open water efficiency	$\eta_O = 0.5287$
diameter	$D = 5.8046\ \text{m}$
pitch–diameter ratio	$P/D = 0.7764$
advance coefficient	$J = 0.4378$
expanded area ratio	$A_E/A_0 = 0.5337$

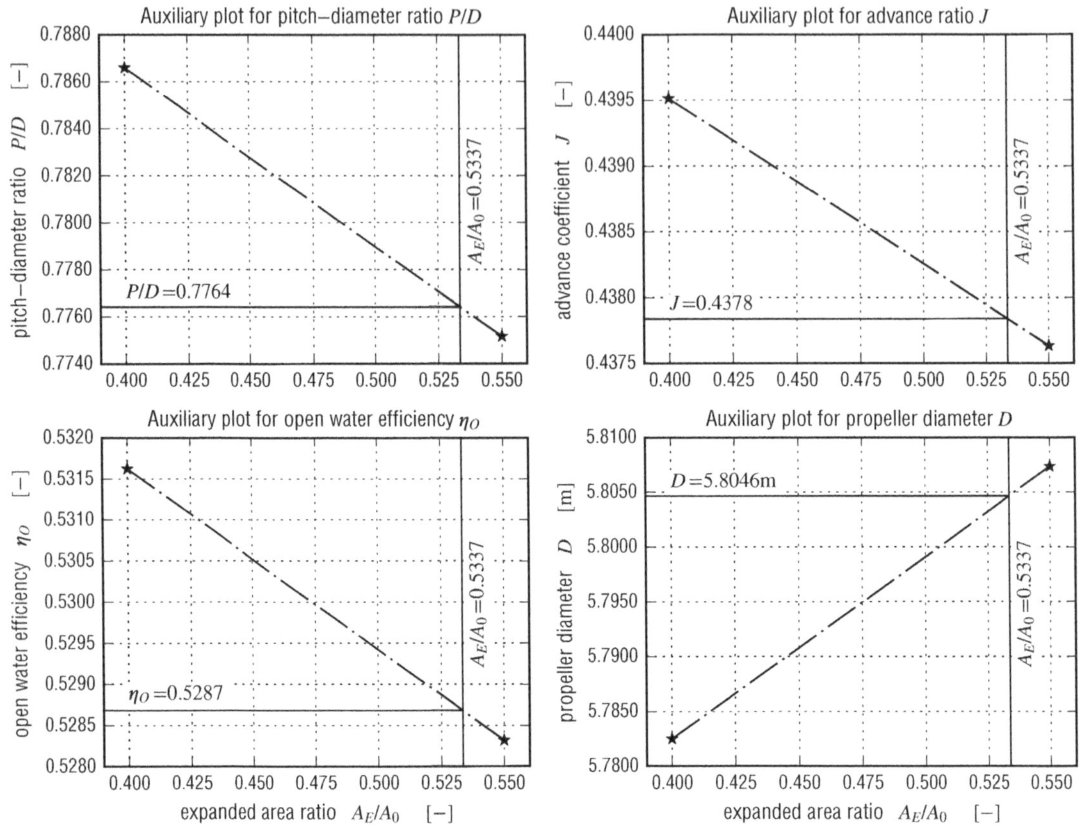

Figure 49.8 Auxiliary plots to determine final optimum propeller characteristics. The ★ mark results of first and second iterations

This propeller provides a thrust of 807.09 kN. Checking with the Burrill chart for a third time yields a required expanded area ratio of $A_E/A_0 = 0.5340$, which confirms our final selection.

49.2.7 Attainable speed check

Check of attainable speed

Propeller design tasks 2 and 4 use the required thrust as input. This ensures that the propeller is capable of bringing the vessel up to its design speed, assuming that the engine is producing the necessary torque. Propeller design tasks 1 and 3, however, use the available delivered power (or torque) as input. We selected a propeller which absorbs the available torque with high efficiency. Unfortunately, this does not guarantee that the propeller is producing sufficient thrust to attain the desired design speed. Therefore, a check of the achievable speed is required for design tasks 1 and 3. At this point, a resistance estimate is necessary like the one shown in Table 49.3.

Available vs. required thrust

Figure 49.9 shows the ship speed as a function of required thrust in service. T_{req} has been computed from the resistance data in Table 49.3 and Equation (49.33). For

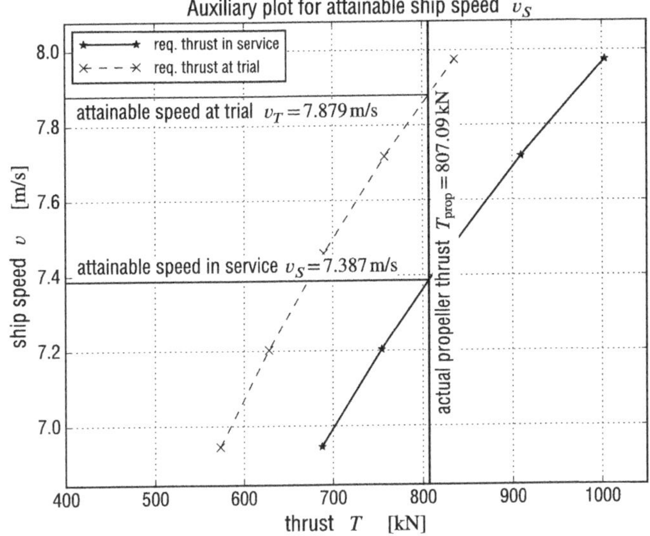

Figure 49.9 Auxiliary plot to determine the attainable ship speed

comparison, the curve of thrust for trial conditions is shown as well. In order to find the attainable speed, we mark the available propeller thrust $T_{prop} = 807.09$ kN in the graph by a vertical line. The intersection with the required thrust curve reveals the actual attainable ship speed.

Attainable velocity

In this example, a speed of $v_S = 7.387$ m/s or 14.36 kn is achieved, which is 99% of the target design speed of 14.5 kn. Note that the same thrust would suffice for a speed of about 15.3 kn at trial. If the lower speed is unacceptable to the owner, the following remedies may be pursued:

- The rating or power output of diesel engines may be adjusted within certain limits. If the specified brake power of $P_B = 7650$ kW is not already the maximum range of the engine model, the rating may be increased. In the example, an increase of brake power by about 3% at the same rate of revolution would be sufficient to attain the desired design speed.
- In some cases, it may be possible to achieve the same power output at a lower rate of revolution. In the example above, a propeller optimized for $n = 104$ rpm would drive the ship at design speed with the stated brake power of 7650 kW. A combination of power increase and lower rate of revolution may be used if the engine layout diagram permits it.
- If an adjustment of engine rating is impossible, design measures have to be taken to improve the overall propulsive efficiency. Optimizations of bulbous bow and hull shape in the run may lower resistance and/or improve hull efficiency.

The propeller selection process must be repeated once a higher brake power, lower rate of revolution, lower resistance, or improved hull–propeller interaction parameters are known.

Process automation

Obviously, the whole propeller selection process may be implemented as a program, eliminating the need to read values off charts. In fact, both examples here were solved by a computer program which also creates the data plots. Plotting or using the charts is a good way to check for errors in the input and for finding ways to improve the design.

References

Burrill, L. and Emerson, A. (1963). Propeller cavitation: further tests on 16in propeller models in the King's College cavitation tunnel. *Transactions of the North East Coast Institution of Engineers and Shipbuilders (NECIES)*, 79:295–320.

Harvald, S. (1983). *Resistance and propulsion of ships.* John Wiley & Sons, New York, NY.

ITTC (2011). *Fresh water and seawater properties.* International Towing Tank Conference, Recommended Procedures and Guidelines 7.5-02-01-03. Revision 2.

Kristensen, H. and Lützen, M. (2012). Prediction of resistance and propulsion power of ships. Technical Report Project No. 2010-56, Work Package 2, Report No. 04, University of Southern Denmark and Technical University of Denmark.

Lewis, E., editor (1988). *Principles of naval architecture,* volume II – Resistance, propulsion and vibration. The Society of Naval Architects and Marine Engineers, Jersey City, NJ, second edition.

Oosterveld, M. and van Oossanen, P. (1975). Further computer-analyzed data of the Wageningen B-screw series. *International Shipbuilding Progress,* 22:251–261.

50

Holtrop and Mennen's Method

In the late 1970s and early 1980s, J. Holtrop and G.G.J. Mennen developed a resistance and propulsion prediction method based on the regression analysis of model tests and trial data of MARIN, the model basin in Wageningen, The Netherlands (Holtrop, 1977; Holtrop and Mennen, 1978, 1982; Holtrop, 1984, 1988). All necessary equations are presented in their newest published form and an example data set with intermediate results allows readers to check their own implementations.

Learning Objectives

At the end of this chapter students will be able to:

- understand applicability and limitations of Holtrop and Mennen's resistance and propulsion estimate
- implement the method as a computer program or spreadsheet
- apply the method in design projects

50.1 Overview of the Method

Holtrop and Mennen's method is arguably the most popular method to estimate resistance and powering of displacement type ships. It is based on the regression analysis of a vast range of model tests and trial data which give it a wide applicability. It is – to the best of the author's knowledge – the only method that adopted the use of the ITTC form factor k. Resistance is calculated as a dimensional force. The method also provides formulas to estimate the hull–propeller interaction parameters thrust deduction, full scale wake fraction, and relative rotative efficiency.

50.1.1 Applicability

Range of application

Not much information is provided in the publications about the range of application of the method. Holtrop (1977) provides a table with ranges for prismatic coefficient, length to beam ratio, and Froude number of the ship types considered in the original

Fundamentals of Ship Hydrodynamics: Fluid Mechanics, Ship Resistance and Propulsion, First Edition.
Lothar Birk.

Companion website: www.wiley.com/go/birk/hydrodynamics

Table 50.1 Required and optional input parameters for Holtrop and Mennen's method

Parameter	Symbol	Remarks
	required parameters	
length in waterline	L_{WL}	
molded beam	B	
molded mean draft	T	typically $T = \frac{1}{2}(T_A + T_F)$
molded draft at aft perpendicular	T_A	
molded draft at forward perpendicular	T_F	
volumetric displacement (molded)	V	alternatively use the block coefficient as $C_B = V/(BT\,L_{WL})$
prismatic coefficient (based on L_{WL})	C_P	
midship section coefficient	C_M	or use $C_M = C_B/C_P$
waterplane area coefficient	C_{WP}	may have to be estimated in early design stages
longitudinal center of buoyancy	ℓ_{CB}	positive forward; with respect to $L_{WL}/2$ in percent of L_{WL}
area of ship and cargo above waterline	A_V	projected in direction of v_S
immersed transom area	A_T	measured at rest
transverse area of bulbous bow	A_{BT}	measured at forward perpendicular
height of center of A_{BT} above basis	h_B	has to be smaller than $0.6T_F$
propeller diameter	D	
propeller expanded area ratio	A_E/A_0	
stern shape parameter	C_{stern}	typical values provided with Equation (50.14) below
	optional parameters	
wetted surface (hull)	S	
wetted surface of appendages	S_{APP_i}	bilge keels, stabilizer fins, etc.
half angle of waterline entrance	i_E	
diameter of bow thruster tunnel	d_{TH}	

regression analysis. Summarizing the numbers, reasonable estimates can be expected for cases that fit the following conditions:

$$\begin{aligned} Fr &\leq 0.45 \\ 0.55 \leq C_P &\leq 0.85 \\ 3.9 \leq \frac{L}{B} &\leq 9.5 \end{aligned} \tag{50.1}$$

50.1.2 Required input

Input to Holtrop and Mennen's method consists of principal dimensions and a few basic hull form parameters. The necessary parameters are listed in Table 50.1.

Note that the method uses the length in waterline L_{WL} as the characteristic length of the vessel. Do not forget to recompute the block coefficient C_B and prismatic coefficient C_P based on L_{WL}.

Characteristic length

$$C_{B_{(\text{w.r.t } L_{WL})}} = \left(\frac{L_{PP}}{L_{WL}}\right) C_{B_{(\text{w.r.t } L_{PP})}} \tag{50.2}$$

Do the same for the prismatic coefficient, if it is provided on the basis of length between perpendiculars L_{PP}.

In early design stages some input parameters may still be unknown. They may be initially derived from design formulas and later replaced with the actual values.

Missing parameters

If, for instance, the longitudinal center of buoyancy is not yet known, the following suggestion for the optimum LCB location by Guldhammer and Harvald (1974) may be used.

Longitudinal center of buoyancy

$$\ell_{CB} = -\left(0.44\, Fr_{\text{design}} - 0.094\right) \tag{50.3}$$

It is important to enter ℓ_{CB} as a percentage. Say a ship with a waterline length of $L_{WL} = 182$ m has its longitudinal center of buoyancy $L_{CB} = 1.5$ m aft of $L_{WL}/2$. Thus ℓ_{CB} will be negative (aft of $L_{WL}/2$) and

$$\ell_{CB} = -\frac{L_{CB}}{L_{WL}} \cdot 100 = -\frac{1.5\text{ m}}{182\text{ m}} \cdot 100\% = -0.8242\%$$

Only the value −0.8242 will be entered into the formulas. If LCB is given with respect to $L_{PP}/2$, it first has to be referenced to $L_{WL}/2$.

A regression equation based on a graph by Jensen (1994) may help with the midship section coefficient:

Midship section coefficient

$$C_M = \frac{1}{1 + (1 - C_B)^{3.5}} \tag{50.4}$$

The waterplane area coefficient C_{WP} is often unknown until the lines plan is completed. Bertram and Wobig (1999) offer this formula for initial estimates:

Waterplane area coefficient

$$C_{WP} = \begin{cases} 0.763\left(C_P + 0.34\right) & \text{for tanker, bulk carrier, general cargo with } 0.56 < C_P < 0.87 \\ 3.226\left(C_P - 0.36\right) & \text{for container ships with } 0.57 < C_P < 0.62 \end{cases} \tag{50.5}$$

Similar formulas may be found in Watson (1998) and Papanikolaou (2014).

The wetted surface of the hull at rest S may be estimated by the following formula, if it is not yet known from hydrostatic analysis (Holtrop and Mennen, 1982).

Wetted surface

$$S = c_{23}\, L_{WL}\,(2T + B)\sqrt{C_M} + 2.38\,\frac{A_{BT}}{C_B} \qquad \text{see also Equation (50.8)} \tag{50.6}$$

with the factor

$$c_{23} = \left[0.453 + 0.4425\, C_B - 0.2862\, C_M - 0.003467\,\frac{B}{T} + 0.3696\, C_{WP}\right] \tag{50.7}$$

Equation (50.6) was later updated to provide a more accurate prediction, especially for slender hull forms with small midship section coefficients (Holtrop, 1988).

$$S = L_{WL}(2T + B)\sqrt{C_M}\left[0.615989\, c_{23} + 0.111439\, C_M^3 + 0.000571111\, C_{\text{stern}} + 0.245357\frac{c_{23}}{C_M}\right] + 3.45538\, A_T + \frac{A_{BT}}{C_B}\left(1.4660538 + \frac{0.5839497}{C_M}\right) \tag{50.8}$$

Waterline entrance angle For the half angle of the waterline entrance, Holtrop and Mennen (1982) provide the following formula:

$$i_E = 1 + 89\, \mathrm{e}^a \tag{50.9}$$

The necessary exponent a is found by evaluating

$$a = -\left[\left(\frac{L_{WL}}{B}\right)^{0.80856}(1 - C_{WP})^{0.30484}\left[1 - C_P - 0.0225\,\ell_{CB}\right]^{0.6367}\left(\frac{L_R}{B}\right)^{0.34574}\left(\frac{100\,V}{L_{WL}{}^3}\right)^{0.16302}\right] \tag{50.10}$$

Note that the angle i_E is returned in degrees like it is needed in formulas below.

Length of run The length of the run L_R is required as an additional input for Equation (50.10). It may be estimated by the following formula (Holtrop, 1984):

$$L_R = L_{WL}\left(\frac{1 - C_P + 0.06\, C_P\, \ell_{CB}}{4\, C_P - 1}\right) \tag{50.11}$$

50.2 Procedure

The Holtrop (1984) method computes a dimensional total resistance which is broken down into several components:

- frictional resistance R_F,
- appendage resistance R_{APP},
- wave resistance R_W,
- resistance due to bulbous bow near the water surface R_B,
- pressure resistance due to immersed transom R_{TR},
- model–ship correlation resistance R_A, and
- air resistance R_{AA}.

As mentioned, Holtrop and Mennen's method is the only early design estimate for resistance and propulsion that has adopted the ITTC form factor approach. Since the use of a form factor affects the estimate of the residuary resistance, or, in this case, the wave resistance, the method should *not be used without* a form factor.

Form factor is used

Resistance components will be computed as functions of Froude and Reynolds numbers for the range of speeds in question.

Froude and Reynolds number

$$Fr = \frac{v_S}{\sqrt{g\,L_{WL}}} \qquad Re = \frac{v_S\,L_{WL}}{\nu} \tag{50.12}$$

Note that both are based on the length in waterline L_{WL} in the context of this method.

50.2.1 Resistance components

The frictional resistance R_F is computed on the basis of the ITTC 1957 model–ship correlation line coefficient C_F (2.18) as the resistance of a flat plate with wetted surface S.

Frictional resistance

$$R_F = \frac{1}{2}\rho\,v_S^2\,S\,C_F \tag{50.13}$$

Do not forget to convert density ρ, ship speed v_S, and wetted surface S into a consistent set of units.

The flat plate resistance R_F is later augmented by a form factor k when the resistance components are assembled into the total resistance. First, a constant labeled c_{14} has to be found which captures the influence of the aft body shape.

Form factor

$$c_{14} = 1.0 + 0.011\,C_{\text{stern}} \quad \text{with} \tag{50.14}$$

Aft body shape	C_{stern}
pram with gondola	−25
V-shaped sections	−10
normal sections	0
U-shaped sections	+10

Using the constant c_{14}, the length of run L_R (50.11), and input values from Table 50.1, the hull form factor k may be estimated (named k_1 in Holtrop, 1984).

$$k = -0.07 + 0.487118\,c_{14}\left[\left(\frac{B}{L_{WL}}\right)^{1.06806}\left(\frac{T}{L_{WL}}\right)^{0.46106}\left(\frac{L_{WL}}{L_R}\right)^{0.121563}\left(\frac{L_{WL}^3}{V}\right)^{0.36486}\left(1-C_P\right)^{-0.604247}\right] \tag{50.15}$$

The original formula for the form factor provided by Holtrop (1984) computes $(1 + k)$. In a later update, Holtrop (1988) introduced a speed dependent correction for the form factor. However, this only seems to yield improvements for very fast vessels ($Fr > 0.5$).

A method to estimate the appendage resistance R_{APP} is provided in Holtrop and Mennen (1982). The form factor values for appendages in Table 50.2 have been updated in Holtrop (1988). Appendages mostly affect the viscous resistance. As discussed at length,

Appendage resistance

Table 50.2 Approximate values for appendage form factors k_{2i} according to Holtrop (1988)

Appendage	k_{2i} value
rudder behind skeg	0.2–0.5
rudder behind stern	0.5
twin screw rudder (slender)	1.5
twin screw rudder (thick)	2.5
shaft brackets	2.0–4.0
skeg	0.5–1.0
strut bossing	2.0–3.0
hull bossing	1.0
exposed shafts (angle with buttocks about 10 degrees)	1.0
(angle with buttocks about 20 degrees)	4.0
stabilizer fins	1.8
dome	1.7
bilge keels	0.4

Reynolds numbers are considerably smaller in model tests than at full scale. Consequently, model tests are not best suited to quantify appendage resistance. Unfortunately, they are often the only option. In addition, effects of appendages are usually registered as a whole and not individually. Practice has shown that reasonable estimates can be made based on the individual form factors listed in Table 50.2. A single, equivalent form factor is determined for the calculation of appendage resistance (50.18).

$$(1 + k_2)_{\text{eq}} = \frac{\sum_i (1 + k_{2i}) S_{APP_i}}{\sum_i S_{APP_i}} \tag{50.16}$$

The resistance due to a bow thruster tunnel opening may be computed according to

$$R_{TH} = \rho\, v_S^2\, \pi\, d_{TH}^2\, C_{D_{TH}} \tag{50.17}$$

The drag coefficient $C_{D_{TH}}$ for the thruster tunnel takes values between 0.003 and 0.012. The smaller values are for thrusters which are in the cylindrical part of the bulbous bow, i.e. the rim of the opening is fairly parallel to the midship plane. See also Equation (51.27).

The appendage resistance is calculated as the sum of thruster resistance and all considered appendages.

$$R_{APP} = \frac{1}{2}\rho\, v_S^2\, (1 + k_2)_{\text{eq}}\, C_F \sum_i S_{APP_i} + \sum R_{TH} \tag{50.18}$$

C_F is the ITTC 1957 model–ship correlation line coefficient which was already computed for the frictional resistance R_F.

Wave resistance Wave resistance is a function of Froude number Fr. For the estimate of R_W, Holtrop

(1984) subdivided the range of Froude numbers into three sections.

$$R_W(Fr) = \begin{cases} R_{Wa}(Fr) \ \text{Equation (50.20)} & \text{if } Fr \leq 0.4 \\ \text{interpolation, see Equation (50.22)} & \text{if } 0.4 < Fr \leq 0.55 \\ R_{Wb}(Fr) \ \text{Equation (50.21)} & \text{if } Fr > 0.55 \end{cases} \quad (50.19)$$

The wave resistance for Froude numbers $Fr < 0.4$ is computed from (Holtrop, 1984)

$$R_{Wa}(Fr) = c_1\, c_2\, c_5\, \rho\, g\, V\, \exp\left[m_1\, Fr^d + m_4 \cos\left(\lambda\, Fr^{-2}\right)\right] \quad (50.20)$$

The expression $\exp[x]$ is used here for the function e^x to make the formula more readable. Equations for the coefficients in Equation (50.20) are provided in Table 50.3.

The wave resistance for Froude numbers $Fr > 0.55$ is computed from (Holtrop, 1984)

$$R_{Wb}(Fr) = c_{17}\, c_2\, c_5\, \rho\, g\, V\, \exp\left[m_3\, Fr^d + m_4 \cos\left(\lambda\, Fr^{-2}\right)\right] \quad (50.21)$$

Formulas for the coefficients in Equation (50.21) are provided in Tables 50.3 and 50.4.

For the remaining range of Froude numbers $0.4 > Fr \leq 0.55$, a linear blending between Equations (50.20) and (50.21) is performed (Holtrop, 1984). Ships typically do not operate in this uneconomical speed range for prolonged periods of time.

$$R_W(Fr) = R_{Wa}(0.4) + \frac{(20\, Fr - 8)}{3}\left[R_{Wb}(0.55) - R_{Wa}(0.4)\right] \quad (50.22)$$

The expressions $R_{Wa}(0.4)$ and $R_{Wb}(0.55)$ mean that the wave resistance is evaluated with the equation for R_{Wa} or R_{Wb} at Froude number $Fr = 0.4$ or $Fr = 0.55$ respectively. Do not forget to recompute the factor m_4 with the respective Froude numbers $Fr = 0.4$ or $Fr = 0.55$.

Holtrop (1988) presents additional reduction factors to capture the effect of bulbous bow and transom on the ship's wave resistance. However, only few data sets were available where models had been tested with and without bulbous bow or transom. For early design purposes one should venture on the side of caution not to underpredict the wave resistance.

Resistance of bulbous bow

The formula for the pressure resistance of a bulbous bow close to the water surface was updated in Holtrop (1988) to include the effects of forward sinkage h_F and local wave height at the bow h_W.

$$h_F = C_P\, C_M\, \frac{B\, T}{L_{WL}} \left(136 - 316.3\, Fr\right) Fr^3 \quad \text{but } h_F \geq -0.01\, L_{WL} \quad (50.23)$$

$$h_W = \frac{i_E\, v_S^2}{400\, g} \quad \text{but at most } h_W \leq 0.01\, L_{WL} \quad (50.24)$$

The values of h_F and h_W are used in the definition of an immersion Froude number Fr_i for the bulbous bow.

$$Fr_i = \frac{v_S}{\sqrt{g\left(T_F - h_B - 0.25\sqrt{A_{BT}} + h_F + h_W\right)}} \quad (50.25)$$

Table 50.3 Coefficients for the wave resistance computation in Equation (50.20) if $Fr \leq 0.4$ (Holtrop, 1984)

$$c_7 = \begin{cases} 0.229577 \left(\frac{B}{L_{WL}}\right)^{(1/3)} & \text{if } B/L_{WL} \leq 0.11 \\ \frac{B}{L_{WL}} & \text{if } 0.11 < B/L_{WL} \leq 0.25 \\ 0.5 - 0.0625 \frac{L_{WL}}{B} & \text{if } B/L_{WL} > 0.25 \end{cases}$$

$$c_1 = 2223105\, c_7^{3.78613} \left(\frac{T}{B}\right)^{1.07961} \left(90 - i_E\right)^{-1.37565}$$

$$c_3 = 0.56 \frac{A_{BT}^{1.5}}{\left[BT\left(0.31\sqrt{A_{BT}} + T_F - h_B\right)\right]}$$

$$c_2 = e^{(-1.89\sqrt{c_3})}$$

$$c_5 = 1 - 0.8 \frac{A_T}{B\, T\, C_M}$$

$$c_{15} = \begin{cases} -1.69385 & \text{if } \frac{L_{WL}^{\;3}}{V} \leq 512 \\ -1.69385 + \frac{\frac{L_{WL}}{V^{(1/3)}} - 8}{2.36} & \text{if } 512 < \frac{L_{WL}^{\;3}}{V} \leq 1726.91 \\ 0 & \text{if } \frac{L_{WL}^{\;3}}{V} > 1726.91 \end{cases}$$

$$c_{16} = \begin{cases} 8.07981\, C_P - 13.8673\, C_P^2 + 6.984388\, C_P^3 & \text{if } C_P \leq 0.8 \\ 1.73014 - 0.7067\, C_P & \text{if } C_P > 0.8 \end{cases}$$

$$d = -0.9$$

$$\lambda = \begin{cases} 1.446\, C_P - 0.03 \frac{L_{WL}}{B} & \text{if } \frac{L_{WL}}{B} \leq 12 \\ 1.446\, C_P - 0.36 & \text{if } \frac{L_{WL}}{B} > 12 \end{cases}$$

$$m_1 = 0.0140407 \frac{L_{WL}}{T} - 1.75254 \frac{V^{(1/3)}}{L_{WL}} - 4.79323 \frac{B}{L_{WL}} - c_{16}$$

$$m_4 = 0.4\, c_{15}\, e^{(-0.034\, Fr^{-3.29})}$$

Table 50.4 Additional coefficients for the wave resistance computation in Equation (50.21) if $Fr > 0.55$ (Holtrop, 1984)

$$c_5 = 1 - 0.8 \frac{A_T}{B\,T\,C_M}$$

$$c_{17} = 6919.3\,C_M^{-1.3346} \left(\frac{V}{{L_{WL}}^3}\right)^{2.00977} \left(\frac{L_{WL}}{B} - 2.0\right)^{1.40692}$$

$$m_3 = -7.2035 \left(\frac{B}{L_{WL}}\right)^{0.326869} \left(\frac{T}{B}\right)^{0.605375}$$

$$m_4 = 0.4\,c_{15}\,\mathrm{e}^{(-0.034\,Fr^{-3.29})}$$

The parameter P_B quantifies the emergence of the bulb from the still water line.

$$P_B = 0.56 \frac{\sqrt{A_{BT}}}{T_F - 1.5\,h_B + h_F} \tag{50.26}$$

The additional resistance R_B is computed with (Holtrop, 1984):

$$R_B = 0.11\,\rho\,g\left(\sqrt{A_{BT}}\right)^3 \frac{Fr_i^3}{1 + Fr_i^2}\,\mathrm{e}^{(-3.0\,P_B^{-2})} \tag{50.27}$$

Again, these formulas should be applied with caution since they are based on a limited number of data points (Holtrop, 1984).

The effects of a bulbous bow are only captured in a very broad way. One should refrain from optimizing the bulbous bow based on the presented formulas. This task should be done with a nonlinear wave resistance code in later design stages (see for instance Raven, 1996).

Transom resistance

An immersed transom may cause an additional pressure resistance R_{TR}. It is a function of a depth Froude number Fr_T which considers the immersion of the transom. If a transom area $A_T > 0$ is given as input, Fr_T is defined by

$$Fr_T = \frac{v_S}{\sqrt{\dfrac{2\,g\,A_T}{(B + B\,C_{WP})}}} \tag{50.28}$$

The expression $A_T/(B + B\,C_{WP})$ is a measure for the average draft of the transom. If this average draft is small compared to the speed, the flow will separate cleanly at the transom edge and the additional transom drag vanishes. This is expressed with the coefficient c_6.

$$c_6 = \begin{cases} 0.2\,(1 - 0.2\,Fr_T) & \text{if } Fr_T < 5 \\ 0 & \text{if } Fr_T > 5 \end{cases} \tag{50.29}$$

The transom drag R_{TR} is deemed negligible above the limit $Fr_T = 5$.

$$R_{TR} = \frac{1}{2}\rho v_S^2 A_T c_6 \qquad (50.30)$$

Correlation allowance resistance

The correlation allowance considered here includes effects of roughness and additional phenomena not captured in other resistance components. Note that correlation allowance and roughness effects have been separated in the current ITTC performance prediction procedure (ITTC, 2014). First, we need the additional coefficient c_4.

$$c_4 = \begin{cases} \dfrac{T_F}{L_{WL}} & \text{if } T_F/L_{WL} \leq 0.04 \\ 0.04 & \text{if } T_F/L_{WL} > 0.04 \end{cases} \qquad (50.31)$$

T_F is the draft at the forward perpendicular.

Then, the correlation allowance coefficient follows from Equation (50.32) below (Holtrop, 1988).

$$C_A = 0.00546\left(L_{WL} + 100\right)^{-0.16} - 0.002 + 0.003\sqrt{\frac{L_{WL}}{7.5}}\, C_B^4\, c_2\, (0.04 - c_4) \qquad (50.32)$$

The coefficient c_2 is found in Table 50.3. Holtrop (1988) states that with modern hull coatings, values of C_A may be achieved that are $0.1 \cdot 10^{-3}$ lower than predicted. However, this will not make a significant difference for early design estimates. The effect of surface roughness higher than the standard value of $k_S = 150\ \mu$m may be estimated by an addition to C_A (Holtrop and Mennen, 1982).

$$\Delta C_A = \begin{cases} 0 & \text{if } k_S = 150\ \mu\text{m} \\ \dfrac{0.105\, k_S^{(1/3)} - 0.005579}{L_{WL}^{(1/3)}} & \text{if } k_S > 150\ \mu\text{m} \end{cases} \qquad (50.33)$$

In contrast to other formulas in Holtrop and Mennen's method, Equation (50.33) is not dimensionless. k_S and L_{WL} have to be entered in meters to obtain correct results.

The correlation resistance is then given by

$$R_A = \frac{1}{2}\rho v_S^2 \left(C_A + \Delta C_A\right)\left[S + \sum S_{APP}\right] \qquad (50.34)$$

Air resistance

The resistance caused by moving the ship above the waterplane through air at rest is calculated according to the standard ITTC procedure.

$$R_{AA} = \frac{1}{2}\rho_A v_S^2 C_{DA} A_V \qquad (50.35)$$

A_V is the area of the lengthwise projection of hull and superstructure above the waterline. The density of air is $\rho_A = 1.225$ kg/m^3 for standard atmospheric pressure and a temperature of 15 °C. The default air drag coefficient is $C_{DA} = 0.8$. Note that C_{DA} itself cannot simply be added to the other ship resistance coefficients since it is not based on the wetted surface.

50.2.2 Total resistance

The resistance components from the previous subsection are assembled into the total resistance.

$$R_T = (1+k)\,R_F + R_{APP} + R_A + R_W + R_B + R_{TR} + R_{AA} \tag{50.36}$$

This calm water resistance should be augmented by a service margin before it is used to select an optimum propeller for the ship.

50.2.3 Hull–propeller interaction parameters

Single screw vessels

Holtrop (1984) provides different sets of formulas of the propulsion parameters for single and twin screw vessels. The formulas for single screw vessels are more complex and are discussed first.

Viscous resistance coefficient

In both cases a viscous resistance coefficient C_V is needed, which combines all friction related components of the resistance and the correlation resistance from Equation (50.34).

$$C_V = \frac{(1+k)\,R_F + R_{APP} + R_A}{\frac{1}{2}\,\rho\, v_S^2\,(S + \sum_i S_{APP_i})} \tag{50.37}$$

Full scale wake fraction

The effect of the hull onto propeller inflow is expressed as a wake fraction w_S for the full scale vessel.

$$w_S = c_9\, c_{20}\, C_V\, \frac{L_{WL}}{T_A}\left[0.050776 + 0.93405\,\frac{c_{11}\, C_V}{1 - C_{P1}}\right] + 0.27915\, c_{20}\sqrt{\frac{B}{L_{WL}\,(1 - C_{P1})}} + c_{19}\, c_{20} \tag{50.38}$$

Formulas for the necessary coefficients are listed in Table 50.5.

Thrust deduction fraction

Thrust deduction fraction for single screw vessels is estimated as

$$t = \frac{0.25014\left(\frac{B}{L_{WL}}\right)^{0.28956}\left(\frac{\sqrt{B\,T}}{D}\right)^{0.2624}}{\left(1 - C_P + 0.0225\,\ell_{CB}\right)^{0.01762}} + 0.0015\, C_{\text{stern}} \tag{50.39}$$

Relative rotative efficiency

Finally, Holtrop (1984) states the following formula for the relative rotative efficiency of single screw vessels.

$$\eta_R = 0.9922 - 0.05908\,\frac{A_E}{A_0} + 0.07424\left(C_P - 0.0225\,\ell_{CB}\right) \tag{50.40}$$

Table 50.5 Coefficients for the full scale wake fraction of single screw vessels in Equation (50.38) (Holtrop, 1984)

$$c_8 = \begin{cases} \dfrac{S}{L_{WL}\,D}\left(\dfrac{B}{T_A}\right) & \text{if } B/T_A \leq 5 \\[2ex] \dfrac{S\left(7\dfrac{B}{T_A} - 25\right)}{L_{WL}\,D\left(\dfrac{B}{T_A} - 3\right)} & \text{if } B/T_A > 5 \end{cases}$$

$$c_9 = \begin{cases} c_8 & \text{if } c_8 \leq 28 \\ 32 - \dfrac{16}{c_8 - 24} & \text{if } c_8 > 28 \end{cases}$$

$$c_{11} = \begin{cases} \dfrac{T_A}{D} & \text{if } T_A/D \leq 2 \\ 0.0833333\left(\dfrac{T_A}{D}\right)^3 + 1.33333 & \text{if } T_A/D > 2 \end{cases}$$

$$c_{19} = \begin{cases} \dfrac{0.12997}{(0.95 - C_B)} - \dfrac{0.11056}{(0.95 - C_P)} & \text{if } C_P \leq 0.7 \\[2ex] \dfrac{0.18567}{(1.3571 - C_M)} - 0.71276 + 0.38648\,C_P & \text{if } C_P > 0.7 \end{cases}$$

$$c_{20} = 1 + 0.015\,C_{\text{stern}}$$

$$C_{P1} = 1.45\,C_P - 0.315 - 0.0225\,\ell_{CB}$$

Twin screw vessels

The respective formulas for twin screw vessels are:

$$w_S = 0.3095\,C_B + 10\,C_V\,C_B - 0.23\,\frac{D}{\sqrt{BT}} \tag{50.41}$$

$$t = 0.325\,C_B - 0.1885\,\frac{D}{\sqrt{BT}} \quad \text{twin screw vessels} \tag{50.42}$$

$$\eta_R = 0.9737 + 0.111\,(C_P - 0.0225\,\ell_{CB}) - 0.06325\,\frac{P}{D} \tag{50.43}$$

With the stated estimates for resistance and hull–propeller interaction parameters, a power prediction can be completed following the procedure outlined in Section 44.2.

50.3 Example

As an example, we repeat the resistance estimate for the container ship from Section 31.3 and add an estimate for the powering requirements.

50.3.1 Completion of input parameters

The basic input parameters from Table 31.5 are used again to facilitate a comparison. However, block coefficient and prismatic coefficient have to be recomputed with respect to the length in waterline L_{WL} as explained in Section 50.1.2.

Length in waterline

The waterline length is taken as the sum of length between perpendiculars and the aft overhang, assuming that the waterline reaches all the way to the unwetted transom ($A_T = 0\ \text{m}^2$).

$$L_{WL} = L_{PP} + L_{\text{aft}} = 145\ \text{m} + 2.7\ \text{m} = 147.7\ \text{m} \tag{50.44}$$

Block and prismatic coefficient

Based on this length in waterline, we obtain for block and prismatic coefficients:

$$C_B = \frac{V}{L_{WL}BT} = 0.6492 \tag{50.45}$$

$$C_P = \frac{V}{L_{WL}A_M} = 0.6659 \tag{50.46}$$

***LCB* position**

The position of the longitudinal center of buoyancy also has to be stated as a fraction of length in waterline. The given LCB_0 position of 0.4% $L_{PP}/2$ before midships translates into

$$\ell_{CB} = 1.3067\% \quad \text{forward of } L_{WL}/2. \tag{50.47}$$

Waterplane area coefficient

Furthermore, Holtrop and Mennen's method requires the waterplane area coefficient C_{WP}. With the help of the first case in Equation (50.5), we estimate:

$$C_{WP} = 0.7675 \tag{50.48}$$

Wetted surface

For consistency, the presented results are based on the given wetted surface of $S = 4400\ \text{m}^2$. Equation (50.8) yields a slightly smaller value of $S = 4380.5947\ \text{m}^2$, but we rather employ given data when it is available.

Air drag input

For the air resistance estimate, a transverse area above the waterline of $A_V = 383.76\ \text{m}^2$ has been used. The air drag coefficient is $C_{DA} = 0.8$, and air density is $1.225\ \text{kg/m}^3$.

Half angle of entrance

With the provided data, Equations (50.9) and (50.11) predict a half angle of the waterline entrance and length of run of $i_E = 19.231$ degree and $L_R = 53.982$ meters.

Propeller data

The propulsion estimate is based on the propeller data from Table 50.6 and open water characteristics according to the K_T and K_Q polynomials presented in Section 47.2.

50.3.2 Resistance estimate

Computational coefficients

A resistance and powering estimate is executed using the equations presented in this

Table 50.6 Propeller data for powering estimate; see also Table 31.5

Propeller data		
number of blades	Z	5
diameter	D	4.9000 m
pitch–diameter ratio	P/D	0.9924
expanded area ratio	A_E/A_0	0.7692

Table 50.7 Holtrop and Mennen resistance and powering estimate example; speed independent procedural coefficients

c_1=2.045485	c_7= 0.162492	c_{15}=−1.693850	d =−0.900000
c_2=0.707301	c_8 =17.793996	c_{16}= 1.293587	λ = 0.778264
c_3=0.033572	c_9 =17.793996	c_{17}= 1.731558	m_1 =−2.135451
c_4=0.040000	c_{11}= 1.673469	c_{19}= 0.042998	m_3 =−2.076076
c_5=1.000000	c_{14}= 1.000000	c_{20}= 1.000000	P_B = 2.555278
			C_{P1}= 0.621151

Table 50.8 Holtrop and Mennen resistance and powering estimate example; speed dependent procedural coefficients

v_S	Fr	Fr_i	Fr_T	c_6	m_2	$m_3 Fr^d$	m_4	$m_4 \cos(\lambda/Fr^2)$
[kn]	[−]	[−]	[−]	[−]	[−]	[−]	[−]	[−]
15.00	0.20275	1.36572	0.00	0.20	−0.06596	−8.72909	−0.00104	−0.00104
15.50	0.20951	1.39805	0.00	0.20	−0.07697	−8.47525	−0.00202	−0.00088
16.00	0.21627	1.42948	0.00	0.20	−0.08855	−8.23651	−0.00359	0.00214
16.50	0.22303	1.46001	0.00	0.20	−0.10060	−8.01153	−0.00595	0.00594
17.00	0.22979	1.48968	0.00	0.20	−0.11303	−7.79915	−0.00926	0.00524
17.50	0.23655	1.51848	0.00	0.20	−0.12576	−7.59831	−0.01369	−0.00310
18.00	0.24331	1.54644	0.00	0.20	−0.13869	−7.40808	−0.01933	−0.01617
18.50	0.25006	1.57358	0.00	0.20	−0.15177	−7.22764	−0.02627	−0.02608
19.00	0.25682	1.59991	0.00	0.20	−0.16491	−7.05623	−0.03452	−0.02485

chapter. Table 50.7 lists values for all procedural coefficients that do not depend on ship speed. Table 50.8 presents intermediate coefficients which are a function of ship speed.

Form factors Equation (50.15) yields an ITTC form factor of $k_1 = 0.193743$. The appendage resistance is based on the surface of the bilge keels $S_{APP} = 52\,\mathrm{m}^2$ and a form factor of $k_2 = 0.4$.

Correlation coefficient The estimate for the model–ship correlation coefficient (50.32) results in $10^3 C_A = 0.433837$.

Resistance components Based on the input and the coefficients listed in Tables 50.7and 50.8, resistance components and total resistance have been computed for trial conditions. Holtrop and Mennen's method computes dimensional resistance values rather than resistance coefficients. Table 50.9 states resistance values for the same set of Froude numbers that has been used in Section 31.3. The values lie about 10% above the estimate from Section 31.3 but the gap closes for higher speeds.

Table 50.9 Holtrop and Mennen resistance and powering estimate example; resistance components and total resistance

v_S [kn]	Fr [–]	R_F [kN]	R_A [kN]	R_W [kN]	R_B [kN]	R_{APP} [kN]	R_{AA} [kN]	R_T [kN]
15.00	0.20275	206.82	58.31	34.60	32.56	3.42	11.20	386.98
15.50	0.20951	219.94	62.27	44.93	33.87	3.64	11.96	419.21
16.00	0.21627	233.44	66.35	57.61	35.15	3.86	12.74	454.37
16.50	0.22303	247.31	70.56	72.88	36.39	4.09	13.55	492.70
17.00	0.22979	261.56	74.90	90.61	37.61	4.33	14.38	534.07
17.50	0.23655	276.18	79.37	110.48	38.79	4.57	15.24	578.14
18.00	0.24331	291.18	83.97	132.61	39.93	4.82	16.12	625.06
18.50	0.25006	306.55	88.70	158.08	41.05	5.07	17.03	675.88
19.00	0.25682	322.29	93.56	188.80	42.13	5.33	17.97	732.51

Table 50.10 Holtrop and Mennen resistance and powering estimate example; wake fraction and self propulsion point analysis

v_S [kn]	v_S [m/s]	Fr [–]	w_{TS} [–]	v_A [m/s]	T_{req} [kN]	C_S [–]	J_{TS} [–]	K_{TS} [–]	$10K_{QTS}$ [–]
15.0	7.717	0.2028	0.2696	5.636	482.69	0.61677	0.6108	0.2301	0.3790
15.5	7.974	0.2095	0.2694	5.826	522.89	0.62544	0.6082	0.2314	0.3807
16.0	8.231	0.2163	0.2693	6.015	566.75	0.63591	0.6052	0.2329	0.3827
16.5	8.488	0.2230	0.2691	6.204	614.56	0.64813	0.6016	0.2346	0.3850
17.0	8.746	0.2298	0.2690	6.393	666.15	0.66156	0.5978	0.2365	0.3874
17.5	9.003	0.2365	0.2688	6.583	721.13	0.67555	0.5940	0.2383	0.3899
18.0	9.260	0.2433	0.2687	6.772	779.64	0.69009	0.5900	0.2403	0.3924
18.5	9.517	0.2501	0.2685	6.961	843.04	0.70616	0.5858	0.2423	0.3952
19.0	9.774	0.2568	0.2684	7.151	913.68	0.72532	0.5808	0.2447	0.3983

50.3.3 Powering estimate

Constant hull-propeller interaction parameters

The powering estimate starts with the computation of thrust deduction fraction and relative rotative efficiency. Both are treated as constants in this method.

thrust deduction fraction	$t = 0.1983$
relative rotative efficiency	$\eta_R = 0.9940$

With the thrust deduction fraction, the required thrust can be computed via Equation (44.1). Once the estimate for the wake fraction is completed, the procedure in Section 44.2 is followed to predict the necessary delivered power for trial conditions.

Delivered power

Table 50.10 shows the results for wake fraction estimates and the self propulsion point analysis based on the propeller characteristics provided in Table 50.6. Table 50.11 summarizes propulsive efficiencies, the propeller rate of revolution, and the necessary delivered power P_D.

Table 50.11 Holtrop and Mennen resistance and powering estimate example; efficiencies, propeller rate of revolution, and delivered power

v_S [kn]	v_S [m/s]	Fr [–]	η_H [–]	η_O [–]	η_D [–]	n [1/s]	n [rpm]	P_D [kW]
15.0	7.717	0.2028	1.0976	0.5902	0.6440	1.883	112.990	4637.23
15.5	7.974	0.2095	1.0974	0.5884	0.6418	1.955	117.277	5208.37
16.0	8.231	0.2163	1.0971	0.5861	0.6392	2.028	121.700	5850.65
16.5	8.488	0.2230	1.0969	0.5835	0.6363	2.104	126.269	6573.49
17.0	8.746	0.2298	1.0967	0.5808	0.6331	2.182	130.949	7378.31
17.5	9.003	0.2365	1.0965	0.5780	0.6299	2.262	135.703	8263.98
18.0	9.260	0.2433	1.0963	0.5750	0.6265	2.342	140.542	9239.76
18.5	9.517	0.2501	1.0961	0.5717	0.6229	2.425	145.520	10328.13
19.0	9.774	0.2568	1.0959	0.5680	0.6187	2.512	150.747	11573.40

References

Bertram, V. and Wobig, M. (1999). Simple empirical formulae to estimate main form parameters of ships. *Schiff & Hafen*, (11):118–121.

Guldhammer, H. and Harvald, S. (1974). *Ship resistance – effect of form and principal dimensions (revised).* Akademisk Forlag, Copenhagen.

Holtrop, J. (1977). A statistical analysis of performance test results. *International Shipbuilding Progress*, 24(270):23–28.

Holtrop, J. (1984). A statistical re-analysis of resistance and propulsion data. *International Shipbuilding Progress*, 31(363):272–276.

Holtrop, J. (1988). A statistical resistance prediction method with a speed dependent form factor. In *Scientific and Methodological Seminar on Ship Hydrodynamics (SMSSH '88)*, Varna, Bulgaria.

Holtrop, J. and Mennen, G. (1978). A statistical power prediction method. *International Shipbuilding Progress*, 25(290):253–256.

Holtrop, J. and Mennen, G. (1982). An approximate power prediction method. *International Shipbuilding Progress*, 29(335):166–170.

ITTC (2014). *1978 ITTC performance prediction method.* International Towing Tank Conference, Recommended Procedures and Guidelines 7.5-02-03-01.4. Revision 3.

Jensen, G. (1994). Moderne Schiffslinien. In Keil, H., editor, *Handbuch der Werften*, volume XXII, pages 93–110. Schiffahrts-Verlag Hansa.

Papanikolaou, A. (2014). *Ship design – Methodologies of preliminary design.* Springer, Dordrecht, The Netherlands.

Raven, H. (1996). *A solution method for the nonlinear ship wave resistance problem.* PhD thesis, Technical University Delft, Delft, The Netherlands.

Watson, D. (1998). *Practical ship design.* Elsevier Ocean Engineering Book Series. Elsevier Science, Oxford, United Kingdom.

Self Study Problems

1. Discuss the principal difference between Guldhammer and Harvald's method for the resistance estimate and Holtrop and Mennen's method.

2. For a ship design project, the following data is provided:

Ship data	
length between perpendiculars	$L_{PP} = 145.00$ m
length in waterline	$L_{WL} = 149.80$ m
molded beam	$B = 22.30$ m
molded draft	$T = 8.26$ m
block coefficient (based on L_{PP})	$C_B = 0.66$
prismatic coefficient (based on L_{PP})	$C_P = 0.68$

   Compute the input values for block coefficient C_B and prismatic coefficient C_P for Holtrop and Mennen's method.

3. Implement the Holtrop and Mennen method as a program in Python, Matlab, or similar, and test it with the data presented in the last section.

51

Hollenbach's Method

Hollenbach's method for resistance and powering estimates of single and twin screw vessels was developed in the 1990s based on test data of the Schiffbau Versuchsanstalt in Vienna, Austria. This method has a narrower range of applicability than Holtrop and Mennen's method, but seems to provide more reliable results, especially for twin screw vessels. In addition to the equations for a resistance prediction, formulas provide estimates of upper and lower bound of the resistance. An example and comparison with results from Sections 31.3 and 50.3 close out the discussion of resistance and propulsion.

Learning Objectives

At the end of this chapter students will be able to:

- understand applicability and limitations of Hollenbach's resistance and powering estimate
- implement the method as a computer program or spreadsheet
- apply the method in design projects

51.1 Overview of the method

Origin In the 1990s, U. Hollenbach evaluated test data of the Vienna Model Basin (www.sva.at) and developed regression formulas for resistance and propulsion estimates of single and twin screw vessels. The formulas were tested against model test data of the Hamburg Ship Model Basin (www.hsva.de) and, in general, showed less standard deviation than other methods. This is an indication that Hollenbach's method is more reliable than previous methods (Hollenbach, 1996). The improvement is particularly visible for twin screw vessels.

Underlying data The formulas are based on experiments for 433 displacement type vessels. Since some of the vessels had variants and were tested on different drafts, a total of 793 resistance and 1103 propulsion tests form the basis for Hollenbach's method. The developed formulas have been published in Hollenbach (1997a) and a series of papers with updates to the

Fundamentals of Ship Hydrodynamics: Fluid Mechanics, Ship Resistance and Propulsion, First Edition.
Lothar Birk.

Companion website: www.wiley.com/go/birk/hydrodynamics

Table 51.1 Recommended limits for principal dimensions and form parameters of single screw vessels on design draft

Permissible range of main dimensions			
Parameter	Symbol	Unit	Range
ship speed	v_S	[m/s]	displacement type vessel
length between perpendicular	L_{PP}	[m]	42.0 ... 205.0
length–beam ratio	L_{PP}/B	[–]	4.71 ... 7.11
beam–draft ratio	B/T	[–]	1.99 ... 4.00
propeller diameter–draft ratio	D/T_A	[–]	0.43 ... 0.84
block coefficient (based on L_{PP})	C_B	[–]	$0.60^{(*)}$ ... 0.83
length–displacement ratio	$L_{PP}/\sqrt[3]{V}$	[–]	4.49 ... 6.01

(*) The resistance estimate for vessels on design draft was later expanded to smaller values of the block coefficient (see Equation (51.10))

method (Hollenbach, 1996, 1997b, 1998a,b, 1999). In addition to the mean resistance at trial conditions, the method provides estimates for the minimum resistance (best case scenario) and for the maximum resistance (worst case scenario). Formulas for the resistance in ballast condition are presented as well. A complete propulsion estimate is possible based on estimates for the hull efficiency, thrust deduction fraction, and relative rotative efficiency, if open water propeller characteristics are known.

51.1.1 Applicability

Range of application

Based on the dimensions of the vessels used for the regression analysis, Hollenbach recommends staying within the limits for the main dimensions listed in Table 51.1 for single screw vessels. Similar limiting values for single screw ships in ballast condition and twin screw vessels are provided in Hollenbach (1999).

$V = C_B L_{PP} B T$ is the volumetric displacement. All other parameters are explained below in Section 51.1.2. Results are expected to be less reliable for vessels that fall outside these ranges. In addition to the restrictions on geometry, the method cd provides estimates for the range of Froude numbers for which the predictions are considered valid.

51.1.2 Required input

Input data

One of the advantages of Hollenbach's method is that it requires only a few basic hull form parameters. The parameters are listed in Table 51.2.

If the waterline length is not known, it can be set to 1.035 L_{PP}.

Computational length

A calculation length L_c is used in some of the formulas. It is determined on the basis of length between perpendiculars L_{PP} and length over wetted surface L_{OS}.

$$L_c = \begin{cases} L_{OS} & \text{if } L_{OS} < L_{PP} \\ L_{PP} + \frac{2}{3}\left(L_{OS} - L_{PP}\right) & \text{if } L_{PP} < L_{OS} < 1.1 L_{PP} \\ 1.0667 L_{PP} & \text{if } L_{OS} > 1.1 L_{PP} \end{cases} \tag{51.1}$$

Table 51.2 Required and optional input parameters for Hollenbach's method

Parameter	Symbol	Remarks
	required parameters	
length between perpendiculars	L_{PP}	
length in waterline	L_{WL}	check definition for ballast condition
length over wetted surface	L_{OS}	check definition for ballast condition
molded beam	B	
molded draft at aft perpendicular	T_A	
molded draft at forward perpendicular	T_F	
propeller diameter	D	
block coefficient (based on L_{PP})	C_B	
transverse vertical area above waterline	A_V	for air resistance
number of rudders	N_{rudders}	1 or 2 (for twin screw vessels)
number of shaft brackets	N_{brackets}	0, 1 or 2 (for twin screw vessels)
number of shaft bossings	N_{bossings}	0, 1 or 2 (for twin screw vessels)
number of side thrusters	$N_{\text{thrusters}}$	between 0 and 4
	optional parameters	
wetted surface (hull)	S	
wetted surface of appendages	S_{APP_i}	bilge keels, stabilizer fins, etc.
diameter(s) of transverse thruster tunnels	d_{TH}	for appendage resistance fins, etc.

The draft T of the vessel midships is taken as mean value of the drafts at fore and aft perpendiculars.

$$T = \frac{T_A + T_F}{2} \tag{51.2}$$

Wetted surface In early design stages the wetted surface S is often not yet known. Hollenbach (1997a) estimates S in two steps: First, a shape factor k is computed according to the formula:

$$\begin{aligned} k = {} & s_0 + s_1 \left(\frac{L_{OS}}{L_{WL}}\right) + s_2 \left(\frac{L_{WL}}{L_{PP}}\right) + s_3 C_B + s_4 \left(\frac{L_{PP}}{B}\right) \\ & + s_5 \left(\frac{B}{T}\right) + s_6 \left(\frac{L_{PP}}{T}\right) + s_7 \left(\frac{T_A - T_F}{L_{PP}}\right) + s_8 \left(\frac{D}{T}\right) \\ & + k_{\text{rudders}} N_{\text{rudders}} + k_{\text{brackets}} N_{\text{brackets}} + k_{\text{bossings}} N_{\text{bossings}} \end{aligned} \tag{51.3}$$

Then, the actual wetted surface of the hull is approximated by

$$S = k\, L_{PP} (B + 2T) \tag{51.4}$$

The shape factor should not be confused with the form factor used in the ITTC performance prediction method. Obviously, the factors k_{rudders}, k_{brackets}, and k_{bossings} are only used for twin screw vessels. Possible values for number of rudders N_{rudders}, brackets N_{brackets}, and bossings N_{bossings} are listed in Table 51.2. The estimated wetted surface includes rudder and skeg for single screw vessels.

Table 51.3 Coefficients for an estimate of the wetted surface

	single propeller vessels		twin propeller vessels	
	design draft	ballast condition	design draft with bulb	design draft w/o bulb
s_0	−0.6837	0.8037	−0.4319	−0.0887
s_1	0.2771	0.2726	0.1685	0.0000
s_2	0.6542	0.7133	0.5637	0.5192
s_3	0.6422	0.6699	0.5891	0.5839
s_4	0.0075	0.0243	0.0033	−0.0130
s_5	0.0275	0.0265	0.0134	0.0050
s_6	−0.0045	−0.0061	−0.0005	−0.0007
s_7	−0.4798	0.2349	−2.7932	−0.9486
s_8	0.0376	0.0131	0.0072	0.0506
k_{rudders}	—	—	0.0131	0.0076
k_{brackets}	—	—	−0.0030	0.0036
k_{bossings}	—	—	0.0061	0.0049

Wetted surface of appendages

If other formulas are used for the wetted surface of the hull, the wetted area of appendages S_{APP} for twin screw vessels may be approximated based on the following formula (Hollenbach, 1997a):

$$S_{APP} = S\left(p_1 + p_2 \exp\left[-\frac{L_{PP}\,T}{1000\text{m}^2}\right]\right) \tag{51.5}$$

with the following properties and constants

$$S = \text{wetted surface of hull without appendages}$$

$$p_1 = \begin{cases} 0.0325 \text{ as upper limit} \\ 0.0280 \text{ as lower limit} \end{cases}$$

$$p_2 = \begin{cases} 0.045 \text{ as upper limit} \\ 0.010 \text{ as lower limit} \end{cases}$$

51.2 Resistance Estimate

Hollenbach's method divides the resistance into five components.

1. frictional resistance R_F without a form factor,
2. residuary resistance R_R,
3. correlation allowance R_A,
4. appendage resistance R_{APP}, and an

5. environmental resistance R_{env}.

A dimensionless coefficient is computed for each component.

51.2.1 Frictional resistance coefficient

Frictional resistance without form factor

The ITTC 1957 model–ship correlation line is used for the frictional resistance coefficient.

$$C_F = \frac{0.075}{\left[\log_{10}(Re) - 2\right]^2} \tag{2.18}$$

The Reynolds number *Re* is computed based on the computational length L_c (51.1) and the kinematic viscosity of seawater $\nu = 1.1892 \cdot 10^{-6}$ m/s^2 at a temperature of 15 °C.

$$Re = \frac{v_S\, L_c}{\nu} \tag{51.6}$$

51.2.2 Mean residuary resistance coefficient

Mean residuary resistance

The central part of almost any resistance estimate is the determination of the residuary resistance or wave resistance. Since Hollenbach's method does not employ a form factor, the residuary resistance will encompass wave resistance and the major part of the viscous pressure resistance.

In his regression analysis, Hollenbach (1997a) found that a better correlation between formula and data could be found if the residuary resistance coefficient is taken to a basis of beam times draft ($B\,T$) instead of the usual wetted surface S. Therefore, Hollenbach's residuary resistance coefficient is defined as

$$C_{R_{BT}} = \frac{10\, R_R}{\frac{1}{2}\,\rho\, v_S^2\, BT} \tag{51.7}$$

The additional scaling factor 10 adjusts the decimals so that values of $C_{R_{BT}}$ usually range from 0.05 to 0.3.

Formulas for $C_{R_{BT}}$ have been provided for several cases:

1. minimum and mean residuary resistance of single screw vessels at design draft
2. mean residuary resistance of single screw vessels at ballast draft
3. minimum and mean residuary resistance of twin screw vessels at design draft

The computation of $C_{R_{BT}}$ is performed in two steps. First, a standard value $\tilde{C}_{R_{std}}$ is determined and then multiplied with several correction factors.

C_R for standard hull form

The standard value is computed with a parabolic polynomial.

$$\begin{aligned} \tilde{C}_{R_{std}} = {} & b_{11} + b_{12} F_N + b_{13} F_N^2 \\ & + \left(b_{21} + b_{22} F_N + b_{23} F_N^2\right) C_B \\ & + \left(b_{31} + b_{32} F_N + b_{33} F_N^2\right) C_B^2 \end{aligned} \tag{51.8}$$

Table 51.4 Coefficients for computation of the standard residuary resistance coefficient in Hollenbach's method

	mean residuary resistance			minimum residuary resistance	
	single screw		twin screw	single screw	twin screw
	design draft	ballast draft	design draft	design draft	design draft
b_{11}	−0.57424	−1.50162	−5.3475	−0.91424	3.27279
b_{12}	13.3893	12.9678	55.6532	13.3893	−44.1138
b_{13}	90.596	−36.7985	−114.905	90.596	171.692
b_{21}	4.6614	5.55536	19.2714	4.6614	−11.5012
b_{22}	−39.721	−45.8815	−192.388	−39.721	166.559
b_{23}	−351.483	121.82	388.333	−351.483	−644.456
b_{31}	−1.14215	−4.33571	−14.3571	−1.14215	12.4626
b_{32}	−12.3296	36.0782	142.738	−12.3296	−179.505
b_{33}	459.254	−85.3741	−254.762	459.254	680.921

The polynomial uses the block coefficient C_B as a parameter and is a function of the special Froude number

$$F_N = \frac{v_S}{\sqrt{g L_c}} \tag{51.9}$$

Note that the ITTC recommended length in waterline has been replaced by the computational length L_c from Equation (51.1).

The necessary coefficients b_{ij} in (51.8) are listed in Table 51.4 for all cases. In the case of single screw vessels at design draft, the coefficient b_{11} for the mean residuary resistance is corrected for vessels with block coefficients smaller than 0.6 (Hollenbach, 1998a).

$$b_{11} = \begin{cases} -0.87674 & \text{if } C_B < 0.49 \\ -0.57424 - 25(0.6 - C_B)^2 & \text{if } 0.49 \leq C_B < 0.6 \\ -0.57424 & \text{see Table 51.4} \end{cases} \tag{51.10}$$

Note that this correction applies only to the mean residuary resistance of single screw vessels on design draft. In all other cases, the b_{11} values of Table 51.4 apply.

Correction factors

The standard value is then multiplied with eight factors representing additional influences on residual resistance.

$$C_{R_{BT}} = \tilde{C}_{R_{\text{std}}} \cdot k_{Fr} \cdot k_L \cdot k_{BT} \cdot k_{LB} \cdot k_{LL} \cdot k_{AO} \cdot k_{Tr} \cdot k_{Pr} \cdot N_{\text{rudders}}^{a_7} \cdot N_{\text{brackets}}^{a_8} \cdot N_{\text{bossings}}^{a_9} \cdot N_{\text{thrusters}}^{a_{10}} \tag{51.11}$$

The correction factors are:

1. High Froude number factor k_{Fr}

   For ship speeds exceeding a critical Froude number Fr_{crit}, the wave resistance is growing faster than predicted by formula (51.8). With the estimate for the critical Froude number

   $$Fr_{\text{crit}} = d_1 + d_2 C_B + d_3 C_B^2 \tag{51.12}$$

the high Froude number factor is defined as

$$k_{Fr} = \begin{cases} 1 & \text{if } F_N < Fr_{\text{crit}} \\ \left(\dfrac{F_N}{Fr_{\text{crit}}}\right)^{c_1} & \text{if } F_N \geq Fr_{\text{crit}} \end{cases} \tag{51.13}$$

2. Length factor k_L

   The residuary resistance is still dependent on length

$$k_L = e_1 \left(\frac{L_{PP}}{[\text{m}]}\right)^{e_2} \tag{51.14}$$

   k_L obviously has to be dimensionless. Thus, the equation is valid only if the length between perpendiculars is entered in meters.

3. Beam–draft ratio factor k_{BT}

$$k_{BT} = \begin{cases} (1.99)^{a_1} & \text{if } \dfrac{B}{T} < 1.99 \\ \left(\dfrac{B}{T}\right)^{a_1} & \text{if } \dfrac{B}{T} \geq 1.99 \end{cases} \tag{51.15}$$

4. Length–beam ratio factor k_{LB}

$$k_{LB} = \begin{cases} \left(\dfrac{L_{PP}}{B}\right)^{a_2} & \text{if } \dfrac{L_{PP}}{B} \leq 7.11 \\ (7.11)^{a_2} & \text{if } \dfrac{L_{PP}}{B} > 7.11 \end{cases} \tag{51.16}$$

5. Wetted length ratio factor k_{LL}

$$k_{LL} = \begin{cases} \left(\dfrac{L_{OS}}{L_{WL}}\right)^{a_3} & \text{if } \dfrac{L_{OS}}{L_{WL}} \leq 1.05 \\ (1.05)^{a_3} & \text{if } \dfrac{L_{OS}}{L_{WL}} > 1.05 \end{cases} \tag{51.17}$$

6. Aft overhang ratio factor k_{AO}

$$k_{AO} = \begin{cases} \left(\dfrac{L_{WL}}{L_{PP}}\right)^{a_4} & \text{if } \dfrac{L_{WL}}{L_{PP}} \leq 1.06 \\ (1.06)^{a_4} & \text{if } \dfrac{L_{WL}}{L_{PP}} > 1.06 \end{cases} \tag{51.18}$$

7. Trim correction factor k_{Tr}

$$k_{Tr} = \left[1 + \frac{T_A - T_F}{L_{PP}}\right]^{a_5} \tag{51.19}$$

8. Propeller factor k_{Pr}

$$k_{Pr} = \begin{cases} (0.43)^{a_6} & \text{if } \dfrac{D}{T_A} < 0.43 \\ \left(\dfrac{D}{T_A}\right)^{a_6} & \text{if } 0.43 \leq \dfrac{D}{T_A} \leq 0.84 \\ (0.84)^{a_6} & \text{if } \dfrac{D}{T_A} > 0.84 \end{cases} \tag{51.20}$$

The additional constants a_1 through e_2 are listed in Table 51.5. Note that the coefficients a_1 and a_2 have different signs compared with Hollenbach's papers because the corresponding ratios in Equations (51.15) and (51.16) have been inverted to simplify notation.

Range of Froude numbers

The formulas for the residuary resistance are valid in specific Froude number ranges $Fr_{\min} \leq F_N \leq Fr_{\max}$ (Hollenbach, 1999).

$$\text{lower limit} \quad Fr_{\min} = \min\left(f_1,\ f_1 + f_2(f_3 - C_B)\right) \tag{51.21}$$

$$\text{upper limit} \quad Fr_{\max} = g_1 + g_2 C_B + g_3 C_B^2 \tag{51.22}$$

Predictions for Froude numbers outside of the indicated range should be interpreted with great care and confirmed by other methods.

51.2.3 Minimum residuary resistance coefficient

Minimum residuary resistance

In essence, the best case scenario resistance estimate relies on the same formulas (51.8) and (51.11) for $\tilde{C}_{R_{\text{std}}}$ and $C_{R_{BT}}$. However, the following modifications to the coefficients are made:

1. The coefficients b_{ij} change; see Table 51.4. In the case of the single screw vessel, only b_{11} changes. All coefficients b_{ij} change for the minimum residuary resistance of twin screw vessels.

2. Coefficients a_5 and a_6 vanish, i.e. no corrections for trim and propeller are employed.

3. No correction is made for higher Froude numbers, the factor k_{Fr} is equal to one: $k_{Fr} = 1$.

4. Also the length factor is set to $k_L = 1$ ($e_1 = 1$ and $e_2 = 0$ respectively).

The lower and upper limits for the Froude number are calculated with the same formulas but with changed coefficients (see Table 51.6). For Froude numbers $F_N < Fr_{\min}$ the lower limit $Fr_{\min}$ should be used in the formula for $\tilde{C}_{R_{\text{std}}}$.

Table 51.5 Coefficients for correction factors of the standard residuary resistance coefficient in Hollenbach's method

	mean residuary resistance			minimum residuary resistance	
	single screw		twin screw	single screw	twin screw
	design draft	ballast draft	design draft	design draft	design draft
a_1	0.3382	0.7139	0.2748	0.3382	0.2748
a_2	−0.8086	−0.2558	−0.5747	−0.8086	−0.5747
a_3	−6.0258	−1.1606	−6.761	−6.0258	−6.761
a_4	−3.5632	0.4534	−4.3834	−3.5632	−4.3834
a_5	9.4405	11.222	8.8158	0.0	0.0
a_6	0.0146	0.4524	−0.1418	0.0	0.0
a_7	0.0	0.0	−0.1258	0.0	0.0
a_8	0.0	0.0	0.0481	0.0	0.0
a_9	0.0	0.0	0.1699	0.0	0.0
a_{10}	0.0	0.0	0.0728	0.0	0.0
c_1	$\left(\frac{F_N}{Fr_{\text{crit}}}\right)$	$10\,C_B\left(\frac{F_N}{Fr_{\text{crit}}}-1\right)$	$\left(\frac{F_N}{Fr_{\text{crit}}}\right)$	0.0	0.0
d_1	0.854	0.032	0.897	0.0	0.0
d_2	−1.228	0.803	−1.457	0.0	0.0
d_3	0.497	0.739	0.767	0.0	0.0
e_1	2.1701	1.9994	1.8319	1.0	1.0
e_2	−0.1602	−0.1446	−0.1237	0.0	0.0

Table 51.6 Factors for lower and upper limit formulas of the range of Froude numbers in which the C_R formulas are valid

	mean residuary resistance			minimum residuary resistance	
	single screw		twin screw	single screw	twin screw
	design draft	ballast draft	design draft	design draft	design draft
f_1	0.17	0.15	0.16	0.17	0.15
f_2	0.20	0.10	0.24	0.20	0.0
f_3	0.60	0.50	0.60	0.60	0.0
g_1	0.642	0.42	0.83	0.614	0.952
g_2	−0.635	−0.20	−0.66	−0.717	−1.406
g_3	0.150	0.0	0.0	0.261	0.643

51.2.4 Residuary resistance coefficient

As mentioned above, Hollenbach's residuary resistance coefficients are based on the unusual reference surface $BT/10$. Therefore, the coefficients $C_{R_{BT}}$ for mean and minimum residuary resistance have to be rescaled based on the wetted surface S before they can be added to the other resistance coefficients:

Residuary resistance coefficient

$$C_R = \frac{R_R}{\frac{1}{2}\rho v_S^2 S} = C_{R_{BT}} \frac{BT}{10\,S} \tag{51.23}$$

51.2.5 Correlation allowance

A formula developed at the Schiffbautechnische Versuchsanstalt Wien (Vienna Model Basin) is used for the correlation allowance.

Correlation allowance

$$C_A = \begin{cases} (0.35 - 0.002 L_{PP})\,10^{-3} & \text{if } L_{PP} < 175\,\text{m, } L_{PP} \text{ in meter} \\ 0 & \text{if } L_{PP} \geq 175\,\text{m} \end{cases} \tag{51.24}$$

The formula yields negative values for ships with a length between perpendiculars larger than 175 m. For these cases, we set the correlation allowance to zero.

51.2.6 Appendage resistance

Hollenbach's method considers two distinct appendage resistance categories:

1. frictional and form resistance of bilge keels, stabilizer fins, and exposed shafts, shaft brackets and bossings especially for twin screw vessels, and
2. the resistance of transverse thrusters if present.

The formulas for both resistance components are borrowed from Holtrop and Mennen (1982).

The viscous resistance of appendages is estimated to

Viscous resistance of appendages

$$R_{APP} = \frac{1}{2}\rho v_S^2 S_{APP}(1 + k_2)_{eq} C_F \tag{51.25}$$

The equivalent factor $(1 + k_2)_{eq}$ is derived from the weighted sum of individual appendage contributions and the total wetted area $S_{APP} = \sum S_{APP_i}$ of all appendages:

$$(1 + k_2)_{eq} = \frac{\sum_i (1 + k_{2i}) S_{APP_i}}{\sum_i S_{APP_i}} \tag{51.26}$$

Suitable form factors k_{2i} for streamlined and flow-oriented appendages are also provided in Holtrop (1988) and repeated in Table 50.2 for completeness. Again, rudder(s) and skegs are included in the hull surface estimate for Hollenbach's method. Consequently, they are not considered part of the appendage resistance. Equation (51.25) uses the same ITTC 1957 model–ship correlation coefficient C_F (2.18) as the hull frictional resistance.

Resistance due to thruster tunnels

The resistance from transverse thrusters may be approximated by Holtrop's method (Holtrop and Mennen, 1982):

$$R_{TH} = \rho v_S^2 \pi d_{TH}^2 C_{D_{TH}} \tag{50.17}$$

Resistance coefficients for bow thruster openings $C_{D_{TH}}$ are in the range 0.003 to 0.012. Hollenbach suggests the following estimate:

$$C_{D_{TH}} = 0.003 + 0.003 \left(\frac{10 d_{TH}}{T} - 1 \right) \tag{51.27}$$

Finally, the appendage resistance coefficient is derived from the formula:

$$C_{APP} = \frac{R_{APP} + R_{TH}}{\frac{1}{2} \rho v_S^2 S} \tag{51.28}$$

51.2.7 Environmental resistance

Air resistance

For ship trial conditions, the resistance of a ship moving through air at rest is estimated according to ITTC recommended procedures (ITTC, 2017).

$$C_{AAS} = C_{DA} \frac{\rho_{\text{air}} A_{VS}}{\rho S} \tag{51.29}$$

A standard value of $C_{DA} = 0.8$ can be used for the air drag coefficient, if wind tunnel test data are unavailable.

Additional resistance contributions from wind and waves may be included here as well for estimates of resistance in service conditions.

$$C_{\text{env}} = C_{AAS} \left[+ C_{\text{wind}} + C_{\text{wave}} \right] \tag{51.30}$$

51.2.8 Total resistance

Mean, minimum, and maximum resistance

The total resistance coefficient for the full scale vessel is given as the sum of the five component coefficients:

$$\begin{aligned} C_{T_{\text{mean}}} &= C_F + C_{R_{\text{mean}}} + C_A + C_{APP} + C_{\text{env}} \\ C_{T_{\text{min}}} &= C_F + C_{R_{\text{min}}} + C_A + C_{APP} + C_{\text{env}} \\ C_{T_{\text{max}}} &= h_1 C_{T_{\text{mean}}} \end{aligned} \tag{51.31}$$

Make sure the residuary coefficient has been converted to its standard form based on the wetted surface S. Obviously, we employ the mean residuary resistance coefficient to form the mean total resistance and use the minimum residuary resistance coefficient for the minimum total resistance. An estimate for the total maximum resistance coefficient

Vessel type	Draft	Average η_R
single screw	design	1.009
single screw	ballast	1.000
twin screw	design	0.981

Table 51.7 Suggested values for the relative rotative efficiency η_R, if the propulsion estimate is based on Wageningen B-Series propeller data (Hollenbach, 1999)

is obtained by multiplying the mean total resistance coefficient with the factor $h_1 = 1.204$ for a single screw vessel on design draft, $h_1 = 1.194$ for a single screw vessel on ballast draft, and $h_1 = 1.206$ for a twin screw vessel.

The total calm water resistance is computed in the usual way using the respective total resistance coefficient.

$$R_T = \frac{1}{2}\rho v_S^2 \, S \, C_T \tag{51.32}$$

The limiting minimum and maximum resistance values will not be exceeded with a probability of 95%.

51.3 Hull–Propeller Interaction Parameters

Keep in mind that the estimated resistance is for trial conditions and for a vessel without a propulsor. In order to combine the results of the resistance estimate with data from open water tests (propeller without ship), we have to specify the changes in hull and propeller properties if the propeller is working behind the hull. The parameters used to describe this hull–propeller interaction are:

1. relative rotative efficiency η_R,
2. thrust deduction fraction t, and
3. wake fraction w.

Hollenbach found that relative rotative efficiency η_R and thrust deduction fraction t showed very little correlation to the main dimensions of the vessel. They depend on shape details of the stern of a vessel.

51.3.1 Relative rotative efficiency

Relative rotative efficiency

Due to the lack of correlation between relative rotative efficiency and the ship's principal dimensions, Hollenbach proposed constant values for η_R based on ship type and draft. Hollenbach (1997a) notes that the efficiency of stock propellers used in the propulsion tests influence the result. If the stock propeller has less efficiency than a comparable Wageningen B-Series propeller, relative rotative efficiency becomes larger and vice versa.

In case the propulsion estimate is based on a Wageningen B-Series propeller, the values of Table 51.7 can be used for the relative rotative efficiency.

Table 51.8 Suggested values for the thrust deduction fraction t (Hollenbach, 1999)

Vessel type	Condition	Appendage arrangement	t	Std. deviation
single screw	design	–	0.190	20.4%
single screw	ballast	–	0.195	26.1%
twin screw	design	twin rudder, shaft brackets	0.150	29.1%
twin screw	design	twin rudder, twin skegs	0.186	22.4%
twin screw	design	single rudder, shaft brackets	0.130	36.0%
twin screw	design	single rudder, shaft bossings	0.113	34.5%

51.3.2 Thrust deduction fraction

Thrust deduction fraction

Since no correlation between main dimensions and the thrust deduction fraction could be found from the model test results, Hollenbach (1999) suggests using a constant value for t in early design stages according to Table 51.8.

Since the thrust deduction fraction does not appear to show significant viscous effects, the value of t is applied to the full scale vessel without corrections. The stated standard deviations for the thrust deduction fraction indicate that there is considerable uncertainty in the values. However, this by no means indicates that more complex formulas – like the ones presented by Holtrop (1984) – are more accurate.

51.3.3 Wake fraction

Hull efficiency of model

Hollenbach does not include explicit formulas for the wake fraction w. Instead, he developed formulas for the hull efficiency η_{HM} at model scale. The hull efficiency combines effects of wake fraction w_{TM} and thrust deduction fraction t.

$$\eta_{HM} = \frac{1-t}{1-w_{TM}} \tag{51.33}$$

According to Hollenbach's method, hull efficiency of models for single screw vessels on design draft is given by

$$\eta_{HM} = 0.948\, C_B^{0.3977} \left(\frac{R_{T\text{mean}}}{R_T}\right)^{-0.58} \left(\frac{B}{T}\right)^{0.1727} \left(\frac{D^2}{BT}\right)^{-0.1334} \quad \text{(design)} \tag{51.34}$$

and on ballast condition by:

$$\eta_{HM} = 1.055\, C_B^{1.0099} \left(\frac{L_{PP}}{B}\right)^{0.2991} \left(\frac{L_{WL}}{L_{PP}}\right)^{-3.2806} \left(\frac{D}{T}\right)^{-0.2317} \quad \text{(ballast)} \tag{51.35}$$

For twin screw vessels in design condition, the following formula should be used.

$$\eta_{HM} = C\, C_B^{0.1202} \left(\frac{D}{T}\right)^{-0.0285} \quad \text{(twin screw)} \tag{51.36}$$

In the formulas above, B is the ship molded beam, D the propeller diameter, L_{PP} the length between perpendiculars, L_{WL} the length in the water line, and T the mean molded draft.

Table 51.9 Suggested values for constant C for the hull efficiency of twin screw vessel models (Hollenbach, 1999)

Vessel type	Condition	Appendage arrangement	C value
twin screw	design	twin rudder, shaft brackets	1.125
twin screw	design	twin rudder, twin skegs	1.224
twin screw	design	single rudder, shaft brackets	1.086
twin screw	design	single rudder, shaft bossings	1.096

The factor $R_{T\text{mean}}/R_T$ in the hull efficiency estimate for the single screw vessels is meant to adjust the propulsion prediction for varying confidence in the resistance estimate. If there is an indication (by comparing with similar vessels or other methods) that the actual resistance is closer to $R_{T\text{minimum}}$ rather than the estimate $R_{T\text{mean}}$, the factor becomes $R_{T\text{mean}}/R_{T\text{minimum}} > 1$. If you think $R_{T\text{mean}}$ is too optimistic, the factor could be $R_{T\text{mean}}/R_{T\text{maximum}} < 1$ or anything in between. If no additional data is available, use $R_{T\text{mean}}/R_T = 1$.

The factor C in the η_{HM} formula for twin screw vessels should be taken from Table 51.9.

Wake fraction of model

Once the hull efficiency of the model has been determined, we compute the wake fraction w_{TM} of the model. Solving Equation (51.33) for the wake fraction yields

$$w_{TM} = 1 - \frac{1-t}{\eta_{HM}} \tag{51.37}$$

The additional subscript $_T$ for the wake fraction is a reference to thrust identity which is commonly assumed for the assessment of self-propulsion tests.

Wake fraction of full scale ship

In contrast to the thrust deduction fraction, wake fraction is affected by the viscous boundary layer of the ship hull. Therefore, the model value has to be adjusted for the full scale vessel. Following the formula provided by the ITTC 1978 Performance Prediction Method, the full scale wake fraction w_{TS} is given as

$$w_{TS} = (t+0.04) + \left(w_{TM} - (t+0.04)\right)\left[\frac{C_{FS}+C_A}{C_{FM}}\right] \tag{51.38}$$

The thrust deduction fraction is used to represent the potential wake fraction w_p. The summand 0.04 accounts for the influence of a rudder onto the wake (ITTC suggested value). The difference $\left(w_{TM} - (t+0.04)\right)$ is an approximation of the frictional wake w_f which is scaled by the ratio of frictional forces $\frac{C_{FS}+C_A}{C_{FM}}$ of full scale vessel and model.

ITTC 1957 model–ship correlation line

Obviously, we need a Reynolds number Re_M for the model to compute the ITTC 1957 model–ship correlation coefficient C_{FM} for the model. According to Hollenbach, the Reynolds number is based on length in waterline and kinematic viscosity of fresh water at 15 °C temperature.

$$C_{FM} = \frac{0.075}{\left(\log_{10}(Re_M)-2\right)^2} \qquad \text{with } Re_M = \frac{v_M\, L_{WL_M}}{\nu_M} \tag{51.39}$$

Hollenbach (1999) states that the equations (51.34), (51.35), and (51.36) are for models of length $L_{PP\,M} = 6.5$ m.

Using the actual length between perpendiculars L_{PP} of our ship project, we derive the scale of our virtual model

$$\lambda_{\text{virtual}} = \frac{L_{PP}}{L_{PP_M}} = \frac{L_{PP}}{6.5\ \text{m}} \tag{51.40}$$

Based on the virtual scale, we obtain length in waterline and speed of the virtual model.

$$L_{WL_M} = L_{WL}/\lambda_{\text{virtual}} \tag{51.41}$$

$$v_M = v_S/\sqrt{\lambda_{\text{virtual}}} \tag{51.42}$$

Assuming fresh water at a temperature of 15 °C, we take the kinematic viscosity of the ITTC water property tables and compute Re_M, C_{FM}, and the full scale wake fraction w_{TS} (51.38).

Hull efficiency of full scale vessel From the full scale wake fraction follows the full scale hull efficiency.

$$\eta_{HS} = \frac{1-t}{1-w_{TS}} \tag{51.43}$$

51.4 Resistance and Propulsion Estimate Example

Finally, we repeat the resistance and propulsion estimate for the 1000 TEU container ship from Sections 31.3 and 50.3 using the formulas presented in this chapter.

51.4.1 Completion of input parameters

Lengths The principal dimensions for the example are listed in Tables 31.5 and 50.6. As stated in Equation (50.44) in the previous chapter, length over water line is $L_{WL} = 147.7$ m. Length over wetted surface is $L_{OS} = 151.0$ m. The calculation length L_c (51.1) is $L_c = 149.0$ m and is used to compute the Froude number F_N (51.9).

Dimensionless parameters The block coefficient of the vessel based on its length between perpendiculars L_{PP} is

$$C_B = \frac{V}{L_{PP}BT} = 0.6613 \tag{51.44}$$

The following dimensionless ratios apply:

$$\frac{L_{PP}}{B} = 6.0417 \qquad \frac{B}{T} = 2.9268$$
$$\frac{D}{T_A} = 0.5972 \qquad \frac{L_{PP}}{\sqrt[3]{V}} = 5.5668 \tag{51.45}$$

In its design condition the ship floats on even keel; therefore, $T_A = T$. The resulting ratios fall within the limits stated in Table 51.1.

Table 51.10 Residuary resistance coefficients for minimum and mean resistance cases

			Constant factors					
		k_L	k_{BT}	k_{BL}	k_{LL}	k_{AO}	k_{Pr}	k_{Tr}
Minimum $C_{R_{BT}}$ factors:		1.0000	1.4379	0.2335	0.8753	0.9364	1.0000	1.0000
Mean $C_{R_{BT}}$ factors:		0.9778	1.4379	0.2335	0.8753	0.9364	0.9925	1.0000

		Speed dependent results							
		minimum residuary resistance				mean residuary resistance			
v_S	F_N	$C_{R_{std}}$	k_{Fr}	$C_{R_{BT}}$	$C_{R_{min}}$	$C_{R_{std}}$	k_{Fr}	$C_{R_{BT}}$	C_R
[kn]	[–]	[–]	[–]	[–]	$[10^{-3}]$	[–]	[–]	[–]	$[10^{-3}]$
15.0	0.2019	0.3851	1.0000	0.1060	0.4741	0.7251	1.0000	0.1937	0.8663
15.5	0.2086	0.4251	1.0000	0.1170	0.5234	0.7651	1.0000	0.2044	0.9141
16.0	0.2153	0.4705	1.0000	0.1295	0.5792	0.8105	1.0000	0.2165	0.9683
16.5	0.2221	0.5212	1.0000	0.1435	0.6417	0.8612	1.0000	0.2300	1.0289
17.0	0.2288	0.5773	1.0000	0.1589	0.7107	0.9173	1.0000	0.2450	1.0959
17.5	0.2355	0.6387	1.0000	0.1758	0.7863	0.9787	1.0000	0.2614	1.1692
18.0	0.2422	0.7055	1.0000	0.1942	0.8685	1.0455	1.0000	0.2792	1.2490
18.5	0.2490	0.7775	1.0000	0.2140	0.9572	1.1175	1.0000	0.2985	1.3351
19.0	0.2557	0.8550	1.0000	0.2353	1.0526	1.1950	1.0000	0.3192	1.4276

Surfaces

For the calculations we use the given wetted surface of $S = 4400.0\ \text{m}^2$. The wetted surface estimate according to Equations (51.3) and (51.4) yields:

$$k = 0.75938 \qquad S = 4448.453\ \text{m}^2 \quad \text{(estimate)} \tag{51.46}$$

The latter deviates from the actual value by just 1.1%. For the transverse area above the water line, we use again $A_V = 383.76\ \text{m}^2$.

51.4.2 Powering estimate

Residuary resistance

The resistance estimate starts with the computation of residuary resistance coefficients. Unlike other procedures, Hollenbach's method provides values for a best scenario minimum and a mean residuary resistance coefficient. The necessary formulas and coefficients are stated in Sections 51.2.2, 51.2.3, and 51.2.4. The Froude numbers F_N stay within the limits provided by Equations (51.21) and (51.22) for the mean residuary resistance estimate: $0.1577 < F_N < 0.2877$. In the best case scenario (minimum) C_R, the highest Froude number exceeds the respective upper limit by less than 1%. We will ignore this because it applies only to values above the design speed. Table 51.10 lists the constant and speed dependent factors and final residuary resistance coefficient values for the example ship.

Correlation allowance and environmental resistance

Equations (51.24) and (51.29) are employed to estimate correlation allowance and the air resistance.

$$C_A = 0.06 \cdot 10^{-3} \qquad C_{env} = C_{AAS} = 0.0833 \cdot 10^{-3} \tag{51.47}$$

Table 51.11 Resistance coefficients for example vessel by Hollenbach's method

v_S [kn]	v_S [m/s]	Fr [–]	$C_{R_{min}}$ [10^{-3}]	C_R [10^{-3}]	C_F [10^{-3}]	C_{APP} [10^{-3}]	$C_{T_{min}}$ [10^{-3}]	C_T [10^{-3}]	$C_{T_{max}}$ [10^{-3}]
15.0	7.717	0.2019	0.4741	0.8663	1.5370	0.0254	2.1799	2.5720	3.0967
15.5	7.974	0.2086	0.5234	0.9141	1.5308	0.0253	2.2228	2.6135	3.1467
16.0	8.231	0.2153	0.5792	0.9683	1.5248	0.0252	2.2725	2.6616	3.2046
16.5	8.488	0.2221	0.6417	1.0289	1.5190	0.0251	2.3291	2.7163	3.2704
17.0	8.746	0.2288	0.7107	1.0959	1.5134	0.0250	2.3924	2.7776	3.3442
17.5	9.003	0.2355	0.7863	1.1692	1.5080	0.0250	2.4625	2.8455	3.4260
18.0	9.260	0.2422	0.8685	1.2490	1.5028	0.0249	2.5394	2.9199	3.5156
18.5	9.517	0.2490	0.9572	1.3351	1.4977	0.0248	2.6230	3.0009	3.6131
19.0	9.774	0.2557	1.0526	1.4276	1.4928	0.0247	2.7134	3.0885	3.7185

Total resistance coefficients include constant $C_{env} = 0.0833 \cdot 10^{-3}$.

Table 51.12 Comparison of total calm water resistance estimates

Hollenbach						Guldhammer and Harvald Table 31.13		Holtrop and Mennen Table 50.9	
v_S [kn]	Fr [–]	Re [–]	$R_{T_{min}}$ [kN]	R_T [kN]	$R_{T_{max}}$ [kN]	Fr [–]	R_T [kN]	Fr [–]	R_T [kN]
15.0	0.2019	966854468.0	293.00	345.71	416.24	0.2005	335.62	0.2028	386.98
15.5	0.2086	999082950.3	319.02	375.10	451.62	0.2072	362.25	0.2095	419.21
16.0	0.2153	1031311432.5	347.54	407.04	490.08	0.2139	391.03	0.2163	454.37
16.5	0.2221	1063539914.8	378.80	441.78	531.90	0.2205	422.36	0.2230	492.70
17.0	0.2288	1095768397.1	413.04	479.54	577.37	0.2272	456.82	0.2298	534.07
17.5	0.2355	1127996879.3	450.52	520.58	626.78	0.2339	495.13	0.2366	578.14
18.0	0.2422	1160225361.6	491.51	565.16	680.45	0.2406	538.28	0.2433	625.06
18.5	0.2490	1192453843.9	536.29	613.56	738.72	0.2473	592.72	0.2501	675.88
19.0	0.2557	1224682326.1	585.16	666.05	801.92	0.2540	662.39	0.2568	732.51

Both are assumed to be independent of ship speed. An appendage resistance component has been added for the bilge keels. For the formula (50.18) provided by Holtrop and Mennen (1982), an area of $S_{APP} = 52.0\ \text{m}^2$ and a form factor of $k_2 = 0.4$ has been used. Subsequently, the dimensional appendage resistance is converted into a dimensionless coefficient with Equation (51.28).

Resistance coefficients The estimated resistance coefficients are listed in Table 51.11. The maximum total resistance coefficient has been obtained by multiplying the mean total resistance coefficient with the factor $h_1 = 1.204$ (see Section 51.2.8).

Comparison of resistance estimates Table 51.12 and Figure 51.1 compare the total resistance estimates derived with the three methods presented in this book: Hollenbach, Guldhammer and Harvald (Chapter 31), and Holtrop and Mennen (Chapter 50). The Froude numbers vary slightly because each method uses a different reference length. All estimates fall within the range defined by the minimum and the maximum resistance predicted by Hollenbach's method. Holtrop

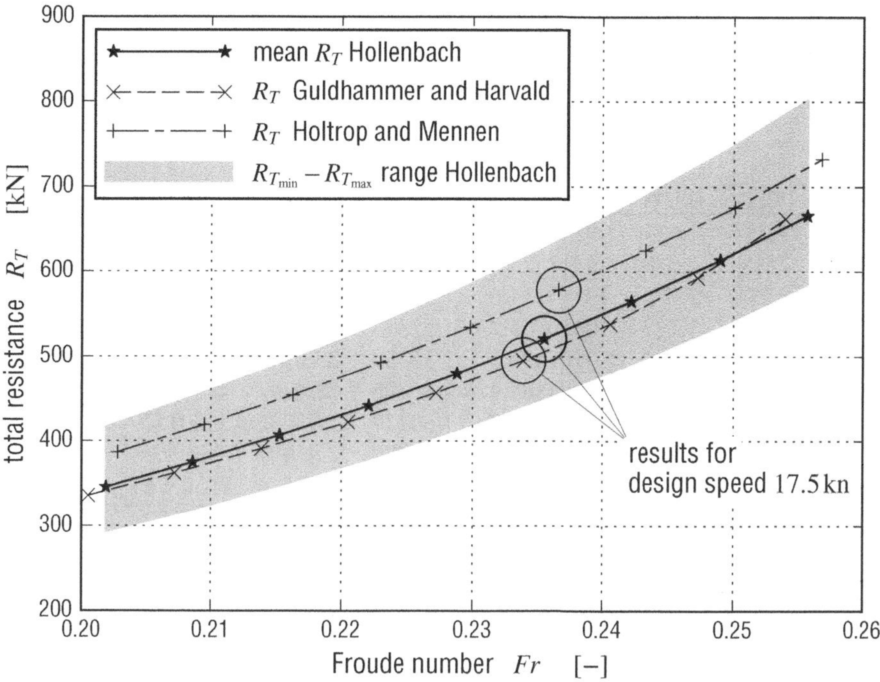

Figure 51.1 Comparison of total resistance estimates for the methods by Hollenbach, Guldhammer and Harvald, and Holtrop and Mennen

and Mennen's estimate is higher than the mean total resistance by Hollenbach. Guldhammer and Harvald's estimate is close to Hollenbach's but features a steeper increase at the end of the Froude number range. At the design speed of 17.5 kn, Guldhammer and Harvald's estimate is actually 4.9% lower than Hollenbach's. However, due to the longer reference length and, consequently, slightly lower Froude numbers, Guldhammer and Harvald's resistance curve appears to lie above Hollenbach's mean resistance curve. Holtrop and Mennen's resistance estimate lies about 11% above Hollenbach's for the design speed. These deviations are fairly typical for different methods.

Thrust deduction fraction, relative rotative efficiency

Hollenbach's method defines constant values for thrust deduction fraction t and relative rotative efficiency η_R. From Tables 51.7 and 51.8 we retrieve

$$\eta_R = 1.009 \qquad \text{and} \qquad t = 0.19$$

The powering estimate with Holtrop and Mennen's method resulted in a relative rotative efficiency of $\eta_R = 0.9940$ and a thrust deduction fraction of $t = 0.1983$ (Section 50.3.3). Andersen and Guldhammer (1986) use a relative rotative efficiency of $\eta_R = 1.0$ and with Equation (34.15) follows a thrust deduction fraction of $t = 0.2019$. Although, the values of the thrust deduction vary quite a bit between the three methods, the resulting factor $1/(1-t)$ for the required thrust varies only by about 1.5%.

Full scale wake fraction

The length of water line of a virtual model is needed for the estimate of the full scale wake fraction. According to Equation (51.40), a virtual model scale $\lambda_{\text{virtual}} = 22.3077$ is derived for a length between perpendiculars of $L_{PP} = 145$ m. Then, the waterline

Table 51.13 Prediction of model and full scale wake fraction

Hollenbach							H and M Table 50.10
v_S [kn]	v_S [m/s]	v_M [m/s]	Re_M [–]	C_F+C_A $[10^{-3}]$	C_{FM} $[10^{-3}]$	w_{TS} [–]	w_{TS} [–]
15.0	7.717	1.634	9500732.4	1.59703	3.02687	0.30285	0.26959
15.5	7.974	1.688	9817423.5	1.59079	3.00963	0.30298	0.26942
16.0	8.231	1.743	10134114.6	1.58477	2.99307	0.30311	0.26926
16.5	8.488	1.797	10450805.7	1.57898	2.97715	0.30323	0.26911
17.0	8.746	1.852	10767496.8	1.57339	2.96183	0.30335	0.26896
17.5	9.003	1.906	11084187.9	1.56799	2.94707	0.30346	0.26882
18.0	9.260	1.961	11400878.9	1.56277	2.93282	0.30357	0.26868
18.5	9.517	2.015	11717570.0	1.55772	2.91907	0.30368	0.26854
19.0	9.774	2.069	12034261.1	1.55283	2.90578	0.30378	0.26841

length of the virtual model is $L_{WL_M} = 6.6210$ m. Model speeds v_M that correspond to the investigated full scale speeds follow from Equation (51.42). Based on L_{WL_M} and v_M, Reynolds numbers for the model Re_M are computed, followed by the friction coefficient C_{FM} for the model (51.39). Once C_{FM} is known, the model wake fraction is corrected for the full scale vessel via Equation (51.38).

The hull efficiency of the model is treated as a speed independent constant as long as the factor $R_{T\text{mean}}/R_T$ is constant. Lacking other guidance, we use $R_{T\text{mean}}/R_T = 1$. The estimated hull efficiency and wake fraction at model scale are:

$$\eta_{HM} = 1.28179 \qquad w_{TM} = 0.36807$$

Table 51.13 summarizes the remaining values for the full scale wake fraction prediction based on Equation (51.38). For comparison, the last column in Table 51.13 repeats the result from Table 50.10 obtained by Holtrop and Mennen's method. Equation (34.8) by Andersen and Guldhammer (1986) yields a considerably smaller full scale wake fraction of $w_S = 0.21661$ if the model value is converted with the suggested factor of 0.7 which applies to the ship trial condition.

Self propulsion point In Hollenbach's method the full scale power propulsion analysis of Section 44.2 is repeated three times because the required thrust will be different for minimum, mean, and maximum resistance cases. Detailed results are presented for the mean resistance case only. Table 51.14 summarizes the data for the self propulsion point analysis based on the estimated mean total resistance. Characteristics of a Wageningen B-Series propeller are assumed, and its principal dimensions are listed in Table 50.6. The self propulsion points change if the analysis is based on the minimum or maximum total resistance curves.

Delivered power Once the self propulsion points are known, Equations (44.6) through (44.10) yield rate of revolution, torque, delivered power, and efficiencies. Table 51.15 shows the values of torque, rate of revolution, and delivered power for the self propulsion points from Table 51.14. Table 51.16 shows the corresponding system efficiencies. A quasi-propulsive

Table 51.14 Self propulsion point based on mean resistance curve

v_S [kn]	v_S [m/s]	Fr [–]	w_{TS} [–]	v_A [m/s]	C_S [–]	J_{TS} [–]	K_{TS} [–]	$10K_{QTS}$ [–]
15.0	7.717	0.2019	0.3028	5.380	0.59864	0.6163	0.2274	0.3754
15.5	7.974	0.2086	0.3030	5.558	0.60852	0.6133	0.2289	0.3774
16.0	8.231	0.2153	0.3031	5.736	0.61995	0.6099	0.2306	0.3796
16.5	8.488	0.2221	0.3032	5.914	0.63291	0.6061	0.2325	0.3821
17.0	8.746	0.2288	0.3033	6.093	0.64741	0.6019	0.2345	0.3848
17.5	9.003	0.2355	0.3035	6.271	0.66345	0.5973	0.2367	0.3878
18.0	9.260	0.2422	0.3036	6.449	0.68103	0.5925	0.2391	0.3909
18.5	9.517	0.2490	0.3037	6.627	0.70014	0.5874	0.2416	0.3942
19.0	9.774	0.2557	0.3038	6.805	0.72078	0.5820	0.2442	0.3976

Table 51.15 Prediction of rate of revolution and delivered power for trial condition based on mean resistance curve

v_S [kn]	v_S [m/s]	Fr [–]	T [kN]	Q [kNm]	n [1/s]	n [rpm]	P_D [kW]
15.0	7.717	0.2019	426.80	342.15	1.781	106.877	3829.36
15.5	7.974	0.2086	463.08	370.76	1.849	110.964	4308.25
16.0	8.231	0.2153	502.52	401.76	1.919	115.169	4845.40
16.5	8.488	0.2221	545.40	435.36	1.992	119.497	5447.98
17.0	8.746	0.2288	592.02	471.78	2.066	123.955	6123.89
17.5	9.003	0.2355	642.69	511.23	2.142	128.546	6881.85
18.0	9.260	0.2422	697.73	553.96	2.221	133.276	7731.49
18.5	9.517	0.2490	757.48	600.22	2.302	138.149	8683.37
19.0	9.774	0.2557	822.28	650.25	2.386	143.170	9749.06

Table 51.16 Predicted efficiencies based on mean resistance curve

v [kn]	v [m/s]	Fr [–]	η_O [–]	η_B [–]	η_H [–]	η_D [–]
15.0	7.717	0.2019	0.5942	0.5996	1.1619	0.6967
15.5	7.974	0.2086	0.5921	0.5974	1.1621	0.6942
16.0	8.231	0.2153	0.5896	0.5949	1.1623	0.6915
16.5	8.488	0.2221	0.5868	0.5921	1.1625	0.6883
17.0	8.746	0.2288	0.5837	0.5890	1.1627	0.6848
17.5	9.003	0.2355	0.5804	0.5856	1.1629	0.6810
18.0	9.260	0.2422	0.5768	0.5820	1.1631	0.6769
18.5	9.517	0.2490	0.5729	0.5781	1.1633	0.6725
19.0	9.774	0.2557	0.5689	0.5740	1.1634	0.6678

efficiency of 63.86% is achieved at the design speed of 17.5 kn. This could possibly be improved by designing a wake adapted propeller using lifting line theory and other methods.

Table 51.17 Comparison of predicted rate of revolution and delivered power

	Hollenbach						Guldhammer and Harvald		Holtrop and Mennen Table 50.11	
v_S [kn]	n_{min} [rpm]	$P_{D_{min}}$ [kW]	n_{mean} [rpm]	$P_{D_{mean}}$ [kW]	n_{max} [rpm]	$P_{D_{max}}$ [kW]	n [rpm]	P_D [kW]	n [rpm]	P_D [kW]
15.0	101.84	3134.85	106.88	3829.36	113.18	4816.33	111.81	4067.26	112.99	4637.23
15.5	105.81	3540.31	110.96	4308.25	117.54	5420.96	115.88	4545.39	117.28	5208.37
16.0	109.91	3998.65	115.17	4845.40	122.03	6099.78	120.05	5077.00	121.70	5850.65
16.5	114.14	4516.78	119.50	5447.98	126.66	6861.99	124.34	5671.91	126.27	6573.49
17.0	118.52	5102.39	123.96	6123.89	131.44	7717.78	128.79	6343.20	130.95	7378.31
17.5	123.04	5764.01	128.55	6881.85	136.36	8678.38	133.42	7108.52	135.70	8263.98
18.0	127.71	6511.05	133.28	7731.49	141.44	9756.18	138.31	7992.05	140.54	9239.76
18.5	132.53	7353.89	138.15	8683.37	146.68	10964.81	143.86	9121.81	145.52	10328.13
19.0	137.52	8303.96	143.17	9749.06	152.08	12319.22	150.28	10598.40	150.75	11573.40

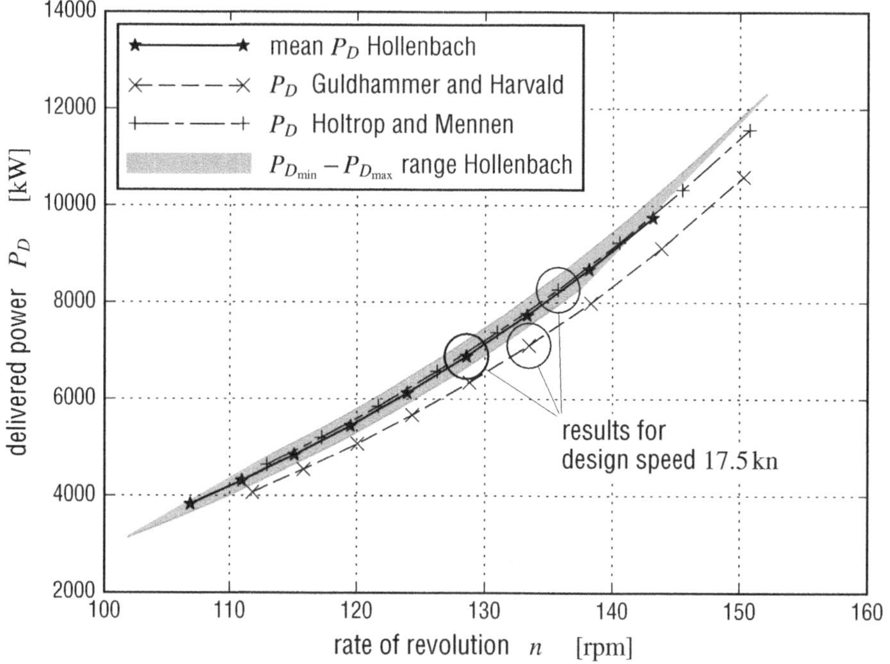

Figure 51.2 Comparison of predicted rate of revolution and delivered power for the methods by Hollenbach, Guldhammer and Harvald, and Holtrop and Mennen

Comparison Table 51.17 and Figure 51.2 present a comparison of the estimated rate of revolution and delivered power for the prediction methods discussed in this book. Although the curves are close together there are differences, especially in the predicted rate of revolution for each speed. Circles mark the values predicted for the design speed. Holtrop and Mennen's method predicts the highest delivered power of $P_D = 9414.4$ kW at a rate of revolution of $n = 170.53$ rpm. Hollenbach's estimate is the most optimistic with a delivered power of 7747.2 kW at 161.51 rpm.

Engine selection

An engine may be selected based on the powering prediction. The delivered power is converted into the engine brake power P_B via Equations (32.16) and (32.15). Proper sea and engine margins have to be added to the brake power predicted for trial conditions. The final combination of rate of revolution and brake power is matched with the engine layout diagram. This is a marine engineering rather than a hydrodynamic problem. The reader can find details in the engine selection guides published by engine manufacturers.

Conclusion

The spread of the predicted power values is an indication of the uncertainty intrinsic to resistance and propulsion estimates used in early design phases. Better results can hardly be expected since only a few form parameters are used to describe the hull shape. Too much of the flow patterns depends on details of the hull geometry, which will not be known until the lines plan is completed. Once the lines are faired, a model may be manufactured and tested. The hull geometry may also serve as the starting point for a CFD analysis if computational resources and expertise are available.

References

Andersen, P. and Guldhammer, H. (1986). A computer-oriented power prediction procedure. In *Proc. of Int. Conf. on Computer Aided Design, Manufacture, and Operation in the Marine and Offshore Industries (CADMO '86)*, Washington, DC, USA.

Hollenbach, K. (1996). Verfahren zur Abschätzung von Widerstand und Propulsion von Ein- und Zweischraubenschiffen im Vorentwurf. In *Jahrbuch der Schiffbautechnischen Gesellschaft*, volume 90, pages 119–133. Schiffbautechnische Gesellschaft (STG).

Hollenbach, K. (1997a). *Beitrag zur Abschätzung von Widerstand und Propulsion von Ein- und Zweischraubenschiffen im Vorentwurf.* PhD thesis, Institut für Schiffbau, Universität Hamburg, Hamburg, Germany.

Hollenbach, K. (1997b). Beitrag zur Abschätzung von Widerstand und Propulsion von Ein- und Zweischraubenschiffen im Vorentwurf. IfS Report 588, Institut für Schiffbau, Universität Hamburg, Hamburg, Germany.

Hollenbach, K. (1998a). Estimating resistance and propulsion for single-screw and twin-screw ships. *Schiffstechnik/Ship Technology Research*, 45(2):72–76.

Hollenbach, K. (1998b). Weiterentwicklung eines verfahrens zur Abschätzung von Widerstand und Propulsion von Ein- und Zweischraubenschiffen im Vorentwurf. In *Jahrbuch der Schiffbautechnischen Gesellschaft*, volume 92, pages 91–104. Berlin.

Hollenbach, K. (1999). Estimating resistance and propulsion for single-screw and twin-screw ships in the preliminary design. In *Proc. of 10th Int. Conference on Computer Applications in Shipbuilding (ICCAS '99)*.

Holtrop, J. (1984). A statistical re-analysis of resistance and propulsion data. *International Shipbuilding Progress*, 31(363):272–276.

Holtrop, J. (1988). A statistical resistance prediction method with a speed dependent form factor. In *Scientific and Methodological Seminar on Ship Hydrodynamics (SMSSH '88)*, Varna, Bulgaria.

Holtrop, J. and Mennen, G. (1982). An approximate power prediction method. *International Shipbuilding Progress*, 29(335):166–170.

ITTC (2017). *1978 ITTC performance prediction method.* International Towing Tank Conference, Recommended Procedures and Guidelines 7.5-02-03-01.4. Revision 4.

Self Study Problems

1. Discuss the principal differences and similarities between the resistance estimates based on Guldhammer and Harvald's method, Holtrop and Mennen's method, and Hollenbach's method.

2. For a ship design project the following data is provided.

Ship data	
length between perpendiculars	$L_{PP} = 145.00$ m
length in waterline	$L_{WL} = 149.80$ m
molded beam	$B = 22.30$ m
molded draft	$T = 8.26$ m
block coefficient (based on L_{PP})	$C_B = 0.66$
prismatic coefficient (based on L_{PP})	$C_P = 0.68$

Compute the input values for block coefficient C_B and prismatic coefficient C_P for Hollenbach's method. Compare the results with the values from the corresponding problem at the end of Chapter 50.

3. Implement Hollenbach's resistance and propulsion estimate as a program in Python, Matlab, or similar, and test it with the data presented in the last section.

Index

R

Printed and bound by CPI Group (UK) Ltd, Croydon, CR0 4YY

16/04/2025

14658562-0004